Springer Proceedings in Energy

The series Springer Proceedings in Energy covers a broad range of multidisciplinary subjects in those research fields closely related to present and future forms of energy as a resource for human societies. Typically based on material presented at conferences, workshops and similar scientific meetings, volumes published in this series will constitute comprehensive state-of-the-art references on energy-related science and technology studies. The subjects of these conferences will fall typically within these broad categories:

- Energy Efficiency
- Fossil Fuels
- Nuclear Energy
- Policy, Economics, Management & Transport
- Renewable and Green Energy
- Systems, Storage and Harvesting
- Materials for Energy

eBook Volumes in the Springer Proceedings in Energy will be available online in the world's most extensive eBook collection, as part of the Springer Energy eBook Collection. To submit a proposal or for further inquiries, please contact the Springer Editor in your region:

Kamiya Khatter (India)
Email: kamiya.khatter@springer.com
Loyola D'Silva (All other countries)
Email: loyola.dsilva@springer.com

M. Ziya Sogut • T. Hikmet Karakoc •
Omer Secgin • Alper Dalkiran
Editors

Proceedings of the 2022 International Symposium on Energy Management and Sustainability

ISEMAS 2022

Springer

Editors
M. Ziya Sogut
Maritime Faculty
Piri Reis University
Istanbul, Türkiye

Omer Secgin
Technology Faculty
Mechanical Engineering Department

Sakarya University of Applied Sciences
Sakarya, Türkiye

T. Hikmet Karakoc
Faculty of Aeronautics and Astronautics
Eskisehir Technical University
Eskisehir, Türkiye

Information Technology Research
and Application Centre
Istanbul Ticaret University
İstanbul, Türkiye

Alper Dalkiran
School of Aviation
Süleyman Demirel University
Keciborlu, Isparta, Türkiye

ISSN 2352-2534 ISSN 2352-2542 (electronic)
Springer Proceedings in Energy
ISBN 978-3-031-30170-4 ISBN 978-3-031-30171-1 (eBook)
https://doi.org/10.1007/978-3-031-30171-1

This Springer imprint is published by the registered company Springer Nature Switzerland AG
The registered company address is: Gewerbestrasse 11, 6330 Cham, Switzerland

Preface

The continued development of energy and environmental sustainability is critical as a global issue. Today, two prominent issues in the development strategies of all nations are energy management and sustainability. These concepts continue to impact many fields of study in various scientific disciplines. Developed in this context, the International Symposium on Energy Management and Sustainability (ISEMAS) event examines research on the manageability of energy that improves social orientation and environmental sustainability and attempts to develop solutions – the objective is to be a platform that supports development. Energy efficiency is critical due to the social awareness created in all sectoral components by global climate change. The manageability of energy is a multidimensional issue related to many variables and scientific criteria. ISEMAS aims to deal with all aspects of energy management, which has a multifaceted discipline process, and to examine the effect of sustainability on energy management processes. ISEMAS-22 was carried out for this purpose. The symposium presented papers from 29 countries, including distinguished keynote and invited speakers. The full texts of these papers are published in this book.

This symposium focused on awareness as an environmental identity and its dissemination. For this purpose, to contribute to ecological sustainability, the symposium organization has planted a new tree for each participant who presented a paper at the symposium. ISEMAS will continue as an international event focusing on the multidisciplinary study of energy management and sustainable development and will continue to contribute to the advancement of science in the coming years.

Istanbul, Türkiye M. Ziya Sogut

Contents

Chapter 1
Energy Efficiency in the Drilling of Hollow Parts: A Sample Application

Omer Secgin, M. Ziya Sogut, and İbrahim Özsert

1.1 Introduction

In manufacturing industries, energy is a very important cost item. Reducing these energy costs provides huge gains to the company. Energy efficiency is defined as the ratio between energy output and input. It includes methods to increase energy savings, measures to reduce losses, studies that reduce operating costs, and meeting investment costs as soon as possible (Pamir 2003).

Studies on energy savings and efficiency, which is one of the most important cost items, are increasing. Distribution of electrical energy demand Turkey in 2015; in industry 25.7%, for residences and public institutions 19.1%, for business 47.6%, for lighting 1.9% and in other sectors 5.7% (TUİK n.d.). However, the rate of meeting the total demand for 2010 with domestic production is 28%, and 12% of our total electrical energy consumption is due to losses and leakages (Meral et al. 2009).

In Turkey, most of the electrical energy is consumed in industry. Therefore, it is necessary to focus on energy efficiency in the industry. It is understood from the statistical data that our industrialists, who have to compete in the international arena, can use energy more efficiently (Kavak 2005). As a result of the energy-saving studies carried out in Turkey, serious developments have been made in terms of energy efficiency in various organizations and enterprises (Narin and Akdemir 2006).

Söğüt et al. (2011) conducted an energy survey in a tomato paste factory based on production and energy consumption data. First of all, they determined the energy

O. Secgin (✉) · İ. Özsert
Sakarya University of Applied Sciences, Technology Faculty, Mechanical Engineering Department, Sakarya, Türkiye
e-mail: omersecgin@subu.edu.tr; ozsert@subu.edu.tr

M. Z. Sogut
Maritime Faculty, Piri Reis University, İstanbul, Türkiye

M. Z. Sogut et al. (eds.), *Proceedings of the 2022 International Symposium on Energy Management and Sustainability*, Springer Proceedings in Energy,
https://doi.org/10.1007/978-3-031-30171-1_1

sources used, production data and energy consumption values of the factory, and unit energy costs. Based on these data, they developed energy consumption relations, then calculated the target energy consumptions, and revealed the energy-saving potentials of the enterprise.

Bayındır et al. (Bayındır et al. 2008) designed and implemented energy monitoring system hardware and software for a business. The energy monitoring system consists of monitoring, archiving, and transferring the energy parameters measured by the energy analyzer from the critical point in the enterprise to the graphical environment.

The most important step to be taken regarding energy efficiency is energy saving. In this study, energy saving in hole drilling operation of hollow parts was investigated.

This study is a continuation of the study titled "A Parametric Program for Productivity of Drilling Process of Hollow Parts" by Seçgin and Özsert (2016). Energy measurements could not be made at the time of the study. Later, this study was prepared by making energy measurements by the authors.

1.2 Manufacturing Process and Operations

The drilling process is one of the important machining methods that is frequently used in the industry, especially in the aviation field where screw connections are used a lot (El-Sonbaty et al. 2004). In addition to teaching theoretical knowledge in the field of energy saving, the most important thing is to reveal the behavior of energy saving. Many people know why they need to save energy but cannot turn this knowledge into behavior. To be successful in energy saving, behavior improvement studies are needed (Erten 2006). In this study, the contribution of a macro program to energy efficiency by shortening the machining time in the drilling operation of hollow parts (pipe, internal channel, etc.) in previously developed CNC machines was investigated. With this program, it is aimed to increase productivity by shortening the drilling times of workpieces with voids (Seçgin and Özsert 2016). Examples of hollow parts, which are the subject of this study, are given in Fig. 1.1.

Canned cycles (G81, G82, and G83) are generally used in drilling operations on CNC machines. Thanks to these ready-made cycles, holes are defined and drilled in a very practical and easy way. In Fig. 1.2, the working logic of the G81 canned cycle is given.

While the abovementioned canned cycles are very useful for drilling filled parts, they do not provide the desired efficiency in the drilling of hollow parts. As explained in Fig. 1.2, the G81 cycle works from the beginning of the hole down to the bottom of the hole with a constant cutting speed. After the cutting tool pierces the solid area at the top of the piece, it is wasted. It continues to move with the cutting feedrate even though it does not remove the chip. Then, it reaches the filled area at the bottom and pierces that part as well.

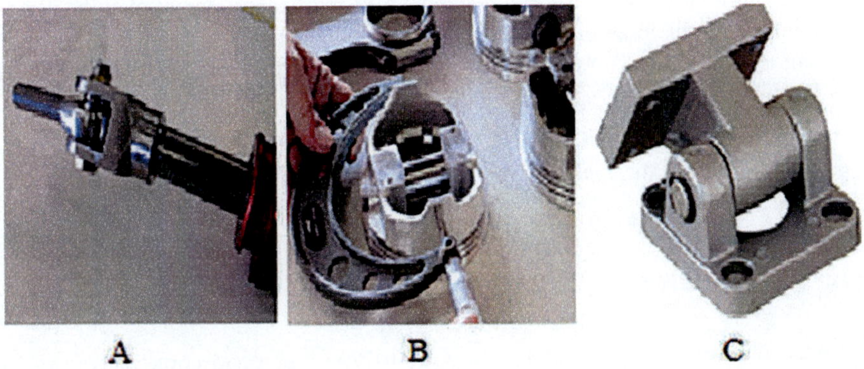

Fig. 1.1 (**a**) Flexible joint (Web Page n.d.-a). (**b**) Piston (Web Page n.d.-b). (**c**) Joint connection. (Web Page n.d.-c)

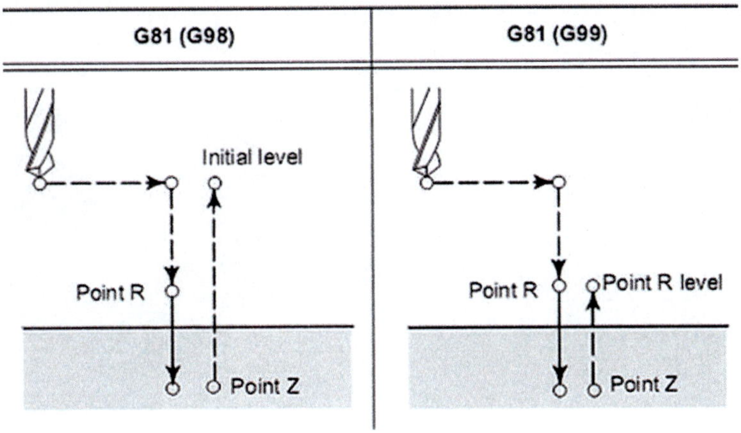

Fig. 1.2 G81 cycle. (FANUC n.d.)

The movement of the cutting tool with the cutting feedrate in the cavity causes a waste of time. Therefore, the G81 cycle reduces the machining efficiency. In the drilling process of the mentioned parts, a cycle is needed that will enable the cutting tool to move with a rapid feedrate in the cavity part.

In the macro program presented in this study, the operation is performed as follows:

- The cutting tool, with the rapid feedrate of the machine, goes to the center of the hole in the X and Y directions.
- The cutting tool, with the rapid feedrate of the machine, goes to the safe approach position of the part in the Z axis.
- The cutting tool, with the cutting feedrate, pierces the upper solid part of the part in the Z axis and exits to the hollow part.

- The cutting tool, with the rapid feedrate of the workbench, comes over the part to be drilled at the bottom in the Z axis.
- The cutting tool, with the cutting feedrate, pierces the bottom solid part in the Z axis.

With the rapid feedrate of the machine, the cutting tool comes out of the part by moving upward in the Z axis.

1.3 Materials and Method

Within the scope of this study, the CM-10125 joint connection piece, whose dimensions were given in Fig. 1.3 and Table 1.1, was drilled on a DAHLIH MCV 860 BSD milling machine. The rapid speed of the machine is 30,000 mm/min.

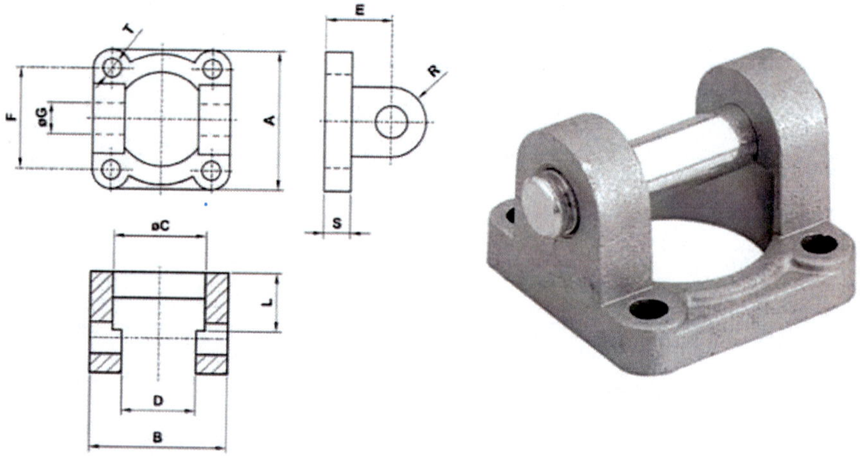

Fig. 1.3 Example part: Pneumatic joint connection. (Web Page n.d.-c)

Table 1.1 Dimensions of the pneumatic joint connection (Seçgin and Özsert 2016)

CODE	A	B	C	D	E	F	G	R	S	T
CM-10032	45	45	30	26	22	32.5	10	11	10	7
CM-10040	52	52	35	28	25	38	12	13	10	7
CM-10050	65	62	40	32	27	46.5	12	13	12	9
CM-10063	75	70	45	40	32	56.5	16	17	12	9
CM-10080	95	90	45	50	36	72	16	17	16	11
CM-10100	115	110	55	60	41	89	20	21	16	11
CM-10125	140	130	60	70	50	110	25	26	20	14
CM-10160	180	170	65	90	55	140	30	31	20	18
CM-10200	220	170	75	90	60	175	30	31	25	18

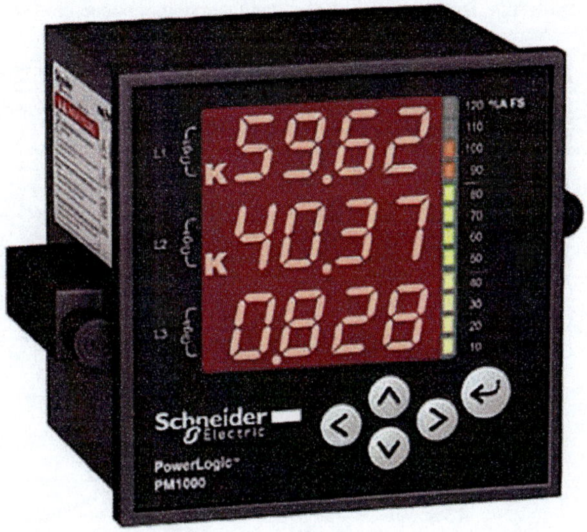

Fig. 1.4 Schneider Powerlogic PM1000 energy analyzer

The energy consumed during the experiment was measured using the Schneider Powerlogic PM1000 energy analyzer. The analyzer in question is given in Fig. 1.4.

1.4 Results

Drilling times of one pneumatic joint connection are given in Table 1.2. In Fig. 1.5, increased efficiency in terms of drilling time is given by drilling a pneumatic joint connection with the macro program. Increased efficiency by drilling one pneumatic joint connection with the macro program was given in Fig. 1.5.

When ten pieces are attached to the bench table at once, the total drilling time with the G81 cycle takes 571 s. With the developed special macro program, this process takes 341 s. In this case, 1384 watts of energy are consumed in the drilling of ten pieces with the G81 cycle, while 848 watts of energy are consumed in the drilling with the specially developed macro program. The amount of energy consumed according to both methods is given in Fig. 1.6. It has been observed that 38% energy saving is achieved by using the developed special macro program.

1.5 Conclusions

Canned cycles are simple and easy programming methods used in axis control of CNC machines. However, existing canned cycles can lead to unnecessary time losses in operations such as drilling operations of hollow parts (pipes, channels, etc.). In this study, the effectiveness of the macro program, which prevents waste in drilling hollow parts, was investigated.

Table 1.2 Drilling times of one pneumatic joint connection (Seçgin and Özsert 2016)

Part code	Drilling time with G81 (seconds)	Drilling time with macro (seconds)	Gain (seconds)	Efficiency (%)
CM-10032	19.698	10.942	8.756	44.45
CM-10040	22.7532	13.44	8.756	44.45
CM-10050	25.9692	15.064	10.9052	41.99
CM-10063	16.86	16.86	13.6116	44.67
CM-10080	38.5116	20.92	17.5916	45.68
CM-10100	47.034	25.94	21.094	44.85
CM-10125	55.677	31.2	24.477	44.00
CM-10160	72.36	40.52	31.84	44.00
CM-10200	72.36	40.52	31.84	44.00

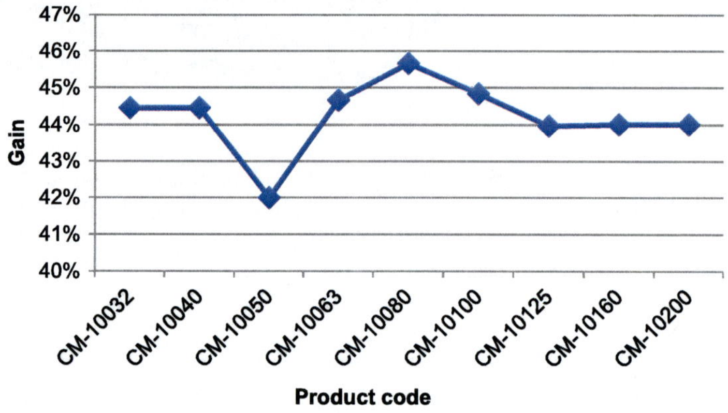

Fig. 1.5 Increased efficiency by drilling one pneumatic joint connection with the macro program. (Seçgin and Özsert 2016)

Fig. 1.6 Electricity consumption in the drilling of ten pieces

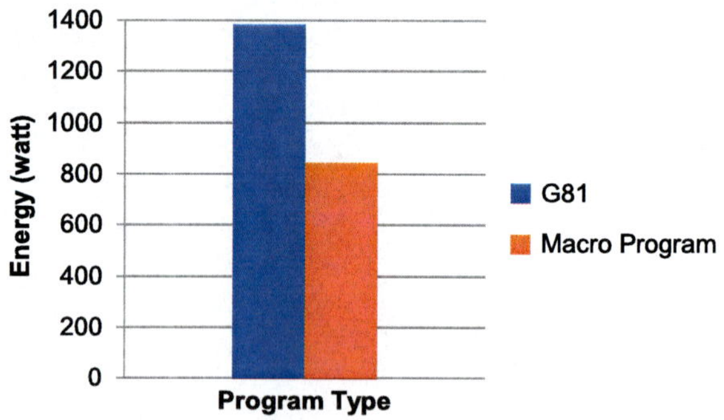

In this study, in which a developed macro program was used, ten pneumatic joint connection parts with a single hole were drilled in CNC milling. In the realized application, it was observed that while the total drilling time was improved by 40%, energy savings of 38% were achieved. For other drilling applications where the program in question can be used, the increase in efficiency is undoubtedly closely related to the void ratio in the part.

References

Bayındır, R., Demirbaş, Ş., Bektaş, A., & Çolak, İ. (2008). Observation of Electric Energy at an Industrial Plant. Erciyes Üniversitesi Fen Bilimleri Enstitüsü Dergisi, 24, 154–164.

El-Sonbaty, I., Khashaba, U. A., & Machaly, T. (2004). Factors affecting the machinability of GFR/epoxy composites. Composite Structures, 63(3–4), 329–338. https://doi.org/10.1016/S0263-8223(03)00181-8

Erten, S. (2006). Enerji Tasarrufu Davranışında Ortaya Çıkabilecek Psikolojik Ve Sosyolojik Engeller. 25. Enerji Verimliliği Konferansı, 23–28.

FANUC. (n.d.). FANUC Series 16 18 160 180 – Model B for Machining Center Operator's Manual.

Kavak, K. (2005). Energy Efficiency in the World and Turkey and Investigation of Energy Efficiency in Turkish Industry. Uzmanlık Tezi, İktisadi Sektörler ve Koordinasyon Genel Müdürlüğü.

Meral, M. E., Teke, A., & Tümay, M. (2009). Energy Efficiency in Electrical Utilities. Uludağ University Journal of The Faculty of Engineering and Architecture, 14, 31–37.

Narin, M., & Akdemir, S. (2006). Energy Efficiency and Turkey. UEK-TEK Uluslararası Ekonomi Konferansı, Türkiye Ekonomi Kurumu.

Pamir, A. N. (2003). Energy in the World and in Turkey, Turkey's Energy Resources and Energy Policies. DESEM, İzmir, 134, 2–5.

Seçgin, Ö., & Özsert, İ. (2016). A Parametric Program for Productivity of Drilling Process of Hollow Parts. 7th International Symposium On Machining, 432–442.

Söğüt, Z., İlten, N., & Oktay, Z. (2011). Examination of Energy Saving Potential Depending on Energy Audit in Tomato Paste Factory. International Advanced Technologies Symposium (IATS'11), May, 16–18.

TUİK. (n.d.). TUİK 2015 enerji istatistiği. www.tuik.gov.tr/PreIstatistikTablo.do?istab_id=1579

Web Page. (n.d.-a). Retrieved January 1, 2016, from http://www.utvguide.net/news/uploaded_images/Teryx-BilletKingRack-Clevis-1-756442.jpg

Web Page. (n.d.-b). Retrieved January 1, 2016, from http://www.mustangandfords.com/techarticles/engine/mufp_0601_ford_460_engine_build/photo_18.html

Web Page. (n.d.-c). Retrieved January 1, 2016, from http://www.univer.com.au/cylinder-mountings-cm10.htm

Chapter 2
An Energy Productivity Analysis in the Scope of the Pressure Collector System that Has Been Designed with the Hydrophore System

Fatih Uysal and Mehmet Masum Olmuştur

Nomenclature

CHS	Classic hydrophore system
DMA	District meter areas
HS	Hydrophore system
IPHS	Initial pressure hydrophore system
N_2	Nitrogen gas
WDN	Water distribution network

2.1 Introduction

In order to meet today's increasing energy crisis, renewable energy sources are being sought. In addition, scientific researches are still being conducted for the determination and consumption of new energy sources and will keep increasing as the number of people that moves on to modern living results in the increasement of energy consumption (Arun Shankar et al. 2016; Halkos and Gkampoura 2021). Therefore, it is not sufficient to find new energy sources and put them into use; it is important to examine the existing systems that consume energy and to enable them to do the same work with less energy consumption. The lamps used for lighting and automobile engines' ability to do more work despite the decrease in energy

F. Uysal (✉) · M. M. Olmuştur
Faculty of Technology, Mechanical Engineering, Sakarya University of Applied Sciences, Sakarya, Türkiye
e-mail: fatihuysal@subu.edu.tr

M. Z. Sogut et al. (eds.), *Proceedings of the 2022 International Symposium on Energy Management and Sustainability*, Springer Proceedings in Energy,
https://doi.org/10.1007/978-3-031-30171-1_2

consumption are examples of these developments. While 46% of the electrical energy produced in the world is consumed by electric motors, 22% of it is used from pump systems (Arun Shankar et al. 2016). By planning the pump operation depending on the water demand, both the electrical energy used by the pumps and the resulting carbon emissions can be reduced (Menke et al. 2016).

Advanced technologies have always been used in the supply and delivery of water, which is an indispensable substance for human life. With the modernizing world, the spread of living spaces all over the world and the increase in building heights, it has become necessary to use different technologies in water supply to houses. Energy use can increase by 15–50% if the water supply system is operated according to the maximum water use that also takes into account future uses (Nikolenko and Shvagirev 2018). Today, organizations/firms responsible for water supply bring water to the entrances of buildings at a pressure between 0.3 and 0.8 MPa. In order to reduce network failures in regions with high water pressure, district meter areas (DMA) are established to reduce the pressure when water usage decreases (KILIÇ 2021), as well as in networks with low pressure, pressure is increased at times of increased use with hydrophore systems installed on the network (Narayanan et al. 2012). Nevertheless, a minimum pressure of 0.2 MPa is needed to use water comfortably. However, the elevation differences in the landforms where the water distribution network (WDN) is located and the increase in building heights reduce the pressure of the water at the end destination. Therefore, a hydrophore system (HS) is being used to increase the water pressure. Due to the incompressibility of water in HS, a compressible element is needed. Thus, while the pressure fluctuations and pressure-induced pulses in the installation are reduced, a certain amount of domestic water can be supplied to the installation while the pump is stopped between certain pressures in start-stop (standing at minimum pressure and operating at maximum pressure) hydrophore systems. Thus, continuous operation of the pump under load is prevented. In the first hydrophore systems, the air was pumped to the booster tank by means of a compressor; hydrophore systems that have been developed thereafter have been developed in such a way that it is able to absorb the air itself. Due to the dissolution of air in water over time and the decrease in the amount of air in the tank, membrane booster systems have been developed with advances in rubber technology. Operating the pump at a minimum pressure and stopping at a maximum pressure causes it to operate in the pressure-flow ranges before and after the maximum operating efficiency, which reduces the overall pump efficiency (Fig. 2.3). In order to reduce the efficiency loss here, electronically controlled hydrophore systems that can operate the pump at variable speed at constant pressure values have been developed with the developing technology (Arslan and N 2006; Gevorkov et al. 2016). Some of the pressure in the hydrophore systems, which can provide the water at the pressure required for comfortable use, compensates for the friction losses in the pipes that increase depending on the flow, and the stopping pressure of the system is usually adjusted to meet the losses at maximum consumption (Cabrera et al. 2015). However, water use is often not at the maximum level. By changing the set pressure of the hydrophore system, energy consumption can be reduced by 27% (Diaz et al. 2017a, 2017b). However, in these systems, the response time of the pressure change should be evaluated so that the comfort does not deteriorate. In hydrophore systems, the speed control of the pump and the adjustment of the pressure and speed cause the pump to operate with low efficiency, especially because the high fan speed, which should be at high pressure values, cannot be achieved. In this case, the energy consumption per unit water volume of the pump increases. Therefore, studies are carried out to keep the pressure of the hydrophore tank constant and to examine the energy efficiency accordingly (Latchooomun et al. 2019). Optimization of different types of pumps used in booster systems, can be easily done by MATLAB SIMSCAPE etc. programs (Nikolenko and Ryzhakov 2020).

In this study, instead of the classic hydrophore system (CHS), which consists of a tank, pump and booster tank and resets the pressure of the water in the tank (at the pump inlet) and repressurizes it, an initial pressure hydrophore system (IPHS) is proposed, with the addition of a pressure collector to the same system, which transfers the pressure in the WDN into the pump without resetting it. Models of both systems were created in the MATLAB SIMSCAPE program, and the energy consumption during the first filling of the hydrophore tank was compared.

2.2 The Modelling of the Hydrophore Systems

Within the scope of the CHS, the water deriving from the WDN is being emptied within the water tank (2), being sucked up via an electric pump (5) and being pumped into the facility. In case the tank is full, the water is closed by the floater (1). The accumulator is mounted in order to benefit from the compression feature of the gases over the discharge line (8). This way, a certain volume of water can be stored between certain pressures by utilizing the compression feature of the gas in the accumulator (generally N_2 or air is being used). As a result, the HS does not need to operate constantly. The operation and stopping of the electric pump at the set pressure is provided by the control unit (7). In case of a breakdown in the hydrophore system, WDN and the installation are connected to each other via check valves (3) so that low pressure water can be supplied to the installation with WDN pressure. This way, water can come to the installation from the side where the pressure is high. In any case of a breakdown, valves (4) are being shut down and accumulators (8) are being each individually maintained (Fig. 2.1).

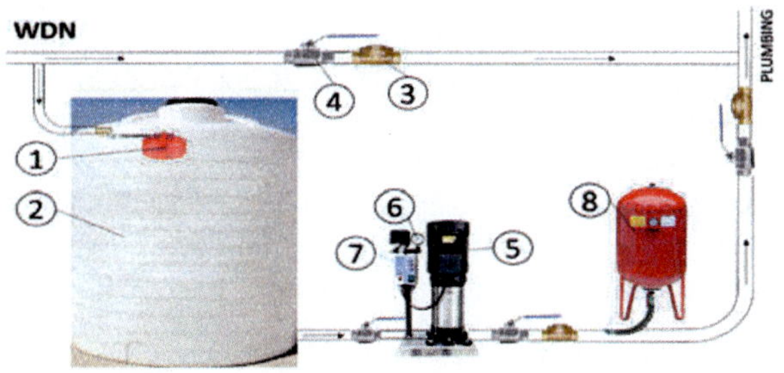

Fig. 2.1 CHS working principle

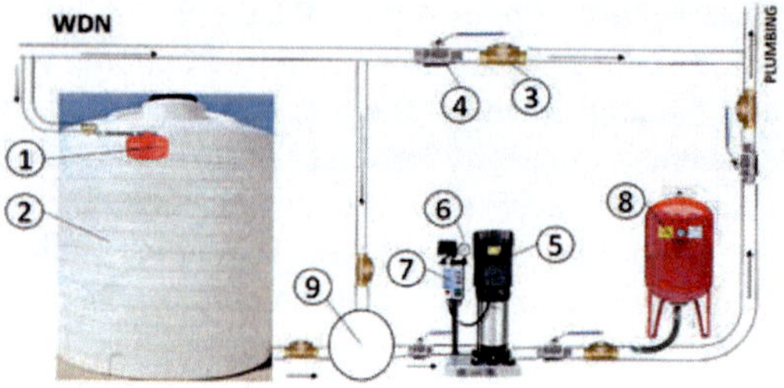

Fig. 2.2 The working principle of the IPHS

HS components:

1. Floater
2. Tank
3. Check valve
4. Valve
5. Electric pump
6. Manometer
7. Control unit
8. Accumulator (hydrophore tank)
9. Pressured collector (for Fig. 2.2)

Pressure collector has been added additionally in IPHS (9) (Fig. 2.2). This way, pressured water deriving from the WDN is being transferred directly to the electric pump without losing it in the water tank first whereby the water pressure is being transferred to the pump entrance. As a result, the pressured water of the network and the electric pump will be able to work as a series pump (Çengel and Cimbala 2008), and the pressure energy will be evaluated by HS without resetting it.

The centrifugal pump graphics are usually existing out of three parts. The first is the head-flow, the second is the efficiency-flow, and the third is the engine power-flow curves (Fig. 2.3). Pressure-flow curves start at a maximum working pressure (pressure) and increase as pressure decreases. In the engine power-flow graph, the increase in engine power continues as the flow rate increases. The efficiency-flow curve emerges in a completely different way from these two curves. When the flow is on zero level (the maximum delivery head), the efficiency will be on zero level as well and the pressure will decrease whereby the flow will increase and the pump feat will parabolic increase as well. After the efficiency reaches its peak, it starts to decrease and decreases parabolic. Therefore, the basic parameter in determining the working places of the pumps is efficiency, and pumps should be selected with the highest efficiency they can operate (Çengel and Cimbala 2008; Yalcin 1998). Pump measurements should not be made when the flow level is below a specific level. Other pumps should be chosen when the flow level will be below such level. It is easier to choose the right pump when the suck and pressure heights are constant. However, it is necessary to consider energy efficiency in systems where the head is variable, such as the hydrophore system. But it is a difficult process to

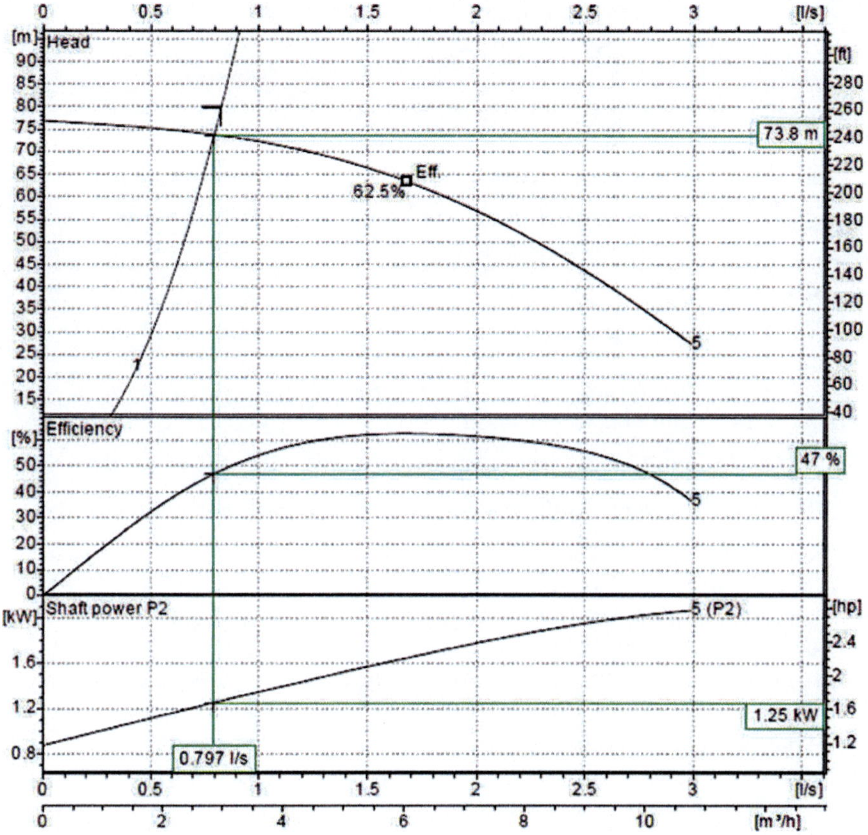

Fig. 2.3 The curve of a centrifugal pump; pressure relative to flow (H_m), efficiency (η) and power (P) curves ("Pump Selection" 2015)

mathematically establish and solve the energy equation of a system where the head, flow and efficiency vary depending on each other. Therefore, the computer programme MATLAB SIMSCAPE has been used to conduct the necessary analysis.

A centrifugal pump for the modelling of the CHS within the computer programme MATLAB SIMSCAPE has been used. An asynchronous motor has been used for the drive of the centrifugal pump. In addition, in order to provide water to the centrifugal pump, a 10-ton tank has been used. Consequently, an accumulator has been mounted at the exit of the pump centrifuge and in order to avoid any leak back to the pump, and a check valve has been placed between the centrifugal pump and the accumulator. This way the CHS model has been designed within the computer programme MATLAB SIMSCAPE (Fig. 2.4). Since the study was on the analysis of energy efficiency during the first filling of the accumulator, no accumulator output was created. Since the elements used are thermal liquid elements, the heat outputs are isolated.

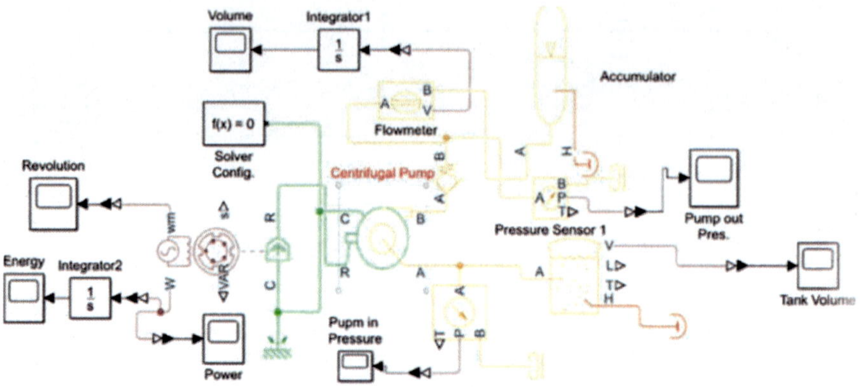

Fig. 2.4 CHS MATLAB SIMSCAPE model

The parameters within MATLAB SIMSCAPE have been regulated as follows:

- Pump parameterization: 1D tabulated data head and brake power vs. capacity.
- Mechanical orientation: positive.
- Capacity vector: [0, 12, 16.67, 30, 40, 53.33, 66.67, 83.33] Lpm.
- Head vector: [65, 60, 50, 47.5, 45, 40, 35, 25] m.
- Brake power vector: [0.4, 0.5, 0.6, 0.75, 0.85, 0.93, 0.98, 0.99] kW.
- Reference density: 998 kg/m^3
- Reference angular speed: 2900 rpm
- Angular speed threshold for flow reversal: default.
- Cross-sectional area of connecting pipe: 0.01 m^2.

The accumulator tanks' parameters were as follows:

- Total accumulator volume: 70 L.
- Minimum gas volume: 10 L.
- Pre-charge pressure (gauge): 0.1 MPa.
- Specific heat ratio, hard-stop stiffness coefficient and hard-stop damping coefficient: default.
- Cross-sectional area of port A: 0.01 m^2.

The parameters of the electric motor are as follows:

- Model parameterization: by motor ratings.
- Rated mechanical power: 2000 W.
- Rated speed: 2900 rpm.
- Rated RMS line-to-line voltage: 400 V.
- Rated supply frequency: 50 Hz.
- Rated RMS line current 3.5 A.
- R1 parameterization and star connections: default.
- Number of pole pairs: 1.
- Number of phases: 3.

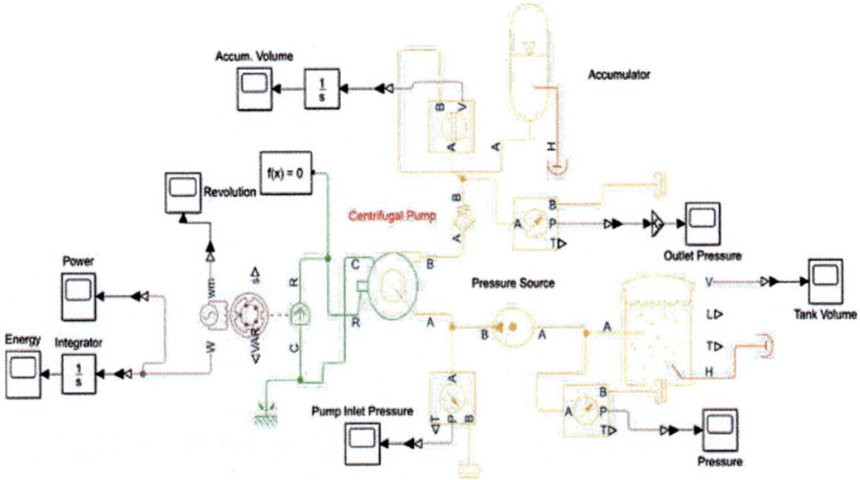

Fig. 2.5 IPHS MATLAB SIMSCAPE model

The measurements were taken from the normal booster system created as follows:

- The inlet and outlet pressures of the centrifugal pump with pressure sensors.
- The tank volume.
- The accumulator volume.
- The feat of an ordinary asynchronous motor, motor power and the energy consumption volume of the motor.

As the water level of the accumulator and the energy consumption of the motor could not directly be measured, it has been monitored by using integral components.

In the IPHS, the same measurements were made in the CHS. However, a pressure source symbolizing the pressure of the pressure collector has been added to this system. The pressure source in the MATLAB SIMSCAPE model provides the system at the set pressure and the pressure required by the system without loss (Fig. 2.5). To understand whether the pressure source is working well, a pressure measurement has been made at the entrance and exit of the pump by mounting a 0.2 MPa pressure at the entrance of the pump.

2.3 Result and Discussion

Within the models that have been created in MATLAB SIMSCAPE, the CHS model has been simulated with 40 s and the IPHS has been simulated with 20 s. Within the time frame of both models, both systems have stored water in the accumulator at the same pressure and flow. This way the energy consumption within both the systems has been compared.

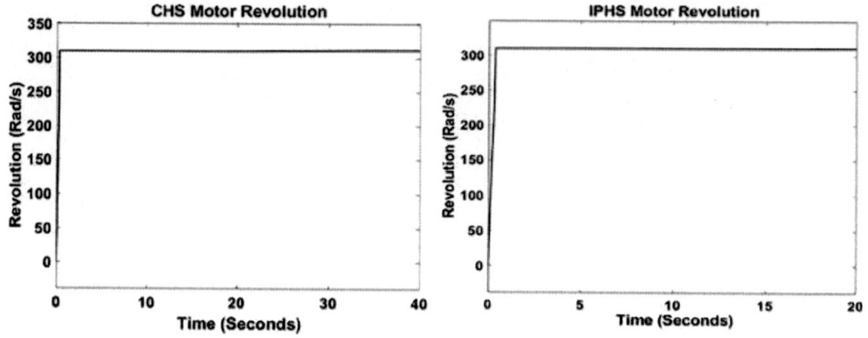

Fig. 2.6 Motor revolution of the CHS and IPHS

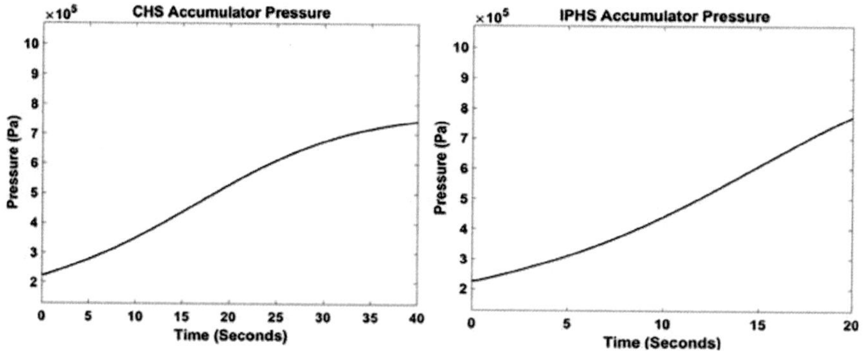

Fig. 2.7 Accumulator pressure of the CHS and IPHS

One of the indicators that is showing the fact that the simulation is working correctly is the speed indicator of the asynchronous motor as that one is at the nominal speed. If the engine encounters a load higher than its normal load, there will be differences in its speed. Therefore, when the operating speed of the simulation system is being checked, it is seen that after a very short time at the beginning of the simulation, the motor reaches its nominal speed of 305 rad/s and continues to operate at this speed (Fig. 2.6).

HS is operated at a certain pressure and stopped at its maximum pressure. Since the energy consumption during the first filling of the accumulator was investigated in the simulation, it was ensured that both hydrophore systems were stopped at approximately the same pressure values. While the normal hydrophore system reaches a pressure value of 0.75 MPa in 40 s, the recommended system achieves the same pressure value in approximately 20 s (Fig. 2.7). Due to the structure of the accumulator, it has been expected that the same pressure within the system will store the same volume of water. When the curves are examined, it is seen that the pressure

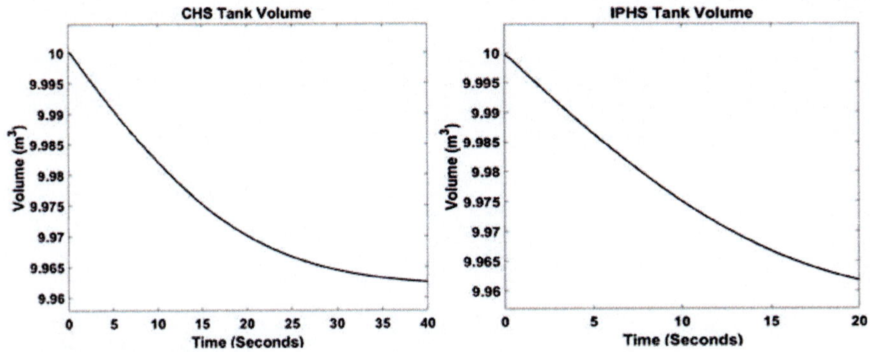

Fig. 2.8 Tank volume of the CHS and IPHS

curve of IPHS is more linear (Fig. 2.8), the curve of CHS is linear in the first 20 s but later becomes horizontal and the pressure residue decreases over time. The reason for this situation is that the flow rate of the pump operating in CHS decreases as the pressure rises and the flow approaches zero as it approaches the maximum pressure. In IPHS, on the other hand, due to the pre-pressure of 0.2 MPa, the pump-operating pressure reaches the maximum level, and therefore a pressure of 0.75 MP is reached before the pressure approaches zero. This is in line with the operating curve of the pump (Fig. 2.3) and confirms the increase in energy efficiency (Yalcin 1998).

Although there is no need for a 10-ton tank to test the hydrophore system, such a tank has been chosen because the hydrophore systems are considered as a backup water source. This situation is compatible with normal hydrophore applications. The decrease in the tank level should be the same as the amount of water pumped into the accumulator. Thus, the accuracy of the established model is shown. The amount of water withdrawn in 40 s in CHS is 35.5 litres, which is approximately the same as the amount of water drawn in 20 s in the recommended hydrophore system. The former results have revealed that both the systems are being simulated under the same conditions. The amount of water withdrawn in the first 20 s in CHS is approximately 30 litres, and there is a vertical decrease in the tank level until this time. After 20 s, the decrease in the tank level starts to become horizontal and the amount of water withdrawn in the remaining time is only 5.5 litres. In IPHS, on the other hand, the decrease in tank level continues vertically for 20 s (Fig. 2.8). This is in line with the operating curve of the pump. According to the results, it is revealed that for HS pumps to work efficiently, they must be stopped at a pressure level slightly away from the maximum pressure levels.

The water that was filling up the accumulator was harmonious with the volume of water that was decreasing in both models. While the water within the CHS has filled up with 30 litres at 20 s, the volume of water was 5.5 litres during the remaining 20 s. In the IPHS the water volume within the accumulator was 35.5 litres around 20 s. This proves that the established models work correctly (Fig. 2.9).

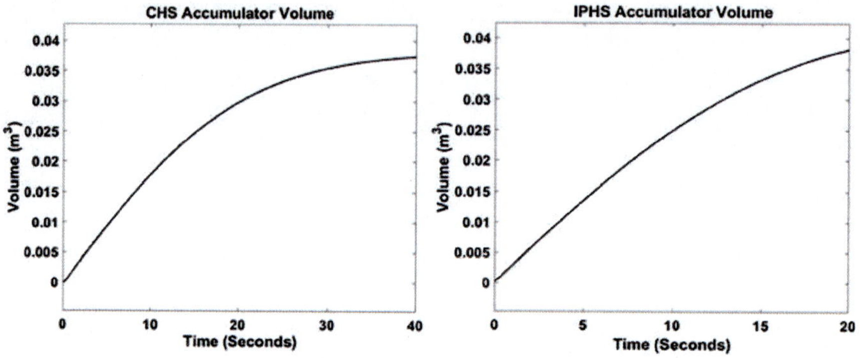

Fig. 2.9 The accumulator volume of the CHS and IPHS

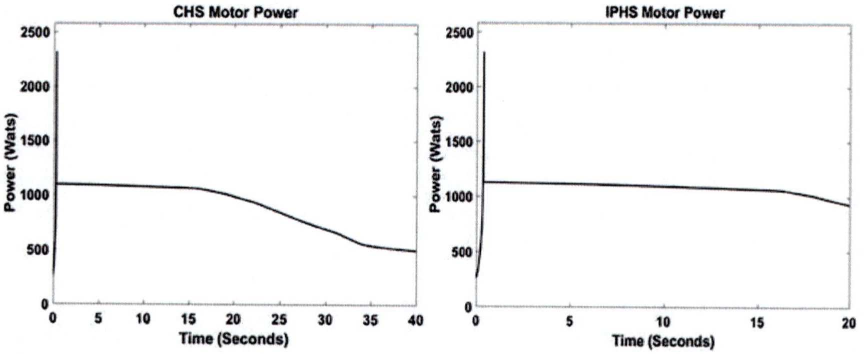

Fig. 2.10 CHS and IPHS motor power

Asynchronous motors can draw up to five times their rated current at start-up. This is clearly seen in the graphs with a maximum power of 2.3 kW (Fig. 2.10). In a short time, the engine power reaches the nominal power of 1 kW and decreases over time. Since the engine speed comes to normal in a very short time, the initial overcurrent does not affect the energy consumption much. However, considering that the hydrophore systems start and stop very frequently depending on the volume of the accumulator tanks, it is a fact that high starting currents will also increase energy consumption in long-term use. Therefore, it reveals that the volume of accumulators should be optimized in terms of cost-accumulator volume according to the installed system. Since the flow is high and the pressure is low at the beginning of the graph, the power is also high, and as the pressure rises, the flow and engine power decrease (Fig. 2.10). This is in line with the operating curve of the pump (Fig. 2.3).

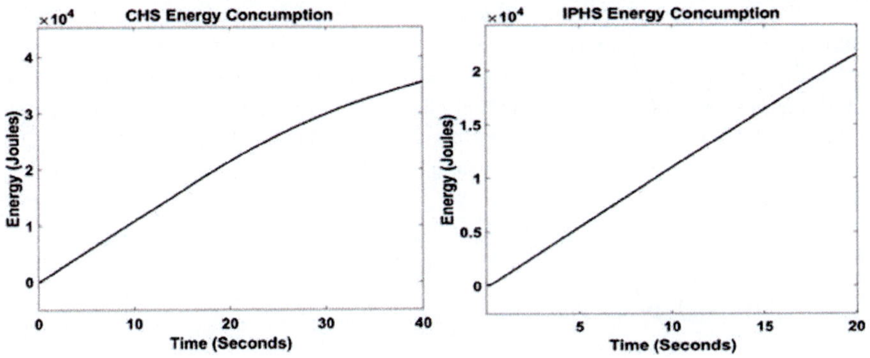

Fig. 2.11 CHS and IPHS energy consumption graphics

It is expected that the initial pressure will contribute to the starting of the engine and the starting power will be lower. However, there is no such thing in the graphic. This means that the simulation programme has not taken the initial pump pressure into consideration. In addition, the initial pressure in the accumulator does not have any effect on engine power. Therefore, it is being expected that in real practice the results of both the CHS and the IPHS will be different. Engine starting power can be expected to be higher in CHS and lower in IPHS. This situation can be investigated with an experimental study.

The total energy consumption of the HS is being realized by the electric motor. In the MATLAB SIMSCAPE programme, the time-dependent power graph is integrated to show the energy consumption (Figs. 2.4 and 2.5). Thus, a graph of time-dependent energy consumption was obtained. Since the duration of the excessive power at the start of the motor is very short, it is not reflected in the energy graph as instantaneous consumption. This situation can be expected to be different from the actual application. The results of both systems have shown the same amount of energy consumption around in 20 s. However, since the normal hydrophore system did not reach the desired pressure, it continued to operate and therefore the energy consumption continued. As a result, the CHS had to consume 45% more energy to have the same pressure as the IPHS (Fig. 2.11). Although CHS runs twice as long as IPHS, its energy consumption has not doubled. The reason for this is that as the pressure in the normal hydrophore system increases, the flow rate of the pump decreases and accordingly the power drawn from the engine decreases. The reason for the pressure difference is the 0.2 MPa inlet pressure in the proposed system. As this pressure had worked just as the pump series (Çengel and Cimbala 2008), it had stored the water within the accumulator with less pressure and higher flow and therefore the energy consumption has decreased. This is compatible with the operating characteristic of the pump (Fig. 2.3).

Although the energy consumed by both systems at the end of 20 s is close to each other, the pressure of CHS is 0.5 MPa, the pressure of IPHS is 0.76 MPa, the volume of water pumped by CHS is 30 litres and the amount of water pumped by IPHS is 38 litres. It took 40 s for the pressure and flow values of the CHS to reach the values of the IPHS. This increases the energy consumption of the CHS to 3.45 kJ during the initial charging of the accumulator, while the energy consumption of the IPHS remains at 2.15 kJ.

2.4 Conclusion

In the study, a normal water booster system was pre-pressurized with a pressurized collector and the energy consumption of the accumulator was analysed in the MATLAB SIMSCAPE programme.

The results have been as follows:

- The IPHS was consumed 45% less energy than the CHS.
- When the pressure had been defined as energy per unit volume, the energy that enters the pump decreased the pump's energy consumption.
- Another reason for the reduction of the energy consumption was the fact that the pump has operated in a more efficient area on the pump's operating curve. Therefore, in the selection of electric pumps for HS, the efficiency of the HS between operating and stopping pressures should have been considered without considering a point in the operating curve of the pump.
- It was expected that the initial pressure given to the pump inlet would reduce the consumed energy during the first engine start-up. However, this is not seen in the power and energy consumption graphics. Experimental examination of this situation may yield different results.

The research has revealed that analyses of the energy consumption at the places where the HS has been installed should be made in accordance with the operating parameters of the system. In today's conditions, where energy costs are increasing rapidly, making this evaluation will contribute to the reduction of energy costs.

In the next study, the results obtained by applying IPHS on a CHS will be compared.

References

Arslan Ş, N EÜİ. Analysis of a hydrophore system based on mechatronics. Sak Univ J Sci. 2006;10: 73–80.

Arun Shankar VK, Umashankar S, Paramasivam S, Hanigovszki N. A comprehensive review on energy efficiency enhancement initiatives in centrifugal pumping system. Appl Energy [Internet]. Elsevier Ltd; 2016;181:495–513. Available from: https://doi.org/10.1016/j.apenergy.2016.08.070

Cabrera E, Gómez E, Cabrera E, Soriano J, Espert V. Energy Assessment of Pressurized Water Systems. J Water Resour Plan Manag. 2015;141:04014095.

Çengel YA, Cimbala JM. Fluid Mechanics: Fundamentals and Applications. Third. Stenquist B, Lorrine B, editors. New York: McGraw-Hill; 2008.

Diaz C, Ruiz F, Patino D. Analysis of water booster pressure systems as dispatchable loads in smart-grids. 2017 IEEE PES Innov Smart Grid Technol Conf Eur ISGT-Europe 2017a - Proc. 2017;2018-Janua:1–6.

Diaz C, Ruiz F, Patino D. Modeling and control of water booster pressure systems as flexible loads for demand response. Appl Energy [Internet]. Elsevier Ltd; 2017b;204:106–16. Available from: http://dx.doi.org/https://doi.org/10.1016/j.apenergy.2017.06.094

Gevorkov L, Vodovozov V, Lehtla T, Raud Z. Hardware-in-the-loop simulator of a flow control system for centrifugal pumps. Proc – 2016 10th Int Conf Compat Power Electron Power Eng CPE-POWERENG 2016. IEEE; 2016;472–7.

Halkos GE, Gkampoura EC. Evaluating the effect of economic crisis on energy poverty in Europe. Renew Sustain Energy Rev [Internet]. Elsevier Ltd; 2021;144:110981. Available from: https://doi.org/https://doi.org/10.1016/j.rser.2021.110981

KILIÇ R. Effectıve management of district meters areas to reduce physıcal losses. Eur J Sci Technol. 2021;306–15.

Latchooomun L, Sockalingum T, Poulle KV, King RTFA, Busawon KK, Barbot JP. Design of a water pressure boosting system for pressure-driven demand in a distribution network. Proc 2018 5th Int Symp Environ Energies Appl EFEA 2018. 2019;

Menke R, Abraham E, Parpas P, Stoianov I. Demonstrating demand response from water distribution system through pump scheduling. Appl Energy [Internet]. Elsevier Ltd; 2016;170:377–87. Available from: http://dx.doi.org/https://doi.org/10.1016/j.apenergy.2016.02.136

Narayanan I, Sarangan V, Vasan A, Srinivasan A, Sivasubramaniam A, Murt BS, et al. Efficient booster pump placement in water networks using graph theoretic principles. 2012 Int Green Comput Conf IGCC 2012. 2012;

Nikolenko I, Shvagirev P. Analysıs Of Methods Of Improvıng Energy Effıcıency Of Pumpıng Statıons Power Unıts Of Water Supply Systems. Internatıonal Q J [Internet]. 2018;7:67–77. Available from: https://bibliotekanauki.pl/articles/410678

Nikolenko I V., Ryzhakov AN. Parallel Operation Mode Optimization of Different-Type Centrifugal Pumps of a Water Supply Booster Pumping Station. 2020 Int Conf Dyn Vibroacoustics Mach DVM 2020. 2020;

Pump Selection [Internet]. Standart. 2015. Available from: https://sps.standartpompa.com/bin/standartpompaV3.dll?RQID=DF60310F556B42AD9FC9B037DA61B7DD

Yalcin K (Trakya U. Volumetric and Centrifugal Pumps. First. İstanbul: Caglayan; 1998.

Chapter 3
Industrial Energy Efficiency Lessons from Past Experience for Today and Tomorrow

Levent Kılıç, Gürhan Dural, Hilal Bilgin, and Bora Tüzer

Nomenclature

KPI	Key performance indicators
SCADA	Supervisory control and data acquisition

3.1 Introduction

Each energy conversion is basically carried out in a certain way, reducing the associated loss in each step to increase the overall energy efficiency (ISO 50001 2018; ISO 50006 2014; BS EN 17267 2019).

Although technologies are diverse not only in new investments but also in effective and efficient management, proven and valid principles must be followed. The basic criteria can be divided into four components: economic, technological, environmental and social. While modelling, to show them in sub-criteria as seen in Table 3.1 will be much appropriate (Andrei et al. 2022).

Effective energy management means that the users are ensured to receive the required energy, which should be supplied at the lowest cost with the desired quality.

L. Kılıç (✉) · B. Tüzer
Şişecam R & D Energy Efficiency Unit, Şişecam R & D Center, Kocaeli, Turkey
e-mail: lkilic@sisecam.com; btuzer@sisecam.com

G. Dural
7Cbasalia, Istanbul, Turkey
e-mail: gurhan.dural@7cbasalia.com

H. Bilgin
Yahyakaptan Mah. G1 8B Kocaeli, Izmit, Turkey

M. Z. Sogut et al. (eds.), *Proceedings of the 2022 International Symposium on Energy Management and Sustainability*, Springer Proceedings in Energy,
https://doi.org/10.1007/978-3-031-30171-1_3

Table 3.1 Main and sub-criteria (Andrei et al. 2022)

Economical	Technological	Environmental	Social
Investment costs	Economically lifetime	Area occupied	Employment
Operational costs	Capacity usage	CO_2 emissions reducing	
Production costs	Recycling time	Reducing wastes	

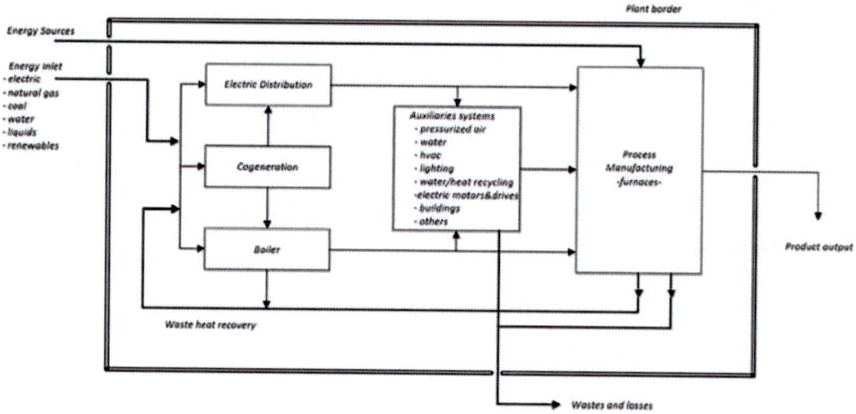

Fig. 3.1 Typical plant layout for energy flow

Of course, all of these goals need to be achieved while properly protecting both production and environmental needs.

The energy management programs must include four main steps: (1) analysis of historical data, (2) energy audits and accounting, (3) engineering analyses and investment proposals based on feasibility studies and (4) staff training and information.

In order to control energy efficiency in each stage, several key performance indicators (KPIs) should be defined before and after taking each action. Common components across all key success indicators are shown as follows:

- Profit.
- Time.
- Technological efficiency.

The main steps and indicators listed above need to be carefully implemented and correlated to achieve the best results.

Industrial facilities with high energy density and consumption, seen in Fig. 3.1, try to solve the efficiency problem by strengthening their processes and auxiliary facilities, seen in Fig. 3.2 (Wei et al. 2016). That carrying out studies within the scope of energy efficiency standards in all components, starting from the factory boundary conditions in the industrial facilities, is important.

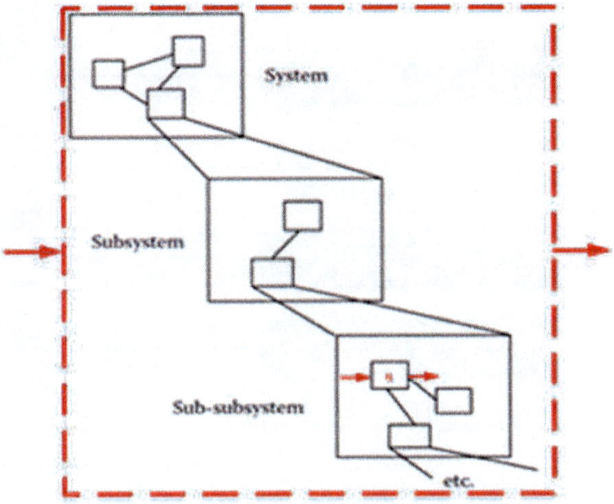

Fig. 3.2 Multiple efficiency levels of systems. (Wei et al. 2016)

The purpose of this article is to provide an overview of energy conversion and management by monitoring the energy flow of some systems and equipment within the field boundaries. Energy efficiency can be done at any point and level. For this, all issues should be reduced to the core by presenting a few basic formulas and data (Giacone and Manco 2012).

Energy crises and fluctuations in energy, which seriously affect the industrial age from time to time, have made companies and final consumers aware that the energy problem exists and will always exist. Therefore, the choice of energy-saving technologies will significantly affect not only the system it serves but also the entire facility.

3.2 Industrial Systems and Equipment

3.2.1 Compressed Air Systems

The formula for energy efficiency can simply be defined as the ratio of output to input. On the basis of compressors, since the output will be the produced air and the input will be the electrical energy consumed, the calculation will be electrical energy consumed per unit of air produced (Kluczek and Olszewski 2017; Nehler 2018).

As other technical criteria for compressors:

- Capacity (providing the specified flow according to atmospheric data under operating conditions ([Nm^3/h]).
- Energy consumption/cost/efficiency (amount of electrical energy used to produce 1 unit of air [kWh/Nm^3]).
- 1-, 3- and 10-year maintenance costs

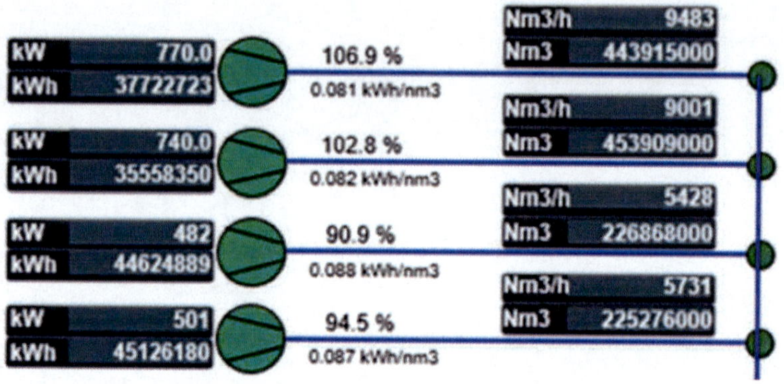

Fig. 3.3 Provided four-bar KPI levels (min. 25% better)

- Field experience on brand/product and ease of operation and common use of existing spare parts in the field.
- Domestic/international service, supply availability and response time.
- Other priority operating conditions (heat exchanger pipe type, etc.)

On average, compressors at different pressure levels consume up to 40–50% of industrial electricity consumption depending on the melting process. The influencing factors in Fig. 3.3 show the 25% energy efficiency achieved on a four-bar line.

Another energy efficiency method is to check and fix leak points. Measured 7061 points show that the main problems of leaks are:

- Pneumatic fittings → 43%
- Mechanical fittings → 10%
- Regulators → 13%
- Pipes → 9%
- Piston and valves → 17%
- Others (manometers, bellows, etc.) → 8%

3.2.2 *Electrical Machines*

Apart from mandatory criteria such as power range, speed, insulation level, etc., other criteria to be considered when purchasing an engine are:

- Energy efficiency.
- Initial purchase price.
- Operating costs such as maintenance, spare parts, storage, etc.

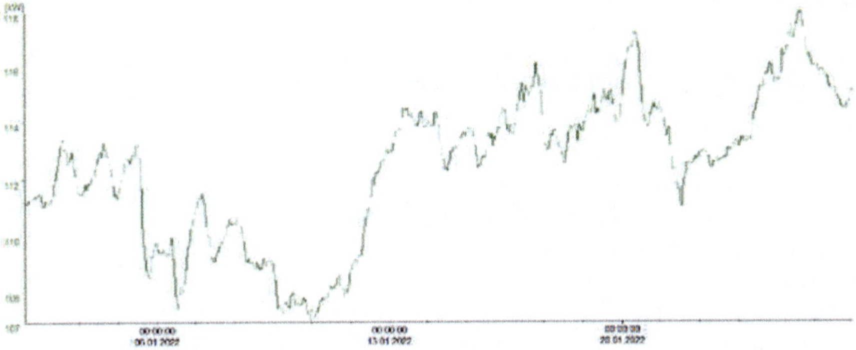

Fig. 3.4 Variable loading of random electrical machine

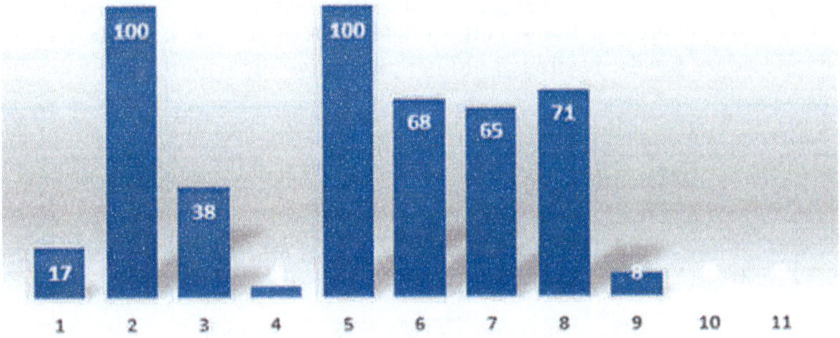

Fig. 3.5 Pilot project by 10 plants for 30,000+ electrical motor replacement

	electrical machine efficiency			
yearly running hours	≤ IE1 (EFF2)	IE2	IE3	IE4
4500	replace by IE3 - IE4	winding once	rewinding	rewinding
6000	replace by IE3 - IE4	winding once or replace	rewinding	rewinding
8760	replace by IE3 - IE4	replace by IE3 - IE4	rewinding	rewinding

Fig. 3.6 When to rewind or replace of electrical motors

Selection and efficiency numerical data from the energy monitoring system supervisory control and data acquisition (SCADA) can be used in real-time analysis.

- Capacity utilization is between 40% and 80%.
- Especially, the intersection points of the fans and pumps are effective on the efficiency class of the electrical machine (Fig. 3.4).

Variable loading of random electrical machine is shown in Fig. 3.4.

Data from the pilot projects in Figs. 3.5 and 3.6 provide technical information on how to decide on new installations and/or rewinding or replacing electrical motors.

3.2.3 Lighting

It is important to pay attention to the following criteria in lighting:

- Net lumens.
- Power.
- Total current harmonic distortion (THDI%).
- Total voltage harmonic distortion (THDV%).
- Power factor (PF) and CosΦ.
- VA.
- Colour rendering.
- Actually measured colour temperature.
- Drive, led brand and model.

After the above qualification is achieved, max $\left(\frac{Lm}{W}\right)$ ratio will be decisive for the best performance. It is a criterion that all systems and equipment should be not only efficient but also economical and comply with the conditions in Table 3.1.

Compiled data for the reference old lamps is given in Table 3.2.

Table 3.2 Compiled data for the reference old lamps

Brand	Power [W]	Ballast loss [W]	Total power [W]	Lumen	Lamp efficiency	Lamp lumen	Life	System efficiency
Mercury vapour	125	14	139	6300	65%	4095	8000	29.46%
Mercury vapour	250	24	274	13,000	60%	7800	8000	28.48%
Mercury vapour	400	35.4	435.4	22,000	60%	13,200	8000	30.31%
Metal halide	150	17	167	12,000	75%	9000	6000	53.89%
Metal halide	250	25.2	275.2	19,000	60%	11,400	6000	41.42%
Metal halide	400	30	430	31,000	60%	18,600	6000	43.25%
Metal halide	1000	72	1072	85,000	85%	72,250	4000	67.39%
Sodium vapour	70	12	82	5600	50%	2800	12,000	34.14%
Sodium vapour	150	25	175	17,500	85%	14,875	12,000	85.00%
Sodium vapour	250	30	280	33,200	85%	28,220	12,000	100.00%
Fluorescent	18	2.5	20.5	1350	65%	878	15,000	42.82%
Fluorescent	36		36	3350	70%	2345	15,000	65.13%
Fluorescent	58		58	5200	70%	3640	15,000	62.75%

3.2.4 *Power Quality*

The energy quality problems caused by the failure of the energy distribution systems affect not only the energy quality problems but also the energy efficiency of the facility. When it comes to energy efficiency in industrial plants, power quality events create negative sequence components that cause power loss in conductors and electric motors and fail like relays in all systems. Moreover, they can damage system components. The criteria are clearly defined in the TS EN 50160 standard (TS EN 50160 2011; Karthick et al. 2021).

The power quality statistics of 2021 for the connection points of 12 plants located in different industrial zones to the grid shows that there is much to be done for improvement as follows:

- Average 3.94 power outages with min.→ 0 and max.→ 12
- Average 144.54 times voltage drop with min.→ 44 and max.→ 382
- Average 123.17 times voltage increase with min.→ 0 and max.→ 425

All of them have a heavy impact not only on electrical equipment but also on energy efficiency.

3.3 Conclusion

In this study, the key success indicators that should be standardized and implemented in order to increase energy efficiency in industrial facilities have been tried to be given with methodological and numerical examples. It is important to take not only technological but also economic, environmental and social criteria as a reference in product/equipment selection.

Real-time monitoring will increase energy efficiency by tapping KPIs quickly. Energy efficiency will ensure the effective use of the limited resources of the country with engineering analysis, as well as the purpose of existence of the facilities that provide financial gain.

References

Andrei M, Thollander P, Sannö A (2022) Knowledge demands for energy management in manufacturing industry – A systematic literature review. Renewable and Sustainable Energy Reviews. https://doi.org/10.1016/j.rser.2022.112168

British Standards Institute (2019) Energy measurement and monitoring plan for organisations – Design and implementation (BS EN Standard No. 17267:2019, Final Draft).

Türk Standardları Enstitüsü (2011) Genel elektrik şebekeleri tarafından sağlanan elektriğin gerilim karakteristikleri (TS EN Standard No. 50160). https://intweb.tse.org.tr/Standard/Standard/Standard.aspx?081118051115108051104119110104055047105102120088111043113104073100047113049066103069117067088079

Giacone E, Manco S (2012) Energy efficiency measurement in industrial processes. Energy 38(1): 331–345. https://doi.org/10.1016/j.energy.2011.11.054

International Organization for Standardization (2018) Energy management systems – Requirements with guidance for use (ISO Standard No. 50001:2018). https://www.iso.org/standard/69426.html

International Organization for Standardization (2014) Energy management systems – Measuring energy performance using energy baselines (EnB) and energy performance indicators (EnPI) – General principles and guidance (ISO Standard No. 50006:2014). https://www.iso.org/standard/51869.html

Karthick T, Charles Raja S, Jeslin Drusila Nesamalar J, Chandrasekaran K (2021) Design of IoT based smart compact energy meter for monitoring and controlling the usage of energy and power quality issues with demand side management for a commercial building. Sustainable Energy, Grids and Networks. https://doi.org/10.1016/j.segan.2021.100454

Kluczek A, Olszewski P (2017) Energy audits in industrial processes. Journal of Cleaner Production 142:3437-3453. https://doi.org/10.1016/j.jclepro.2016.10.123

Nehler T (2018) Linking energy efficiency measures in industrial compressed air systems with non-benefits – A review. Renewable and Sustainable Energy Reviews 89:72-87. https://doi.org/10.1016/j.rser.2018.02.018

Wei M, Hong SH, Alam M (2016) An IoT-based energy management platform for industrial facilities. Applied Energy 164:607-619. https://doi.org/10.1016/j.apenergy.2015.11.107

Chapter 4
A Brief Comparison of Risk Analysis Methods for Fuel Cell Ships

Omer Berkehan Inal, Muhammed Fatih Gulen, Cengiz Deniz, and Murat Mert Tekeli

Nomenclature

AHP	Analytic hierarchy process
ETA	Event tree analysis
FMEA	Failure mode and effect analysis
FRAM	Functional resonance analysis method
FTA	Fault tree analysis
GHG	Greenhouse gas
HAZOP	Hazard and operability analysis
IMO	International Maritime Organization
MCFC	Molten carbonate fuel cell
PEMFC	Proton exchange membrane fuel cell
SOFC	Solid oxide fuel cell

4.1 Introduction

Ship-sourced harmful emissions are being limited by International Maritime Organization (IMO) in order to slow down global warming. Greenhouse gas emissions from shipping are mainly from the massive usage of fossil fuels. According to IMO,

O. B. Inal (✉) · C. Deniz
Marine Engineering Department, Maritime Faculty, Istanbul Technical University, Tuzla/Istanbul, Turkey
e-mail: inalo@itu.edu.tr

M. F. Gulen · M. M. Tekeli
Department of Maritime Transportation and Management Engineering, Maritime Faculty, Istanbul Technical University, Tuzla/Istanbul, Turkey

M. Z. Sogut et al. (eds.), *Proceedings of the 2022 International Symposium on Energy Management and Sustainability*, Springer Proceedings in Energy,
https://doi.org/10.1007/978-3-031-30171-1_4

worldwide shipping consumes 300 million tons of fossil fuel annually, approximately (Zincir and Deniz 2021). This consumption is principally separated into three different fuel types: 72% heavy fuel oil, 26% marine diesel oil, and 2% liquefied natural gas (Inal et al. 2022). This consumed fossil fuel amount causes an increase in the CO_2 from 713 mt in 2012 to 755 mt in 2018 according to the fourth IMO GHG Study. The aim of reducing greenhouse gas emissions by at least 50% by 2050 is a real struggle in the short-term and mid-term for the maritime industry. Therefore, more ecofriendly power generation sources are being sought by maritime stakeholders. In this perspective, marine alternative fuels such as methanol, ammonia, or hydrogen and alternative power generators such as fuel cells, batteries, and solar panels are gaining importance to reduce emissions.

Zero-emission electric propulsion by batteries and supercapacitors offers a good solution, but the limitations on the energy storage side cause a problem when the typical oceangoing ships are considered. Hybridization of energy storage devices with conventional marine diesel engines is another efficient solution; however, while the system complexity increases, energy management strategy becomes vital. For this reason, to determine the typical operation profile of the case, ship is essential to establish a proper energy management system. Furthermore, it is impossible to achieve the ultimate zero-emission goal with hybridization using conventional marine diesel engines, since they use fossil fuels. On the other hand, hydrogen fuel cells seem as strong options considering their lower GHG emissions and higher efficiency (Inal and Deniz 2018). However, since they are a quite new technology for the shipping industry, preliminary assessments on the onboard usage should be carried out. Although several types of research and development projects have been examined (van Biert et al. 2016), risk analysis is fundamental to establishing a commercial and fleet-based approach (Aarskog et al. 2020).

For this purpose, five qualitative and five quantitative risk assessment methods have been chosen to understand which combination fits better to evaluate the risks of a fuel cell ship. Onboard hydrogen storage, proton exchange membrane (PEM) fuel cell operational properties, hydrogen bunkering, and system reliability (high- and low-pressure piping) have been considered as the root causes of the fire or explosion risks.

As a nature of the system, each risk analysis method has different strong and weak sides. However, globally, some perform better among them. The paper focuses on the risk analysis selection for fuel cell-powered ships.

4.2 PEM Fuel Cells and Risk Assessment Methods

Fuel cells are electrochemical devices that convert the chemical energy of the fuel directly into electrical power. In both literature and industry, fuel cells are named by their electrolyte types, such as proton exchange membrane fuel cell, solid oxide fuel cell, molten carbonate fuel cell, alkaline fuel cell, and phosphoric acid fuel cells. Among different types, PEM fuel cells are one of the most promising fuel cell types with their large range of applications (Tronstad et al. 2017). They can find different

Table 4.1 Evaluated risk analysis methods

Risk analysis methods	
Qualitative	Quantitative
Fault tree analysis (FTA)	Artificial neural network
Event tree analysis (ETA)	Bayesian network
Hazard and operability analysis (HAZOP)	Analytic hierarchy process (AHP)
Functional resonance analysis method (FRAM)	Monte Carlo analysis
Bow-tie analysis	Failure mode and effect analysis (FMEA)

roles in the industry from fuel cell electric cars to stationary power supply units (Daud et al. 2017; Inal and Deniz 2020). PEM fuel cells use hydrogen in high purity as fuel and their efficiency is around 50–60% (Tronstad et al. 2017). Although the high-temperature working fuel cell types (MCFC and SOFC) can use hydrocarbons as fuel, the purity of the hydrogen for PEMs is critical.

Table 4.1 gives the selected risk analysis methods to evaluate in this paper. Fault tree analysis (FTA), event tree analysis (ETA), hazard and operability analysis (HAZOP), functional resonance analysis method (FRAM), and bow-tie analysis methods form the selected risk analysis methods, respectively. For the quantitative methods, artificial neural network (ANN), Bayesian network (BN), analytic hierarchy process (AHP), Monte Carlo analysis, and failure mode and effect analysis (FMEA) are evaluated.

The details of the methods are given briefly as follows: FTA is a deductive technique that can be used to classify the instrumental relationships leading to a specific failure mode. It is a top-down approach and is a graphical representation of the relationships between the failure modes (Whiteley et al. 2016). An event tree is a graphical representation of the possible outcomes of an incident that results from a selected initiating event. The analysis considers the responses of operators and of safety systems to the initiating event as well as random effects such as the chance of ignition of a flammable mixture when determining the potential outcomes. Event trees are often used with a qualitative assessment of the effects in a qualified risk assessment (Crawley 2020). HAZOP is a formal, qualitative, systematic, and rigorous examination of a plant, process, or operation, to identify credible deviations from the design intent in the context of the complete system, which can contribute to the realization of hazards or operability problems, by applying the experience, judgment, and imagination, stimulated by keywords, of a team (Brazier et al. 2021). FRAM is a systemic risk assessment and accident analysis method that is generated from the fundamentals of resilience engineering (Hollnagel and Goteman 2004). The FRAM describes system failures (unforeseen events) as the outcome of a functional resonance arising from the variability of normal performance. The resonance principle is called to explain that small or even insignificant variabilities can trigger each other to cause disproportionately considerable effects (Hollnagel 2013). A bow-tie diagram is a type of diagram used to model and visualize risk management and preparedness. It is a graphical tool to illustrate an accident scenario, starting from causes of the accidents and ending with its consequences (Khakzad et al. 2012). ANNs are efficient data-driven modeling tools widely used for nonlinear system dynamic modeling and identification, due to their universal approximation capabilities and flexible structure that allow capturing complex nonlinear behaviors (Shokry and Espuña 2018). A Bayesian network (BN) is a probabilistic graphical model for representing knowledge about an uncertain domain where each node corresponds to a random variable and each edge represents the conditional probability for the corresponding random variables (Ben-Gal et al. 2005). The analytic hierarchy process (AHP) is a mathematical methodology of measurement through pairwise comparisons and relies on the judgments of experts to derive priority scales, which one measures subjectivity in a mathematically objective way (França et al. 2019). Monte Carlo methods, or Monte Carlo experiments, are a broad class of computational algorithms that rely on repeated random sampling to obtain numerical results. The underlying concept is to use randomness to solve problems that might be deterministic in principle. It is frequently used to provide a quantitative approach to qualitative risk analysis methods (Patriarca et al. 2017). FMEA is a bottom-up approach to analyzing equipment, or a system, with relation to its failure events. The technique is systematic scrutiny of all of the individual ways in which a component or piece of equipment can fail and the effect of that failure on the overall system's operation. It is widely used in industry as a means to identify, rank, and mitigate against the component failure modes (Whiteley et al. 2016).

4.3 Discussion

Typical hydrogen fuel cell-powered ships comprise similar subsystems such as onboard hydrogen storage, hydrogen transfer systems and piping, hydrogen bunkering, and electrical components. Therefore, Fig. 4.1 gives the probable major risk sources that are considered during the assessment system selection.

The risks considered in this study are of both social and technical origin. Hydrogen refueling and storage hazards can occur due to equipment and human reliability. Traditional cause-effect risk assessment models are insufficient for socio-technical systems due to the complex interactions between causes (Hirose and Sawaragi 2020). Therefore, qualitative and quantitative approaches are weak when applied alone. So, combining assessment methods from both perspectives would be more effective.

The bow-tie analyzing method becomes more prominent with its ability to involve FTA and ETA. It is highly effective in analyzing both cause and consequence effect relationships in many risky scenarios. Furthermore, it is a comprehensive risk analysis method, thanks to including preventive safety barriers in bow-tie diagrams that are used to reduce the severity and likelihood of hazards. The combination of the bow-tie method with AHP or Monte Carlo simulation would be more effective since the analysis becomes semiquantitative (Elidolu et al. 2022).

Secondly, FRAM is also an effective technique to analyze the risk from socio-technical systems. It is useful for understanding the complex dynamics that arise from the interactions between the various functions that form the system (Hollnagel and Goteman 2004; Hollnagel 2013). In the case of a fuel cell-powered ship, the risk

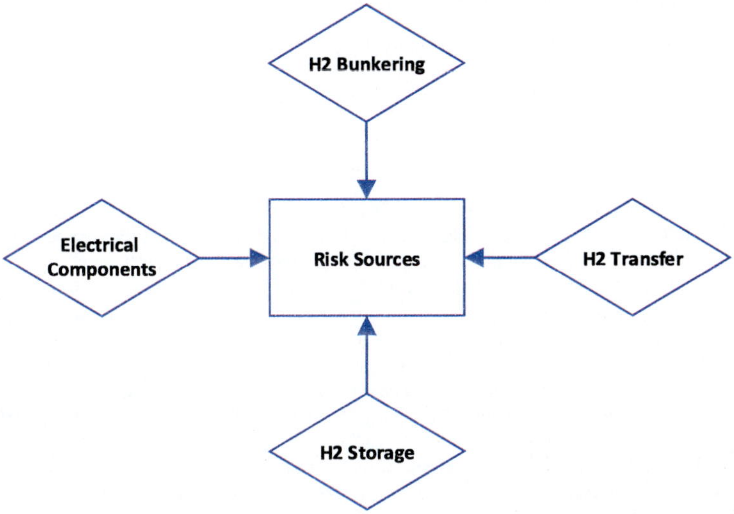

Fig. 4.1 Major risk sources

functions are variables, and it directly affects the other functions in terms of resonance. So, this brings variability in the whole system that causes a failure. By the way, FRAM is an efficient tool in case of high variability in terms of determining the function's interrelations. However, like the bow-tie analysis and other qualitative methods, supporting FRAM with quantitative methods like AHP or Monte Carlo simulation would give more effective solutions.

4.4 Conclusion

The result of this study will enlighten the operational side of the fuel cell ships from the risk assessment perspective. This chapter serves as an overview of the selection strategy of the risk assessment methods for hydrogen PEM fuel cell-powered ships. Hydrogen bunkering, storage, transfer, and electrical components are accepted as the major potential risk sources. Both can be affected by human error and also technical causes. Therefore, the system is considered a socio-technical and one-way approach; neither qualitative nor quantitative is enough. According to the preliminary analysis of the risk assessment strategies, the combination of the functional resonance analysis method (FRAM) and the bow-tie analysis as qualitative together with quantitative methods like AHP or Monte Carlo simulation seems like the most suitable risk analysis methods in both qualitative and quantitative ways. In further studies, case-based analysis can be carried out by using the selected risk analysis method.

Acknowledgments This research did not receive any specific grant from funding agencies in the public, commercial, or not-for-profit sectors.

References

Aarskog, F. G., Hansen, O. R., Strømgren, T., & Ulleberg, Ø. (2020). Concept risk assessment of a hydrogen-driven high speed passenger ferry. International Journal of Hydrogen Energy, 45(2), 1359–1372. https://doi.org/https://doi.org/10.1016/j.ijhydene.2019.05.128

Ben-Gal, I., Shani, A., Gohr, A., Grau, J., Arviv, S., Shmilovici, A., Posch, S., & Grosse, I. (2005). Identification of transcription factor binding sites with variable-order Bayesian networks. Bioinformatics, 21(11), 2657–2666. https://doi.org/https://doi.org/10.1093/BIOINFORMATICS/BTI410

Brazier, A., Edwards, D., Macleod, F., Skinner, C., & Vince, I. (2021). Hazard and operability (HAZOP) analysis. Trevor Kletz Compendium, 9–45. https://doi.org/https://doi.org/10.1016/B978-0-12-819447-8.00001-8

Crawley, F. (2020). Event tree analysis. In A Guide to Hazard Identification Methods (pp. 125–130). Elsevier. https://doi.org/10.1016/b978-0-12-819543-7.00014-8

Daud, W. R. W., Rosli, R. E., Majlan, E. H., Hamid, S. A. A., Mohamed, R., & Husaini, T. (2017). PEM fuel cell system control: A review. In Renewable Energy (Vol. 113, pp. 620–638). https://doi.org/10.1016/j.renene.2017.06.027

Elidolu, G., Akyuz, E., Arslan, O., & Arslanoğlu, Y. (2022). Quantitative failure analysis for static electricity-related explosion and fire accidents on tanker vessels under fuzzy bow-tie CREAM approach. Engineering Failure Analysis, 131, 105917. https://doi.org/https://doi.org/10.1016/J.ENGFAILANAL.2021.105917

França, J. E. M., Hollnagel, E., dos Santos, I. J. A. L., & Haddad, A. N. (2019). FRAM AHP approach to analyse offshore oil well drilling and construction focused on human factors. Cognition, Technology & Work 2019 22:3, 22(3), 653–665. https://doi.org/10.1007/S10111-019-00594-Z

Hirose, T., & Sawaragi, T. (2020). Extended FRAM model based on cellular automaton to clarify the complexity of socio-technical systems and improve their safety. Safety Science, 123.

Hollnagel, E. (2013). An Application of the Functional Resonance Analysis Method (FRAM) to Risk Assessment of Organisational Ch. Swedish Radiation Safety Authority, 1–81. www.stralsakerhetsmyndigheten.se

Hollnagel, E., & Goteman, Ö. (2004). The Functional Resonance Accident Model. Proceedings of Cognitive System Engineering in Process Plant, 155–161. http://82.94.179.196/bookshelf/books/403.pdf

Inal, O. B., Charpentier, J. F., & Deniz, C. (2022). Hybrid power and propulsion systems for ships: Current status and future challenges. Renewable and Sustainable Energy Reviews, 156. https://doi.org/https://doi.org/10.1016/j.rser.2021.111965

Inal, O. B., & Deniz, C. (2020). Assessment of fuel cell types for ships: Based on multi-criteria decision analysis. Journal of Cleaner Production, 265, 121734. https://doi.org/https://doi.org/10.1016/j.jclepro.2020.121734

Inal, O. B., & Deniz, C. (2018). Fuel Cell Availability for Merchant Ships. 3rd International Naval Architecture and Maritime Symposium, May.

Khakzad, N., Khan, F., & Amyotte, P. (2012). Dynamic risk analysis using the bow-tie approach. Reliability Engineering & System Safety, 104, 36–44. https://doi.org/https://doi.org/10.1016/J.RESS.2012.04.003

Patriarca, R., Di Gravio, G., & Costantino, F. (2017). A Monte Carlo evolution of the Functional Resonance Analysis Method (FRAM) to assess performance variability in complex systems. Safety Science, 91, 49–60. https://doi.org/https://doi.org/10.1016/J.SSCI.2016.07.016

Shokry, A., & Espuña, A. (2018). The Ordinary Kriging in Multivariate Dynamic Modelling and Multistep-Ahead Prediction. Computer Aided Chemical Engineering, 43, 265–270. https://doi.org/https://doi.org/10.1016/B978-0-444-64235-6.50047-4

Tronstad, T., Åstrand, H. H., Haugom, G. P., & Langfeldt, L. (2017). Study on the use of Fuel Cells in Shipping. 1–108.

van Biert, L., Godjevac, M., Visser, K., & Aravind, P. V. (2016). A review of fuel cell systems for maritime applications. Journal of Power Sources, 327(X), 345–364. https://doi.org/https://doi.org/10.1016/j.jpowsour.2016.07.007

Whiteley, M., Dunnett, S., & Jackson, L. (2016). Failure Mode and Effect Analysis, and Fault Tree Analysis of Polymer Electrolyte Membrane Fuel Cells. International Journal of Hydrogen Energy, 41(2), 1187–1202. https://doi.org/https://doi.org/10.1016/J.IJHYDENE.2015.11.007

Zincir, B., & Deniz, C. (2021). Methanol as a Fuel for Marine Diesel Engines. In P. C. Shukla, G. Belgiorno, G. Di Blasio, & A. K. Agarwal (Eds.), Alcohol as an Alternative Fuel for Internal Combustion Engines (pp. 45–85). Springer Singapore. https://doi.org/https://doi.org/10.1007/978-981-16-0931-2_4

Chapter 5
Investigation of the Gap Between the Predicted Mean Vote (PMV) and the Actual Vote (AMV) of the Students in CSB Climate Zone

Şevval Örfioğlu, Aydın Ege Çeter, Mehmet Furkan Özbey, and Cihan Turhan

Nomenclature

AMV	Actual mean vote
ASHRAE	American Society of Heating, Refrigerating and Air-Conditioning Engineers
ISO	International Organization for Standardization
PMV	Predicted mean vote

5.1 Introduction

Occupants' thermal comfort in an indoor environment has an important impact on their health and efficiency. Thermal comfort is defined as that condition of mind that expresses satisfaction with the thermal environment (ISO 2005). According to the

Ş. Örfioğlu · C. Turhan (✉)
Faculty of Engineering, Energy Systems Engineering, Atılım University, Ankara, Turkey
e-mail: cihan.turhan@atilim.edu.tr

A. E. Çeter
School of Natural and Applied Science, Mechanical Engineering, Atılım University, Ankara, Turkey

M. F. Özbey
School of Natural and Applied Science, Mechanical Engineering, Atılım University, Ankara, Turkey

Faculty of Engineering, Mechanical Engineering, Atılım University, Ankara, Turkey
e-mail: furkan.ozbey@atilim.edu.tr

M. Z. Sogut et al. (eds.), *Proceedings of the 2022 International Symposium on Energy Management and Sustainability*, Springer Proceedings in Energy,
https://doi.org/10.1007/978-3-031-30171-1_5

definition, the sensation of thermal comfort is highly subjective and depends on factors such as health, climate, culture, living conditions, behaviour, mood, age, even weather season, etc. (Castilla et al. 2010; Cruz et al. 2009; Battle et al. 2020).

The impact of thermal comfort is more critical for students, as students spend 30% of their total time in classrooms; furthermore, thermal comfort conditions directly affect students' efficiency with respect to attention, understanding and learning capacity (De Giuli et al. 2012). According to Jiang et al. (2018), thermal discomfort affects the learning performance of students negatively. To this aim, providing good environmental conditions for educational buildings is very important to prevent thermal discomfort on students. To assess thermal comfort conditions, predicted mean vote (PMV) is an index that describes the thermal comfort value of the environment, accepted by the most prominent and widely used American Society of Heating, Refrigerating, Air-Conditioning Engineers (ASHRAE) (2017) and the International Standard, ISO 7730:2005 (ISO 2005). The PMV index depends on indoor environmental parameters that mean radiant temperature, indoor temperature, air velocity and relative humidity besides personal parameters such as occupants' metabolic rate and clothing levels (ASHRAE 2017; Dyvia and Arif 2021). These parameters also impact occupants' actual thermal sensation that was derived from actual mean vote (AMV) (Kuru and Calis 2017), which is the mean value of occupants' actual thermal sensation. In the same environment, everyone feels the thermal conditions of the environment differently, and there is a gap that occurs between the AMV and PMV. Therefore, to ensure the thermal comfort of occupants literally, analysing the relationship between the PMV and AMV values is significant. To determine this gap, a study was conducted for secondary classrooms, which were located in the Mediterranean climate (Pereira et al. 2014). According to the results, the sensation of occupants was either much cooler or much warmer when compared with the sense proposed by the PMV model (Kuru and Calis 2017). Because the PMV values and the AMV of occupants might not match with each other, further studies are important to figure out the correlation of these values (Kuru and Calis 2017; Cheong et al. 2007).

The PMV index is in a range from -3 to +3 (Marinakis et al. 2017). According to the ASHRAE scale, (3) cold, (2) cool, (1) slightly cool, (0) neutral, (+1) slightly warm, (+2) warm and (+3) hot (ASHRAE 2017), the AMV has the same ranges as the PMV. For thermal comfort standards, the PMV value of the environment should be 0 (neutral) with a tolerance of ± 0.5, which means that an environment provides fully thermal comfort (Balbis-Morejon et al. 2020).

This study aims to investigate and understand the gap between the PMV and AMV of the students in a temperate climate zone. The following section of the paper explains the methodology. Then, results and discussion and conclusions parts are presented.

5.2 Methodology

In this study, a university study hall, which was located in the Engineering Faculty of a university in Ankara, Turkey (39.81°N 32.72°E), was determined as an experimental site. Under the Köppen climate classification, Ankara is included in the Csb-type climate zone (Weatherbase 2022).

The total area of the experimental site was 365 m^2. For the north direction, the site had a large continuous window with one external wall on the north and five internal adiabatic walls. The building's external wall had consisted of pumice concrete, cement plastering and cement screed; however, the window frame was aluminium with double glazing window (13 mm air gap). Furthermore, the site was a mixed-mode building without mechanical ventilation, and radiators were used for just heating purposes. Occupants provide ventilation by opening windows/doors. The heating system in the study hall was fixed at 22 °C in the winter season. The study was conducted between 15th of August and 15th of December 2021. For this study, a thermal comfort sensor (Fig. 5.1) was placed in the study hall of the Engineering Faculty of a university in Ankara, Turkey, namely, Delta Ohm HD32.3TCA (Delta Ohm 2021). By this sensor, environmental parameters of the study hall (air velocity, relative humidity, indoor air temperature and mean radiant temperature) were collected. Experiments were conducted in the 10 m^2 area in the middle of the study hall to reduce the effect of natural ventilation from doors and windows. It is worth noting that the study hall (Fig. 5.2) was generally used by engineering faculty students.

The sensors in the inner area were positioned at a height of 1.1 m, which is recommended by the ASHRAE 55 Standard. In addition, a sensor used in outdoor temperature and relative humidity measurements is at a height of 1.7 m that is determined by the standard (ASHRAE 55 2017). Data was taken every 10 min and converted to hourly averages. The locations of the sensors were shown in Fig. 5.3. After sensor placement to the study hall, the process of surveying the

Fig. 5.1 Delta Ohm thermal comfort measurement device

Fig. 5.2 Snapshot of the study hall

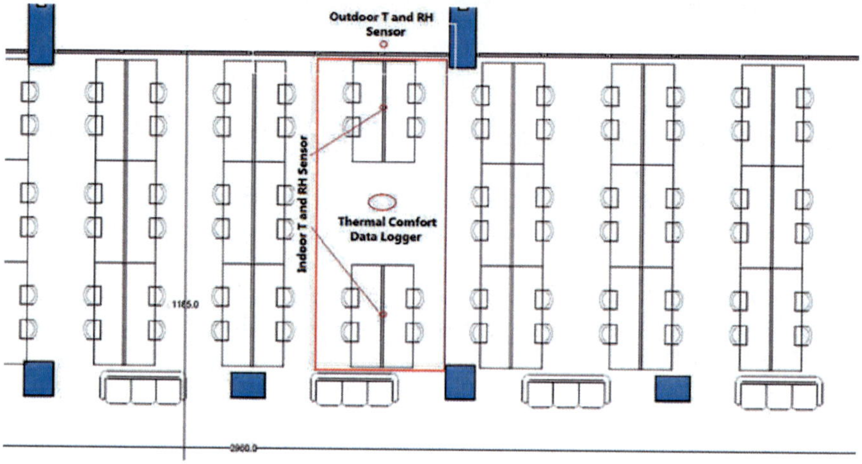

Fig. 5.3 Sensor locations in the study hall

students was started. To conduct these surveys, a mobile application was developed for smartphones in order to get the feedback of the occupants. Through the surveys, students' personal information (such as gender, weight and height), thermal sensation votes and garments were collected, and during the data collection process, surveys were done with students whose age range was 18–26. The thermal sensation votes of the students in that environment were obtained from surveys when data was taken from the thermal comfort sensor as soon as the students started the surveys.

Before starting surveys, participants were asked whether they have any metabolic disease and whether they have nicotine, caffeine, etc. at least 12 h before studies. Moreover, the students were requested to rest for 15 min to ensure accurate results. Filling out the questionnaires lasted approximately 10 min and all data was anonymous. Instead of the classic 7-scaled thermal comfort scale, 13-scale was used in the study in order to increase the accuracy and precision of the results. The scale for thermal sensation included hot (3), between warm and hot (2.5), warm (2), between slightly warm and warm (1.5), slightly warm (1), between neutral and slightly warm (0.5), neutral (0), between slightly cool and neutral (−0.5), slightly cool (−1), between cool and slightly cool (−1.5), cool (−2), between cold and cool (−2.5) and cold (3). Participants were requested to define their thermal sensation on this 13-point scale. The answers to this thermal scale were used to compare the AMV and PMV values. A total of 1296 data (Total of the PMV + AMV data) was collected in this study.

5.3 Results and Discussion

The number of students used for the study was 648. All participants were in the range of 18–26; 69.4% of participants were male, and 30.6% of participants were female. A comparison of the AMV and PMV was made for each participant as indicated in Table 5.1 as an example.

A linear relationship was obtained between the values. This relationship was shown by Eq. 5.1.

$$y = 0.8788x + 0.2905 \tag{5.1}$$

where x is the PMV and y represents the AMV values of the students.

Figure 5.4 shows the differences between PMV and AMV of the occupants. As a result of the analysis, 22.4% of the participants felt colder, 77.3% of the participants felt warmer and 0.33% of participants felt the same with the PMV. This result shows that occupants generally felt warmer than measured PMV values in Csb-type climate zones.

Other studies conducted in different climate zones in order to establish differences between the PMV and AMV observed similar results (Gallardo et al. 2016; Kuru and Calis 2017).

Table 5.1 Comparison of the PMV and AMV

Subject number	PMV	AMV	PMV – AMV
1	−0.24	−0.5	0.26
2	−0.62	0	−0.62
3	−0.31	0	−0.31
4	−0.32	0	−0.32
5	−0.3	0.5	−0.8
6	−0.26	0	−0.26
7	−0.47	0.5	−0.97
8	−0.1	0.5	−0.6
9	−0.91	−0.5	−0.41
10	−0.77	−1	0.23

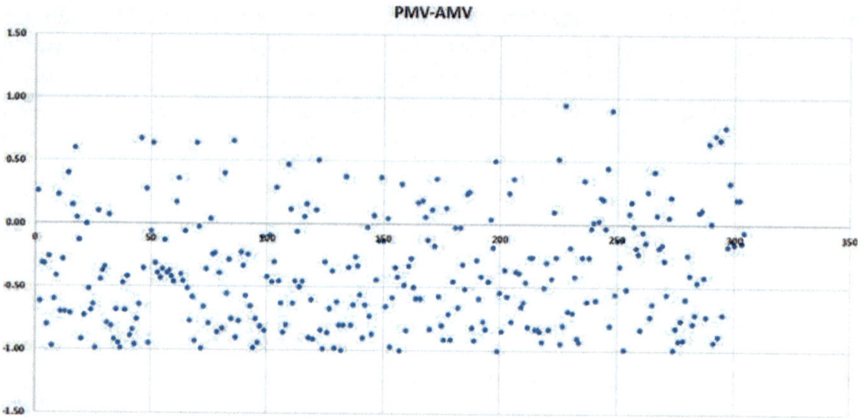

Fig. 5.4 Differences of the PMV and AMV

This equation is important to understand the significant difference between the PMV and AMV of students in Csb type climate zone. Also, since it was concluded that behavioural, psychological and adaptation of psychological were three types of supposed reasons for the inconsistency between the PMV and AMV (De Dear and Brager 1998; Auliciems 1981), this gap could give information about the thermal behaviours of the engineering faculty students in this zone.

There were three limitations of this study. Firstly, since there was a COVID-19 pandemic period during the experiments, all occupants used face masks. Secondly, the experiments were carried out in the student study hall since students used this hall in order to study their exams and lessons, which means that their stress levels were high, and high-stress levels might cause the sensation of the thermal comfort to be warmer. Finally, since the study hall was an engineering faculty, the number of male participants was higher.

5.4 Conclusions

In this study, thermal comfort parameters in a university's engineering faculty study hall were collected through objective sensors. To obtain the PMV value, indoor air temperature, mean radiant temperature, relative humidity and air velocity were measured by sensors, while the thermal sensation of occupants (AMV) was collected through surveys, which were conducted by a developed mobile application. Then, according to these values, linear equality between the PMV and AMV of occupants, in a university study hall located in the Csb type climate zone, was presented. The main findings of this study can be summarised as follows:

- The university students in Csb type climate zone felt generally much warmer; the AMV values of occupants were much higher than the PMV values. This situation showed the cultural options and features of the students who were in Csb type climate zone.
- Linear regression derived for the students in the study hall can be used in Csb type climate zone in order to develop energy-efficient devices for heating and/or cooling purposes.

Acknowledgement The Scientific and Technological Research Council of Turkey (TÜBİTAK) funded this research, and their contribution is gratefully acknowledged (Project Number: 120M890).

References

American Society of Heating, Refrigerating and Air-Conditioning Engineers. 2017. Thermal Environment Conditions for Human Occupancy (Standard 55). Atlanta, GA: ASHRAE.

Auliciems, A (1981) Towards a psycho-physiological model of thermal perception. International Journal of Biometeorology 25: 109–122. https://doi.org/10.1007/BF02184458

Balbis-Morejón M, Rey-Hernández J, Amaris-Castilla C, Velasco-Gómez E, San José Alonso JF, Rey-Martínez FJ (2020) Experimental study and analysis of thermal comfort in a university campus building in tropical climate. Sustainability 12:8886. https://doi.org/10.3390/su12218886

Battle EAO, Palacio JCE, Lora EES, Reyes AMM, Moreno MM, Morejón MB (2020) A methodology to estimate baseline energy use and quantify savings in electrical energy consumption in higher education institution buildings: Case study, Federal University of Itajubá (UNIFEI). Journal of Cleaner Production 244: 118551. https://doi.org/10.1016/j.jclepro.2019.118551

Castilla, M., Álvarez JD, Berenguel M, Pérez M, Rodríguez F, Guzmán JL (2010) Técnicas de control del confort en edificios. Revista Iberoamericana de Automática e Informática Industrial RIAI 7: 5–24. https://doi.org/10.1016/S1697-7912(10)70038-8

Cheong KWD, Yu WJ, Sekhar SC, Tham KW, Kosonen R (2007) Local thermal sensation and comfort study in a field environment chamber served by displacement ventilation system in the tropics. Building and Environment 42: 525–533. https://doi.org/10.1016/j.buildenv.2005.09.008

Cruz EMG, Claret G, Moreles M (2009) About thermal comfort: Neutral temperatures in the humid tropic. Revista Iberoamericana de Automática e Informática Industrial. Journal of Scientific Research 4: 33–38.

de Dear RJ, Brager GS (1998) Developing an Adaptive Model of Thermal Comfort and Preference. ASHRAE Transactions 104: 145–167

De Giuli V, Da Pos O, De Carli M (2012) Indoor environmental quality and pupil perception in Italian primary schools. Building and Environment 56: 335–345. https://doi.org/10.1016/j.buildenv.2012.03.024

DELTA OHM HD32.3TCA, Microclimate Measurement Device, Delta OHM, Italy.

Dyvia HA, Arif C (2021) Analysis of thermal comfort with predicted mean vote (PMV) index using artificial neural network. IOP Conference Series Earth and Environmental Science 622: 012019. doi:https://doi.org/10.1088/1755-1315/622/1/012019

Gallardo A, Palme M, Lobato-Cordero A, Beltran RD, Gaona G (2016) Evaluating Thermal Comfort in a Naturally Conditioned Office in a Temperate Climate Zone. Buildings 6: 27. https://doi.org/10.3390/buildings6030027

https://www.weatherbase.com/weather/weathersummary.php3?s=82171&cityname=Ankara, +Turkey, accessed on January 20, 2022.

ISO, 2005, ISO Standard 7730-2005, Ergonomics of the Thermal Environment. Analytical Determination and Interpretation of Thermal Comfort Using Calculation of the PMV and PPD Indices and Local Thermal Comfort Criteria, ISO: Geneva.

Jiang J, Wang D, Liu Y, Xu Y, Liu J (2018) A study on pupils' learning performance and thermal comfort of primary schools in China. Building and Environment 134: 102–113. https://doi.org/10.1016/j.buildenv.2018.02.036

Kuru M, Calis G (2017) Understanding the Relationship between Indoor Environmental Parameters and Thermal Sensation of users Via Statistical Analysis. Procedia Engineering 196: 808–815. https://doi.org/10.1016/j.proeng.2017.08.011

Marinakis V, Doukas H, Spiliotis E, Papastamatiou I (2017) Decision support for intelligent energy management in buildings using the thermal comfort model. International Journal of Computational Intelligence Systems 10: 882–893. https://doi.org/10.2991/ijcis.2017.10.1.59

Pereira LD, Raimondo D, Corgnati SP, da Silva MG (2014) Assessment of indoor air quality and thermal comfort in Portuguese secondary classrooms: Methodology and results. Building and Environment. 81: 69–80 https://doi.org/10.1016/j.buildenv.2014.06.008

Chapter 6
Development of an Energy Efficiency Project for a Glass Production Plant

Hakan Orhon and Levent Kılıç

6.1 Introduction

Glass production is an energy-intensive industry. Every year vast amounts of electricity and natural gas are being used to melt glass and shape it. Since Turkey does not produce its own natural gas, importing gas is a significant burden on the economy and its budget. Therefore, increasing the efficiency of the industry, hence decreasing consumption, holds high importance. Since reducing consumptions would also mean fewer emissions, increasing efficiency increases savings and reduces the burden on the environment. When compared to electricity, natural gas plays a much larger role in the energy consumed, although it is mostly used within furnaces and therefore has a very little room for improvement. On the other hand, electric motors and lightning could be improved without much effort.

With the economic development in Turkey in the 1980s, the energy production and consumption balances have changed greatly and the increase in energy imports has become mandatory. Turkey is in the position of an energy-importing country due to the inability of its energy production capacity to meet the energy demand in general. As a natural consequence of this, the efficiency of use of energy and energy resources has become very important. The General Directorate of Electrical Works Survey Administration (EIEI) has been working for a long time under the Ministry of Energy; in order to increase energy efficiency, studies, training, awareness raising, statistics, evaluation and legislation development activities continue.

H. Orhon (✉) · L. Kılıç
Sisecam R & D Batch & Mechanical Systems Department, Sisecam R & D Center, Koceli, Turkey

Sisecam R & D Energy Efficiency Unit, Sisecam R & D Center, Koceli, Turkey
e-mail: haorhon@sisecam.com

M. Z. Sogut et al. (eds.), *Proceedings of the 2022 International Symposium on Energy Management and Sustainability*, Springer Proceedings in Energy,
https://doi.org/10.1007/978-3-031-30171-1_6

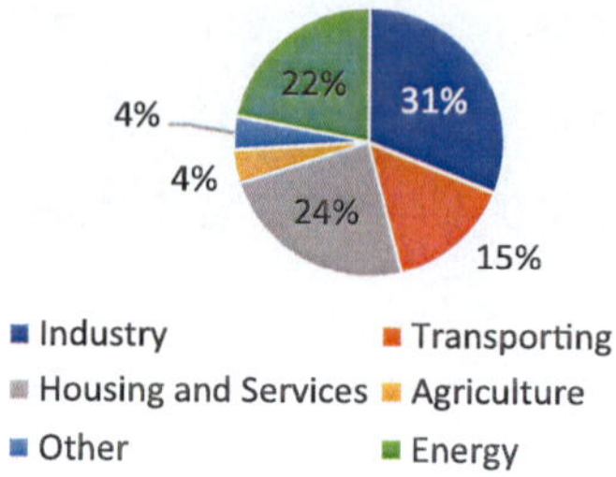

Fig. 6.1 Primary energy consumptions by sectors in Turkey (ijareeie.com, www.energysage.com)

For this purpose, studies, training, awareness raising, statistics, evaluation and legislation development are carried out. In order to increase energy efficiency in the industry, a project study was carried out with the Japanese Technical Cooperation Organization (JICA) (Heperkan and Olgun 2008). These studies reflected in the legislation with the regulations made recently and the regulations that are still in progress. The legislation that entered into force are listed below, in the order of publication dates (Energy Efficiency Law, Law No. 5627).

1. 2 May 2007 Energy Efficiency Law Law No. 5627
2. 9 October 2008 Regulation on Thermal Insulation in Buildings No: 27019
3. 25 October 2008 Regulation on Increasing Efficiency in the Use of Energy Resources and Energy No: 27035
4. 5 December 2008 Energy Performance Regulation in Buildings No: 27075
5. 6 February 2009 Energy No. 5627 Communiqué on the Procedures and Principles to be Applied on Authorizations, Certifications, Reporting and Projects to be Made within the Scope of the Efficiency Law No: 27133 (Fig. 6.1)

In accordance with the Regulation on Increasing Efficiency in Energy Consumption of Industrial Organizations, which was published in the Official Gazette dated 11.11.1995 and numbered 22460, which was prepared with the aim of providing the necessary regulations to increase the energy efficiency in the industrial sector with high energy consumption in Turkey. All factories with an annual total energy consumption of 2000 Tons Equivalent Petrol and above are obliged to appoint an Energy Manager. In accordance with the regulation, Energy Manager courses are organized by the General Directorate of EIEI for engineers working in industrial establishments.

The "Energy Efficiency Law" No. 5627 was published in the Official Gazette dated 02 May 2007 with the aim of using energy resources and energy effectively, to prevent waste, to alleviate the burden of energy costs on the economy and to protect the environment (Energy Efficiency Law, Law No. 5627).

Energy Efficiency Coordination Board was established with the participation of relevant ministries, undersecretariats, professional chambers and unions. With the approval of this board, authorization certificates can be given to universities and professional chambers by the General Directorate of Electrical Power Resources Survey and Development Administration for the purpose of providing applied training. The validity of the certificate is 5 years, and it must be renewed at the end of it. Authorization with a validity of 3 years can be given to companies to carry out training, energy audit, consultancy and implementation activities by organizations that have the authority to give the certificate. With this law, the framework of authority of institutions and organizations has been determined, and the concept of energy manager has been revealed and determined (Gupta 2015).

The purpose of this regulation, which was published in the Official Gazette in 25.10.2008 with the number 27035, is to regulate the principles and procedures for increasing efficiency in the use of energy resources and to use energy effectively, to prevent energy waste, to alleviate the burden of energy costs on the economy and to protect the environment (Regulation on Increasing Efficiency in the Use of Energy Resources and Energy 2008). The regulation regulates the authorization and supervision of institutions and companies, the establishment of an energy management unit at the facilities, training and certifications, supporting efficiency-enhancing projects, demand-side management and practices for increasing energy efficiency in electricity generation, transmission and distribution.

With the regulation, the limit of having an energy manager in industrial enterprises has been determined as 1 per 1000 Tons of Equivalent Petroleum (TEP) per year. In the calculation of this limit, the average of the total energy consumption of the last 3 years was taken as a basis.

6.2 Methodology and Results

6.2.1 Renovation of Lightning System

The lighting system throughout the plant was examined and an inventory list was prepared. In general, it was determined that the level of illumination intensity was **low, e.g. the mean lux values** were measured at the end of cooling at 80 lux, loading areas 25, warehouses 50 and oven elevation 145. Analyses were carried out with the Dialux program to ensure that the existing luminaires provide illumination level in accordance with the standards and a better lightning system was elected using LED luminaires in a way that would ensure standards/efficiency.

A new-generation LED luminaire application has been considered, replacing the existing luminaires. With the new luminaires, the lighting quality will be improved, and the total consumption will be reduced.

The list of luminaires was taken from the maintenance crew and the performance of the luminaires was measured using a lux meter. A total of 2.816 lamps were measured. A set of new generation luminaires was proposed, totalling 2.695 lamps. If the current luminaires were to be replaced:

- 912.660 kWh/year energy savings
- 400.475 TL yearly savings
- 532.1 tons/year CO_2 reductions

The cost of renovating the lightning system is 1.1 million TL with a return of investment of 2.8 years.

6.2.2 Renovation of Electric Motors

The list of electric motors was taken from the maintenance department. A set of new generation of electric motors were proposed. If the current motors were to be replaced with the new 24 efficient models:

- 132.720 kWh of energy savings
- 58.235 TL yearly savings
- 156,1 tons/year CO_2 reductions

The cost of replacing these motors will be 212.325 TL with a return of investment of 3.65 years.

6.2.3 Renovation of Pumps

Energy savings can be achieved by replacing the pumps with more efficient pumps suitable for the needs of purification tank, A-C plunger, demineralized, mould water and technological water pumps. The pumps are without speed drive and the existing speed drives of technological water pumps can be used. Speed drives are a step forward for digitalization and must be implemented for further increase in efficiency.

The list of pumps was taken from the maintenance crew. A set of new generation of pumps were proposed, replacing a total of seven pumps.

- 267.760 kWh of energy savings
- 117.278 TL yearly savings
- 156,1 ton/year CO_2 reductions

The cost of replacing these pumps will be 224.600 TL with a return of investment of 1.92 years (European Green Deal 2019).

6.2.4 Renovation of Cooling Fans

The fans for which efficiency analysis was carried out within the scope of the study are listed below. The main electricity consumer fans in the business are the fans called oven fans.

Examined fans are speed driven and set manually according to the recipe, and speed control is not made via a feedback signal.

It is possible to obtain 75% efficiency under operating conditions in furnace fans whose efficiency was examined. In this context, it is recommended to replace the fans with an efficiency value below 60% with high efficiency models.

The list of cooling fans was taken from the maintenance crew. A set of new generation of fans were proposed, replacing a total of seven fans:

- 914.088 kWh of energy savings
- 401.116 TL yearly savings
- 532,93 tons/year CO_2 reductions

The cost of replacing these fans will be 489.8 TL with a return of investment of 1.22 years (Paşabahçe Kırklareli (Pk) Fabrikası. Detaylı Enerji Etüdü Raporu 2020).

6.3 Conclusions

There are a lot of equipment that are working inefficiently, but due to working conditions and work required, it is not always possible to make the machinery run at its full potential. Every year new models of machinery are coming in to market, but it is not feasible to renew every single machine in a plant. Therefore the ones that are working over a threshold must work even if there are better models in the market. There can always be more efficient models; it is just a matter of investment and time.

Sooner or later these investments must take place, and Green Deal will force EU's trade partners to reduce carbon footprint. There will be extra taxation for companies who are producing their product without the means of renewable energy. Companies who are using vast amounts of fossil fuels will require every bit of energy savings it can in order to stay competitive amongst its rivals in EU market.

References

A European Green Deal. (2019) Striving to be the first climate-neutral continent. Available at: https://ec.europa.eu/info/strategy/priorities-2019-2024/european-green-deal_en

Gupta, Anupama., A Review on Energy Management and Audit (2015) . International Journal of Advanced Research in Electrical, Electronics and Instrumentation Engineering. Vol. 4, Issue 2, February 2015. Available from: https://www.ijareeie.com/upload/2015/february/54_A%20Review.pdf

Escon Enerji Sistemleri Ve Cihazları San. Tic. A.Ş., Paşabahçe Kırklareli (Pk) Fabrikası. Detaylı Enerji Etüdü Raporu. 17.01.2020

Enerji Kaynaklarının ve Enerjinin Kullanımında Verimliliğin Artırılmasına İlişkin Yönetmelik, Sayı: 27035, Resmi Gazete, 25.10.2008

Enerji Verimliliği Kanunu, Kanun No. 5627, Resmi Gazete, 2 Mayıs 2007

Heperkan, H. A., Olgun, B., "Enerji Verimliliği ve Türkiye'deki Mevzuat", Isıtma, Soğutma, Klima, Havalandırma, Yalıtım, Pompa, Vana, Tesisat, Su Arıtma ve Güneş Enerjisi Sistemleri Dergisi, Eylül-Ekim 2008

Chapter 7
Design of a Sustainable Combined Power Plant with sCO_2–BC and Ejector Cooling System Driven by Solar Energy

Fatih Yılmaz, Murat Ozurk, and Resat Selbas

Nomenclature

AC	Air compressor
BC	Brayton cycle
des	Destruction rate
e	Exit
ex	Exergy flow rate
$\dot{E}$	Energy rate
h	Specific enthalpy
in	Inlet
PDSC	Parabolic dish solar collector
TIT	Turbine inlet temperature
$\dot{Q}$	Heat rate
$\dot{W}$	Work rate
0	Reference point

7.1 Introduction

Environmental problems are one of the most challenging topics for humans and the globe. Currently, for more than 80% of energy demand, still, carbon-based fuels are widely used (Temiz and Dincer 2021; IEA 2018). As a result of the usage of fossil

F. Yılmaz (✉) · M. Ozurk · R. Selbas
Department of Mechatronics Engineering, Faculty of Technology, Isparta University of Applied Sciences, Isparta, Turkey
e-mail: muratozturk@isparta.edu.tr; resatselbas@isparta.edu.tr

M. Z. Sogut et al. (eds.), *Proceedings of the 2022 International Symposium on Energy Management and Sustainability*, Springer Proceedings in Energy,
https://doi.org/10.1007/978-3-031-30171-1_7

fuels, environmental complications, e.g., acid rain, global warming, and ozone depletion, arise, which are expressed as environmental problems. On the other hand, one of the most important methods to tackle these difficulties is the use of nonfossil-based energy sources. Moreover, solar energy can be called the basis of renewable energy sources. In recent years, researchers' interest in solar-assisted generation systems has increased linearly. Also, it is concluded that solar energy-integrated plants that are cogeneration, trigeneration, and multigeneration are more attractive and have more high performance, in the open literature.

A comprehensive review study of solar-driven ORC systems has been conducted by Loni et al. (2021). They examined concentrating and non-concentrating solar energy plants. Finally, their investigated study results show that trigeneration and polygeneration plant driven by solar energy and ORC is a promising technology. Tukenmez et al. (2022) proposed a thermodynamic examination of the multigeneration plant for hydrogen production that utilizes solar energy. Their suggested model energetic and exergetic efficiencies are 56.48% and 54.06%, respectively. Pourmoghadam and Mehrpooya (2021) conducted dynamic modeling and analysis of a solar collector and thermochemical cycle integrated plant. Based on their results, the main irreversibility rate is seen in the dish collector part. Georgousis et al. (2022) examined a solar-assisted polygeneration plant with carbon dioxide fluid in terms of multiobjective optimization. The system is designed to generate a few MW of power. The total efficiency of their modeled plant is 55%. The authors, Altayib and Dincer (2022), examined the geothermal energy-based ORC cycle with a solar dish collector for the heating upgrade. They aimed to raise the temperature of the low-temperature geothermal water by using a solar dish at the turbine inlets. According to the examination outcomes, the exergetic efficiency of the entire cycle varies between 40% and 68%. Chen et al. (2022) examined the thermal performance of the new design sCO_2 solar receiver. The result of their proposed study states that thermal efficiency and thermal stress decrease as the PDC inlet temperature increases. In recent years, Yang et al. (2022) have done a study on the integrated design and off-design working of solar energy towers into the sCO_2 Brayton cycle.

As coming abovementioned the literature survey, briefly, it should be noted that the importance of solar energy-based integrated systems is increasing. Therefore, in the coming years, solar-based combined plants will gain more significance. For this aim, the examined plant is investigated as the parabolic dish (PDSC) collector-based plant consisting of a sCO_2-BC and ejector cooling unit for obtaining hydrogen, power, cooling, and hot water, in the proposed book chapter. The main motivation and difference of this study are to utilizing of CO_2 refrigerant in BC and also ejector cooling system (ECS), which is ammonia fluid examined in terms of a thermodynamic approach. To explore how the influence impact of some significant factors on the demonstrated plant's efficiency, a parametric study is also executed. Also, use the CO_2 refrigerant in the systems because of its environmental properties.

7.2 System Description

In this chapter, a PDSC-motivated plant that is seen in Fig. 7.1 is modeled and proposed to generate hydrogen, power, cooling, and hot water. This plant consists of a solar dish collector, $sCO_2 - BC$, an ejector cooling unit, and a PEM electrolyzer. A solar dish collector is used as a thermal energy source in this model. While power generation is realized with the $sCO_2 - BC$, cooling is obtained with the ejector cooling system (ECS). Ammonia is preferred as the refrigerant in the ECS system. The entire system is thermally integrated, as shown in Fig. 7.1.

At point 7, the CO_2 fluid entering BC_Com in the saturated-liquid phase is removed to the supercritical phase here and enters HX1 at point 8. At the HX1 outlet, the CO_2 fluid reaches its high temperature and goes to the BC_Turbine. By reason of the expansion in the BC_Turbin, electricity generation occurs. BC continues to operate as a closed loop. At point 26, water at 80 °C enters the PEM electrolysis, where it decomposes into hydrogen and oxygen molecules. Hydrogen production takes place in this way. Moreover, hot water at point 25 is preferred for heating purposes.

7.3 Mathematical Modeling

In this part, a general mathematical model is addressed to examine the thermodynamic efficiency of the entire system. To perform thermodynamic modeling of any thermal plant, the first four balance equivalences can be inscribed as (Cengel and Boles 2007; Dincer and Rosen 2012):

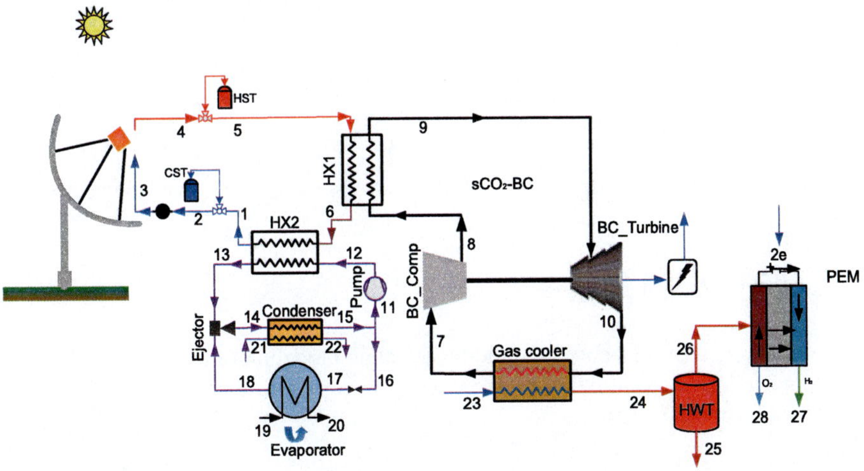

Fig. 7.1 Illustrative figure of the integrated plant

$$\dot{m}_i = \dot{m}_e \tag{7.1}$$

$$\dot{Q}_i + \dot{W}_i + \sum \dot{m}_i h_i = \dot{Q}_e + \dot{W}_e + \sum \dot{m}_e h_e \tag{7.2}$$

$$\sum \left(\frac{\dot{Q}}{T}\right)_{in} + \sum \dot{m}_i s_i + \dot{S}_{gen} = \sum \left(\frac{\dot{Q}}{T}\right)_e + \sum \dot{m}_e s_e \tag{7.3}$$

$$\dot{Ex}^{\dot{Q}_{in}} + \dot{Ex}^{\dot{W}_{in}} + \sum \dot{m}_i ex_i = \dot{Ex}^{\dot{Q}_e} + \dot{Ex}^{\dot{W}_e} + \sum \dot{m}_e ex_e + \dot{Ex}_{des} \tag{7.4}$$

Also, work and heat exergy rates can be modeled as:

$$\dot{Ex}^{\dot{Q}} = \left(1 - \frac{T_0}{T_c}\right)\dot{Q} \tag{7.5}$$

$$\dot{Ex}^{\dot{W}} = \dot{W} \tag{7.6}$$

In Eq. (7.4), the specific exergy term, after disregarding the kinetic and potential exergies, can be cleared:

$$ex = (h - h_0) - T_0(s - s_0) \tag{7.7}$$

The thermodynamic balance equation of the examined plant's components is given in Table 7.1. The following formulations are written for each system component and then analyzed with the EES (Klein 2021) package program.

Finally, the following equations are the energetic and exergetic mathematical formulation of the entire model:

$$\eta_{overall} = \frac{\dot{W}_{net} + (\dot{Q}_{cooling}) + (\dot{m}_{H_2} LHV_{H_2}) + (\dot{Q}_{hotwater})}{(\dot{Q}_{solar})} \tag{7.8}$$

$$\Psi_{overall} = \frac{\dot{W}_{net} + \left(\dot{Ex}^{\dot{Q}}_{cooling}\right) + (\dot{m}_{H_2} ex_{H_2}) + \left(\dot{Ex}^{\dot{Q}}_{hotwater}\right)}{\left(\dot{Ex}^{\dot{Q}}_{solar}\right)} \tag{7.9}$$

7.4 Results and Discussion

The modeled paper introduces the thermodynamic modeling of the combined cycle for beneficial products utilizing solar energy. To address the thermodynamic modeling, some important aspects are presented as follows and also are tabulated in Table 7.2:

Table 7.1 Thermodynamic balance formulation of the purposes system components

Comp.	Mass balance	Energy balance	Exergy balance
PDSC	$\dot{m}_3 = \dot{m}_4$	$\dot{m}_3 h_3 + \dot{Q}_{solar} = \dot{m}_4 h_4$	$\dot{Ex}_3 + \dot{Ex}^{\dot{Q}}_{solar} = \dot{Ex}_4 + \dot{Ex}_{Des,PDC}$
HX1	$\dot{m}_5 + \dot{m}_8 = \dot{m}_6 + \dot{m}_9$	$\dot{m}_5 h_5 + \dot{m}_8 h_8 = \dot{m}_6 h_6 + \dot{m}_9 h_9$	$\dot{Ex}_5 + \dot{Ex}_8 = \dot{Ex}_6 + \dot{Ex}_9 + \dot{Ex}_{Des,HX1}$
HX2	$\dot{m}_6 + \dot{m}_{12} = \dot{m}_1 + \dot{m}_{13}$	$\dot{m}_6 h_6 + \dot{m}_{12} h_{12} = \dot{m}_1 h_1 + \dot{m}_{13} h_{13}$	$\dot{Ex}_6 + \dot{Ex}_{12} = \dot{Ex}_1 + \dot{Ex}_{13} + \dot{Ex}_{Des,HX2}$
Condenser	$\dot{m}_{21} + \dot{m}_{14} = \dot{m}_{15} + \dot{m}_{22}$	$\dot{m}_{21} h_{21} + \dot{m}_{14} h_{14} = \dot{m}_{15} h_{15} + \dot{m}_{22} h_{22}$	$\dot{Ex}_{21} + \dot{Ex}_{14} = \dot{Ex}_{15} + \dot{Ex}_{22} + \dot{Ex}_{Des,con}$
Evaporator	$\dot{m}_{17} + \dot{m}_{19} = \dot{m}_{18} + \dot{m}_{20}$	$\dot{m}_{17} h_{17} + \dot{m}_{19} h_{19} = \dot{m}_{18} h_{18} + \dot{m}_{20} h_{20}$	$\dot{Ex}_{17} + \dot{Ex}_{19} = \dot{Ex}_{18} + \dot{Ex}_{20} + \dot{Ex}_{Des,evap}$
Ejector	$\dot{m}_{13} + \dot{m}_{18} = \dot{m}_{14}$	$\dot{m}_{13} h_{13} + \dot{m}_{18} h_{18} = \dot{m}_{14} h_{14}$	$\dot{Ex}_{13} + \dot{Ex}_{18} = \dot{Ex}_{14} + \dot{Ex}_{Des,ejec}$
Pump	$\dot{m}_{11} = \dot{m}_{12}$	$\dot{m}_{11} h_{11} + \dot{W}_{Pump} = \dot{m}_{12} h_{12}$	$\dot{Ex}_{11} + \dot{W}_{Pump} = \dot{Ex}_{12} + \dot{Ex}_{Des,pump}$
BC_Comp.	$\dot{m}_7 = \dot{m}_8$	$\dot{m}_7 h_7 + \dot{W}_{BC\_comp} = \dot{m}_8 h_8$	$\dot{Ex}_7 + \dot{W}_{BC\_comp} = \dot{Ex}_8 + \dot{Ex}_{Des,BC\_Comp}$
BC_Turbine	$\dot{m}_9 = \dot{m}_{10}$	$\dot{m}_9 h_9 = \dot{m}_{10} h_{10} + \dot{W}_{BC\_Turbine}$	$\dot{Ex}_9 = \dot{Ex}_{10} + \dot{W}_{BC\_turbine} + \dot{Ex}_{Des,BC\_Turbine}$
Gas cooler	$\dot{m}_{10} + \dot{m}_{23} = \dot{m}_{24} + \dot{m}_7$	$\dot{m}_{10} h_{10} + \dot{m}_{23} h_{23} = \dot{m}_{24} h_{24} + \dot{m}_7 h_7$	$\dot{Ex}_{10} + \dot{Ex}_{23} = \dot{Ex}_{24} + \dot{Ex}_7 + \dot{Ex}_{Des,gas\ cooler}$
PEM	$\dot{m}_{26} = \dot{m}_{27} + \dot{m}_{28}$	$\dot{m}_{26} h_{26} + \dot{W}_{PEM} = \dot{m}_{27} h_{27} + \dot{m}_{28} h_{28}$	$\dot{Ex}_{26} + \dot{W}_{PEM} = \dot{Ex}_{27} + \dot{Ex}_{28} + \dot{Ex}_{Des,PEM}$

Table 7.2 Modeling system assumption and input parameters

Parameters	Value
Receiver emissivity	90%
Concentrate ratio, C	330
Solar system's working fluid	$60NaNO_3\_40KNO_3$
Solar flux	800 W/m2
BC compressor inlet pressure	$1.7{*}P_{critic}$
Compressor compression rate	2.5
Compressor isentropic efficiency	90%
Pinch point temperature	5–30 °C
Working fluid of ejector cooling system	NH_3
Reference temperature	25 °C
Reference pressure	101.325 kPa

Table 7.3 Thermodynamic analysis results of the modeled plant

Parameters	Value
Total power rate	204.8 kW
Hydrogen production rate	0.0006602 kgs^{-1}
Cooling load capacity	703.8 kW
COP_{en}	0.3876
COP_{ex}	0.1491
η_{BC}	15.14%
ψ_{BC}	30.93%
$\eta_{overall}$	43.35%
$\psi_{overall}$	22.10%

- Kinetic and potential energy changes are neglected.
- Pressure variations between components are neglected.
- Pumps and turbines also have isentropic efficiency.

Thermodynamic modeling is carried out in this proposed study, accompanied by the acceptance in Table 7.2, and the results are offered in Table 7.3. Looking at this table, the power and hydrogen generation capacity of the proposed system are computed as 204.8 kW and 0.0006602 kgs^{-1}, respectively. Accordingly, the energy and exergy performance of the entire system is determined as 43.35% and 22.10%.

Referring to the parametric analysis consequences, to begin with, Fig. 7.2 presents the effect of solar flux on the overall plant beneficial crops. As expected, the amounts of beneficial products are increasing with the growth in solar flux. By growing the solar flux from 700 W/m^2 to 900 W/m^2, the net power rate, cooling capacity, and hydrogen generation rate increase.

Also, Fig. 7.3 clearly shows the rise in plant performance with the growth of turbine inlet temperature (TIT). As the temperature at the turbine inlet rises from 250 to 350 °C, both the energetic and exergetic performance of the total plant increase linearly. The main reason for this increasing behavior is the higher temperature (higher enthalpy) fluid enters the BC system at point 9, and the performance

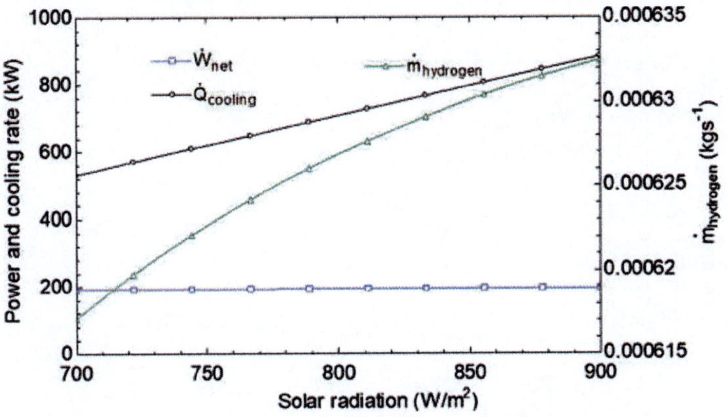

Fig. 7.2 Influence of solar radiation on the beneficial products from the system

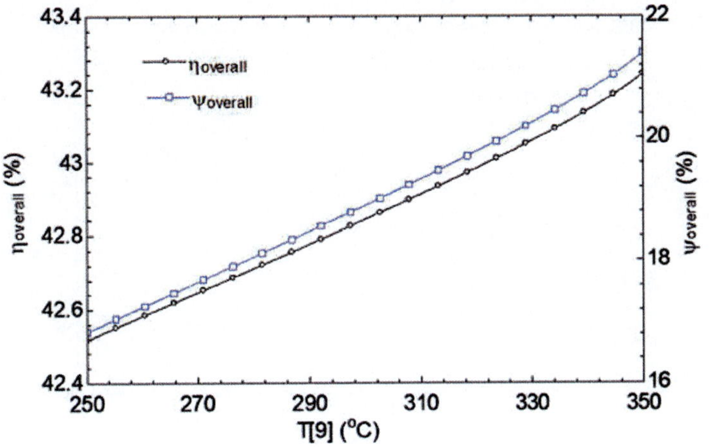

Fig. 7.3 Variation of the effectivenes of the entire system vs TIT

goes up as a result of higher temperature entry in the subsystems. Meanwhile, Fig. 7.4 displays the variation of the useful outputs obtained with the change in turbine inlet temperature. The reason for seeing a linear increase in useful output is that more power is available from the system at higher temperatures.

The impact of the last parameter change, the air compressor's compression ratio (r_{AC}) on the plant performance, and the beneficial crops obtained are examined and presented in Figs. 7.5 and 7.6, respectively. For both figures, both performance and useful outputs increased, excluding cooling capacity. The decline in cooling capacity is the decline in the mass flow rate of the refrigerant circulating in the cooling system as a result of rising to high temperatures. As r_{AC} increases from 2.5 to 4.5, the amount of hydrogen obtained from the proposed system increases linearly. The reason for this is more electrical power generation.

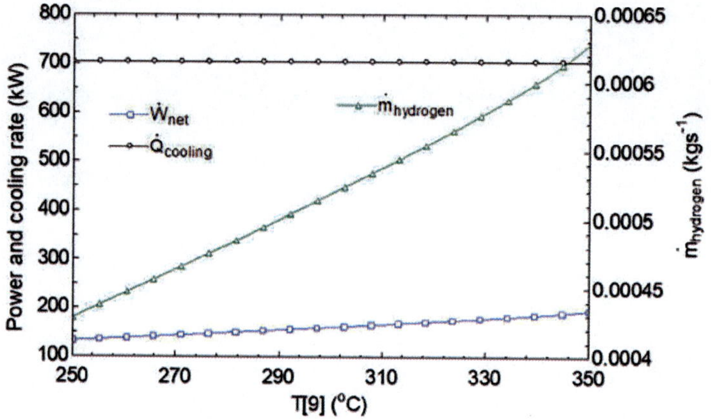

Fig. 7.4 Effect of TIT on the valuable productions from the overall plant

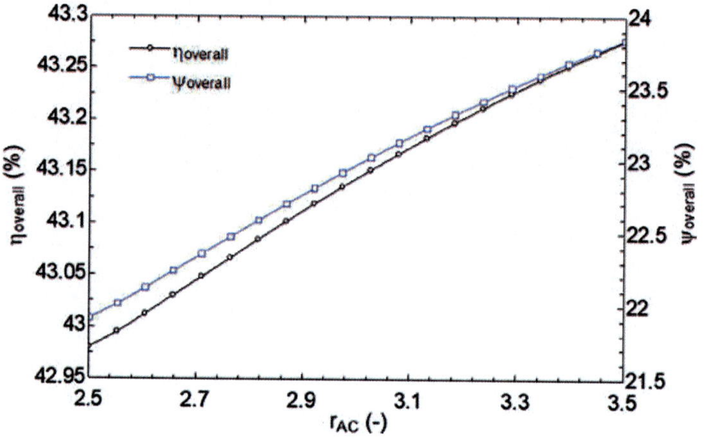

Fig. 7.5 Influence of r_{AC} on the performance rate of the total cycle

For the book chapter, the final figure, Fig. 7.7, offers the irreversibility of the major units in the proposed cycle. What happens here is that the most irreversibility is determined in the PDC subsystem with 2388 kW and then the HXs. The main reason for these high values is that the heat transfer mechanisms that take place here are high. It may be possible to obtain more efficient systems by reducing the irreversibility of these components.

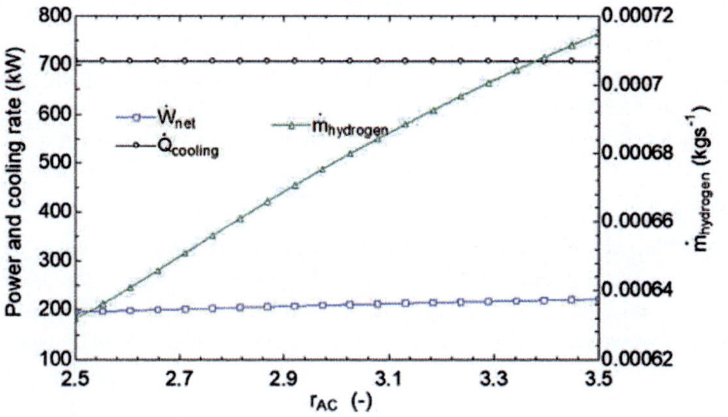

Fig. 7.6 Effect of r_{AC} on the valuable products from the overall cycle

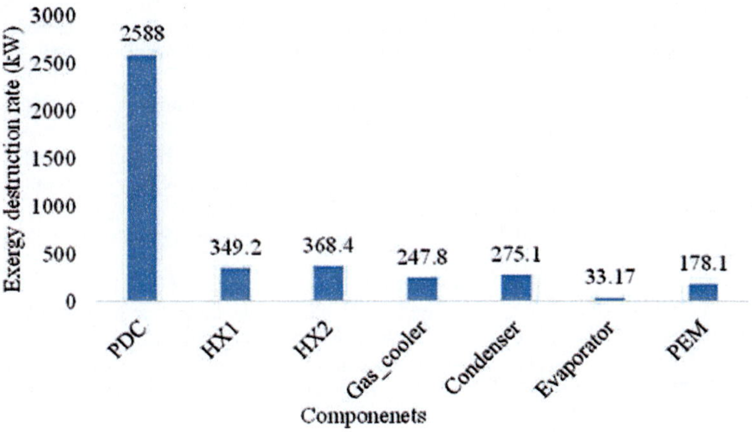

Fig. 7.7 Exergy destruction rates of the designed system components

7.5 Conclusion

In the advised and examined book chapter, thermodynamic modeling of the PDSC integrated unified plant is conducted to determine the performance and beneficial output rates. This model comprises the sCO_2 – BC, solar dish unit, a PEM electrolyzer, and ejector cooling system. Energy and exergy effectiveness, as well as irreversibility examination, are applied in detail for the entire system. Looking at the examination outcomes, the main conclusions of this paper can be written as:

1. The quantities of the net power and hydrogen of the modeled system are 204.8 kW and 0.0006602 kg/s.
2. The advised cooling system's cooling capacity is found as 703.8 kW.

3. Performance coefficients for energetic and exergetic methods of the ejector cooling plant are found as 0.38 and 0.14, respectively.
4. Overall energy and exergy performances of the modeled plant are figured as 43.35% and 22.10%, respectively.
5. Based on parametric analysis results, increment in the solar flux, in turbine inlet temperature, and the r_{AC} have an optimistic impact on the overall system's performance and beneficial products.

The developed study deals with the development of the combined sCO_2 – BC plant driven by solar energy that generates hydrogen, power, hot water, and cooling. It must be noted that in the coming years, to mitigate and finish the environmental detriments, renewable energy-based energy generation plants should be widely used.

References

Altayib, K., & Dincer, I. (2022). Development of a geothermal-flash organic Rankine cycle-based combined system with solar heat upgrade. Energy Conversion and Management, 252, 115,120. https://doi.org/10.1016/j.enconman.2021.115120

Cengel, Y. A., & Boles, M. A. (2007). Thermodynamics: An Engineering Approach sixth Editon (SI Units). The McGraw-Hill Companies, Inc., New York.

Chen, Y., Wang, D., Zou, C., Gao, W., & Zhang, Y. (2022). Thermal performance and thermal stress analysis of a supercritical CO2 solar conical receiver under different flow directions. Energy, 123,344. https://doi.org/10.1016/j.energy.2022.123344

Dincer, I., & Rosen, M. A. (2012). Exergy: energy, environment and sustainable development. Newnes, Elsevier.

Georgousis, N., Lykas, P., Bellos, E., & Tzivanidis, C. (2022). Multi-objective optimization of a solar-driven polygeneration system based on CO2 working fluid. Energy Conversion and Management, 252, 115,136. https://doi.org/10.1016/j.enconman.2021.115136

IEA, World Energy Outlook 2018. OECD; 2018. https://doi.org/10.1787/weo-2018-en.

Loni, R., Mahian, O., Markides, C. N., Bellos, E., le Roux, W. G., Kasaeian, A., ... & Rajaee, F. (2021). A review of solar-driven organic Rankine cycles: Recent challenges and future outlook. Renewable and Sustainable Energy Reviews, 150, 111,410. https://doi.org/10.1016/j.rser.2021.111410

Pourmoghadam, P., & Mehrpooya, M. (2021). Dynamic modeling and analysis of transient behavior of an integrated parabolic solar dish collector and thermochemical energy storage power plant. Journal of Energy Storage, 42, 103,121. https://doi.org/10.1016/j.est.2021.103121

Klein, S., 2021, Engineering equation solver (EES), AcademicCommercial, V11.199. 2021. Madison, USA, F-chart software.

Temiz, M., & Dincer, I. (2021). Enhancement of solar energy use by an integrated system for five useful outputs: System assessment. Sustainable Energy Technologies and Assessments, 43, 100,952. https://doi.org/10.1016/j.seta.2020.100952

Tukenmez, N., Yilmaz, F., & Ozturk, M. (2022). Parametric analysis of a solar energy based multigeneration plant with SOFC for hydrogen generation. International Journal of Hydrogen Energy, 47(5), 3266–3283. https://doi.org/10.1016/j.ijhydene.2021.01.131

Yang, J., Yang, Z., & Duan, Y. (2022). A review on integrated design and off-design operation of solar power tower system with S–CO2 Brayton cycle. Energy, 123,348. https://doi.org/10.1016/j.energy.2022.123348

Chapter 8
Evaluation of the Thermodynamic Performance Analysis of Geothermal Energy-Assisted Combined Cycle for Power, Heating, and Hydrogen Generation

Fatih Yılmaz, Murat Ozurk, and Resat Selbas

Nomenclature

$\dot{E}$	Energy rate
ex	Exergy rate
h	Specific enthalpy
s	Specific entropy
$\dot{Q}$	Heat rate
$\dot{W}$	Work rate
PEM	Proton exchange membrane
KC	Kalina cycle
in	Inlet
e	Exit
0	Reference point

8.1 Introduction

Globalization and the increase in human population worldwide have increased the demand for clean and sustainable energy in parallel with this situation. In 2021, electricity demand in major emerging economies, for example, China, is projected to increase by 4.5%, according to a projection from the IEA (IEA 2021). Fossil

F. Yılmaz (✉) · M. Ozurk · R. Selbas
Depermant of Mechatronics Engineering, Faculty of Technology, Isparta University of Applied Sciences, Isparta, Turkey
e-mail: muratozturk@isparta.edu.tr; resatselbas@isparta.edu.tr

M. Z. Sogut et al. (eds.), *Proceedings of the 2022 International Symposium on Energy Management and Sustainability*, Springer Proceedings in Energy,
https://doi.org/10.1007/978-3-031-30171-1_8

resources still have serious potential in meeting these increasing energy demands and cause environmental issues, for example, global warming, acid rain, melting of glaciers, ozone depletion, etc. At this point, to tackle these environmental detriments, the utilization of renewable energy sources is more important. In addition, another important factor in reducing environmental impacts is the efficient use of energy resources. In this context, cogeneration, trigeneration, and multigeneration systems, that is, integrated systems, which generate different useful outputs with single energy input, offer many advantages, especially solar or geothermal supported.

When investigating the open literature survey, there are many papers about renewable energy (i.e., geothermal)-based combined plants for various beneficial outputs. Tukenmez et al. (2021) modeled a multigeneration plant that utilizes geothermal energy to generate liquid hydrogen. The study integrated various sub-plants thermally, and then entire energy and exergy effectiveness is computed as 25.62% and 52.69%, respectively. Bicer and Dincer (2016) investigated a novel solar-geothermal energy-supported combined cycle that the energetic and exergetic effectivenees can reach up to 10.8% and 46.3%. Yilmaz (2021) developed and analyzed thermodynamic modeling of the integrated cycle utilizing geothermal energy. In his planned work, totally energetic and exergetic efficiency is determined as 63.28% and 55.99%. In addition, a book chapter on hydrogen production from geothermal energy was suggested by Ozturk and Dincer (2021). In this book chapter, a comprehensive investigation is conducted on the hydrogen generation methods from geothermal sources. Javadi et al. (2021) observed the thermodynamic and economic examination of the new combined cycle. In the proposed study, the cost of maximum hydrogen production capacity is 5 \$/kg. Altayib and Dincer (2022) proposed a geothermal and ORC-based combined plant that is integrated with solar energy. They investigated a flash-ORC system in terms of thermodynamic viewpoints. Finally, their modeled plant's exergy efficiency varies between 40% and 68%.

In 2022, Zinsalo et al. (2022) advised a combined system that is comprised of an ORC and also motivated by geothermal energy. They conducted energy, exergy, and economic analysis to determine the modeled plant performance and economic viewpoints. According to their results, the best energy efficiency is R12333zd (E) fluid with 19.2–9.32%. Furthermore, in the same year, Gürbüz et al. (2022) examined a new two-stage ORC system driven by a real geothermal power plant. They conducted an enhanced exergo-environmental examination.

In light of the literature review, there are numerous research papers of the various designs to enhance the thermal efficiency of the geothermal energy-based integrated cycle. When examined by years, it can be remarked that the subject of this research study is up-to-date. In this regard, the main aim of this suggested cycle is to design and examine the low-temperature geothermal energy-based combined cycle to generate heating, power, and hydrogen. The property of parameters, e.g., reservoir temperature, reservoir outlet mass flow rate, and quality, which are important parameters affecting system performance, are parametrically investigated. Meanwhile, the prominent novelties of this book chapter can be written as follows:

- To generate power from steam turbine in Kalina cycle (KC) NH_3-H_2O working fluid are used in this paper.
- To compare the performance of single and multiple production configurations.
- To investigate hydrogen production capacity by PEM electrolysis.
- To conduct comprehensive thermodynamic modeling of this paper.

8.2 Modeling System Description

As coming to modeled system description, briefly, as charted in Fig. 8.1, this integrated plant comprises a steam turbine, a Kalina cycle, a PEM electrolyzer, and a flashing process. The geothermal water inlet at state 1, first, goes in the flashing part and then goes up the separator 1 (with steam-liquid phase). Here, the steam-liquid mixture fluid separates into saturated liquid and steam phases. Subsequently, the saturated steam ($x = 1$) enters the steam turbine (T1), where expansion results in power generation. Subsequently, the saturated liquid at state 3 enters HEX 1, and after providing the thermal energy demand for the KC, HEX 2 is also used in hot water production to heating applications. Hot water at point 23 can be used in low-temperature studies for heating purposes.

The geothermal water exiting at state 7 enters HEX 4 and transfers its heat to the water at state 24. On the other hand, water comes in the PEM unit at 80 °C at ambient pressure, that is, 101.325 kPa. The excess electricity obtained is used to generate hydrogen in PEM. Finally, the geothermal fluid returns to the ground at about 60 °C at point 9.

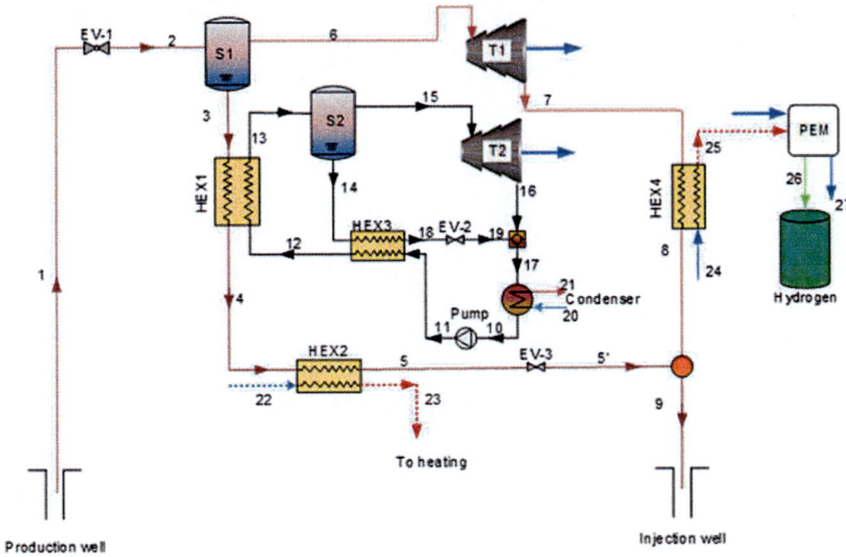

Fig. 8.1 Presentation chart of the modeled cycle

8.3 Analysis

This study deals with a comprehensive thermodynamic analysis of the geothermal energy-supported integrated cycle to produce hot water, power, and hydrogen. In general, four balance equations, which are mass, energy, entropy, and exergy equilibriums, are used to conduct the basic thermodynamic modeling of any thermal plant. These formulations based on the first and second laws of thermodynamics can be modeled as below (Dincer and Rosen 2012; Cengel and Boles 2015; Moran et al. 2010):

$$\sum \dot{m}_{in} = \sum \dot{m}_{out} \tag{8.1}$$

$$\sum \dot{Q} + \sum \dot{m}_i h_i = \sum \dot{W} + \sum \dot{m}_{out} h_{out} \tag{8.2}$$

$$\sum \left(\frac{\dot{Q}}{T}\right)_{in} + \sum \dot{m}_{in} s_{in} + \dot{S}_{gen} = \sum \left(\frac{\dot{Q}}{T}\right)_{out} + \sum \dot{m}_{out} s_{out} \tag{8.3}$$

$$\dot{Ex}^{\dot{Q}_{in}} + \dot{Ex}^{\dot{W}_{in}} + \sum \dot{m}_i ex_i = \dot{Ex}^{\dot{Q}_e} + \dot{Ex}^{\dot{W}_e} + \sum \dot{m}_e ex_e + \dot{Ex}_{des} \tag{8.4}$$

Meanwhile, the terms of the heat and work exergy can be formulated as:

$$\dot{Ex}^{\dot{Q}} = \left(1 - \frac{T_0}{T_c}\right)\dot{Q} \tag{8.5}$$

$$\dot{Ex}^{\dot{W}} = \dot{W} \tag{8.6}$$

After disregarding the kinetic and potential exergy, each state flow exergy can be also described as follows:

$$ex = (h - h_0) - T_0(s - s_0) + ex_{ch} \tag{8.7}$$

The following equations are the energetic and exergetic mathematical formulation of the entire cycle:

$$\eta_{overall} = \frac{\dot{W}_{net} + (\dot{Q}_{heating}) + (\dot{m}_{H_2} LHV_{H_2})}{\dot{m}_1 (h_1 - h_9)} \tag{8.8}$$

$$\psi_{overall} = \frac{\dot{W}_{net} + \left(\dot{Ex}^{\dot{Q}}_{heating}\right) + (\dot{m}_{H_2} ex_{H_2})}{\dot{m}_1 (ex_1 - ex_9)} \tag{8.9}$$

Table 8.1 Assumptions for the thermodynamic modeling

Parameters	Unit	Value
Geothermal mass glow rate	kg/s	50
Geothermal source temperature	°C	180
Pressure ratio of expansion valve	–	4
Pump inlet pressure	kPa	900
Pressure ratio of pump	–	2.5
Isentropic efficiency of pump	%	85
Isentropic efficiency of turbines	%	88
Working fluid	–	NH_3-H_2O
Reinjection well temperature	°C	70
Ambient temperature	°C	25
Ambient pressure	kPa	101.325

For modeling the thermodynamic analysis, some assumptions and input indicators are employed and presented in Table 8.1. To perform all the mathematical calculations, Engineering Equation Solver (EES) (Klein 2021) program is used.

Table 8.2 presents the mass, energy, and exergy formulations of the system's major units designed according to the thermodynamic equilibrium equations.

8.4 Results and Discussion

In this part of the chapter, an extensive thermodynamic and parametric analysis is conducted and results are presented. For this aim, in light of the present in Table 8.1 assumptions, examination results are tabulated in Table 8.3. This table presents also a comparison of the single generation (SG) and multigeneration (MG) options, based on performance rate. It is stated that the MG plant has higher energy and exergy performance than the SG plant. The total hydrogen generation capacity of this plant is 0.0087 kg/s. While the energy efficiency for SG is 3.70%, the energy efficiency for MG is calculated as 28.47%. This result shows that the efficiency of MG systems is higher.

Looking at parametric analysis results, Fig. 8.2 presents the impact of geothermal water temperature (T[1]) on the beneficial outputs from the proposed plant. With an increase of the T[1] by 160–200 °C, net power, heating, and hydrogen generation capacities increase. It should be noted that this parameter has a constructive outcome on the system. Furthermore, the performance rate of this modeled plant increased by expanding T[1] that as seen in Fig. 8.3. The key motive for this increase is that the system performance increases as a result of the rise in the beneficial crops obtained in the cycle with the escalation in temperature.

Table 8.2 Mass, energy, and exergy balance equation of the examined system's components

Comp.	Mass balance	Energy balance	Exergy balance
S1	$\dot{m}_2 = \dot{m}_3 + \dot{m}_6$	$\dot{m}_2 h_2 = \dot{m}_3 h_3 + \dot{m}_6 h_6$	$\dot{E}x_2 = \dot{E}x_3 + \dot{E}x_6 + \dot{E}x_{Des,S1}$
S2	$\dot{m}_{13} = \dot{m}_{14} + \dot{m}_{15}$	$\dot{m}_{13} h_{13} = \dot{m}_{14} h_{14} + \dot{m}_{15} h_{15}$	$\dot{E}x_{13} = \dot{E}x_{14} + \dot{E}x_{15} + \dot{E}x_{Des,S2}$
HEX1	$\dot{m}_3 + \dot{m}_{12} = \dot{m}_4 + \dot{m}_{13}$	$\dot{m}_3 h_3 + \dot{m}_{12} h_{12} = \dot{m}_4 h_4 + \dot{m}_{13} h_{13}$	$\dot{E}x_3 + \dot{E}x_{12} = \dot{E}x_4 + \dot{E}x_{13} + \dot{E}x_{Des,HEX1}$
HEX2	$\dot{m}_4 + \dot{m}_{22} = \dot{m}_5 + \dot{m}_{23}$	$\dot{m}_4 h_4 + \dot{m}_{22} h_{22} = \dot{m}_5 h_5 + \dot{m}_{23} h_{23}$	$\dot{E}x_4 + \dot{E}x_{22} = \dot{E}x_5 + \dot{E}x_{23} + \dot{E}x_{Des,HEX2}$
HEX3	$\dot{m}_{11} + \dot{m}_{14} = \dot{m}_{12} + \dot{m}_{18}$	$\dot{m}_{11} h_{11} + \dot{m}_{14} h_{14} = \dot{m}_{12} h_{12} + \dot{m}_{18} h_{18}$	$\dot{E}x_{11} + \dot{E}x_{14} = \dot{E}x_{12} + \dot{E}x_{18} + \dot{E}x_{Des,HEX3}$
HEX4	$\dot{m}_7 + \dot{m}_{24} = \dot{m}_8 + \dot{m}_{25}$	$\dot{m}_7 h_7 + \dot{m}_{24} h_{24} = \dot{m}_8 h_8 + \dot{m}_{25} h_{25}$	$\dot{E}x_7 + \dot{E}x_{24} = \dot{E}x_8 + \dot{E}x_{25} + \dot{E}x_{Des,HEX4}$
T1	$\dot{m}_6 = \dot{m}_7$	$\dot{m}_6 h_6 = \dot{m}_7 h_7 + \dot{W}_{T1}$	$\dot{E}x_6 = \dot{E}x_7 + \dot{W}_{T1} + \dot{E}x_{Des,T1}$
T2	$\dot{m}_{15} = \dot{m}_{16}$	$\dot{m}_{15} h_{15} = \dot{m}_{16} h_{16} + \dot{W}_{T1}$	$\dot{E}x_{15} = \dot{E}x_{16} + \dot{W}_{T2} + \dot{E}x_{Des,T2}$
Pump	$\dot{m}_{10} = \dot{m}_{11}$	$\dot{m}_{10} h_{10} + \dot{W}_P = \dot{m}_{11} h_{11}$	$\dot{E}x_{10} + \dot{W}_P = 1 + \dot{E}x_{Des,P}$
Condenser	$\dot{m}_{17} + \dot{m}_{20} = \dot{m}_{10} + \dot{m}_{21}$	$\dot{m}_{17} h_{17} + \dot{m}_{20} h_{20} = \dot{m}_{10} h_{10} + \dot{m}_{21} h_{21}$	$\dot{E}x_{17} + \dot{E}x_{20} = \dot{E}x_{10} + \dot{E}x_{21} + \dot{E}x_{Des,Con}$
PEM	$\dot{m}_{25} = \dot{m}_{26} + \dot{m}_{27}$	$\dot{m}_{25} h_{25} + \dot{W}_{PEM} = \dot{m}_{26} h_{26} + \dot{m}_{27} h_{27}$	$\dot{E}x_{25} + \dot{W}_{PEM} = \dot{E}x_{26} + \dot{E}x_{27} + \dot{E}x_{Des,PEM}$

Table 8.3 Analysis results of the combined cycle

Parameters		Value
Energy efficiency	SG	3.70
Exergy efficiency		14.91
Energy efficiency	MG	28.47
Exergy efficiency		24.82
Net power rate		873.9 kW
Heating capacity		4600 kW
Hydrogen production capacity		0.0087 kg/s

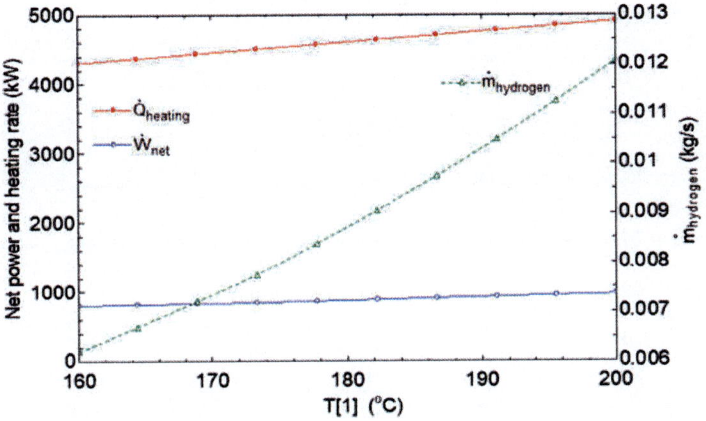

Fig. 8.2 Useful products of the examined cycle vs. reservoir temperature

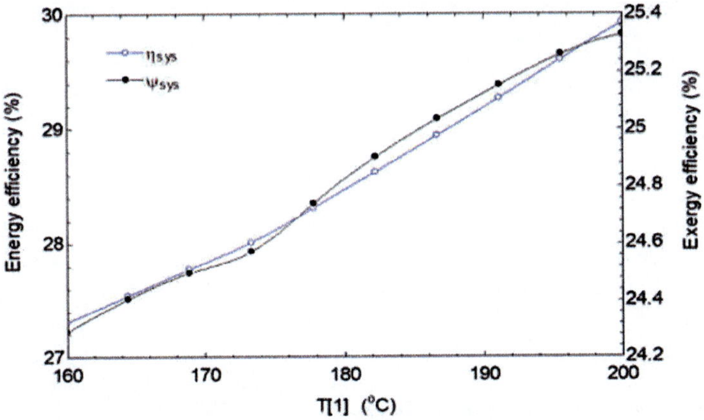

Fig. 8.3 Performance rate of the examined cycle vs. reservoir temperature

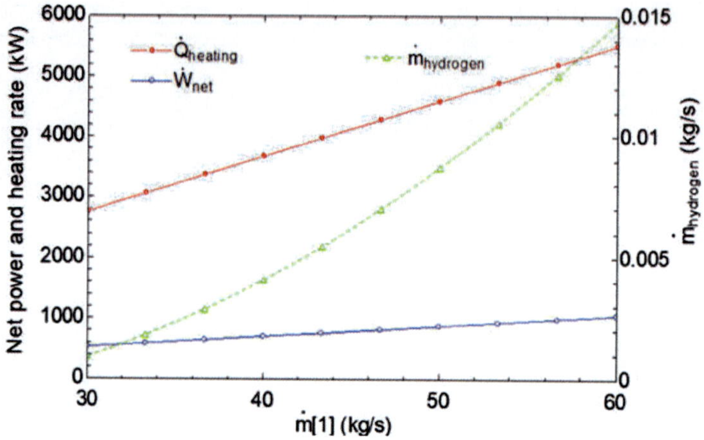

Fig. 8.4 Beneficial products of the advised cycle vs. geothermal flow mass rate

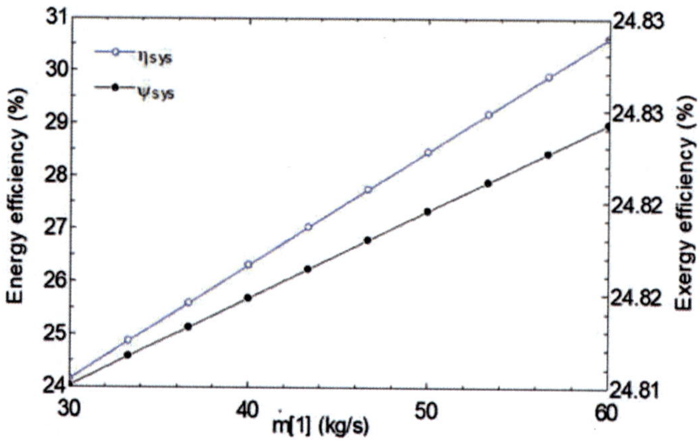

Fig. 8.5 Performance ratios of the examined cycle with geothermal flow mass rate

Another significant variable is the change in the mass flow rate of the geothermal fluid ($\dot{m}_1$). With the growth of $\dot{m}_1$, as revealed in Figs. 8.4 and 8.5, both the beneficial outputs obtained from the model and the increase in system performance are seen. The increasing trend in these graphs is consistent with many studies in the literature. The reason for this rise is that the amount of energy inflowing the cycle increases and more power is obtained with the increase of the mass stream rate to the turbines.

In geothermal power generation systems, the flashing process is an important factor and is directly related to efficiency. Therefore, Fig. 8.6 presents the impact of the change in the quality of state (X_2) on the power available from the system after flashing. With the increase of X_2 from 0 to 0.4, the power obtained from the steam turbine (T1) increases, while the power obtained from the KC turbine (T2) decreases.

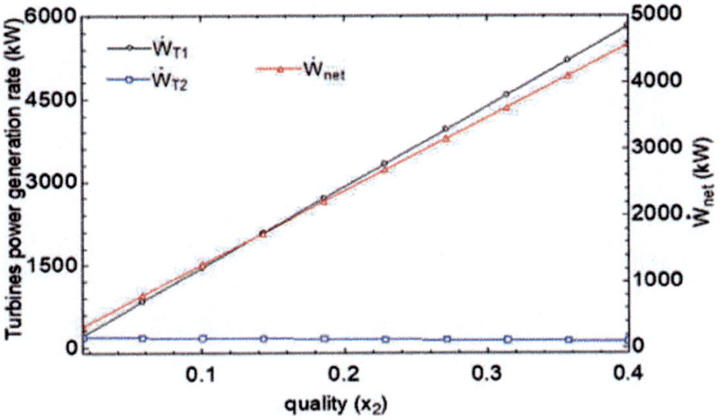

Fig. 8.6 Electrical power rate of the turbines and overall system vs. state 2′ quality

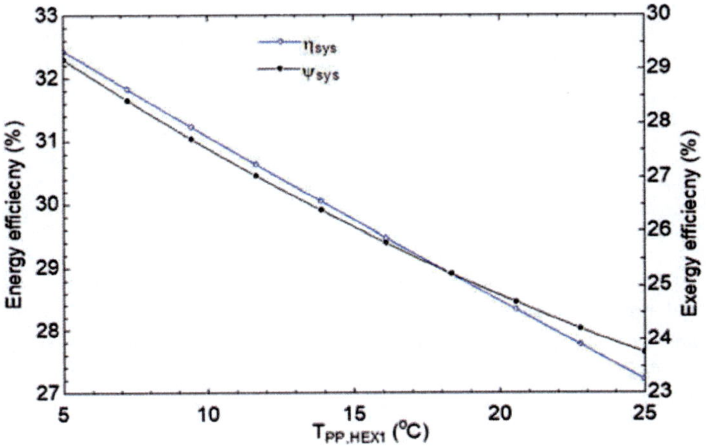

Fig. 8.7 Energetic and exergetic effectiveness of the entire system according to different $T_{PP,\ HEX\ 1}$

However, overall net power generation increases. The cause for this is that while the mass flow to the steam turbine increases at state 2, the mass rate at state 3 decreases.

Figure 8.7 illustrates the efficiency change of the whole cycle with the increase in pinch point temperature of HEX 1 ($T_{PP,\ HEX\ 1}$) from 5 to 20 °C. It is understood from the chart that both energetic and exergy performances decrease linearly with this increase. What happens here is that with the increase of $T_{PP,\ HEX\ 1}$, the lower amount of energy goes to the subsystems, and thus the efficiency of the whole system is negatively labeled. Therefore, $T_{PP,\ HEX\ 1}$ design is an important parameter for thermal systems.

The last chart of this study compares the performance of SG and MG energy generation schemes, as noticed in Fig. 8.8. In the conclusion, it is concluded that

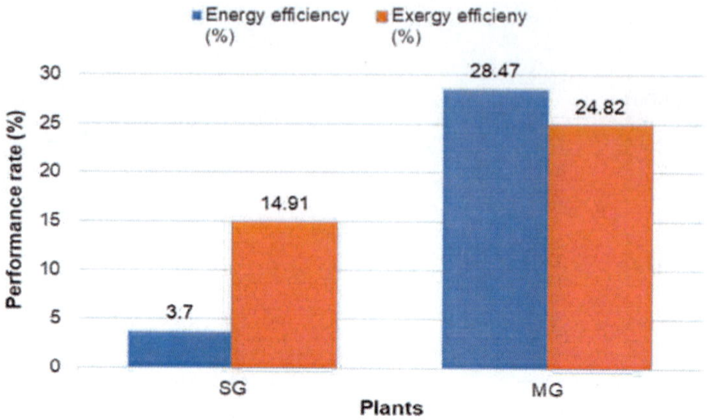

Fig. 8.8 Performance comparison of the SG and MG plants

increasing the beneficial outputs of the plant rises in plant performance. For this reason, to design the higher performance system, various beneficial outputs should be taken into consideration.

8.5 Conclusion

In this proposed work, the thermodynamic analysis of a combined system with geothermal energy-assisted KC and PEM electrolysis is modeled for hydrogen, power, and heating generation. The main purpose of the proposed system is to search for clean and sustainable methods to obtain useful products. In this framework, the effects of the change of factors, for instance, geothermal water and mass flow on the cycle, are investigated by performing a parametric study. Some prominent fundamental analysis results can be summarized as follows:

- Net power generation and hydrogen content from the proposed cycle is 873.9 kW and 0.0087 kg/h, respectively.
- The heating capacity of the water heated during the recycling of geothermal water was calculated as 4600 kW.
- The energy and exergy efficiencies of the entire cycle are 28.47% and 24.82%, respectively.
- Increasing the temperature and mass flow rate of the geothermal source has a positive effect on the system's performance. From another perspective, it is observed that the energetic performance goes up from 24% to 30% with the increase of the $\dot{m}_1$ rate from 30 to 60 kg/s.

As a result, it can be concluded that it is very important to obtain different useful outputs from geothermal energy supported systems, both in terms of the environment and for using the systems more efficiently. These systems will become more significant in the coming years.

References

Altayib, K., & Dincer, I. (2022). Development of a geothermal-flash organic Rankine cycle-based combined system with solar heat upgrade. *Energy Conversion and Management*, *252*, 115120. https://doi.org/10.1016/j.enconman.2021.115120

Bicer, Y., & Dincer, I. (2016). Development of a new solar and geothermal based combined system for hydrogen production. *Solar Energy*, *127*, 269-284. https://doi.org/10.1016/j.solener.2016.01.031

Çengel, Yunus A. Boles, Michael A. (2015). Thermodynamics: an engineering approach. *New York: McGraw-Hil l Education.*

Dincer, I., & Rosen, M. A. (2012). *Exergy: energy, environment and sustainable development.* Newnes. Elsevier

IEA, 2021, https://www.iea.org/fuels-and-technologies/electricity (acc. Date: 29.12.2021)

Gürbüz, E. Y., Güler, O. V., & Keçebaş, A. (2022). Environmental impact assessment of a real geothermal driven power plant with two-stage ORC using enhanced exergo-environmental analysis. Renewable Energy, 185, 1110-1123. https://doi.org/10.1016/j.renene.2021.12.097

Javadi, M. A., Abhari, M. K., Ghasemiasl, R., & Ghomashi, H. (2021). Energy, exergy and exergy-economic analysis of a new multigeneration system based on double-flash geothermal power plant and solar power tower. *Sustainable Energy Technologies and Assessments*, *47*, 101536. https://doi.org/10.1016/j.seta.2021.101536

Moran, M. J., Shapiro, H. N., Boettner, D. D., & Bailey, M. B. (2010). *Fundamentals of engineering thermodynamics*. John Wiley & Sons.

Ozturk, M., & Dincer, I. (2021). Hydrogen production from geothermal power plants. In *Thermodynamic Analysis and Optimization of Geothermal Power Plants* (pp. 207–223). Elsevier.

Klein, S.A. (2021) Engineering equation solver, F-Chart Software, Madison, WI (2021), p. 1

Tukenmez, N., Yilmaz, F., & Ozturk, M. (2021). Thermodynamic performance assessment of a geothermal energy assisted combined system for liquid hydrogen generation. *International Journal of Hydrogen Energy*, *46*(57), 28995-29011. https://doi.org/10.1016/j.ijhydene.2020.12.012

Yilmaz, F. (2021). Performance and environmental impact assessment of a geothermal-assisted combined plant for multi-generation products. *Sustainable Energy Technologies and Assessments*, *46*, 101291. https://doi.org/10.1016/j.seta.2021.101291

Zinsalo, J. M., Lamarche, L., & Raymond, J. (2022). Performance analysis and working fluid selection of an Organic Rankine Cycle Power Plant coupled to an Enhanced Geothermal System. Energy, 123259. https://doi.org/10.1016/j.energy.2022.123259

Chapter 9
A Bow-Tie Analysis for the Navigational Safety and Environmental Sustainability on the 1915 Çanakkale Bridge

Muhammed Fatih Gulen, Murat Mert Tekeli, Omer Berkehan Inal, Ozcan Arslan, and Muhsin Kadioglu

Nomenclature

ENV-C	Environmental consequences
ENV-T	Environmental threats
HR-C	Human-related consequences
MET-T	Meteorological threats
NAV-T	Navigational threats
OP-C	Operational consequences
OP-T	Operational threats
PROP-C	Property consequences
TGDCS	Turkish General Directorate of Coastal Safety
VTS	Vessel traffic services

9.1 Introduction

The Çanakkale Strait is an important shipping route that provides a connection between the Mediterranean Sea and the Black Sea. Thousands of ships pass through this narrow waterway every year. There is also heavy local traffic on the strait.

M. F. Gulen (✉) · M. M. Tekeli · O. Arslan · M. Kadioglu
Maritime Faculty, Maritime Transportation Management Engineering Department, Istanbul Technical University, Tuzla/Istanbul, Turkey
e-mail: gulenmu@itu.edu.tr

O. B. Inal
Department of Marine Engineering, Maritime Faculty, Istanbul Technical University, Istanbul, Türkiye

M. Z. Sogut et al. (eds.), *Proceedings of the 2022 International Symposium on Energy Management and Sustainability*, Springer Proceedings in Energy,
https://doi.org/10.1007/978-3-031-30171-1_9

Table 9.1 Some ship accidents in history

Ship	Accident	Year
SS Esso Maracaibo	General Rafael Urdaneta bridge collision	1964
Star clipper	Tjörn bridge collision	1980
Exxon Valdez	Grounding	1989

This strait, where approximately 540,000 people live on its shores, is also on the migration route of hundreds of fish and bird species. In addition, due to the continuous currents, the strait has a wide geographical impact area, especially the tourism regions in the Aegean Sea.

The 1915 Çanakkale Bridge will be the suspension bridge with the longest middle span in the world with 2023 m. The piers of this huge bridge are within the waterways. Considering the past accidents in the region, the bridge is located in one of the riskiest areas of the strait (Ilgar 2015). In this case, it is of great importance to identify and manage navigational risks (Svensson 2009). On the other hand, the absence of any protective measures to reduce the severity of collisions at the piers of the bridges will lead to the growth of negative consequences in possible collisions (Gulen 2021).

When the accidents in history (Table 9.1) are examined, it is possible to have an idea about the disasters that a possible accident will cause.

Recent ship collisions and near misses in the Çanakkale Strait are also important for us to understand the extent of the navigational risks in the region. On December 7, 2021, the 225-m-long wheat-laden ship "Fortune Trader", passing through the Çanakkale Strait, had a mechanical failure. Vessel traffic services (VTS) promptly directed the pilot and two powerful tugboats to the area and stopped the ship traffic in both directions as a precaution. The ship was drifting at a speed of 4.2 knots and a distance of about 1.5 miles from the steel bridge piers. Luckily, the tugs were able to catch up with the ship without any collision with the bridge piers (Tenker 2021).

According to the Turkish General Directorate of Coastal Safety (TGDCS) data, about 20% of the ships passing through the Çanakkale Strait carry dangerous cargo (TGDCS 2021). An oil spill, explosion, fire, etc. that may occur as a result of a possible accident may cause great environmental destruction.

In this study, a risk analysis was carried out by using the bow-tie analysis method. At every stage of the analysis, the opinions of the experts consisting of captains, pilots, academics, and VTS operators were used. This study offers applicable measures to increase navigational safety and protect environmental sustainability in the 1915 Çanakkale Bridge area.

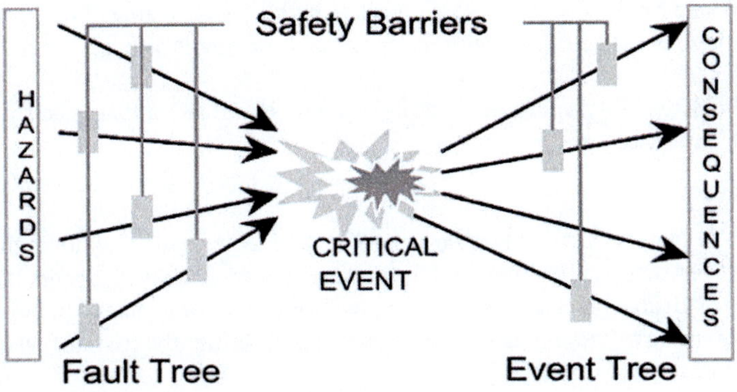

Fig. 9.1 A simplified bow-tie representation. (De Dianous and Fiévez 2006)

9.2 Methodology

Bow-tie analysis is an effective and popular risk assessment method frequently used in high-risk scenarios. It is a graphical tool to illustrate an accident scenario, starting from accident causes and ending with its consequences (Khakzad et al. 2012). Basically, the bow-tie can be considered as an approach that has both proactive and reactive elements and that systematically works through the hazard and its management (Jacinto and Silva 2010).

A bow-tie diagram contains hazards on one side and consequences on the other, with the critical event in the center (Fig. 9.1) (Shahriar et al. 2012). The diagram includes safety barriers to reduce the likelihood of hazards creating a critical event and to reduce the severity of the consequences. Considering safety barriers is also essential in this method to avoid accidents (Elidolu et al. 2022).

The identification of critical event is the first step of the risk analysis. In this study, the bow-tie analysis was applied to evaluate "the ship-bridge collision risk" at the 1915 Çanakkale Bridge. The critical event is defined as "collision with the bridge piers".

The second step of the analysis is to define the potential hazards. The question is "Which hazards/threats cause the critical event to happen?". All hazards were identified by the experts (Gulen et al. 2021a). As a result, hazards that may cause a ship-bridge collision on the 1915 Çanakkale Bridge are grouped into four categories: meteorological threats (MET-T), environmental threats (ENV-T), navigational threats (NAV-T), and operational threats (OP-T).

Meteorological threats:

- MET-T1 Strong current: there is a continuous current from the Marmara Sea to the Aegean Sea in the Çanakkale Strait. The strong current in the strait adversely affects the navigational safety of ships.
- MET-T2 Strong wind: the wind flowing into the İstanbul Strait blows especially from the north/northeast direction due to the topographical features of the

Çanakkale Strait. The wind is compressed in the entrance area of the strait where the bridge is located, increasing its speed and making it difficult for the ships to navigate.
- MET-T3 Visibility: reduced visibility is an important factor that makes it difficult for ships to navigate.

Environmental threats:

- ENV-T1 Narrow navigable waterway: The navigable channel width in the bridge area is very narrow (1600 m) and the distance between the bridge piers is 2023 m.
- ENV-T2 Bridge piers in waterway: As the bridge piers are located in the sea, they further narrow the navigable waterways and bring the risk of ship-bridge collision.

Navigational threats:

- NAV-T1 Heavy passing traffic: There is very heavy traffic in the strait. Thousands of merchant ships pass through the Çanakkale Strait every year. A total of 43.342 ships passed through the strait in 2021 (TGDCS 2021).
- NAV-T2 Heavy local traffic: The ferries that provide transportation between the two lands and the fishing boats in the region create heavy local traffic.
- NAV-T3 Communication traffic: As a result of traffic density, there is intense communication traffic in the region.
- NAV-T4 Human error factor: In a risky waterway where such maximum navigational attention is required, the possibility of marine accidents is substantial. Considering the causes of the 118 marine accidents that occurred between 2004 and 2017 at the Çanakkale Strait, it is seen that human error takes the first place (Tasan 2019).

Operational threats:

- OP-T1 Optional pilotage: According to the Montreux Convention, merchant vessels enjoy the freedom of harmless passage through the Turkish Straits. These vessels, which pass without a pilot, pose a risk in terms of navigational safety in the region.
- OP-T2 Over-workload of VTS: Sector Gelibolu fulfils all of its duties as a single sector such as receiving reports before the strait entrances from ships, determining the strait entry times, and ensuring coordination between the pilots and the ships. In addition to all these tasks, it keeps watching the safe passage of ships through the 1915 Çanakkale Bridge. This over-workload situation can increase the probability of errors.

The third step of the analysis is to evaluate potential consequences. The question is "What consequences may arise if the critical event happens?". Possible consequences are grouped into four main categories: environmental (ENV-C), property (PROP-C), operational (OP-C), and human-related (HR-C). Accordingly, the following catastrophic consequences may occur in a possible ship-bridge accident:

- Sea pollution (ENV-C1).
- Air pollution (ENV-C2).
- Damage to animals (ENV-C3).
- Damage to the ship (PROP-C1).
- Damage to the bridge (PROP-C2).
- Damage to the land vehicles on the bridge (PROP-C3).
- Disruption of land traffic (OP-C1).
- Disruption of sea traffic (OP-C2).
- Deaths (HR-C1).
- Injuries (HR-C2).
- Pollution-related health issues (HR-C3).

The created bow-tie diagram is shown in Fig. 9.2.

9.3 Results and Discussions

The consequences of a ship colliding with a bridge pier can cause effects that will last for many years. This effect will not be limited only in the strait, but will have the potential to spread to a wide area both by air and by sea. In particular, the effects of sea and air pollution can last for years apart from instant results such as deaths, injuries, property damage, and traffic disruption. For this reason, both preventive barriers that will reduce the likelihood and protective barriers that will reduce the severity in case of a collision are of great importance.

Many preventive barriers are used to increase navigational safety in the bridge area: pilotage services, tugboat services, local rules (speed restriction, closure of passage in bad weather, one-way traffic application, and overtaking prohibition near the bridge, etc.), navigational aids, and vessel traffic services. These measures are mostly positive and sufficient; however, improvements can still be made (Gulen 2021). The 1915 Çanakkale Bridge is located within the borders of Sector Gelibolu, which is a part of Çanakkale VTS. However, this sector also covers a very large area in the Sea of Marmara. Sector Gelibolu fulfils all of its duties as a single sector. Additionally, it continuously monitors the safe passage of ships through the 1915 Çanakkale Bridge. In this sense, the over-workload on Sector Gelibolu creates an environment that is tending to mistakes. As a suggestion, a separate VTS sector should be established that will only serve in the bridge area to monitor the ships with the utmost care during the bridge passages. According to experts, the existing infrastructure of Çanakkale VTS is sufficient for a new sector. Thus, the new sector can be put into practice in a short time by equipping sufficient personnel and determining sector boundaries.

In addition, it should be considered as an additional arrangement that at least one tugboat should always be standby on both piers of the bridge to intervene in any emergency immediately.

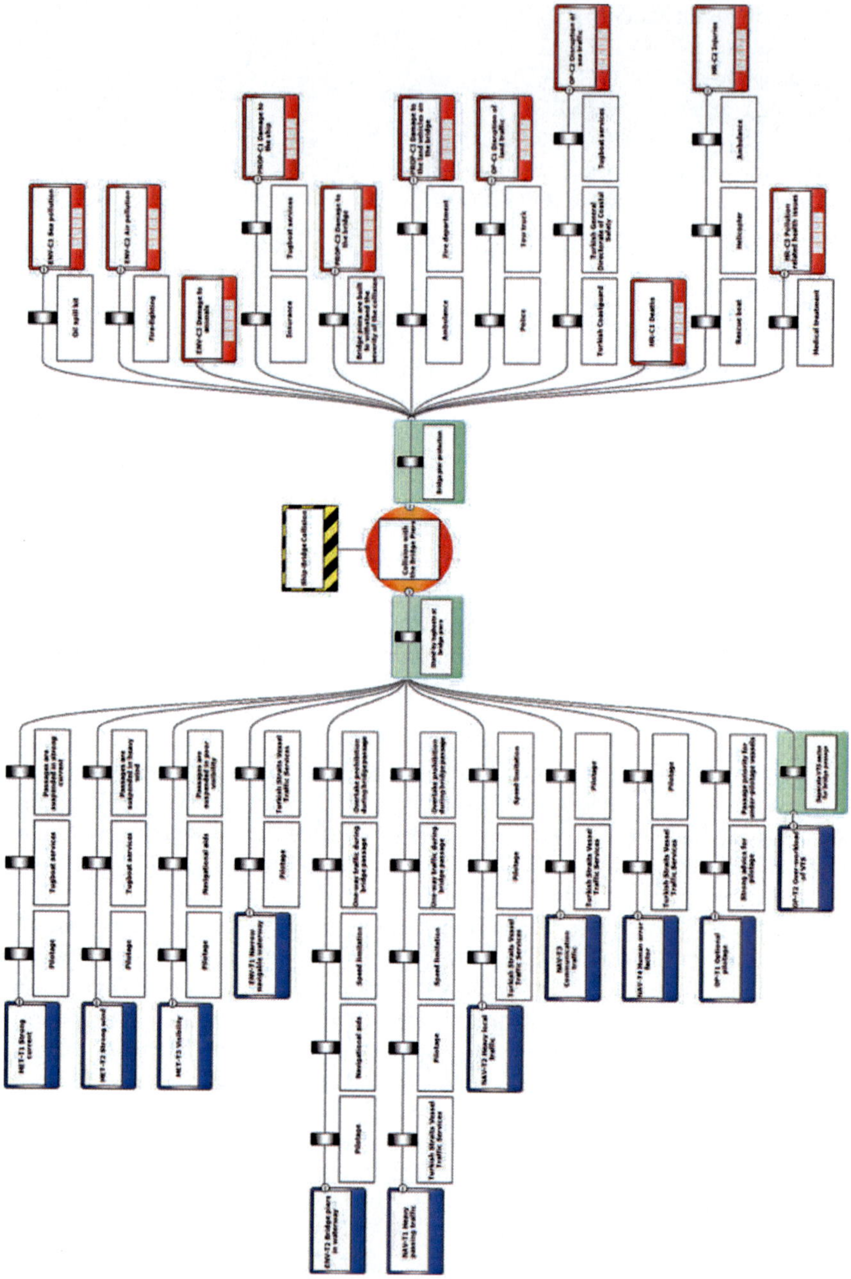

Fig. 9.2 Bow-tie diagram of ship-bridge collision risk at the 1915 Çanakkale Bridge (Green barriers are not in practice, they represent recommendations)

Even if all necessary preventive measures are taken, there is still a risk of ships colliding with bridge piers. No matter how durable the bridge piers are, additional protective measures are needed to reduce the severity of the collision, especially to prevent possible environmental disasters. Unfortunately, there are no protective measures on the piers of the 1915 Çanakkale Bridge. Considering the characteristics of the bridge and the natural conditions, fender protection seems to be the most viable protection measure (Gulen et al. 2021b). However, it is of great importance that the fenders are produced in appropriate sizes and from suitable materials. Engineers should be encouraged to work on this issue.

All recommendations are based on expert opinions. In addition, the recommendations are shown in green in the bow-tie diagram in Fig. 9.2.

9.4 Conclusion

A ship-bridge collision in the Çanakkale Strait has the potential to create catastrophic consequences, primarily in terms of environmental sustainability but also in terms of economy, tourism, and health. For this reason, all relevant stakeholders such as the government, academics, maritime authorities, and engineers should exchange opinions and work together to take all necessary measures to manage and control the current risk.

Acknowledgement This conference paper is derived from the author's MSc thesis.

References

De Dianous V, Fiévez C (2006) ARAMIS project: A more explicit demonstration of risk control through the use of bow–tie diagrams and the evaluation of safety barrier performance. Journal of Hazardous Materials 130(3):220–233. https://doi.org/10.1016/J.JHAZMAT.2005.07.010

Elidolu G, Akyuz E, Arslan O, Arslanoğlu Y (2022) Quantitative failure analysis for static electricity-related explosion and fire accidents on tanker vessels under fuzzy bow-tie CREAM approach. Engineering Failure Analysis 131:105917. https://doi.org/10.1016/J.ENGFAILANAL.2021.105917

Gulen MF (2021) Measures for ship-bridge collisions and establishment of fuzzy-logic based risk assessment model - case study: 1915 Çanakkale Bridge. In Master Thesis. İstanbul Technical University.

Gulen MF, Kadioglu M, Arslan O (2021a) SWOT analysis for safer ship navigation at 1915 Çanakkale Bridge. 4th Global Conference on Innovation in Marine Technology and the Future of Maritime Transportation.

Gulen MF, Kadioglu M, Arslan O (2021b) Evaluation of Preventive and Protective Measures Against Collision Risk to Bridge Piers in 1915 Çanakkale Bridge. Proceedings of the 1st International Conference on the Stability and Safety of Ships and Ocean Vehicles.

Ilgar R (2015) Determination of ship mobility and accident risk map in the Canakkale Strait. Turkey Geographic Magazine 65:1–10

Jacinto C, Silva C (2010) A semi-quantitative assessment of occupational risks using bow-tie representation. Safety Science 48(8):973–979. https://doi.org/10.1016/J.SSCI.2009.08.008

Khakzad N, Khan F, Amyotte P (2012) Dynamic risk analysis using bow-tie approach. Reliability Engineering & System Safety 104:36–44. https://doi.org/10.1016/J.RESS.2012.04.003

Shahriar A, Sadiq R, Tesfamariam S (2012) Risk analysis for oil & gas pipelines: A sustainability assessment approach using fuzzy based bow-tie analysis. Journal of Loss Prevention in the Process Industries 25(3):505–523. https://doi.org/10.1016/J.JLP.2011.12.007

Svensson H (2009) Protection of bridge piers against ship collision. Steel Construction: Design and Research 2(1):21–32

Tasan M (2019) Analyzing Passages and Passage Times of Ships from Turkish Straits. In Master Thesis. Istanbul Technical University.

Tenker S (2021) Meğer felaketi ucuz atlatmışız. https://odatv4.com/guncel/meger-felaketi-ucuz-atlatmisiz-223448

TGDCS (2021). The Statistics Summary Of Vessels Passed Canakkale Strait 2021. https://denizcilikistatistikleri.uab.gov.tr/turk-bogazlari-gemi-gecis-istatistikleri.

Chapter 10
Evaluation of Combustion Characteristics in a Common Rail Diesel Engine Fueled Butanol/N-Heptane/Diesel Blends

Mustafa Vargün, Ahmet Yapmaz, Ilker Turgut Yılmaz, and Cenk Sayın

10.1 Introduction

Due to the increasing energy demand since the beginning of the second half of the nineteenth century, there has been a remarkable increase in carbon dioxide (CO_2) emissions worldwide. This situation raises concerns about global warming and climate change. According to the statistical data published by the International Energy Agency, the transportation sector is directly responsible for 24% of the CO_2 emissions emitted worldwide. The emission of CO_2 occurs as a result of the combustion of hydrocarbon-based fuel used as an energy source (2022b). As a result of determined efforts in European Union countries, approximately 23% improvement was achieved in greenhouse gas emissions in the last 30 years. The use of developing technology, energy efficiency, and renewable and alternative energy sources has an important place in reducing greenhouse gas emissions (2022a). Due to the studies, it is expected that the use of biofuels in the transportation sector will increase by 3% annually until 2026. However, this increase does not provide the necessary conditions for the implementation of the International Energy Agency's sustainability strategies. The International Energy Agency stated that the use of biofuels should increase by at least 10% annually (2022c).

Butanol has a high potential for use because it is an alcohol fuel that can be produced by fermentation of plant materials and has properties close to fossil fuels (Doğan 2011). Butanol has significant advantages over other alcohol fuels. The lower heating value of butanol provides a better engine performance due to its lower heating value higher than that of methanol and ethanol (Kumar et al. 2013), and it

M. Vargün (✉) · A. Yapmaz · I. T. Yılmaz · C. Sayın
Faculty of Technology, Mechanical Engineering, Marmara University, Istanbul, Turkey

M. Z. Sogut et al. (eds.), *Proceedings of the 2022 International Symposium on Energy Management and Sustainability*, Springer Proceedings in Energy,
https://doi.org/10.1007/978-3-031-30171-1_10

can be transported more safely due to the high flash point of butanol (Giakoumis et al. 2013). In addition, butanol shows very good miscibility with fossil diesel fuel (Sarathy et al. 2014).

N-heptane fuel, which is shown with the formula C_7H_{16} and has a straight chemical chain, is used as a reference fuel because its octane number is zero. Since its chemical properties are close to fossil-based diesel fuel, it has a high potential for use in compression ignition engines. It has properties close to fossil-based diesel fuel in terms of combustion characteristics and exhaust emission formation. It was stated that n-heptane fuel showed better combustion performance, especially in low operating conditions (Kolaitis and Founti 2009; ÇeliK et al. 2016).

The aim of this study is to examine the combustion characteristics of a four-cylinder, common rail diesel engine using different fuel mixtures at 80 Nm engine load and three different engine speeds (1500 rpm, 1600 rpm and 1700 rpm). The data obtained as a result of the experiments were compared with the results obtained with fossil-based diesel fuel, which is the reference fuel.

10.2 Another First-Level Paragraph

In this study, a four-stroke, four-cylinder, turbocharged diesel engine with common rail fuel injection system was used. As the main parts of the test system, it consists of test engine, eddy current dynamometer, common rail fuel injection system, cylinder gas pressure measurement system and exhaust emission measurement system. The schematic view of the experimental system is given in Fig. 10.1.

Engine tests were carried out using three different fuel types (D100, B20, H20) at 80 Nm constant engine load and different engine loads (1500 rpm, 1600 rpm and 1700 rpm), and their effects on combustion characteristics were investigated. The characteristics of the test engine used in the experiments are given in Table 10.1. In addition, the amount of air taken into the engine was determined by mass with the help of an air flow meter. A type K temperature sensor is used to measure intake air temperature, engine oil temperature, fuel temperature, exhaust gas temperature and cooling water inlet and outlet temperatures. In the engine tests, eddy current dynamometer of Cusson-P8602 type, maximum power of 150 kW, maximum engine speed of 8000 rpm and maximum torque of 475 Nm was used.

Three different fuel types were used in the experiments. As the reference fuel, the fuel containing 100% fossil-based diesel fuel and named as D100 was used. Other fuel mixtures were prepared by mixing it with D100 fuel at a rate of 20% by volume. The fuel mixture containing 20% butanol + 80% D100 fuel by volume is named as B20. The fuel mixture containing 20% n-heptane + 80% D100 fuel by volume is named as H20. After the fuel mixtures were prepared, observations were made for a certain period of time and no phase separation was observed as a result of the observations.

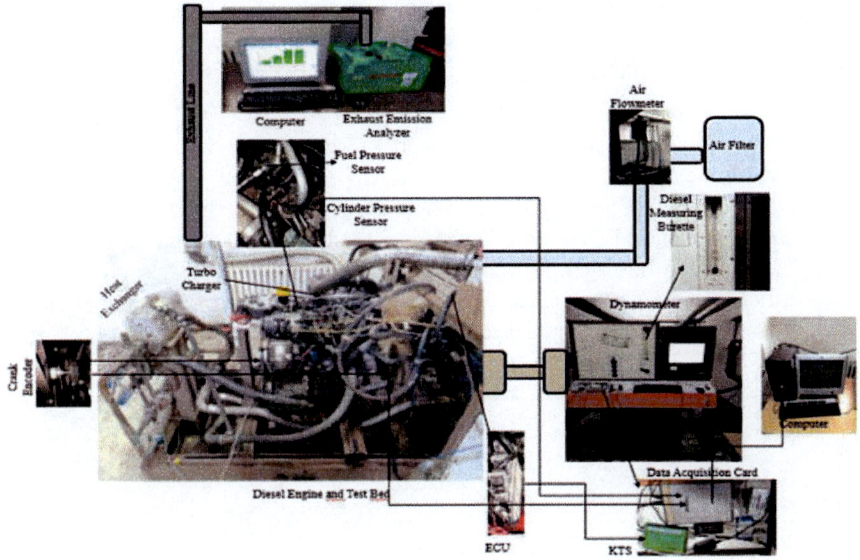

Fig. 10.1 Schematic view of the experimental set-up

Table 10.1 Specifications of the test engine

Engine type	In-line
Cylinder volume	1461 cm^3
Bore	76 mm
Stroke	80.5 mm
Cylinder number	4
Valves	8
Compression rate	18.25:1
Maximum power (4000d/d)	48 kW (65 hp)
Maximum torque (1750d/d)	160 nm
Fuel system	Common rail

Before starting the experiments, the engine coolant temperature was expected to reach 80–90 °C for stable operation of the engine. For combustion analysis, 200 cycles were averaged in the experiments. As a result of the engine tests, cylinder gas pressure, ignition delay, combustion time and NO emission values were measured. The results obtained were compared with the results obtained with the reference fuel D100 fuel.

10.3 Result and Discussion

10.3.1 Cylinder Gas Pressure

In Fig. 10.2, the effects of using different fuel mixtures on cylinder gas pressure at different engine speeds are given. In the tests performed at 1500 and 1600 rpm engine speeds, the highest maximum cylinder gas pressure values were obtained in the use of B20 fuel, 111.5 bar and 111.6 bar, respectively. It is thought that due to the oxygen content in B20 fuel, it improves combustion in the cylinder and causes an increase in cylinder gas pressure. On the other hand, the lowest maximum cylinder gas pressure values at all engine speeds were determined as 108.5 bar at 1500 rpm,

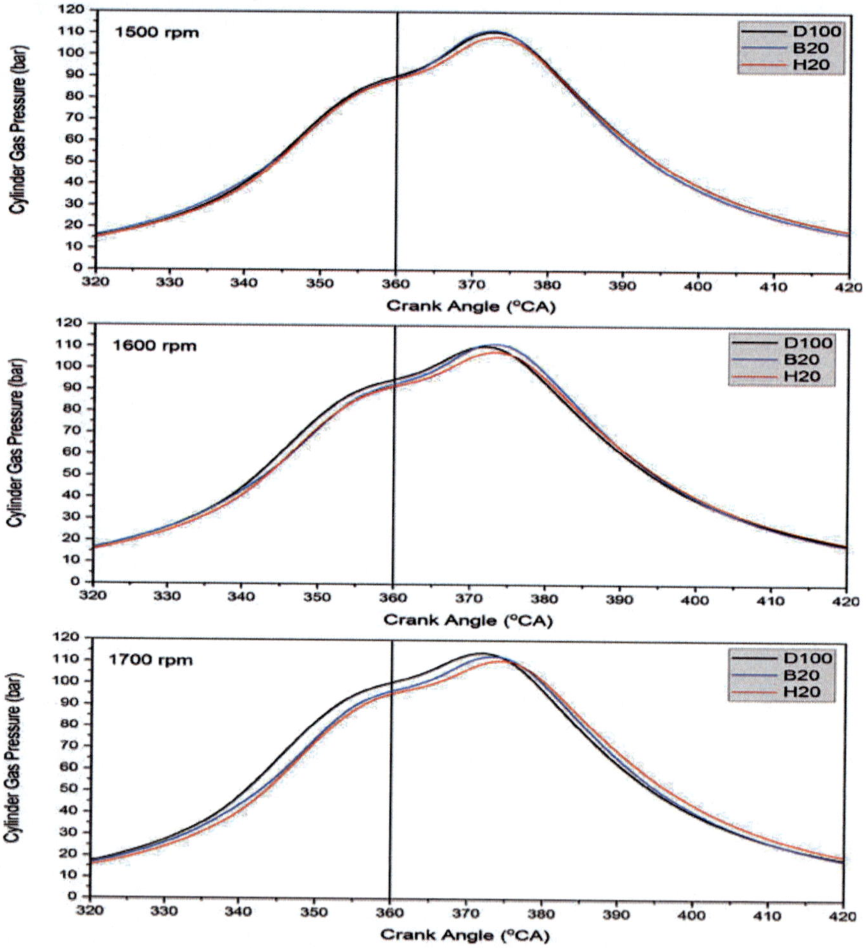

Fig. 10.2 Effect of fuel blends on cylinder gas pressure at different engine speed

108.1 bar at 1600 rpm and 110.8 bar at 1700 rpm with H20 fuel. In all test conditions, the highest maximum cylinder gas pressure was obtained with D100 fuel at 1700 rpm and 114.1 bar at 372 crank angle (°CA). The highest maximum cylinder gas pressure values for all fuel types were determined at 1700 rpm. As a result of the increase in engine speed, a more homogeneous mixture is obtained due to the increase in air-fuel movements in the cylinder; it is thought that there is an increase in cylinder gas pressures.

10.3.2 Ignition Delay and Combustion Duration

In Fig. 10.3, the effects of using different fuel mixtures on ignition delay and combustion duration at different engine speeds are given. The highest ignition delay values were observed in the use of B20 fuel at all engine speeds tested. The highest ignition delay times were obtained with B20 fuel as 17°CA at 1500 rpm and 1600 rpm engine speeds. Due to the low cetane number and high latent heat of vaporization of butanol, it is thought that the ignition delay times increase with the use of B20 fuel. On the other hand, when compared to D100 fuel, the ignition delay times were reduced as a result of the use of H20 fuel. The lowest ignition delay time was obtained at 1600 rpm and 1700 rpm with H20 fuel as 10°CA. It is thought that shorter ignition delay times are seen with the use of H20 fuel, since n-heptane fuel

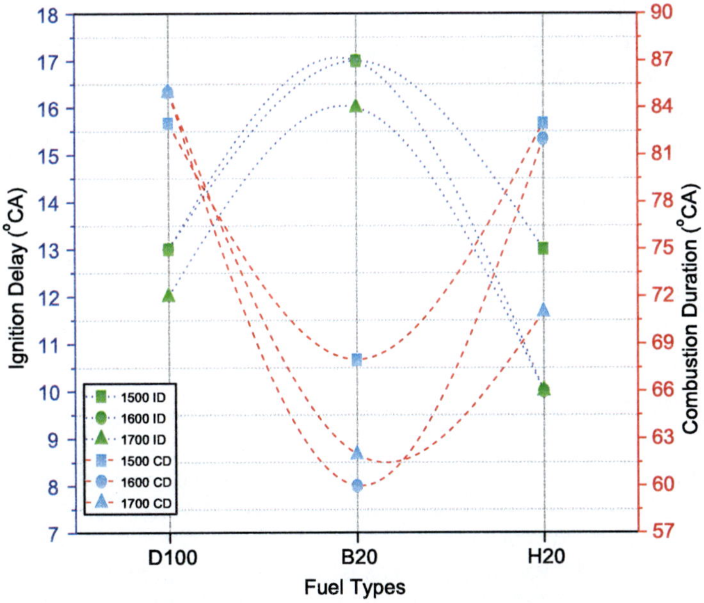

Fig. 10.3 Effect of fuel blends on CD and ID at different engine speed

does not show much resistance to self-ignition due to the high cetane number of the fuel and ignites more easily. In addition, it has been observed that the increase in engine speed in the use of D100 fuel causes a shortening of the ignition delay times. In the use of D100 fuel, the shortest ignition delay was obtained as 12°CA at 1700 rpm engine speed.

When the changes in combustion duration in the use of D100, B20 and H20 fuels were examined, it was seen that the use of B20 and H20 fuels caused a decrease in combustion durations. At the tested engine speeds, the shortest combustion durations were determined in the use of B20 fuel. The shortest combustion time was obtained as 60°CA at 1600 rpm. It is thought that the reason for the shortening of the combustion durations as a result of the use of B20 fuel is that the oxygen content of butanol improves the combustion in the cylinder and provides a faster combustion. In the use of D100 fuel, the combustion time at 1500 rpm was 83°CA, while the combustion times at 1600 rpm and 1700 rpm were 85°CA. In the use of H20 fuel, the shortest combustion time was obtained as 71°CA at 1700 rpm engine speed, while the longest combustion time was 83°CA at 1500 rpm.

10.3.3 NO Emission

In Fig. 10.4, the effects of using different fuel mixtures on NO emissions at different engine speeds are given. In all test conditions, the highest NO emission values were obtained with the use of B20 fuel, while the lowest NO emission values were obtained with H20 fuel. The highest NO emission was determined as 5.6 g/kWh with the use of B20 fuel at 1500 rpm, while it was determined as 5.18 g/kWh with D100 fuel and 4.3 g/kWh with the use of H20 fuel at the same engine speed. In all test conditions, the lowest NO emission was observed as 4.22 g/kWh at 1600 rpm as a result of the use of H20 fuel. In the tests performed at 1700 rpm engine speed, the highest NO emission was observed as 5.27 g/kWh with B20 fuel, while 4.95 g/kWh and 4.27 g/kWh were obtained with the use of D100 and H20 fuels. Due to the oxygen content of B20 fuel, combustion is improved in the cylinder and high temperatures are achieved in the cylinder. Therefore, it is thought that there is an increase in NO emissions.

10.4 Conclusion

In this study, the effects of using D100, B20 and H20 fuels on combustion characteristics at different engine speeds (1500, 1600 and 1700 rpm) in a common rail diesel engine were investigated.

Compared to D100 fuel, the cylinder gas pressure values improved with the use of B20 fuel, while a slight decrease was detected as a result of the use of H20 fuel.

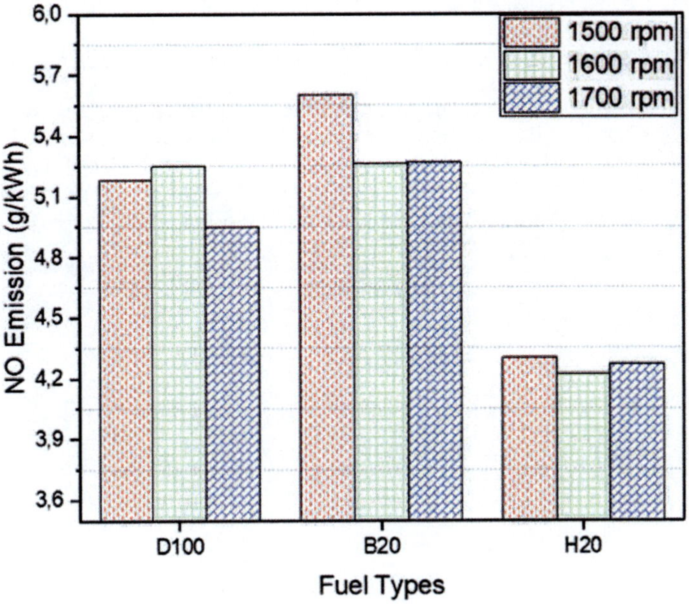

Fig. 10.4 Effect of fuel blends on NO emission at different engine speed

When compared to D100 fuel at all engine speeds tested, approximately 20% reduction was observed in combustion duration as a result of the use of B20 fuel, while an increase of approximately 20% was observed in the ignition delay times. With the use of H20 fuel, an increase in ignition delay was observed by more than 10%, while combustion durations were found to be close to D100 fuel.

Compared to D100 fuel, in NO emission, approximately 7% increase was observed in the use of B20 fuel, while the use of H20 fuel was decreased up to 10% in NO emission.

References

ÇelİK, M. et al. (2016) 'Examination Of The Effects Of N-Heptane Addition To The Canola Methyl Ester On Engine Performance And Combustion Characteristics', J. of Thermal Science and Technology, 36(1), pp. 9–16.

Doğan, O. (2011) 'The influence of n-butanol/diesel fuel blends utilization on a small diesel engine performance and emissions', Fuel, 90(7), pp. 2467–2472. doi:https://doi.org/10.1016/j.fuel.2011.02.033.

Giakoumis, E.G. et al. (2013) 'Exhaust emissions with ethanol or n-butanol diesel fuel blends during transient operation: A review', Renewable and Sustainable Energy Reviews, 17, pp. 170–190. doi:https://doi.org/10.1016/j.rser.2012.09.017.

IEA (2020) (2022a) 'European Union 2020'. IEA. Available at: https://www.iea.org/reports/european-union-2020.

IEA (2020), (2022b) 'Tracking Transport 2020'. IEA. Available at: https://www.iea.org/reports/tracking-transport-2020.

IEA (2020) (2022c) 'Transport Biofuels'. IEA. Available at: https://www.iea.org/reports/transport-biofuels.

Kolaitis, D.I. and Founti, M.A. (2009) 'On the assumption of using n-heptane as a "surrogate fuel" for the description of the cool flame oxidation of diesel oil', Proceedings of the Combustion Institute, 32(2), pp. 3197–3205. doi:https://doi.org/10.1016/j.proci.2008.06.073.

Kumar, S. et al. (2013) 'Advances in diesel–alcohol blends and their effects on the performance and emissions of diesel engines', Renewable and Sustainable Energy Reviews, 22, pp. 46–72. doi: https://doi.org/10.1016/j.rser.2013.01.017.

Sarathy, S.M. et al. (2014) 'Alcohol combustion chemistry', Progress in Energy and Combustion Science, 44, pp. 40–102. doi:https://doi.org/10.1016/j.pecs.2014.04.003.

Chapter 11
Co-combustion of Sewage Sludge with Eco-friendly Fuels to Reduce CO_2 Emissions in Flue Gas

Kubilay Bayramoğlu, Can Coskun, and Zuhal Oktay

Nomenclature

BAC	Boeing Airplane Company
LCC	Lockheed California Company
IATA	International Air Transport Association
ACI	Airport Council International

11.1 Introduction

Sewage sludge is a waste containing organic and heavy metals in its structure. The main technique used in the removal of these wastes is incineration with fuel (Fedorov et al. 2021). Sewage sludge wastes have been widely co-combustion with fossil fuels such as natural gas and coal in recent years (Wilk et al. 2021). In recent years, the use of cleaner energy sources has been made compulsory by many countries due to the emissions resulting from the combustion of fossil fuels. Hydrogen fuel is one of the main fuels used in heat and power applications due to its high combustion efficiency and containing no carbon. However, since hydrogen fuel is not directly found in free form in nature, it is obtained by different techniques (Paparao and Murugan 2021). One of the most common methods used in hydrogen

K. Bayramoğlu
Maritime Faculty, Bulent Ecevit University, Zonguldak, Turkey
e-mail: kubilay.bayramoglu@beun.edu.tr

C. Coskun (✉) · Z. Oktay
Engineering Faculty, Izmir Democracy University, Izmir, Turkey
e-mail: can.coskun@idu.edu.tr; zuhal.oktay@deu.edu.tr

M. Z. Sogut et al. (eds.), *Proceedings of the 2022 International Symposium on Energy Management and Sustainability*, Springer Proceedings in Energy,
https://doi.org/10.1007/978-3-031-30171-1_11

production is electrolysis using solar energy. Electrolysis is an environment-friendly hydrogen production technique that produces hydrogen using only water. In addition to hydrogen, oxygen, which is one of the necessary substances for combustion, is also formed by electrolysis (Cao et al. 2022). Another fuel type that has increased in importance in recent years and is used in combustion processes is ammonia. Ammonia is used instead of fossil fuels because of its no carbon content and its ability to be synthesized with renewable energy sources (Chai et al. 2021).

Sewage sludge is a waste with high water content. The water in its component evaporates with the effect of temperature, and the dry sludge sewage is co-combustion with natural gas. In this study, the co-combustion of hydrogen and ammonia together with sewage sludge as an alternative to natural gas was investigated. Considering the thermodynamic properties of hydrogen, ammonia and natural gas, the amount of fuel giving the same heat energy was determined for each fuel. Hydrogen production was achieved by electrolysis using solar energy. Solar panel and system cost calculations were made for the hydrogen fuel used in the power plant. In addition, to provide energy balance, exhaust gas temperatures and possible emission rates have been determined for each fuel. Emissions were determined by chemical equilibrium constants.

11.2 Methodology

In this study, modelling was carried out for a power plant producing 1 MW of power. In the combustion chamber, the sewage sludge and fuel are co-combustion. As shown in Fig. 11.1, flue gas has been formed in this process.

Sewage sludge containing 76% water by mass was used as waste. The components and mass properties of sewage sludge and natural gas co-combustion are presented in Table 11.1 (Coskun et al. 2020). Outdoor temperature conditions have a crucial impact on the combustion chamber. Energy efficiency of the combustion chamber increases with air temperature (Coskun et al. 2014).

Specific heat capacity and enthalpy for the exhaust gas components formed as a result of combustion were calculated by curve fitting in accordance with the equations given in Eqs. (11.1) and (11.2) (Ferguson 1985).

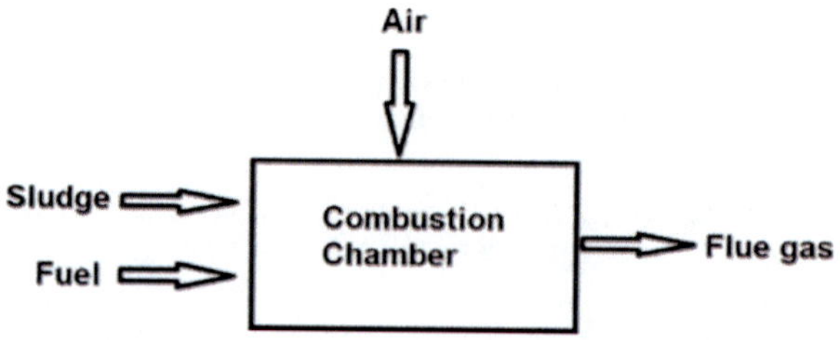

Fig. 11.1 Schematic diagram of combustion chamber

Table 11.1 Fuel properties

Atoms	Units	Sludge	Natural gas
C	kg/s	0.224328	0.01785
H	kg/s	0.024907	0.00595
O	kg/s	0.108279	–
S	kg/s	0.005510	–
N	kg/s	0.0042540	–
Ash	kg/s	0.145475	–
H_2O	kg/s	1.74496	–
LHV	KJ/kg	1121.312	54630.488

$$\frac{C_p}{R} = a_1 + a_2T + a_3T^2 + a_4T^3 + a_5T^4 \tag{11.1}$$

$$\frac{h}{RT} = a_1 + \frac{a_2}{2}T + \frac{a_3}{3}T^2 + \frac{a_4}{4}T^3 + \frac{a_5}{5}T^4 + \frac{a_6}{T} \tag{11.2}$$

In the given equations, C_p is the specific heat capacity, R is the ideal gas constant and T is the temperature. Total flue gas-specific heat, enthalpy of each products and total enthalpy are calculated as Eqs. (11.3), (11.4) and (11.5), respectively (McAllister et al. 2011):

$$C_{P_{\text{flue gas}}} = \sum_{i=1}^{n} v_i C_{P_i} \tag{11.3}$$

$$h_i = \int C_{P_i} dT \tag{11.4}$$

$$h_{\text{flue gas}} = \sum_{i=1}^{n} v_i h_i \tag{11.5}$$

In addition, the possible ash $C_{P_{\text{ash}}}$ value in the flue gas and sewage can be calculated with Eq. (11.6) (Coskun et al. 2020).

$$C_{P_{\text{ash}}} = \frac{-197.633 + 1.164 \cdot T}{1 + 0.838 \cdot T + 1.115 \cdot 10^{-5} \cdot T^2} \tag{11.6}$$

The co-combustion of sewage sludge with alternative fuels such as hydrogen and ammonia was carried out numerically according to the chemical equilibrium using the equilibrium constant. The general form of the expression of co-combustion of sewage sludge with alternative fuel is given in Eq. (11.7). In the given equations, it is assumed that 12 species are formed in the exhaust compositions (Kayadelen and Ust 2013).

$$\text{Sewage sludge} + \text{Alternative fuel} + \frac{a_s}{\phi}(O_2 + 3.76N_2) \rightarrow n_1CO_2 + n_2H_2O + n_3N_2 + n_4O_2 + n_5CO + n_6H_2 + n_7H + n_8O + n_9OH + n_{10}NO + n_{11}SO + n_{12}\text{ash} \quad (11.7)$$

where a_s is stochiometric air fuel ratio, ϕ is equivalence ratio and n_1, n_2,. . . . n_{10} are mole of flue gas. The equivalence ratio is described as in Eq. (11.8) and mole fraction of the products is obtained as in Eq. (11.9).

$$a_s = n_c + \frac{n_H}{4} - \frac{n_O}{2} \quad (11.8)$$

$$v_i = \frac{n_i}{n_T} \quad (11.9)$$

where n_c, n_H and n_O are atom number of carbon, hydrogen and oxygen in the fuel, respectively. Also, n_i is mole of the i. species in the flue gas and n_T is the total mole of flue gas.

11.2.1 Chemical Equilibrium Method

The chemical equilibrium constants method is a numerical method proposed by Olikara and Borman (1975), which provides the estimation of the combustion products in equilibrium with the help of thermodynamic properties. Equation (11.10) shows the combustion of any hydrocarbon fuel and the possible types that can occur (Ferguson 1985).

$$C_aH_bO_cN_d + \frac{a_s}{\phi}(O_2 + 3.76N_2) \rightarrow n_1CO_2 + n_2H_2O + n_3N_2 + n_4O_2 + n_5CO + n_6H_2 + n_7H + n_8O + n_9OH + n_{10}NO \quad (11.10)$$

The given expression is also solved with ten unknown equations. Atom balance equations can be written in Eqs. (11.11)–(11.14)

$$C : a = (y_1 + y_5)N \quad (11.11)$$

$$H : b = (2y_2 + 2y_6 + y_7 + y_9)N \quad (11.12)$$

$$O : c + 2a_s/\phi = (2y_1 + y_2 + 2y_4 + y_5 + y_8 + y_9 + y_{10})N \quad (11.13)$$

$$N : d + 7.52a_s/\phi = (2y_3 + y_{10})N \quad (11.14)$$

where N is the total number of moles. It can be expressed as Eq. (11.15) so that the sum of the mole fractions equals 1.

$$\sum_{i=1}^{10} y_i = 1 \tag{11.15}$$

The three equations obtained from the given expressions are presented in Eqs. (11.16), (11.17), and (11.18) (Kökkülünk et al. 2014).

$$d_1 = \frac{b}{a} \tag{11.16}$$

$$d_2 = \frac{c}{a} + 2\frac{a_s}{\phi a} \tag{11.17}$$

$$d_3 = \frac{d}{a} + \frac{7.52a_s}{\phi a} \tag{11.18}$$

In addition to the given equations, six gas-phase equilibrium reactions are presented between Eqs. (11.19), (11.20), (11.21), (11.22), (11.23), and (11.24), including components such as hydrogen, oxygen and water (Gonca 2015).

$$\frac{1}{2}H_2 \Leftrightarrow H \qquad K_1 = \frac{y_7 P^{1/2}}{{y_6}^{1/2}} \tag{11.19}$$

$$\frac{1}{2}O_2 \Leftrightarrow O \qquad K_2 = \frac{y_8 P^{1/2}}{{y_4}^{1/2}} \tag{11.20}$$

$$\frac{1}{2}H_2 + \frac{1}{2}O_2 \Leftrightarrow OH \qquad K_3 = \frac{y_9}{{y_4}^{1/2}{y_6}^{1/2}} \tag{11.21}$$

$$\frac{1}{2}O_2 + \frac{1}{2}N_2 \Leftrightarrow NO \qquad K_4 = \frac{y_{10}}{{y_4}^{1/2}{y_3}^{1/2}} \tag{11.22}$$

$$\frac{1}{2}H_2 + \frac{1}{2}O_2 \Leftrightarrow H_2O \qquad K_5 = \frac{y_2}{{y_4}^{1/2}{y_6}^{1/2}p^{1/2}} \tag{11.23}$$

$$CO + \frac{1}{2}O_2 \Leftrightarrow CO_2 \qquad K_6 = \frac{y_2}{{y_4}^{1/2}{y_5}^{1/2}p^{1/2}} \tag{11.24}$$

Olikara ve Borman (1975) expresses the equilibrium equations between 600 and 4000 K with curve fitting as Eq. 11.25.

$$\log_{10} K_i(T) = A_i \ln\left(\frac{T}{1000}\right) + \frac{B_i}{T} + C_i + D_i T + E_i T^2 \tag{11.25}$$

where T is the Kelvin and K_i is the curve-fit coefficient represented in Table 11.2.

Table 11.2 Equilibrium constant K_i curve-fit coefficients

K_i	A_i	B_i	C_i	D_i	E_i
K_1	+0.432168E+00	−0.112464E +05	+0.267269E +01	−0.745744E−04	+0.242484E−08
K_2	+0.310805E+00	−0.129540E +05	+0.321779E +01	−0.738336E−04	+0.344645E−08
K_3	−0.141784E+00	−0.213308E +04	+0.853461E +00	+0.355015E−04	−0.310227E−08
K_4	+0.150879E−01	−0.470959E +04	+0.646096E +00	+0.272805E-05	-0.154444E-08
K_5	-0.752364E+00	+0.124210E +05	-0.260286E +01	+0.259556E-03	-0.162687E-07
K_6	-0.415302E-02	+0.148627E +05	-0.475746E +01	+0.124699E-03	-0.900227E-08

11.2.2 System Definition

The chemical formula of sewage sludge used in the given equation was obtained from atomic balance as $C_{4.32}H_{8.58}O_2N_{0.61}S_{0.04}$ (Coskun et al. 2020). The mass flow rate of ammonia (NH_3) and hydrogen (H_2) used as alternative fuels are given in Eq. (11.26).

$$\dot{m}_{NG} \cdot LHV_{NG} = \dot{m}_{H_2} \cdot LHV_{H_2} = \dot{m}_{NH_3} \cdot LHV_{NH_3} \tag{11.26}$$

In the given equations, LHV_{NG}, LHV_{H_2} and LHV_{NH_3} is lower heating value for natural gas, hydrogen and ammonia, respectively. In this study, LHV is taken as 120 MJ/kg and 18 MJ/kg for hydrogen and ammonia, respectively. The hydrogen used in the co-combustion of sludge sewage and hydrogen was produced by proton exchange membrane (PEM) electrolysis. In PEM electrolysis, water is consumed and hydrogen and oxygen (O_2) are formed (Akyuz et al. 2011). The chemical equation of PEM electrolysis is presented in Eq. (11.27).

$$H_2O \rightarrow H_2 + \frac{1}{2}O_2 \tag{11.27}$$

In the given equation, 1 mole of hydrogen and 0.5 mole of oxygen are produced for 1 mole of water consumption. Produced hydrogen will be added to the system as fuel and oxygen will be added to the system as excess air. Thus, the nitrogen (N_2) rate in the air supplied to the system will be reduced. Combustion chamber energy balance is calculated as Eq. (11.28).

$$\dot{E}_{sluge} + \dot{E}_{fuel} + \dot{E}_{air} = \dot{E}_{flue\ gas} \tag{11.28}$$

where $\dot{E}_{sludge}$ is sludge energy rate, $\dot{E}_{fuel}$ is fuel energy rate, $\dot{E}_{air}$ is air energy rate and $\dot{E}_{flue\ gas}$

is flue gas energy rate. $\dot{E}_{sludge}$ energy rate can be determined as Eq. (11.29) (Coskun et al. 2020).

$$\dot{E}_{sludge} = h_{sludge}(T) \cdot \dot{m}_{sludge} \tag{11.29}$$

In the given expression, $h_{sludge}(T)$ represents the enthalpy of sewage sludge. It is accepted that water and ash are in the composition of sewage sludge.

Eqs. (11.30), (11.31) and (11.32) show the co-combustion of natural gas, hydrogen and ammonia with sewage sludge, respectively.

$$C_{4.32}H_{8.58}O_2N_{0.61}S_{0.04} + CH_4 + \frac{a_s}{\phi}(O_2 + 3.76N_2) \rightarrow F \tag{11.30}$$

$$C_{4.32}H_{8.58}O_2N_{0.61}S_{0.04} + H_2 + \frac{a_s}{\phi}(aO_2 + 3.76N_2) \rightarrow F \tag{11.31}$$

$$C_{4.32}H_{8.58}O_2N_{0.61}S_{0.04} + NH_3 + \frac{a_s}{\phi}(O_2 + 3.76N_2) \rightarrow F \tag{11.32}$$

In the given equations, the coefficient a denotes the oxygen addition has been formed by PEM electrolysis, and F denotes the flue gas. In addition, the coefficient of ϕ for each equation is taken as 0.8.

11.3 Results and Discussion

Combustion processes of three fuels have been examined separately with sewage sludge. The total energy for each fuel has been assumed to be the same for 1 MW electricity production. The compressed inlet air temperature, which is introduced with blower to the combustion chamber, has been taken as 905 K each combustion model. In addition, natural gas, hydrogen and ammonia used as a fuel temperature have been assumed to be 288 K. The enthalpy values used in the energy balance of the fuel given to the system are taken as LHV values.

The total energy of sewage sludge, natural gas and air calculated according to the temperature values are given in Table 11.3.

Table 11.4 expresses the mass flow rates, enthalpy and energy flow rates of sewage sludge, hydrogen, combustion air and flue gas. The hydrogen used in the combustion of the sludge was produced from water by the PEM electrolysis method.

Table 11.3 Specification of fuel for CH_4

Specification	Enthalpy [kJ/kg]	Mass flow rate [kg/s]	Energy flow rate [kW]
Sludge + water	2943	2.296	6757
Natural gas	55,500	0.0238	1320.9
Air	939.6	4.998	4735.58
Flue gas	1792	7.32	12,813

Table 11.4 Specification of fuel for H_2

Specification		Enthalpy [kJ/kg]	Mass flow rate [kg/s]	Energy flow rate [kW]
Sludge + water		2943	2.296	6757
Hydrogen		120,000	0.011	1320.9
Air	O_2	939.6	1.26	4830.3
	N_2		3.875	
Flue gas		1733.169	7.44	12908.5

Table 11.5 Specification of fuel for NH_3

Specification	Enthalpy [kJ/kg]	Mass flow rate [kg/s]	Energy flow rate [kW]
Sludge+water	2943	2.296	6757
Ammonia	18,900	0.0699	1320.9
Air	939.6	5.48	5149
Flue gas	1687.1	7.84	13226.9

Table 11.6 Some thermodynamic parameters

ν_i	Products	CH_4	H_2	NH_3
ν_1	CO_2	0.077936	0.07546	0.071809
ν_2	H_2O	0.399317	0.423155	0.40602
ν_3	N_2	0.489966	0.467703	0.489823
ν_4	O_2	0.025689	0.026375	0.025363
ν_5	CO	1.84E-08	3.87E-09	3.88E-09
ν_6	H_2	1.06E-08	3.13E-09	3.2E-09
ν_7	H	0	0	0
ν_8	O	6.52E-09	1.67E-09	1.7E-09
ν_9	OH	3.01E-06	1.35E-06	1.34E-06
ν_{10}	NO	0.000153	9.34E-05	9.44E-05
ν_{11}	SO	0.000743	0.000772	0.000738
ν_{12}	Ash	0.006194	0.00644	0.006152
T [K]		1339	1295	1264
$Cp_{flue\ gas}$ (kJ/kg·K)		1.339	1.34	1.334
$h_{flue\ gas}$ (kJ/kg)		1792.921	1735.3	1686.176

Oxygen formed together with hydrogen by PEM electrolysis was added to the system as combustion air. Thus, the oxygen concentration in the combustion air increased. The *a* given in Eq. 11.28 has been determined as 1.08 with the addition of excess oxygen. The value given in Eq. (11.28) was 1.08 with the addition of excess oxygen. The mass flow rate of hydrogen used as fuel has been determined to give the same energy as natural gas.

The mass flow rate of fuel and air to be supplied to the combustion chamber using ammonia instead of natural gas as fuel and the mass flow rate of flue gas formed are given in Table 11.5.

For the general combustion equation given in Eq. 11.7, the mole fractions of each combustion product are given in Table 11.6 for different fuels. The obtained data

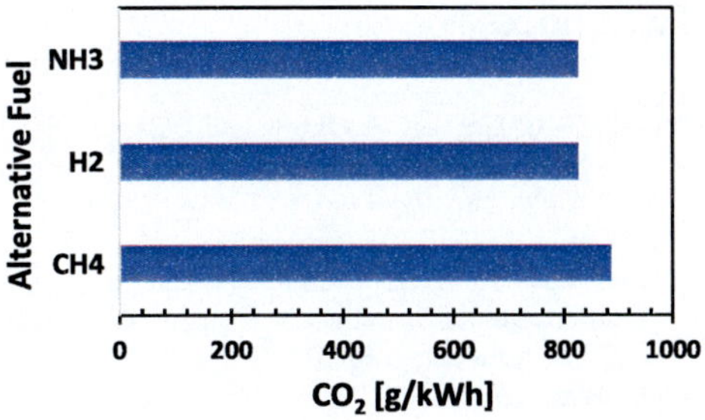

Fig. 11.2 Amount of CO_2 for alternative fuels

were obtained from five atomic equilibrium and seven equilibrium constants equations. Thus, 12 unknowns have been solved by linearized 12 equilibrium equations using the Gaussian method. The total enthalpy, temperature and specific heat values of the products formed because of the co-combustion of sludge and eco-friendly fuels are the thermodynamic parameters presented in Table 11.6.

Another subject examined within the scope of the study is the change in the amount of carbon dioxide according to the combustion of natural gas, hydrogen and ammonia. Figure 11.2 shows the change of three fuels CO_2 emissions.

The findings show that carbon dioxide emissions resulting from the co-combustion of hydrogen and ammonia fuel, which do not contain carbon atoms, with sewage sludge are reduced by 60.56 g/kWh compared to the co-combustion of natural gas.

11.4 Conclusion

The use of eco-friendly fuels in combustion processes plays an important role in reducing emissions. In this study, the co-combustion process of hydrogen and ammonia with sewage sludge was investigated and compared with natural gas. As a result of the study, the end products of combustion, thermodynamic properties and the amount of CO_2 emission were determined.

The amounts of hydrogen and ammonia and the ratio of air used in combustion were determined to provide the same energy, and it was ensured that the flue gas molar fraction was determined by using the method of chemical equilibrium constants.

Another result obtained in the study is the determination of CO_2 emissions, which are important for sustainability. While the amount of CO_2 formed because of co-combustion sewage sludge with natural gas was 888 g/kWh, it was found as 827.44 g/kWh in the use of hydrogen and ammonia. The emission amounts given show that CO_2 emissions resulting from sustainable hydrogen and ammonia combustion have decreased by 6.82% compared to natural gas combustion. Therefore, the use of environmentally friendly fuels in combustion processes plays an important role in sustainability and reducing emissions.

References

Akyuz, E., C. Coskun, Z. Oktay, and I. Dincer. 2011. "Hydrogen Production Probability Distributions for a PV-Electrolyser System." International Journal of Hydrogen Energy 36(17): 11292–99. https://doi.org/10.1016/j.ijhydene.2010.11.125.

Cao, Yan et al. 2022. "Hydrogen Production Using Solar Energy and Injection into a Solid Oxide Fuel Cell for CO2 Emission Reduction; Thermoeconomic Assessment and Tri-Objective Optimization." Sustainable Energy Technologies and Assessments 50(November 2021): 101767. https://doi.org/10.1016/j.seta.2021.101767.

Chai, Wai Siong et al. 2021. "A Review on Ammonia, Ammonia-Hydrogen and Ammonia-Methane Fuels." Renewable and Sustainable Energy Reviews 147(January): 111254. https://doi.org/10.1016/j.rser.2021.111254.

Coskun, Can, Zuhal Oktay, Tunc Koksal, and Bahadır Birecikli. 2020. "Co-Combustion of Municipal Dewatered Sewage Sludge and Natural Gas in an Actual Power Plant." Energy 211.

Coskun, Can, Ertürk, Mustafa, Oktay, Zuhal, Hepbasli, Arif. 2014. "A new approach to determine the outdoor temperature distributions for building energy calculations." Energy Conversion and Management 78: 165–172

Fedorov, A. V. et al. 2021. "Combustion of Sewage Sludge in a Fluidized Bed of Catalyst: ASPEN PLUS Model." Journal of Hazardous Materials 405(August 2020): 124196. https://doi.org/10.1016/j.jhazmat.2020.124196.

Ferguson, C R. 1985. International Combustion Engines; Applied Thermosciences. United States: John Wiley and Sons,New York, NY. https://www.osti.gov/biblio/5822574.

Gonca, Guven. 2015. "Investigation of the Influences of Steam Injection on the Equilibrium Combustion Products and Thermodynamic Properties of Bio Fuels (Biodiesels and Alcohols)." Fuel 144: 244–58. https://doi.org/10.1016/j.fuel.2014.12.032.

Kayadelen, Hasan Kayhan, and Yasin Ust. 2013. "Prediction of Equilibrium Products and Thermodynamic Properties in H 2O Injected Combustion for CαH BOγNδ Type Fuels." Fuel 113(x): 389–401. https://doi.org/10.1016/j.fuel.2013.05.095.

Kökkülünk, Görkem et al. 2014. "Theoretical and Experimental Investigation of Steam Injected Diesel Engine with EGR." Energy 74(C): 331–39.

McAllister, Sara, Jyh-Yuan Chen, and A. Carlos Fernandez-Pello. 2011. Fundamentals of Combustion Processes. http://link.springer.com/10.1007/978-1-4419-7943-8.

Olikara, Cherian, and Gary L. Borman. 1975. "A Computer Program for Calculating Properties of Equilibrium Combustion Products with Some Applications to I.C. Engines." Automotive Engineering Congress and Exposition: 23.

Paparao, Jami, and S. Murugan. 2021. "Oxy-Hydrogen Gas as an Alternative Fuel for Heat and Power Generation Applications - A Review." International Journal of Hydrogen Energy 46(76): 37705–35. https://doi.org/10.1016/j.ijhydene.2021.09.069.

Wilk, Małgorzata, Maciej Śliz, and Bogusław Lubieniecki. 2021. "Hydrothermal Co-Carbonization of Sewage Sludge and Fuel Additives: Combustion Performance of Hydrochar." Renewable Energy 178: 1046–56.

Chapter 12
Energy Usage in Glass Industry: Past, Today, and Tomorrow

Onur Kodak, Farshid Sadeghi-Khaneghah, Alp Er Ş. Konukman, Levent Kılıç, Neşet Arzan, and Gürhan Dural

Nomenclature

SPP	Solar power plant
PGC	Power grid connection
NGF	Natural gas furnace
PV	Photovoltaic

12.1 Introduction

In our world where technology is accelerating day by day, the glass industry has an important place. A series of products such as flat glass, container glass, household glassware, fiberglass, etc. are generally produced in the glass industry. The glass sector is an energy-intensive industrial sector and the industrial energy demand of this sector is increasing day by day. From past to present, many different types of energy sources (biomass, coal, oil, natural gas, renewable energy) have been used in the glass industry. The glass industry, which is among the industries involving high-temperature processes, is of great importance worldwide (Conradt 2019). A large

O. Kodak · F. Sadeghi-Khaneghah · A. E. Ş. Konukman (✉)
Faculty of Engineering, Gebze Technical University, Kocaeli, Türkiye
e-mail: o.kodak2020@gtu.edu.tr; f.sadeghi2021@gtu.edu.tr; konukman@gtu.edu.tr

L. Kılıç · N. Arzan
Şişecam R&D Center, Kocaeli, Türkiye
e-mail: lkilic@sisecam.com; narzan@sisecam.com

G. Dural
7Cbasalia Global, Istanbul, Türkiye
e-mail: gurhan.dural@7cbasalia.com

M. Z. Sogut et al. (eds.), *Proceedings of the 2022 International Symposium on Energy Management and Sustainability*, Springer Proceedings in Energy,
https://doi.org/10.1007/978-3-031-30171-1_12

part of the energy consumed in glass furnaces is used for melting the glass raw material at high temperatures (Zier et al. 2021). Container glass (bottles and jars) represents approximately 60% of the entire glass industry and 30% for the construction and automotive industries. Ten percent of the glass production is made of glass fiber and special glasses (Griffin et al. 2021; Sardeshpande et al. 2007). Product outputs in the glass industry are container glass, flat glass, household glass, special glass continuous filament fiberglass, and glass wool (Zier et al. 2021). As fossil fuels pollute the environment, increasing energy efficiency in systems is of great importance for the future of the world (Mirzaei et al. 2018) (Sardeshpande et al. 2007).

12.1.1 Past

Historically, wood, coal, natural gas, and electricity have been used as energy sources in glass production (Griffin et al. 2021). Since the outbreak of the oil crisis in the last century, the need to reduce energy consumption per unit product has become one of the key factors in industrial furnace designs (Weber et al. 2020). Glass products are widely used in buildings, automobiles, solar cells and display devices (Li et al. 2020). Glass furnaces generally operate 8760 h/year and its lifetime is approximately 10 years. When the term of lifetimes expires, the furnace destroys and rebuilt (Danieli et al. 2019). While the life of a glass furnace was 1–1.5 years in 1930, the life of the furnace exceeding 10 years has now become a standard (Conradt 2019).

12.1.2 Today

Today, glass melting is an energy-intensive process that requires high temperatures and is mostly derived from the combustion of natural gas (Giuffrida et al. 2018). When fuel oil, coal, or natural gas is used as the main energy source in the production system, approximately 1.2798, 0.6250 and 0.4498 t_{CO_2}/t_{glass} emissions occur, respectively (Hu et al. 2018). The Chinese glass industry meets its energy needs with fuel oil (13.1%), natural gas (15.5%), coal (44.3%), electricity, and other sources (27.1%). On the other hand, the USA and Europe use natural gas as an energy source in the glass industries with a share of 80% and 90%, respectively (Zier et al. 2021). Today, energy conversions have to be evaluated not only for the variability of unit costs but also for the energy gain of spaces such as roof space, its direct impact on the reduction of emissions and waste, and efficient resource use.

12.1.3 Tomorrow

Natural gas will continue to be the main fuel for glass production until 2050 (Griffin et al. 2021). But in the future, countries are planning to use renewable energy sources such as hydrogen, nitrogen, biomass, solar energy and wind energy instead of carbon-based fossil fuels.

Hydrogen gas is accepted as the cleanest energy source that can be obtained from different production technologies (El-Shafie et al. 2021). The energy content of hydrogen is approximately 2.5 times higher than conventional fuels (El-Shafie et al. 2021).

Oxygen and nitrogen gas are used in many industrial processes. A computer model of a nitrogen generator was created and analyzed to enrich the oxygen content of combustion air in a furnace for energy-saving purposes. They further stated that the optimum level of oxygen content in the air is between 21% and 30% and that more than 20% fuel savings can be achieved with oxygen enrichment. In this range (21–30%), there is an optimum point between energy savings and increased pollutant emissions. The disadvantage of increasing the oxygen concentration in the combustion air is higher pollutant emissions (Ming et al. 2013).

Biomass constitutes approximately 14% of the total energy consumption in the world and is classified as the fourth energy source in the world (Abuelnuor et al. 2014). Biomass energy is generally obtained from wood and wood waste (64%), solid waste (24%), agricultural waste (5%), and landfill gas (5%) (Abuelnuor et al. 2014). Biomass energy reduces the dependence of humanity on petroleum products, natural gas, and coal (Abuelnuor et al. 2014). Biomass material absorbs carbon dioxide during growth and releases energy during combustion by releasing the carbon dioxide it contains (Abuelnuor et al. 2014).

12.2 Energy Consumption in Glass Furnaces

Generally, glass furnaces consume about 80% of the total energy consumed in a glass factory (Sardeshpande et al. 2011; Sardeshpande et al. 2007; Zier et al. 2021). The theoretical energy requirements for glass production are endothermic heat for glass reaction, sensible heat for glass heating, and sensible heat for intermittent gases (gases from the glass reaction) (Sardeshpande et al. 2007). Even the most efficient furnace has an energy consumption of about 70% higher than the theoretical minimum consumption (2275 kJ/kg) (Sardeshpande et al. 2007). If the cullet is not used, soda-lime glass (container glass) needs 2671 kJ/kg of energy. If the glass furnace is working with 100% cullet, the theoretical minimum energy requirement is 1886 kJ/kg. The reason for this decrease is due to the decrease in the heat of reaction (487 kJ/kg) and the sensible heat (298 kJ/kg) of intermittent gases (Sardeshpande et al. 2007).

Recovery of waste heat increases energy efficiency and reduces the consumption of fossil energy sources. Therefore, it can reduce greenhouse gas emissions (CO_2 and NO_x and SO_x, etc.) (Ren and Wang 2017). In addition, efforts to increase energy efficiency in glass furnaces are accelerating day by day due to reasons such as greenhouse gas emissions from carbon fuels, global warming, and fuel savings.

Although electric furnaces are not used much today, due to the expensiveness of electric energy (up to 85%), it is estimated that these furnaces will be used more widely in the future. One of the main advantages of electric melting compared to fossil fuel melting is its high efficiency (Zier et al. 2021). In many countries, electricity prices are higher than natural gas prices. Electric melting is a highly energy-efficient method for glass melting production. Electrical energy is generally used in furnaces either as full electric or as an electrical supplement. In the electric booster furnace, the electric booster system is incorporated into the fuel-fired glass furnace (Li et al. 2019).

12.3 Feasibility of Using Hydrogen or Electric Energy Instead of Natural Gas

For improving the existing furnace's energy efficiency and alternative for prospects, the feasibility of using other energy sources alongside natural gas for melting the glass is investigated. For this, four different cases of glass melting heat sources are considered for a glass production factory with four furnaces. These cases for melting heat sources are as follows and illustrated in Fig. 12.1:

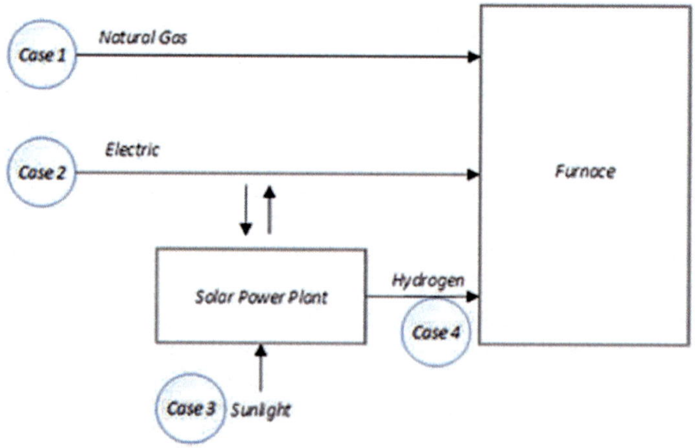

Fig. 12.1 System schematic for four cases

Table 12.1 General information

Parameters	Symbols	Unit	Values
Furnace production capacity	$\dot{m}_r$	$\left[\frac{ton_{glass}}{day}\right]$	400
Total glass production	$\dot{m}_{r,m}$	$\left[\frac{ton_{glass}}{day}\right]$	1600
Glass minimum theoretical melting energy	e_{min}	$\left[\frac{kJ}{kg}\right]$	1563 (Sardeshpande et al. 2007)
Natural gas thermal energy input per glass mass	$e_{n,f}$	$\left[\frac{kJ}{kg}\right]$	3838 (Sardeshpande et al. 2007)
Hydrogen thermal energy input per glass mass	e_{H_2f}	$\left[\frac{kJ}{kg}\right]$	3473
Electric melting thermal energy input per glass mass	$e_{e,f}$	$\left[\frac{kJ}{kg}\right]$	1839
Natural gas combustion furnace thermal efficiency	$\eta_{n,f}$	[%]	42 (Li et al. 2020)
Hydrogen combustion furnace thermal efficiency (≈)	η_{H_2f}	[%]	45
Electrically melting furnace thermal efficiency	$\eta_{e,f}$	[%]	85 (Zier et al. 2021)
Natural gas low heating value	LHV_{NG}	$\left[\frac{MJ}{Nm^3}\right]$	35.450
Hydrogen low heating value	LHV_{H_2}	$\left[\frac{MJ}{Nm^3}\right]$	10.75
Electrolyze power consumption	H_{H_2}	$\left[\frac{kWh}{Nm^3.H_2}\right]$	4.3 (SinoHy Energy 2022)

- Case 1: 100% from the combustion of natural gas (reference and common case)
- Case 2: 40% from electric energy (obtained from grid connection) and 60% from the combustion of natural gas (electrically boosted furnaces)
- Case 3: 40% from electric energy (obtained from the solar power plant and from grid connection) and 60% from the combustion of natural gas (renewable integrated electrically boosted furnaces)
- Case 4: 40% from the combustion of hydrogen (produced by an electrolyzing process using electric energy obtained from the solar power plant and from grid connection) and 60% from the combustion of natural gas

All four furnaces are the same dimensions and the modifications are considered. Information about furnaces, factory, and solar power plant are presented in Table 12.1:

Some of the most important information are presented in Table 12.1, including general information regarding the furnaces, natural gas, and hydrogen heating properties and efficiencies of natural gas and hydrogen combustion furnaces.

12.3.1 Case 1: Full Natural Gas

For melting 4 × 400 tons (1600 tons) glass per day using natural gas combustion, the total required natural gas energy $E_{tot,1}$ (Eq. 12.1) based on the data from Table 12.1 and using the equation below is:

$$\begin{aligned} E_{\text{tot},1} = \dot{m}_{r,m} \cdot e_{n,f} = 1{,}600{,}000\left(\frac{\text{kg}}{\text{day}}\right) \cdot 3838\left(\frac{\text{kJ}}{\text{kg}}\right) \cdot \frac{1}{3600}\left(\frac{\text{kWh}}{\text{kJ}}\right) \\ = 1{,}705{,}777\left(\frac{\text{kWh}}{\text{day}}\right) \end{aligned} \tag{12.1}$$

The required natural gas volume V_{NG} based on Eq. 12.2 is:

$$\begin{aligned} V_{\text{NG}} = \dot{m}_{r,m} \cdot e_{n,f} \cdot \frac{1}{\text{LHV}_{\text{NG}}} = 1{,}600{,}000\left(\frac{\text{kg}}{\text{day}}\right) \cdot 3838\left(\frac{\text{kJ}}{\text{kg}}\right) \cdot \frac{1}{35{,}450}\left(\frac{\text{Nm}^3}{\text{kJ}}\right) \\ = 173{,}224\left(\frac{\text{Nm}^3}{\text{day}}\right) \end{aligned} \tag{12.2}$$

where $\dot{m}_{r,m}$ is the total glass production capacity, $e_{n,f}$ is the natural gas thermal energy input per glass mass, and LHV_{NG} is the natural gas low heating value.

12.3.2 Case 2: Electric Boosting Natural Gas Furnace (40% Electric)

In electric boosting furnaces, electrodes are submerged in glass melt to assist the melting process. Using specific heat from Table 12.1, the energy required for melting 40% of glass by electrodes E_{elec} based on Eq. 12.3 is:

$$\begin{aligned} E_{\text{elec}} = 0.4 \cdot \dot{m}_{r,m} \cdot e_{e,f} = 0.4 \cdot 1{,}600{,}000\left(\frac{\text{kg}}{\text{day}}\right) \cdot 1839\left(\frac{\text{kJ}}{\text{kg}}\right) \cdot \frac{1}{3600}\left(\frac{\text{kWh}}{\text{kJ}}\right) \\ = 326{,}933\left(\frac{\text{kWh}}{\text{day}}\right) \end{aligned} \tag{12.3}$$

where $\dot{m}_{r,m}$ is the total glass production capacity and $e_{n,f}$ is the electric melting thermal energy input per glass mass.

The energy required for melting 60% of glass by combustion of natural gas energy E_{NG} based on Eq. 12.4 is:

$$E_{\mathrm{NG}} = 0.6 \cdot E_{\mathrm{tot,1}} = 1,023,466\left(\frac{\mathrm{kWh}}{\mathrm{day}}\right) \tag{12.4}$$

The required natural gas volume V_{NG} is (Eq. 12.5):

$$\begin{aligned} V_{\mathrm{NG}} &= 0.6 \cdot 1,600,000\left(\frac{\mathrm{kg}}{\mathrm{day}}\right) \cdot 3838\left(\frac{\mathrm{kJ}}{\mathrm{kg}}\right) \cdot \frac{1}{35,450}\left(\frac{\mathrm{Nm}^3}{\mathrm{kJ}}\right) \\ &= 103,934\left(\frac{\mathrm{Nm}^3}{\mathrm{day}}\right) \end{aligned} \tag{12.5}$$

The total required energy in electric boosting furnaces $E_{tot,\,2}$ is (Eq. 12.6):

$$E_{\mathrm{tot,2}} = E_{\mathrm{elec}} + E_{\mathrm{NG}} = 1,350,399\left(\frac{\mathrm{kWh}}{\mathrm{day}}\right) \tag{12.6}$$

12.3.3 *Case 3: Electric Boosting (Solar Power Plant and Power Grid Connection) Natural Gas Furnace (40% SPP-PGC, 60% NGF)*

In this case, part of the required electric energy for melting glass using electrodes is obtained from a renewable source. This renewable source is solar panels installed on the roof of the factory. Based on data provided from Şişecam Co. for the Turkey/Eskişehir-Polatlı factory, based on Table 12.2, solar panels can produce an average daily photovoltaic energy $E_{\mathrm{elec},\,pv}$ of 41,737 (kWh/day), a minimum of 21,548 (kWh/day) of energy for December and 69,676 (kWh/day) for July.

Therefore, in the maximum capacity, the solar power plant can supply up to 21.31% of the electricity demand of electrodes, and the rest should be supplied by a grid connection. The energy used by the power grid $E_{\mathrm{elec},\,\mathrm{grid}}$ is (Eq. 12.7):

Table 12.2 Solar power plant information

Parameters	Unit	Values
Factory roof area	[m^2]	100,000
Total panel area	[m^2]	66,666
Photovoltaic panel efficiency (≈)	[%]	20
Installation capacity (peak)	[kW$_p$]	13,333

$$E_{\text{elec,grid}} = E_{\text{elec}} - E_{\text{elec,pv}} = 326,933\left(\frac{\text{kWh}}{\text{day}}\right) - 41,737\left(\frac{\text{kWh}}{\text{day}}\right) = 285,196\left(\frac{\text{kWh}}{\text{day}}\right) \quad (12.7)$$

where E_{elec} is the energy required for melting 40% of glass by electrodes and $E_{\text{elec, pv}}$ is the average daily photovoltaic energy.

The total required energy in electric boosting (photovoltaic energy and grid energy) furnaces $E_{\text{tot, 3}}$ based on Eq. 12.8 is:

$$E_{tot,3} = E_{\text{elec,grid}} + E_{\text{elec,pv}} + E_{\text{NG}} = 1,350,399\left(\frac{\text{kWh}}{\text{day}}\right) \quad (12.8)$$

12.3.4 Case 4: Hydrogen and Natural Gas Furnace (40% Hydrogen)

Using specific heat from Table 12.1, the energy required for melting 40% of glass by hydrogen combustion energy E_{H_2} based on Eq. 12.9 is:

$$E_{\text{H}_2} = 0.4 \cdot \dot{m}_{r,m} \cdot e_{\text{H}_2,f} = 0.4 \cdot 1,600,000\left(\frac{\text{kg}}{\text{day}}\right) \cdot 3473\left(\frac{\text{kJ}}{\text{kg}}\right) \cdot \frac{1}{3600}\left(\frac{\text{kWh}}{\text{kJ}}\right) = 617,422\left(\frac{\text{kWh}}{\text{day}}\right) \quad (12.9)$$

where $e_{\text{H}_2,f}$ is the hydrogen thermal energy input per glass mass. The required hydrogen gas volume V_{H_2} (Eq. 12.10) is:

$$V_{\text{H}_2} = 0.4 \cdot \dot{m}_{r,m} \cdot e_{H_2,f} \cdot \frac{1}{LHV_{H_2}} = 0.4 \cdot 1,600,000\left(\frac{\text{kg}}{\text{day}}\right) \cdot 3473\left(\frac{\text{kJ}}{\text{kg}}\right) \cdot \frac{1}{10,750}\left(\frac{\text{Nm}^3}{\text{kJ}}\right) = 206,764\left(\frac{\text{Nm}^3}{\text{day}}\right) \quad (12.10)$$

where $e_{\text{H}_2,f}$ is the hydrogen gas thermal energy input per glass mass and LHV_{H_2} is the hydrogen gas low heating value.

For producing this amount of hydrogen from water by electrolyzing process, according to Table 12.1, the required energy $E_{\text{electrolyze}}$ based on Eq. 12.11 is:

$$E_{\text{electrolyze}} = H_{H_2} \cdot V_{H_2} = 4.3\left(\frac{\text{kWh}}{\text{Nm}^3}\right) \cdot 206,764\left(\frac{\text{Nm}^3}{\text{day}}\right) = 889,088\left(\frac{\text{kWh}}{\text{day}}\right) \quad (12.11)$$

where H_{H_2} is the electrolyze power consumption per cubic meter of hydrogen and V_{H_2} (Eq. 12.12) is the required hydrogen gas volume.

$$E_{\text{elec,grid}} = E_{\text{electrolyze}} - E_{\text{elec,pv}} = 889,088 - 41,737 = 857,351\left(\frac{\text{kWh}}{\text{day}}\right) \quad (12.12)$$

In the maximum capacity, the solar power plant can supply only up to 7.83% of the electricity demand of hydrogen production (16, 189 $\left(\frac{\text{Nm}^3}{\text{day}}\right)$), and the rest should be supplied by a grid connection. In this case, the amount of energy needed from the power grid $E_{\text{tot, 4}}$ based on Eq. 12.13 is:

$$E_{\text{tot,4}} = E_{\text{electrolyze}} + E_{\text{NG}} = 1,922,554\left(\frac{\text{kWh}}{\text{day}}\right) \quad (12.13)$$

12.4 Conclusion

In this study, a brief introduction to the glass industry and many aspects of it through time is discussed. Also, three alternative cases suggested for melting 40% of daily melting capacity instead of natural gas combustion, which gives us four cases, and one of these cases is a reference case for comparison. These cases are as follows: using electric boosting for melting glass all obtained from grid connection, supplying a part of this electric energy from rooftop PV panels, and finally melting 40% of glass by hydrogen burners. All these four cases were investigated analytically and the results are presented in Table 12.3. In the cases examined, it has been shown computationally that a conventional system working with natural gas can be

Table 12.3 Energy obtained from conventional sources (1600 ($\text{ton}_{\text{glass}}$/day) pull rate)

Case	Natural gas combustion energy E_{NG} $\left(\frac{\text{kWh}}{\text{day}}\right)$	Electric energy from the grid $E_{\text{elec, grid}}$ $\left(\frac{\text{kWh}}{\text{day}}\right)$	Electric energy from the photovoltaic panel $E_{\text{elec, pv}}$ $\left(\frac{\text{kWh}}{\text{day}}\right)$	Total energy E_{tot} $\left(\frac{\text{kWh}}{\text{day}}\right)$
Case 1	1,705,777	0	0	1,705,777
Case 2	1,023,466	326,933	0	1,350,399
Case 3	1,023,466	285,196	41,737	1,350,399
Case 4	1,023,466	857,351	41,737	1,922,554

hybridized with electrical energy obtained from the sun, which can be met as a renewable energy source. It has also been conceptually shown that the hybrid structure can be further diversified by hydrogen energy production and use and can be freed from dependence on only one energy source.

As we see from Table 12.3, hydrogen production is an energy-consuming process and rooftop PVs cannot respond to its demand; therefore, most of the required energy should be supplied from grid connection. Also, electric melting uses the lowest energy compared to natural gas and hydrogen combustion melting. Finally, we can say that in case of feasibility, the best option is to use electric boosting since it has higher efficiency compared to combustion melting, even compared to hydrogen combustion melting. In addition, using electrical melting reduces greenhouse gas (CO_2 and NO_x and SO_x, etc.). It is better we supply electric melting required energy from renewable sources such as solar power plant. But, the required space for producing this much of energy from solar panels is too much and the area designated for the PV panels in the factory of interest is low.

Acknowledgment This study was supported by the Scientific and Technological Research Council of Türkiye (TÜBİTAK), (2244 Industrial Ph.D. Fellowship Program, Project No: 119C189).

References

Abuelnuor AAA, Wahid MA, Hosseini SE, et al. (2014) Characteristics of biomass in flameless combustion: A review. Renewable and Sustainable Energy Reviews 33:363–370. https://doi.org/10.1016/J.RSER.2014.01.079

Conradt R (2019) Prospects and physical limits of processes and technologies in glass melting. Journal of Asian Ceramic Societies 7:377–396. https://doi.org/10.1080/21870764.2019.1656360

Danieli P, Rech S, Lazzaretto A (2019) Supercritical CO2 and air Brayton-Joule versus ORC systems for heat recovery from glass furnaces: Performance and economic evaluation. Energy 168:295–309. https://doi.org/10.1016/j.energy.2018.11.089

El-Shafie M, Kambara S, Hayakawa Y (2021) Energy and exergy analysis of hydrogen production from ammonia decomposition systems using non-thermal plasma. International Journal of Hydrogen Energy 46:29361–29375. https://doi.org/10.1016/J.IJHYDENE.2020.08.249

Giuffrida A, Chiesa P, Drago F, Mastropasqua L (2018) Integration of oxygen transport membranes in glass melting furnaces. In: Energy Procedia. Elsevier Ltd, pp 599–606. https://doi.org/10.1016/j.egypro.2018.08.147

Griffin PW, Hammond GP, McKenna RC (2021) Industrial energy use and decarbonisation in the glass sector: A UK perspective. Advances in Applied Energy 3:100037. https://doi.org/10.1016/J.ADAPEN.2021.100037

Hu P, Li Y, Zhang X, et al. (2018) CO2 emission from container glass in China, and emission reduction strategy analysis. Carbon Management 9:303–310. https://doi.org/10.1080/17583004.2018.1457929

Li L, Han J, Lin H-J, et al. (2020) Simulation of glass furnace with increased production by increasing fuel supply and introducing electric boosting. International Journal of Applied Glass Science 11:170–184. https://doi.org/10.1111/IJAG.13907

Li L, Lin H-J, Han J, et al. (2019) Three-Dimensional Glass Furnace Model of Combustion Space and Glass Tank with Electric Boosting. Materials Transactions 60:1034–1043. https://doi.org/10.2320/MATERTRANS.M2019044

Ming X, Borgnakke DS, Campos MA, et al. (2013) Possibility of combustion furnace operation with oxygen-enriched gas from nitrogen generator. University of Michigan, ACEEE Summer Study on Energy Efficiency in Industry.

Mirzaei M, Ahmadi MH, Mobin M, et al. (2018) Energy, exergy and economics analysis of an ORC working with several fluids and utilizes smelting furnace gases as heat source. Thermal Science and Engineering Progress 5:230–237. https://doi.org/10.1016/J.TSEP.2017.11.011

Ren L, Wang H (2017) Parameter optimization and performance comparison of several power cycles for waste heat recovery from moderate temperature flue gas. Energy Procedia 142:1333–1339. https://doi.org/10.1016/J.EGYPRO.2017.12.516

Sardeshpande V, Anthony R, Gaitonde UN, Banerjee R (2011) Performance analysis for glass furnace regenerator. Applied Energy 88:4451–4458. https://doi.org/10.1016/J.APENERGY.2011.05.028

Sardeshpande V, Gaitonde UN, Banerjee R (2007) Model based energy benchmarking for glass furnace. Energy Conversion and Management 48:2718–2738. https://doi.org/10.1016/J.ENCONMAN.2007.04.013

SinoHy Energy (2022) 25m^3/h Alkaline Water Electrolysis Hydrogen Generation Equipment. https://www.sinohyenergy.com/25m3-h-alkaline-water-electrolysis-hydrogen-generation-equipment/. Accessed 28 Feb 2022

Weber R, Gupta AK, Mochida S (2020) High temperature air combustion (HiTAC): How it all started for applications in industrial furnaces and future prospects. Applied Energy 278:. https://doi.org/10.1016/J.APENERGY.2020.115551

Zier M, Stenzel P, Kotzur L, Stolten D (2021) A review of decarbonization options for the glass industry. Energy Conversion and Management: X 10:. https://doi.org/10.1016/J.ECMX.2021.100083

Chapter 13
A Multi-criteria Evaluation Framework for Prioritizing the Geothermal Power Plant Site Selection Factors by Fuzzy AHP

Ertugrul Ayyildiz and Alev Taskin

Nomenclature

AHP	Analytic hierarchical process
MCDM	Multi-criteria decision-making

13.1 Introduction

Energy is one of the basic inputs necessary for social and economic development (Toman and Jemelkova 2003). In parallel with the increasing population, urbanization, industrialization, the development of technology and the increase in the level of welfare, energy consumption is increasing in an unstoppable way. Approximately 90% of the world's energy consumption is met by coal, oil and natural gas, which are called fossil fuels (Abas et al. 2015). The fact that fossil energy sources create pollution due to the high rate of carbon dioxide given to the air makes it necessary to use alternative energy sources (Long et al. 2015). On the other hand, considering that fossil fuel reserves and coal will run out in 200 years and oil in 40 years (Shafiee and Topal 2009), it becomes clear how important it is to replace them with new energy sources. This situation requires efficient use of energy and benefiting from renewable energy sources. For these reasons, studies on the research and utilization

E. Ayyildiz (✉)
Engineering Faculty, Industrial Engineering, Karadeniz Technical University, Trabzon, Turkiye
e-mail: ertugrulayyildiz@ktu.edu.tr

A. Taskin
Engineering Faculty, Industrial Engineering, Karadeniz Technical University, Trabzon, Turkiye

Faculty of Machinery, Industrial Engineering, Yildiz Technical University, Istanbul, Turkiye

M. Z. Sogut et al. (eds.), *Proceedings of the 2022 International Symposium on Energy Management and Sustainability*, Springer Proceedings in Energy,
https://doi.org/10.1007/978-3-031-30171-1_13

of alternative energy sources that can replace fossil fuels have accelerated in recent years.

Renewable energy refers to the energy source that can maintain its current state in the evolution of nature (Litvinenko 2020). Renewable energy sources can be classified as solar energy, wind energy, biomass energy, hydrogen energy, hydraulic energy, geothermal energy and water-power energies obtained from wave energy and fusion energy (Yücenur and Ipekçi 2021).

Geothermal energy is an important alternative energy source because it does not cause air pollution with its low carbon dioxide emission rate and it is renewable (Adedoyin et al. 2021). Compared to renewable energy sources such as solar and wind, geothermal energy has an advantageous position as it is uninterrupted (Choudhary et al. 2022). Geothermal resource is defined as hot water and steam, which is created by the heat accumulated in various depths of the earth, whose temperature values are constantly above the atmospheric average temperature in the region and which may contain more various minerals, salts and gases than the normal surface and underground waters around it (Lee 2001). Geothermal energy includes all kinds of indirect and direct benefits obtained from these sources. Depending on its temperature, geothermal energy is used in various fields in industry, especially for electricity generation, heating and treatment purposes (Lund et al. 2011). It is possible to integrate a high-temperature geothermal fluid in many areas.

Turkey has a great geothermal potential due to the abundance of orogenic, magmatic and volcanic activities in connection with its location on the Alpine-Himalayan orogenic belt (Akpinar et al. 2008). Depending on active faults and volcanism in Turkey, there are more than 600 geothermal springs in the Aegean, Northwest, Central Anatolia and East and Southeast Anatolia regions (Hepbasli and Ozgener 2004). Turkey is the country with the 7th largest geothermal energy potential in the world (Akpinar et al. 2008).

Geothermal energy, which is one of the most important types of sustainable and renewable energy, requires a high amount of investment to be obtained. The amount of electricity produced in these power plants, which cause minimal damage to the environment, varies according to the conditions of the region. Although the power plant location is selected by taking some criteria into account, no study has been carried out to bring a comprehensive set of criteria. In this study, we prioritize the factors to evaluate the geothermal power plant location problem by fuzzy analytic hierarchy process (AHP).

13.2 Fuzzy AHP

AHP is a decision-making method used in solving complex multi-criteria decision-making (MCDM) problems and applied by assigning relative importance degrees to decision criteria and alternatives. AHP was first developed by Myers and Alpert in 1968, and then it was modelled by Saaty with a comprehensive study and started to

Table 13.1 Linguistic terms and corresponding fuzzy numbers

Linguistic term	Triangular fuzzy numbers
Absolutely low important – ALI	(1/9, 1/9, 1/9)
Very low important – VLI	(1/8, 1/7, 1/6)
Low important – LI	(1/6, 1/5, 1/4)
Slightly low important – SLI	(1/4, 1/3, 1/2)
Equal important – EI	(1, 1, 1)
Slightly high important – SHI	(2, 3, 4)
High important – HI	(4, 5, 6)
Very high important - VHI	(6, 7, 8)
Absolutely more important - AMI	(9, 9, 9)

be used in solving complex decision-making problems. AHP can be defined as a decision-making process based on the evaluation of decision alternatives in terms of criteria affecting the decision to be made by establishing a hierarchical decision structure (Yildiz et al. 2020).

The hierarchical structure of AHP enables to analyse complex decision-making problems by simplifying them (Solangi et al. 2019). One of the advantages of AHP is that it allows comparison of decision criteria used in both qualitative and quantitative situations. Due to such features, AHP is frequently used in decision-making processes (Dos Santos et al. 2019).

The emergence of fuzzy logic has become inevitable due to the inadequacy of the modern logic's structure, which has only one interpretation, rejects ambiguity and consists of right and wrong. Fuzzy logic is a logic that is modelled and processed based on fuzzy sets. With fuzzy sets, it is easy to model situations that are frequently encountered in real life without uncertainty and ambiguity. With the models made using linguistic variables in fuzzy set theory, results very close to real life can be obtained (Ayyildiz et al. 2021). Fuzzy sets are defined by assigning a membership degree to expressions containing uncertainty.

Fuzzy set theory has been applied to many problems and many fuzzy AHP applications have been made in the literature, since many decisions cannot be represented with precise numerical expressions or there are decisions that contain uncertainty (Yildiz et al. 2022). The fact that fuzzy logic can be applied to both qualitative and quantitative decision problems (Ayyildiz 2021) due to the use of linguistic variables has been one of the biggest reasons for applying fuzzy AHP. In fuzzy AHP, decision-makers can make more accurate evaluations, thanks to linguistic variables, while making pairwise comparisons of their criteria. In this study, fuzzy AHP proposed by Buckley is utilized to determine criteria weights. The application steps are given below (Buckley 1985):

Step 1: Criteria are determined for the specified decision-making problem.

Step 2: Pairwise comparison matrix is constructed using the linguistic terms given in Table 13.1 to determine the weights of the criteria.

Let a_{ij} be the pairwise comparison of criterion i with criterion j and n is the number of criteria.

$$A = \begin{bmatrix} 1 & a_{12} & \cdots & a_{1n} \\ a_{21} & 1 & \cdots & a_{2n} \\ \vdots & \vdots & \ddots & \vdots \\ a_{n1} & a_{n2} & \cdots & a_{nn} \end{bmatrix} \tag{13.1}$$

Step 3: The linguistic expressions are converted to triangular fuzzy numbers using Table 13.1.

$$\tilde{C} = \begin{bmatrix} 1 & \tilde{c}_{12} & \cdots & \tilde{c}_{1n} \\ \tilde{c}_{21} & 1 & \cdots & \tilde{c}_{2n} \\ \vdots & \vdots & \ddots & \vdots \\ \tilde{c}_{n1} & \tilde{c}_{n2} & \cdots & \tilde{c}_{nn} \end{bmatrix} \tag{13.2}$$

where $\tilde{c}_{ij}$ is the fuzzy comparison value of a_{ij}.

Step 4: Comparison matrix is tested for consistency.

$$\text{CI} = \frac{\lambda_{max} - n}{n - 1} \tag{13.3}$$

$$\text{CR} = \frac{\text{Cl}}{\text{RI}} \tag{13.4}$$

RI presents the random index and determined from Saaty's table (Saaty 1977). If consistency ratio (CR) is determined smaller than 0.1, then the matrix is determined as consistent.

Step 5: Geometric mean of fuzzy value is calculated for each criterion.

$$\tilde{r}_i = \left(\tilde{c}_{i1} \bigotimes \tilde{c}_{i2} \bigotimes \ldots \bigotimes \tilde{c}_{in}\right)^{1/n} \tag{13.5}$$

Step 6: Fuzzy weight matrix is calculated.

$$\tilde{w}_i = \left(\tilde{r}_i \bigotimes (\tilde{r}_1 + \tilde{r}_2 + \ldots + \tilde{r}_n)^{-1}\right. \tag{13.6}$$

Step 7: Fuzzy weights are converted to crisp weight (Gumus et al. 2013).

$$v_i = \frac{(u\tilde{w}_{ii} - l\tilde{w}_{ii}) + (m\tilde{w}_{ii} - l\tilde{w}_{ii})}{3 + l\tilde{w}_{ii}} \tag{13.7}$$

where $l\tilde{w}_i$, $m\tilde{w}_i$ and $u\tilde{w}_i$ present the lower, medium and upper value of v_i.

Step 8: Crisp weights are normalized.

$$w_i = \frac{v_i}{\sum_{i=1}^{n} v_i} \tag{13.8}$$

13.3 Application of Fuzzy AHP

The priorities of the criteria to select the best location for geothermal power plant are determined by fuzzy AHP in this section. For this purpose, criteria weights are determined by applying the fuzzy AHP steps described in detail in Sect. 13.2. Application phase is explained step by step below.

Step 1: Firstly, five criteria are determined based on the literature review and expert opinions as given in Fig. 13.1

Step 2: A five-person expert group is consulted to get their opinions about criteria. Two academicians from different universities, two managers from different energy firms and one manager from local government express their opinions about criteria. Table 13.2 presents the pairwise comparison matrix for each expert.

Step 3: Linguistic terms are converted to fuzzy numbers based on Table 13.1.

Step 4: After that, matrices are tested for consistency and determined as consistent.

To aggregate opinions from five different experts, geometric mean of fuzzy evaluation of each pairwise comparison is calculated, and aggregated decision matrix is constructed as given in Table 13.3.

Step 5: Geometric means for criteria are calculated by Eq. 13.5 as presented in Table 13.4.

Step 6: Fuzzy weight matrix is calculated by Eq. 13.6 and given in Table 13.5.

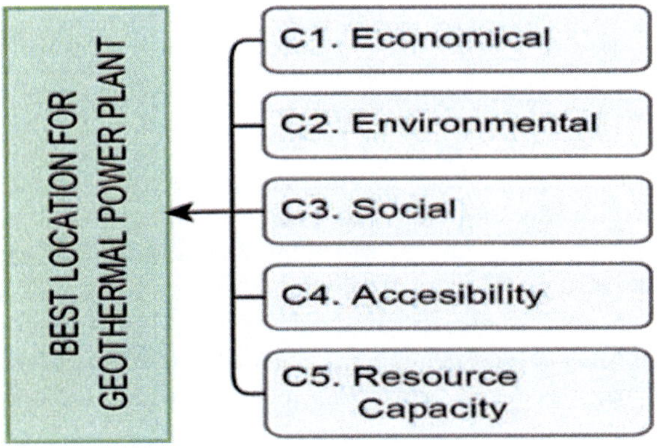

Fig. 13.1 Criteria for geothermal power plant location selection

Table 13.2 Pairwise comparison matrices

E-1	C1	C2	C3	C4	C5	E-4	C1	C2	C3	C4	C5
C1	EI	LI	SHI	SLI	SHI	C1	EI	SHI	AMI	VHI	HI
C2	HI	EI	AMI	SHI	HI	C2	SLI	EI	VHI	HI	SHI
C3	SLI	ALI	EI	LI	SHI	C3	ALI	VLI	EI	LI	LI
C4	SHI	SLI	HI	EI	VHI	C4	VLI	LI	HI	EI	SLI
C5	SLI	LI	SLI	VLI	EI	C5	LI	SLI	HI	SHI	EI
E-2	C1	C2	C3	C4	C5	E-5	C1	C2	C3	C4	C5
C1	EI	SHI	SHI	SHI	VHI	C1	EI	EI	HI	SHI	SLI
C2	SLI	EI	LI	SLI	SHI	C2	EI	EI	VHI	HI	SLI
C3	SLI	HI	EI	SHI	SHI	C3	LI	VLI	EI	EI	LI
C4	SLI	SHI	SLI	EI	SHI	C4	SLI	LI	EI	EI	VLI
C5	VLI	SLI	SLI	SLI	EI	C5	SHI	SHI	HI	VHI	EI
E-3	C1	C2	C3	C4	C5						
C1	EI	LI	SHI	AMI	SHI						
C2	HI	EI	AMI	AMI	AMI						
C3	SLI	ALI	EI	SHI	SLI						
C4	ALI	ALI	SLI	EI	SLI						
C5	SLI	ALI	SHI	SHI	EI						

Step 7: Unnormalized criteria weights are calculated by Eq. 13.7, as given in Table 13.6.

Step 8: Lastly, final criteria weights are determined by Eq. 13.8. Figure 13.2 shows the final weights.

According to criteria weights presented in Fig. 13.2, "C2. Environmental" is the most significant criterion for determining the best location to construct geothermal power plant. The criterion has the weight of 0.35. So, it can be said that decision-makers should consider environmental effect of the plant. They should develop environment-friendly strategies, especially for the waste discharge process. "C1. Economical" is determined as the second most important criterion. The installation, operation and other costs of the plant are important for decision-makers.

13.4 Conclusion

The energy problem in Turkey is growing day by day as well as in the rest of the world. Especially for Turkey, data on energy production and supply clearly reveals this situation. Turkey, which imports more than its own energy production, should develop its energy resources and thus increase its production capacity. In this situation, where fossil fuel reserves are about to run out and consumption is constantly increasing, but production is insufficient, it is necessary to use renewable energy potential efficiently in order to reduce Turkey's dependence on foreign sources.

Table 13.3 Aggregated decision matrix

	C1			C2			C3			C4			C5		
C1	1.00	1.00	1.00	0.64	0.82	1.00	3.10	4.14	5.10	2.22	2.85	3.57	1.89	2.54	3.29
C2	1.00	1.23	1.55	1.00	1.00	1.00	3.45	3.80	4.19	2.35	2.95	3.65	2.05	2.67	3.37
C3	0.20	0.24	0.32	0.24	0.26	0.29	1.00	1.00	1.00	0.64	0.82	1.00	0.49	0.65	0.87
C4	0.28	0.35	0.45	0.27	0.34	0.43	1.00	1.23	1.55	1.00	1.00	1.00	0.62	0.80	1.06
C5	0.30	0.39	0.53	0.30	0.37	0.49	1.15	1.53	2.05	0.94	1.25	1.61	1.00	1.00	1.00

Table 13.4 Geometric means of criteria

Criteria	Geometric means
C1: Economical	1.53, 1.89, 2.27
C2: Environmental	1.75, 2.06, 2.40
C3: Social	0.43, 0.51, 0.61
C4: Accessibility	0.54, 0.65, 0.79
C5: Resource capacity	0.63, 0.78, 0.97

Table 13.5 Fuzzy weights of criteria

Criteria	Fuzzy weights
C1: Economical	0.22, 0.32, 0.46
C2: Environmental	0.25, 0.35, 0.49
C3: Social	0.06, 0.09, 0.12
C4: Accessibility	0.08, 0.11, 0.16
C5: Resource capacity	0.09, 0.13, 0.20

Table 13.6 Unnormalized weights of criteria

Criteria	v_i
C1: Economical	0.334
C2: Environmental	0.363
C3: Social	0.090
C4: Accessibility	0.117
C5: Resource capacity	0.140

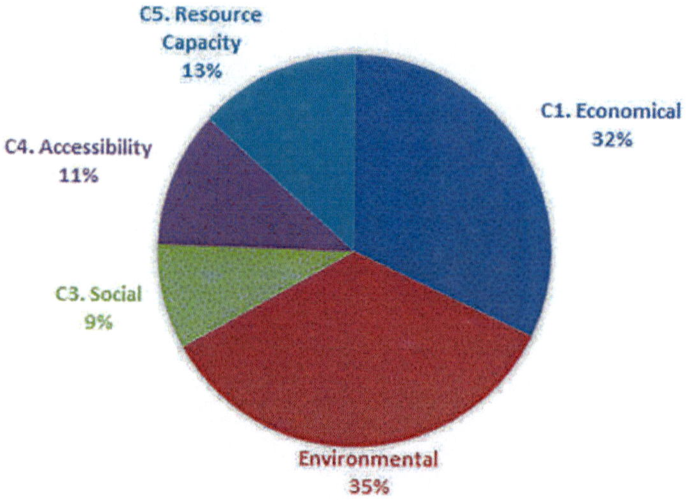

Fig. 13.2 Final criteria weights for geothermal power plant location selection

Turkey has a very good geographical position in terms of renewable energy potential. Being able to evaluate this potential correctly is extremely important in terms of the amount of energy to be obtained. Turkey has a very important potential in terms of geothermal energy source due to its location. Turkey is located in the Alpine-Himalayan orogenic belt. When the characteristics of the belt it is located in are examined, it is an area where orogenic, magmatic and volcanic activities are abundant. Therefore, there are many geothermal resource areas in the Aegean, Northwest, Central Anatolia and Southeast and East Anatolia regions where active faults and volcanism are located in the Alpine-Himalayan orogenic belt.

In this chapter, the criteria for determining the best location for geothermal power plant are evaluated. A multi-criteria decision-making analysis is carried out by adopting fuzzy AHP, where many criteria can be used effectively at the same time for the geothermal power plant location selection problem. As a result of the fuzzy AHP calculation, the most significant criterion is determined as the "Environmental".

As future suggestions, hybrid or integrated multi-criteria decision-making methods can be utilized to validate the results. The candidate locations can be evaluated with respect to these criteria.

References

Abas N, Kalair A, Khan N (2015) Review of fossil fuels and future energy technologies. Futures 69: 31–49. https://doi.org/10.1016/j.futures.2015.03.003

Adedoyin FF, Alola AA, Bekun FV (2021) The alternative energy utilization and common regional trade outlook in EU-27: Evidence from common correlated effects. Renew Sustain Energy Rev 145:111092. https://doi.org/10.1016/J.RSER.2021.111092

Akpinar A, Kömürcü MI, Önsoy H, Kaygusuz K (2008) Status of geothermal energy amongst Turkey's energy sources. Renew Sustain Energy Rev 12:1148–1161. https://doi.org/10.1016/J.RSER.2006.10.016

Ayyildiz E (2021) Interval valued intuitionistic fuzzy analytic hierarchy process-based green supply chain resilience evaluation methodology in post COVID-19 era. Environ Sci Pollut Res 2021 1: 1–19. https://doi.org/10.1007/S11356-021-16972-Y

Ayyildiz E, Erdogan M, Taskin Gumus A (2021) A Pythagorean fuzzy number-based integration of AHP and WASPAS methods for refugee camp location selection problem: a real case study for Istanbul, Turkey. Neural Comput Appl 1–18. https://doi.org/10.1007/s00521-021-06195-0

Buckley JJ (1985) Fuzzy hierarchical analysis. Fuzzy Sets Syst 17:233–247. https://doi.org/10.1016/0165-0114(85)90090-9

Choudhary A, Majumdar R, Saha SK (2022) Hybridisation of geothermal source with ORC-based load loop for uninterrupted generation of steady power. Int J Sustain Energy 41:58–84. https://doi.org/10.1080/14786451.2021.1895779

Dos Santos PH, Neves SM, Sant'Anna DO, et al. (2019) The analytic hierarchy process supporting decision making for sustainable development: An overview of applications. J Clean Prod 212: 119–138. https://doi.org/10.1016/J.JCLEPRO.2018.11.270

Gumus AT, Yesim Yayla A, Çelik E, Yildiz A (2013) A Combined Fuzzy-AHP and Fuzzy-GRA Methodology for Hydrogen Energy Storage Method Selection in Turkey. Energies 2013, Vol 6, Pages 3017-3032 6:3017–3032. https://doi.org/10.3390/EN6063017

Hepbasli A, Ozgener L (2004) Development of geothermal energy utilization in Turkey: a review. Renew Sustain Energy Rev 8:433–460. https://doi.org/10.1016/J.RSER.2003.12.004

Lee KC (2001) Classification of geothermal resources by exergy. Geothermics 30:431–442. https://doi.org/10.1016/S0375-6505(00)00056-0

Litvinenko V (2020) The Role of Hydrocarbons in the Global Energy Agenda: The Focus on Liquefied Natural Gas. Resour 2020, Vol 9, Page 59 9:59. https://doi.org/10.3390/RESOURCES9050059

Long X, Naminse EY, Du J, Zhuang J (2015) Nonrenewable energy, renewable energy, carbon dioxide emissions and economic growth in China from 1952 to 2012. Renew. Sustain. Energy Rev. 52:680–688

Lund JW, Freeston DH, Boyd TL (2011) Direct utilization of geothermal energy 2010 worldwide review. Geothermics 40:159–180. https://doi.org/10.1016/J.GEOTHERMICS.2011.07.004

Myers JH, Alpert MI (1968) Determinant Buying Attitudes: Meaning and Measurement. J Mark 32: 13–20. https://doi.org/10.1177/002224296803200404

Saaty TL (1977) A scaling method for priorities in hierarchical structures. J Math Psychol 15:234–281. https://doi.org/10.1016/0022-2496(77)90033-5

Shafiee S, Topal E (2009) When will fossil fuel reserves be diminished? Energy Policy 37:181–189. https://doi.org/10.1016/j.enpol.2008.08.016

Solangi YA, Tan Q, Mirjat NH, et al. (2019) An integrated Delphi-AHP and fuzzy TOPSIS approach toward ranking and selection of renewable energy resources in Pakistan. Processes 7:. https://doi.org/10.3390/pr7020118

Toman MA, Jemelkova B (2003) Energy and Economic Development: An Assessment of the State of Knowledge. Energy J 24:93–112. https://doi.org/10.5547/ISSN0195-6574-EJ-VOL24-NO4-5

Yildiz A, Ayyildiz E, Gumus AT, Ozkan C (2020) A Modified Balanced Scorecard Based Hybrid Pythagorean Fuzzy AHP-Topsis Methodology for ATM Site Selection Problem. Int J Inf Technol Decis Mak 19:365–384. https://doi.org/10.1142/S0219622020500017

Yildiz A, Ayyildiz E, Taskin Gumus A, Ozkan C (2022) Evaluation of quality expectations for intercity bus firms by interval type-2 trapezoidal fuzzy AHP and firm selection. J Fac Eng Archit Gazi Univ 37:757–770. https://doi.org/10.17341/GAZIMMFD.625921

Yücenur GN, Ipekçi A (2021) SWARA/WASPAS methods for a marine current energy plant location selection problem. Renew Energy 163:1287–1298. https://doi.org/10.1016/J.RENENE.2020.08.131

Chapter 14
Necessity of Ecological Efficiency Indicator Modal of Air Pollutants and Emissions from Ships in Maritime Transportation: Policy Perspective

Ufuk Yakup Çalışkan and Burak Zincir

Nomenclature

BC	Black carbon
CO_2	Carbon dioxide
CH_4	Methane
CNG	Compressed natural gas
DWT	Deadweight
EEDI	Energy Efficiency Design Index
GHG	Greenhouse gases
GT	Gross tonnage
GTP	Global Temperature-change Potential
GWP	Global Warming Potential
HPDF	High-pressure dual fuel
IFO	Intermediate fuel oil
IFO 380	Intermediate fuel oil with maximum viscosity of 380 centistokes
IMO	International Maritime Organization
IPCC	Intergovernmental Panel on Climate Change
LFO	Light fuel oil
LNG	Liquefied natural gas
LPDF	Low-pressure dual fuel

U. Y. Çalışkan (✉)
Division of Transportation Services, Bartın University, Bartin, Turkey
e-mail: ucaliskan@bartin.edu.tr

B. Zincir
Maritime Faculty, Istanbul Technical University, Istanbul, Turkey
e-mail: bzincir@itu.edu.tr

M. Z. Sogut et al. (eds.), *Proceedings of the 2022 International Symposium on Energy Management and Sustainability*, Springer Proceedings in Energy,
https://doi.org/10.1007/978-3-031-30171-1_14

MDO	Marine diesel oil
MGO	Marine gas oil
NO	Nitrogen monoxide
NO_2	Nitrogen dioxide
NO_x	Nitrogen oxides
PM	Particulate matter
RF	Radiative forcing
SO	Sulfur monoxide
SO_2	Sulfur dioxide
SO_x	Sulfur oxides
VLS	Very low sulfur
VOC	Volatile organic compound

14.1 Introduction

Air pollution is one of the main concerns that the shipping community is keen on reducing along with greenhouse gases. Mostly, ships release two types of gases: GHGs, such as CO_2 and CH_4, and air pollutants, such as SOx and NOx (Kontovas 2020). NOx is a generic term to describe mono-nitrogen oxides, NO, and NO_2. PM or fine particulate matter is a term to explain the suspension of solid and liquid particles with a diameter of 2.5 μm and 10 μm suspended in the air, abbreviated as $PM_{2.5}$ and PM_{10}, respectively. SOx is to refer sulfur and oxygen-containing components such as SO and SO_2 (European Environment Agency 2021).

Emission metrics are the tools to scale different compounds' impacts on the climate. The metrics can be more than one dimensional to serve the only purpose. International agreements, emission trading schemes, consideration of a potential trade-off in emissions, and comparing means of transport or industrial activities are some of the application areas of emission metrics. Compounds that are non-CO_2 have been left out in many policy negotiations in the past. For instance, the Kyoto Protocol does not cover short-lived substances that maritime transportation emits, such as BC.

There are undeniable difficulties in developing new metrics such as finding an appropriate structure for the metric, associated value judgments for the selected period, and quantification of input parameters. For the short-lived species, the urgency of the climate policies is getting to dangerous levels. Apart from gases included in the Kyoto Protocol, quantitative evaluation of most polluters and GHGs is not consistent since the place of emission, altitude of emission, regional distribution, and distance to the shore can be the direct reasons to alter the value of metrics. In implementing climate policy, GWPs and GTPs as an alternative metric are the two most known metrics to calculate relative and absolute contributions of different substances to climate change, including region to region and sector differences. GWP is the principal policy-making tool that the Kyoto protocol uses that integrates with RF of a substance over a period relative to that from CO_2. GTP is the ratio of change in mean surface temperature at a chosen point in time from the substance of interest relative to that from CO_2. The choice of metric is dependent on the particular application, thus relevancy (Shine et al. 2005). RF and lifetimes from the literature to derive GWPs and GTPs for maritime transportation-related emissions are evaluated (Fuglestvedt et al. 2010).

There are limited works on the estimation of shipping gaseous and particle emissions in emission metrics. Yet, existing ones are not focusing on the unique nature of the maritime transportation's policy environment. The paper discusses an approach to assess an indicator of primary ship-borne emissions. The aim is to keep climate change and air quality indicators altogether. Derivatives of emission estimation methods are discussed to target the issue with more accuracy in ship-borne emissions.

14.2 Influences of Policies on Emissions

Seventy-five percent of the direct and indirect aerosol effects from shipping, which forms low marine clouds to increase albedo, are related to the fuel sulfur content (Lauer et al. 2007). Regardless of the metrics used, the decrease in sulfur content is causing a net warming effect (Kontovas 2020; Lauer et al. 2009). Lindstad et al. (2015) suggested burning high-sulfur content fuels on the high seas. In Fig. 14.1, changes in the fuel preferences of the maritime transportation industry could explain the nonintegrated approach.

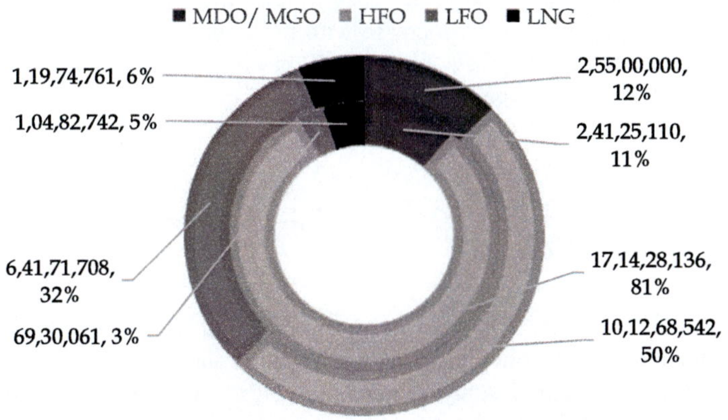

Fig. 14.1 Consumption of common marine fuels (99.91% in 2020) in 2019 (inner circle) and 2020 (outer circle) for ships of 5.000 GT and above by EEDI ship type and EEDI size category, including others and passenger ship categories for ships not subject to EEDI (authors curation based on IMO 2020, 2021a)

Table 14.1 World LNG fleet by fuel type as of January 1, 2021

	Ships	Ships %[a]	GT	GT %[a]	DWT	DWT %[a]
LNG – VLS IFO	373	0.37	36.964.811	2.57	30.159.817	1.42
LNG – VLS MDO	168	0.17	10.814.060	0.75	8.190.743	0.39
LNG – VLS MGO	37	0.04	424.846	0.03	430.662	0.02
LNG	32	0.03	459.380	0.03	139.039	0.01
Biofuel – LNG	4	>0.00	43.851	>0.00	3.907	>0.00
CNG – VLS MDO	3	>0.00	111.058	0.01	105.325	>0.00
IFO 380 – LNG	2	>0.00	251.144	0.02	18.400	>0.00
CNG – VLS MGO	1	>0.00	30.742	>0.00	31.473	>0.00
LNG – MDO	1	>0.00	65.314	>0.00	22.437	>0.00

Authors curation based on United Nations Conference on Trade and Development (2021)
[a]"Ships %*, GT %*, and DWT %*" denotes the respective unit of value in percentage of total world maritime fleet

Marine fuels shown in Fig. 14.1 represent 99.91% of marine fuels in 2020. There was a 1.84% increase in total reporting ships but 4.91% fuel consumption loss in total. Share of LNG in total consumption rose from 4.92% to 5.9%. LNG is the only cargo permitted as fuel by the IMO (Dobrota et al. 2013). Switching a dual-fuel vessel from diesel fuel to natural gas mitigates NOx, $PM_{2.5}$, CO_2, and BC emissions by 92%, 93%, 18%, and 97%, respectively (Peng et al. 2020). However, due to the technical imperfection of marine engines, methane slip is unavoidable. The most popular LNG engine technology is LPDF engine with four-stroke and medium speed technology. The second most popular is HPDF. In terms of GHG emissions, LPDF and HPDF emit 70% and 4% more emissions, respectively, against MGO in life cycle analyses (Pavlenko et al. 2020). Since near-term climate forcers are evaluated less impactful in the 100-year GWP scale (GWP-100 hereafter), possible climate benefits of reducing CH_4 emissions are not included in the maritime transportation agenda even though the sharpest increase with more than 2.5 times is in methane emissions (Sofiev et al. 2018; IMO 2021b). In Table 14.1 existing LNG-fueled ships are separated by their fuel type.

14.3 Emission Metrics and True Impact of Maritime Transportation

This study is in line with UN IPCC Assessment Report 5. Table 14.2 shows the climatic impact of various pollutants that mostly ships emit.

IMO's (2021b) inventory study values align with GWP-100. The only way to calculate emissions or the true impact of maritime transportation is to weigh emissions of different gases in their lifetimes and according to effects in the atmosphere. GWP and GTP values of NOx are strongly negative in any given time horizon. The lower-altitude nature of maritime transportation is making O_3 emissions strongly

Table 14.2 Pollutants' climate impact on different emission metrics and timescales

Pollutant	Lifetime (years)	GWP-20	GWP-100	GTP-20	GTP-100
CO_2	Variable[a]	1	1	1	1
CH_4	12.4[a]	84	28	67	4
N_2O	121[a]	264	265	277	234
NOx	2.5–10 h (NO) 4 h (NO_2)	19	–11	–87	–2.9
SO_2	13–78 h	−150	−43	−44	−6.1
BC	Few days to 2 weeks	1600	460	470	64
BC (total effect)	Few days to 1 month	3200	900	920	130
CO	2 months	6–9.3	2–3.3	3.7–6.1	0.29–0.55
OC	2–3 months	–240	–69	–71	–10
VOCs	30 min to months	14	4.5	7.5	0.66

CO_2 emissions: IPCC (2013). NOx: Fuglestvedt et al. (2010); based on Wild et al. (2001); Eyring et al. (2010); based on Song et al. (2003); Von Glasow et al. (2003); SO_2: Fuglestvedt et al. (2010); Lee et al. (2011); BC: Bond et al. (2013); Brewer (2021); CO: Fuglestvedt et al. (2010); OC: Fuglestvedt et al. (2010); VOC: Khalil and Rasmussen (1990); Williams and Koppmann (2007); Fuglestvedt et al. (2010) based on Collins et al. (2002); IPCC (2013)

[a]The GHG's atmospheric lifetimes are perturbation lifetime, not the lifetime of the atmospheric burden.

Table 14.3 CO_2 emission factors of main marine fuels

Heavy fuel oil	Marine gas oil – marine diesel oil (distillate marine fuels)	Liquefied natural gas
3,114 g CO_2/g fuel	3,206 g CO_2/g fuel	2,750 g CO_2/g fuel

IPCC (2007)

positive. SO_2 emissions are another solid driver for the negative RF values. The direct GWP-100 for shipping was ranging in between -11 and -43 (Fuglestvedt et al. 2010; Mhyre et al. 2013). According to Eyring et al. (2010), Fuglestvedt et al. (2009), and Lauer et al. (2007), shipping has negative RF values.

Estimation of emission is done most conveniently by multiplying fuel with the respective emission factor of the fuel (Kontovas and Psaraftis 2016). CO_2 equivalent (CO_2e) indicators are to compare emissions of various gases to CO_2. GWP-100 values are set as 1, 28, and 265 for CO_2, CH_4, and N_2O, respectively, in the 4th GHG Study of IMO. CO_2 emission factors of main marine fuels are given to set an example in Table 14.3 (IPCC 2006, 2013; IMO 2014, 2021b).

In contrast, Peng et al. (2020) calculated the average GWP-20 for natural gas and diesel oil on the same scale as 1515 CO_2e g/kWh and 725 CO2e g/kWh, respectively. The ratio of natural gas in GWP-20 to diesel is 2.10, while the GTP-20 calculation is slightly lower, with 1.96. Considering CH_4 has most of its impact in the short term, and LNG is seen as a transition fuel for zero emission, it would be dangerous to compare all the compounds in ship plume on the same scale.

Around 70% of emissions of oceangoing ships occur within 400 km of land (Eyring et al. 2010). Therefore, air quality should be prioritized as much as climate change in policy-making. Cardu and Baica (2001) set out an energy ecologic magnitude model for flue gases emitted by thermopower plants. CO_2e values and pollution indicator are given below in Eqs. 14.1, 14.2, and 14.3.

$$(CO_2)_e = (CO_2) + 700(SO_2) + 1000\,(NO_x) \tag{14.1}$$

$$700(SO_2) = (SO_2)_e \text{ and } 1000\,(NO_x) = (NO_x)_e \tag{14.2}$$

$$\Pi_g = \frac{(CO_2)_e}{Q_i}\,(\text{kg/MJ}) \tag{14.3}$$

Π_g is the pollution indicator of the flue gases, and Q_i refers to the lower heating value of the fuel. The mass unit refers to (kg) resulting $(CO_2)_e$ quantity from burning 1 kg of fuel. Inspired by their work, Lora and Salomon (2005) broadened with PM emissions. In the same approach, Coronado et al. (2009) used the same coefficient for CO_2e, a hypothetical pollutant concentration factor determined by Eqs. 14.4 and 14.5. Apart from CO_2, the rest of the gases' effects are taken from World Health Organization's (WHO) air quality standards. The multiplication factor of the gases other than CO_2 has been calculated with a maximal acceptable limit of μg/m^3 when minimal concentration that patients of asthma and particulate matters depend on the ashes of the fuels selected (WHO 2006).

$$(CO_2)_e = (CO_2) + 80(SO_2) + 50\,(NO_x) + 67(PM) \tag{14.4}$$

$$(SO_x)_e = 80\,(SO_x)(NO_x)_e = 50(NO_x)\;(PM)_e = 67(PM) \tag{14.5}$$

14.4 Discussions and Results

Endres et al. (2018) state that international law treats the atmosphere and the ocean very differently, even though they are tightly linked via chemical, physical, and biological processes. Deposition and dissolution of gases in the atmosphere affect the ocean's surface (Sèbe et al. 2022). Releasing pollutants to the ocean affects the atmosphere via outgassing and aerosol production (Brévière and the SOLAS Scientific Steering Committee 2016).

A delayed policy movement on CH_4 emissions cannot be fully compensated for later in the century since the most dramatically increasing emission rates belong to those emissions (Van Den Berg et al. 2015; IMO 2021b). As of right now, there are no known regulations on CH_4 emissions, but there are strict regulations that get tighter each year on SO_2 emissions in the maritime transportation industry. This approach is far from holistic.

Since it is advised to design the metric suitable for policy-makers, we suggest combining air quality and the climate change coefficients in a holistic approach (Shine 2009; Tanaka et al. 2013). Therefore, targeted emission should include well-mixed GHGs; CO_2, CH_4, and N_2O, near-term climate forcers; tropospheric O_3, aerosols (BC), and aerosol precursors (NO_x). Due to playing a significant role in increasing the low marine cloud's albedo, SO_x emission controls should not be this strict (Lauer et al. 2007, 2009; Fuglestvedt et al. 2009; Mhyre et al. 2013; Jensen et al. 2016).

14.5 Conclusion

This study reflects the emission impacts and contradictions in the trade-off of policies controlling air quality and climate change within the maritime transport industry. It is aimed to pinpoint target emissions and the issues in preparing an emission metric to use in policy-making for maritime transportation. It is advised to use the metric selective, for instance, GTP depending on the atmospheric lifetime, including precursor effects of the compounds. The holistic approach should be adopted to calculate CO_2e emissions considering warming and pollution. The policy-making stage could be more transparent via using the given CO_2e values for each compound. This work is the first development stage of the maritime transportation industry policy-making metric for emissions. Further development and the study of the metric will be concluded.

References

Bond TC, Doherty SJ, Fahey DW, Forster PM et al. (2013) Bounding the role of black carbon in the climate system: A scientific assessment. Journal of geophysical research: Atmospheres, 118(11), 5380–5552. https://doi.org/10.1002/jgrd.50171

Brévière E, and the SOLAS Scientific Steering Committee (eds) (2016) SOLAS 2015–2025: Science Plan and Organisation. Kiel. solas-int.org

Brewer, T (2021) Introduction: Problems, Policies and Technologies. In Transportation Air Pollutants (pp. 1–11). Springer, Cham. https://doi.org/10.1007/978-3-030-59691-0_1

Cârdu M, and Baica M (2001) A seismic vision regarding a methodology to estimate globally the energy-ecologic efficiency of thermopower plants (TPPs). Energy conversion and management, 42(11), 1317–1325. https://doi.org/10.1016/S0196-8904(00)00138-2

Collins WJ, Derwent RG, Johnson CE, and Stevenson DS (2002) The oxidation of organic compounds in the troposphere and their global warming potentials. Climatic Change, 52(4), 453–479. https://doi.org/10.1023/A:1014221225434

Coronado CR, de Carvalho Jr JA, Yoshioka JT, and Silveira JL (2009) Determination of ecological efficiency in internal combustion engines: The use of biodiesel. Applied Thermal Engineering, 29(10), 1887–1892. https://doi.org/10.1016/j.applthermaleng.2008.10.012

Dobrota D, Lalić B, and Komar I (2013) Problem of boil-off in LNG supply chain. Transactions on Maritime Science, 2(2),91–100. https://doi.org/10.7225/toms.v02.n02.001

Endres S, Maes F, Hopkins F, Houghton K et al. (2018) A New Perspective at the Ship-Air-Sea-Interface: The Environmental Impacts of Exhaust Gas Scrubber Discharge. Front. Mar. Sci. 5: 139. https://doi.org/10.3389/fmars.2018.00139

European Environment Agency (2021) Emissions of air pollutants from transport. eea.europa.eu/data-and-maps/indicators/

Eyring V, Isaksen, IS, Berntsen T, Collins WJ et al. (2010) Transport impacts on atmosphere and climate: Shipping. Atmospheric Environment, 44(37), 4735–4771. https://doi.org/10.1016/j.atmosenv.2009.04.059

Fuglestvedt J, Berntsen T, Eyring V, Isaksen I et al. (2009) Shipping Emissions: From Cooling to Warming of Climate—and Reducing Impacts on Health. Environmental Science & Technology 2009 43 (24), 9057–9062. https://doi.org/10.1021/es901944r

Fuglestvedt JS, Shine KP, Berntsen T, Cook J et al. (2010) Transport impacts on atmosphere and climate: Metrics. Atmospheric Environment, 44(37), 4648–4677. https://doi.org/10.1016/j.atmosenv.2009.04.044

IMO (2014) Third IMO GHG study in: T.W.P. Smith, J.P. Jalkanen, B.A. Anderson, J.J. Corbett, J. Faber, S. Hanayama, E. O'Keeffe, S. Parker, L. Johansson, L. Aldous, C. Raucci, M. Traut, S. Ettinger, D. Nelissen, D.S. Lee, S. Ng, A. Agrawal, J. J. Winebrake, M. Hoen, S. Chesworth, A. Pandey (Eds.), International Maritime Organization (IMO) London, UK, June. imo.org/en/OurWork/Environment/Pages/Greenhouse-Gas-Studies-2014.aspx

IMO (2020) Report of fuel oil consumption data submitted to the IMO Ship Fuel Oil Consumption Database in GISIS (Reporting year: 2019) MEPC 76/6/1; IMO: London, UK. https://ibia.net/wp-content/uploads/2021/03/MEPC-76-6-1-Report-Of-Fuel-Oil-Consumption-Data-Submitted-To-The-Imo-Ship-Fuel-Oil-Consumption-Databas...-Secretariat.pdf

IMO (2021a) Report of fuel oil consumption data submitted to the IMO Ship Fuel Oil Consumption Database in GISIS (Reporting year: 2020) MEPC 77/6/1; IMO: London, UK. imo.org/

IMO (2021b) Fourth IMO GHG Study. International Maritime Organisation (IMO) Faber, J., Hanayama, S., Zhang, S., Pereda, P., Comer, B., Hauherhof, H., van der Loeff, W., Smith, T., Zhang, Y., Kosaka, H., Adachi, M., Bonello, J., Galbarith, C., Gong, Z., Hirata, K., Hummels, D., Klejin, A., Lee, D.S., Liu, Y., Lucchesi, A., Mao, X., Muraoka, E., Osipova, L., Qian, H., Rutherford, D., Suarez de la Fuante, S., Yuan, H., Perico, V., Wu, L., Sun, D., Yoo, D. and Xing, H., 4 Albert Embankment, London SE1 7SR. imo.org/en/OurWork/Environment/Pages/Fourth-IMO-Greenhouse-Gas-Study-2020.aspx

IPCC (2006) Chapter 3: Mobile Combustion, Volume 2: Energy. In: S. Eggleston, et al. (eds) IPCC Guidelines for National Greenhouse Gas Inventories. Hayama: Intergovernmental Panel on Climate Change (IPCC), pp. 3.1–3.78.

IPCC (2007) Climate Change 2007 – the Physical Science Basis: Working Group I Contribution to the Fourth Assessment Report of the IPCC. ipcc.ch/report/ar4/wg1/

IPCC (2013) Climate Change the Physical Science Basis. Working Group I Contribution to the IPCC Fifth Assessment Report (Ar5). Final Draft Underlying Scientific-Technical Assessment. Working Group I – Twelfth Session. Stockholm, 23–26 September. ipcc.ch/report/ar5/wg1/

Jensen M, Wang J, and Wood R (2016) Atmospheric System Research Marine Low Clouds Workshop Report (No. DOE/SC-ASR-16-001). ARM Data Center, Oak Ridge National Laboratory (ORNL), Oak Ridge, TN (United States). asr.science.energy.gov/

Khalil MAK, and Rasmussen RA (1990) The global cycle of carbon monoxide: Trends and mass balance. Chemosphere, 20(1–2), 227–242. https://doi.org/10.1016/0045-6535(90)90098-E

Kontovas, CA (2020). Integration of air quality and climate change policies in shipping: The case of sulphur emissions regulation. Marine Policy, 113, 103815. https://doi.org/10.1016/j.marpol.2020.103815

Kontovas CA, and Psaraftis HN (eds) (2016) Transportation emissions: Some basics. In Green transportation logistics (pp. 41–79). Springer, Cham. https://doi.org/10.1007/978-3-319-17175-3_2

Lauer A, Eyring V, Hendricks J, Jöckel, P et al. (2007) Global model simulations of the impact of ocean-going ships on aerosols, clouds, and the radiation budget. Atmospheric Chemistry and Physics, 7(19), 5061–5079. https://doi.org/10.5194/acp-7-5061-2007

Lauer A, Eyring V, Corbett J J, Wang C et al. (2009) Assessment of Near-Future Policy Instruments for Oceangoing Shipping: Impact on Atmospheric Aerosol Burdens and the Earth's Radiation Budget. Environmental Science & Technology 43 (15), 5592–5598 https://doi.org/10.1021/es900922h

Lee C, Martin RV, van Donkelaar A, Lee H et al. (2011) SO2 emissions and lifetimes: Estimates from inverse modeling using in situ and global, space-based (SCIAMACHY and OMI) observations. Journal of Geophysical Research: Atmospheres, 116(D6). https://doi.org/10.1029/2010JD014758

Lindstad H, Eskeland G S, Psaraftis HN, Sandaas I, et al. (2015) Maritime shipping and emissions: A three-layered, damage-based approach. Ocean Engineering, 110, 94–101. https://doi.org/10.1016/j.oceaneng.2015.09.029

Lora EES, and Salomon KR (2005) Estimate of ecological efficiency for thermal power plants in Brazil. Energy Conversion and Management, 46(7–8), 1293–1303. https://doi.org/10.1016/j.enconman.2004.06.028

Myhre G, Shindell D, Bréon F-M, Collins W et al. (2013) Anthropogenic and Natural Radiative Forcing. In: Climate Change 2013: The Physical Science Basis. Contribution of Working Group I to the Fifth Assessment Report of the Intergovernmental Panel on Climate Change. Cambridge University Press, Cambridge, United Kingdom and New York, NY, USA. https://doi.org/10.1017/CBO9781107415324.018

Pavlenko N, Comer B, Zhou Y, Clark N et al. (2020) The climate implications of using LNG as a marine fuel. International Council on Clean Transportation: Berlin, Germany. theicct.org/publication/the-climate-implications-of-using-lng-as-a-marine-fuel/

Peng W, Yang J, Corbin J, Trivanovic U et al. (2020) Comprehensive analysis of the air quality impacts of switching a marine vessel from diesel fuel to natural gas. Environmental Pollution, 266, 115404. https://doi.org/10.1016/j.envpol.2020.115404

Sèbe M, Scemama P, Choquet A, Jung JL et al. (2022) Maritime transportation: Let's slow down a bit. Science of the Total Environment, 811, 152262. https://doi.org/10.1016/j.scitotenv.2021.152262

Shine K.P (2009) The global warming potential—the need for an interdisciplinary retrial Climatic Change 96(4), 467–472. https://doi.org/10.1007/s10584-009-9647-6

Shine KP, Fuglestvedt JS, Hailemariam K, and Stuber N (2005) Alternatives to the global warming potential for comparing climate impacts of emissions of greenhouse gases. Climatic Change, 68(3), 281–302. https://doi.org/10.1007/s10584-005-1146-9

Sofiev M, Winebrake JJ, Johansson L, Carr EW et al. (2018) Cleaner fuels for ships provide public health benefits with climate tradeoffs. Nature communications, 9(1), 1–12. https://doi.org/10.1038/s41467-017-02774-9

Song CH, Chen G, Hanna SR, Crawford J et al. (2003) Dispersion and chemical evolution of ship plumes in the marine boundary layer: Investigation of O3/NOy/HOx chemistry. Journal of Geophysical Research: Atmospheres, 108(D4). https://doi.org/10.1029/2002JD002216

Tanaka K, Johansson DJA, O'Neill BC, and Fuglestvedt JS (2013) Emission metrics under the 2 °C climate stabilization target Climatic Change 117(4), 933–941. https://doi.org/10.1007/s10584-013-0693-8

United Nations Conference on Trade and Development (2021) Review of Maritime Transport 2021 Report. Geneva November 2021. https://unctad.org/publications

Van Den Berg M, Hof AF, Van Vliet J, and Van Vuuren DP (2015) Impact of the choice of emission metric on greenhouse gas abatement and costs. Environmental Research Letters, 10(2), 024001. https://doi.org/10.1088/1748-9326/10/2/024001

Von Glasow R, Lawrence MG, Sander R, and Crutzen PJ (2003) Modeling the chemical effects of ship exhaust in the cloud-free marine boundary layer. Atmospheric Chemistry and Physics, 3(1), 233–250. https://doi.org/10.5194/acp-3-233-2003

WHO (2006) Air Quality Guidelines for Particulate Matter, Ozone, Nitrogen Dioxide and Sulfur Dioxide, Summary of Risk Assessment. World Health Organisation, apps.who.int/iris/handle/10665/69477

Wild O, Prather MJ, and Akimoto H (2001) Indirect long-term global radiative cooling from NOx emissions. Geophysical Research Letters, 28(9), 1719–1722. https://doi.org/10.1029/2000GL012573

Williams J, and Koppmann R (2007) Volatile organic compounds in the atmosphere: an overview. Volatile Organic Compounds in the Atmosphere, 1.

Chapter 15
Determination of Electric Vehicle Battery Cell Optimal Spacing Using the Intersection of Asymptotes Method

Sahin Gungor

Nomenclature

A	Area, m^2
CP	Specific heat, J/kg°C
Dh	Hydraulic diameter, m
f	Friction factor
h	Heat transfer coefficient, W/m^{2}°C
I	Circuit current, A
H	Battery cell height, m
k	Thermal conductivity, W/m°C
$\dot{m}$	Mass flow rate, kg/s
n	Number of battery cells
Nu	Nusselt number
P	Wetted perimeter, m
Pr	Prandtl number
ΔP	Pressure difference, Pa
$\dot{q}$	Heat transfer rate, W
Re	Reynolds number
s	Battery cell spacing, m
t	Stack thickness, m
T	Temperature, K
u	Velocity, m/s

S. Gungor (✉)
Department of Mechanical Engineering, Izmir Katip Celebi University, Izmir, Turkiye

Department of Mechanical Engineering, Villanova University, Villanova, PA, USA
e-mail: sahin.gungor@ikc.edu.tr; sahin.gungor@villanova.edu

M. Z. Sogut et al. (eds.), *Proceedings of the 2022 International Symposium on Energy Management and Sustainability*, Springer Proceedings in Energy,
https://doi.org/10.1007/978-3-031-30171-1_15

V	Voltage, V
w	Battery cell width, m

Greek Letters

α	Thermal diffusivity, m^2/s
ρ	Density, kg/m^3
μ	Dynamic viscosity, Pa.s
τ	Shear stress, N/m^2
ν	Kinematic viscosity, m^2/s

Subscripts

c	Circuit
f	Friction
o	Open circuit
w	Wall region
∞	Free stream

15.1 Introduction

Countries have recently attended the United Nations climate change conference of COP26 to come up with solutions on how to eliminate the carbon footprint resulting from industrial processes, transportation, and any human activities. Targeted outputs of the COP26 conference are to reduce the national basis harmful emissions and promote zero-carbon and carbon-negative technologies (UN-Climate Change Conference 2021). At this point, electric vehicles (EVs) are the most suitable candidate for near-future transportation, especially if the electricity is generated via renewable energy sources such as wind, solar, or biomass. Today, all EV manufacturers prefer lithium-ion (Li-ion) battery technologies due to their superiorities in energy density, lifetime, and recyclability (Fan et al. 2020). Yet, overheating within the EV battery pack, milage range limit, and the scant number of charging stations are the main lack of the EV industry.

Li-ion batteries have nonuniform temperature distribution caused by nonhomogeneous heat generation (Ghalkhani et al. 2017). The generated heat mainly results from the electrochemical reactions, Joule heating, and electrode side kinetics. Determination of battery heat generation characteristics is also essential to achieve successful thermal management for electric vehicles (Chen et al. 2014). Therefore, understanding the heat generation and thermal characteristic phenomena for each type of battery cell is the key issue in the literature (Rizk et al. 2018). The general form of the heat generation equation is given by (Karimi and Li 2013):

$$\dot{q} = I(V_0 - V_C) - IT\frac{dV_0}{dT} \tag{15.1}$$

On the other hand, charging/discharging current rate (C-rate) is the crucial parameter in EV battery pack systems directly affecting the heat generation and temperature levels (Damay et al. 2015). Once the battery temperature level exceeds the optimal ranges, batteries may expose to layer degradation, solid-electrode interface (SEI) formation (Kolzenberg et al. 2020), and even thermal runaway at higher temperature levels (Feng et al. 2019). Here the thermal runaway is the most critical issue causing many safety risks such as explosion, fire, and harmful emissions (Lei et al. 2019: Liao et al. 2019). Battery thermal management systems (BTMS) have emerged to prevent these safety risk and performance losses. BTMS should ensure removal of the generated heat within the EV battery pack systems (Dincer et al. 2017: Kim et al. 2019) and maintain the temperature levels during charge/discharge operations. Furthermore, these systems should satisfy the optimal operating temperature range during charging and discharging operations (Pesaran 2001, 2002).

Note that many types of battery thermal management techniques have been proposed by the R&D researchers and EV industry thermal designers (Mali et al. 2021). The main BTMS techniques can be classified into various categories, yet the most appropriate one is energy consumption. Active thermal management techniques contain forced air, forced liquid, multiphase systems, immersion cooling, and thermoelectric strategy, while the passive ones include natural convection, heat pipe, and phase change material (PCM) strategies (Kim et al. 2019). Although there are numerous discrete or integrated thermal management techniques, EV industry dominantly prefers forced-air and forced liquid strategies. Main reasons behind this choice are application simplicity and cost advantage of the air-based techniques, effectiveness of liquid-cooling strategies, and thermal stability. On the other hand, leakage problem of the liquid medium, low heat capacity of the air coolant, and high-level pumping power requirements are the lack of these thermal management techniques.

This study mainly focuses on forced-air cooling strategy satisfying the thermal management requirements of an EV battery pack. Note that a manifold covering the battery pack was assumed to uniformly distribute the airflow between the battery cells. As the cell spacing between the batteries dominantly varies the heat transfer and pressure drop, this study investigates the optimal cell spacing with the light of thermal boundary layer approach and method of the intersection of asymptotes. Proposed technique ensures the determination of optimal battery cell spacing satisfying better thermal performance with less pressure drop. The findings are applicable to all kinds of air-cooled battery models.

15.2 Methodology

In this study, the optimal battery cell spacing is determined via analytical and similarity solutions of the heat transfer and pressure drop along the battery surfaces. Figure 15.1 presents the investigated battery pack domain with forced-air cooling. Here the battery cells are assumed as very thin (pouch type) in order to eliminate the local pressure losses just above the battery cells. The airflow having uniform flow rate among the battery cells can be considered as distributed by a fan and manifold system. Furthermore, the cell spacing (s) is maintained constant for all the gaps. On the other hand, the battery cells and spacings were considered as having periodic thermal and fluid flow characteristics for designing an adequate and realistic electric vehicle thermal management system.

Consider the geometries given in Fig. 15.2, in which the airflow has free stream velocity of u_∞ and inlet temperature of T_∞. Here heat transfer rate and pressure drop of each investigated configuration are calculated analytically. Here two different conditions of small spacings and large spacings are examined to determine the heat transfer limits (Fig. 15.2a, b). In the small spacing investigations, the cell spacing is considered as $s \rightarrow 0$. The channels are assumed as slender ensuring the flow is fully developed along the channel height (H). It is obvious that the inlet temperature reaches the battery wall temperature of T_{wall} in this case. Once the heat transfer and pressure drop of small spacing case are calculated, this trend is used in the asymptotic analysis to obtain the optimal battery cell spacing. In this study, both the graphical trend of the heat transfer limits and the mathematical form of the heat transfer rates are documented, and determination of the optimal point is discussed in detail.

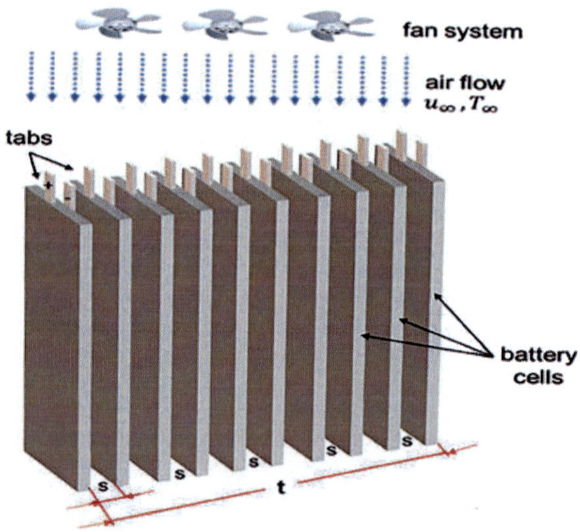

Fig. 15.1 Investigated air-cooled battery module configuration

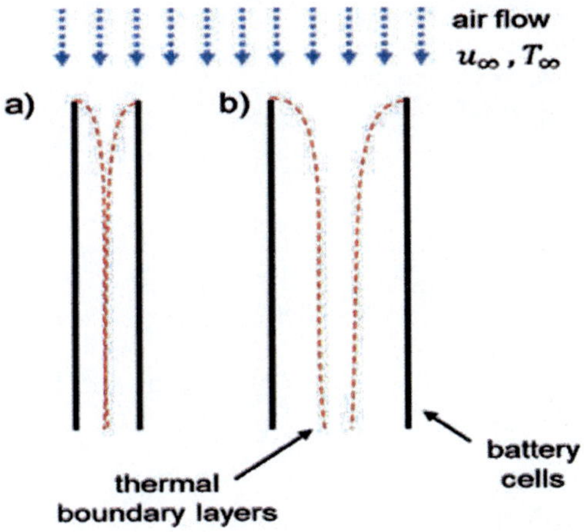

Fig. 15.2 Configurations for (**a**) small spacing limit and (**b**) large spacing scenario

Another investigated case is related to large spacing limits. Note that both cases are examined in the same boundary conditions and geometrical aspects. We can consider the cell spacing as $s \to \infty$ that causes the formation of distinct thermal boundary layers. The flow is also fully developed in this part, yet the free stream velocity (u_∞) is required to the heat transfer limit of the large spacing case.

15.3 Results and Discussion

Here the heat transfer and pressure drop limits of an air-cooled battery thermal management system are determined for two extreme conditions: small battery cell spacing and large one. First, consider the cell spacing limit $s \to 0$. The pressure drop and mean velocity among the Li-ion battery cells can be calculated as follows (Bejan 2013):

$$\frac{dP}{dy} = \mu \frac{d^2u}{dx^2} = \text{constant} \tag{15.2}$$

$$u = \frac{3}{2} U \left[1 - \left(\frac{x}{s/2} \right)^2 \right] \tag{15.3}$$

$$U = \frac{s^2}{12\mu} \left(-\frac{dP}{dy} \right) \tag{15.4}$$

where pressure (P) is a function of vertical axis (y). Here no slip boundary conditions of u = 0 at y= ±s/2 were applied to obtain mean velocity given in Eq. 15.3. The mass flow rate of the system is:

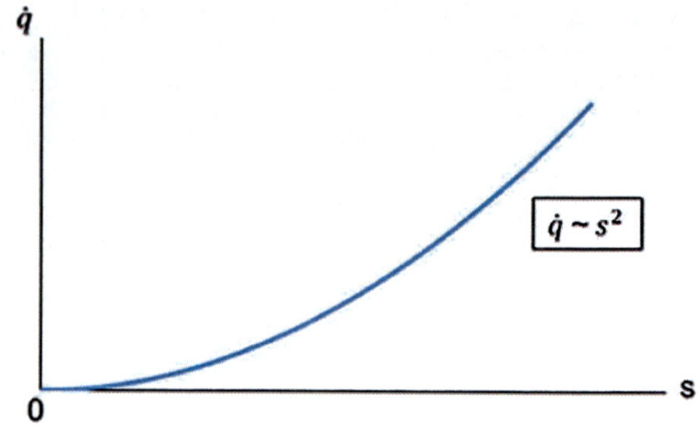

Fig. 15.3 Scale analysis result among the heat transfer rate and small battery cell spacing

$$\dot{m} = \rho_{\mathrm{air}} u\,(tL) \tag{15.5}$$

in which n is the number of battery channel, ρ is the air density, t is the stack thickness, and L denotes the unit length in the perpendicular direction to Fig. 15.2. Furthermore, the total heat transfer rate of small spacing case is given as follows:

$$\dot{q} = \dot{m}\, C_p(T_{\mathrm{wall}} - T_\infty) \tag{15.6}$$

$$\dot{q} = \rho_{\mathrm{air}}\, tL \frac{\Delta P\, s^2}{12\mu H} C_p(T_{\mathrm{wall}} - T_\infty) \tag{15.7}$$

Here the heat transfer rate from the battery surfaces is proportional to s^2 (Fig. 15.3) for the investigated small spacing limit.

On the other hand, large battery spacing is the opposite extreme in which $s \to \infty$. In this case, the boundary layers (see Fig. 15.2b) become distinct. Each channel is like the entrance region to a parallel plate duct. The force balance is given by (Bejan 2013):

$$t\,\Delta P = n\, 2\overline{\tau}_w H \tag{15.8}$$

where n is the number of battery channels and $\overline{\tau}_w$ corresponds the wall shear stress. Furthermore, skin friction coefficient (C_f) and wall shear stress can be calculated as follows:

$$C_{f,y} = 0.664\, Re_H^{-0.5} \tag{15.9}$$

$$\overline{\tau}_w = 1.328\, Re_H^{-0.5} \left(\frac{1}{2}\rho u_\infty^2\right) \tag{15.10}$$

Note that Reynolds number based on the height of the Li-ion battery cells should be double-checked as these calculations are valid under laminar flow conditions.

$$u_\infty = \left(0.753 \frac{t\,\Delta P}{n\rho\sqrt{\nu H}}\right)^{2/3} \tag{15.11}$$

where ν is the kinematic viscosity of the air coolant. We need Nusselt correlations satisfying the laminar flow conditions and Pr > 0.5 in order to calculate the heat transfer rate when the battery cells have large spacings. The overall Nusselt number of this case can be calculated via the correlation given below (Bejan and Lorente 2008):

$$Nu_H = 0.664\,Re_H^{1/2} Pr^{1/3} \tag{15.12}$$

$$Nu_H = \frac{\overline{h}\,H}{k_{\text{air}}} = \frac{\dot{q}''\,H}{(T_{\text{wall}} - T_\infty)\,k_{\text{air}}} \tag{15.13}$$

Here $\overline{h}$ is the average convective heat transfer coefficient, k_{air} is the thermal conductivity of the air media, and $\dot{q}''$denotes the heat flux.

$$\dot{q} = 2ntH\,\dot{q}'' \tag{15.14}$$

$$\dot{q} = 2nt\left(0.664\,Re_H^{1/2} Pr^{1/3}\right)(T_{\text{wall}} - T_\infty)\,k_{air} \tag{15.15}$$

and if we insert Eq. 15.11 into Eq.15.15:

$$\dot{q} = 1.2081\,tk_{\text{air}}\left[\frac{\Delta P^{1/3} H^{1/3} Pr^{1/3}}{\nu^{2/3}\rho^{1/3}s^{2/3}}\right](T_{\text{wall}} - T_\infty) \tag{15.16}$$

Equation 15.16 presents the total heat transfer limit of the large battery cell spacing case. It is obvious that the total heat transfer rate depends on both thermophysical properties of the coolant and the geometrical aspects such as battery cell height, total thickness of the battery module, and the gap between the Li-ion battery cells. In the large spacing case, heat transfer rate from the battery surfaces is proportional to $s^{-2/3}$. Once the heat transfer limits are determined for the small and large battery cell spacing cases, the intersection of asymptotes method is applied for detecting the optimal battery cell spacing satisfying better thermal performance with the same pressure drop. Figure 15.4 presents the air-cooled battery cell spacing application of asymptotic analyses.

Intersection of asymptotes method yields us an order of magnitude for the optimal spacing ensuring better thermal performance. Figure 15.4 shows the graphical presentation of the result, yet the equation of optimal point is as follows (Bejan 2013: Kim et al. 2007):

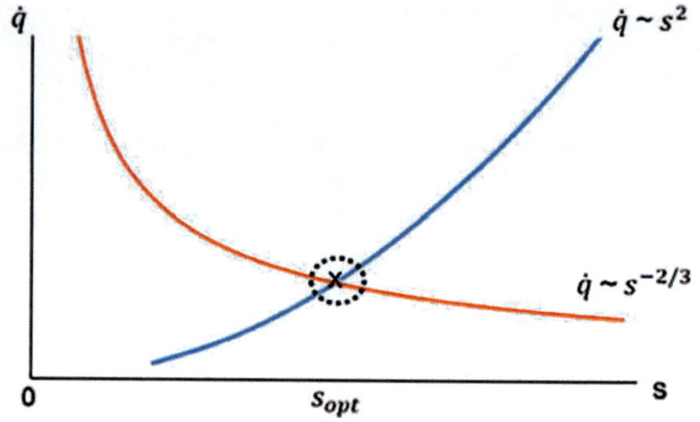

Fig. 15.4 Intersection of asymptotes method application for air-cooled battery thermal management problem

$$s_{\mathrm{opt}} \simeq 2.73H\left(\frac{\Delta P\, H^2}{\alpha\mu}\right)^{-0.25} \tag{15.17}$$

where α is the thermal diffusivity depending on thermal conductivity, density, and specific heat of the air coolant. Note that the dimensionless pressure drop within the parenthesis is termed as the Bejan number. Here the pressure drop between the battery cells can be calculated as (Fox et al. 2014):

$$\Delta P = f\frac{PH}{A}\left(\frac{1}{2}\rho u^2\right) \tag{15.18}$$

where f corresponds the friction factor experienced by the airflow, P is the wetted perimeter, and A is the cross section of the flow channel. Note that the battery cell spacing (s) is the same dimension with the thickness of flow channel. Equation 15.18 can be rewritten with respect to hydraulic diameter (D_h) as given below:

$$\Delta P = 4f\frac{H\rho u^2}{D_h} \tag{15.19}$$

Note that battery width (ww) is comparatively larger than battery cell spacing in a realistic electric vehicle battery module system, i.e., for $w \gg s$:

$$\Delta P = 96\frac{\dot{m}\nu H}{wD_h^3} \tag{15.20}$$

According to all these analytical calculations and scale analyses, Table 15.1 presents the results of some optimal battery cell spacing scenarios calculated at an approximate film temperature. Calculations were performed via Eqs. 15.17 and 15.20, and air velocity at the inlet was assumed as 1 ms^{-1} corresponding to Reynolds numbers of 1164 and 1219 for the investigated cross sections.

Table 15.1 Optimal battery cell spacing for air-cooled electric vehicle battery packs

$w \times H$ (mm)	$\Delta P_{analytical}$ (Pa)	s_{opt} (mm)
100×100	0.37	4.21
100×200	0.74	5.95
200×200	0.67	6.09
200×300	1.02	6.74

15.4 Conclusion

This study covers the analytical calculation steps for optimal battery cell spacing of air-cooled electric vehicle battery module. Heat transfer performance for the small and large spacing limits is investigated to obtain the optimal battery cell spacing satisfying enhanced thermal performance under the same energy consumption level. Optimal battery cell spacing is obtained via thermal boundary layer evolution and intersection of asymptotes method. Furthermore, optimal battery cell spacing trials were documented for some geometrically available Li-ion battery cells:

- The results indicate that optimal cell spacing is directly dependent on battery cell dimensions.
- Optimal cell spacing increases in parallel with the pressure drop along airflow channel.
- The findings show that 6 mm cell spacing can be assumed as global optimal at about Re $\approx$ 1200.

This approach can be implemented in real-case air-cooled battery module/pack applications and will be compared with numerical analyses in the near future.

Acknowledgment I would like to express my deepest appreciation to Prof. Sylvie Lorente (Villanova University, USA) as she inspired me to apply this strategy into an air-cooled battery thermal management system.

References

Bejan A, Lorente S (2008) Design with Constructal Theory. 1st Edition, WILEY.

Bejan A (2013) Convection Heat Transfer. 4th Edition, WILEY.

Chen K, Unsworth G, Li X (2014) Measurements of heat generation in prismatic Li-ion batteries. Journal of Power Sources 261:28–37. https://doi.org/10.1016/j.jpowsour.2014.03.037

Damay N, Forgez C, Bichat M, Friedrich G (2015) Thermal modelling of large prismatic LiFePO4/graphite battery. Coupled thermal and heat generation models for characterization and simulation. Journal of Power Sources 283:37–45. https://doi.org/10.1016/j.jpowsour.2015.02.091

Dincer I, Hamut HS, Javani N (2017) Thermal management of electric vehicle battery Systems. Automotive Series. WILEY.

Fan E, Li L, Wang Z, Lin J, Huang Y, Yao Y, Chen R, Feng W (2020) Sustainable recycling technology for Li-ion batteries and beyond: Challenges and future prospects. Chemical Reviews 120:14:7020–7063. https://doi.org/10.1021/acs.chemrev.9b00535

Feng X, Zheng S, Ren D, He X, Wang L, Cui H, Ouyang M (2019) Investigating the thermal runaway mechanisms of Li-ion batteries based on thermal analysis database. Applied Energy 246:53–64. https://doi.org/10.1016/j.apenergy.2019.04.009

Fox RW, McDonald AT, Pritchard PJ, Mitchell JW (2014) Fluid Mechanics. 9th Edition, WILEY.

Ghalkhani M, Bahiraei F, Nazri G, Saif M (2017) Electrochemical–thermal model of pouch-type Lithium-ion batteries. Electrochimica Acta 247:569–587. https://doi.org/10.1016/j.electacta.2017.06.164

Karimi G, Li X (2013) Thermal management of lithium-ion batteries for electric vehicles. International Journal of Energy Research 37:13–24. https://doi.org/10.1002/er.1956

Kim J, Oh J, Lee H (2019) Review on battery thermal management system for electric vehicles. Applied Thermal Engineering 149:192–212. https://doi.org/10.1016/j.applthermaleng.2018.12.020

Kim S, Lorente S, Bejan A (2007) Vascularized materials with heating from one side and coolant forced from the other side. International Journal of Heat and Mass Transfer 50:17:3498-3506. https://doi.org/10.1016/j.ijheatmasstransfer.2007.01.020

Kolzenberg L, Latz A, Horstmann B (2020) Solid–electrolyte interphase during battery cycling: Theory of growth regimes. Chem Sus Chem 13:3901-3910. https://doi.org/10.1002/cssc.202000867

Lei Z, Maotao Z, Xiaoming X, Junkui G (2019) Thermal runaway characteristics on NCM lithium-ion batteries triggered by local heating under different heat dissipation conditions. Applied Thermal Engineering 159:113847. https://doi.org/10.1016/j.applthermaleng.2019.113847

Liao Z, Zhang S, Li K, Zhang G, Habetler TG (2019) A survey of methods for monitoring and detecting thermal runaway of lithium-ion batteries. Journal of Power Sources 436:226879. https://doi.org/10.1016/j.jpowsour.2019.226879

Mali V, Saxena R, Kumar K, Kalam A, Tripathi B (2021) Review on battery thermal management systems for energy-efficient electric vehicles. Renewable and Sustainable Energy Reviews 151: 111611. https://doi.org/10.1016/j.rser.2021.111611

Pesaran AA (2001) Battery thermal management in EV and HEVs: issues and solutions. Advanced Automotive Battery Conference 43:34-49.

Pesaran AA (2002) Battery thermal models for hybrid vehicle simulations. Journal of Power Sources 110:377-382. https://doi.org/10.1016/S0378-7753(02)00200-8

Rizk R, Louahlia H, Gualous H, Schaetzel P (2018) Experimental analysis and transient thermal modelling of a high-capacity prismatic lithium-ion battery. International Communications in Heat and Mass Transfer 94:115–125. https://doi.org/10.1016/j.icheatmasstransfer.2018.03.018

United Nations (2021) Climate Change Conference (COP26) https://ukcop26.org/, Accessed: July 16th, 2022.

Chapter 16
Investigating the Drying Kinetics of Pineapple Dried in Passive Indirect Mode Solar Dryer: Comparative Analysis With and Without Thermal Energy Storage System

Mulatu C. Gilago, Vishnuvardhan Reddy Mugi, and V. P. Chandramohan

Nomenclature

A	Area, m^2
cp	Specific heat, J/kg °C
h	Heat transfer coefficient, W/m^2K
m	Mass of moisture removed, kg
η	Efficiency, %
d	Dryer
i	Initial
f	Final
e	Effective
w	Water removed
ISD	Indirect solar dryer

M. C. Gilago
Department of Mechanical Engineering, National Institute of Technology Warangal, Warangal, Telangana, India

Mechanical Engineering Department, Wachemo University, Hosaena City, Ethiopia
e-mail: mulatucg@student.nitw.ac.in

V. R. Mugi · V. P. Chandramohan (✉)
Department of Mechanical Engineering, National Institute of Technology Warangal, Warangal, Telangana, India
e-mail: vpcm80@nitw.ac.in

M. Z. Sogut et al. (eds.), *Proceedings of the 2022 International Symposium on Energy Management and Sustainability*, Springer Proceedings in Energy,
https://doi.org/10.1007/978-3-031-30171-1_16

16.1 Introduction

Solar drying can be considered as one of the ancient styles of drying and it involves concurrent mass and heat transports. Drying agricultural herbs in an indirect solar dryer (ISD) is advantageous over other methods in that it preserves nutritional elements. Also, the final dried products are clean and free of dust contamination. During the drying of agriproducts, the method of drying should be carefully selected and controlled so that the required quality of the product can be achieved; evaluating the drying kinetics is very important to optimize the drying parameters for a specific product during any drying experiment (Babu et al. 2018).

Tagnamas et al. (2021) performed an experiment on the investigation of drying kinetics of carob seeds (*Ceratonia siliqua* L.) in a forced convection ISD. They reported that the activation energy was 41.46 kJ/mol, and the drying efficiency and the diffusivity were in the range of 2.6–4.2% and 1.1968×10^{-9}–4.1482×10^{-9} m^2/s, respectively.

Generally, from the available literature, few studies investigated on drying kinetics and the importance of herbal products drying in solar dryer (Bhardwaj et al. 2021). But study is reported on the kinetics of drying for pineapple in ISD supported with thermal energy storage (TES) system.

Some studies reported on the influence of adding TES in an ISD and the merits and demerits of drying agricultural products in solar dryers (Alimohammadi et al. 2020). But there is no data on natural convection ISD and on the comparative evaluation of the setups without and with TES (setup-1 and setup-2) during drying pineapple. There are limited data on drying kinetics such as coefficients of surface transfer (Wang and Brennan 1995), drying rate (Hidalgo et al. 2021), actual heat supplied (Babu et al. 2018) and diffusion coefficient (Hidalgo et al. 2021) in drying agricultural produces. But no study was found on the kinetics of drying for pineapple dried in natural convection ISD setup-1 and setup-2. Few studies reported on the parameters of drying performance such as η_c, η_d and activation energy (E_a) (Vijayan et al. 2016), specific energy consumption (SEC) and specific moisture extraction rate (SMER) (Hidalgo et al. 2021). But study is found on the drying performance parameters for natural convection ISD setup-2 during drying pineapple.

Based on the figured-out gaps, below-listed goals were planned and achieved in this study: (i) to execute drying experiments of pineapple in setup-1 and setup-2; (ii) to assess the drying performance parameters like collector efficiency, actual heat supply and drying efficiency; (iii) to evaluate the drying kinetics of pineapple: effective diffusivity, drying rate and h and h_m; (iii) to estimate the E_{ac}, SEC and SMER for both setups; and (iv) to perform a comparative analysis of both setups based on drying kinetics and overall performances.

16.2 Materials and Methods

A passive ISD has been employed to dry pineapple that was sliced in 5 mm thickness. After the test was done in a setup without TES system (setup-1), a rectangular framed TES was integrated to modify the existing setup (setup-2). The setup-2's materials and specification details are reported in the earlier work of the authors (Gilago and Chandramohan, 2022). Its working principle is diagrammatically represented in Fig. 16.1, and the components with their details are reported in previous study by the same authors (Gilago and Chandramohan 2021). A total of 0.8 kg (0.2 kg × 4) of pineapple was placed on four vertically parallel trays. The mass variation and drying parameter data were recorded carefully, and the parameters of performance for the dryer and drying kinetics of the sample were determined for both cases.

After 24 h of drying at 105 °C, 5 randomly chosen slices out of 12 samples were used to determine initial moisture content (MC_i). The MC_i was 7.9112 db that is determined by:

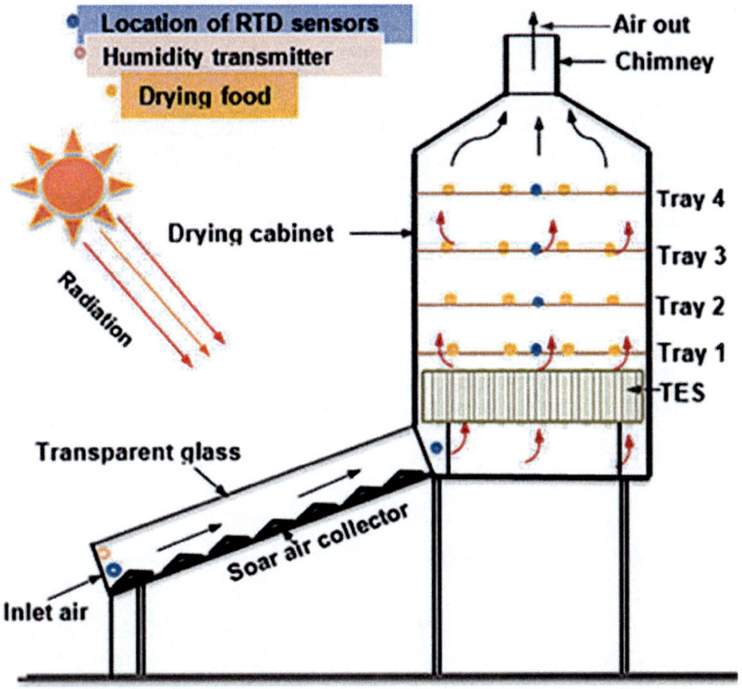

Fig. 16.1 Working principle of NCISD supported with TES

$$\text{MC} = \frac{m_i - m_f}{m_{i/f}} \times 100 \quad (16.1)$$

16.2.1 Evaluation of Energy

An explanation of drying energy was presented by the application of mass conservation (Chandramohan and Talukdar 2010):

$$\sum \dot{m}_i = \sum \dot{m}_o \quad (16.2)$$

Energy of the solar air heater (SAH) is calculated using

$$Q_{ic} = I_s A_c \quad (16.3)$$

$$Q_{oc} = \dot{m} c_p (T_{co} - T_{ci}) \quad (16.4)$$

where Q_{ic} is the heat input and Q_{oc} (Q_{ac}) is the useful heat output for the collector. I_s (W/m^2) represents intensity of solar radiation and A_{c} is collector's area.

$$\eta_c = \frac{Q_{oc}}{Q_{ic}} = \frac{\dot{m} c_p (T_{co} - T_{ci})}{I_s A_{SAC}} \quad (16.5)$$

To estimate the efficiency of the collector (η_c), Eq. (16.5) has been employed. The energy inside the drying compartment is calculated as:

$$E_{in} = I_a A_s t_d \quad (16.6)$$

$$\eta_d = \frac{m_w L_w}{I_a A_s} \quad (16.7)$$

where E_{in} is the input energy (kWh), I_a is the average solar radiation for the total drying time (kW/m^2), A_s is the area of the system where the solar flux reaches (m^2), t_d is the total drying time (h), m_w is the mass of water removed (kg) and Lw is the latent heat of water (kJ/kg).

16.2.2 Analysing the Drying Kinetics

Based on experimental data of mass loss, D_e was calculated (Lingayat et al. 2020):

$$ln(MR) = ln\left(\frac{8}{\pi^2}\right) - \pi^2 \frac{D_e}{4L^2} t_d \tag{16.8}$$

The activation energy (E_{ac}) was calculated using experimental data by applying Arrhenius equation (Goud et al. 2019), where t_d (h) is the drying time and L (m) is the thickness of the sample.

$$D_e = D_o \, exp\left(-\frac{E_{ac}}{RT}\right) \tag{16.9}$$

Assume D_o (m^2/s) is the pre-exponential factor and R (J/mol K) is the universal gas constant. Moisture ratio (MR) can be estimated by using:

$$MR = \frac{MC_t}{MC_i} \tag{16.10}$$

where MC_t (db or wb) is MC at an instant of time.

$$DR = \frac{MC_t - MC_{t+dt}}{dt} \tag{16.11}$$

Time and change in time are represented by the subscripts t and dt, respectively. Coefficient (h_m) of mass transfer was evaluated using (Goud et al. 2019):

$$h_m = \frac{V}{A_t} ln(MR) \tag{16.12}$$

where the sample's volume is V (m^3), its surface area is A_t (m^2) and its thickness is L (m).

$$h = h_m \frac{k}{D_{ab} Le^{1/3}} \tag{16.13}$$

$$Le = \frac{\alpha}{D_{ab}} \tag{16.14}$$

According to Wang and Brennan (1995), the coefficient of heat transfer (h) can be calculated using Eq. (16.13), where D_{ab} (m^2/s) corresponds to the water moisture diffusivity in the air (0.282×10^{-4} m^2/s), k (W/m K) corresponds to the thermal conductivity of air, and Le corresponds to the Lewis number that describes the relationship between thermal and concentration boundary lines (Mugi and Chandramohan 2021).

SEC (kWh/kg) and SMER (kg / kWh) were estimated (Bhardwaj et al. 2021) using:

$$SEC = \frac{E_{in}}{m_w} \tag{16.15}$$

$$SMER = \frac{m_w}{E_{in}} \tag{16.16}$$

16.3 Result Analysis and Discussion

16.3.1 Data of Solar Radiation

Data of solar radiation was recorded during drying pineapple in a passive ISD without (setup-1) and with TES system (setup-2) and displayed in Fig. 16.3. It was recorded from morning 8:00 am to 6:00 pm for setup-1, while from 8:00 am to midnight for setup-2. The average solar radiation was recorded to be 659.5 and 623.6 W, for setup-1 and setup-2, respectively.

16.3.2 Evaluating Heat Supplied to Drying Section

The instantaneous actual heat supply (Q_{ac}) with time during drying pineapple in setup-1 and setup-2 is displayed in Fig. 16.2. The average Q_{ac} for setup-1 and setup-2 was calculated to be 717.6 and 812.9 W, respectively.

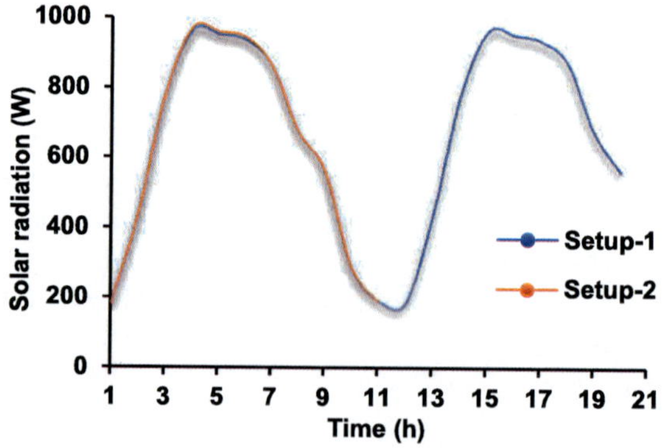

Fig. 16.2 Instantaneous solar radiation for the drying test days of setup-1 and setup-2

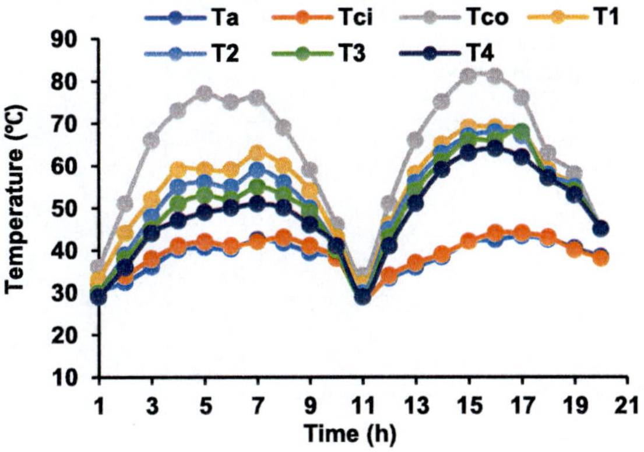

Fig. 16.3 Temperature data recorded during drying of pineapple in setup-1

16.3.3 Collector Efficiency

The efficiency of the SAC (η_c) has been evaluated from recorded data during drying pineapple in setup-1 and setup-2. The average and maximum η_c for setup-1 and setup-2 were estimated to be 63.6 and 61.3% and 70.6 and 92.9%, respectively.

16.3.4 Drying Efficiency

The drying efficiency (η_d) of setup-1 and setup-2 has been evaluated during drying pineapple. The average values of η_d for setup-1 and setup-2 were 6.4 and 10.7%, respectively. Setup-2 improved drying efficiency by 67.2% compared to setup-1.

16.3.5 Temperature Distribution

The temperature data for the experiment dates during drying pineapple in an ISD without a TES unit has been recorded and reported in Fig. 16.3. In Fig. 16.3, Ta, Tci, Tc, T1, T2, T3 and T4 are denoting ambient air, collector inlet, collector outlet, and trays 1, 2, 3 and 4, temperature, respectively. The average values of the same were 38.5, 39.1, 62.9, 54.6, 52.2, 50.5 and 48.4 °C, respectively. The maximum for the same was 43.4, 44, 81, 69, 68, 68 and 64 °C, respectively.

Similarly, the temperature distribution on setup-2 is demonstrated in Fig. 16.4. From Fig. 16.4, the average values of Ta, Tci, Tc, T1, T2, T3 and T4 in setup-2 were 33.2, 33.5, 42, 407, 39.4, 39.1 and 38.2°C, respectively. The maximum for the same was 40.2, 40.2, 66.5, 55, 53, 51 and 50°C, respectively.

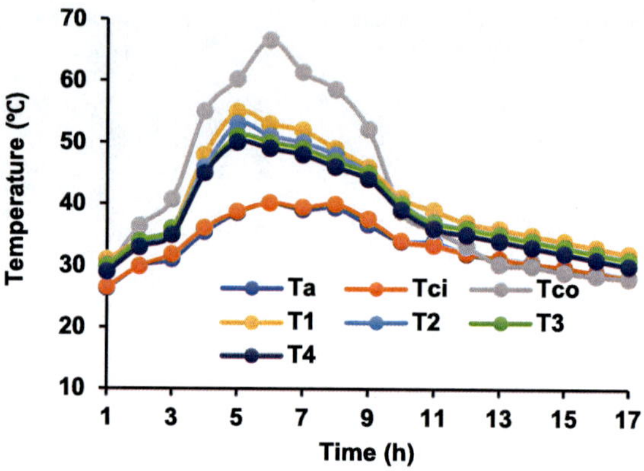

Fig. 16.4 Temperature data recorded during drying of pineapple in setup-2

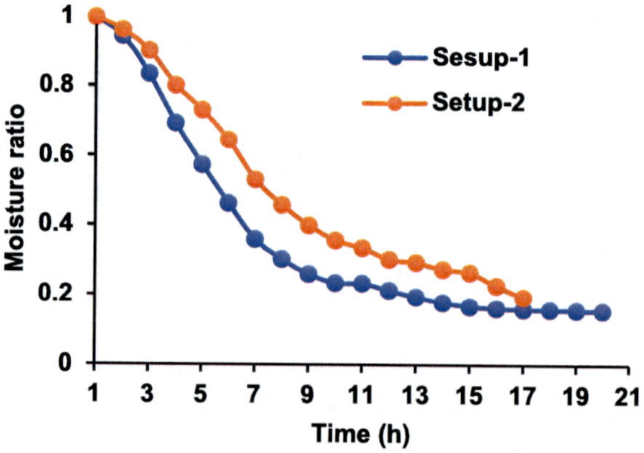

Fig. 16.5 Instantaneous moisture ratio of pineapple in during drying setup-1 and setup-2

16.3.6 Drying Kinetics

16.3.6.1 Moisture Ratio

The ratio of moisture present in the drying sample at an instant to the MC_i of the drying sample is the moisture ratio (MR). MR is evaluated by using Eq. (16.10) and described in Fig. 16.5. The characteristic graph of the MC with time is similar to the sketch of MC. MR is a function of MC, which is the reason for the similarity of the trends.

16.3.6.2 Rate of Drying

The rate of drying (DR) of pineapple during drying in setup-1 and setup-2 has been investigated. The instantaneous DR with time is demonstrated in Fig. 16.6. The DR increased at higher rate up to its maximum value at noon. From the maximum values

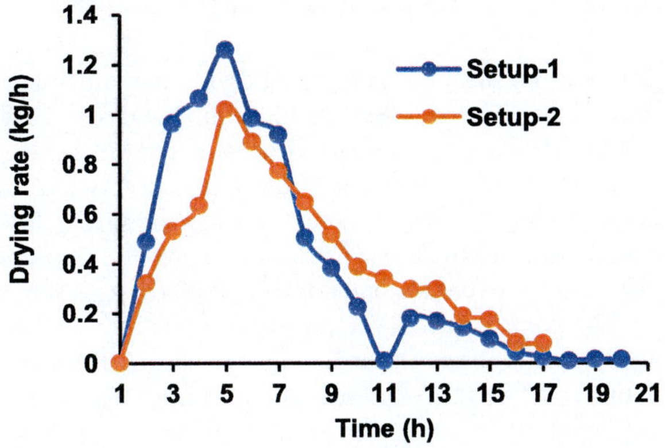

Fig. 16.6 Drying rate vs time during drying pineapple in setup-1 and setup-2

onwards, the DR decreased with a fast rate for setup-1 than setup-2. As there was TES in setup-2 which could maintain the temperature inside the drying section, the DR was seen higher than setup-1. The average and maximum DR were evaluated to be 0.375 and 0.420 kg/h and 1.26 and 1.018 kg/h, respectively, for setup-1 and setup-2. Setup-2 improved DR by 12% compared to setup-1.

16.3.6.3 Effective Moisture Diffusion Coefficient

Effective moisture diffusivity (D_e) has been evaluated for pineapple in setup-1 and setup-2. The average D_e for setup-1 and setup-2 was calculated to be 7.30703×10^{-09} and 5.22919×10^{-09} m^2/s, while their values were in the range of 2.28596×10^{-09}–1.07343×10^{-08} m^2/s and 2.28596×10^{-09}–9.00441×10^{-09} m^2/s, respectively.

16.3.6.4 Coefficient of Heat Transfer

Equation (16.12) was applied to evaluate the coefficient of heat transfer (h) of pineapple slices dried in setup-1 and setup-2. The mean values of it for setup-1 and setup-2 were 9.0 and 5.3 W/m^2K, respectively.

16.3.6.5 Coefficient of Mass Transfer

Coefficient of mass transfer (h_m) of pineapple while drying in setup-1 and setup-2 has been investigated. The average h_m for setup-1 and setup-2 was 0.007842 and 0.004597 m/s, respectively. It was evaluated in the range of 0–0.013371912 m/s for setup-1 and 0–15.40436494 m/s for setup-2.

16.3.6.6 Activation Energy, SMER and SEC

The SEC for the pineapple during drying in setup-1 and setup-2 was evaluated using Eq. (16.15). The average SEC for setup-1 and setup-2 was estimated to be 4.843 and 0.3053 kWh/kg, respectively. The estimated values of SEC for the same were in the range of 0.2441–31.43 and 0.2348–0.5112 kWh/kg, respectively. Comparatively, setup-2 showed improvement by minimizing the SEC by 93.7%. Similarly, the SMER for pineapple during drying in setup-1 and setup-2 was calculated to be in the range of 0–0.0041 and 0.00196–0.00426 kg/kWh, respectively.

The average SMER for setup-1 and setup-2 was 0.207 and 3.28 kg/kWh, respectively. Setup-2 shows an improvement of 3.073 kg/kWh of SMER compared to setup-1. Moreover, the activation energy (Eacv) for the drying experiment of pineapple has been evaluated for both setups. Its average for setup-1 and setup-2 was 41.25 and 36.76 kJ/mol, respectively. There was 10.88% improvement of Eacv in case of setup-2 compared to setup-1.

16.4 Conclusion

An experimental assessment of the drying kinetics of pineapple during drying in a passive solar dryer (ISD), which is indirect type, comparing without thermal energy storage (TES) (setup-1) and with TES (setup-2) has been performed. The drying performances of the two setups have been comparatively analysed. Accordingly, the following main conclusion points are inferred:

- With the recorded average solar data of setup-1 and setup-2 drying days being 659.5 and 623.6 W, the average collector efficiency was 63.6 and 61.3%, respectively. The mean heat supplied to the drying section for the same was 717.6 and 812.9 W, respectively. Similarly, the mean values of drying efficiency of setup-1 and setup-2 were 6.4 and 10.7%, respectively. Setup-2 improved drying efficiency by 67.2% compared to setup-1.
- The estimated average drying rate was 0.375 and 0.420 kg/h, respectively, for setup-1 and setup-2. There was 12% increment in drying kinetics in setup-2 in comparison with setup-1.
- The effective moisture diffusion coefficient was 7.30703×10^{-09} and 5.22919×10^{-09} m^2/s in setup-1 and setup-2, respectively. The average heat transfer coefficient for setup-1 and setup-2 were 9.0 and 5.3 W/m^2K, respectively. Similarly, the average mass transfer coefficient for the same was 0.007842 and 0.004597 m/s, respectively.
- The average activation energy for setup-1 and setup-2 was 41.25 and 36.76 kJ/mol, respectively. There was a 10.88% reduction of activation energy in the case of setup-2 compared to setup-1.

- The mean specific energy consumption (SEC) in setup-1 and setup-2 was 4.843 and 0.3053 kWh/kg, respectively. Comparatively, setup-2 showed an improvement by reducing the SEC by 93.7%.
- The average unit rate of moisture extraction (SMER) for pineapple during drying in setup-1 and setup-2 was 0.207 and 3.28 kg/kWh, respectively. Setup-2 showed an improvement of 3.073 kg/kWh of SMER compared to setup-1.
- Even though it took 16 h for both setups for drying the sample from the moisture content of 7.911 to 0.4169 (db), setup-2 performed well as the drying in it was completed in 1 day with better efficiency.

References

Alimohammadi Z, Samimi Akhijahani H, Salami P (2020) Thermal analysis of a solar dryer equipped with PTSC and PCM using experimental and numerical methods. Solar Energy 201: 157–177. https://doi.org/10.1016/j.solener.2020.02.079

Babu AK, Kumaresan G, Raj VAA, Velraj R (2018) Review of leaf drying: Mechanism and influencing parameters, drying methods, nutrient preservation, and mathematical models. Renewable and Sustainable Energy Reviews 90:536–556. https://doi.org/10.1016/j.rser.2018.04.002

Bhardwaj AK, Kumar R, Kumar S, Goel B, Chauhan R (2021) Energy and exergy analyses of drying medicinal herb in a novel forced convection solar dryer integrated with SHSM and PCM. Sustainable Energy Technologies and Assessments 45:101119. https://doi.org/10.1016/j.seta.2021.101119

Chandramohan VP, Talukdar P (2010) Three dimensional numerical modeling of simultaneous heat and moisture transfer in a moist object subjected to convective drying. International Journal of Heat and Mass Transfer 53:4638–4650. https://doi.org/10.1016/j.ijheatmasstransfer.2010.06.029

Gilago MC, Chandramohan VP (2021) Performance evaluation of natural and forced convection indirect type solar dryers during drying ivy gourd: An experimental study. Renewable Energy 182. https://doi.org/10.1016/j.renene.2021.11.038.

Gilago M C, Chandramohan VP (2022) Effect of phase change materials on the performance of natural convection indirect type solar dryer during drying ivy gourd. Heat Transfer Engineering 44 (7). https://doi.org/10.1080/01457632.2022.2079045.

Goud M, Reddy MVV, CVP, SS (2019) A novel indirect solar dryer with inlet fans powered by solar PV panels: Drying kinetics of Capsicum Annum and Abelmoschus esculentus with dryer performance. Solar Energy 194:871–885. https://doi.org/10.1016/j.solener.2019.11.031.

Hidalgo LF, Candido MN, Nishioka K, Freire JT, Vieira GNA (2021) Natural and forced air convection operation in a direct solar dryer assisted by photovoltaic module for drying of green onion. Solar Energy 220: 24–34. https://doi.org/10.1016/j.solener.2021.02.061

Lingayat A, Chandramohan VP, Raju VRK, Kumar A (2020) Development of indirect type solar dryer and experiments for estimation of drying parameters of apple and watermelon: Indirect type solar dryer for drying apple and watermelon. Thermal Science and Engineering Progress 16. https://doi.org/10.1016/j.tsep.2020.100477.

Mugi VR, Chandramohan VP (2021) Energy, exergy and economic analysis of an indirect type solar dryer using green chilli: A comparative assessment of forced and natural convection. Thermal Science and Engineering Progress 24:100950. https://doi.org/10.1016/j.tsep.2021.100950.

Tagnamas Z, Kouhila M, Bahammou Y, Lamsyehe H, Moussaoui H, Idlimam A, Lamharrar A (2021) Drying kinetics and energy analysis of carob seeds (Ceratonia siliqua L.) convective solar drying. Journal of Thermal Analysis and Calorimetry 146:16-24. https://doi.org/10.1007/s10973-021-10632-6

Vijayan S, Arjunan TV, Kumar A (2016) Mathematical modeling and performance analysis of thin layer drying of bitter gourd in sensible storage based indirect solar dryer. Innovative Food Science and Emerging Technologies 36:59–67. DOI: https://doi.org/10.1016/j.ifset.2016.05.014

Wang N, Brennan J G (1995) A mathematical model of simultaneous heat and moisture transfer during drying of potato. Journal of Food Engineering 24:47–60. https://doi.org/10.1016/0260-8774(94)P1607-Y

Chapter 17
An Exergetic Investigation of a Marine Diesel Engine

Turgay Köroğlu and Arif Savaş

Nomenclature

IMO International Marine Organization

17.1 Introduction

The energy need of the world has increased rapidly due to population growth, developments in industry, and the modernization of human life (Sarıkoç et al. 2020). Various sectors are at the forefront of energy consumption. The transportation sector accounts not only for approximately 20% of the world's energy consumption but also for approximately 23% of total CO_2 emissions (Kalghatgi 2018).The sector is developing day by day, and the amount of energy it requires is constantly increasing. One of the most important transportation subsectors is maritime transportation. In comparison with 1990, the maritime sector has increased approximately 2.5 times and reached 10 billion tons of goods capacity; hence it dominates more than 80% of the total transportation (Unctad 2015).

A large part of energy production is obtained from internal combustion engines (Wang et al. 2021). Nevertheless, there are basically two problems, namely, depletion of fossil fuels and their emissions to the environment. Declared by the International Marine Organization (IMO), Tier III stricts 80% less nitrogen oxide (NOx) emissions than Tier I (Ni et al. 2020). To overcome this issue, emission reduction studies in diesel engines are conducting to increase the efficiency of the systems

T. Köroğlu (✉) · A. Savaş
Maritime Faculty, Bandirma Onyedi Eylul University, Balıkesir, Turkey
e-mail: tkoroglu@bandirma.edu.tr; asavas@bandirma.edu.tr

M. Z. Sogut et al. (eds.), *Proceedings of the 2022 International Symposium on Energy Management and Sustainability*, Springer Proceedings in Energy,
https://doi.org/10.1007/978-3-031-30171-1_17

(Karvounis et al. 2018; Yao et al. 2009). Baldi et al. applied energy and exergy analyses on cruise ships based on 1-year operational activities of a ship in the Baltic Sea. They determined that the driving force uses 46% of the total energy and the rest in electricity production and heat. On the other hand, it was determined that 76% of the total exergy was destroyed in all processes involving combustion (Baldi et al. 2018). Baldi et al. also performed energy and exergy analyses on a chemical tanker. Results yielded that the propulsion occupied 70% of the energy onboard and power and heat were left with the rest. In addition, the recovery of exhaust gas heat is important in terms of exergy, because they showed that it represents 18% of the engine power output (Baldi et al. 2014).

Yesilyurt and Arslan investigated the energy and exergy efficiency of biodiesel, which they obtained by adding waste cook and canola oil at the same rate, at different injection pressures. Diesel fuel usage showed that lower fuel pressure brings high energy and exergy efficiencies; however, biodiesel needs more pressure to achieve its high efficiencies, which are still slightly lower than diesel fuel with lower exergy destructions (Yesilyurt and Arslan 2019). Panigrahi et al. performed exergy analysis on a four-stroke engine by mixing alternative fuel and diesel fuel. It has been determined that the energy carried in the exhaust gases of diesel fuel is higher than that of biodiesel. They determined that the efficiency in biodiesel is higher than diesel fuel (Panigrahi et al. 2014).

Murugapoopathi and Vasudevan conducted energy and exergy analyses by experimenting different biodiesel ratios with different compression ratios. As a result of the study, it was observed that rubber seed oil mixed with diesel in variable compression ratio engine gave a better performance in the thermodynamic analysis of methyl esters (Murugapoopathi and Vasudevan 2019).

In this study, an eight-cylinder two-stroke turbocharged diesel engine of a crude oil carrier is investigated by applying energy and exergy analyses. The results reveal the potential recovery of exergy to produce more power and the improvement potential of the engine efficiency to lead a better environmental and economic return.

17.2 Method and Material

Exergy is defined as the maximum potential work that could be harvested when a system interacts with its environment. It consists of different parts; however mostly, engineers deal with the physical part of the exergy of a stream as (Gozmen Şanli et al. 2019):

$$\dot{E}x \;= \dot{m}[(\mathrm{h} - \mathrm{h0}) - \mathrm{T0}(\mathrm{s} - \mathrm{s0})] \tag{17.1}$$

where $\dot{m}$ is the mass flow rate, h is the specific enthalpy, T_0 is the environmental temperature, and s is the specific entropy. Moreover, exergy of a heat transfer could be described as (Şanli and Uludamar 2020):

$$\dot{Ex}_Q = (1 - T_0/T)\dot{Q} \tag{17.2}$$

Exergy balance is being used to reveal the exergy destruction ($\dot{Ex}_D$) of any system goes under a process, which leads the engineer to determine the inefficiencies (Panigrahi et al. 2014).

$$\dot{Ex}_{in} - \dot{Ex}_{out} - \dot{Ex}_D = \mathrm{d}\,Ex_{sys}/dt \tag{17.3}$$

where $\dot{Ex}_{in}$ and $\dot{Ex}_{out}$ are inlet and outlet exergies of investigated system, respectively, while d Ex_{sys}/dt is the exergy change of the system, which is zero for the steady-state steady-flow systems. Fuel energy $\dot{E}$ fuel is:

$$\dot{E}_{fuel} = \dot{m}_{fuel}\,H_u \tag{17.4}$$

where $\dot{m}$ fuel and H_u are mass flow rate and heating value of the fuel (42700 kj/kg), respectively. The exergy of the fuel could be calculated as (Koroglu and Sogut 2018):

$$\dot{Ex}_{fuel} = 1,07\dot{E}_{fuel} \tag{17.5}$$

While making the calculations, considering that the amount of the fuel is very low in the fuel-air ratio and that the nitrogen in the exhaust gas is dominant, the exhaust gas is assumed as air (Çavuş et al. 2021). Moreover, exergies of lubrication oil and jacket water heats are calculated by the assumption of average temperature with respect to inlet and outlet temperatures (Şanli and Uludamar 2020) (Fig. 17.1).

Energy and exergy analyses of the MAN B & W 8S40ME-C9.5-HPSCR onboard of a crude oil carrier, which was designed to produce 9080 kW at 146 RPM of power under the nominal maximum continuous conditions, were carried out. In the study, the ambient temperature and pressure were assumed as 25 °C and 1 atm, respectively. Under these conditions, the propulsion requirement of the ship was to have an 8500 kW of power as MCR at 140 RPM at 100% load (Man B & W 2022).

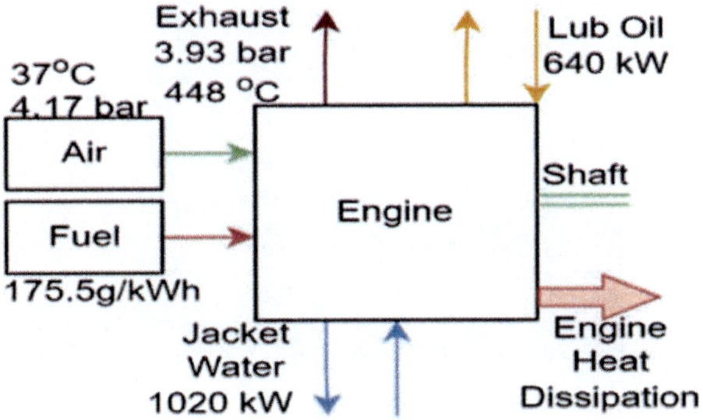

Fig. 17.1 System diagram

17.3 Result and Discussion

The energy and exergy values and the mass flow rates of the streams are given in Table 17.1.

It is obvious that the inlet air has a high amount of energy; however its exergy is less than half of its energy due to its low temperature. Exhaust carries the highest amount of the energy as 55.6% that is given by the inlet streams and could be seen on Fig. 17.2. Power output follows the exhaust energy. That is, it is required to harvest energy from the exhaust, which could be done by annexing a turbocharger and may be a power turbine to produce more power. The energy share of the lubrication oil and the jacket water is low in comparison.

On the other hand, exergy analysis reveals that the highest exergy belongs to the output power as expected. Energy analysis misleads as it can be seen in Fig. 17.3. It is shown that, even, the highest amount of energy is in the exhaust, and its potential to produce maximum work is limited to 23% of the overall exergy inlet. Moreover, it seems impractical to focus primarily on jacket water or lubrication oil exergy, while it is important to recover exhaust exergy. As mentioned above, turbocharger could

Table 17.1 System data and results

	$\dot{m}$ (kg/s)	Exergy, $\dot{E}x$ (kW)	Energy, $\dot{E}$ (kW)
Air	16,9	2048	5237,31
Fuel	0,4	18915	17677,8
Exhaust	17,3	4897	12744,91
Jacket Water	N/A	91,05	1020
Lub oil	N/A	73,08	640
Power	N/A	8500	8500
Loss	N/A	7401.87	10.2

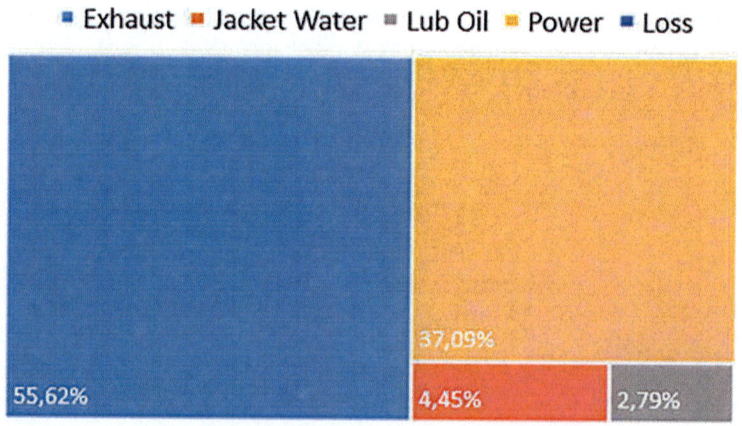

Fig. 17.2 Distribution of total energy

Fig. 17.3 Distribution of total exergy

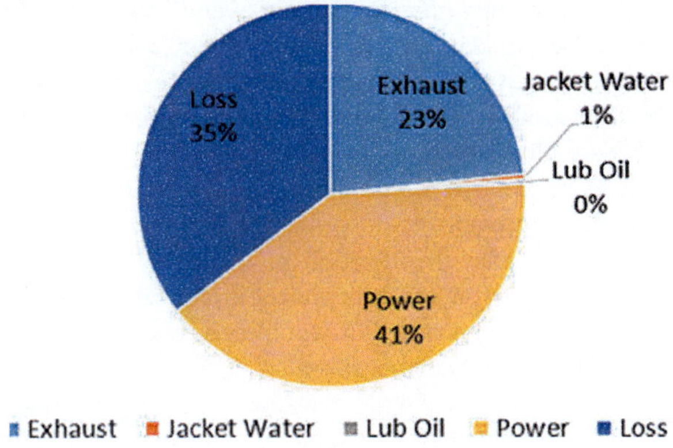

be annexed with a consequent organic Rankine cycle instead of power turbine if the outlet state of the exhaust is low to drive the power turbine. It also could be utilized to produce domestic steam and hot water.

Exergy destruction, represented as "Loss" in Table 17.1 and Fig. 17.3, has the second highest share within the system as 35%. That is a huge amount in comparison. The investigated engine would have friction, heat loss, incomplete combustion, pressure drops, leaks, etc. Therefore, it would be better to lubricate the system to produce less friction; in fact, poly-tetrafluoroethylene coating could be considered. Moreover, combustion efficiency could improve with additives. Better o- rings as well as oil filming would stop leaks and therefore pressure and temperature drops. Lastly, insulation could prevent heat loss.

Lubrication oil exergy is the lowest due to its nature to work in less temperature state, and its first goal is lubricate instead of cooling. In contrast, the second lowest exergy belongs to the jacket water, but its exergy could be utilized to produce water from flash desalination process as well as domestic hot water.

17.4 Conclusion

In this study, an eight-cylinder two-stroke marine diesel engine is investigated by applying both energy and exergy analyses. Energy analysis gives results that would lead the improvement efforts into one direction; however, it is the exergy analysis to reveal the potential of improvement. That could be saidby the results of the exhaust: The results of the energy analysis showed that the exhaust gas carries 55% of the overall energy, while exergy analysis resulted the share of the exhaust gas is only 23% of the overall exergy. Moreover, exergy destruction is the highest after the power output. Therefore, the reasons of the exergy destruction may be eliminated with respect to given advices. It is important to note that the investigation resulted in a low heat dissipation. That could mean the jacket water is taking the heat out efficiently. Lubrication oil also carries the heat, but it is a lower amount as expected, which is still quite high in comparison with the heat dissipation.

Lastly, energy efficiency of the overall system is 37%, while exergy efficiency is 41%. That could be concluded even if less amount of energy is converted into power, it is converted efficiently. It would be necessary to point out that the engine needs to be improved. It would be done internally or externally. The internal measures are discussed before as to create less friction or improvement of the combustion. However, these steps could bring more economical burden. Therefore, considering external improvements to create more power such as annexing new devices to the exhaust and useful utilization of the jacket water and/or the lubrication oil heat would increase the overall efficiency. One could not say there would be less cost for the external applications. It would be investigated, evaluated, and compared to the internal improvement steps.

The future work should consider the mentioned measures as well as the inclusion of the turbocharge system not only in the exergy point of view but also in economic considerations.

References

Baldi, F., Ahlgren, F., Nguyen, T. van, Thern, M., & Andersson, K. (2018). Energy and exergy analysis of a cruise ship. Energies, 11(10). https://doi.org/10.3390/en11102508

Baldi, F., Johnson, H., Gabrielii, C., & Andersson, K. (2014). Energy and exergy analysis of ship energy systems - The case study of a chemical tanker. Proceedings of the 27th International Conference on Efficiency, Cost, Optimization, Simulation and Environmental Impact of Energy Systems, ECOS 2014. https://doi.org/10.5541/ijot.70299

Çavuş, İ., Bayer, M. U., & Köroğlu, T. (2021). Performing Exergy Analysis On A Marine Engine Waste Heat To Recovery Systems Increase Fuel Consumption Efficiency. Proceedings of the 2nd International Congress on Ship and Marine Technology, 603– 611.

Gozmen Şanli, B., Uludamar, E., & Özcanli, M. (2019). Evaluation of energetic-exergetic and sustainability parameters of biodiesel fuels produced from palm oil and opium poppy oil as alternative fuels in diesel engines. Fuel, 258. https://doi.org/10.1016/j.fuel.2019.116116

Kalghatgi, G. (2018). Is it really the end of internal combustion engines and petroleum in transport? In Applied Energy (Vol. 225, pp. 965–974). Elsevier Ltd. https://doi.org/10.1016/j.apenergy.2018.05.076

Karvounis, N., Pang, K. M., Mayer, S., & Walther, J. H. (2018). Numerical simulation of condensation of sulfuric acid and water in a large two-stroke marine diesel engine. Applied Energy, 211, 1009–1020. https://doi.org/10.1016/j.apenergy.2017.11.085

Koroglu, T., & Sogut, O. S. (2018). Conventional and advanced exergy analyses of a marine steam power plant. Energy, 163, 392–403. https://doi.org/10.1016/j.energy.2018.08.119

Man B & W. (2022). CEAS Engine Data report 8S40ME- C9.5-HPSCR with scrubber.

Murugapoopathi, S., & Vasudevan, D. (2019). Energy and exergy analysis on variable compression ratio multi-fuel engine. Journal of Thermal Analysis and Calorimetry, 136(1), 255–266. https://doi.org/10.1007/s10973-018-7761-2

Ni, P., Wang, X., & Li, H. (2020). A review on regulations, current status, effects and reduction strategies of emissions for marine diesel engines. In Fuel (Vol. 279). Elsevier Ltd. https://doi.org/10.1016/j.fuel.2020.118477

Panigrahi, N., Mohanty, M. K., Mishra, S. R., & Mohanty, R. C. (2014). Performance, Emission, Energy, and Exergy Analysis of a C.I. Engine Using Mahua Biodiesel Blends with Diesel. International Scholarly Research Notices, 2014, 1–13. https://doi.org/10.1155/2014/207465

Şanli, B. G., & Uludamar, E. (2020). Energy and exergy analysis of a diesel engine fuelled with diesel and biodiesel fuels at various engine speeds. Energy Sources, Part A: Recovery, Utilization and Environmental Effects, 42(11), 1299–1313. https://doi.org/10.1080/15567036.2019.1635229

Sarıkoç, S., Örs, İ., & Ünalan, S. (2020). An experimental study on energy-exergy analysis and sustainability index in a diesel engine with direct injection diesel- biodiesel-butanol fuel blends. Fuel, 268. https://doi.org/10.1016/j.fuel.2020.117321

Unctad. (2015). Review of Maritime Transport 2015. In Unctad (Issue October). http://unctad.org/rmt

Wang, P., Hu, Z., Shi, L., Tang, X., Liu, Y., & Deng, K. (2021). Experimental investigation of the effects of Miller timing on performance, energy and exergy characteristics of two-stage turbocharged marine diesel engine. Fuel, 292. https://doi.org/10.1016/j.fuel.2021.120252

Yao, M., Zheng, Z., & Liu, H. (2009). Progress and recent trends in homogeneous charge compression ignition (HCCI) engines. In Progress in Energy and Combustion Science (Vol. 35, Issue 5, pp. 398–437). https://doi.org/10.1016/j.pecs.2009.05.001

Yesilyurt, M. K., & Arslan, M. (2019). Analysis of the fuel injection pressure effects on energy and exergy efficiencies of a diesel engine operating with biodiesel. Biofuels, 10(5), 643–655. https://doi.org/10.1080/17597269.2018.1489674

Chapter 18
Bibliometric Analysis of Alternative Fuel in Marine

Arif Savaş, Muhammed Umar Bayer, İrfan Çavuş, and Tolga Berkay Şirin

18.1 Introduction

There has been an increase in the need for energy due to the rapid development of industry in the world, the increase in population and the modernization of human life (Sarıkoç et al. 2020). Energy is consumed in various sectors. One of the most important of these sectors is the transportation sector. While the transportation sector consumes approximately 20% of the world's energy, it produces approximately 23% of the total CO2 emissions depending on this consumption (Kalghatgi 2018; Prussi et al. 2021). In the transportation sector, maritime transportation, which is one of the most important types of transportation and currently constitutes approximately 80% of the total transportation in the world, comes. Maritime transport is increasing rapidly. Compared to 1990, the total transportation increased 2.5 times and reached the level of approximately 10 billion tons (Unctad 2015). Depending on this increase, energy consumption increases. In addition to increasing energy efficiency, there are also studies to reduce emissions. In maritime transport, emission studies are carried out according to Tier III declared by the International Maritime Organization (IMO) (Ni et al. 2020; Ovaska et al. 2019).

Most of the energy production is produced by burning fossil fuels on the side of internal combustion engines. Considering the limited fossil fuels and emission effects, studies are carried out to obtain energy and reduce emissions by burning new fuels in engines (Wang et al. 2021; Yilmaz et al. 2015). Alternative fuel usage area in marine is limited due to the high required energy. These are generally biogas,

A. Savaş (✉) · M. U. Bayer · İ. Çavuş · T. B. Şirin
Maritime Faculty, Bandirma Onyedi Eylul University, Balıkesir, Turkey
e-mail: asavas@badirma.edu.tr; mbayer@bandirma.edu.tr; icavus@bandirma.edu.tr; tsirin@bandirma.edu.tr

M. Z. Sogut et al. (eds.), *Proceedings of the 2022 International Symposium on Energy Management and Sustainability*, Springer Proceedings in Energy,
https://doi.org/10.1007/978-3-031-30171-1_18

dimethyl ether, ethanol, liquefied natural gas, liquefied petroleum gas, methanol, biodiesel, and ammonia (Bilgili 2020; Mohd Noor et al. 2018).

Bibliometric analysis is the study of many fields, including the general study of a field and the analysis of prominent authors with their work on the field. These review topics can be authors, number of publications, publishing countries, and keywords used in publications (Bonilla et al. 2015a, b; del Giudice et al. 2021).

Roskilly et al. Emission analysis was carried out using biodiesel in two small boat diesel engines. As a result of the tests, it was observed that more fuel was burned in order to obtain the same power in the motor. A decrease in NOx emission was observed in the use of biodiesel. In addition, a decrease in CO emission was observed in the engine operating with biodiesel fuel at high loads (Roskilly et al. 2008). Murillo et al. Engine performance and emission emission analyzes were made by using used cooking oil as biodiesel. Studies have determined that there is a decrease of approximately 12% in CO emissions but an increase of up to 20% in NOx emissions. It has been determined that there is an increase in specific fuel consumption, but according to the authors, this is at an acceptable level considering the exhaust emissions and biodiesel is usable (Murillo et al. 2007). Gabiña et al. compared two alternative fuels based on waste oil by testing. In this study, better combustion was obtained due to the improvement in injection timing. As a result of the study, they found a significant reduction in carbon emissions in both fuels. NOx emission in distillate fuel was higher than in alternative fuel. According to the authors, the fuel is suitable for feasibility in boats such as fishing boats due to the performance of alternative fuel (Gabiña et al. 2016). There are also researches on various subjects on bibliometric analysis. There are publications on a variety of topics, such as biotechnology (Dalpé 2002), economics (Bonilla et al., 2015), transport and emissions (Tian et al. 2018), marine litter (Renzi et al. 2020), and smart cities (Mora et al. 2017).

In this study, a bibliometric analysis of the studies on the use of alternative fuels in maritime transport was made. During the study, Web of Science (WoS) and VOSviewer package program were used. A study has been made on the change of authors, countries, keywords, and publications according to the years of all studies carried out to date.

18.2 Material and Method

Web of Science was used in this study, since most publications are available. The results were obtained by searching "Alternative fuel in marine" on the WoS search engine. The results were analyzed with the help of the VOSviewer package program.

As a result of the examination, a total of 1226 publications were reached in 1990.

Figure 18.1 shows the annual change in the publications made in 2010 and after. Publications made in 2010 and later constitute approximately 85% of the total publications. Publications continued to increase after 2010. The types of publications made can be seen in Table 18.1. According to the table, approximately 76% of the publications are in the type of research articles.

Fig. 18.1 Publication years

Table 18.1 Document types

Document types	Record count	%
Articles	927	75,61
Proceedings Papers	189	15,42
Review articles	141	11,50
Book chapters	17	1,39
Early access	13	1,06
Editorial Materials	3	0,24

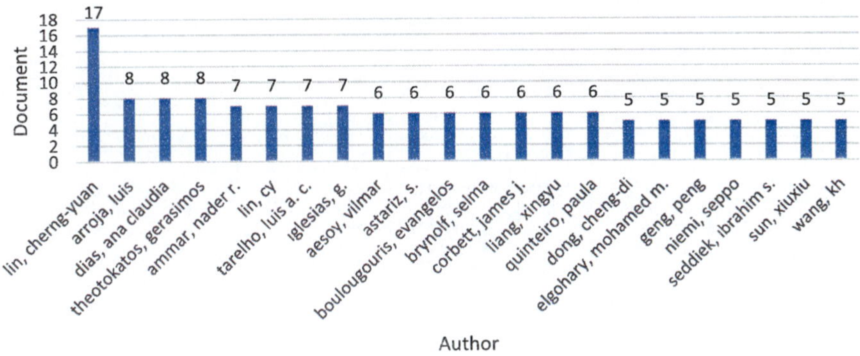

Fig. 18.2 Authors

18.3 Result and Discussion

18.3.1 Author

In the study, it is seen that a total of 4570 different authors published. Among these authors, when the authors with 5 or more publications in total were filtered, it was determined that there were 22 authors in total.

Figure 18.2 shows the authors who have published five or more publications in total. Cherng-yuan is the most published author with 17 publications in total. This author is followed by Ana Claudia, Luis, and Gerasimos with eight publications each.

18.3.2 Keywords

In the examinations made, 3555 different keywords were found. A total of 105 words with 5 or more words were identified. The words used in 20 or more of these words are shown in Fig. 18.3 The relations of these words with each other can be seen bibliometrically in Fig. 18.4.

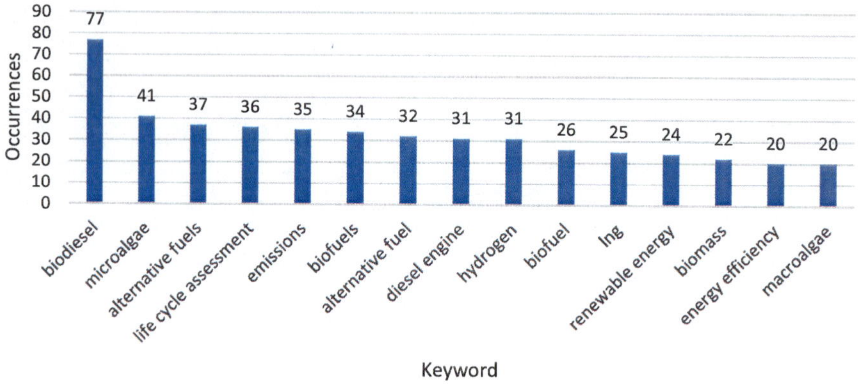

Fig. 18.3 Keywords

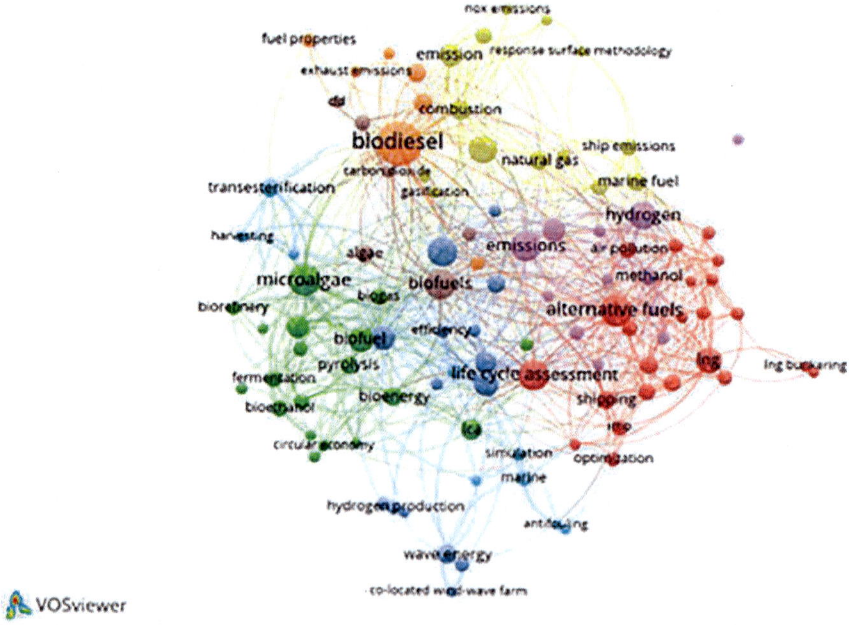

Fig. 18.4 Map of keywords

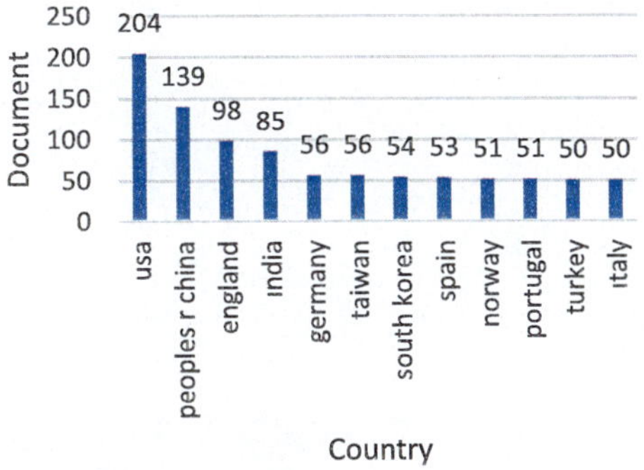

Fig. 18.5 Country

As seen in Fig. 18.4, colors represent a cluster in bibliometric analysis. The size of the circles represented by the words is proportional to the number of uses. In addition, it is understood that the more the word is in the center and connected with other words, the more the interaction of the word is. The word "biodiesel," which was used 77 times in this study, was the most used keyword. The second place is the word "microalgae," which is used 41 times.

18.3.3 Country

In the study, it was determined that a total of 88 countries broadcast. When the countries that broadcast at least 5 or more in these countries are classified, there are 52 countries. Countries with 40 or more broadcasts are shown in Fig. 18.5.

While the USA is the country with the most publications with 204 publications, China follows the USA with 139 publications. The relationship between the publications of the countries is seen in the bibliometric map given in Fig. 18.6.

As seen in Fig. 18.6, each color represents a cluster. It is understood that the larger the circle the countries are represented, the more publications they have. In addition, the more ties countries have with other countries, the more their publications are related to each other. As seen in Fig. 18.6, the USA and China are in the middle. In addition, these two countries are connected with other countries. On the other hand, countries such as Egypt, which appear far and small in the figure, are countries with fewer publications and less interaction with their publications.

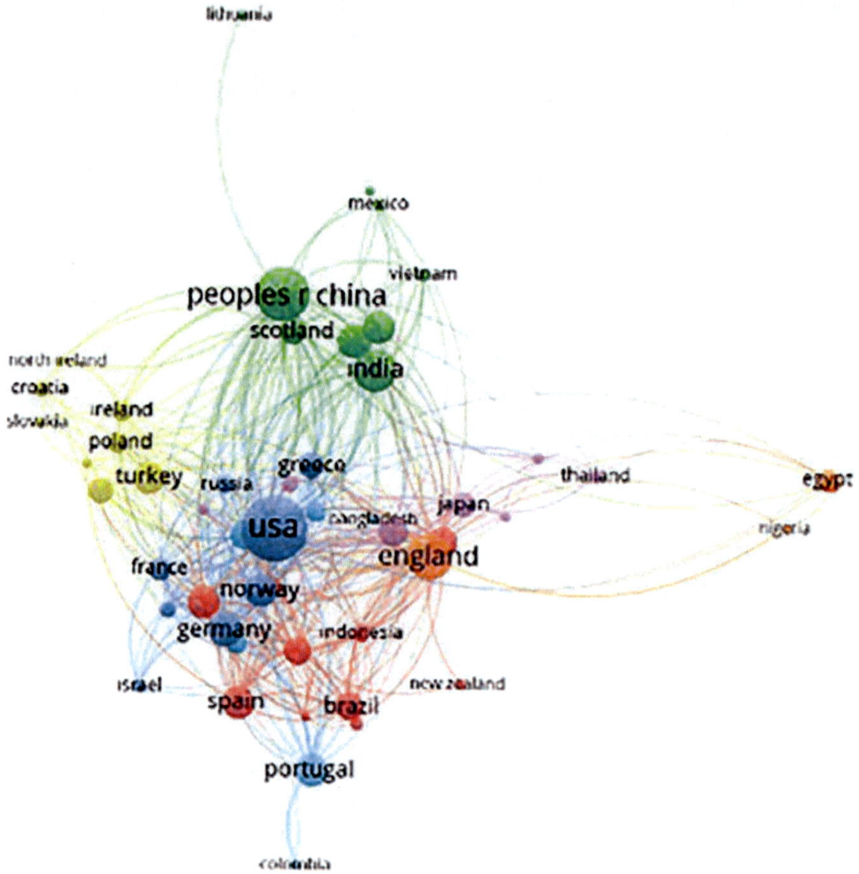

Fig. 18.6 Map of country

18.4 Conclusion

In the study, bibliometric analysis of "Alternative fuel in marine" was made. A total of 1226 publications made in 1900 until today have been accessed. Looking at the distribution of publications by years, it is seen that approximately 85% of the total publications were made in 2010 and later. From this, it can be deduced that the studies are new and will continue to be studied.

Approximately 76% of these publications are research articles. While a total of 88 different countries broadcast in the broadcasts, the number of countries that broadcast at least 5 or more is 52 in total. The USA is the most broadcasting country with 204 publications. China follows the USA.

When examined in terms of the author, it is seen that there are 4570 different authors in the study. The number of those who published 5 or more publications among these authors are 22. Cherng-yuan is the most published author with 17 publications.

When the keywords were examined, it was determined that a total of 3555 keywords were used. The number of words used 5 or more in these keywords is 105. The most used word among the words was the word biodiesel, which is an alternative fuel type.

References

Bilgili, L. (2020). Comparative assessment of alternative marine fuels in life cycle perspective. Renewable and Sustainable Energy Reviews, 144(April). 10.32388/yfb30w

Bonilla, C. A., Merigó, J. M., & Torres-Abad, C. (2015a). Economics in Latin America: a bibliometric analysis. Scientometrics, 105(2), 1239–1252. https://doi.org/10.1007/s11192-015-1747-7

Bonilla, C. A., Merigó, J. M., & Torres-Abad, C. (2015b). Economics in Latin America: a bibliometric analysis. Scientometrics, 105(2), 1239–1252. https://doi.org/10.1007/s11192-015-1747-7

Dalpé, R. (2002). Bibliometric analysis of biotechnology. In Budapest Scientometrics (Vol. 55, Issue 2). Kluwer Academic Publishers.

del Giudice, M., di Vaio, A., Hassan, R., & Palladino, R. (2021). Digitalization and new technologies for sustainable business models at the ship–port interface: a bibliometric analysis. Maritime Policy and Management. https://doi.org/10.1080/03088839.2021.1903600

Gabiña, G., Martin, L., Basurko, O. C., Clemente, M., Aldekoa, S., & Uriondo, Z. (2016). Waste oil-based alternative fuels for marine diesel engines. Fuel Processing Technology, 153, 28–36. https://doi.org/10.1016/j.fuproc.2016.07.024

Kalghatgi, G. (2018). Is it really the end of internal combustion engines and petroleum in transport? In Applied Energy (Vol. 225, pp. 965–974). Elsevier Ltd. https://doi.org/10.1016/j.apenergy.2018.05.076

Mohd Noor, C. W., Noor, M. M., & Mamat, R. (2018). Biodiesel as alternative fuel for marine diesel engine applications: A review. Renewable and Sustainable Energy Reviews, 94(February 2017), 127–142. https://doi.org/10.1016/j.rser.2018.05.031

Mora, L., Bolici, R., & Deakin, M. (2017). The First Two Decades of Smart-City Research: A Bibliometric Analysis. Journal of Urban Technology, 24(1), 3–27. https://doi.org/10.1080/10630732.2017.1285123

Murillo, S., Míguez, J. L., Porteiro, J., Granada, E., & Morán, J. C. (2007). Performance and exhaust emissions in the use of biodiesel in outboard diesel engines. Fuel, 86(12–13), 1765–1771. https://doi.org/10.1016/j.fuel.2006.11.031

Ni, P., Wang, X., & Li, H. (2020). A review on regulations, current status, effects and reduction strategies of emissions for marine diesel engines. In Fuel (Vol. 279). Elsevier Ltd. https://doi.org/10.1016/j.fuel.2020.118477

Ovaska, T., Niemi, S., Sirviö, K., Heikkilä, S., Portin, K., & Asplund, T. (2019). Effect of alternative liquid fuels on the exhaust particle size distributions of a medium-speed diesel engine. Energies, 12(11), 1168–1176. https://doi.org/10.3390/en12112050

Prussi, M., Scarlat, N., Acciaro, M., & Kosmas, V. (2021). Potential and limiting factors in the use of alternative fuels in the European maritime sector. Journal of Cleaner Production, 291. https://doi.org/10.1016/j.jclepro.2021.125849

Renzi, M., Pauna, V. H., Provenza, F., Munari, C., & Mistri, M. (2020). Marine litter in transitional water ecosystems: State of the art review based on a bibliometric analysis. Water (Switzerland), 12(2). https://doi.org/10.3390/w12020612

Roskilly, A. P., Nanda, S. K., Wang, Y. D., & Chirkowski, J. (2008). The performance and the gaseous emissions of two small marine craft diesel engines fuelled with biodiesel. Applied Thermal Engineering, 28(8–9), 872–880. https://doi.org/10.1016/j.applthermaleng.2007.07.007

Sarıkoç, S., Örs, İ., & Ünalan, S. (2020). An experimental study on energy-exergy analysis and sustainability index in a diesel engine with direct injection diesel-biodiesel-butanol fuel blends. Fuel, 268. https://doi.org/10.1016/j.fuel.2020.117321

Tian, X., Geng, Y., Zhong, S., Wilson, J., Gao, C., Chen, W., Yu, Z., & Hao, H. (2018). A bibliometric analysis on trends and characters of carbon emissions from transport sector. Transportation Research Part D: Transport and Environment, 59, 1–10. https://doi.org/10.1016/j.trd.2017.12.009

Unctad. (2015). Review of Maritime Transport 2015. In Unctad (Issue October). http://unctad.org/rmt

Wang, C., Ju, Y., & Fu, Y. (2021). Comparative life cycle cost analysis of low pressure fuel gas supply systems for LNG fueled ships. Energy, 218. https://doi.org/10.1016/j.energy.2020.119541

Yilmaz, N., Vigil, F. M., Benalil, K., Davis, S. M., & Calva, A. (2015). Erratum: Effect of biodiesel-butanol fuel blends on emissions and performance characteristics of a diesel engine (Fuel (2014) (46-50)). Fuel, 139, 781. https://doi.org/10.1016/j.fuel.2014.09.019

Chapter 19
Investigation of Different Raw Material Needs of the Energy Sector and Future Prospects

Tolga Berkay Şirin, Muhammed Umar Bayer, İrfan Çavuş, and Arif Savaş

19.1 Introduction

Energy meets many needs that become inevitable for the continuation of human life, such as household needs (such as heating, lighting, water supply), agricultural uses (such as irrigation, harvest), industrial production, transportation, health, education and communication (Asif and Muneer 2007; Kaygusuz 2011). Furthermore, it is a critical component of economic growth and socio-economic development (Herbert et al. 2007).

Global energy demand has increased throughout history due to many factors such as human population, urbanization and modernization, and fossil fuels have been used to meet these increasing energy needs (Asif and Muneer 2007). Today, fossil fuels such as crude oil, natural gas and coal still make up about 80% of the energy supply. Nuclear energy, which can be called a relatively cleaner energy source, is used only at a rate of 4.3%. However, fossil fuels cause air pollution and climate change, and nuclear energy causes technical safety problems, high rate of radioactive atomic waste production and negative economy (Barbir et al. 1990; Stache 2021). Therefore, as a result of changes in global energy policies, the energy sector has shifted to cleaner renewable energy sources.

Renewable energy is clean energy gathered from renewable sources such as sunlight, wind, rain, tides, waves and geothermal heat, which are naturally occurring sources in nature (Sattar et al. 2020). In order to obtain energy from these sources and use it efficiently, tools, machines and structures such as turbines for the wind, solar panels for the sun and energy converters for the waves are produced. This not

T. B. Şirin (✉) · M. U. Bayer · İ. Çavuş · A. Savaş
Maritime Faculty, Bandirma Onyedi Eylül University, Balıkesir, Turkey
e-mail: tsirin@bandirma.edu.tr; mbayer@bandirma.edu.tr; icavus@bandirma.edu.tr; asavas@bandirma.edu.tr

M. Z. Sogut et al. (eds.), *Proceedings of the 2022 International Symposium on Energy Management and Sustainability*, Springer Proceedings in Energy,
https://doi.org/10.1007/978-3-031-30171-1_19

only helps to reduce emissions, which is of great importance to the environment today, but also helps to immensely eliminate other environmental problems (Bencs et al. 2020).

The shift of countries to renewable energy sources has led to the spread and popularity of studies in this field. Although the development of wind turbines and solar panels is important, one of the most critical factors to be considered in this regard is the raw materials or minerals, namely, mines, of which we manufacture engineering components. The transition to zero-emission energy sources also contains vital elements for the research of raw materials. In this sense, it can be summarized as the availability, extraction, processing and proper functioning of the supply chain of materials used in the production of turbines, panels or energy converters. This paper presents only a preliminary study that covers these issues in general terms.

19.2 Raw Materials in Energy Industry

The periodic table currently contains 118 elements (Curry et al. 2021). These elements are used alone or together in the energy sector for many purposes. In the transition to renewable energy, raw materials used in traditional energy systems are replaced by other raw materials. Figure 19.1 shows the increase in elements used in the energy sector when considering the elements discovered over the years. For instance, instead of fossil fuels such as coal, oil and natural gas, lithium, indium, and molybdenum used in batteries, solar panels and wind turbines will be preferred, respectively.

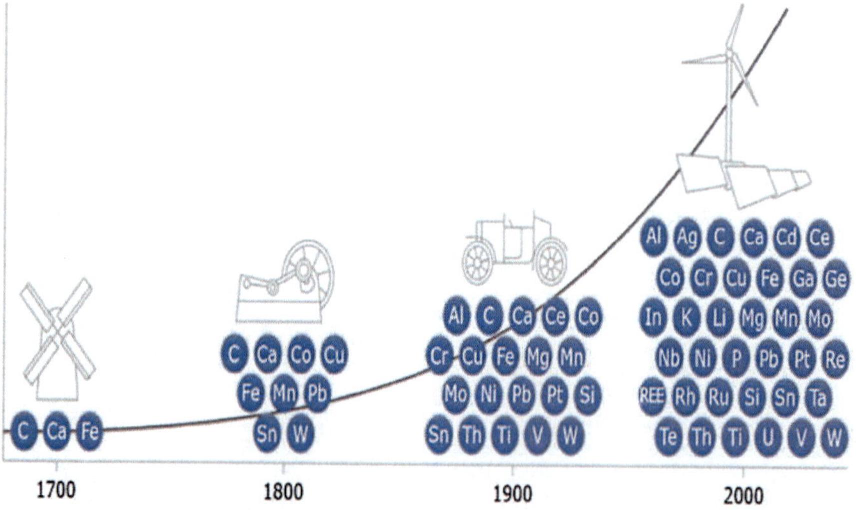

Fig. 19.1 Raw materials used along the timeline in the energy sector. (Wellmer et al. 2019)

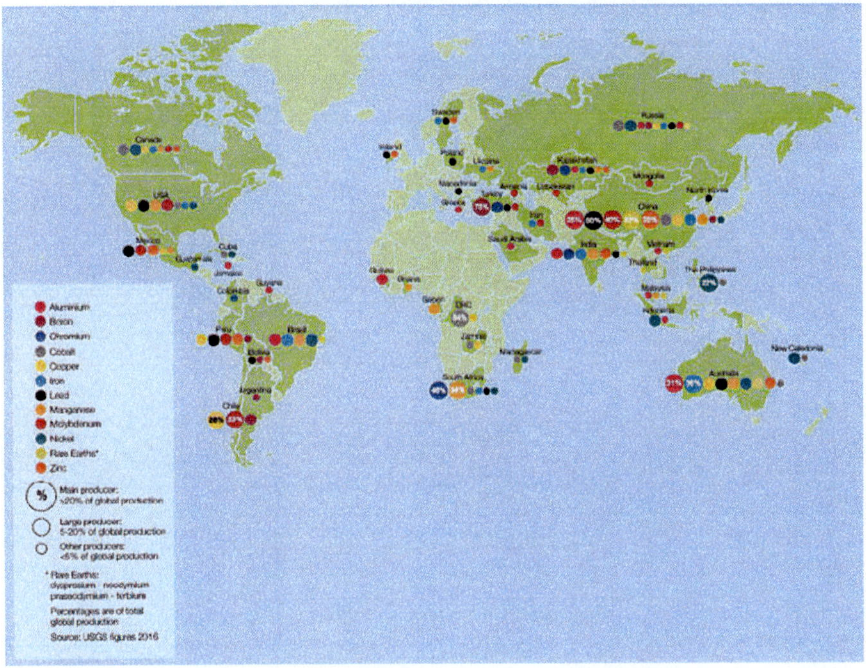

Fig. 19.2 Producers of raw materials used in wind turbines. (Carrara et al. 2020)

Clean energy technologies are more raw material-intensive than fossil fuel technologies. However, there are many different raw materials available in the territory of each country and renewable energy technologies using these raw materials. Figure 19.2 presents a simple example of this map for the raw materials required for wind turbines. Therefore, many factors need to be researched and analysed by countries that switch to renewable energy technologies and need access to the raw materials required for these technologies. Some of those are as follows:

- Investment in mining and ore processing technologies
- Meeting the water requirement for mining
- Development of recycling and energy storage systems
- Good management of the supply chain
- Risking the degradation of land and settlement
- Political situations, etc.

If the mentioned factors are met, renewable energy technologies can be used. However, the disadvantages such as weather dependency, high installation costs, high noise (for wind), fluctuation (for solar) and intermittency (for wind) should not be overlooked (Ayaz et al. 2017).

Table 19.1 Change predictions of minerals for 2050

Mineral	2020 Production (thousand tonnes)	2050 Annual projected demand (thousand tonnes)	2050 Demand as a % of 2020 production
Lithium	82	415	506%
Cobalt	140	644	460%
Graphite	1,100	4,590	417%
Indium	0.9	1.73	192%
Vanadium	86	138	161%
Nickel	2,500	2,268	91%
Silver	25	15	60%
Lead	4,400	781	18%
Molybdenum	300	33	11%
Aluminium	65,200	5,583	9%
Copper	20,000	1,378	7%
Manganese	18,500	694	4%
Chromium	40,000	366	0.92%
Iron	1,500,000	7,584	0.51%
Titanium	8,200	3.44	0.04%

Bhutada (2021)

Energy sources utilize numerous raw materials that offer different properties and functionalities. Examples can be multiplied such as providing both high temperature and pressure resistance by using titanium-steel alloys in geothermal power plants, obtaining high conductivity by using silver in solar panels and increasing hardness and corrosion resistance by using chromium-steel alloys in hydroelectric power plants. The demand for these energy technologies and raw materials will increase with our energy needs. Some of the raw materials that are expected to see increasing demand from energy technologies until 2050 according to current production levels are presented in Table 19.1, and the estimation of increasing raw material needs according to some renewable technologies is presented in Fig. 19.3.

According to the data, due to the increasing interest in electric vehicle (EV) batteries (lithium-ion batteries), it is estimated that the key ingredients of these batteries, lithium, graphite and cobalt will create the highest demand. Considering that this increase is only for these batteries and is used for other purposes, it can be thought that the demand may be much higher.

When we take a look at the bottom row of the list, we see that there are indium and vanadium raw materials. These raw materials, which most of us do not know where they are used, are used in solar energy technologies and energy storage technologies (vanadium redox flow batteries) and can be considered as second level raw materials where the greatest increases will be experienced. Figure 19.4 shows assembly lines containing raw materials from which existing components in solar and wind systems (most promising) are produced.

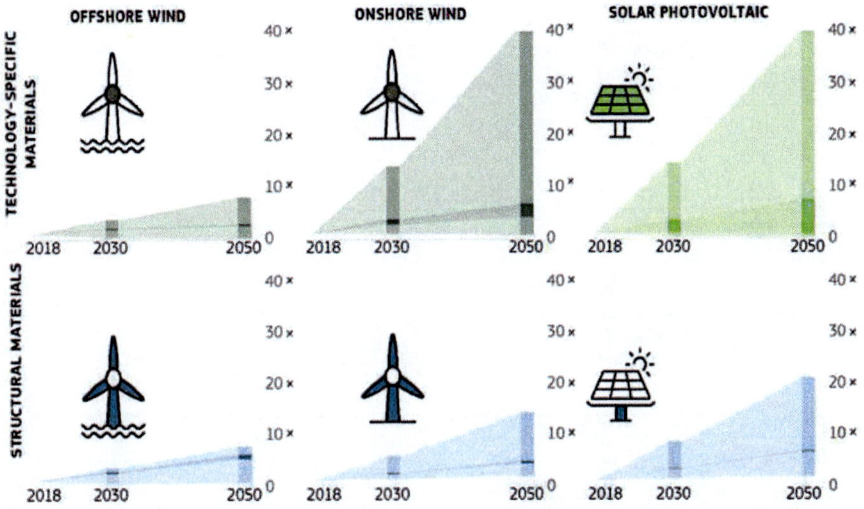

Fig. 19.3 Raw material demand forecast. (Carrara et al. 2020)

Despite these increases in demand, the need for more traditional and abundant raw materials in the earth's crust, which we all know, can be seen to decrease significantly in the transition to renewable energies. Table 19.2 shows the raw materials used in different clean energy technologies and their current status.

Increasing demand for raw materials does not necessarily mean that this supply can be sustained, because each raw material is found in certain quantities in the world and is in the territory of different countries. The creation of a continuous and healthy supply chain is another issue that will play a critical role in the transition to renewable energy.

There are many reasons why the raw materials in the table are considered critical. For example, the limited supply of dysprosium and neodymium minerals required for wind power or the high political risks and limited supply flexibility of indium, gallium and tellurium minerals due to the Chinese supply chain are the reasons behind these types of minerals being critical. These examples can be extended, and future measures can be taken with the increase of studies and analyses for the problems mentioned.

19.3 Conclusion

This paper is a superficial review of raw material changes in the energy sector. Estimated analyses include that the consumption of raw materials will gradually increase and concerns about meeting the increasing demand. Studies on renewable

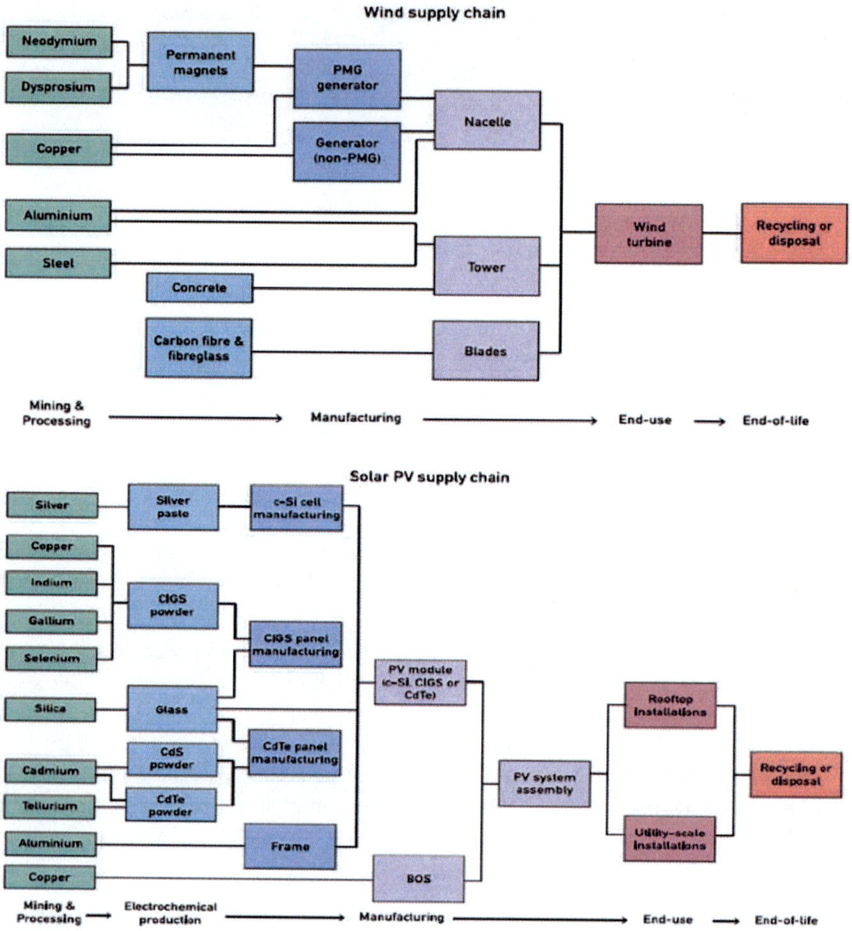

Fig. 19.4 Raw material assembly lines for wind and solar PV materials. (Giurco et al. 2019)

and clean energy are limited. Despite this, the rapid increase in interest today shows that new developments will emerge in this field. Although current analyses show that there will be no problems in many raw materials in the perspective of 2050, the intensification of studies in this field and the acceleration of the transition may reverse the effects.

Table 19.2 Potentially critical raw materials used in some clean energy technologies

Technology	Potentially critical raw materials
Batteries	
Lead acid	Lead
Nickel metal hydride	Lanthanum, nickel
Lithium ion	Cobalt, graphite, lithium
Emerging	Magnesium, vanadium
Solar power	
Silicon based	Nickel, silver, tin
Copper-indium-gallium selenide	Gallium, indium, selenium
Cadmium telluride	Tellurium
Magnets (for high efficiency motors)	
Alnico	Cobalt, nickel
Samarium-cobalt	Cobalt, samarium
Neodymium-iron-boron	Dysprosium, neodymium, praseodymium, terbium
Lighting	
Fluorescent	Cerium, europium, lanthanum, manganese, terbium, yttrium
Light-emitting diodes (LEDs)	Cerium, europium, gallium, germanium, indium, lanthanum, nickel, silver, terbium, tin, yttrium
Catalysts	Cerium, lanthanum, palladium, platinum, rhodium
Fuel cells	Cerium, cobalt, gadolinium, lanthanum, palladium, platinum, rhodium, yttrium
Nuclear	Cobalt, gadolinium, hafnium, indium
Hydrogen electrolysis	Palladium, platinum, rhodium

Eggert (2017)

References

Asif M, Muneer T (2007) Energy supply, its demand and security issues for developed and emerging economies. Renewable and sustainable energy reviews 11(7):1388-1413. https://doi.org/10.1016/j.rser.2005.12.004

Ayaz K, Sulemani MS, Ahmed N (2017) Efficient Energy Performance within Smart Grid. Smart Grid and Renewable Energy 8(3):75-86. https://doi.org/10.4236/sgre.2017.83005

Barbir F, Veziroğlu T, Plass JR H (1990) Environmental damage due to fossil fuels use. International journal of hydrogen energy 15(10):739-749. https://doi.org/10.1016/0360-3199(90)90005-J

Bencs P, Al-Ktranee M, Mészáros KM (2020) Effects of solar panels on electrical networks. Analecta Technica Szegedinensia 14(1):50-60. https://doi.org/10.14232/analecta.2020.1.50-60

Bhutada G (2021) The Raw Material Needs of Energy Technologies https://elements.visualcapitalist.com/the-raw-material-needs-of-energy-technologies/.

Carrara S, Alves Dias P, Plazzotta B, Pavel C (2020) Raw materials demand for wind and solar PV technologies in the transition towards a decarbonised energy system. Luxembourg: Publications Office of the European Union 68. https://doi.org/10.2760/160859

Curry E, Metzger A, Zillner S, Pazzaglia J-C et al. (2021) The Elements of Big Data Value: Foundations of the Research and Innovation Ecosystem. Springer Nature. p.i-xxiii. https://doi.org/10.1007/978-3-030-68176-0

Eggert R (2017) Materials, critical materials and clean-energy technologies. EPJ Web of Conferences, EDP Sciences 00003. https://doi.org/10.1051/epjconf/201714800003

Giurco D, Dominish E, Florin N, Watari T et al. (2019) Requirements for Minerals and Metals for 100% Renewable Scenarios. In: Teske S. (eds) Achieving the Paris Climate Agreement Goals. Springer Cham. https://doi.org/10.1007/978-3-030-05843-2_11

Herbert G J, Iniyan S, Sreevalsan E, Rajapandian S (2007) A review of wind energy technologies. Renewable and sustainable energy Reviews 11(6):1117-1145. https://doi.org/10.1016/j.rser.2005.08.004

Kaygusuz K (2011) Energy services and energy poverty for sustainable rural development. Renewable and sustainable energy reviews 15(2):936-947. https://doi.org/10.1016/j.rser.2010.11.003

Sattar M A, Sameeroddin M, Deshmukh M K G, Sami M A (2020) Renewable energy and its industrial applications. Int. Res. J. Eng. Technol. (IRJET 2020) 7(6):6042-6046. https://www.irjet.net/archives/V7/i6/IRJET-V7I61130.pdf

Stache C (2021) Germany shuts down half of its remaining nuclear plants. Aljazeera, 12.31.2021 https://www.aljazeera.com/news/2021/12/31/germany-shuts-down-half-of-its-remaining-nuclear-plants.

Wellmer F-W, Buchholz P, Gutzmer J, Hagelüken C et al. (2019) Fundamentals. In: Raw Materials for Future Energy Supply. Springer Cham. https://doi.org/10.1007/978-3-319-91229-5_2

Chapter 20
Investigation of Main Engine Turbocharger Fouling Effects on Fuel Oil Consumption by Using Engine Room Simulator

Bulut Ozan Ceylan and Yasin Arslanoğlu

Nomenclature

ERS	Engine room simulator
F/O	Fuel oil
IMO	International Maritime Organization
kW	Kilowatt
M/E	Main engine
P_{comp}	Compression pressure
P_{max}	Maximum pressure
RPM	Revolutions per minute
TEMP	Temperature

B. O. Ceylan (✉)
Department of Maritime Transportation and Management Engineering, Istanbul Technical University, Istanbul, Turkey

Department of Marine Engineering, Bandirma Onyedi Eylül University, Balikesir, Turkey
e-mail: bceylan@bandirma.edu.tr

Y. Arslanoğlu
Department of Maritime Transportation and Management Engineering, Istanbul Technical University, Istanbul, Turkey
e-mail: arslanoglu@itu.edu.tr

M. Z. Sogut et al. (eds.), *Proceedings of the 2022 International Symposium on Energy Management and Sustainability*, Springer Proceedings in Energy,
https://doi.org/10.1007/978-3-031-30171-1_20

20.1 Introduction

A significant percentage of commercial ships are powered by two-stroke marine diesel engines (Ryu et al. 2016). Since two-stroke marine diesel engines are more efficient and have a lower specific fuel consumption than other varieties of internal combustion engines, they are the pioneer option for marine applications (Vera-García et al. 2020; Ünver et al. 2021). Additionally, this type of diesel engine that is placed as the main engine of ships has been designed and manufactured in large sizes, high power, and complex structure including various engine systems (Cai et al. 2017). Two-stroke marine diesel engine cross-section and engine parts are demonstrated in Fig. 20.1 (MAN 2021).

The main engine has diversified types of auxiliary components such as cooling, heating, turbocharging starting air, fuel, and lubricating systems (Ceylan et al. 2021). Between these subsystems, turbocharging is one of the most critical auxiliary system since it provides pressurized air to the main engine scavenge manifold for the cylinder combustion process. Thanks to the compressed air supply, effective combustion takes place in the engine cylinders, and the power of the engine increases.

According to the literature, researchers have shared different studies about T/C. Feneley et al. (2017) establish a review of variable geometry turbocharger improvements for heat energy recovery. With a similar approach, Lee et al. (2016) carry out an overview of electric turbochargers for downsized internal combustion engines. In the maritime field, Wei et al. (2020) use an unsupervised algorithm for fault diagnosis of marine turbocharger systems. Knežević et al. (2020) suggest risk assessment and failure diagnosis of marine diesel engine turbocharger system. Additionally, Cui et al. (2018) offer a performance evaluation approach for a turbocharger in a diesel engine. Lin et al. (2021) carry out a reliability analysis of marine low-speed diesel engine turbocharger turbine. On the other side, Liu et al. (2020) investigate the effects of turbocharger blade roughness in a marine diesel engine. Finally, Ntonas et al. (2020) suggest an integrated simulation context for assessing turbocharger malfunction effects on diesel engine performance parameters.

In this study, the effects of T/C exhaust side fouling problem on the fuel consumption of marine diesel engines were investigated. The application was carried out on a full mission engine room simulator, and a two-stroke MAN B & W 6S50 MC-C engine was used in this session. Within this scope, the paper is organized as follows. This section gives the motivation behind the research. The second section introduces the material and method of the study. Section three includes a T/C fouling problem application on a full-mission simulator. Finally, section four concludes the research and advises future studies.

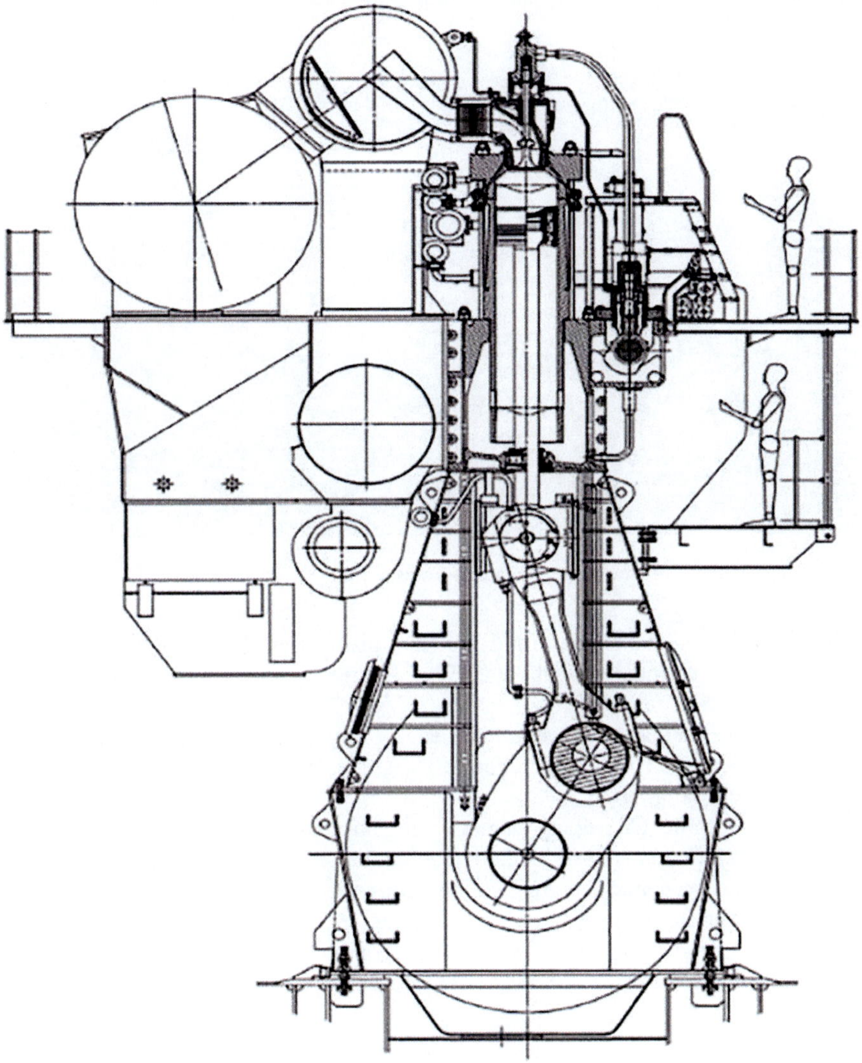

Fig. 20.1 Two-stroke marine diesel engine cross-section. (MAN 2021)

20.2 Material and Method

In this investigation, a realistic simulation session was conducted using the Transas Engine Room Simulator (ERS) 5000. The application was carried out on ten integrated workstations with the intent of an interactive and quick responsive simulation process. As the main engine of product tanker ship, MAN B & W 6S50 MC-C, two strokes, single acting, direct reversible, crosshead type marine diesel

Fig. 20.2 Simulated product tanker. (Wärtsilä-Transas 2021)

with six cylinders (500 mm bore and 2000 mm stroke length) was selected. The main engine generates a maximum of 8600 kW at 127 rpm. A simulated product tanker is seen in Fig. 20.2.

During the simulation process, main engine commands and related engine rpm values are controlled in the engine room. The simulated engine command console is demonstrated in Fig. 20.3.

According to the ERS 5000, engine commands and rpm set points are shown in Table 20.1. The engine room simulator (ERS) is an essential instrument for maritime education and academic research. In numerous academic investigations involving operations that are difficult to test in real-time, engine room simulator is utilized. Moreover, according to Wartsila, the engine simulator can produce results that closely resemble real-time operating data (Wärtsilä-Transas 2021).

20.3 Application

In this study, the main engine turbocharger fouling problem was tested in Transas 5000 ERS by using the simulator's malfunction interface. After the initial parameter's record, the main engine T/C exhaust side fouling malfunction was gradually added to the session. It was initiated at 2% and increased to 25% gradually by the instructor. Under these conditions, cylinder exhaust outlet temp (°C), M/E F/O consumption (L/h), No. 1 cylinder P_{comp} (bar), and No. 1 cylinder P_{max} (bar) parameters are recorded. The recorded T/C speed and main engine F/O consumption data is seen in Figs. 20.4 and 20.5.

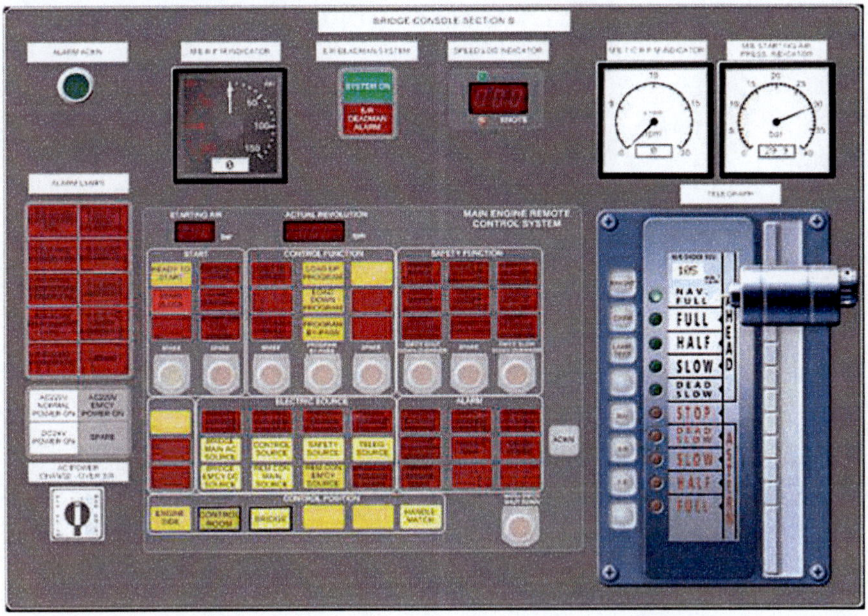

Fig. 20.3 Simulated engine room command console. (Wärtsilä-Transas 2021)

Table 20.1 Technical description of the engine commands

Command	Revolution per minute
Navigation Full ahead	89/128 rpm
Full ahead	76/80 rpm
Half ahead	68/71 rpm
Slow ahead	56/58 rpm
Dead slow ahead	32/43 rpm
Stop	0 rpm
Dead slow astern	−38 rpm
Slow astern	−58 rpm
Half astern	−75 rpm
Full astern	−85/90 rpm

20.4 Conclusion

Ships are massive and complex engineering structures subjected to the harsh conditions of the seas. The main engine plays a significant role in this structure as it powers the ship's propulsion. On the other side, high-power marine diesel engines require several pieces of auxiliary. T/C is one of the essential auxiliary equipment that provides compressed air to the main engine. However, this equipment is subject to fouling phenomenon over time because it also faces harsh sea conditions. Additionally, this pollution may have a substantial impact on the main engine's

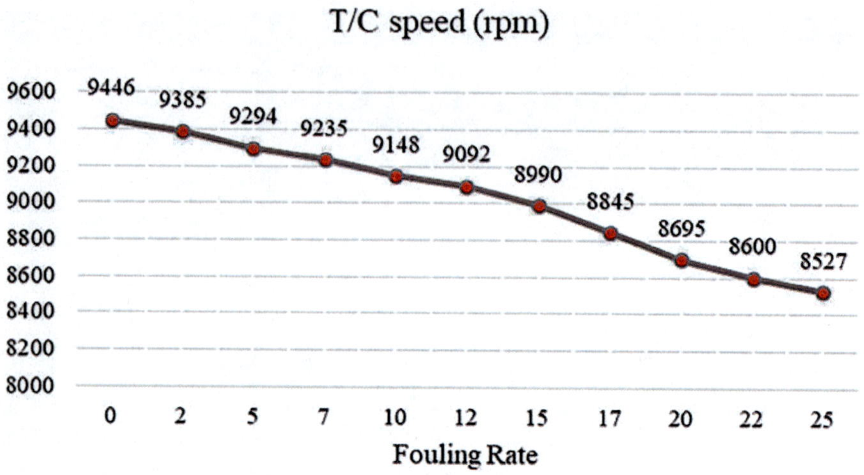

Fig. 20.4 Record of T/C speed (rpm) data of the main engine

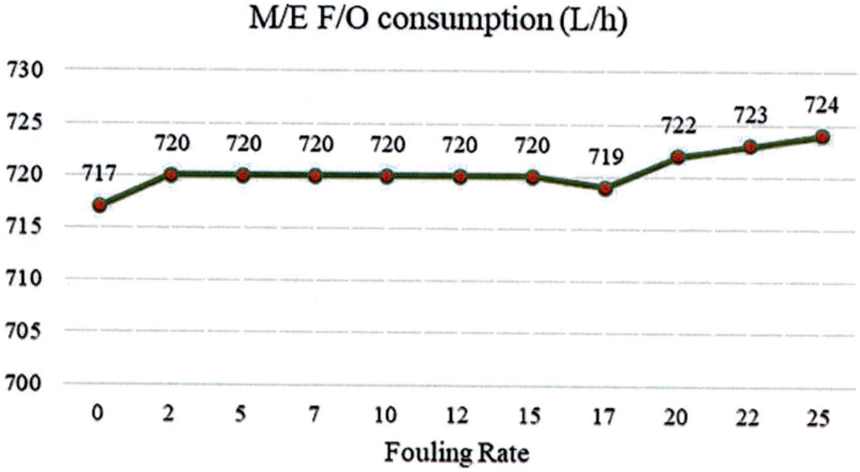

Fig. 20.5 Record of F/O consumption (L/h) data of the main engine

parameters. In this sense, full mission engine simulation of a two-stroke MAN B & W 6S50 MC-C marine diesel engine was used to examine the effects of T/C exhaust side fouling in this study. In the simulation process, fouling was added gradually (% 2–25) to the session by the instructor. Related engine parameters and main engine fuel oil consumption were recorded. According to the results, the fouling in the exhaust side of T/C decreased the rotation speed of the equipment, the reduced speed caused less air compression, and the combustion process of the engine was adversely affected. In addition, the maximum and minimum pressures of the cylinders decreased. Finally, the main engine fuel consumption has increased due to the poor combustion of the cylinders.

Acknowledgment The authors would like to thank Bandirma Onyedi Eylül University for support of this research.

References

Cai, C., Weng, X., & Zhang, C. (2017). A novel approach for marine diesel engine fault diagnosis. Cluster computing, 20(2), 1691-1702.

Ceylan, B. O., Akyuz, E., & Arslan, O. (2021). Systems-Theoretic Accident Model and Processes (STAMP) approach to analyse socio-technical systems of ship allision in narrow waters. Ocean Engineering, 239, 109804.

Cui, X., Yang, C., Serrano, J. R., & Shi, M. (2018). A performance degradation evaluation method for a turbocharger in a diesel engine. Royal Society open science, 5(11), 181093.

Feneley, A. J., Pesiridis, A., & Andwari, A. M. (2017). Variable geometry turbocharger technologies for exhaust energy recovery and boosting-A Review. Renewable and sustainable energy reviews, 71, 959-975.

Knežević, V., Orović, J., Stazić, L., & Čulin, J. (2020). Fault tree analysis and failure diagnosis of marine diesel engine turbocharger system. Journal of Marine Science and Engineering, 8(12), 1004.

Lee, W., Schubert, E., Li, Y., Li, S., Bobba, D., & Sarlioglu, B. (2016). Overview of electric turbocharger and supercharger for downsized internal combustion engines. IEEE Transactions on Transportation Electrification, 3(1), 36-47.

Lin, L. E. I., Ming-ze, D. I. N. G., Hong-wei, H. U., Yun-xiao, G. A. O., Hai-lin, X. I. O. N. G., & Wei, W. A. N. G. (2021). Reliability analysis and optimization of turbocharger turbine of marine low speed diesel engine under complex load. In IOP Conference Series: Materials Science and Engineering (Vol. 1043, No. 3, p. 032055). IOP Publishing.

Liu, C., Cao, Y., Ding, S., Zhang, W., Cai, Y., & Lin, A. (2020). Effects of blade surface roughness on compressor performance and tonal noise emission in a marine diesel engine turbocharger. Proceedings of the Institution of Mechanical Engineers, Part D: Journal of Automobile Engineering, 234(14), 3476-3490.

MAN, 2021. MAN B & W S60MC-C8.2-TII Project Guide. https://man-es.com/applications/projectguides/2stroke/content/epub/S60MC-C8_2.pdf (Accessed 06 January 2021).

Ntonas, K., Aretakis, N., Roumeliotis, I., Pariotis, E., Paraskevopoulos, Y., & Zannis, T. (2020). Integrated simulation framework for assessing turbocharger fault effects on diesel-engine performance and operability. Journal of Energy Engineering, 146(4), 04020023.

Ryu, Y., Lee, Y., & Nam, J. (2016). Performance and emission characteristics of additives-enhanced heavy fuel oil in large two-stroke marine diesel engine. Fuel, 182, 850-856.

Ünver, B., Altın, İ., & Gürgen, S. (2021). Risk ranking of maintenance activities in a two-stroke marine diesel engine via fuzzy AHP method. Applied Ocean Research, 111, 102648.

Vera-García, F., Pagán Rubio, J. A., Hernández Grau, J., & Albaladejo Hernández, D. (2020). Improvements of a failure database for marine diesel engines using the RCM and simulations. Energies, 13(1), 104.

Wartsila Transas, 2021. ERS 5000 Engine Room Simulator [WWW Document]. URL https://www.transas.com/products/simulation/engine-room-and-cargo-handling-simulators/ERS5000

Wei, Y., Liu, H., Chen, G., & Ye, J. (2020). Fault diagnosis of marine turbocharger system based on an unsupervised algorithm. Journal of Electrical Engineering & Technology, 15(3), 1331-1343.

Chapter 21
Optimization of Tilt Angle and Maximization of Solar Radiation for Fixed and Tracking Surfaces: A Case Study for Gaziantep, Turkey

Batur Alp Akgül, Fatih Alisinanoğlu, and Mustafa Sadettin Özyazıcı

Nomenclature

TMS Turkish Meteorological Service
NASA National Aeronautics and Space Administration

21.1 Introduction

As the gap between production and consumption widens in conventional energy sources, solar energy has become one of the most significant energy sources on the planet. Solar radiation data has a crucial role in solar systems for optimal design because accurate information on the availability of solar radiation is required for the installation of solar systems. The effectiveness of solar radiation is greatly influenced by meteorological circumstances, seasonal fluctuations, geographic location, and day of the year (Gunerhan and Hepbasli 2007; Ulgen 2007). The total amount of average solar radiation value is the most important indicator to determine energy generation.

The solar radiation value is changed according to a region in the world due to the geoid shape of the earth. Gaziantep Province is located in the Southeastern Anatolia Region in Turkey, and the case simulation study and measurement data are conducted in Gaziantep. The aim of this research is to find the optimal tilt angles for Gaziantep region and to investigate the maximum solar radiation values monthly,

B. A. Akgül (✉) · F. Alisinanoğlu · M. S. Özyazıcı
Electrical & Electronics Engineering, Hasan Kalyoncu University, Gaziantep, Turkey
e-mail: fatih.alisinanoglu@hku.edu.tr; msadettin.ozyazici@hku.edu.tr

M. Z. Sogut et al. (eds.), *Proceedings of the 2022 International Symposium on Energy Management and Sustainability*, Springer Proceedings in Energy,
https://doi.org/10.1007/978-3-031-30171-1_21

seasonally, and yearly of solar surfaces set to optimal tilt angles. It is also aimed to investigate the performance of one-axis and two-axis solar tracking systems and compare them to fixed-axis systems.

21.2 Related Works in Turkey

Researchers in Turkey have performed some studies in the literature regarding optimizing tilt angle and calculating solar radiation for improving solar energy system efficiency. The studies carried out optimization of tilt angles and maximization of solar radiation falling on the tilted surface for the different locations contributed to Turkey on this field. Researchers have contributed to this field with studies on tilt angle optimization and maximization of solar radiation falling on the tilted surface at various locations (Akgül et al. 2022; Kaygusuz 2020; Yolcan and Köse 2020; Akyürek et al. 2019; Bakırcı and Kaltakkıran 2018). Also, there are other studies in the literature that provide maximum radiation with solar tracking systems and compare radiation values for Turkey with fixed systems (Aydemir and Toslak 2019; Eke and Senturk 2012; Kıvrak et al. 2012).

21.3 Materials and Methods

In this section, materials and techniques such as the case study region, used data, description of the calculation model, and so on are described.

21.3.1 Case Study Region

The case study region is Gaziantep with the scope of this study, the global radiation distribution of Gaziantep Province is given in Fig. 21.1 provided by the TMS, and as can be seen from the radiation values of Fig. 21.1, the region has a good solar radiation potential to plant solar systems.

21.3.2 Used Data and Calculation Model

The daily total solar radiation on the horizontal surfaces has to be provided before to reach optimal tilt angle and the maximal solar radiation values falling on the solar surfaces. Various estimation methods and simulation programs are used by researchers to find the optimal tilt angle and to obtain maximal solar radiation. The RETScreen model calculates using satellite and ground station interactive

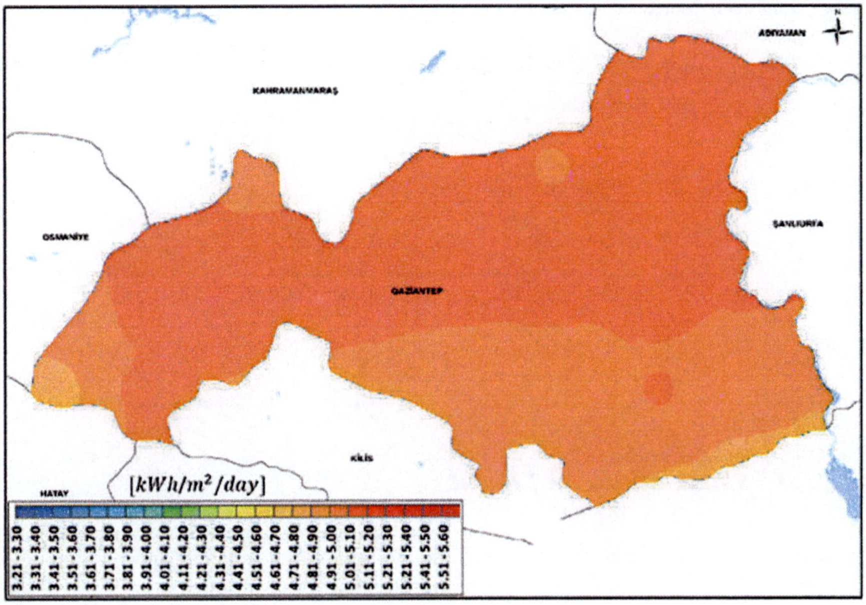

Fig. 21.1 Solar radiation distribution of Gaziantep. (TMS 2022)

meteorological data provided by NASA. It simulates the optimal hourly, daily, monthly, and yearly solar radiation values in different options, depending on the geographical region. Therefore, the RETScreen model is preferred in this research to have to make realistic analyses.

The model uses an approach similar to that proposed model by Klein and Theilacker (Duffie and Beckman 1991), and the algorithm is extended to tracking systems to investigate maximal solar radiation in the fixed-axis, one-axis, and two-axis constructed solar surfaces (Braun and Mitchell 1983). Daily and hourly radiation values are obtained for fixed and tracking systems. The calculation is performed with 1-h step covering 365 days of the year.

21.3.3 Horizontal Solar Radiation Models

Because the amount of radiation can be varied according to the region, the monthly daily average radiation values falling on a horizontal surface are the most important parameter to optimize the tilt angles and to obtain maximum solar radiation. The monthly average daily solar radiation values provided by NASA are presented in Table 21.2.

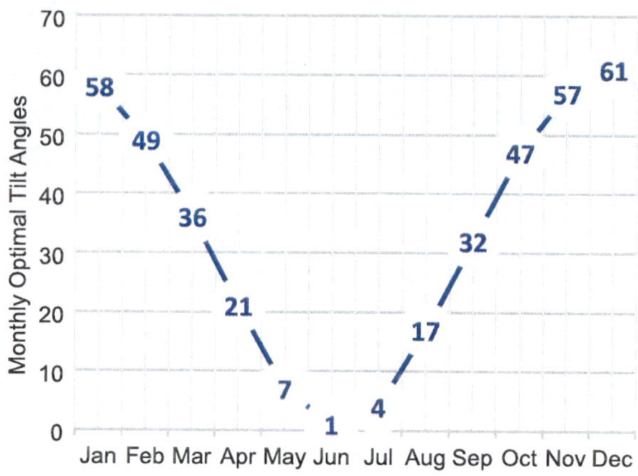

Fig. 21.2 Yearly optimal tilt angle values of Gaziantep

21.3.4 Calculations and Results

Calculations are carried out in this section, and the findings such as optimal tilt angle values and solar radiation gains are reported as well as a comparison of fixed and tracking systems.

21.3.4.1 Optimal Tilt Angle Results of the Region

If the correct tilt angle values are taken into account, the maximum solar radiation values can be obtained. Gaziantep Province's optimal tilt angles for solar systems are determined, and the comparison of monthly optimal tilt angle values is given in Fig. 21.2. The tilt angles at which the amount of radiation falling on the solar surface is maximum are determined as the optimal angle. In other words, the optimal tilt angle for each month of the year when solar radiation is at its peak is shown in Fig. 21.2. Calculations are done for all tilt angles between 0° and 90° to calculate the solar radiation gathered by the optimally tilted surface monthly for Gaziantep Province.

By using climatic and latitude data provided from NASA for Gaziantep, optimal tilt angles of the solar surfaces are calculated for a specific period. For Gaziantep Province, the monthly, seasonal, and annual optimal tilt angle values are presented in Table 21.1, and the monthly average daily solar radiation gain at optimal tilt angles is presented in Table 21.2. The changes in the quantity of radiation gain as a result of the tilt angle effect can be seen from Table 21.2.

It is seen that the optimal tilt angle values are determined between 1° in June and 61° in December throughout the year. The yearly optimal tilt angle is determined to be 28° for a south-facing solar surface. The lowest optimal tilt angle is found to be between 1° in June and 4° in July, whereas the maximum monthly average daily

Table 21.1 Monthly, seasonal, and yearly optimal tilt angle values of Gaziantep

Month	Monthly OPT	Seasonal OPT	Annual OPT
Dec	58	56	28
Jan	49		
Feb	36		
Mar	21	21	
Apr	7		
May	1		
Jun	4	7	
Jul	17		
Aug	32		
Sep	47	44	
Oct	57		
Nov	61		

Table 21.2 Average daily solar radiation gain at optimal tilt angles

Month	Horizontal	Monthly [kWh/m^2]	Seasonal [kWh/m^2]	Annual [kWh/m^2]
Dec	1.94	3.30	3.63	2.88
Jan	2.19	3.55		3.16
Feb	2.98	4.06		3.85
Mar	4.15	4.83	5.57	4.79
Apr	5.22	5.47		5.44
May	6.6	6.64		6.37
Jun	7.82	7.82	7.60	7.24
Jul	7.92	7.93		7.46
Aug	6.94	7.15		7.05
Sep	5.71	6.51	5.17	6.49
Oct	3.82	5.09		4.87
Nov	2.57	4.10		3.68

solar radiation levels are found in the same months. The optimal tilt angle for the winter season is 56° with the lowest monthly average daily solar radiation levels in December and January. For the summer season, the optimal tilt angle is 7°.

21.3.4.2 Radiation Values Collected by the Fixed Surfaces

The total radiation values have been calculated by changing the tilt of the south-oriented solar surfaces with steps between 0° and 90°. It is assumed that the azimuth angle of the solar surface is fixed to 0° for directly facing the equator. When the solar radiation values falling on the solar surfaces with different tilt angles are compared in Table 21.2, the monthly average daily solar radiation values range from 2.88 kWh/m^2 in December to 7.46 kWh/m^2 in July if the solar surface set the yearly optimal tilt angle.

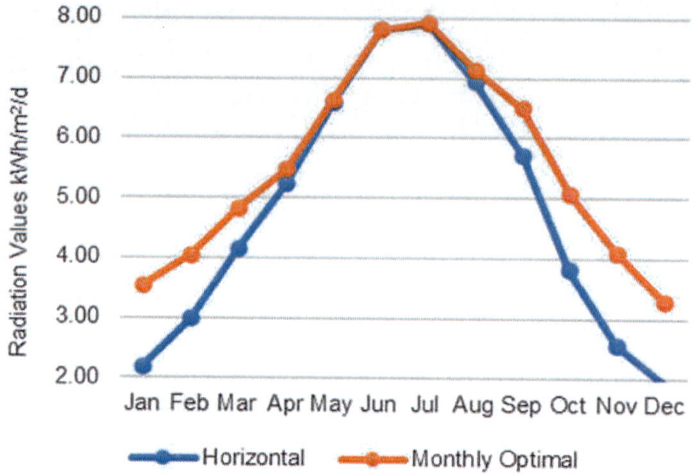

Fig. 21.3 Radiation gain obtained at monthly optimal tilts

In the winter season (December, January, and February), the optimal tilt angle is found as 56°, and the amount of monthly average daily radiation is calculated between 3.30 and 4.06 kWh/m^2. In the spring season (March, April, and May), the optimal tilt angle is found as 21°, and the amount of monthly average daily radiation is calculated between 4.83 and 6.64 kWh/m^2. In the summer season (June, July, and August), the optimal tilt angle is found as 7°, and the amount of monthly average daily radiation is calculated between 7.15 and 7.82 kWh/m^2. In the autumn season (September, October, and November), the optimal tilt angle is found as 44°, and the amount of monthly average daily radiation is calculated between 4.10 and 6.51 kWh/m^2.

Based on the findings, it has been calculated that changing the tilt angle every month in Gaziantep can result in a significant increase in the amount of radiation landing on the solar surfaces as compared to an annual constant tilt angle. This can provide a major efficiency impact, especially for large-capacity solar systems. The maximal solar radiation gain graph obtained at the monthly optimal tilt angles is given in Fig. 21.3. As can be seen from the graph in Fig. 21.3, if the optimal tilt angle for each month is chosen, the maximum amount of radiation may be obtained from the solar surfaces. As a result, solar surfaces should be tilted at optimal angles to maximize the solar radiation collected by the solar surfaces.

21.3.4.3 Radiation Gain for the Solar Tracking Systems

Gaziantep Province's tracking systems are useful tools for solar systems that require improved solar radiation efficiency. The monthly average daily solar radiation values are investigated of the desired region, and the results are presented in Table 21.3 for one-axis and two-axis tracking systems. It is clear from Table 21.3

Table 21.3 Radiation gain for the solar tracking systems of the region

Month	One-axis [kWh/m^2]	Two-axis [kWh/m^2]
Jan	3.87	4.12
Feb	4.66	4.78
Mar	6.01	6.01
Apr	6.69	6.74
May	8.47	8.74
Jun	9.98	10.49
Jul	10.46	10.89
Aug	9.89	10.02
Sep	8.59	8.59
Oct	6.36	6.47
Nov	4.55	4.83
Dec	3.52	3.81
AVG	6.93	7.14

that Gaziantep is one of the most productive regions in Southeastern Anatolia in terms of solar radiation efficiency when tracking systems are used. In Gaziantep, the maximum monthly average daily radiation values are 10.46 kWh/m^2 for one-axis and 10.89 kWh/m^2 for two-axis. Also, the minimum monthly average daily solar radiation values are 3.52 Wh/m^2 for one-axis and 3.81 kWh/m^2 for two-axis. The monthly optimal tilt angles provided in Fig. 21.2 can be utilized for Gaziantep Province in circumstances when it is not possible to establish the tilt angles of the surfaces or there is no solar tracking equipment.

In addition, the yearly total solar radiation values falling on the solar surfaces in the case of setting yearly fixed optimal tilt angles and in the case of using solar tracking systems are presented in Fig. 21.4. It demonstrates that tracking systems make a significant contribution to energy efficiency. It is calculated that optimally angled surfaces provide approximately 10% more solar radiation yearly compared to surfaces that are not angled at all for the fixed surface. It is also calculated that one-axis tracking systems provide approximately 31% more solar radiation and two-axis tracking systems provide approximately 35% more solar radiation than the fixed-axis systems in Gaziantep Province.

Because solar systems have a 25–30-year life span, proper tilting of the surfaces should be taken into consideration. In solar systems, a one-axis tracking system is highly recommended. Solar tracking systems, on the other hand, have a high operating and maintenance cost and are not always practicable due to a lack of available area. If changing the tilt angles of the solar surface is not possible monthly or seasonally, the annual fixed optimal tilt angle values can be used for solar energy systems in Gaziantep Province.

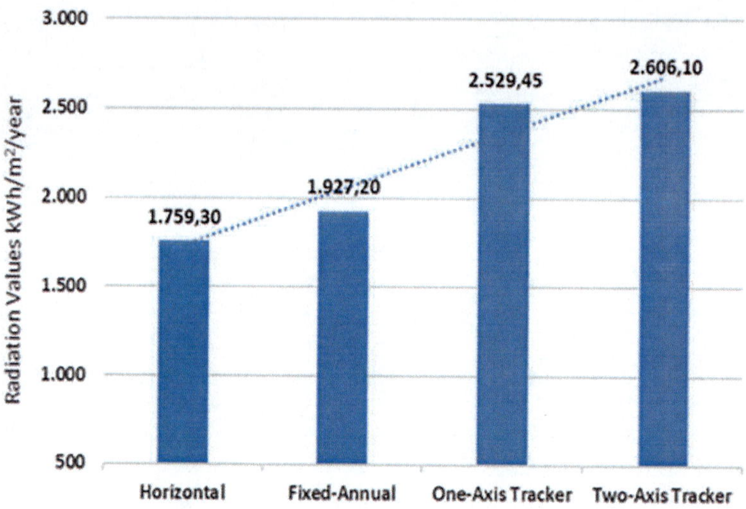

Fig. 21.4 Comparison of yearly total solar radiation values

21.4 Conclusions

In Gaziantep Province, this research contributes to determining the optimal tilt angle of solar surfaces and measuring total radiation falling on the tilted surface. The research also adds to the estimate of radiation values obtained by solar surfaces when tracking systems are used to assess the presence of solar potential in Gaziantep. In this research, it is determined that the yearly optimal tilt angle for Gaziantep is calculated at 28°, and it is also observed that the optimal tilt angles during the year varied between 1° and 61°. It has been found that the amount of radiation falling on the tilted solar surfaces of the tracking systems is always higher than in fixed systems. It is observed that the one-axis tracking system harvests approximately 31% higher solar radiation and two-axis tracking system harvests approximately 35% than the fixed-axis system in Gaziantep. It is observed that optimally angled surfaces harvest approximately 10% more solar radiation yearly compared to surfaces that are not angled at all for the fixed surface. The tilt angles of solar surfaces that are not structured in a solar tracking system can be changed monthly and seasonally to maximize the amount of radiation. It has also been discovered that tracking systems greatly increase the amount of radiation collected. As a result, using tracking devices to achieve high solar radiation efficiency is critical. It is anticipated that this research can help solar engineers and designers to improve the efficiency of solar surfaces, properly set optimal tilt angles, and choose the right methodology for installing solar systems in Gaziantep. This research is expected to aid solar engineers and designers in improving the effectiveness of solar surfaces, properly setting optimal angles, and selecting the best methodology for solar system installation in Gaziantep.

References

Akgül BA, Alisinanoğlu F, Özyazici MS (2022) Obtaining Maximum Radiation by Determining Optimum Tilt Angles in Large-Scale Grid-Connected PV Plants, Efficiency Analysis of Solar Tracking Systems: A Case Study for Sanliurfa, Turkey. ISER International Conference, Los Angeles. 21–24.

Akyürek Z, Akyüz AÖ, Güngör A (2019) Optimizing the Tilt Angle of Solar Panels to Reduce Carbon Footprint: Case for the West Mediterranean Region of Turkey. International Journal of Engineering, Design, and Technology. 1(1):10–15

Aydemir E, Toslak F (2019) Efficiency Comparison of The Fixed Axis System with Double Axis Movement Solar Tracking System. IV. ECOCEE Conf. 1170–1174.

Bakırcı K, Kaltakkıran G (2018) Determination of Tilt Angles of Solar Panels: A Study on Gaziantep. 3rd International Energy & Engineering Congress Proceedings Book. 393–401.

Braun JE, Mitchell JC (1983) Solar Geometry for Fixed and Tracking Surfaces. Solar Energy. 31(5):439–444.

Duffie JA Beckman WA (1991) Solar Engineering of Thermal Processes. New York, Wiley, and Sons, USA.

Eke R, Senturk A (2012) Performance Comparison of a Double-Axis Sun Tracking Versus Fixed PV System. Solar Energy. 86(1):2665–2672.

Gunerhan H, Hepbasli A (2007) Determination of the Optimum Tilt Angle of Solar Collectors for Building Applications. Building and Environment. 42(2):779–783.

Kaygusuz K (2020) Calculation of Solar Radiation Data on Horizontal and Tilted Surfaces for Trabzon, Turkey. Journal of Engineering Research and Applied Science. 9(2):1471–1476.

Kıvrak S, Gunduzalp M, Dincer F (2012) Theoretical and Experimental Performance Investigation of a Two-Axis Tracker Under the Climatic Condition of Denizli, Turkey. PRZEGLĄD ELEKTROTECHNICZNY (Electrical Review). 332–336.

Turkish Meteorological Service (TMS) (2022) Turkey Global Solar Radiation Long-Year Average (2004-2021) HELIOSAT Model Products. https://www.mgm.gov.tr/kurumici/radyasyon_iller.aspx?il=gaziantep

Ulgen K (2007) Optimum Tilt Angle for Solar Collectors. Energy Sources, Part A: Recovery, Utilization, and Environmental Effects. 28(13):1171–1180.

Yolcan OO, Köse R (2020) Finding Optimum Tilt Angles of Photovoltaic Panels: Kütahya Case Study. International Journal of Energy Studies. 5(2):89–105.

Chapter 22
Applied Time Series Regression by Using Random Forest Algorithm for Forecasting of Electricity Consumption on a Daily Basis

Khalid Alhashemi and O. Tolga Altinoz

Nomenclature

RF	Random forest
PCC	Pearson correlation coefficient
RMSE	Root mean squared error
MAPE	Mean absolute percentage error

22.1 Introduction

Knowing the instantaneous amount of electricity that is required to be generated has always been an important part of avoiding overflowing power generation losses and preventing any power outages. This is where energy consumption forecasting becomes necessary to obtain. Energy consumption is affected by many factors over time; however, these factors may have a trivial or significant effect as happened during the period of lockdown due to the outbreak of the COVID-19. Precise prediction also has benefits in estimating the cost of power generation and even in estimating the profits of utility companies as well. Short-term forecasting can provide a good prediction due to the short period of the forecasting, and it has a low chance of being influenced by force majeure factors; on the contrary, long-term forecasting would be a major benefit in addition to what was mentioned above, e.g., it can help countries to set their financial budgets for which have a synchronous grid with each other.

K. Alhashemi (✉) · O. T. Altinoz
Department of Electrical and Electronics Engineering, Ankara University, Ankara, Turkey
e-mail: kwhalhashemi@ankara.edu.tr; taltinoz@ankara.edu.tr

M. Z. Sogut et al. (eds.), *Proceedings of the 2022 International Symposium on Energy Management and Sustainability*, Springer Proceedings in Energy,
https://doi.org/10.1007/978-3-031-30171-1_22

The main objective of this chapter is to apply the concept of RF algorithm to an enhanced model to predict virtual future values of electric energy consumption based on a time series index on a daily basis; the period being predicted already has real values in the dataset, and we chose to predict in this period to assess the results that we earned by the actual data in the dataset. Consequently, the values of the dataset for the period that we want to predict will not be included in the code as training sets. The most effective and appropriate variable features that make the forecasting approach successful were selected. The proposed method will use Python with a dataset of energy demand in Spain between 2014 and 2018.

The concept of the RF came into light by Ho (1995); he proposed a method of building multiple trees in a random selection of feature space. RF algorithm has been extended and registered as a trademark by Breiman (2001).

Improving the prediction performance by the relevant and informative features can be done by using the ranking criterion and the dynamic feature elimination strategy (Nguyen et al. 2006). Electricity consumption predicting has generally been divided by researchers into two approaches: statistical and artificial intelligence approaches. The statistical approach is basic and easy to use, but it can only handle samples that have a limited number of features, necessitating high sample data stationarity, and it has low prediction accuracy, while the artificial intelligence approach has high prediction accuracy and controllable generalization errors and can handle a huge number of features with different types of data (Xiong et al. 2021).

In general, using RF logarithm provides a highly accurate prediction since its results are taken from applying a multiple decision tree algorithm (which belongs to the family of supervised learning algorithms) and with no overfitting or high variance occurring in training dataset when there will be more trees (Stephan et al. 2015).

22.2 Random Forest Algorithm

RF is an ensemble learning model method of machine learning algorithms that is based on the decision tree algorithm. It is widely used in both of classification and regression models of the machine learning algorithm. The decision of a RF algorithm is made by creating a large number of decision trees and consulting them all at once. The suggested name of "forest" refers to the numerous trees that have been built inside the model, and the suggested name of "random" indicates the random distribution of trees in the forest because both the sampling of the data frame and the selection of the features are done at random. For obtaining final decision for any applied model, the dataset must contain sets of rows that represent observations (it refers to the number of time indexes in time series forecasting) and also contain sets of columns that represent the predictors or attributes and their related features (Nallathambi and Ramasamy 2017).

RF in general has two parameters which should be tuned:

1. Number of trees to grow
2. Number of features that are sampled randomly in each split as candidates

The frequent use of RF came as a result of the distinguished benefits: especially it has high accuracy, it can deal with large datasets, pruning trees is unnecessary, overfitting is not a problem, and it can apply to datasets that have errors and/or a lot of missing values. For example, let us take a decision tree algorithm as a compassion with RF; decision tree has a high variance, but when we mix a bunch of trees in parallel, the final variance will be low since each decision tree is perfectly trained on a particular sample data; as a result, the outcome will be dependent on numerous decision trees rather than one. In a classification model, the outcome is obtained by using the majority voting classifier, and in a regression model, it will take the mean of all the outcomes, even though the data samples that have irregular regression will increase overfitting, but the effect of extreme values will not be efficient.

Bootstrap aggregating, which is a prevalent algorithm, is used with most machine learning algorithms to make the outcome more stable and accurate; it also helps to avoid the overfitting and to lowering the variance that occurs in a noisy dataset. For a training sample $D = \{(X_i, Y_i), i = 1, 2, ..., n\}$, to estimate any regression function, it must predict the response of Y, which is associated with the random feature of X; the independent random features are $\theta_1, \theta_2, \theta_3, ..., \theta_M$, and the predicted value at point x in the jth tree is $m_n(x, \theta_j)$, where M is the number of decision trees that were generated in the training sample; the decision tree becomes $f(x, D, \theta)$, and each decision tree deals with the regression to generate the projected value of y_i. As a result, the arithmetic mean of the decision is calculated for each tree and gives the final decision (Mahrukh et al. 2020). The schematic of the RF is shown in Fig. 22.1.

RF is an aggregating algorithm model that forecasts by integrating all decision tree sequences with the help of the subsampling of data. Aside from that, all model bases in a decision tree are configured individually. Moreover, bagging plays an important role in the random selection of ideal equally spaced splitting in each data sample and in the convergence of the relationships between the features of the data.

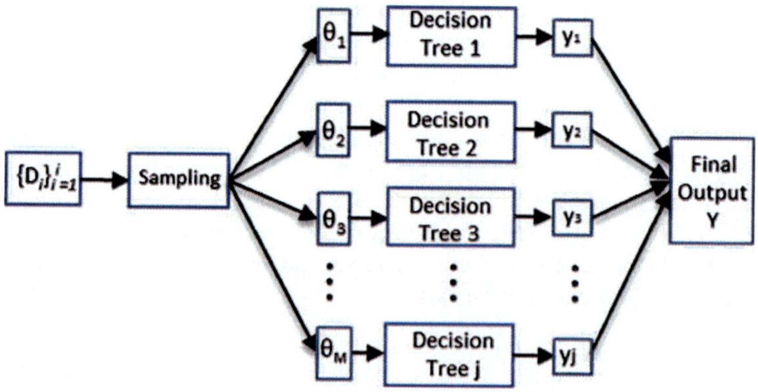

Fig. 22.1 Schematic diagram of RF algorithm (Mahrukh et al. 2020)

22.3 Methodology

22.3.1 *Building a Random Forest Prediction Model*

The proposed approach in this paper is to apply the concept of the RF algorithm to an enhanced model to predict future values of electric energy consumption based on a time series index on a daily basis. Many related features of the dataset are candidate to be generated as a new input index. Pearson correlation analysis has been used to evaluate the corresponding correlations by determining the relevance of each significant feature and the relationships between them (Liu et al. 2021), the dataset has been split into training and testing sets, and the splitting ratio has been tested to give the best prediction results.

Some parameters must be chosen to fit with a specific model and its dataset, e.g., the number of estimators (trees), the values of the maximum features, and random state. The structure of the regression model has been established, and the errors of different predicted periods have been measured to determine the accuracy of the predictions.

22.3.2 *Evaluation of the Dataset*

One of the biggest obstacles facing researchers during the development of the forecasting model is finding a good dataset that is free of wrong typing, missing values, and overfitting. Therefore, some statistical analysis must be applied to the dataset to know whether it is effective at giving a precise prediction. There are many different analyses that can be applied to the dataset, e.g., testing the normality distribution by calculating the skewness and the kurtosis, measuring the volatility to know how the data is dispersed, and discovering interesting patterns based on some subjective information by using sequence mining analysis.

22.3.3 *Parameter Setting of the Model*

Statistical analyses are used on datasets that have one or few outputs to create new or several inputs; these new inputs are evaluated to find values that are closely related to the output. In this chapter, a distributed lag model and moving average analysis or also known as a rolling mean and both maximum and minimum moving average values and moving standard deviation have been used to create new inputs. Distributed lag is used to predict the present values of a dependent variable using both present and past lag values of a related regression variable. Moving average is used for smoothing the variations and exposing the trends of the dataset. The concept of making an analysis in time series forecasting models is to capture the most recent changes in energy consumption that will have an effect on the near-future consumption.

PCC is used in this paper to examine the direction and strength of the linear relationship between each feature that we created. PCC is calculated as the product of the mean differences, and its square summation between two variables, when PCC applied, is generally represented by p and may be referred to as the correlation coefficient of the product difference (Liu et al. 2021). PCC can be written as the following equation:

$$p = \frac{\text{cov}(x,\ y)}{\sigma_x \sigma_y} = \frac{E\,(x - \mu_x)(y - \mu_y)}{\sigma_x \sigma_y} \tag{22.1}$$

where cov(x, y) denotes the covariance of the two variables x and y, μ_x and μ_y denote the average values of variables x and y, and σ_x and σ_y denote the standard deviations of variables x and y. The correlation coefficient p has a value within the range of [−1, 1]. A value of p close to −1 or 1 refers to a strong linear relationship between two variables, when $p = 0$, and there is no linear correlation between the two variables, even if they are related in other ways.

22.3.4 Evaluation of the Predicted Results

The comparison and evaluation of the predicted results of any forecasting model can be measured as long as actual data is available. Measuring the differences between actual and predicted values will decide the accuracy in both choosing the logarithm that is suitable for a certain dataset and choosing the parameters that have a huge impact on improving the results. The parameters could be, for example, the number of trees and their depths in a RF algorithm or the values of epsilon and C parameter in a super vector machine algorithm.

RMSE and MAPE are used to measure the accuracy of the forecasting model, and it is obvious that these parameters are measuring the differences between the predicted values and the actual data that is used to build a model with its training and testing sets. The results of RMSE and MAPS obtained by using predictive modeling based on applying a machine learning algorithm can express the quality of building a particular model. The error components of RMSE and MAPE can be represented by Eqs. 22.2 and 22.3, respectively:

$$\text{RMSE} = \sqrt{\frac{1}{n}\sum\nolimits_{i=1}^{n}\left(yi - \widehat{yi}\right)^2} \tag{22.2}$$

$$\text{MAPE} = \left(\frac{1}{n}\sum\nolimits_{i=1}^{n}\frac{\left(yi - \widehat{yi}\right)}{yi}\right) * 100 \tag{22.3}$$

where y_i denotes the actual data value, $\widehat{yi}$ denotes the forecasting value based on previous values, and n denotes the nth observation.

22.4 Results and Discussion

The right prediction process starts with choosing a good dataset. For time series prediction, it cannot be accurately predicted unless the dataset has a time component. In this paper, the dataset used is a daily record of the energy demand in megawatt-hour of Spain (2014–1018), it contains 1825 observations, and it has no missing values. As shown in Fig. 22.2, the graphical representation of the entire dataset shows a little upward trend stationary, but this visual may not always provide accurate impression, so it is preferable to use statistical tests to confirm the validity of the dataset that gives the best results. The dataset is available in Excel format file at: https://www.esios.ree.es/en.

Before entering into the building process, statistical analysis of the dataset has been done, normal distribution and normal curve of observations have been plotted as seen in Fig. 22.3, and the null-hypothesis significance has been tested by calculating the p-value that refers to the probability of occurrence of the observed difference in random chance. A dataset with a p-value less than 0.05 is considered to have a null hypothesis (Dallal 1999). The dataset used in this paper has a p-value of 2.54×10^{-10}.

The approach to enhancing the model's predicting accuracy is to create new features to be included in the training and test sets. The new features that have been created are the past distributed lags from 1 to 30 days, the sets of the moving average with its maximum and minimum values, and the moving standard deviation for (2, 3, 7, 30) days. All of these features are due to the next output value.

PCC between new features mentioned and the next day's value was calculated using Eq. 22.1. Figure 22.4 shows the correlation matrix of Pearson coefficient between some selected features, and it is clear that the distributed lag of the past 6 days has a strong linear correlation ($p = 0.8$) with the value of the next day.

The values of the dataset of the year 2018 will not be selected to be contained in the training set because the goal of this work is to predict different periods in year 2018, and the reason behind choosing this year is to make an evaluation of the prediction process by calculating its errors since the actual values are available in the

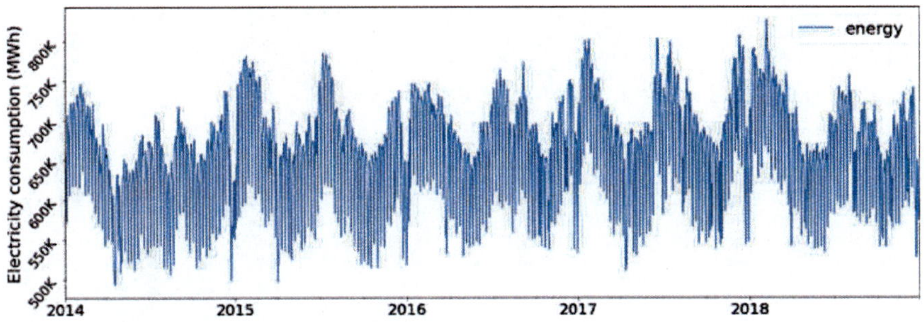

Fig. 22.2 Graphical representation of the dataset

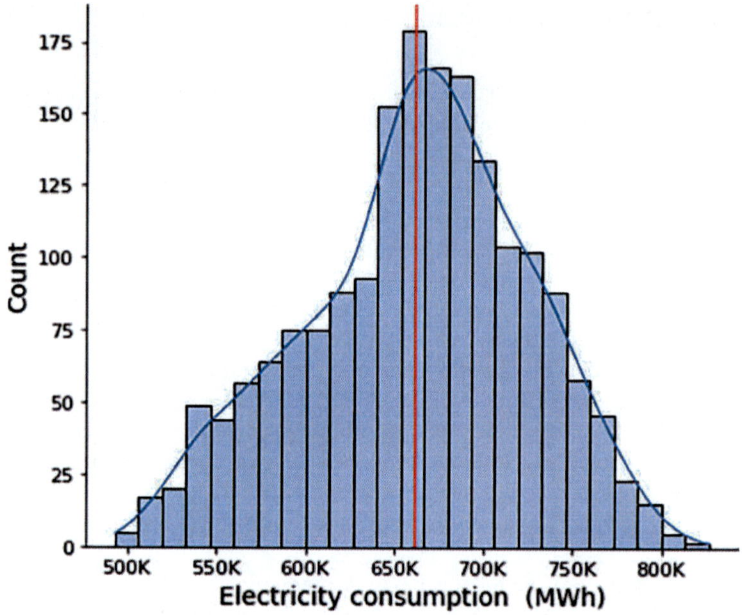

Fig. 22.3 Bell-shaped curve of the normal distribution

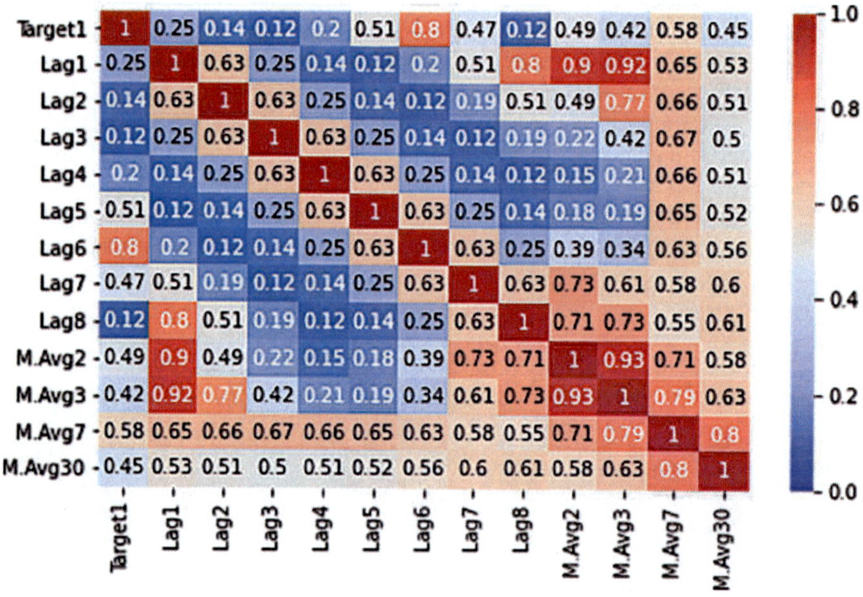

Fig. 22.4 Heatmap of the correlation matrix

original dataset. Afterward, with a good predicted results and a good choice of parameters, the model can be used to predict future values that do not exist in the original dataset.

Collecting more data and features and choosing the highest affected ones has a huge impact to accurate the predication of the model; therefore all parameters that used to build a RF model must be met to achieve the highest results with less error. The optimization tool (Scikit-Learn) of Python programming language has been used to determine the best tune parameters; it gives the optimized values of maximum depth and maximum feature for each individual tree by calculating the less RMSE of the fold cross-validation among many variables, with each iteration of hyperparameter optimization varying the number of trees and calculating the MAPE for predicting the power consumed for the next day, week, and month by using Eq. 22.3. Different values were tried for the maximum features and maximum depth, and each iteration gave a different number of the best values.

As mentioned above, the data split into the training set is from the years 2014 to 2017. For the daily recorded dataset with some deleted data due the distributed lags, the observations will be 1765 rows. The more rows in the data, the more trees are needed, and the best performance is obtained with tuning the number of trees and the other parameters.

Different numbers of trees were tried and the results are tightly closed, as shown in Table 22.1. But when trees = 300, it gives the fairly best results.

The results obtained in this paper by training the model of RF regression gave a prediction of the energy consumed in different periods of year 2018 in Spain. For 1 day ahead of prediction with the use of the best parameter selection, the MAPE was measured as 2.05%. After performing the prediction of a period of time ahead, the model uses the actual data as training data and the prediction process of this period will start over. Figure 22.5 gives a good visual of the prediction for 1 day ahead. Note that there is an error rate in the prediction during the first week of February, which may be due to a sudden drop in temperature that occurred at that time.

Table 22.1 Evaluation results

	MAPE of prediction in (%)				
No. of trees	1 day	7 days	30 days	Max features	Max depth
10	2.4	3.53	4.54	16	30
50	2.13	3.29	4.2	16	30
100	2.08	3.25	4.16	8	30
200	2.09	3.2	4.1	8	30
250	2.08	3.23	4.12	8	30
300	2.05	3.2	4.08	16	30
350	2.08	3.21	4.1	16	30
400	2.07	3.23	4.12	8	30
500	2.08	3.21	4.1	8	30
1000	2.06	3.2	4.1	16	10
2000	2.06	3.2	4.12	16	30

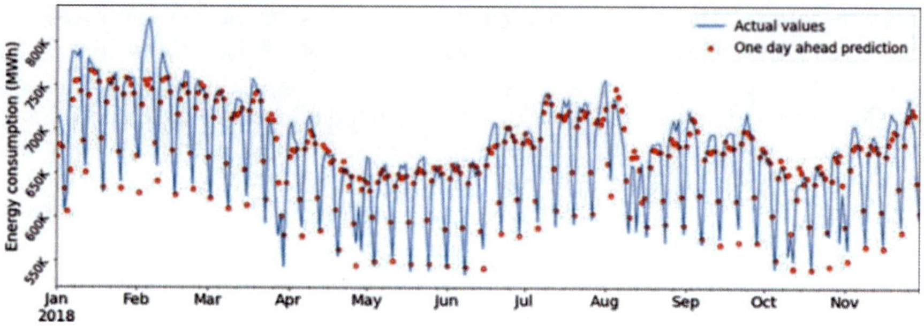

Fig. 22.5 One day ahead prediction vs actual values

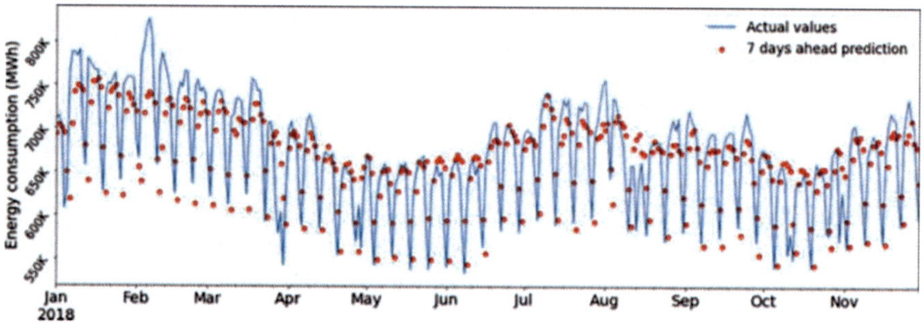

Fig. 22.6 Seven days ahead prediction vs actual values

As it was done in the previous step, the prediction of 7 and 30 days ahead has been performed and their MAPE has been measured (for trees = 300) as 3.2% and 4.08%, respectively. Although there is an increase in MAPE for the 7 days ahead than for the 1 day ahead prediction, however, it can be noted from Fig. 22.6 that the prediction for 7 days ahead is accurate enough as a 1 day ahead prediction with the continuation of the error that occurred for the first week of February, which is observed in Fig. 22.5.

The differences between actual and predicted values that occurred at the beginning of February in 1 day ahead and 7 days ahead predictions, as shown in Figs. 22.5 and 22.6, respectively, have become involved in the whole first quarter of the year in 30 days ahead prediction (see Fig. 22.7).

It turns out that the forecast is better for the shorter period due to changes that occur in energy consumption, such as sudden weather changes that lead to a change in the usual daily consumption.

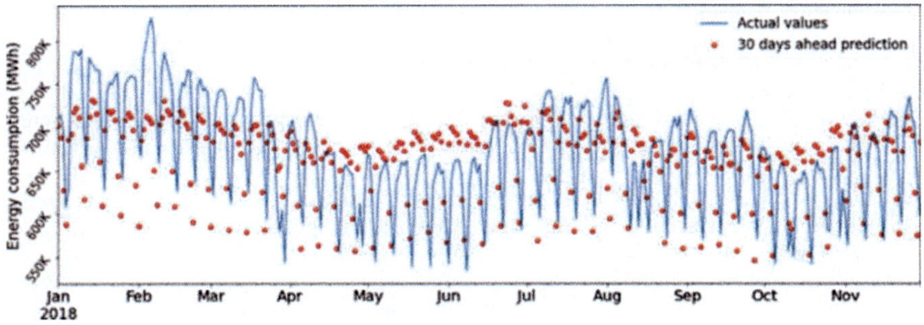

Fig. 22.7 Thirty days ahead prediction vs actual values

22.5 Conclusion

Predicting the need for electrical energy consumption in future time periods is important to optimize the operation of the power plants and reduce their gas emissions, which also has a financial benefit. For datasets with low input, like in time series models, it is indispensable to find more inputs that cooperate and are somehow related together to give the output, which leads to increasing the accuracy of the prediction. In this study, the RF algorithm was applied to predict the energy consumed in Spain for different periods of the year 2018. New inputs were created and tested, and the ones that had an effect on increasing the prediction and had low error rates were chosen. In conclusion, the right choice of the new feathers helps to increase the accuracy of the prediction. The proposed model has a good prediction for short periods of time (up to a month ahead), but for long periods, there are some intervals of time in which the prediction may not be accurate enough.

References

Breiman L (2001) Random forests. Machine Learning 45(1):5–32. https://doi.org/10.1023/a:1010933404324

Dallal GV (1999) The Little Handbook of Statistical Practice. http://www.jerrydallal.com/LHSP/LHSP.htm

Ho TK (1995) Random decision forests. Proceedings of 3rd International Conference on Document Analysis and Recognition 1:278–282. https://doi.org/10.1109/ICDAR.1995.598994

Liu Y, Chen H, Zhang L, Feng Z (2021) Enhancing building energy efficiency using a random forest model: A hybrid prediction approach. Energy Reports 7:5003–5012. https://doi.org/10.1016/j.egyr.2021.07.135

Mahrukh AW, Lian GX, Bin SS (2020) Prediction of power transformer oil chromatography based on LSTM and RF model. 2020 IEEE International Conference on High Voltage Engineering and Application (ICHVE). https://doi.org/10.1109/ICHVE49031.2020.9279968

Nallathambi S, Ramasamy K (2017) Prediction of electricity consumption based on DT and RF: An application on USA Country Power Consumption. 2017 IEEE International Conference on Electrical, Instrumentation and Communication Engineering (ICEICE). https://doi.org/10.1109/iceice.2017.8191939

Nguyen HN, Vu TN, Ohn SY, et al. (2006) Feature elimination approach based on Random Forest for cancer diagnosis. Lecture Notes in Computer Science :532–542. https://doi.org/10.1007/11925231_50

Stephan J, Stegle O, Beyer A (2015) A random forest approach to capture genetic effects in the presence of population structure. Nature Communications 6(1). https://doi.org/10.1038/ncomms8432

Xiong X, Xu Z, Yuan Y (2021) Grey correlation-oriented random forest and particle swarm optimization algorithm for power load forecasting. Journal of Applied Science and Engineering 25:19–30. https://doi.org/10.6180/jase.202202_25(1).0003

Chapter 23
Comparison Between PSO-Based and fmincon-Based Approaches of Optimal Power Flow for a Standard IEEE-30 Bus System

Omar Sagban Al-butti and O. Tolga Altinoz

Nomenclature

OPF	Optimal power flow
FMINCON	Find the minimum of a constrained nonlinear multivariable function
PSO	Particle swarm optimization
ED	Economic dispatch
MATLAB	Matrix laboratory

23.1 Introduction

Optimal power flow (OPF) is used to determine the best operating levels for electric power plants (systems) while considering the limits of the equipment and satisfying the operating constraints. It includes three different objectives, economic dispatch, security constrained optimal power flow (SCOPF), and multi-objective optimization, and improves the performance of the electrical system, hence the possibility to control the output power from the generators and change it to fit the demand within the minimum permissible limits at the lowest possible cost (Abou El Ela et al. 2010).

This study aims to present methods for different solutions, through reliable and efficient approaches, to solve the OPF problem for the standard IEEE 30-bus system, using the fmincon optimization method and particle swarm optimization (PSO)

O. S. Al-butti (✉) · O. T. Altinoz
Department of Electrical and Electronics Engineering, Gazi University, Ankara, Turkey

Department of Electrical and Electronics Engineering, Ankara University, Ankara, Turkey
e-mail: osagban.al-butti@gazi.edu.tr; taltinoz@ankara.edu.tr

M. Z. Sogut et al. (eds.), *Proceedings of the 2022 International Symposium on Energy Management and Sustainability*, Springer Proceedings in Energy,
https://doi.org/10.1007/978-3-031-30171-1_23

method. The results will then be evaluated after comparing both methods. The goal of the OPF is to reduce the total cost function of generating units and transmission losses and maintain the design and performance of the whole power system. It does this while meeting the operational requirements.

Many different types of study have been conducted at OPF, with differing degrees of difficulty. In a power system, OPF is an optimal way of decreasing generation costs by effectively scheduling and updating generation. In the study presented by Vijayvargia et al. (2016), the MATLAB program was developed for three conventional load flow analysis methods: the Gauss-Seidel method, the Newton-Raphson method, and the Fast-Decoupled method. The three load flow methods have been compared based on the number of iterations obtained. Results showed that with the least number of iterations, the Fast-Decoupled approach produces results that are similar to those obtained by the Newton-Raphson and Gauss-Seidel methods. Wankhade et al. (2012) investigated the OPF problem using both continuous and discrete control variables, as well as traditional OPF and genetic algorithm (GA)-based OPF. The results indicate that genetic algorithm outperformed the conventional technique in terms of computational time and generation cost/fuel cost, demonstrating that genetic algorithm is better at finding the optimal global cost than the conventional method. In the study presented by Om and Shukla (2015), MATLAB software was used to compare the genetic algorithm (GA) and PSO based on OPF methods. As a result, among these ways, the PSO method reduces the generating cost and losses of an IEEE-30 bus system more than the GA method. Benhamida (2006) developed the design of the *fmincon* routine, which is part of the MATLAB optimization, which was designed to solve the ED of thermal units, including transmission losses.

The following is a breakdown of the paper's arrangement. Section 23.2 describes the problem formulation of the OPF including the objective functions and constraints. Section 23.3 introduces the solution strategy for the objective function by implementing the *fmincon* and PSO algorithms. Section 23.4 depicts the system and the numerical results. Finally, the conclusion is summarized in Sect. 23.5.

23.2 Methodology

The formulation used to solve the OPF issue in an IEEE-30 bus system, including the objective functions and constraints, is discussed in this section.

23.2.1 Objective Function

The goal of the OPF is to reduce the total cost function of generating units and transmission losses and maintain the design and performance of the whole power system. It does this while meeting the operational requirements (Graa and Benhamida 2020).

The basic equation of OPF systems:

$$\sum_i P_{Gi} = P_{\text{Load}} + P_{\text{Loss}} \tag{23.1}$$

P_{Gi} is the power output of *i*th generator, P_{Loss} is the total transmission line loss, and $P_{\text{Load}}$is the total load demand of the power system.

As shown in Eq. (23.2), the cost function for generators, the most basic of OPF systems, aims to match the generating output to the needed load while attempting to minimize the generation cost. ED is the term for this.

$$\text{Min}(f) = \sum_i \left(a_i + b_i P_{Gi} + c_i P_{Gi}^2\right) \tag{23.2}$$

a_i, b_i, and c_i are the generator cost coefficients.

The power injected into each bus is given by:

$$P_i = \sum_{j=1}^{N} \text{V}_i \mid \text{V}_j \mid \mid \text{Y}_{ij} \mid \cos\left(\delta_i - \delta_j - \theta_i\right) \tag{23.3}$$

$$Q_i = \sum_{j=1}^{N} \text{V}_i \mid \text{V}_j \mid \mid \text{Y}_{ij} \mid \sin\left(\delta_i - \delta_j - \theta_i i\right) \tag{23.4}$$

P_i and Q_i are the active and reactive power outputs of *i*th generator, δ is the voltage angle, and Y_{ij} is the admittance of line between *i*th and *j*th bus.

23.2.2 System Constraints

To make a more effective conventional OPF simulation, the system can be extended to include some additional system limitations.

Equality Constraints

$$P_i = P_{\text{Load}} + P_{\text{Loss}} \tag{23.5}$$

$$Q_i = Q_{\text{Load}} + Q_{\text{Loss}} \tag{23.6}$$

Inequality Constraints The generators are operating within the following maximum and minimum power limits:

$$P_G^{\min} < P_G < P_G^{\max} \tag{23.7}$$

$$Q_G^{\min} < Q_G < Q_G^{\max} \tag{23.8}$$

The following are the system-wide voltage constraints per unit for each bus:

$$V_i^{\min} < V_i < V_j^{\max} \tag{23.9}$$

The capacity of the lines between the buses is as follows:

$$S_{ij}^{\min} < S_{ij} < S_{ij}^{\max} \tag{23.10}$$

23.3 Solution Method

23.3.1 Solution Method by Fmincon

Several different algorithms can be used with the *fmincon* function to iterate through and solve the optimization problem (Jasemi and Abdi 2022). The *fmincon* function was used to find the minimum of a constrained nonlinear multivariable function. It operates by changing a function's value and returning the function's lowest value within the limited constraints.

A constrained nonlinear multivariable function is written as:

Objective function:

$$\min{}_x f(x) \tag{23.11}$$

Constraints:

$$A.x \leq b,\ \ \text{Aeq}.x = \text{beq}, c(x) = 0, \text{ceq}(x) = 0, lb \leq x \leq ub \tag{23.12}$$

A and Aeq are matrices, *b* and beq are vectors, $c(x)$and ceq(x) are functions that are passed x and return vectors, $f(x)$ is the cost function, the variable *lb* is the lower limits of the x vector, and the variable *ub* is the upper limits of the x variable.

The variable x_0 represents the objective function's initial guess. The following variables can be modified to determine the lowest cost of generation at the generators:

1. *The bus voltages.* On the bus, the voltage boundaries have been manually set. The lowest and higher limits of the bus voltages have been set at 0.9 per unit and 1.1 per unit, respectively.
2. *The real power limits.* The generator bounds are directly taken from the entered data in the bus's matrix variable.

The main objective of this program is to build a stable system within the specified parameters and at the lowest possible cost (Mijatovi 2013).

23.3.2 Solution Method by PSO

The main idea of ED is to distribute load demand among generators at the lowest possible cost while remaining within system limits (Khaled et al. 2017).

The OPF problem was solved using the PSO-based technique (Hasan and Mohammed 2020). The problem was framed as an optimization problem with constraints (Abido 2002). The optimization problem can be expressed mathematically as follows:

$$\min{}_x f(x, u) \tag{23.13}$$

$$g(u, x) = 0, h(u.x) = 0 \tag{23.14}$$

f is the objective function, g is the equality constraints, and h is the inequality constraint.

The PSO is developed according to the following:

- *Swarm:* The problem of ED is formulated by Eq. (23.1). For the ED problem, the decision variable is real power generations, and accordingly, real power generations are used to build swarms. The following is a matrix representation of a complete swarm:

$$S = \begin{pmatrix} P_{11} & P_{12} & P_{1M} \\ P_{21} & P_{22} \; \ldots & P_{2M} \\ \vdots & \vdots & \vdots \\ P_{N1} & P_{N2} & P_{NM} \end{pmatrix}. \tag{23.15}$$

  M is a number of generators in the system, the position of the particle would be represented as a vector length NG, and N are particles in the swarm.
- *Initialization:* Each element of the swarm matrix given above is randomly initialized within capacity limitations, dependent on Eq. (23.7). The velocity of the particles is initialized using this inequality.

$$V_{ij}^{\min} < V_{ij} < V_{ij}^{\max} \tag{23.16}$$

  The limit of maximum velocity for the jth dimension is expressed as

$$V_{ij}^{max} = \frac{P_j^{max} - P_j^{min}}{\alpha} \tag{23.17}$$

α is selected a number of intervals.

- *Objective function evaluation*: One of the committed generators is chosen as a slack generator(s) to meet energy limitations. After the slack generator's value has been limited, the penalty factor term is added to the objective function to penalize the fitness value. This function is written as follows:

$$f^j = F\left(P_i^j\right) - \varphi^j \tag{23.18}$$

φ^j is a penalty factor,

$$\varphi^j = \begin{cases} \left(P_s^j - P_s^{min}\right)^2; & P_s^j < P_s^{min} \\ \left(P_s^{max} - P_s^j\right)^2; & P_s^j > P_s^{max} \\ 0; & P_s^{min} \le P_s^j \le P_s^{mix} \end{cases} \tag{23.19}$$

- *Best position:* $p$best and Gbest are essential components of the PSO approach. The position that has the lowest value of the objective function is referred to as $p$best. Among best, Gbest is the best value. $p$best denotes the particle's best position and Gbes denotes the best among of the $p$best.
- *Movement of particle:* For the updated velocities and positions, the particles in the swarm are moved to a new position using the equations below.

$$v_{ij}^{new} = w.v_{ij} + c1 * \text{rand1} * \left(p\text{best} - P_{ij}\right) + c2 * \text{rand2} * \left(G\text{best} - P_{ij}\right) \tag{23.20}$$

$$P_{ij}^{new} = P_{ij} + v_{ij}^{new} \tag{23.21}$$

$c1$ and $c2$ are the acceleration constant, rand1 and rand2 are the uniform random value in the range [0,1], and w is the inertia weight.

- *Updating the best global position:* The values of the objective function evaluate the particles in their new positions. The upgradation of *pbest* of particles takes place in this stage. The Gbest is selected from among the $p$best. Retaining as $f$best is an objective value at $f$best.
- *Stopping:* The stopping criterion is the maximum number of iterations taken for this task. If the stopping condition is not satisfied, the method described above will be repeated with an incremental t value.
- *PSO parameters:*

Case (1) population size = 50, iteration = 100, $c1 = 1.8903$, $c2 = 2.1225$, and $w = 0.1618$.

Case (2) population size = 100, iteration = 100, and $w = 0.1618$; here in this case will be changed the acceleration constant as follows:

(a) $c1 = 1.8903$, $c2 = 2.1225$
(b) $c1 = 2.1225$, $c2 = 1.8903$

(c) $c1 = 2, c2 = 2$
(d) $c1 = 1.7, c2 = 1.7$
(e) $c1 = 1.4, c2 = 1.4$
(f) $c1 = 0.5, c2 = 1.5$
(g) $c1 = 1.5, c2 = 0.5$

Case (3) population size = 200, iteration = 100, $c1$ = 1.8903, $c2$ = 2.1225, and = 0.1618.

23.4 Result and Discussion

Figure 23.1 depicts the IEEE 30-bus standard test system. The line and bus data of the system are available at www.ee.washington.edu/research/pstca/ (Niknam et al. 2011; Alsac and Stott 1974). Six generators are located on buses 1, 2, 5, 8, 11, and 13 in the system and the total demand for the system was 283.4 (MW). Bus number one was considered as the slack busbar. The minimum and maximum magnitudes of

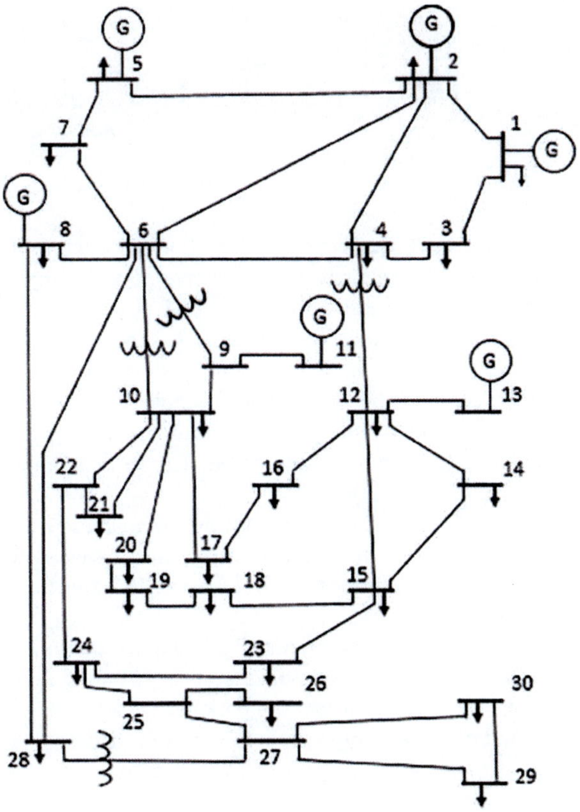

Fig. 23.1 IEEE 30-bus standard test system. (Niknam et al. 2011)

Table 23.1 Cost function and generator limit variables for the IEEE 30-bus test system

No. bus	P_G min	P_G max	Q_G min	Q_G max	Cost function
1	50	200	−20	250	$2P_i + 0.00375P_i^2$
2	20	80	−20	100	$1.75P_i + 0.01750P_i^2$
5	15	50	−15	80	$1.0P_i + 0.0625P_i^2$
8	10	35	−15	60	$3.25P_i + 0.00834P_i^2$
11	10	30	−10	50	$3.0P_i + 0.025P_i^2$
13	12	40	−15	60	$3.0P_i + 0.025P_i^2$

Alsac and Stott (1974)

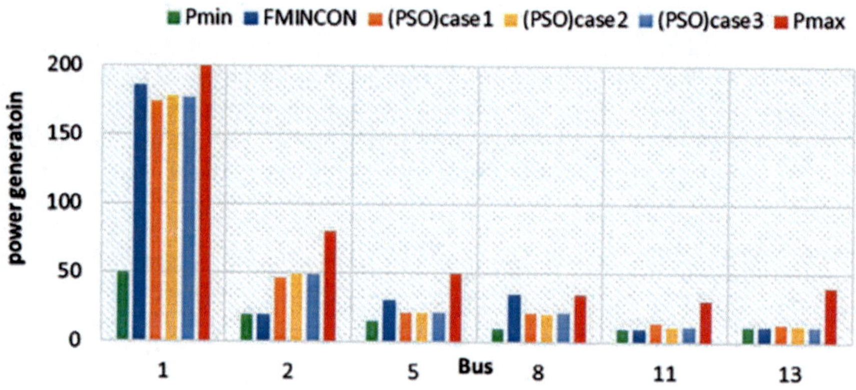

Fig. 23.2 Comparative chart of active power generation using fmincon and PSO

voltage for the buses were considered at 0.95 and 1.05 per unit, respectively. Fuel cost coefficients and active and reactive power outputs were taken from Alsac and Stott (1974) and as shown in Table 23.1.

Figure 23.2 shows a comparative chart of outputs of control variables for active power generation in an IEEE 30-bus system using *fmincon* and PSO.

The PSO and the *fmincon* were applied to the IEEE 30-bus standard and the results of objective functions, power outputs of generator units, and transmission losses of the system were compared. The *fmincon* approach of this study had better performance as in Table 23.2.

Table 23.3 shows the results of the comparison for PSO technique output active power generation, fuel cost, transmission losses, and time when the number of population and iterations are both set at 100 and the acceleration constant values are changed.

The acceleration coefficients *c*1 and *c*2 control the stochastic effect of the personal and societal components on a particle's total velocity. The constants *c*1 and *c*2 are also known as trust parameters, with *c*1 expressing a particle's trust in itself and *c*2 expressing a particle's trust in its neighbors (Ratnaweera et al. 2004).

Table 23.2 Results of PSO and fmincon approaches

IEEE 30 bus	FMINCON	PSO_{case1}	PSO_{case2}	PSO_{case3}
P_1 (MW)	186	174.34	177.38	177.23
P_2 (MW)	20	46.30	48.76	48.95
P_3 (MW)	30	21.76	21.20	21.45
P_4 (MW)	35	21.8	21.02	21.35
P_5 (MW)	10	14.29	11.64	11.56
P_6 (MW)	12	13.72	12.55	12
P_{loss} (MW)	9.70	9.26	9.12	9.13
P_{tot} (MW)	293.01	292.30	292.56	292.53
Cost ($/h)	797.74	801.08	800.87	800.83
Time (s)	11.68	30.90	61.84	138.74

Table 23.3 Results of PSO with different values of $c1$ and $c2$

IEEE 30- bus	Case2.a	Case2.b	Case2.c	Case2.d	Case2.e	Case2.f	Case2.g
P_1 (MW)	177.38	178.41	176.98	176.32	179.81	178.46	162.95
P_2 (MW)	48.76	48.81	48.83	49.08	48.34	49.10	54.17
P_3 (MW)	21.20	21.60	21.36	21.50	22.05	21.09	22.16
P_4 (MW)	21.02	19.63	20.22	20.89	17.14	18.98	19.22
P_5 (MW)	11.64	12.02	13.21	12.79	11.48	13.07	16.62
P_6 (MW)	12.55	12.071	12.01	12.00	14.02	12.00	17.05
P_{loss} (MW)	9.12	9.25	9.22	9.20	9.21	9.29	8.78
P_{tot} (MW)	292.56	292.66	292.56	292.58	292.84	292.69	292.18
Cost ($/h)	800.87	800.90	801.00	800.84	801.03	801.09	805.24
Time (s)	61.84	64.53	63.85	65.21	63.38	63.86	65.51

Several cases have been tested for the acceleration coefficients ($c1$, $c2$) as shown in Table 23.3 and as follows:

1. If $c1 >> c2$, each particle is strongly drawn to its own optimal place, causing excessive roaming. On the other hand, if $c2 >> c1$, particles are more strongly pulled to the global best position, forcing them to rush to the optimal location prematurely.
2. If $c1$ and $c2$ values are low, resulting in smooth particle paths, allowing particles to wander away from the good position to detect before being pulled back toward the good position. High values result in a faster movement toward or past good locations.
3. If $c1 \approx c2$ and $c1 = c2$, particles are most effective when they work together (Del Valle et al. 2008).

23.5 Conclusion

The PSO-based approach and the *fmincon*-based approach were applied to the IEEE 30-bus standard, and the results of the objective functions, power outputs of generator units, and transmission losses in the system were compared. The *fmincon* approach to this study had an overall better performance than PSO.

In the case of PSO. First, it has been shown that with the same number of iterations and an increase in population, the generation fuel cost and losses decrease slightly, but the computational time gets longer. Second, when the population and iterations are fixed, and the acceleration constant is changed, we conclude the particles' strength comes from their ability to work together, and they are most effective when $c1 \approx c2$ and $c1 = c2$; as in case2a, case2b, and case2c, particles are attracted toward the global best position.

References

Abido MA (2002) Optimal power flow using particle swarm optimization. International Journal of Electrical Power & Energy Systems 24(7):563–571. https://doi.org/10.1016/s0142-0615(01)00067-9

Abou El Ela AA, Abido MA, Spea SR (2010) Optimal power flow using differential evolution algorithm. Electric Power Systems Research 80(7):878–885. https://doi.org/10.1016/j.epsr.2009.12.018

Alsac O, Stott B (1974) Optimal load flow with steady-state security. IEEE Transactions on Power Apparatus and Systems PAS-93(20):745–751. https://doi.org/10.1109/tpas.1974.293972

Benhamida F (2006) A New Solution Method to Economic Dispatch using the MATLAB Function (fmincon). Modelling, Measurement and Control A 79:1–13. https://www.researchgate.net/publication/289450902_A_solution_method_to_economic_dispatch_using_the_matlab_function_FMINCON

Del Valle Y, Venayagamoorthy GK, Mohagheghi S, et al (2008) Particle swarm optimization: Basic concepts, variants and applications in Power Systems. IEEE Transactions on Evolutionary Computation 12(2):171-195. https://doi.org/10.1109/tevc.2007.896686

Graa A, Benhamida F (2020) A review on optimization methods applied to Energy Management System. Serbian Journal of Management 15(2):371–382. https://doi.org/10.5937/sjm15-22519

Hasan V, Mohammed F (2020) Optimal Power Flow for a power system under particle swarm optimization (PSO) based. International Journal of Computer Applications177(33):56–62. https://doi.org/10.5120/ijca2020919757

Jasemi A, Abdi H (2022) Probabilistic Multi-Objective Optimal Power Flow in an AC/DC Hybrid Microgrid Considering Emission Cost. Journal of Operation and Automation in Power Engineering 10(1):13–27. https://doi.org/10.22098/joape.2022.8156.1565

Khaled U, Eltamaly AM, Beroual A (2017) Optimal power flow using particle swarm optimization of renewable hybrid distributed generation. Energies 10(7):1013. https://doi.org/10.3390/en10071013

Mijatovic A (2013) Solving optimal power flow with voltage constraints using MATLAB optimization toolbox. School of Engineering and Information Technology. Doctoral dissertation. http://researchrepository.murdoch.edu.au/id/eprint/21660

Niknam T, Narimani Mrasoul, Jabbari M, Malekpour AR (2011) A modified shuffle frog leaping algorithm for multi-objective Optimal Power Flow. Energy 36(11):6420–6432. https://doi.org/10.1016/j.energy.2011.09.027

Om H, Shukla S (2015) Optimal power flow analysis of IEEE-30 bus system using soft computing techniques. International Journal of Engineering Research & Science (IJOER) 1(8):55–60. https://ijoer.com/Paper-November-2015/IJOER-NOV-2015-21.pdf

Ratnaweera A, Halgamuge SK, Watson HC (2004) Self-organizing hierarchical particle swarm optimizer with time-varying acceleration coefficients. IEEE Transactions on Evolutionary Computation 8(3):240–255. https://doi.org/10.1109/tevc.2004.826071

Vijayvargia A, Jain S, Meena S, Gupta V et al. (2016) Comparison between Different Load Flow Methodologies by Analyzing Various Bus Systems. International Journal of Electrical Engineering 9(2):127–138. https://www.ripublication.com/irph/ijee16/ijeev9n2_01.pdf.

Wankhade CM, Saoji BP, Vaidya AP (2012) Comparative Study of GA Based Optimal Power Flow. International Journal on Advanced Electrical and Electronics Engineering ISSN 2278-8948 1(1):20–25. http://www.irdindia.in/journal_ijaeee/pdf/vol1_iss1/5.pdf

Chapter 24
Goal-Oriented Requirements Engineering Approach to Energy Management Systems

Murat Pasa Uysal

Nomenclature

EnMS	Energy management system
EMIS	Energy management information systems
GORE	Goal-oriented requirements engineering
GRL	Goal-oriented requirements language

24.1 Introduction

ISO 50001 defines an energy management system (EnMS) as a system that establishes an energy policy, which includes performance goals, objectives, and energy targets, along with action plans and responsibilities to achieve these goals, objectives, and energy targets (ISO 50001 2018). Therefore, an EnMS enables organizations to take actions for the improvement of energy performance, energy efficiency, energy use, and consumption (Eccleston et al. 2012). On the other hand, implementation, operation, and assessment of an EnMS also necessitate the establishment of an effective energy management information system (EMIS). EMISs collect, record, process, analyse, and present different types of organizational data related to energy use (Uysal and Söğüt 2017). There is a strong bidirectional relationship and dependency; therefore, we combine and call EnMS and EMIS as EnM(I)S. As being similar to other types of information systems (IS), the design and development of EnMS and EMIS require conducting effective requirements engineering

M. P. Uysal (✉)
Department of Management Information Systems, Baskent University, Ankara, Turkey
e-mail: mpuysal@baskent.edu.tr

M. Z. Sogut et al. (eds.), *Proceedings of the 2022 International Symposium on Energy Management and Sustainability*, Springer Proceedings in Energy,
https://doi.org/10.1007/978-3-031-30171-1_24

(RE) processes, which should regard the organizational context, business strategy, and goals (Berenbach et al. 2009; Aurum and Wohlin 2005). This type of RE approach is defined as goal-oriented requirements engineering (GORE) (Lamsweerde 2001).

One of the important factors for unsuccessful IS projects is the deficiency in the RE processes and determining goals, not only for IS but also for EnMIS. In other words, there is a clear need for bridging the gap between EnMS and EMIS in terms of RE. However, the review of literature cannot provide sufficient work presenting the guidelines for GORE of EnMS and EnMIS.

24.2 Background

24.2.1 Energy Management System

As being a continual improvement framework and based on the Plan-Do-Check-Act (PDCA) cyclic method, an EnMS incorporates the energy management of an organization into its existing practices (ISO 50001 2018). PDCA method includes the following cyclic steps:

Plan:

- Understanding the organizational context
- Establishing an energy management team and an energy policy
- Conducting an energy review to identify:
  - Significant energy uses (SEUs)
  - Energy performance indicators (EnPIs)
  - Energy baseline(s) (EnBs)
  - Objectives (goals) and energy targets
- Developing action plans required to deliver results that will improve energy performance in accordance with the energy policy

Do:

- Implementing the action plans
- Ensuring operational and maintenance controls and communication

Check:

- Monitoring, measuring, analysing, evaluating, auditing, and conducting management review(s) of energy performance and the EnMS

Act:

- Taking necessary actions to address nonconformities
- Continually improving energy performance and the EnMS

Energy performance is a measurable result related to energy efficiency, energy use, and energy consumption. In the context of EnMS, objectives are results to be achieved, which are set by the organization and desired to be consistent with the energy policy. Energy targets are specific, quantifiable objectives or events of energy performance improvements. An energy objective, as being a long-term goal, is usually applied across the entire organization; however, energy targets can vary over time and across various organizational functions and activities (Eccleston et al. 2012). There may be three types of objectives. The first type is the improvement of something, which could be the maximization or minimization of something, such as "reduce energy consumption". The second type is maintaining something as in "maintain energy efficiency". The third type is "studying/researching" as in "conducting a feasibility study" (Howel 2014).

24.2.2 Goal-Oriented Requirements Engineering

A goal may be defined as a strategic achievement that has to be accomplished by an IS. Goals can be expressed at different levels of abstraction. Therefore, they are useful for elicitation, conceptualization, modelling, analysing, and identifying alternatives and conflict resolutions during RE processes. As being one of the four types of RE approach, GORE mainly focuses on the strategic context of IS requirements and thus helps elaborate the requirements that support the organizational goals (Lamsweerde 2001). GORE's main concerns are (a) how to achieve organizational goals, (b) how to operationalize the goals into services or products, and (c) how to assign tasks and responsibilities.

Goals are usually modelled according to their fundamental features and attributes and also the relationships with other goals and model elements. Goal modelling supports formal, semi-formal, qualitative, and heuristic reasoning during RE. Therefore, goal-oriented requirements language (GRL) would be very helpful for modelling, communicating, and reasoning the requirements of EnMS (GRL 2021). GRL's main concern is to understand "why" is the system needed? The decision-makers or modellers are not interested in the operational or detailed specifications at the early phases of RE. There are three categories of GRL concepts: intentional elements (IE), intentional relationships (IR), and actors. The IE of GRL are goal, task, softgoal, belief, and resource. Means-ends, contribution, decomposition, dependency, and correlation relationships form the IR and they are depicted by different types of arrows (Fig. 24.1).

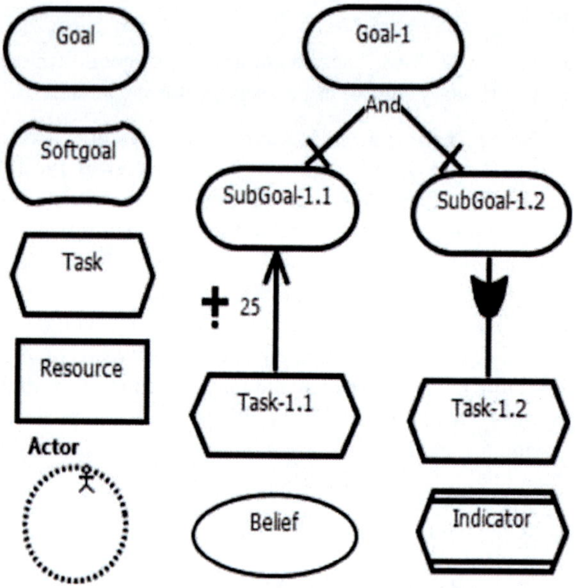

Fig. 24.1 Intentional elements and relationships of GRL

24.3 Method

In this study, we adopt a GORE approach and extend it for the RE processes of EnMS. Eclipse IDE and jUCMNav v7 plug-in are used for the design and development of GRL models. The two-phase GORE approach is as follows:

- Phase 1: Definition of an EnMS's goals and concepts
- Phase 2: Modelling of the EnMS by using the GORE tools and techniques

24.3.1 Limitations

The results of this study are limited to the data and facts retrieved from the online report of a real industrial case study (BCH 2016). However, the detailed processes for goal formulation, verification, and validation could not be conducted due to these limitations.

24.4 Case Study

This section depends on a company's report belonging to an ISO 50001 case study, which is obtained from the company's website (BCH 2016). The company is located in Canada and owns an integrated paper and pulp manufacturing plant. It has two paper machines and two pulp machines to produce newsprint and kraft pulp, which is also used for manufacturing printing, writing, and tissue papers. After the announcement informing that electricity costs would increase by 27% over 5 years, a decision was made to implement ISO 50001. The company estimated that it could reduce its energy expenditure by 3.2% in 3 years. Therefore, it initially obtained an ISO 50001 certificate and implemented its EnMS. This included an action plan with energy reviews, energy baselines, the establishment of energy objectives (goals), and energy targets. Energy performance improvements were focused on thermo-mechanical pulping, self-power generation, the steam system, and boilers. Energy sources involved electricity, natural gas, gasoline, and diesel. Consequently, the company's cumulative energy reduction was 4.8% or 100 giga-watt-hours (GWh) after a 2-year implementation. In other words, it accomplished 4.8% savings in 2 years instead of 3.2% in 3 years.

For the purposes of simplicity and space limitations, only the planning phase is included when GORE is reapplied to the RE processes of the company's EnMS. First, we analysed the case report. Second, we extracted the EnMS-related data, facts, and knowledge. Finally, the GORE modelling process is conducted depending on this knowledge.

24.4.1 Phase 1 (Definition of EnMS Goals and Concepts)

The steps of Phase 1 are given as follows:

- Step a: Energy management team is established.
- Step b: Organizational objectives (goals), energy targets, scope, and boundaries are identified.
- Step c: Performance indicators (EnPIs) and energy baselines (EnB) are measured and determined.
- Step d: Required tasks and resources that support the objectives are identified.

The scope of EnMS is defined as the thermo-mechanical pulping, self-power generation, steam system, and boilers. The main EnMS tasks are energy review, determining EnPIs, EnB, monitoring performance, and energy targets. Electricity, natural gas, gasoline, and diesel are the energy sources. Table 24.1 presents the EnMS goals and concepts that are extracted from the case study report:

Table 24.1 EnMS goals and concepts

Goals (energy objectives)	Softgoals	Energy targets	EnPIs
Reduce energy cost (3.2%)	Integrated management	Reduce GWh	CO
Reduce energy consumption	Long-term stability	CO	CO
Improve energy efficiency	Reduced risk	CO	CO
Fulfil regulatory requirements	Enhanced sustainability	CO	CO

CO Could not be obtained from the report

24.4.2 Phase 2 (GORE Modelling of EnMS)

The steps of Phase 2 are given as follows:

- Step a: Mapping EnMS stakeholders, goals, subgoals, softgoals, and objectives to the IE of GRL
- Step b: Mapping the relationships, which are related to the goals, subgoals, and softgoals, to the IR of GRL
- Step c: Mapping the supporting EnMS concepts to the GRL tasks, resources, key performance indicators (EnPIs), and beliefs
- Step d: Development of GRL model(s) by using IE and IR elements

Step a and Step b: The IE [actors (stakeholders), goals (objectives), subgoals, and softgoals of EnMS] and the IR are modelled in Fig. 24.2:

As can be seen from Fig. 24.2, the actor "Managers and Energy Management Team" carries out the actions to achieve the main goal by exercising its know-how (GRL 2021). The main goal of the company is to reduce energy cost by 3.2%. This

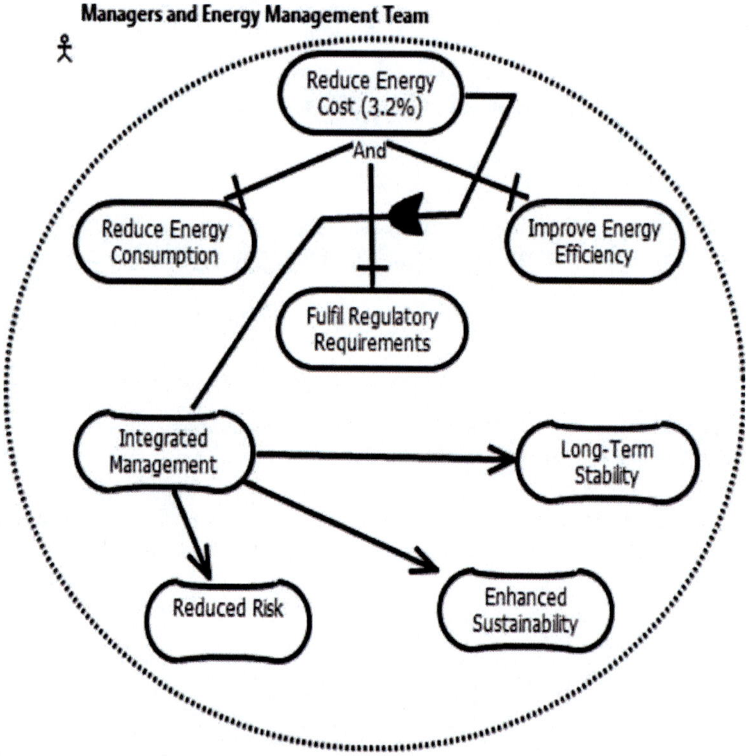

Fig. 24.2 The IE and IR of EnMS

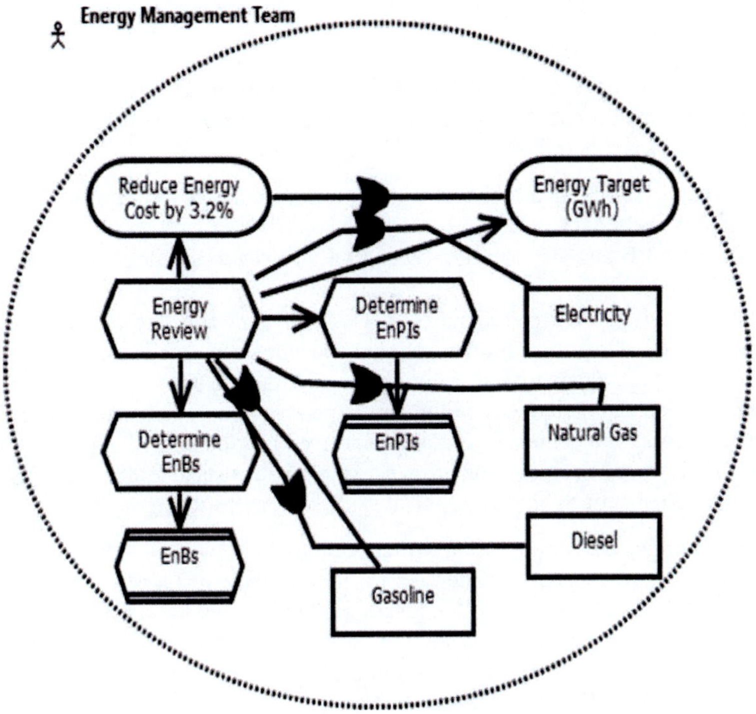

Fig. 24.3 The GRL tasks, resources, EnBs, and EnPIs

goal is decomposed into three subgoals: "reduce energy consumption", "improve energy efficiency", and "fulfil regulatory requirements". The main goal is dependent on the softgoal "integrated management". This softgoal also contributes to the other softgoals: "long-term stability", "enhanced sustainability", and "reduced-risk".

Step c: The GRL tasks, energy resources, key performance indicators (EnPIs), and the IR are modelled as follows (Fig. 24.3):

As can be seen from Fig. 24.3, the actor "energy management team" carries out the actions to achieve the main goal by exercising the core EnMS tasks. For example, the task "energy review" contributes to the "energy target (GWh)" and also to the task "determine EnBs", which eventually determines the EnBs as a GRL indicator. The EnPIs are specified by the task "determine EnPIs". The processes that are included by the task "energy review" are directly dependent on the resources: electricity, natural gas, gasoline, and diesel, respectively.

Step d: The combination GRL models of Phase 1 and Phase 2 are given in the Appendix.

24.5 Conclusion

In this study, we have proposed a GORE approach to the RE of EnMSs and then applied it to a real industrial case study. The main conclusions that may be drawn from this study are as follows:

- Deficiencies in RE have been and will be the main concern not only for IS but also for EnMS and EMIS.
- High-level RE analysis should precede the low-level and detailed requirement specifications.
- GORE can help bridge the gap between the "why" and "how" questions that underline the design and implementation decisions and details of a system.
- GORE supports qualitative or formal reasoning during RE.
- GORE helps elaborate the requirements supporting goals and, thus, provides the criterion for completeness of requirements specification and validation.
- GORE helps communicate the ideas between various stakeholders, ranging from non-technical staff to highly technical staff, such as software engineers.

Consequently, the contributions of this study are twofold: (1) to draw the practitioners' and researchers' attention to the problems of GORE for EnMS and EMIS and (2) to propose a GORE process model for EnMSs. As future work, we plan to extend this approach to another case, which would start from the very beginning of the EnMS project. Therefore, this study can be viewed as an initial step towards the GORE of EnMS and EMIS.

Acknowledgement The author confirms that he has no competing financial and other types of interests or personal and organizational relationships that could have appeared to influence the research study reported in this paper.

Appendix: Integrated GRL Models of Phase 1 and Phase 2

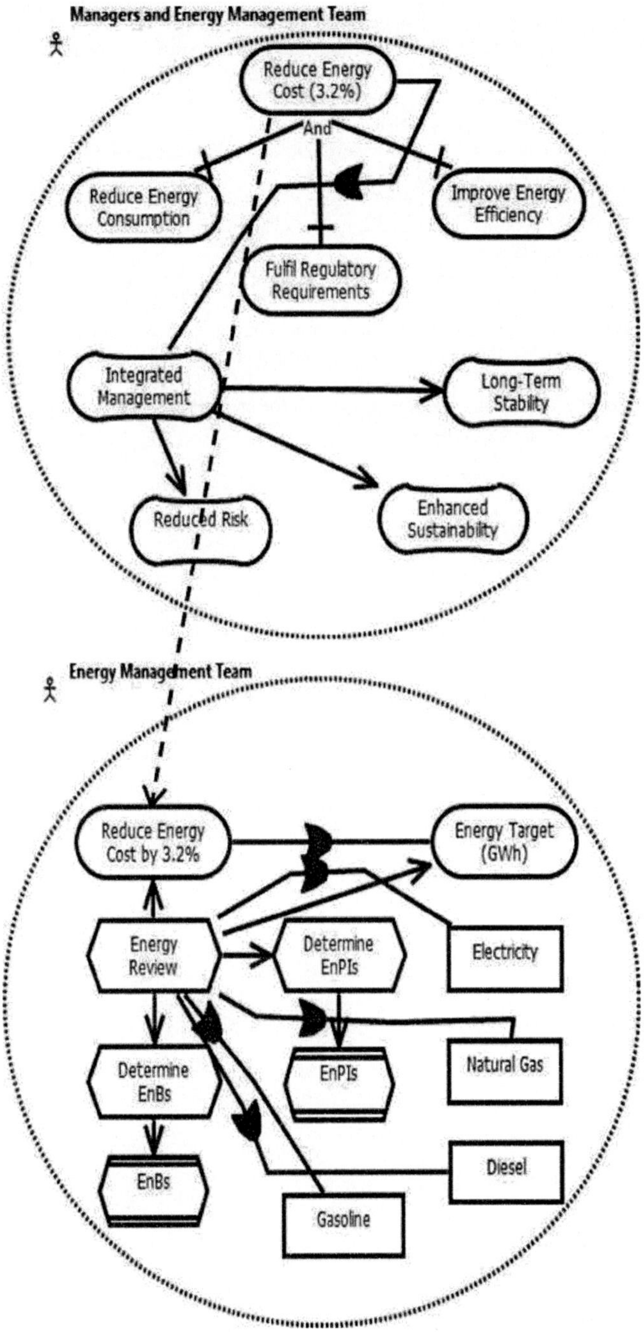

References

Aurum A Wohlin C (2005) Engineering and managing software requirements. Springer, Germany

BCH, (2016), Catalyst Crofton ISO 50001 Case Study, BC Hydro Report. https://www.bchydro.com/index.html.

Berenbach B Paulish D Kazmeier J Rudorfer A (2009) Software & Systems Requirements Engineering: In Practice. McGraw-Hill, USA.

Eccleston C March F Cohen T (2012) Inside Energy: Developing and Managing an ISO 50001 Energy Management System. CRC Press, USA.

GRL, 2021, GRL Web Site, University of Toronto. http://www.cs.toronto.edu/km/GRL.

Howel MT (2014) Effective implementation of an ISO 50001 energy management system (EnMS). ASQ Quality Press, USA.

ISO 50001 (2018) Energy management systems- requirements with guidance for use, International Organization.

Lamsweerde AV (2001) Goal-Oriented requirements engineering: A guided tour, Proceedings of Fifth IEEE International Symposium on Requirements Engineering, 27–31 Aug. 2001, Toronto, Canada.

Uysal MP Söğüt MZ (2017) An integrated research for architecture-based energy management in sustainable airports. Energy:140, 387–1397.

Chapter 25
Decision-Making on Nuclear Power Plant Site Selection in Turkey

Muhammed Sutcu and Ibrahim Tumay Gulbahar

Nomenclature

MAUT	Multi-attribute utility theory
SUF	Single-attribute utility function
Fuzzy DEA	Fuzzy data envelopment analysis
Fuzzy GP	Fuzzy goal programming
MCDM	Multi-criteria decision-making
CCSD	Correlation coefficient and standard deviation

25.1 Introduction

The need for energy is increasing continuously in the world to meet the daily life requirements of people. Population growth, developments in technology, etc. lead to a burst of the energy demand. To meet these demands, different types of energy sources are used such as fossil fuel plants, thermal power plants, dams, wind turbines, and so on. However, more energy requirements direct countries, companies, investors, etc. to demand satisfaction from more adequate resources such as nuclear power plants.

For an investment, criteria must be determined to make it worthwhile. Nevertheless, there exist many aspects that have impacts on decision-making. For the past years, the Turkish government was working on these concerns about nuclear power plant construction. At the end, they decided to build a power plant in the province of

M. Sutcu · I. T. Gulbahar (✉)
Faculty of Engineering, Abdullah Gul University, Kayseri, Türkiye
e-mail: muhammed.sutcu@agu.edu.tr; ibrahim.gulbahar@agu.edu.tr

M. Z. Sogut et al. (eds.), *Proceedings of the 2022 International Symposium on Energy Management and Sustainability*, Springer Proceedings in Energy,
https://doi.org/10.1007/978-3-031-30171-1_25

İçel, and it is announced in 2010 by signing the cooperation agreement between Turkey and Russia (Rosatom 2010).

Facility location is a noteworthy method to have a more efficient and effective site selection for years. Different approaches are used to provide solutions like statistical models, simulations, optimization models, decision-making strategies, etc. (Mahmoudi et al. 2022).

According to Cedolin et al. (2017), facility location selection has strategic importance for investors, which would be governments, companies, etc., because it influences not only manufacturing and transportation costs but also productivity and lead times to a great extent. They propose a solution to the facility location selection problem by applying fuzzy data envelopment analysis (fuzzy DEA) and fuzzy goal programming (fuzzy GP) methods.

Multi-attribute utility theory (MAUT) is another approach to select the best location for a facility among the alternatives. Emeksiz and Yüksel (2022) proposed a novel approach for a site selection for a sustainable bioenergy production facility. They hybridize the multi-criteria decision-making (MCDM) approaches by using MAUT.

Canbolat et al. (2007), for a global manufacturing facility, present a multi-phase approach to selecting a country. For the decision process, an influence diagram is used. Also, a risk profile is generated by analyzing the uncertainties regarding cost with a decision tree. Then, this risk profile is used as a measure in the MAUT model. In addition to these, a survey with the decision-makers has been conducted and the weights are specified. Lastly, one of the five countries is determined to make the investment on.

Another study done by Singh (2016) proposes the multi-attribute utility theory and correlation coefficient and standard deviation (CCSD) to solve real-time allocation problems. The correlation coefficient and the standard deviation are incorporated with multi-attribute utility theory to obtain the best choice from a finite set of alternative facility locations.

As in the examples from the literature, for the investment of the nuclear power plant in Turkey, there were 36 volunteer cities that are willing to provide large lands. This study aims to seek the best city for building the first nuclear plant in Turkey.

The government of Turkey started to build nuclear facilities amid controversy in April 2015. It is being constructed in Akkuyu in İçel on the coast of the Mediterranean. The paper evaluates the decision that İçel is the best location for nuclear power plant. In the solution pathway, MAUT approach has been applied as the methodology.

The remainder of this article is structured as follows: In the methodology section, the used methodology is presented. In the results section, the outputs of the applied methodology and the explanation of them are given. Lastly, the conclusion of the article is provided.

25.2 Methodology

Nuclear energy production is affected by many different factors such as delivery, geological factor, meteorological factor, secure factor, and labor quality. These reasons are the restrictions of the selection process like the health of the community and international agreements.

In the light of studies in the related literature, to solve the facility location decision problem, it is needed to determine several criteria. Therefore, to select the best location, MAUT is applied to the seven different factors. The weights of the measures are determined by a survey. The purpose of the survey is to determine the importance level of each measure. This survey includes 14 different questions. In each question, to determine the level of importance of a measure, the respondents gave a score between 0 and 10. The respondents also added a new measure if they wanted to add it, and they also gave a score for the new measure. There are 21 respondents who joined the survey.

According to the result of the survey, the arithmetic mean of the responses for each measure is calculated. Finally, the scores are normalized (between 0 and 1) as in Table 25.1.

Single-attribute utility functions (SUF) are constructed to observe the optimum conditions of each measure. There exist several types of SUFs like S-shape, exponential, linear, etc. Exponential SUFs indicate the different and the same directional importance of range. On the other hand, S-shape SUFs are the combination of two exponential functions. This indicates the decreasing and increasing of the importance of ranges in the same graph. Constructed SUFs in this study are as follows:

Table 25.1 Normalized weights of measures

Measures	Normalized weights
Proximity to ports	0.179
Proximity to supplier	0.152
Landslide category	0.089
Earthquake category	0.089
Precipitation	0.080
Wind speed	0.080
Population density	0.080
Magnitude earthquake	0.054
University enrollment	0.045
Technical and industrial vocational high school enrollment	0.036
Distance to neighborhood countries	0.036
Labor cost	0.027
Unemployment rate	0.027
Rank in the development industrialization region	0.027

Table 25.2 Types of the constructed SUFs

Function	SUF type
Closeness to suppliers	S-shape
Closeness to ports	S-shape
University enrollment	S-shape
Amount of precipitation	S-shape
Unemployment rate	Exponential
Industrialization	Exponential
Labor cost	Exponential
Population density	Linear
Landslide category	Linear
Earthquake category	Linear
Average wind speed	Linear
Distance to neighbor countries	Linear

- Closeness to suppliers
- Closeness to ports
- University enrollment
- Unemployment rate
- Amount of precipitation
- Population density
- Landslide category
- Industrialization
- Labor cost
- Earthquake category
- Average wind speed
- Distance to neighbor countries

As indicated, constructed SUFs are categorized into three different types: S-shape, exponential, and linear. Classes of those are given in Table 25.2 and examples of the SUFs are shared in Figs. 25.1, 25.2, and 25.3.

25.3 Results

The constructed functions are inserted into the "Logical Decisions" software to solve and get the results. The swing weight method is used to order measures. Weights of the measures are provided by the decision-makers' survey answer averages. After running the model with the given information, it is seen that the İçel is the best option. On the other hand, two other possible options are observed as almost the same grade with the İçel, which are Sinop and Adana as in Fig. 25.4.

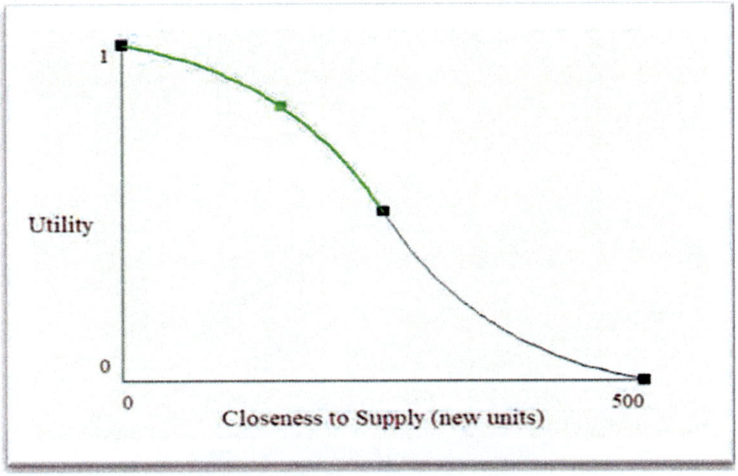

Fig. 25.1 Example of an S-shape SUF (closeness to suppliers)

Fig. 25.2 Example of an exponential SUF (unemployment rate)

25.4 Conclusion

Continuous increases in the energy requirement lead communities to find out new resources for power generation. There exist many alternative resources, but one of the most productive options is nuclear power plants.

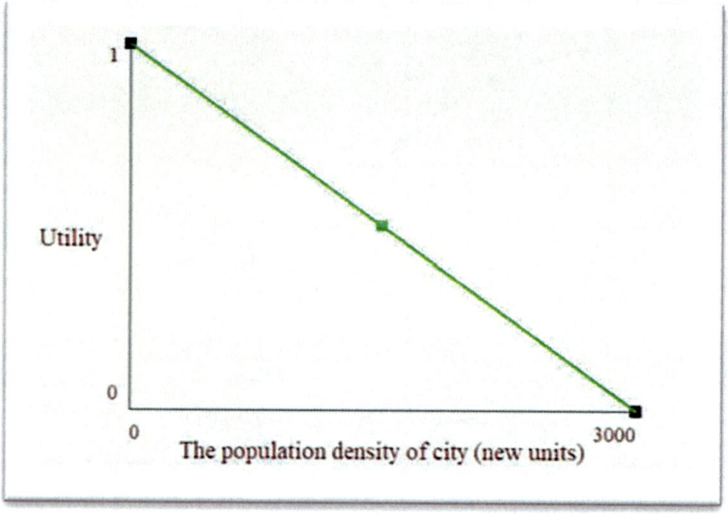

Fig. 25.3 Example of a linear SUF (population density)

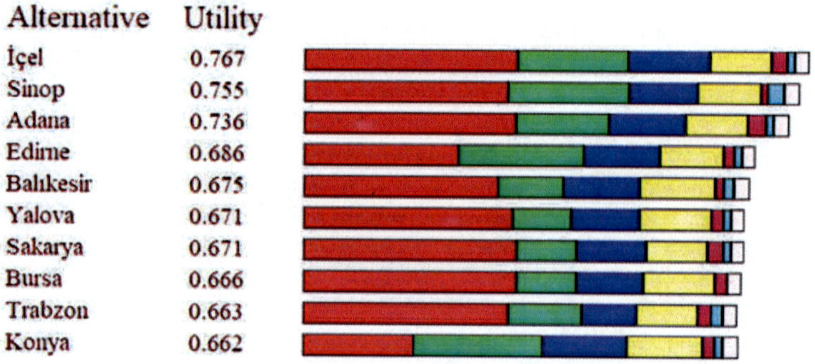

Fig. 25.4 First ten alternative outcomes of the model

Energy generation by nuclear reactors is used by many countries. However, because of the high costs of construction, operations, side industries, and so on, it requires deeply analyzed decisions.

Turkey as a candidate for nuclear power generators decided to build the first nuclear power plant in İçel with a Russian contractor company. On the other hand, there were 36 different candidate cities that are volunteers for this project. Therefore, the decision process must have been handled attentively.

In this study, the decision of the site selection for this nuclear power generation facility is questioned. MAUT modeling is used for analysis and comparisons. Inputs for the model are gathered by a survey, which has 14 questions, from 21 experts. The swing weight method is used to order preferences, and, in the end, it is seen that the decision on the İçel is proper for the construction, although there are alternatives that have almost the same scores.

Results may alter by changing the weights of measures, which means that alternatives other than İçel may become the best option for the facility building.

Acknowledgments The authors would like to thank the anonymous reviewers and editors for their comments and suggestions.

Statement of Conflicts of Interest There is no conflict of interest between the authors.

Statement of Research and Publication Ethics The authors declare that this study complies with Research and Publication Ethics.

References

Akkuyu Nükleer AŞ, (2010) Rosatom. https://www.rosatom.ru/tr/akkuyu-n-kleer/, accessed on November 19, 2021.

Canbolat YB, K Chelst, N Garg, (2007) Combining decision tree and MAUT for selecting a country for a global manufacturing facility. Omega, Volume 35, Issue 3, pp. 312–325. https://doi.org/10.1016/j.omega.2005.07.002

Cedolin M, N Göker, E Dogu, and YE Albayrak, (2017) Facility Location Selection Employing Fuzzy DEA and Fuzzy Goal Programming Techniques. Advances in Fuzzy Logic and Technology 2017. Advances in Intelligent Systems and Computing, vol 641, pp. 466–476. https://doi.org/10.1007/978-3-319-66830-7_42

Emeksiz C, and A Yüksel, (2022) A suitable site selection for sustainable bioenergy production facility by using hybrid multi-criteria decision making approach, case study: Turkey. Fuel, 315, 123214. https://doi.org/10.1016/j.fuel.2022.123214

Mahmoudi M, K Shirzad, V Verter, (2022) Decision support models for managing food aid supply chains: A systematic literature review. Socio-Economic Planning Sciences, 101255. https://doi.org/10.1016/j.seps.2022.101255

Singh RK, (2016) Facility Location Selection Using Maut and CCSD Method. International Journal of Science, Engineering and Technology Research (IJSETR), Volume 5, Issue 2.

Chapter 26
Optimization of CO_2 Conversion and Estimation of Synthetic Methane Production Using Deep Neural Networks

Sercan Yalçın and Münür Sacit Herdem

Nomenclature

LSTM	Long short-term memory
CNN	Convolutional neural network
RMSE	Root-mean-square error
MADE	Mean absolute deviation error
H_2	Hydrogen
CO_2	Carbon dioxide
CH_4	Methane gas

26.1 Introduction

Hydrogen as an energy carrier can be used to decrease carbon dioxide emissions for various sectors. However, one of the most important challenges to using hydrogen for different applications is its storage. Therefore, hydrogen can be converted to alternative fuels such as synthetic methane, methanol, etc. via carbon dioxide hydrogenation.

Carbon dioxide hydrogen production processes produce alternative fuels that have been discussed in terms of different aspects which include catalysts, systems, and techno-economic analysis (Gao et al. 2020; Leonzio 2018; Parigi et al. 2019; Sahebdelfar and Ravanchi 2015).

S. Yalçın (✉) · M. S. Herdem
Faculty of Computer Engineering, Adiyaman University, Adiyaman, Türkiye

Faculty of Mechanical Engineering, Adiyaman University, Adiyaman, Türkiye
e-mail: svancin@adiyaman.edu.tr; herdem@adiyaman.edu.tr

M. Z. Sogut et al. (eds.), *Proceedings of the 2022 International Symposium on Energy Management and Sustainability*, Springer Proceedings in Energy,
https://doi.org/10.1007/978-3-031-30171-1_26

In this conference paper, we use a long short-term memory (LSTM)-convolutional neural network (CNN) hybrid deep architecture to estimate the maximum conversion of carbon dioxide during carbon dioxide hydrogenation at specific pressure and temperature. We also calculate the methane production per used hydrogen for the hydrogenation reaction via the LSTM-CNN-based estimation method.

26.2 Proposed Study

In this work, an LSTM-CNN deep architecture is used to optimize CO_2 conversion and estimation of CH_4 gas. The aim here is to optimize CO_2 conversion and predict the estimation of CH_4 gas methanization with the fewest errors.

LSTM is one of the deep iterative neural networks and is a type of iterative neural network that can memorize past values, inspired by the functions of the human brain (Sarker 2021). In this study, an LSTM architecture was used, as in the following references: Liu et al. (2021) and Gundu and Simon (2021). The mathematical infrastructure of the LSTM model was used in this study. Briefly, there are three gates in the internal structure of an LSTM network: input, forget, and output gates (Shao et al. 2019). An LSTM model architecture is shown in Fig. 26.1.

The CNN part of the study was used for the optimization of the parameters to be given as input to the LSTM block, namely, the CO_2 conversion. The CNN structure in our study was designed based on the study in Wang et al. (2021). The mathematical infrastructure of the CNN model was used in this study. CNNs are deep feedforward neural networks. The architecture of a typical CNN includes a convolution layer, a pool layer, and a fully connected layer (Lu et al. 2020).

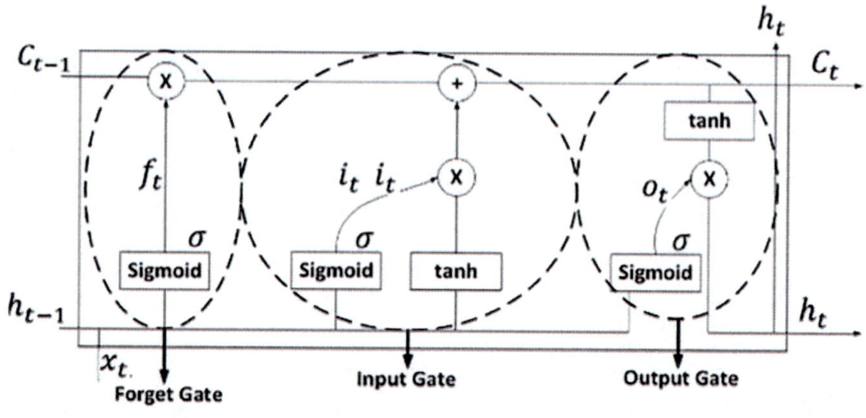

Fig. 26.1 An LSTM architecture model

We propose the LSTM-CNN-based estimation method with six steps:

- Step (1): Input data: The required data are the inputs for the training of LSTM-CNN.
- Step (2): Data standardization: Due to the fact that there is a large gap in the input data, to better train the model, a method with z-score standardization is applied to standardize the input data.
- Step (3): Network initialization: The LSTM-CNN layer's weights and biases are initialized.
- Step (4): Calculation of the CNN layer: The input data are passed through the convolution, pooling, and flattening layers in the CNN layer. This allows us to execute the feature extraction of the input data and obtain the output value.
- Step (5): LSTM layer configuration: For effective climate parameter estimation, it is required to select the appropriate input data. In this study, we design the LSTM layer and the number of neurons to be adaptive to our study.
- Step (6): Output data: At this stage, the estimation of the output data is made, and the error calculation is executed. The proposed model is analyzed based on the root-mean-square error (RMSE) and mean absolute deviation error (MADE) metrics in this work. The RMSE calculation is given as Eq. (26.1). It is expected that the lower the RMSE value, the more successful the data estimation.

$$\mathrm{RMSE} = \sqrt{\frac{\sum_{t=1}^{T} \left(x_t - x_t'\right)^2}{T}} \tag{26.1}$$

where x_t and x_t' are the observed and forecasted values in t times for T samples, respectively. MADE prevents the problem where positive and negative errors in estimation dampen each other. The MADE calculation is presented in Eq. (26.2):

$$\mathrm{MADE} = \frac{1}{T} \sum_{t=1}^{T} \left| x_t - x_t' \right| \tag{26.2}$$

26.3 Experimental Results

To implement the application of this study, the following installations were performed. This study is closely related to open-source software installations. The Python software that comes with Anaconda 3 distribution was utilized. Also, some Python scripts were compiled via PyCharm IDE.

26.3.1 Data

Data for this work was obtained using Aspen Plus. Hydrogen to carbon dioxide ratio, reactor temperature, and pressure were taken as the input parameters, while carbon dioxide conversion and methane production per used hydrogen were the output parameters in the simulator.

26.3.2 Experimental Setup

A function was used to operate the LSTM-CNN TensorFlow Keras model with layers. The model uses a neuron for the output layer because the aim is to estimate a real-valued number. The algorithm utilizes both tanh and ReLU activation function. In addition, the proposed model uses the Adam optimizer. Table 26.1 shows some of the parameters used in the LSTM-CNN-based proposed method.

In this work, the input parameters were modeled, including the H_2/CO_2, reactor temperature (K), and reactor pressure (bar). Ninety percent of the total 7851 data were used for each parameter for training and the remaining 10% for testing.

26.3.3 Optimization of the CO_2 Conversion

In this section, the necessary input parameters for the CO_2 conversion are optimized. Figure 26.2a shows the modeling of H_2/CO_2 parameter. According to the results obtained, the use of H_2/CO_2 between approximately 4.2 and 5.3 can lead to high

Table 26.1 LSTM-CNN parameters used in this study

Parameters	Value
Number of hidden units in LSTM layer	64
Activation function of the LSTM layer	tanh
Convolution layer filters	32
Convolution layer kernel size	1
Activation function of the convolution layer	tanh
Convolution layer padding	Same
Pooling layer pool size	1
Pooling layer padding	Same
Activation function of the pooling layer	ReLU
Time step	12
Batch size	16
Learning rate	0.005
Optimizer	Adam
Loss function	Mean absolute error
Number of epochs	100

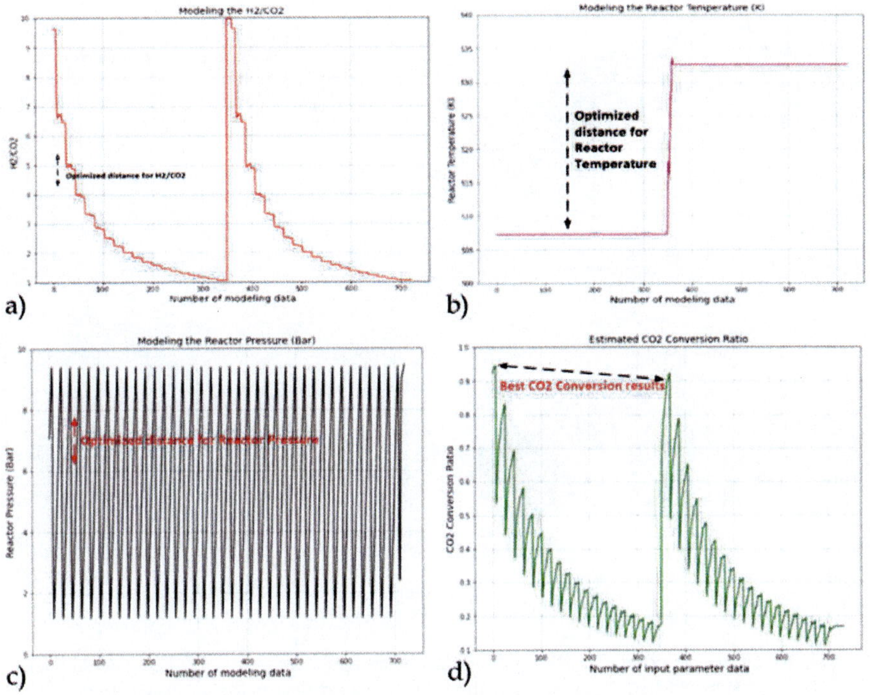

Fig. 26.2 Optimization of the input parameters for CO_2 conversion. (**a**) Modeling the H_2/CO_2. (**b**) Modeling the reactor temperature (K). (**c**) Modeling the reactor pressure (bar). (**d**) Optimization of the CO_2 conversion

carbon dioxide production. The best CO_2 conversion results are obtained as shown in Fig. 26.2d, with reactor temperature of 507 to 532 K from Fig. 26.2b and 6 to 8 bar reactor pressure from Fig. 26.2c.

26.3.4 Estimation of the CH_4/H_2

In this section, an estimation of methane gas production is made. Figure 26.3a shows the determined values of CH_4/H_2 ratios for both training and testing according to the input parameter numbers. The estimated CH_4/H_2 values according to the proposed model are given in Fig. 26.3b. As shown in Fig. 26.3c, the estimated and actual values of methane gas are given together. As can be seen, the actual values and the estimated values are very similar to each other; that is, the prediction error rates are minimized.

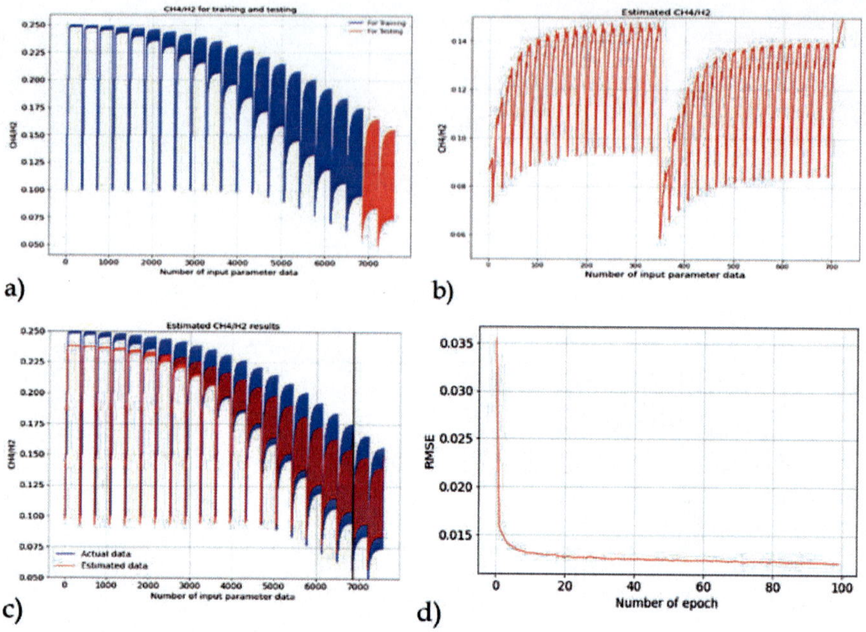

Fig. 26.3 Estimation of the CH_4/H_2. (**a**) CH_4/H_2 for training and testing. (**b**) Estimated/modeled CH_4/H_2. (**c**) Estimated CH_4/H_2 results. (**d**) RMSE values according to the estimated result

It is significant to notice that as the number of epochs increased, the RMSE and MADE metrics decreased, and the estimation performance was enhanced. The reason why the results were successful and produced the fewest errors is that the LSTM-CNN network modeling was designed in accordance with deep network methodologies for the proposed scheme.

RMSE and MADE ratios were calculated to measure the performance of the proposed method. As shown in Fig. 26.3d, the RMSE value is less than 0.015 and the MADE value is about 0.009, and the error rate decreases as the number of epochs increases.

26.4 Conclusion

In this study, we presented an LSTM-CNN deep learning scheme to optimize CO_2 conversion and estimate methane gas. We modeled the H_2/CO_2, reactor temperature (K), and reactor pressure (bar), respectively. As a result, we infer the optimum carbon dioxide ratio to be used. In addition, estimates have been made for the production of CH_4 gas. The proposed algorithm was coded by Python programming. Estimation results were shown by comparing the estimated values obtained with the actual data. The results obtained were quite interesting, and the RMSE and MADE rates were low. In future studies, we will expand this conference paper to develop an estimation tool for the calculation of different outputs for carbon dioxide hydrogenation with variation of different input parameters.

References

Gao R, Zhang C, Lee YJ, Kwak G, Jun KW, Kim SK, Guan G (2020) Sustainable production of methanol using landfill gas via carbon dioxide reforming and hydrogenation: Process development and techno-economic analysis. Journal of Cleaner Production 272: 122552.

Gundu V, and Simon SP (2021) PSO–LSTM for short term forecast of heterogeneous time series electricity price signals, Journal of Ambient Intelligence and Humanized Computing 12:2375–2385.

Leonzio G (2018) State of art and perspectives about the production of methanol, dimethyl ether, and syngas by carbon dioxide hydrogenation. Journal of CO2 Utilization 27: 326–354.

Liu W, Wang Z, Zeng N, Alsaadi FE, and Liu X (2021) A PSO-based deep learning approach to classifying patients from emergency departments, International Journal of Machine Learning and Cybernetics 12: 1939–1948.

Lu W, Li J, Li Y, Sun A, and Wang J (2020) A CNN-LSTM-Based Model to Forecast Stock Prices. Complexity vol. 2020, Article ID 6622927.

Parigi D, Giglio E, Soto A, Santarelli M (2019) Power-to-fuels through carbon dioxide Re-Utilization and high-temperature electrolysis: A technical and economical comparison between synthetic methanol and methane. Journal of Cleaner Production 226: 679–691.

Sarker IH (2021) Deep Learning: A Comprehensive Overview on Techniques, Taxonomy, Applications and Research Directions. SN Computer Science 2:420.

Sahebdelfar S, Ravanchi MT (2015) Carbon dioxide utilization for methane production: A thermodynamic analysis. Journal of Petroleum Science and Engineering 134:14–22.

Shao B, Li M, Zhao Y, and Bian G (2019) Nickel Price Forecast Based on the LSTM Neural Network Problems in Engineering Article ID 1934796.

Wang K, Ma C, Qiaoa Y, Lua X, Hao W, and Dong S (2021) A hybrid deep learning model with 1DCNN-LSTM-Attention networks for short-term traffic flow prediction, Physica A: Statistical Mechanics and its Applications 583: 126293.

Chapter 27
A Comprehensive Review on Sustainability and Energy Management of Seaports

Demir Ali Akyar, Bulut Ozan Ceylan, and Mehmet Serdar Celik

Nomenclature

AES	All-electric ships
AGV	Automated-guided vehicle
ASC	Automated stacking crane
LNG	Liquefied natural gas
RMG	Rail-mounted gantry (RMG) crane
RTG	Rubber-tired gantry (RTG) crane
STS	Ship-to-shore (STS) gantry crane

27.1 Introduction

Maritime transportation and port industry have seen significant business growth rates over the last few decades as the global production and trade soared. So far recently, the need to better understand and monitor the energy-related activities taking place within seaports has become more apparent as a consequence of the growing international trade, increasing energy prices, global risks, public environmental awareness on sustainability, and a greater industrial focus on energy efficiency.

Energy management and sustainability issues were not seen to be particularly urgent until recently. However, considering the current economic and social conjuncture, the changing geography and structure of trade, and greater social awareness and demand for sustainable and green logistics, concepts of energy management,

D. A. Akyar (✉) · B. O. Ceylan · M. S. Celik
Maritime Faculty, Bandirma Onyedi Eylul University, Bandirma, Turkey
e-mail: dakyar@bandirma.edu.tr; bceylan@bandirma.edu.tr; mcelik@bandirma.edu.tr

M. Z. Sogut et al. (eds.), *Proceedings of the 2022 International Symposium on Energy Management and Sustainability*, Springer Proceedings in Energy,
https://doi.org/10.1007/978-3-031-30171-1_27

efficiency, and sustainability have become a priority for the business practitioners and academics (Wilmsmeier and Spengler 2016).

The term so-called sustainability has an interdisciplinary nature. In a general sense, it is "the continuous development that meets the needs of the present, without compromising the ability of future generations to meet their needs." In this sense, the aspects of sustainability cover environmental, economic, and social dimensions (Markley and Davis 2007). In this study, we will touch upon all three aspects in terms of ports' energy management. The performance and competitiveness of seaport terminals depend not only on their level of productivity but also on sustainable activities (Kannika et al. 2019). For this reason, port sustainability stands for "the management of the port business, strategies and activities, resources, information and finance that meet the current and future needs of the port and its stakeholders, while at the same time minimizing the environmental impacts and maximizing social well-being" (Hassini et al. 2012; AAPA 2007).

27.2 Literature Review

Seaports are the meeting points of land and maritime logistics and serve as transfer hubs in the supply chains (Nijdam and Van der Horst 2017). Seaports (from hereafter also referred as ports) can be formed by the union of different terminals handling different types of cargoes, such as liquid bulk, dry bulk, general cargo, and project cargo. A single terminal can theoretically constitute a whole port in case if it controls the entire port area. As a maritime logistics hub, very diverse mechanic and physical activities are performed in seaport terminals, e.g., cargo loading, discharging, stacking, transportation, mode shifting, etc. These physical operations are done through ship to shore cranes, rail-mounted and/or rubber-tyred cranes, and yard equipment, all powered by fossil fuels and electricity. Since such activities are performed perpetually, they account for an enormous consumption of diverse energy resources.

For the purposes of this paper, we define ports as industrial areas where numerous services are given to the vessels and cargo resulting in high demand for energy. As an energy-intensive industry, modern ports consume and generate energy. Some ports have managed to harness the power of the wind and the solar energy. Energy management in ports is one of the sustainable activities that take place in port management.

Energy use in ports, refers to the use of fossil fuels and electricity for port and port-related activities and can be classified into different sub-activities. Most of the energy is consumed by the primary activities in ports, such as cargo handling at terminals, tug services, cargo storage and refrigerating, and port administration buildings and lighting. Other intense activities can be summarized as follows: powering ships (cold ironing), port-related activities such as refineries, gas storage, highway and railway cargo shifting, metal works, and so on (Acciaro et al. 2014).

Due to the growth of global trade, port capacities and infrastructural technology have progressed, making them significant energy consumers. As of 2021, 10.6 billion tons of cargo has been handled in world seaports according to the UNCTAD Review of Maritime Transport report (UNCTAD 2022). Among the cargo handling equipment, we can state ship-to-shore (STS) gantry crane, rubber-tired gantry (RTG) crane, rail-mounted gantry (RMG) crane, container stacking yard trucks, tractors, automated-guided vehicles (AGVs), automated stacking cranes (ASC), straddle carriers, forklifts, reach stackers, pumps for liquid materials, vacuum pneumatic systems, conveyor belts, grabbers, and so on (Alamoush et al. 2021; Bailey and Solomon 2004; IAPH 2008; PIANC 2014; IMO 2018). As for supplementary services, ports use tug and towing boats for nautical services and shuttles for local transfer, storage of dangerous goods and perishables under certain physical conditions, cargo shifting services, etc. All of these operations are carried out by consuming fossil fuels and electricity, generating financial, environmental, and social impacts.

Renewable energy plays an important role in sustainable efforts of modern ports. Some major ports are suitably located in areas where it is possible to efficiently harness the power of the wind (e.g., Rotterdam in the Netherlands, Kitakyushu in Japan), wave (e.g., Port Kembla in Australia and Mutriku in the Basque Country), tide (Dover, UK), and geothermal heat (Hamburg). As a matter of fact, ports have available wide flat surfaces for solar power generation, such as the tops of administration buildings, storage areas, and warehouses (e.g., Tokyo Ohi Terminal and Port of San Diego). Such energy supply by solar power would not be enough to power a huge portion of the energy demand but still can be supportive to the total production. Some ports even have the ability to build wind farms for harnessing the power of the offshore wind energy (Acciaro et al. 2014).

Applications of intelligent, automated, and electric-powered transportation systems are commonly used in container terminals since the handling operations are repetitive tasks. Automated and electric-powered transport systems are utilized in the physical transportation and stacking of containers. These equipments operate through using wireless technology and, improve traceability and ensure the optimum movement of handling in the stacking area and apron thereby achieving energy and time efficiency. On the other hand, benefits of such can be achieved including better utilization of the stacking area, low injury risks, low labor costs, effective data collection/control/transfer, real-time system monitoring, smart electric charging systems, and so on.

Renewable energy sources are becoming more and more widespread especially in shipping business. Vessels powered by LNG are being built, and all-electric ships (AES) are projected in the coming future, including wind and solar power generating vessels (Binti Ahamad et al. 2018).

As the main focus of this chapter is the sustainability and energy management in seaports, Fig. 27.1 summarizes the aspects of different sustainability-related

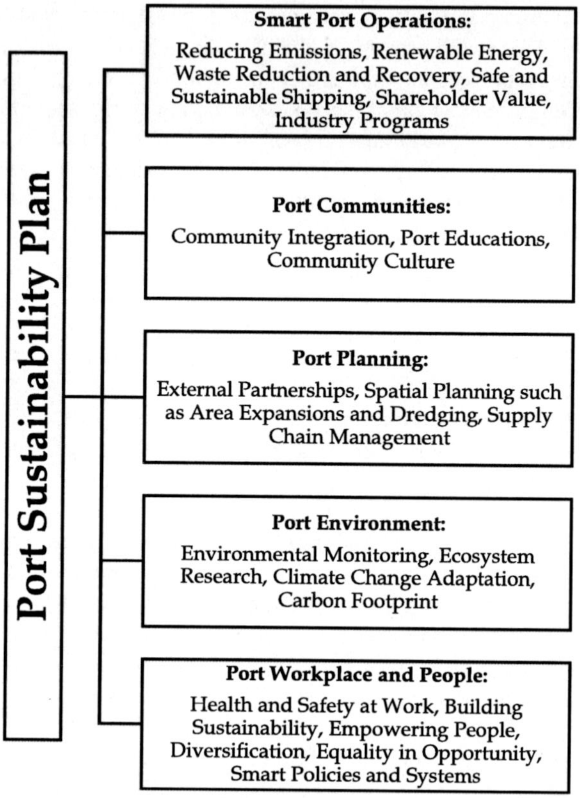

Fig. 27.1 Aspects of port sustainability plans (derived from various port sustainability plan reports)

activities taking place in ports. There are some further measures regarding sustainability including but not limited to:

- Determination of policies and strategies regarding the regulations related to emissions of harmful substances
- Transition from linear economy to self-sustaining circular economy
- Establishment of an energy management system and/or environmental management system to monitor and increase energy efficiency
- A green design of the port-city interaction
- Benchmarking best practices from leading ports in sustainability and environmental issues
- Digitalization and automation of port operations and activities
- Adoption of the term "green growth" to the port strategic plans
- Becoming a part of environmental initiatives such as EcoPorts

Table 27.1 Keywords of the literature review

Number	Search keywords
1	Port (container port or seaport)
2	Energy management (energy policy)
3	Sustainability (and sustainable)
4	Port equipment (and cargo handling)
5	Green port

27.3 Results and Discussion

The purpose of the following literature review is to summarize the existing patterns in research and identify the trends in energy management and sustainability in seaports.

Five keywords have been determined and used to find out the scientific studies related to the relationship between seaports (mainly general cargo), energy management and sustainability (see Table 27.1). Articles published in the Scopus and Web of Science databases are selected within the scope of the research. Searching of the related articles is carried out with the keyword combinations of 1 & 2, 1 & 3, 1 & 4, and 5. Time period is selected as all times until January 2022. The results are listed in Table 27.2, which are sorted in descending order in terms of citations. Three hundred fifty related studies have been found as the searching outcome.

When the studies are examined, it is revealed that the ports have begun to put their interest toward alternative and green energy sources. Port competition, high energy prices, sustainability concerns of ports, social pressures by stakeholders and sectoral legislations require ports to lean toward green energy resources. Wind, solar power, and wave energy converters are used in some cases like the Port of Valencia toward achieving these goals.

Port-city integration and collaborations, smart port, and smart city concepts are some of the fields that are open for progress since ports become highly integrated with the cities surrounding them. During the processes of urban planning, infrastructure planning, transportation network designs, and city-port expansion plans, authorities accordingly pay attention to the energy distribution plans. Government and private sector incentives as well as development projects take place in modern cities toward improving port and city integration, and ports become more active on social responsibility projects.

Monitoring of air, water, sediment quality, and the noise level around the port area, climate change effects and carbon footprints of ports, digitalization and Internet of Things (IoT) as well as equipment automation, LNG storage and fueling ships, and solid/liquid waste management systems are some of the most popular sustainability concerns in the port industry (see Table 27.3).

In the recent past, ports have seen the impacts of the COVID-19 pandemic: trade volumes have fluctuated; there have been disruptions in cargo operations and supply chains due to infections and worker shortages. There have been serious implementations of occupational health and safety measures for seafarers,

Table 27.2 Studies related with sustainable energy management in seaports

Author(s)	Title	Citations
Peris-Mora et al. (2005)	Development of a system of indicators for sustainable port management	131
Acciaro et al. (2014)	Energy management in seaports: A new role for port authorities	119
Frantzeskaki et al. (2014)	The role of partnerships in 'realising' urban sustainability in Rotterdam's City Ports Area, The Netherlands	109
Iris and Lam (2019)	A review of energy efficiency in ports: Operational strategies, technologies and energy management systems	93
Chang and Wang (2012)	Evaluating the effects of green port policy: Case study of Kaohsiung harbor in Taiwan	87
Peris-Mora et al. (2005)	Toward a Smart Sustainable Development of Port Cities/Areas: The Role of the "Historic Urban Landscape" Approach	79
Denktas-Sakar and Karatas-Cetin (2012)	Port Sustainability and Stakeholder Management in Supply Chains: A Framework on Resource Dependence Theory	75
Asgari et al. (2015)	Sustainability ranking of the UK major ports: Methodology and case study	73
Fusco Girard (2013)	Toward a Smart Sustainable Development of Port Cities/Areas: The Role of the "Historic Urban Landscape" Approach	69
Lirn et al. (2013)	Green performance criteria for sustainable ports in Asia	64
Chiu et al. (2014)	Evaluation of Green Port Factors and Performance: A Fuzzy AHP Analysis	58
Yang and Chang (2013)	Impacts of electric rubber-tired gantries on green port performance	56
Parise et al. (2015)	Wise Port and Business Energy Management: Port Facilities, Electrical Power Distribution	44
Martínez-Moya et al. (2019)	Energy efficiency and CO_2 emissions of port container terminal equipment: Evidence from the Port of Valencia	31
Yigit and Acarkan (2018)	A new electrical energy management approach for ships using mixed energy sources to ensure sustainable port cities	27
Cascajo et al. (2019)	Integration of Marine Wave Energy Converters into Seaports: A Case Study in the Port of Valencia	23

stevedores, and shore personnel. For suitable works, home office was established, and severe operational procedures were implemented for physical operations in port facilities (ILO 2021).

Table 27.3 Sustainability measures of seaports

Environmental Management System (EMS) and Environmental Standards	EMS Certification (ISO 14001)
Air, Water, Sediment Quality Monitoring	Corporate Social Responsibility
Sustainability Plans and Reporting	Renewable Energy (RE)
Energy Efficiency and Conservation (EEC)	Shore Power (Cold ironing etc.)
Emissions Monitoring and Reduction Policies	Green Incentives
LNG Storage Facility	Noise Level Monitoring
Digitalization, Internet of Things, Automation in Ports	Green Infrastructure Development
Community Engagement	Port Environmental Review System (PERS)
Environmental Monitoring Report	Research and Development (R & D)
Climate Change Adaptation	Labour and Education
Waste Management Systems	Wildlife Protection

Source: Adopted from Hossain et al. (2021)

27.4 Conclusion

Energy resources and energy management are critical issues in terms of scope, and port authorities and managers should consider them in a comprehensive approach. To conclude, an integrated energy management and operational planning is suggested for ports as large-scale energy end users. All physical work carried out in ports depend on significant amounts of energy. It is extremely important to optimize the energy usage in ports and diversify the sources due to increasing prices and the risk of interruptions. Dependence on a single source of energy provider would diminish the bargaining power of port authority and bring a risk of operational interruptions. When the related literature is examined, it is revealed that the ports have begun to put their interest toward alternative and green energy sources as well as automated and electrically operated port equipment in recent years.

Many alternative scenarios and technological developments have been studied, and yet the biggest but the most critical role here falls to the port managers and port authorities in terms of diversification of energy resources and internalizing green energy resources. Assuring an uninterrupted energy supply and following the globally accepted sustainability measures are of critical importance for port environmental management systems and green transition in ports.

Acknowledgment The authors would like to thank Bandirma Onyedi Eylül University for supporting this research.

References

AAPA (American Association of Port Authorities). (2007). D-11: Embracing the concept of sustainability as a standard practice for ports and the association.

Acciaro, M., Ghiara, H., & Cusano, M. I. (2014). Energy management in seaports: A new role for port authorities. Energy Policy, 71, 4–12.

Alamoush, A. S., Ballini, F., & Ölçer, A. I. (2021). Revisiting port sustainability as a foundation for the implementation of the United Nations Sustainable Development Goals (UN SDGs). Journal of Shipping and Trade, 6(1), 1–40.

Asgari, N., Hassani, A., Jones, D., & Nguye, H. H. (2015). Sustainability ranking of the UK major ports: Methodology and case study. Transportation Research Part E: Logistics and Transportation Review, 78, 19–39.

Bailey D, Solomon G. (2004). Pollution prevention at ports: clearing the air. Environ Impact Assess Rev 24(7–8):749–774.

Cascajo, R., García, E., Quiles, E., Correcher, A., & Morant, F. (2019). Integration of marine wave energy converters into seaports: A case study in the port of Valencia. Energies, 12(5), 787.

Chang, C. C., & Wang, C. M. (2012). Evaluating the effects of green port policy: Case study of Kaohsiung harbor in Taiwan. Transportation Research Part D: Transport and Environment, 17(3), 185–189.

Chiu, R. H., Lin, L. H., & Ting, S. C. (2014). Evaluation of green port factors and performance: a fuzzy AHP analysis. Mathematical problems in engineering, 2014.

Denktas-Sakar, G., & Karatas-Cetin, C. (2012). Port sustainability and stakeholder management in supply chains: A framework on resource dependence theory. The Asian Journal of Shipping and Logistics, 28(3), 301–319.

Frantzeskaki, N., Wittmayer, J., & Loorbach, D. (2014). The role of partnerships in 'realising' urban sustainability in Rotterdam's City Ports Area, The Netherlands. Journal of Cleaner Production, 65, 406–417.

Fusco Girard, L. (2013). Toward a smart sustainable development of port cities/areas: The role of the "Historic Urban Landscape" approach. Sustainability, 5(10), 4329–4348.

Hassini, E., Surti, C., & Searcy, C. (2012). A literature review and a case study of sustainable supply chains with a focus on metrics. International journal of production economics, 140(1), 69–82.

Hossain, T., Adams, M., & Walker, T. R. (2021). Role of sustainability in global seaports. Ocean & Coastal Management, 202, 105435.

IAPH. (2008). IAPH tool box for greenhouse gasses. International Association of Ports & Harbors.

ILO. (2021). International Labour Organization. COVID-19 and the port sector report. https://www.ilo.org/wcmsp5/groups/public/%2D%2D-ed_dialogue/%2D%2D-sector/documents/briefingnote/wcms_810868.pdf. (Accessed 10 January 2021).

IMO. (2018). Port Emission Toolkit Guide No.1: Assessment of Port Emission. GloMeep project coordination unit and the International Maritime Organization, London.

Iris, Ç., & Lam, J. S. L. (2019). A review of energy efficiency in ports: Operational strategies, technologies and energy management systems. Renewable and Sustainable Energy Reviews, 112, 170–182.

Kannika, N., Tan, K. H., & Pawar, K. (2019). Enhancing the Competitiveness of Container Seaports Through Sustainability: A Case Study of Thailand. Procedia Manufacturing, 39, 1587–1596.

Lirn, T. C., Wu, Y. C. J., & Chen, Y. J. (2013). Green performance criteria for sustainable ports in Asia. International Journal of Physical Distribution & Logistics Management.

Markley, M. J., & Davis, L. (2007). Exploring future competitive advantage through sustainable supply chains. International Journal of Physical Distribution & Logistics Management.

Martínez-Moya, J., Vazquez-Paja, B., & Maldonado, J. A. G. (2019). Energy efficiency and CO2 emissions of port container terminal equipment: Evidence from the Port of Valencia. Energy Policy, 131, 312–319.

N. B. Binti Ahamad, J. M. Guerrero, C. -L. Su, J. C. Vasquez and X. Zhaoxia, "Microgrids Technologies in Future Seaports," 2018 IEEE International Conference on Environment and Electrical Engineering and 2018 IEEE Industrial and Commercial Power Systems Europe (EEEIC/I & CPS Europe), 2018, pp. 1–6, https://doi.org/10.1109/EEEIC.2018.8494428.

Nijdam, M., & Van der Horst, M. (2017). Port definition, concepts and the role of ports in supply chains. Ports and Networks: Strategies, Operations, and Perspectives, 9–25.

Parise, G., Parise, L., Martirano, L., Chavdarian, P. B., Su, C. L., & Ferrante, A. (2015). Wise port and business energy management: Port facilities, electrical power distribution. IEEE Transactions on Industry Applications, 52(1), 18-24.

Peris-Mora, E., Orejas, J. D., Subirats, A., Ibáñez, S., & Alvarez, P. (2005). Development of a system of indicators for sustainable port management. Marine pollution bulletin, 50(12), 1649–1660.

PIANC. (2014). Sustainable ports: a guide for port authorities. The World Association for Waterborne Transport Infrastructure. PIANC Maritime Navigation Commission, Brussels.

UNCTAD. (2022). Review of Maritime Transport 2021.

Wilmsmeier, G., & Spengler, T. (2016). Energy consumption and container terminal efficiency.

Yang, Y. C., & Chang, W. M. (2013). Impacts of electric rubber-tired gantries on green port performance. Research in Transportation Business & Management, 8, 67–76.

Yigit, K., & Acarkan, B. (2018). A new electrical energy management approach for ships using mixed energy sources to ensure sustainable port cities. Sustainable cities and society, 40, 126–135.

Chapter 28
Comparative Investigation of the Spray Properties of Ethyl and Methyl Ester-Based Biodiesels

Anılcan Ulu and Güray Yildiz

Nomenclature

CVSC	Constant volume spray chamber
EE	Ethyl ester
ME	Methyl ester
SCA	Spray cone angle
SPL	Spray penetration length

28.1 Introduction

Biodiesels are primarily produced via transesterification, in which methanol is widely utilized as a reactant (Maximo et al. 2018). While methanol is mainly derived from nonrenewable fossil resources such as natural gas and coal (Gui et al. 2009), it is also possible to utilize bio-based ethanol in transesterification. Bioethanol can be derived from various types of lignocellulosic biomass feedstocks (e.g., maize, sugarcane, etc.). The use of bioethanol contributes to a fully renewable biodiesel production process (Verma and Sharma 2016). Biodiesels produced with the use of methanol and ethanol are, respectively, named methyl ester and ethyl ester. Beyond the advantages mentioned above, ethyl esters have further advantages over methyl esters, which makes them preferable for further research. Ethyl esters can have lower iodine values, better oxidation stabilities, and lubricity characteristics than methyl ester counterparts (Reyero et al. 2015). Also, cold weather properties of ethyl esters

A. Ulu (✉) · G. Yildiz
Faculty of Engineering, İzmir Institute of Technology, İzmir, Turkey
e-mail: anilcanulu@iyte.edu.tr; gurayyildiz@iyte.edu.tr

M. Z. Sogut et al. (eds.), *Proceedings of the 2022 International Symposium on Energy Management and Sustainability*, Springer Proceedings in Energy,
https://doi.org/10.1007/978-3-031-30171-1_28

can be better than methyl esters, owing to having lower cloud, pour, and cold filter plugging points (Zanuttini et al. 2014). Besides, some studies showed that ethyl esters can perform similar performance and emission characteristics compared to methyl esters based on the same feedstock (Alptekin et al. 2019; Eliçin et al. 2019).

Biodiesels differ from fossil diesel in terms of their physicochemical properties, such as density, viscosity, calorific value, etc. Considering the fact that such liquid fuels can be mainly used in internal combustion engines, it is important to investigate their spray characteristics to predict the effects of such physicochemical properties on atomization (Tinprabath et al. 2013). Spray characteristics can be evaluated by using different experimental setups such as an optical research engine, constant pressure flow rig, and constant volume spray chamber (CVSC) (Li et al. 2021). Among the available techniques, a CVSC offers the possibility to test a wide range of gas pressures and temperatures (Baert et al. 2009).

Considering the importance of spray investigation and the advantages of ethyl esters, this study aimed to compare the spray characteristics of ethyl ester-type biodiesel with those of its methyl ester counterpart under an injection pressure of 80 MPa and ambient pressure of 1.5 MPa.

28.2 Materials and Methods

Both and ethyl esters were produced from sunflower oil (Kucukbay Oil and Detergent Inc. Trademark: Orkide) via transesterification. The biodiesel production process was the same for both types of biodiesels except for alcohol/lipid ratio and catalyst/lipid ratio. Six moles of methanol and 24 moles of ethanol against 1-mole lipid were, respectively, used to produce the methyl ester and the ethyl ester. KOH was the catalyst, and catalyst/lipid ratios were, respectively, 1 wt.% and 0.1 wt% for methyl and ethyl esters. The transesterification reaction took place at 50 °C for 4 h at a stirring speed of 1100 rpm. The methyl ester and the ethyl ester will be referred to as ME and EE, respectively, in the following. The physical properties of the fuels tested are presented in Table 28.1.

Experiments were conducted in a spray test rig that consisted of a CVSC, fuel injection system, optical setup, gas filling system, and control equipment. The CVSC was filled with nitrogen to obtain a nonreactive environment. Furthermore, the fuel injection system consisted of a fuel tank, fuel pump, fuel filter, common rail, and

Table 28.1 Physical properties of the fuels

Test fuel	EE	ME
Viscosity (mm^2/s) @ 40 °C	5.85	5.17
Density (kg/m^3) @ 15 °C	865.7	880.8
Contact angle (°) with glass @ 25 °C	19.8	20.3

Siemens brand diesel injector. An optical system based on the shadowgraph technique was utilized to detect the spray characteristics of the fuels via a high-speed camera operated at 20,000 fps. In addition, a control system, which contained National Instruments data acquisition cards, and a control program developed in LabVIEW were employed to control the test equipment, to receive the data, and to give the necessary commands.

Tests were performed under an injection pressure of 80 MPa and ambient pressure of 1.5 MPa. The injection duration was 1 ms. During the experiments, the temperature of the chamber was kept at 25 °C. In addition, experiments were repeated three times to ensure the consistency of the work. The average values of these three experiments were calculated, and then, it was found that the deviation of the experiments from the average was lower than 5%.

In the present study, spray characteristics of the fuels were investigated in terms of spray penetration length (SPL) and spray cone angle (SCA). These characteristics were measured by processing the spray images with an image processing algorithm. The image recording process was run synchronously with the injection process; however, the image recording started slightly before the injection to ensure the recording of raw background images. Background images were then subtracted from the spray images; thus, only images containing the fuel spray were obtained. In the next step, these images were exposed to thresholding to obtain binary images for the purpose of determining the edges of the spray. After detecting the spray boundary, spray penetration lengths were measured by using two furthest points on the spray axis. Then, spray cone angle values were determined by using the two outermost points on the spray thickness at 50% of spray penetration length (Xie et al. 2015).

28.3 Results and Discussion

Figure 28.1 shows the spray penetration length and spray cone angle values of the ethyl ester in comparison to those of the methyl ester under the given experimental conditions. Similar spray characteristics between EE and ME were observed in terms of both SPL and SCA. For example, spray penetration length values of ME and EE were, respectively, measured as 53.2 mm and 53.9 mm at 0.6 ms of the injection process. Besides, spray cone angle values for the methyl ester and the ethyl ester were detected as 17.5° and 17.1°, respectively, at 0.6 ms of injection. These values respectively correspond to approximately 1.3% and 2.3% distinctions in terms of SPL and SCA. Considering the repeatability value (±%5), the mentioned differences were not significant. The similarity between the spray properties of both kinds of fuels may be explained by the physical properties of the fuels. Although the physical properties of the fuels were not very different, the ethyl ester had around 13.2% higher viscosity, 1.7% lower density, and 2.5% lesser contact angle than the methyl ester. Generally, higher viscosity and surface tension effects lead to increased spray

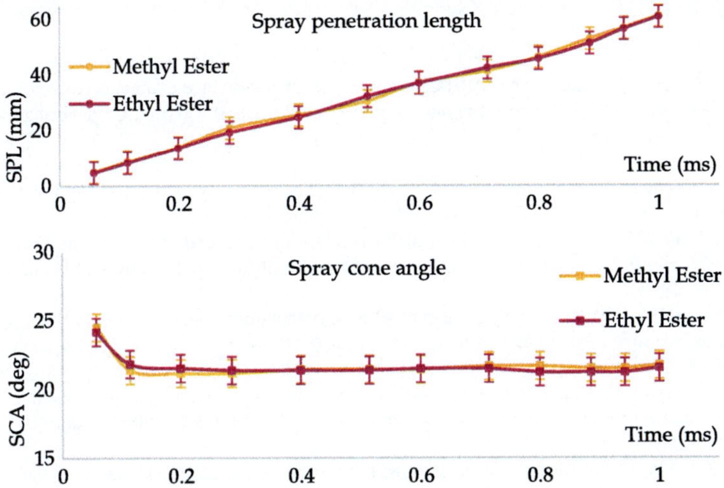

Fig. 28.1 Spray penetration length and spray cone angle curves of the ethyl ester and the methyl ester

penetration length and decreased spray cone angle, and higher density results in raised spray penetration length; however, in this situation, the physical properties may compensate for each other and lead to similar spray properties in terms of both SPL and SCA (Ulu et al. 2022).

28.4 Conclusion

This study aimed at investigating the spray characteristics of two types of biodiesel fuels, i.e., methyl ester and ethyl ester. Experiments were conducted under an injection pressure of 80 MPa, ambient pressure of 1.5 MPa, and chamber temperature of 25 °C. The experimental work showed that the variation between the spray properties, which were spray penetration length and spray cone angle of both types of biodiesels, did not exceed 3%. The differences remained within the repeatability value (±5%). Thence, it could be concluded that the spray characteristics of the ethyl ester were very similar to those of the methyl ester, and thus, ethanol can be a renewable alternative in biodiesel production instead of methanol.

References

Alptekin E, Sanli H, Canakci M (2019) Combustion and performance evaluation of a common rail DI diesel engine fueled with ethyl and methyl esters. Applied Thermal Engineering 149:180–191. https://doi.org/10.1016/j.applthermaleng.2018.12.042

Baert, RSG, Frijters PJM, Somers B, Luijten CCM, de Boer W (2009) Design and Operation of a High Pressure, High Temperature Cell for HD Diesel Spray Diagnostics: Guidelines and Results. SAE Tech. Pap. 2009-01-0649. https://doi.org/10.4271/2009-01-0649

Eliçin A, Öztürk F, Baran MF, Esgici R (2019) The Use of Rapeseed Oil Methyl and Ethyl Esters and of Rapeseed Oil-Diesel Mixtures as Fuels for Diesel Engine. Fresenius Environ. Bull. 28: 7915–7923.

Gui MM, Lee KT, Bhatia S (2009) Supercritical ethanol technology for the production of biodiesel: Process optimization studies. J. Supercritic. Fluids 49:286–292. https://doi.org/10.1016/j.supflu.2008.12.014

Li H, Verschaeren R, Beji T, Verhelst S (2021) Investigation of evaporating sprays in a medium speed marine engine. Exp. Therm. Fluid Sci. 121: 110278. https://doi.org/10.1016/j.expthermflusci.2020.110278

Maximo GJ, Magalhães AMS, Gonçalves MM, Esperança ES et al. (2018) Improving the cold flow behavior of methyl biodiesel by blending it with ethyl esters. Fuel 226:87–92. https://doi.org/10.1016/j.fuel.2018.03.154

Reyero I, Arzamendi G, Zabala S, Gandía M (2015) Kinetics of the NaOH-560 catalyzed transesterification of sunflower oil with ethanol to produce biodiesel. Fuel Proces. Technol. 129: 147–155. https://doi.org/10.1016/j.fuproc.2014.09.008

Tinprabath P, Hespel C, Chanchaona S, Foucher F (2013) Influence of Biodiesel and Diesel Fuel Blends on the Injection Rate and Spray Injection in Non-Vaporizing Conditions. SAE Tech. Pap. 2013-24-0032. https://doi.org/10.4271/2013-24-0032

Ulu A, Yildiz G, Özkol Ü, Rodriguez AD (2022) Experimental investigation of spray characteristics of ethyl esters in a constant volume chamber, Biomass Convers. Biorefin.: 1–18. https://doi.org/10.1007/s13399-022-02476-3

Verma P, Sharma MP (2016) Comparative analysis of effect of methanol and ethanol on Karanja biodiesel production and its optimisation. Fuel 180:164–174. https://doi.org/10.1016/j.fuel.2016.04.035

Xie H, Song L, Xie Y, Pi D, Shao C, Lin Q (2015) An Experimental Study on the Macroscopic Spray Characteristics of Biodiesel and Diesel in a Constant Volume Chamber. Energies 8: 5952–5972. https://doi.org/10.1016/j.fuel.2017.08.070

Zanuttini MS, Pisarello ML, Querini CA (2014) Butia Yatay coconut oil: Process development for biodiesel production and kinetics of esterification with ethanol. Energy Convers. Manag. 85: 407–416. https://doi.org/10.1016/j.enconman.2014.05.080

Chapter 29
A New Solar-Assisted Power, Cooling, and Freshwater Production System Considering the Energy Storage Option

Leyla Khani, Gülden Gökçen Akkurt, and Mousa Mohammadpourfard

Nomenclature

A	Surface area, m^2
c_p	Specific heat capacity, kJ/kgK
d	Diameter, m
$\Delta T_{supheat}$	Superheating degree in the superheater, K
$\dot{E}$	Exergy rate, kW
$\overline{e}^{ch}$	Standard chemical exergy, kJ/kmol
f	Friction factor
gi	Glass cover inner surface
go	Glass cover outer surface
h	Specific enthalpy, kJ/kg
HTF	Heat transfer fluid
k	Thermal conductivity, W/mK
L	Length, m
$\dot{m}$	Mass flow rate, kg/s
P	Pressure, bar

L. Khani
Faculty of Chemical and Petroleum Engineering, University of Tabriz, Tabriz, Iran
e-mail: leyla_khani@tabrizu.ac.ir

G. G. Akkurt (✉)
Department of Energy Systems Engineering, Izmir Institute of Technology, Izmir, Turkey
e-mail: guldengokcen@iyte.edu.tr

M. Mohammadpourfard
Faculty of Chemical and Petroleum Engineering, University of Tabriz, Tabriz, Iran

Department of Energy Systems Engineering, Izmir Institute of Technology, Izmir, Turkey
e-mail: mohammadpour@tabrizu.ac.ir

M. Z. Sogut et al. (eds.), *Proceedings of the 2022 International Symposium on Energy Management and Sustainability*, Springer Proceedings in Energy,
https://doi.org/10.1007/978-3-031-30171-1_29

pi	Receiver pipe inner surface
po	Receiver pipe outer surface
Pr	Prandtl number
$\dot{Q}$	Heat transfer rate, kW
r	Radius, m
R	Universal gas constant, kJ/kgK
Re	Reynolds number
r_p	Pressure ratio
s	Specific entropy, kJ/kgK
T	Temperature, K
t	Time, s
U	Overall heat transfer coefficient, W/m^{2}K
V	Volume, m^3
$\dot{W}$	Power, kW
x	Ammonia concentration
y	Mole fraction

Greek Letters

η	Energy efficiency
σ	Stefan-Boltzmann constant, W/m^{2}K
ε	Exergy efficiency
ρ	Density, kg/m^3

Subscripts

0	Environmental condition
1, ...	State points
a	Air
ch	Chemical
col	Collector
cond	Conduction
conv	Convection
D	Destruction
e	Outlet
f	Fluid
i	Inlet
k	Kth component
ph	Physical
rad	Radiation
S	Sky
ST	Storage tank

29.1 Introduction

The rapid growth of energy utilization rate and the main role of fossil fuels in total energy supply have caused a decrease in fossil fuel resources, an increase in their price, environmental pollution, greenhouse gas emissions, and health issues. Hence, renewable energy sources like solar, wind, and geothermal energy are getting more and more attention because they do not have most of the problems of fossil fuels. Furthermore, the conventional thermal systems are not as efficient as desired. Hence, researchers are looking for new ways to afford the global energy demand effectively (Mossi Idrissa and Goni Boulama 2019). In this way, it seems that combining multigeneration systems with renewable solar energy will be a useful solution (Mirzaee et al. 2019). These systems can achieve high efficiency with lower waste. Moreover, human life depends on water, but the reservoir of drinking water is decreasing in some areas due to drought, pollution of freshwater sources, and a high rate of water consumption. So, scientists have proposed various seawater desalination systems for freshwater production (Brandt et al. 2017).

In this paper, a new solar-based trigeneration system is designed to generate power, freshwater, and cooling. A molten salt heat storage option is added to the solar energy collector to confirm its constant operation during day and night regardless of fluctuations in the solar radiation intensity. Also, the Goswami absorption refrigeration cycle is used to produce electrical power and cooling with receiving solar energy in the superheater and boiler. A multistage flash desalination system is integrated to generate drinking water from seawater. It seems that this proposed system can be a valuable option for electricity, cooling, and potable water production in hot and arid areas that are remote from main power and water networks, but have access to seawater and receive a significant amount of solar radiation. The simplicity, quietness, and flexibility of this system are its other advantages. Necessary equations for the energy and exergy simulation of the system are solved in EES software. The system performance is evaluated under different operating conditions.

29.2 System Description

The proposed system for electrical power, cooling, and potable water generation is shown in Fig. 29.1. According to this figure, solar radiation is the energy source of the system. The first part of the system is the solar energy collector plates with hot and cold molten salt storage tanks to provide the necessary energy of the subsystems at the desired temperature. The second part is the Goswami absorption refrigeration cycle with the ammonia-water solution to generate electricity and refrigeration from solar energy. Finally, the third part is a multistage flash seawater desalination subsystem that receives solar heat and produces freshwater. The brine is emitted to the environment at a justified condition.

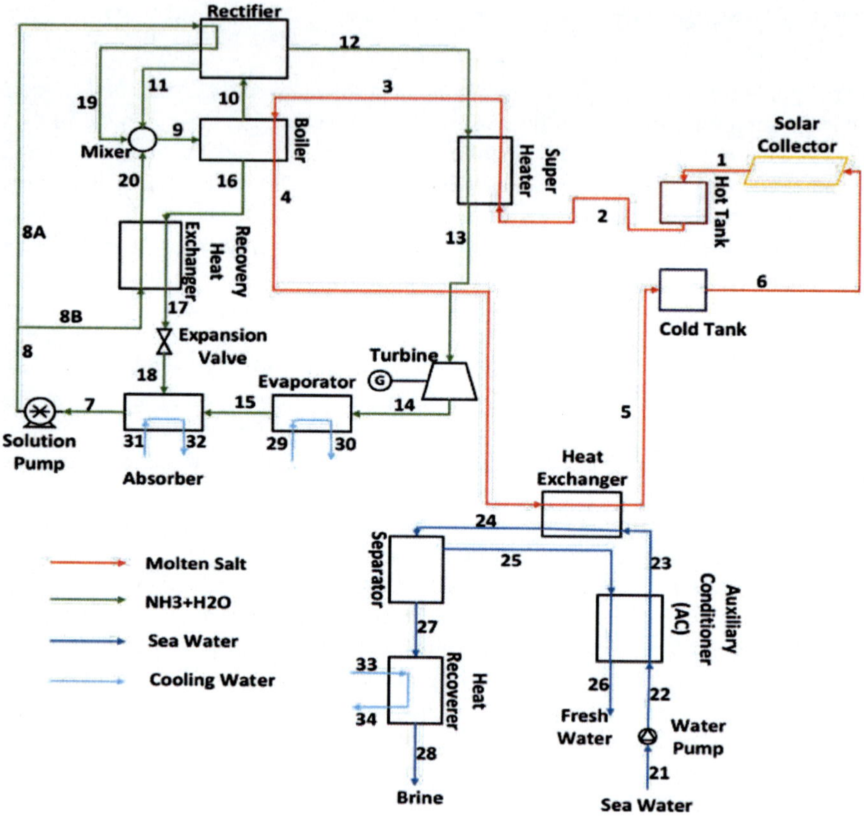

Fig. 29.1 The proposed solar-based power cooling and freshwater trigeneration system

29.3 System Modelling

If changes in potential and kinetic energies are ignored, mass and energy conservation laws along with exergy balance equations are written for each component at a steady state to simulate the thermodynamic performance of the system:

$$\sum \dot{m}_{\mathrm{i}} = \sum \dot{m}_{\mathrm{e}} \tag{29.1}$$

$$\dot{Q}_{\mathrm{k}} - \dot{W}_{\mathrm{k}} = \sum \dot{m}_{\mathrm{e}} h_{\mathrm{e}} - \sum \dot{m}_{\mathrm{i}} h_{\mathrm{i}} \tag{29.2}$$

$$\dot{E}_{\mathrm{D}} = \sum \left(1 - \frac{T_0}{T_{\mathrm{j}}}\right) \dot{Q}_{\mathrm{j}} - \dot{W}_{\mathrm{cv}} + \sum \dot{E}_{\mathrm{i}} - \sum \dot{E}_{\mathrm{e}} \tag{29.3}$$

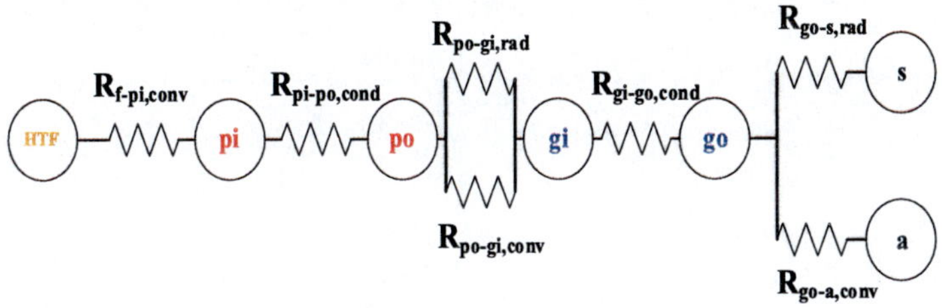

Fig. 29.2 Solar thermal resistance model. (Kalogirou 2012)

In the absence of electrical, magnetic, nuclear, and surface stress effects, and excluding kinetic and potential exergies, the exergy of any stream is the sum of its physical and chemical exergies:

$$\dot{E} = \dot{E}_{\text{ph}} + \dot{E}_{\text{ch}} \tag{29.4}$$

$$\dot{E}_{\text{ph}} = \sum \dot{m}_{\text{i}} \left[\left(h_{\text{i}} - h_{\text{i}}^{0} \right) - T_0 \left(s_{\text{i}} - s_{\text{i}}^{0} \right) \right] \tag{29.5}$$

$$\dot{E}_{\text{ch}} = \sum \dot{m}_{\text{i}} e_{\text{i}}^{\text{ch}} + RT_0 \sum \dot{m}_{\text{i}} \ln y_{\text{i}} \tag{29.6}$$

It should be noted that unlike mass and energy, exergy is preserved only in reversible processes and is destroyed in real processes by the destructions that happen in thermodynamic systems.

In the solar energy collecting system, the solar radiation reaches the parabolic reflector and gets reflected to the receiver tube. Then, it is converted into thermal energy and warms up the molten salt. Thermal resistance method is used to model the thermal operation of the solar energy collecting system, as depicted in Fig. 29.2 (Kalogirou 2012), and its necessary equations are summarized in Table 29.1 (Behar et al. 2015).

Hot and cold molten salt storage tanks are homogeneously mixed (Zolfagharnasab et al. 2020). Therefore, a uniform time-variant temperature profile occurs for each tank and is determined using the following energy balance equations (Ashouri et al. 2015):

$$\left[\left(\rho V c_{\text{p}} \right)_{1} + \left(\rho V c_{\text{p}} \right)_{\text{ST}} \right] \frac{dT}{dt} = \dot{Q}_{\text{in}} - \dot{Q}_{\text{loss}} \tag{29.7}$$

$$\dot{Q}_{\text{loss}} = (UA)_{\text{ST}} (T_{\text{ST}} - T_0) \tag{29.8}$$

$$(UA)_{\text{ST}} = \left(\frac{k_{\text{i}}}{r_{\text{ST1}}} \right) \frac{1}{\ln \left(\frac{r_{\text{ST2}}}{r_{\text{ST1}}} \right)} (2\pi r_{\text{ST1}} L_{\text{ST}}) + \left(\frac{k_{\text{i}}}{\delta_{\text{i}}} \left(2\pi r_{\text{ST1}}^2 \right) \right) \tag{29.9}$$

$$T_{\text{ST,new}} = T_{\text{ST}} + \frac{\Delta t}{\left(\rho V c_{\text{p}} \right)_{1} + \left(\rho V c_{\text{p}} \right)_{\text{ST}}} \left[\dot{Q}_{\text{in}} - \dot{Q}_{\text{loss}} \right] \tag{29.10}$$

Table 29.1 Thermal resistance terms for the solar collector

Term	Equation
$\dot{q}_{\mathrm{HTF-pi,conv}}$	$\frac{\frac{\left[1.82\log\left(\mathrm{Re}_{d_{\mathrm{pi}}}\right)-1.64\right]^{-2}}{8\mathrm{Pr}_f\left(\mathrm{Re}_{d_{\mathrm{pi}}}-1000\right)}}{1+12.7\sqrt{\frac{f_{\mathrm{pi}}}{8}}\left(\mathrm{Pr}_f^{2/3}-1\right)}\left(\frac{\mathrm{Pr}_f}{\mathrm{Pr}_{\mathrm{pi}}}\right)^{0.11}\left(\frac{k_f}{d_{\mathrm{pi}}}\right)\pi d_{\mathrm{pi}}\left(T_{\mathrm{pi}}-T_f\right)$
$\dot{q}_{\mathrm{pi-po,cond}}$	$\frac{2\pi k_{\mathrm{pipe}}\left(T_{\mathrm{pi}}-T_{\mathrm{po}}\right)}{\ln\left(\frac{d_{\mathrm{po}}}{d_{\mathrm{pi}}}\right)}$
$\dot{q}_{\mathrm{po-gi,rad}}$	$\frac{\sigma\pi d_{\mathrm{po}}\left(T_{\mathrm{po}}^4-T_{\mathrm{gi}}^4\right)}{\frac{1}{\varepsilon_{\mathrm{po}}}+\frac{d_{\mathrm{po}}\left(1-\varepsilon_{\mathrm{gi}}\right)}{\varepsilon_{\mathrm{gi}}d_{\mathrm{gi}}}}$
$\dot{q}_{\mathrm{po-gi,conv}}$	$\pi d_{\mathrm{po}}\frac{k_{\mathrm{std}}}{\frac{d_{\mathrm{po}}}{2\ln\left(\frac{d_{\mathrm{gi}}}{d_{\mathrm{po}}}\right)}+b\lambda\left(\frac{d_{\mathrm{po}}}{d_{\mathrm{gi}}}+1\right)}\left(T_{\mathrm{po}}-T_{\mathrm{gi}}\right)$
$\dot{q}_{\mathrm{gi-go,cond}}$	$\frac{2\pi k_g\left(T_{\mathrm{gi}}-T_{\mathrm{go}}\right)}{\ln\left(\frac{d_{\mathrm{go}}}{d_{\mathrm{gi}}}\right)}$
$\dot{q}_{\mathrm{go-}s\mathrm{,rad}}$	$\sigma\varepsilon_{\mathrm{go}}\pi d_{\mathrm{go}}\left(T_{\mathrm{go}}^4-T_s^4\right)$
$\dot{q}_{\mathrm{go-}a\mathrm{,conv}}$	$\pi d_{\mathrm{go}}\frac{k_{\mathrm{a}}}{d_{\mathrm{go}}}C\,\mathrm{Re}_{d_{\mathrm{go}}}^{m}\mathrm{Pr}_{\mathrm{a}}^{n}\left(\frac{\mathrm{Pr}_{\mathrm{a}}}{\mathrm{Pr}_{\mathrm{go}}}\right)^{\frac{1}{4}}\left(T_{\mathrm{go}}-T_{\mathrm{a}}\right)$
$\dot{Q}_u$	$\left(\dot{q}_{\mathrm{po,SolAbs}}+\dot{q}_{\mathrm{go,SolAbs}}-\dot{q}_{\mathrm{heatloss}}\right)L_{\mathrm{col}}$ $\dot{m}_{\mathrm{f}}c_{\mathrm{p,f}}\left(T_{\mathrm{fo}}-T_{\mathrm{fi}}\right)$
$\dot{q}_{\mathrm{heatloss}}$	$\dot{q}_{\mathrm{go-}a\mathrm{,conv}}+\dot{q}_{\mathrm{go-}s\mathrm{,rad}}$

Kalogirou (2012) and Behar et al. (2015)

At last, energy and exergy efficiencies of the system are defined on a daily basis as follows:

$$\eta=\left(\sum_{t=1}^{24}\dot{W}_{\mathrm{net}}+\dot{Q}_{\mathrm{cooling}}+\dot{Q}_{\mathrm{freshwater}}\right)/\sum_{t=1}^{24}\dot{Q}_{\mathrm{solar},t} \tag{29.11}$$

$$\varepsilon=\left(\sum_{t=1}^{24}\dot{W}_{\mathrm{net}}+\dot{E}_{\mathrm{cooling}}+\dot{E}_{\mathrm{freshwater}}\right)/\sum_{t=1}^{24}\dot{E}_{\mathrm{solar},t} \tag{29.12}$$

29.4 Results and Discussion

In this section, the obtained results are presented and discussed. The necessary inputs for simulation of the system are listed in Table 29.2 (Demirkaya et al. 2011; Kalogirou 2012; Behar et al. 2015; Vidal et al. 2006).

Figure 29.3 shows the effect on the system energy and exergy efficiencies of changing Goswami turbine pressure ratio. As evidenced by increasing the pressure ratio in Goswami cycle from 6 to 18, both efficiencies increase by around 3.3%. This trend is due to the increasing generated cooling and freshwater with higher pressure ratios.

Table 29.2 Input data for the system modelling

Parameter	Value
Ambient pressure (bar)	1
Ambient temperature (°C)	25
Pressure ratio for pump and turbine	10
Ammonia concentration at the rectifier exit	0.98
Pump and turbine adiabatic efficiency (%)	85
Superheating degree (°C)	0
Collector apparatus width (m)	5
Collector length (m)	7.8
Receiver outer diameter (m)	0.07
Receiver inner diameter (m)	0.066
Glass cover outer diameter (m)	0.0115
Glass cover inner diameter (m)	0.109
Receiver pipe thermal conductivity (W/mK)	54
Glass cover thermal conductivity (W/mK)	0.78
Glass cover emittance	0.86
Number of collectors	4

Demirkaya et al. (2011), Kalogirou (2012), Behar et al. (2015), and Vidal et al. (2006)

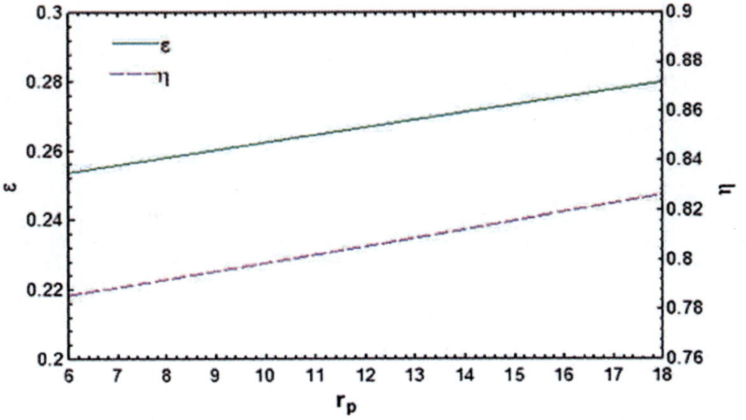

Fig. 29.3 The effect of changing Goswami turbine pressure ratio on the system performance

Figure 29.4 depicts the effects of higher values of the ammonia concentration in the rectifier exit on the system performance. As shown in this figure, the effect of the ammonia concentration on the system energy efficiency is not the same as its effect on the exergy efficiency. As x_{12} increases, the exergy efficiency decreases, but the energy efficiency increases. For energy efficiency, the increasing trend of produced cooling and freshwater is stronger than the decreasing trend of net power output, and thus, energy efficiency increases. However, for the exergy efficiency, the increase in the exergy values of the cooling and freshwater cannot overcome the net power reduction rate, and the exergy efficiency decreases.

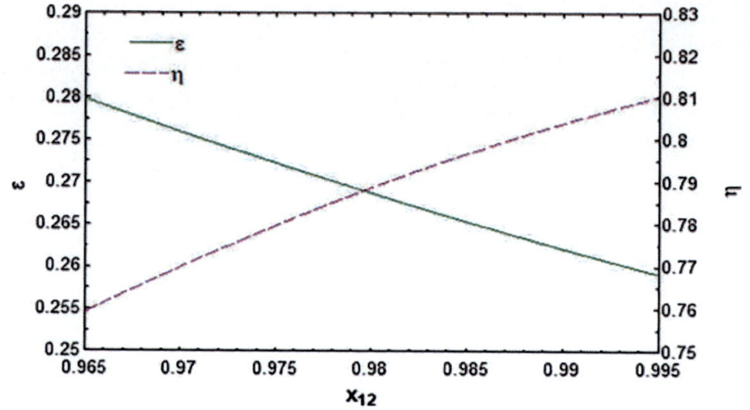

Fig. 29.4 The effect of ammonia concentration in the rectifier outlet point on the system efficiencies

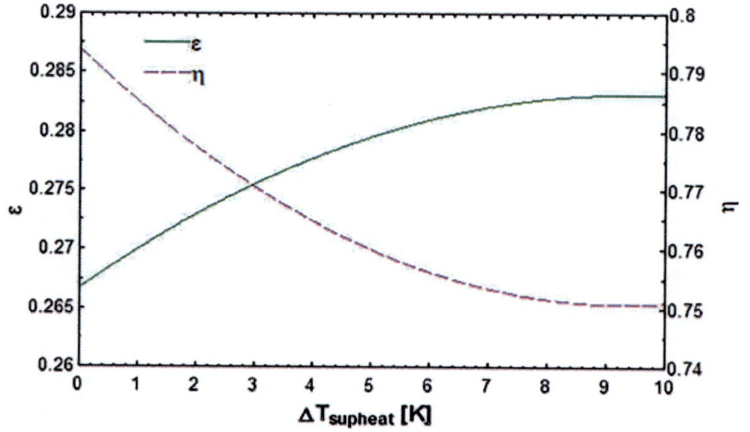

Fig. 29.5 The effect of superheating degree on the system operation

In Fig. 29.5, the effect of superheating degree on the system is illustrated. It shows that as the superheating degree increases from 0 to 10 K, the energy efficiency decreases from 0.795 to 0.75, but the exergy efficiency increases.

29.5 Conclusion

In this work, a new solar-based trigeneration system for electrical power, cooling, and potable water production is proposed and analysed from energy and exergy viewpoints. A molten salt heat storage option is also added to ensure the continuous steady performance of the system. The obtained results show that the energy and exergy efficiencies increase with increasing turbine pressure ratio. Also, higher energy efficiency is achieved if the ammonia concentration at the rectifier exit is increased. However, the exergy efficiency is decreasing in this operating condition. As the superheating degree increases, the system energy efficiency reduces, but the exergy efficiency rises.

References

Ashouri M, Khoshkar Vandani AM, Mehrpooya M, Ahmadi MH, Abdollahpour A (2015) Techno-economic assessment of a Kalina cycle driven by a parabolic Trough solar collector. Energy Conversion and Management 105:1328–39. https://doi.org/10.1016/j.enconman.2015.09.015

Behar O, Khellaf A, Mohammedi K (2015) A novel parabolic trough solar collector model – Validation with experimental data and comparison to Engineering Equation Solver (EES). Energy Conversion and Management 106:268–81. https://doi.org/10.1016/j.enconman.2015.09.045

Brandt MJ, Johnson KM, Elphinston AJ, Ratnayaka DD (2017) Chapter 1 – The Demand for Potable Water. In: Brandt MJ, Johnson KM, Elphinston AJ, Ratnayaka DDBT-TWS (Seventh E, editors., Boston: Butterworth-Heinemann) 1–36. https://doi.org/10.1016/B978-0-08-100025-0.00001-6

Demirkaya G, Vasquez Padilla R, Goswami DY, Stefanakos E, Rahman MM (2011) Analysis of a combined power and cooling cycle for low-grade heat sources. International Journal of Energy Research 35:1145–57. https://doi.org/10.1002/er.1750

Kalogirou SA (2012) A detailed thermal model of a parabolic trough collector receiver. Energy 48: 298–306. https://doi.org/10.1016/j.energy.2012.06.023

Mirzaee M, Zare R, Sadeghzadeh M, Maddah H, Ahmadi MH, Acıkkalp E (2019) Thermodynamic analyses of different scenarios in a CCHP system with micro turbine – Absorption chiller, and heat exchanger. Energy Conversion and Management 198:111919. https://doi.org/10.1016/j.enconman.2019.111919

Mossi Idrissa AK, Goni Boulama K (2019) Advanced exergy analysis of a combined Brayton/Brayton power cycle. Energy 166:724–37. https://doi.org/10.1016/j.energy.2018.10.117

Vidal A, Best R, Rivero R, Cervantes J (2006) Analysis of a combined power and refrigeration cycle by the exergy method. Energy 31:3401–14. https://doi.org/10.1016/j.energy.2006.03.001

Zolfagharnasab MH, Aghanajafi C, Kavian S, Heydarian N, Ahmadi MH (2020) Novel analysis of second law and irreversibility for a solar power plant using heliostat field and molten salt. Energy Science and Engineering 8:4136–53. https://doi.org/10.1002/ese3.802

Chapter 30
Design and Thermodynamic Analysis of a Novel Power, Methanol, and Light Olefins Trigeneration System Fed with Shale Gas

Leyla Khani, Hamidreza Haddadi, Gülden Gökçen Akkurt, and Mousa Mohammadpourfard

Nomenclature

$\dot{E}$	Exergy rate, kW
$\bar{e}^{ch}$	Standard chemical exergy, kJ/kmol
h	Specific enthalpy, kJ/kg
$\dot{m}$	Mass flow rate, kg/s
P	Pressure, bar
$\dot{Q}$	Heat transfer rate, kW
R	Universal gas constant, kJ/kgK
s	Specific entropy, kJ/kgK
T	Temperature, K
$\dot{W}$	Power, kW
y	Mole fraction

Greek Letters

η	Energy efficiency
ψ	Exergy efficiency

L. Khani · H. Haddadi · M. Mohammadpourfard
Faculty of Chemical and Petroleum Engineering, University of Tabriz, Tabriz, Iran
e-mail: leyla_khani@tabrizu.ac.ir; mohammadpour@tabrizu.ac.ir

G. G. Akkurt (✉)
Department of Energy Systems Engineering, Izmir Institute of Technology, Izmir, Turkey
e-mail: guldengokcen@iyte.edu.tr

M. Z. Sogut et al. (eds.), *Proceedings of the 2022 International Symposium on Energy Management and Sustainability*, Springer Proceedings in Energy,
https://doi.org/10.1007/978-3-031-30171-1_30

Subscripts

0	Environmental Condition
ch	Chemical
D	Destruction
e	Outlet
i	Inlet
k	k^{th} component
ph	Physical

30.1 Introduction

With the advancement of global economy and technology, energy has become more important because its supply and demand rates are hardly in balance. Most of the necessary energy is provided by burning fossil fuels. However, environmental problems and the limitations of fossil fuels have urged researchers to find cleaner and more affordable fuels, like shale gas, and modify the energy production processes.

Shale gas is a kind of natural gas, with a different composition of hydrocarbons. The appropriate price of shale gas makes it a good candidate for various utilizations, such as olefin production through methanol-to-olefin (MTO) process (Zendehboudi and Bahadori 2016).

Also, multigeneration systems are getting noticed because they are capable of producing two or more products from one input fuel by recovering waste heat or material. This leads to higher efficiency, lower pollution, more reliability, and less cost (Manesh and Amidpour 2020).

Moreover, chemical looping cycle is one of the promising technologies to reduce carbon dioxide emission. This system consists of two separate reactors, in which solid particles circulate as oxygen carriers. Hence, fuel combustion process occurs with pure oxygen, and there are not any nitrogen or sulfur oxides in the combustion products, so it is possible to separate and store carbon dioxide (Rydén and Lyngfelt 2006). Chemical looping cycle is favorable because of its low environmental pollution, near-zero carbon emission, low cost, variety of fuels, and the ability of carbon capture (Adanez et al. 2012).

In this paper, a new electricity, methanol, and light olefins generation system is designed and evaluated from energy and exergy viewpoints. Shale gas is selected as the fuel of the system. Additionally, the proposed system is an integration of chemical looping reforming and combustion cycles, methanol production unit, methanol-to-light olefins conversion subsystem, and Rankine cycle. Special attention is paid to heat recovery in different components of the system, so the wasted energy is decreased and the system efficiency is increased.

30.2 System Description

The proposed power, methanol, and light olefins trigeneration system is shown in Fig. 30.1. As obvious, it consists of the following subsystems:

- The synthesis gas production from shale gas in chemical looping reforming cycle.
- Methanol production process.
- Methanol-to-light olefins process.
- Heat and power production using thermal recovery, chemical looping combustion cycle, and Rankine cycle.

Shale gas is the main fuel of the system, and electrical power, methanol, ethylene, and propylene are produced. The subsystems and heat exchangers are integrated in such a way that the amount of necessary external heat is reduced significantly.

Iron and titanium composite is used as the oxygen carrier in the chemical looping reforming cycle, and the following reactions occur in that cycle:

$$FeTi_yO_x + CH_4 \rightarrow FeTi_yO_{x-1} + CO + 2H_2 \quad (30.1)$$

$$CH_4 + H_2O \rightarrow CO + 3H_2 \quad (30.2)$$

$$CH_4 + CO_2 \rightarrow 2CO + 2H_2 \quad (30.3)$$

$$FeTi_yO_{x-1} + 0.5O_2 \rightarrow FeTi_yO_x \quad (30.4)$$

Nickel is chosen as the oxygen carrier in the chemical looping combustion cycle for implementing the following reactions (Ishida et al. 2002):

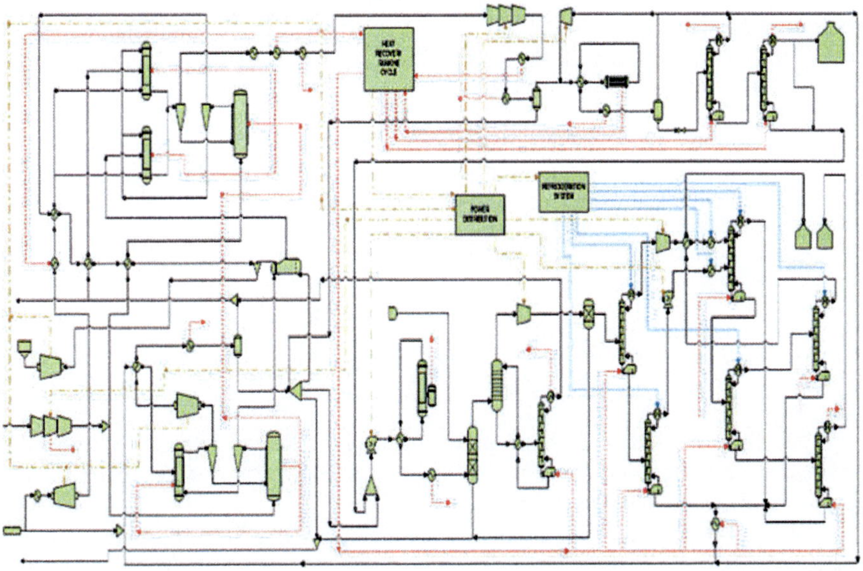

Fig. 30.1 The proposed system for power, methanol, and light olefins generation fed with shale gas

$$C_xH_y + [(4x+y)/2]\ NiO \rightarrow xCO_2 + y/2H_2O + [(4x+y)/2]Ni \quad (30.5)$$

$$CO + NiO \rightarrow CO_2 + Ni \quad (30.6)$$

$$H_2 + NiO \rightarrow H_2O + Ni \quad (30.7)$$

$$2Ni + O_2 \rightarrow 2NiO \quad (30.8)$$

30.3 System Modeling

The proposed system is simulated in Aspen Plus software using the following assumptions:

- The system operates in the steady-state condition.
- Heat loss to the environment is ignorable.
- The ambient temperature and pressure are 25 °C and 1 atm, respectively.
- The air consists of 21% oxygen and 79% nitrogen.
- Compressors, pumps, and turbines have isentropic efficiencies.

If changes in potential and kinetic energies are ignored, mass and energy conservation laws along with exergy balance equations are written for each component at a steady state to simulate the thermodynamic performance of the system:

$$\sum \dot{m}_i = \sum \dot{m}_e \quad (30.9)$$

$$\dot{Q}_k - \dot{W}_k = \sum \dot{m}_e h_e - \sum \dot{m}_i h_i \quad (30.10)$$

$$\dot{E}_D = \sum \left(1 - \frac{T_0}{T_j}\right)\dot{Q}_j - \dot{W}_{cv} + \sum \dot{E}_i - \sum \dot{E}_e \quad (30.11)$$

In the absence of electrical, magnetic, nuclear, and surface stress effects, and excluding kinetic and potential exergies, the exergy of any stream is the sum of its physical and chemical exergies:

$$\dot{E} = \dot{E}_{ph} + \dot{E}_{ch} \quad (30.12)$$

$$\dot{E}_{ph} = \sum \dot{m}_i \left[\left(h_i - h_i^0\right) - T_0\left(s_i - s_i^0\right)\right] \quad (30.13)$$

$$\dot{E}_{ch} = \sum \dot{m}_i e_i^{ch} + RT_0 \sum \dot{m}_i \ln y_i \quad (30.14)$$

It should be noted that unlike mass and energy, exergy is preserved only in reversible processes and is destroyed in real processes by the destructions that happen in thermodynamic systems.

Shale gas composition and properties are listed in Table 30.1 (Zendehboudi and Bahadori 2016).

Table 30.1 Shale gas properties

Parameter	Value
Molar composition	94.3% CH_4 2.7% C_2H_6 1% C_3H_8 1.5% N_2 0.5% CO_2
Pressure	30 bar
Temperature	30 °C

Table 30.2 Validation of the methanol production system

Parameter	Walid et al. (2018)	Present work
T (°C)	256.1	253.4
P (bar)	66.7	69.2
$\dot{m}$ $\left(\frac{kg}{s}\right)$	11284.3	10792.7

The system energy and exergy efficiencies can be defined as follows:

$$\eta = \frac{\dot{W}_{net} + \dot{n}_{methanol}LHV_{methanol} + \dot{n}_{ethylene}LHV_{ethylene} + \dot{n}_{propylene}LHV_{propylene}}{\dot{n}_{shalegas}LHV_{shalegas}} \tag{30.15}$$

$$\psi = \frac{\dot{W}_{net} + \dot{n}_{methanol}\dot{E}_{methanol} + \dot{n}_{ethylene}\dot{E}_{ethylene} + \dot{n}_{propylene}\dot{E}_{propylene}}{\dot{n}_{shalegas}\dot{E}_{shalegas}} \tag{30.16}$$

30.4 Results and Discussion

First, the properties of the methanol generated in the methanol production subsystem of this work are compared to the values of Walid et al. in Table 30.2, proving a good agreement between them.

The exergy destruction ratio of each subsystem is shown in Fig. 30.2. According to this figure, the highest exergy destruction rate, 43%, belongs to the heat and power generation section, which consists of thermal recovery, chemical looping combustion cycle, and Rankine cycle. This can be justified if the high operating temperature of air and fuel reactors and the high compression ratio in this subsystem are considered. On the other hand, the lowest exergy destruction ratio, 6%, is related to the light olefins production process, mainly because of methanol-to-olefin reaction.

The effects on the system energy and exergy efficiencies and the molar flow rate of the output carbon dioxide of changing methanol-to-MTO molar ratio are depicted in Fig. 30.3. As evident in this figure, increasing the methanol-to-MTO process leads to lower values of the energy and exergy efficiencies, but the rate of carbon dioxide output increases. The energy and exergy efficiencies are 71% and 77.2% when no methanol is sent to the MTO process. These efficiencies will be 64.8% and 67.5% if half of the produced methanol is directed to the MTO section. In these cases, carbon dioxide outlet flow rates are 550 and 976.08 kmol/h, respectively.

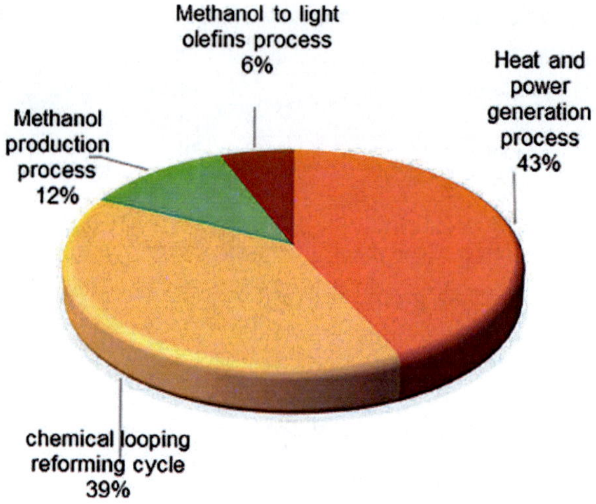

Fig. 30.2 Exergy destruction ratio of the subsystems

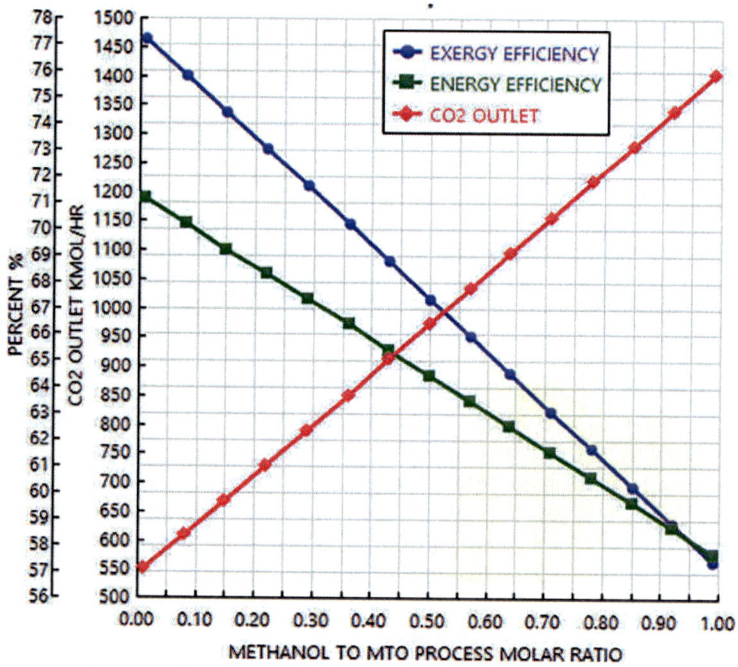

Fig. 30.3 The changes in energy and exergy efficiencies and carbon dioxide outlet flow rate with methanol-to-MTO process ratio

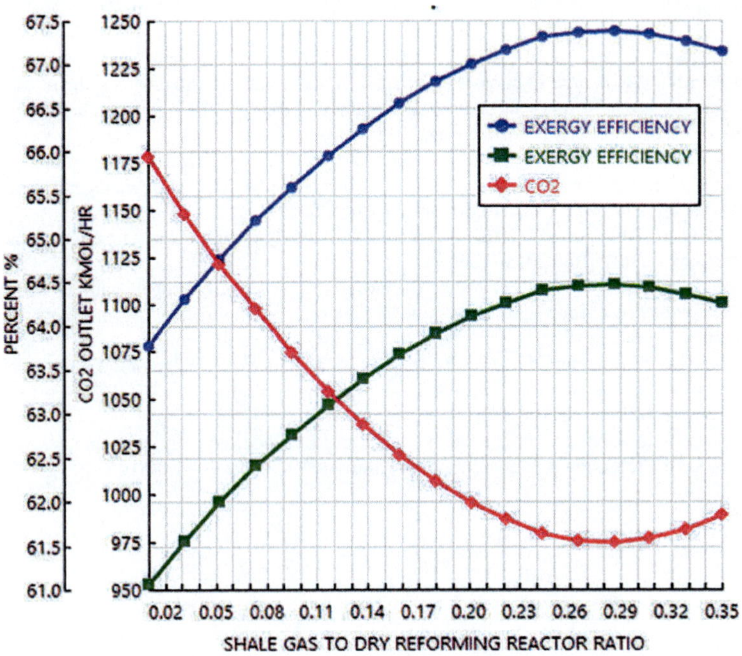

Fig. 30.4 The changes in energy and exergy efficiencies and carbon dioxide outlet flow rate with shale gas to dry reforming reactor ratio

Figure 30.4 illustrates the changes in the energy and exergy efficiencies and carbon dioxide flow rate with varying shale gas to dry reforming reactor ratio. According to this figure, the highest energy and exergy efficiencies, which are 64.5% and 67.4%, respectively, occur when 28.5% of shale gas flow is sent to the dry reforming reactor. Additionally, the lowest carbon dioxide, 975 kmol/h, is achieved at this point.

30.5 Conclusion

A new power, methanol, and light olefins production system has been designed and investigated in this paper. Shale gas is considered as the main fuel of the system. Chemical looping reforming cycle, methanol production subsystem, methanol-to-olefin conversion section, Rankine cycle, thermal recovery component, and chemical looping combustion cycle are the main units of the proposed system. The first and second laws of thermodynamics are written for each component at a steady state. The obtained results show that the heat and power generation sections own the highest exergy destruction rate among other units. Also, when half of the produced methanol is consumed for generating olefins, the system energy and exergy efficiencies are 64.8% and 67.5%, respectively. Finally, if only 28.5% of shale gas is sent to the dry reforming reactor, 975 kmol/h outlet carbon dioxide is achieved.

References

Adanez J, Abad A, Garcia-Labiano F, Gayan P, Luis F (2012) Progress in chemical-looping combustion and reforming technologies. Progress in energy and combustion science 38(2): 215–282. https://doi.org/10.1016/j.pecs.2011.09.001

Ishida M, Yamamoto M, Ohba T (2002) Experimental results of chemical-looping combustion with NiO/NiAl2O4 particle circulation at 1200°C. Energy Conversion and Management 43(9-12): 1469–78. https://doi.org/10.1016/S0196-8904(02)00029-8

Manesh M. H. K, Amidpour M (2020) Cogeneration and Polygeneration Systems. Academic Press. https://doi.org/10.1016/C2018-0-02100-3

Rydén M, Lyngfelt A (2006) Using steam reforming to produce hydrogen with carbon dioxide capture by chemical-looping combustion. International Journal of Hydrogen Energy 31:1271–1283. https://doi.org/10.1016/j.ijhydene.2005.12.003

Walid BA, Hassiba B, Boumediene H, Weifeng S (2018) Improved Design of the Lurgi Reactor for Methanol Synthesis Industry. Chemical Engineering & Technology 41(10):2043–52. https://doi.org/10.1002/ceat.201700551

Zendehboudi S, Bahadori A (2016) Shale oil and gas handbook: theory, technologies, and challenges. Gulf Professional Publishing. https://doi.org/10.1016/C2014-0-01653-X

Chapter 31
Design and Performance Evaluation of a Direct Absorption Solar Collector

Ismail Pacaci, Koray Ulgen, and M. Z. Sogut

Nomenclature

W	Collector wideness, m
C_p	Specific heat, J/kg°C
K	Thermal conductivity, W/mK
ρ	Fluid density, kg/m^3
f	Focal length, m
μ	Viscosity, Ns/m^2
T	Temperature, °C
I	Heat flux, W/m^2
$\varnothing$	Volumetric particle density, m^3
λ	Wavelength
DASC	Direct absorption solar collector
GNP	Graphene nanoparticles
C_g	Geometric concentrating ratio

31.1 Introduction

Energy requirement is increasing parallelly with technological developments in industrial, domestic, and transportation sectors. The increasing use of fossil fuels due to the increase in energy demand jeopardizes the concept of sustainability. To stop these negative effects, new local and global energy policies are created by

I. Pacaci (✉) · K. Ulgen · M. Z. Sogut
Solar Energy Institute, Ege University, Izmir, Turkey

Piri Reis University, İstanbul, Türkiye

M. Z. Sogut et al. (eds.), *Proceedings of the 2022 International Symposium on Energy Management and Sustainability*, Springer Proceedings in Energy,
https://doi.org/10.1007/978-3-031-30171-1_31

governments. According to the new energy policies, the main goal is, firstly, to extend energy generation by using renewable energy sources, thereafter reaching zero carbon footprint. In this sense, solar energy is one of the best technologies to generate clean energy.

Concentrated solar energy collectors currently have no widespread practical application, but they have advantages in high temperature applications, which is promising.

Miller et al. (1951) have made high sensitivity and high surface quality Fresnel lens by plastic raw materials. Thus, Fresnel lenses found an opportunity for a wide area of application. Nelson et al. (1975) used linear focusing lens and reached 143 °C to hot water and steam generate. Harmon (1977) presented an experimental and analytical method to determine concentrated solar energy efficiency of the Fresnel lens used in solar collectors. Akisawa et al. (2007) developed optical design geometry and solar tracking technique with high focusing ability oval non-imaging Fresnel lens. Perini et al. (2017) have done experimental and with technical aspects theoretical analysis of an innovative system capable of bidirectional monitoring for linear focus Fresnel lens.

Various technologies have been considered for solar irradiation on the collector. Fresnel lens is a prominent technology with low dimension and weight specialties. With these advantages of Fresnel lens it is possible to reach high efficiency with low cost (Abbas et al. 2013). There are two types of Fresnel lens used in solar collectors: point focusing lens in high temperature applications and linear focusing lens in solar cooling systems and industrial and steam generation applications.

In this chapter, the design of a direct absorption solar energy collector is a very unique subject in the literature, and its performance is evaluated under the climatic conditions of Izmir province in 2019. Thermal performance of the nanofluid formed by adding graphene oxide nanoparticles in the volume of 0,010 to the water used as a base fluid investigated. In addition, theoretical calculations of two different flow rates of 0,0168 kg/s and 0,0366 kg/s in the system were made, and the effect of the flow rate on the collector system efficiency was revealed.

31.2 Material and Method

Direct absorption solar collectors work with the principle of obtaining high temperature heat by focalizing solar radiation onto the receiver pipe, in which the heat transfer fluid passes through in the most basic form. Therefore, the thermal performance of the heat transfer fluid directly affects the system efficiency.

Solar irradiation directly focuses on the heat transfer fluid, which is circulating in the receiving tube. So, in high temperatures, thermal losses occur in the receiving tube. Thermal losses decrease the system efficiency. Because of this reason a glass cover is added on the receiving tube to minimize thermal losses. Also, the gap between the glass cover and the receiver tube can be vacuumed. This vacuum method increases system efficiency and investment cost. However, unvacuumed tube has low-cost, easy production.

31.2.1 *Fresnel Lens Concentrator*

It is possible to produce high efficiency energy at low cost by Fresnel lenses with small size and low weight. In addition, Fresnel lenses provide optical design ease compared to traditional mirror technologies which are frequently used in solar energy collector applications (Abbas et al. 2013).

Fresnel lenses consist of a chain of prisms. Each prism represents a curve on the lens surface. Fresnel lenses are affected by the deviations that occur in the grooves and prism ends that occur during their production, and they create an image in the focal plane by refracting the lights coming from the source. The schematic representation of traditional Fresnel lens is given in Fig. 31.1.

The focal length (f) that determines the distance between the Fresnel lens and the receiver tube can be calculated with the help of the equation below:

$$f = \frac{d}{\tan\theta} \tag{31.1}$$

In this equation d indicates the radius of the receiver tube and θ is the incidence angle of solar radiation.

The geometric condensation ratio (C_g) of the Fresnel lens is defined in Eq. (31.2). W is the focal point width and b is the width of solar radiation falling on the receiving plane.

$$C_g = \frac{b}{W} \tag{31.2}$$

31.2.2 *Dimension of Solar Collector*

Direct absorption solar collector system efficiency related with collector efficiency, heat transfer coefficient, total heat loses and time constant (Xie et al. 2013). Also, the thermal performance of solar collector increases in direct proportion with the energy transmission ratio on the transfer fluid and absorption capacity. Direct absorption concept occurred to simplify surface absorption design and make direct absorption in the transfer fluid (Gorji and Ranjbar 2015). Adding nanoparticles, which have better optical and thermophysical properties, provides higher energy absorption. Thus, the system efficiency rises.

The collector efficiency can be calculated with the help of the equation given below:

$$\eta = \frac{Q_{\text{use}}}{Q_{\text{inc}}} \times 100 \tag{31.3}$$

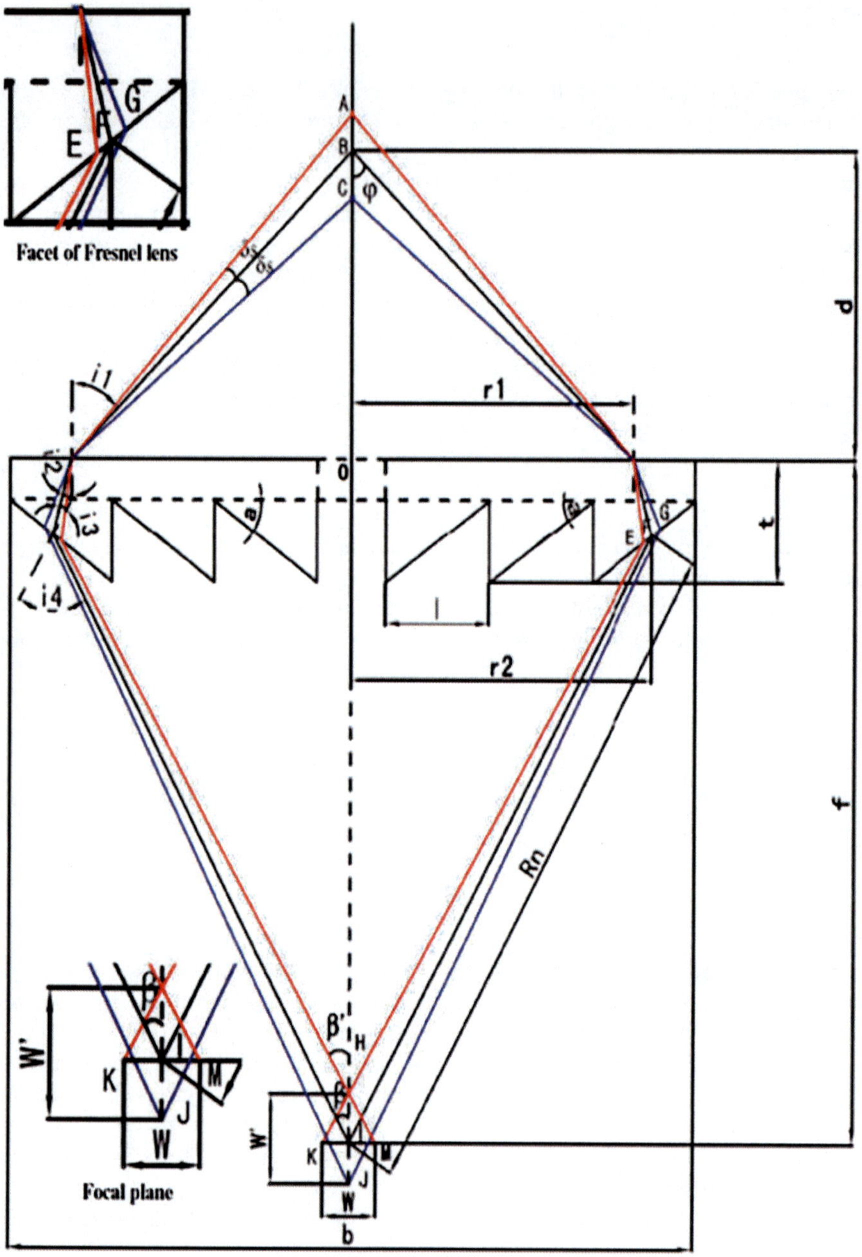

Fig. 31.1 Schematic representation of traditional Fresnel lens. (Xie et al. 2013)

In Eq. (31.3), η is the general collector efficiency, Q_{use} the useful energy at the collector exit, and Q_{inc} the coming energy on the collector surface.

$$Q_{use} = (\rho_{nf} W C_{nf}) \int_{y=0}^{y=H} [u(y)(T_{out}(y) - T_{in})] dy \tag{31.4}$$

In Eq. (31.4), ρ_{nf} indicates the density of nanofluid, W the collector wideness, C_{nf} the thermal capacity of nanofluid, H the film thickness of nanofluid, u the flow rate of nanofluid, T_{in} the inlet temperature, and T_{out} the outlet temperature.

$$Q_{inc} = L\, W \int_{o}^{\infty} I_0(\lambda) d\lambda \tag{31.5}$$

Equation (31.5) represents L as the collector length, λ the wavelength of nanofluid, and I_0 the heat flux coming to the solar collector.

A small-scale test setup was created to determine the thermal performance of direct absorption solar collector for use in high temperature applications (Fig. 31.2). In this system, four units of linear focusing capability Fresnel lens which is able to follow sun in two axes and receiver tube connected to each other series designed. The heat transfer fluid is connected to the solar collector by a flow meter, a reducing valve, and a three-way thermostatic valve to keep the flow rate of fluid 60 l/h and 120 l/h value. Also, heat transfer fluid is stored in the storage tank.

The dimensions of the created test setup are given in Table 31.1. It is assumed that there is a vacuum between the glass cover and the receiver tube.

31.2.3 Nanofluids

Efforts are underway to develop heat transfer fluids that can transfer heat faster in shorter time to reduce energy consumption and provide higher performance for thermal devices. Accordingly, nanofluids became noteworthy to use as a heat transfer fluid.

The new heat transfer fluid, which is formed by suspension of nanoparticles smaller than 100 nm into basic heat transfer fluid, is defined as nanofluid (Li et al. 2020).

Thermal conductivity and erosion value improved with the addition of nanoparticles into the fluid (Kazemi et al. 2020).

Nanofluids have better surface area than basic fluids, so it has increased heat conduction. On the other hand, nanoparticles are of very small size. Therefore, adding nanoparticles ensures reducing erosion, clogging, and pumping power. In this way, nanofluids provide significant energy savings.

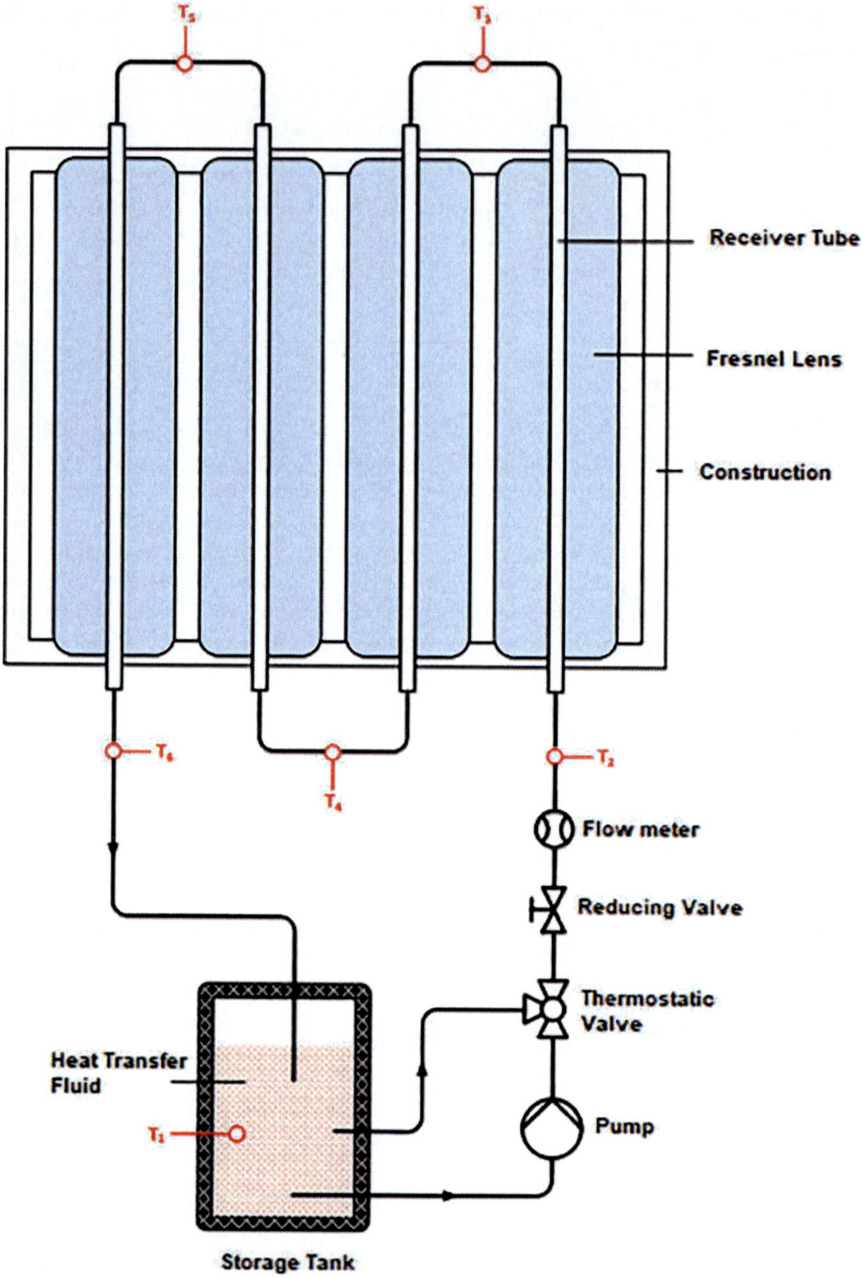

Fig. 31.2 Linear focusing direct absorption solar collector schematic view

Table 31.1 Dimensions of test setup

	Dimensions
Fresnel lens	250 μm bottom film thickness
	130 mm focal length
	180 mm wideness
	550 mm length
Receiver tube	10 mm diameter of inner receiver tube
	23 mm diameter of outer glass cover
	550 mm wideness and set 103 mbar air pressure
	700 mm total length

31.2.4 Thermophysical Properties of Nanofluid

Particle density directly affects the thermophysical properties and chemical structure of nanofluid. Particle density can be calculated by Eq. (31.6):

$$\varnothing = \frac{\frac{W_{np}}{\rho_{np}}}{\frac{W_{np}}{\rho_{np}} + \frac{W_{bf}}{\rho_{bf}}} \tag{31.6}$$

In Eq. (31.7) W_{np} is the nanoparticle mass, ρ_{np} the nanoparticle density, W_{bf} the base fluid mass, and ρ_{bf} the base fluid density (Muraleedharan et al. 2016).

$$\rho_{nf} = \phi_v \rho_{np} + (1 - \phi_v)\rho_f \tag{31.7}$$

It is possible to determine the nanofluid density using the ϕ, which is the particle size distribution, in Eq. (31.6).

The specific heat of nanofluid can be calculated with Eq. (31.8):

$$C_{p,na} = \phi\, C_{p,n} + (1 - \phi)C_{p,a} \tag{31.8}$$

$C_{p,\,n}$ and $C_{p,\,a}$ represent the specific heat of nanoparticle and base fluid, respectively.

The Nusselt (Nu) and Prandtl (Pr) numbers are two important parameters used to analyze the heat transfer capacity of the fluid under laminar and turbulent flow conditions and they vary depending on the thermal conductivity of the fluid. Therefore, in order to investigate the heat transfer potential of the nanofluid, it is necessary to calculate the thermal conductivity (Sezer et al. 2019). It can be defined as the heat-carrying capacity of fluid and can be calculated with the Maxwell method as in Eq. (31.9):

$$k_{nf} = \frac{k_{cp} + 2k_f + 2(k_{cp} - k_f)(1-\beta)^3 f_v}{k_{cp} + 2k_f - (k_{cp} - k_f)(1-\beta)^3 f_v} \tag{31.9}$$

In Eq. (31.9), β is the ratio of interfacial layers between spherical radius, k_{nf} the thermal conductivity of nanofluid, k_f the base fluid thermal conductivity, and k_{cp} the complex particle thermal conductivity.

Viscosity is one of the important flow properties of a heat transfer fluid. The pumping power of the system, pressure drop in laminar flow, and heat transfer by convection area are directly related to the viscosity of the fluid. In almost all of the studies, it has been stated that the viscosity of the nanofluid is higher than the viscosity of the base fluid. It was determined that the viscosity of the nanofluid increases the particle concentration. When the temperature increases, the interaction between the nanoparticles and the fluid weakens, such as molecules of the fluid. For this reason, the viscosity decreases with the increasing temperature (Budak 2016).

$$\mu_{nf} = (1 + 2.5\varnothing + 6.5\varnothing^2)\mu_f \tag{31.10}$$

It is possible to calculate the viscosity of the nanofluid with Eq. (31.10). In there, μ_{nf} represents the viscosity of nanofluid and μ_f the viscosity of the fluid and particle size distribution.

31.2.5 Graphene Added Nanofluids

Graphene particles have higher thermal conductivity value compared to other nanoparticles, the use of graphene added nanofluids consisting of graphene nanoparticles (GNP) increases the thermal conductivity of the fluid. Thus, it can improve the thermal performance of the heat transfer system.

Graphene, which is an allotrope of carbon, has a thermal conductivity value of 3000 W/mK, a specific heat value of 765 J/KgK, and a density of 3600 kg/m^3 at 25 °C. In addition, when graphene is oxidized to graphene oxide, its stability increases, and it does not precipitate in the fluid. Graphene thermal properties at 25 °C are given in Table 31.2.

Table 31.2 Graphene thermal properties at 25 °C

Nanoparticle	Graphene
Density	3600 kg/m^3
Heat capacity	765 J/KgK
Thermal Conductivity	3000 W/mK

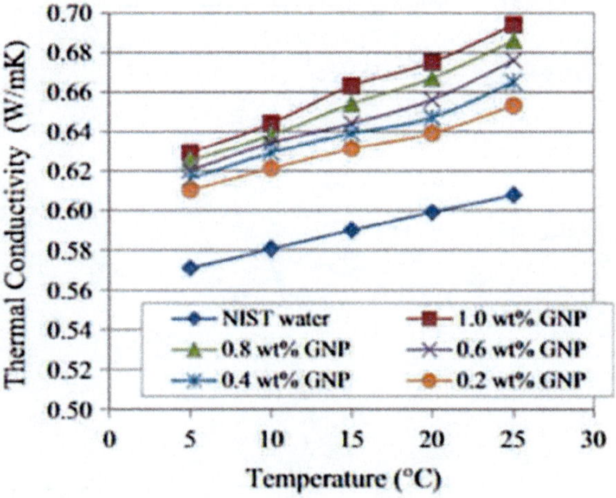

Fig. 31.3 Thermal conductivities of GNP nanofluids with different volumetric concentration. (Wang et al. 2018)

Wang et al. (2018) showed the thermal conductivity values for water and GNP nanofluids in their study. The thermal conductivity values of nanofluids increased linearly with the increase in temperature, as shown in Fig. 31.3.

31.3 Results and Discussions

Direct absorption solar collectors (DASC) are linear focusing systems, and solar radiation is absorbed directly in the heat transfer fluid passing through the receiver tube. Therefore, the performance and efficiency of the solar collector is directly dependent on the solar radiation. The calculation of the direct absorption solar collector thermal performance efficiency has been done in climate condition of Izmir, Turkey, in 2019.

In order to determine performance of DASC for the day representing each month depending on the newly created nanofluid with two different mass flows, the solar radiation values reaching the collector level and the receiver pipe surface were calculated for those days. In this paper 2 days was selected which has solar minimum and maximum radiation values and the thermal performance results are given. Solar radiation coming to a surface following two axes of the sun were calculated by with the hourly data of the province of Izmir.

Based on the obtained result of the calculation, solar radiation on the Fresnel lens, receiver tube and thermal performance changing values are shown in Figs. 31.4, 31.5, 31.6 and 31.7.

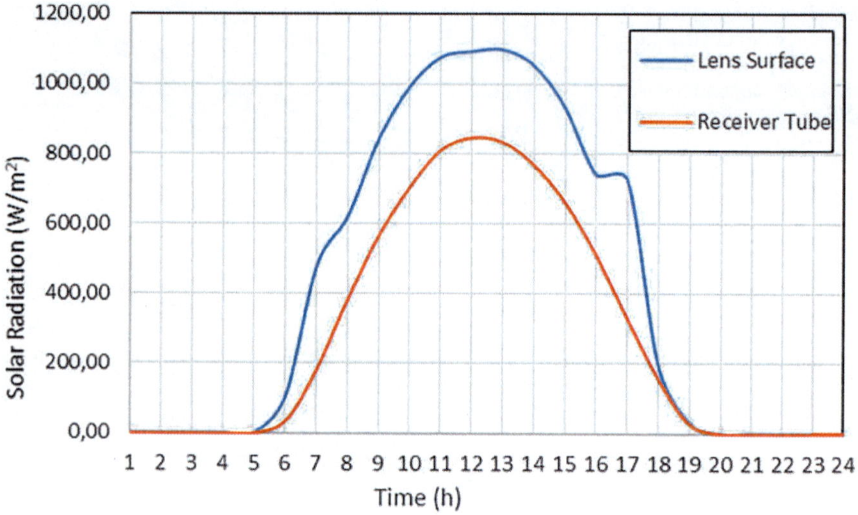

Fig. 31.4 Solar radiation graph of collector in July

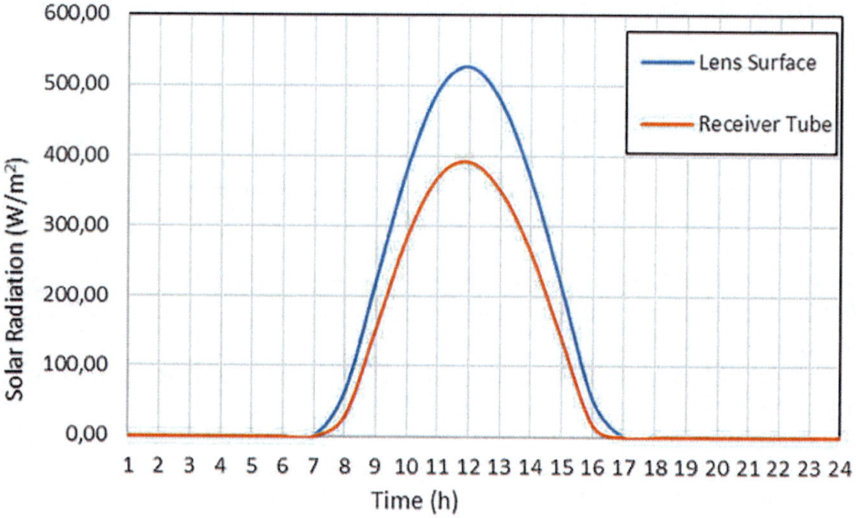

Fig. 31.5 Solar radiation graph of collector in December

These graphs explain to us that the temperature of the collector is directly affected from solar radiation and fluid mass flow. It has been determined that the temperature of the fluid to be obtained from the collector increases when sunlight is high.

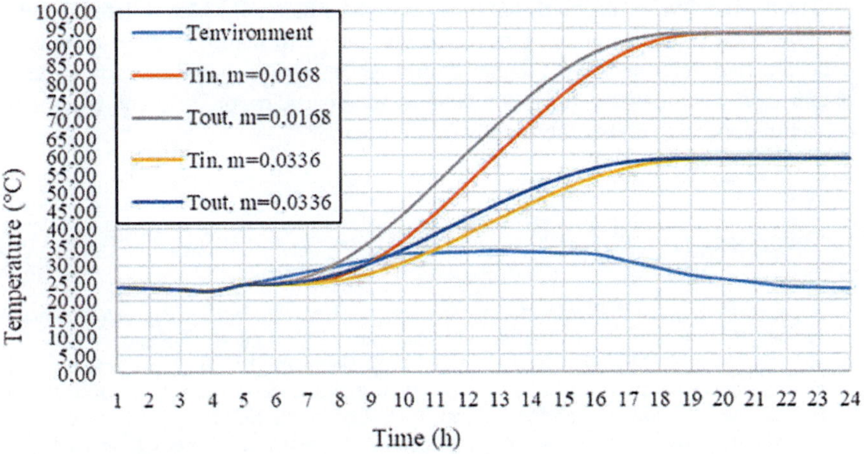

Fig. 31.6 Temperature graph of collector in July

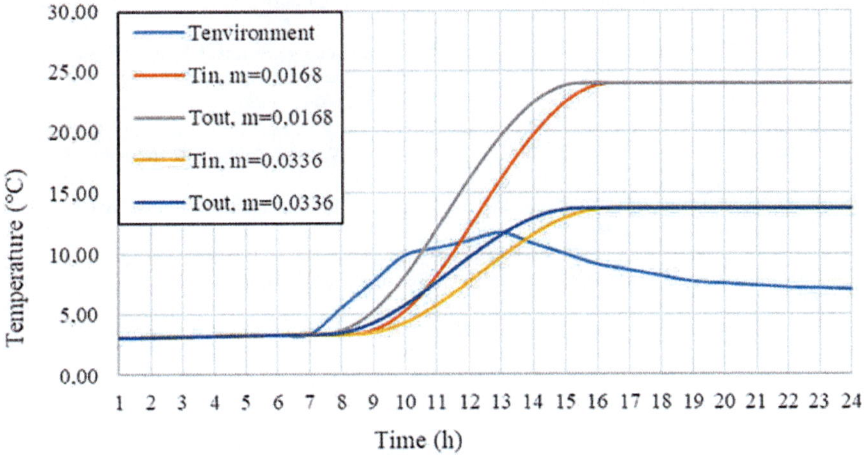

Fig. 31.7 Temperature graph of collector in December

31.4 Conclusion

In this chapter the DASC performance evaluation by design parameters and thermophysical properties is investigated.

Dimensioning of the collector should be optimized according to system requirements because solar radiation absorbing value is dependent on the geometrical dimension of collector.

The thermophysical properties of the heat transfer material which is defined as the working fluid come as the first factor affecting the performance of the solar energy collector. In order to achieve a higher performance, studies should be carried out to obtain new heat transfer fluids by adding nanoparticles of different types and concentrations with higher thermophysical properties.

References

Abbas, R., Muñoz-Antón, J., Valdés, M., & Martínez-Val, J. M. (2013). High concentration linear Fresnel reflectors. Energy Conversion and Management, 72, 60–68. (for journal)

Akisawa, A., Sato, T., Miyazaki, T., Kashiwagi, T., & Hiramatsu, M. (2007, September). High concentration non-imaging Fresnel lens design with flat upper surface. In High and Low Concentration for Solar Electric Applications II (Vol. 6649, pp. 133–140). SPIE.

Budak, N. (2016). Güneş kollektörlerinde nanoakışkan kullanımının ısıl verime etkisinin deneysel incelenmesi

Gorji, T. B., & Ranjbar, A. A. (2015). Geometry optimization of a nanofluid-based direct absorption solar collector using response surface methodology. Solar Energy, 122, 314–325.

Harmon, S. (1977). Solar-optical analyses of a mass-produced plastic circular Fresnel lens. Solar Energy, 19(1), 105–108.

Kazemi, I., Sefid, M., & Afrand, M. (2020). A novel comparative experimental study on rheological behavior of mono & hybrid nanofluids concerned graphene and silica nano-powders: Characterization, stability and viscosity measurements. Powder Technology, 366, 216–229.

Li, X., Chen, W., & Zou, C. (2020). The stability, viscosity and thermal conductivity of carbon nanotubes nanofluids with high particle concentration: A surface modification approach. Powder Technology, 361, 957–967.

Miller, O. E., McLeod, J. H., & Sherwood, W. T. (1951). Thin sheet plastic Fresnel lenses of high aperture. JOSA, 41(11), 807–815.

Muraleedharan, M., Singh, H., Suresh, S., & Udayakumar, M. (2016). Directly absorbing Therminol-Al2O3 nano heat transfer fluid for linear solar concentrating collectors. Solar Energy, 137, 134–142.

Nelson, D. T., Evans, D. L., & Bansal, R. K. (1975). Linear Fresnel lens concentrators. Solar Energy, 17(5), 285–289.

Perini, S., Tonnellier, X., King, P., & Sansom, C. (2017). Theoretical and experimental analysis of an innovative dual-axis tracking linear Fresnel lenses concentrated solar thermal collector. Solar Energy, 153, 679–690.

Sezer, N., Atieh, M. A., & Koç, M. (2019). A comprehensive review on synthesis, stability, thermophysical properties, and characterization of nanofluids. Powder technology, 344, 404–431.

Wang, Y., Al-Saaidi, H. A. I., Kong, M., & Alvarado, J. L. (2018). Thermophysical performance of graphene based aqueous nanofluids. International Journal of Heat and Mass Transfer, 119, 408–417.

Xie, W. T., Dai, Y. J., & Wang, R. Z. (2013). Thermal performance analysis of a line-focus Fresnel lens solar collector using different cavity receivers. Solar Energy, 91, 242–255. (for journal)

Chapter 32
Determination of Combustion Characteristics of Selected Waste Wood Samples and Two Local Lignites by Thermogravimetric Analysis

Kemal Berk Altunkaya, Mihriban Civan, and Sema Yurdakul

Nomenclature

TGA	Thermogravimetric Analysis
MDF	Medium Density Fiberboard
DTG	Derivative Thermogravimetry

32.1 Introduction

The world energy comes from 85% non-renewable energy resources, such as oil, coal, and natural gas. Today, oil reserves are expected to last less than 50 years, while coal reserves would last more than 300 years. In view of the fact that energy needs rise in line with world population, the usage of coal for energy production is expected to gain more importance in the near future. Coal meets about 30% of the world's energy consumption. It is an attractive fuel type because its reserves are spread over a wide area in the world (World coal 2012).

When considered in terms of our country, while fossil fuels have a 90% share in the total primary energy consumption, about 74% of the energy demand is met through imports in Turkey (Turkey's energy strategy 2022). However, this amount,

K. B. Altunkaya · S. Yurdakul (✉)
Environmental Engineering Department, Suleyman Demirel University, Isparta, Turkey
e-mail: semayurdakul@sdu.edu.tr

M. Civan
Environmental Engineering Department, Kocaeli University, Kocaeli, Turkey
e-mail: mihriban.civan@kocaeli.edu.tr

M. Z. Sogut et al. (eds.), *Proceedings of the 2022 International Symposium on Energy Management and Sustainability*, Springer Proceedings in Energy,
https://doi.org/10.1007/978-3-031-30171-1_32

which is a clear indicator of foreign dependency in terms of energy, can be reduced to lower levels by using local resources. The war between Russian and Ukraine shows that the production of national and alternative energy is critical to establishing energy security. Considering the local energy resources, it is seen that Turkey has a significant lignite potential. Lignite coal, with its total (apparent and probable) reserves of 19.32 billion tons, is the energy source that has the largest share in our country's solid fossil fuel reserves (MTA 2022). In this context, the development of proper thermochemical technologies to use lignite potential should be among the priority energy issues of Turkey as the country is mainly dependent on energy import. Biomass is an organic-based fuel that stands out with its low-polluting properties and renewable nature compared to coal. Co-firing of biomass with coals is a practical and cost-effective application. On the other hand, when wood is burned alone, their raw materials are depleted faster, and it leads to some problems such as corrosion due to its high alkalinity content. Therefore, the existence of coal boilers can be used for the burning of biomass-coal mixtures with minor modifications without the need for the design and construction of biomass combustion boilers (Turan 2009).

32.2 Material and Methods

In this study, five waste wood samples were selected to stand for wood processing wastes such as pine wood and four chemically treated wood products. These products are wooden window frame, wooden parquet, and two different furniture made of MDF used for the office furniture from two different factories (representing as f1 and f2). Pine tree samples (*Pinus brutia*) not containing any chemicals and additives were taken from a timber production center in Isparta. In addition to waste wood samples, two different local lignites (viz., Tuncbilek and Soma) were also examined for this study in order to find out the combustion characteristics of Turkish lignites. The sample abbreviation names are listed in Table 32.1.

Twenty-one raw lignite and raw waste wood samples were prepared, and their TG analyses were performed within the scope of the study. TGA results for pine and wooden parquet samples are given in (Fig. 32.1).

Experiments were carried out in the temperature ranging from 30 to 900 °C and three different heating rates: 10 °C/min, 40 °C/min, and 80 °C/min. Volatile matter

Table 32.1 The used abbreviations for samples

Sample	Used abbreviation
Pine	c10[a], c40, c80
Waste furniture 1	f110, f140, f180
Waste furniture 2	f210, f240, f280
Wooden window frame	w10, w40, w80
Wooden parquet	p10, p40, p80

[a]Heating rate as °C min^{-1}

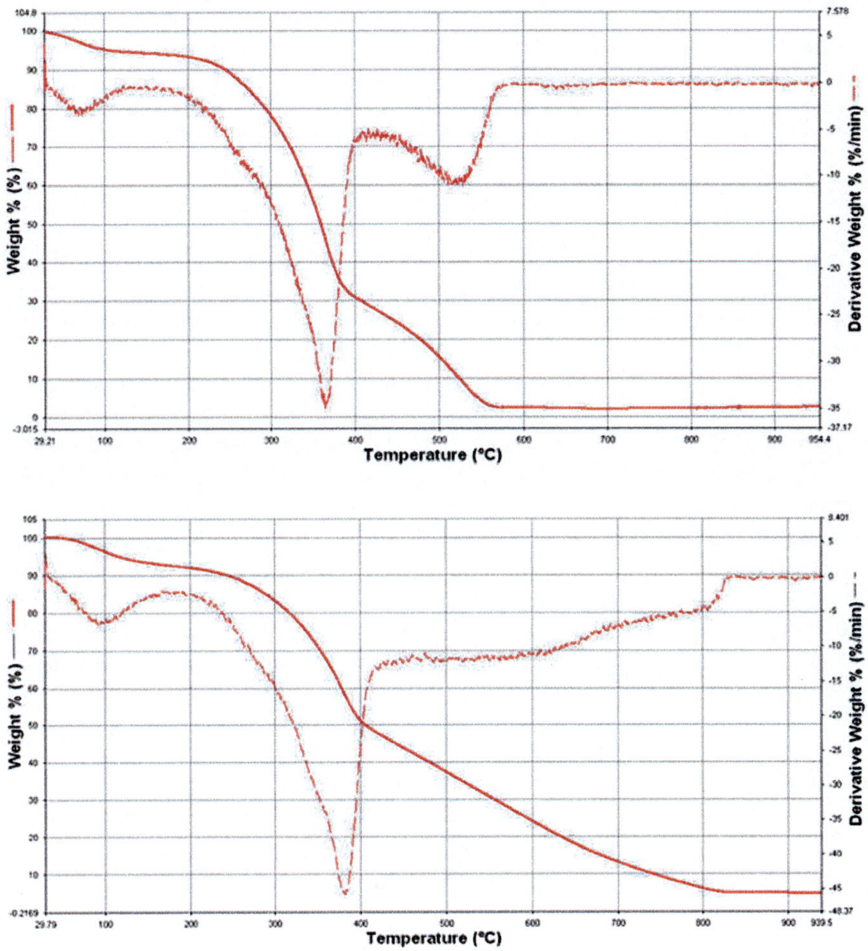

Fig. 32.1 TG and DTG curves of the pine and wooden parquet samples

contents of the samples ranged from 33.32% (Tuncbilek lignite) to 89.90% (pine). The volatile matter contents of the biomass samples were also found to be considerably higher than those of the lignite samples. On the other hand, the ash contents of the lignite samples were found to be quite high compared to the woody biomass, and the ash contents of the samples in the study were found to be between 0.63% (pine) and 42.6% (Soma lignite). The fixed carbon contents of the samples varied between 9.40% (pine) and 24.70% (Tuncbilek lignite). The carbon contents of the samples in this study were found to be between 46.90% (both wooden parquet and f1) and 74.89% (Soma lignite). Although the carbon content of lignite samples is higher than that of biomass, the sulfur content of lignite samples was found to be approximately 0.8%, while no sulfur was found in wood samples. Low nitrogen and sulfur content in any fuels are desirable properties in terms of causing lower nitrogen oxide and sulfur oxide emissions during the use of the fuel in thermochemical processes (Mierzwa-Hersztek et al. 2019).

32.3 Results and Discussion

To examine the combustion characteristics of the selected samples, different parameters such as ignition, burnout, maximum peak temperatures, DTG_{max} values, and the total burnout of the samples as a result of thermal decomposition were investigated.

Ignition temperature and burnout temperature are two of the important properties of the biofuels during the design and operation of thermochemical systems in industrial applications (Bampenrat et al. 2021). The ignition temperature indicates the point on the DTG curve where thermal decomposition begins. The ignition properties of biomass are important in optimizing the combustion behavior of biomass in the combustion units (Cao et al. 2017). The lower the ignition temperature, the easier to ignite the sample (Jia 2021). In this study, the ignition temperatures of the samples were found to be between 131 °C (wooden window frame at 10 °C/min) and 285 °C (Soma lignite at 80 °C/min) (Fig. 32.2). These results indicate that waste wood samples are favorable to combustion as a biomass due to their lower ignition temperatures.

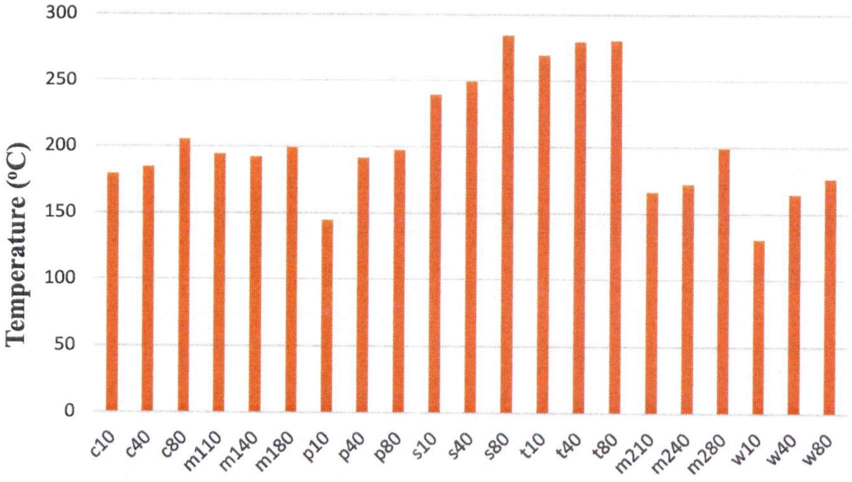

Fig. 32.2 Ignition temperatures of the samples

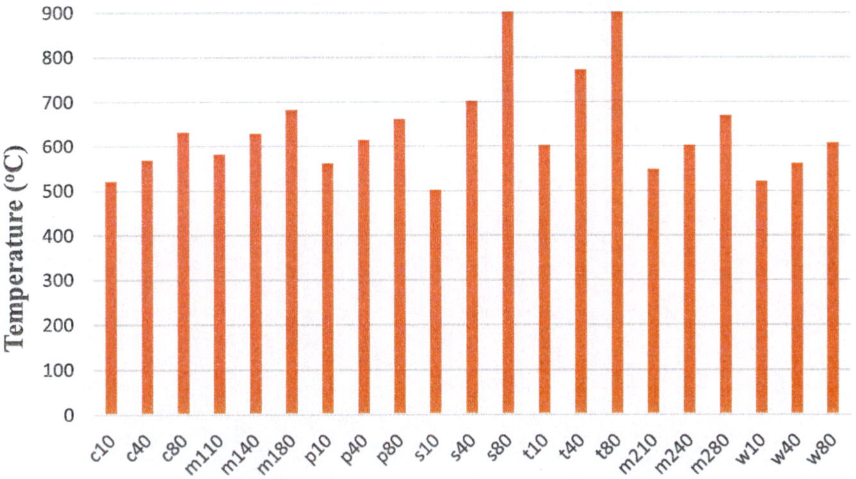

Fig. 32.3 Burnout temperatures of the samples

It was found that the ignition temperatures of waste wood samples were lower by 50–75 °C than that of lignite samples. Therefore, the ignition performance of the biomass can be increased by blending waste wood into lignite (Bampenrat et al. 2021).

Burnout temperature refers to the point where the thermal decomposition is completed. Higher burnout temperatures make samples more difficult to burn. Therefore, higher temperatures and longer residence time are observed at higher burnout temperatures for the complete conversion (Wnorowska et al. 2021). In the study, the burnout temperatures were ranged from 500 °C (Soma lignite at 10 °C/min) to 950 °C (Tuncbilek lignite at 80 °C/min) (Fig. 32.3). The total burnout of the samples also varies from 75 wt.% (Tuncbilek lignite at 80 °C/min) to 99.84 wt.% (f2 at 10 °C/min).

The peak temperature and maximum combustion rate (DTG_{max}) show the points where the weight loss rate is maximum and the weight loss on DTG curve is maximum, respectively (Varol et al. 2010). The peak temperatures of the first peaks (indicating main decomposition) ranged from 331 °C (f1 and pine at 10 °C/min) to 569 °C (Tuncbilek lignite at 80 °C/min) (Fig. 32.4).

The peak temperature is also accepted as an important parameter to evaluate the thermal stability of samples. Higher peak temperature indicates that the sample is thermally more stable (Amit et al. 2021). Therefore, two lignite samples were found to be thermally more stable compared to waste wood samples. Like peak temperatures, higher DTG_{max} values were also obtained in this study since heating rate increased (Fig. 32.5). The DTG_{max} values of the examined samples were found to be between 5.31%/min (Soma at 10 °C/min) and 70.86%/min (pine at 80 °C/min).

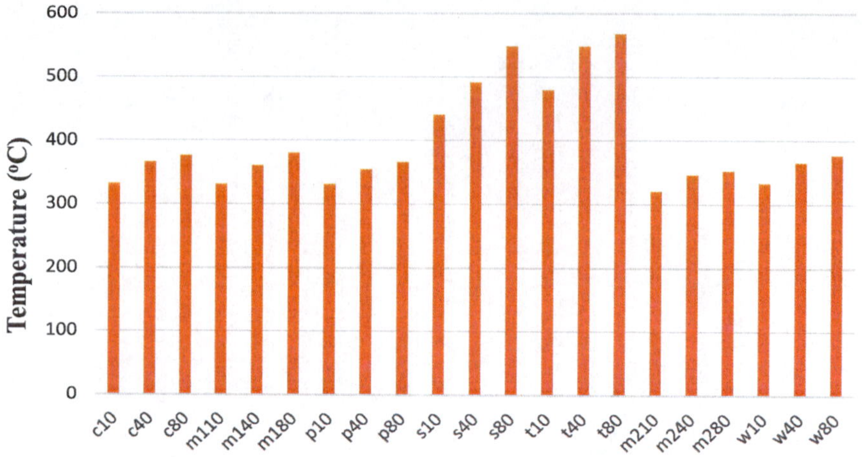

Fig. 32.4 Peak temperatures of the samples

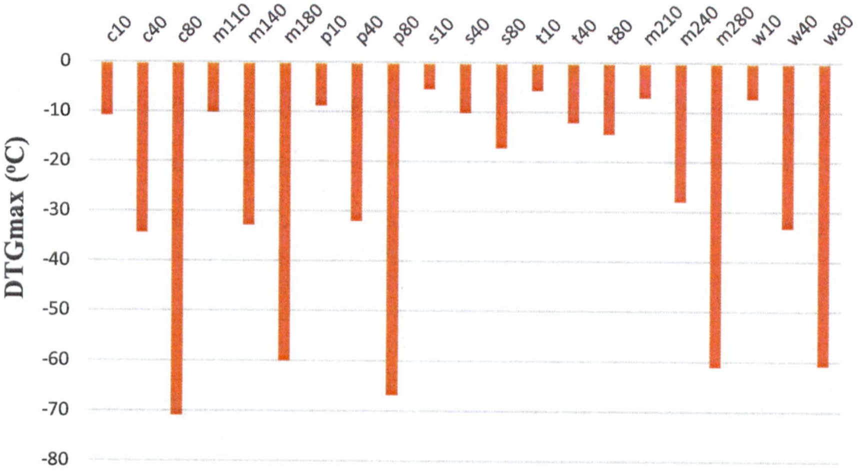

Fig. 32.5 DTG_{max} values of the samples

32.4 Conclusion

- It can be concluded that both DTG_{max} values and ignition points of the waste wood samples were obtained at lower temperatures than that of lignite samples.
- Biomass samples were found to be more reactive than local lignite samples. Accordingly, the blending of lignite with the waste wood in industrial applications and the combustion performance of lignite can be improved in the thermochemical systems.

References

Amit T, Roy R, Raynie DE (2021) Thermal and Structural Characterization of Two Commercially Available Technical Lignins for High-value Applications. https://doi.org/10.26434/chemrxiv.13570790.v1.

Bampenrat A, Sukkathanyawat H, Seangwattana T (2021) Coal/biomass co-combustion investigation by thermogravimetric analysis. In E3S Web of Conferences (Vol. 302, p. 01002). EDP Sciences.

Cao W, Li J, Lue L (2017) Study on the ignition behaviour and kinetics of combustion of biomass, Energy Procedia, 142:136–141. https://doi.org/10.1016/j.egypro.2017.12.022.

Jia G (2021) Combustion Characteristics and Kinetic Analysis of Biomass Pellet Fuel Using Thermogravimetric Analysis, Processes, 9(5):868. https://doi.org/10.3390/pr9050868.

Mierzwa-Hersztek M, Gondek K, Jewiarz M (2019) Assessment of Energy Parameters of Biomass and Biochars, Leachability of Heavy Metals and Phytotoxicity of Their Ashes, J Material Cycles Waste Manage. 21:786–800. https://doi.org/10.1007/s10163-019-00832-6

MTA, 2022. Coal exploration researches, https://www.mta.gov.tr/v3.0/arastirmalar/komur-arama-arastirmalari.

Turan AZ (2009) Linyit biyokütle karışımlarının oksijen ortamında yakılması (Doctoral dissertation, Fen Bilimleri Enstitüsü).

Turkey's energy strategy (2022) https://www.mfa.gov.tr/turkeys-energy-strategy.en.mfa.

Varol M, Atimtay AT, Bay B, Olgun H (2010) Investigation of co-combustion characteristics of low quality lignite coals and biomass with thermogravimetric analysis, Thermochimica acta, 510(1–2):195–201. https://doi.org/10.1016/j.tca.2010.07.014.

Wnorowska J, Ciukaj S, Kalisz S (2021) Thermogravimetric analysis of solid biofuels with additive under air atmosphere, Energies, 14(8):2257. https://doi.org/10.3390/en14082257.

World coal, World coal association (2012) Coal – Energy for Sustainable Development UK. https://sustainabledevelopment.un.org/getWSDoc.php?id=996

Chapter 33
Characterization of Post-consumer Household Plastic Waste: Assessing the Suitability for Hydrocarbon Fuel Production by Pyrolysis

Gulsun Gizem Taylan and Güray Yildiz

Nomenclature

HPW	Household plastic waste
PET	Polyethylene terephthalate
HDPE	High-density Polyethylene
PVC	Polyvinyl chloride
LDPE	Low-density polyethylene
PP	Polypropylene
PS	Polystyrene
PE	Polyethylene
PC	Polycarbonate
PLA	Polylactic acid
ABS	Acrylonitrile butadiene styrene
PA	Polyamide
PBT	Polybutylene terephthalate
ICP-MS	Inductively coupled plasma-mass spectrometry
TGA	Thermogravimetric analysis
DSC	Differential scanning calorimetry

G. G. Taylan
Department of Energy Systems Engineering, Faculty of Engineering, Izmir Institute of Technology, Izmir, Türkiye
e-mail: gulsuntaylan@iyte.edu.tr

G. Yildiz (✉)
Department of Environmental Engineering, Faculty of Engineering, Izmir Institute of Technology, Izmir, Türkiye
e-mail: gurayyildiz@iyte.edu.tr

M. Z. Sogut et al. (eds.), *Proceedings of the 2022 International Symposium on Energy Management and Sustainability*, Springer Proceedings in Energy,
https://doi.org/10.1007/978-3-031-30171-1_33

SEM-EDX Scanning electron microscopy-energy dispersive X-ray spectroscopy
RIC Resin identification codes

33.1 Introduction

Plastics are one of the most utilized materials in various industries such as construction, packaging, automotive, etc. The variety of plastic products has a wide range which includes various types of plastics that could be listed according to the Resin Identification Codes (RIC), as defined by ASTM D7611. These are polyethylene terephthalate (PET, code 1), high-density polyethylene (HDPE, code 2), polyvinyl chloride (PVC, code 3), low-density polyethylene (LDPE, code 4), polypropylene (PP, code 5), polystyrene (PS, code 6), and OTHER (code 7) (Elliott et al. 2020). OTHER is defined as the mixture of miscellaneous plastics, including polyethylene (PE), polycarbonate (PC), polylactic acid (PLA), acrylic, acrylonitrile butadiene styrene (ABS), polyamide (PA), polybutylene terephthalate (PBT), fibreglass, and nylon (Eriksen et al. 2019). Each plastic type demonstrates different features, e.g. PET is lightweight and resistant to pressure and thus is used in water and beverage bottles, whereas HDPE has long polymer chains and is highly crystalline, so the application area consists of detergent, shampoo, and oil containers (Sharuddin et al. 2016). PS is usually used in toys, medical stuff, and electronics due to its light and heat resilience nature and high strength and durability. PP has high resistance to heat and chemicals and high rigidity and, thus, is used in furniture production, carpets, and food packaging (Areeprasert et al. 2017; Fortelny et al. 2004). PVC also has high resistance to temperature and is versatile, so it can be used in the production of credit cards, medical devices, food foil, and packaging. However, the mismanagement of plastic waste has become a global-scale problem, creating serious negative effects on human health and the environment.

Pyrolysis, one of the thermochemical conversion technologies for the conversion of solid waste materials, focuses on the production of intermediate energy carriers at reaction temperatures ca. 500 °C and in the absence of oxygen. The oil produced by the pyrolysis of plastics has a high calorific value; this makes waste plastics a promising feedstock for producing alternative fuels.

HPW is extremely heterogeneous, and hence, its plastic-type distribution plays a critical role in the pyrolysis process. Different plastic types that have varying elemental compositions affect the yield and quality of pyrolysis products. Pyrolysis of plastics having high volatile matter contents results in high liquid product yields. Conversely, high ash containing plastic feedstocks decreases the liquid yield. Among the plastic types, polyolefins (i.e. HDPE, LDPE, and PP) have the highest volatile matter contents of ca. wt. 99%. While conventional fossil-based diesel has a calorific value of 43 MJ/kg, according to the ASTM 1979 standard, the calorific values of produced liquid oil from polyolefins by pyrolysis are higher than 40 MJ/kg (Kumagai and Yoshioka 2020). Polyvinyl chloride (PVC) and PET are problematic feedstocks for pyrolysis due to the high chlorine content in PVC and ester content in

PET. High chlorine content results in HCl formation and chlorinated liquid products. Since HCl release is completed at ca. 310 °C, stepwise systems can be considered suitable for the removal of HCl and improving the quality of the liquid product. Benzoic acid formation due to ester content in PET reduces the pyrolysis oil quality. Besides, PET and PVC have lower C/H ratios that result in lower calorific values compared to that of polyolefins.

The intention is to report our research results concerning the assessment of the suitability of post-consumer HPW as a feedstock for the pyrolysis process. The ultimate goal is to come up with some recommendations and suggestions regarding the requirements for optimal feedstocks for pyrolysis on a commercial/industrial scale.

33.2 Methodology

For this research, a total of 78 kg of HPW was collected over a period of 4 months, specifically between May 2021 and September 2021. The collected material was manually classified according to the Resin Identification Codes (RIC) as defined by ASTM D7611 (Del Rey Castillo et al. 2020). Moreover, unidentified plastic labels were considered a stand-alone category (i.e. label). Figure 33.1 shows the plastic-type distribution of collected HPW. PVC was not determined in the mixture.

Individual HPW samples and HPW mix samples were prepared by considering the weight per cent of plastic-type distribution in total collected plastic waste. HPW MIX 1 represents the constituent distribution of collected HPW (as shown in Fig. 33.1), while HPW MIX 2 consists of the polyolefin distribution percentages in the collected HPW. HPW MIX 2 has constituent percentages of 40.5%, 17.1%, and 42.4% for HDPE, LDPE, and PP, respectively.

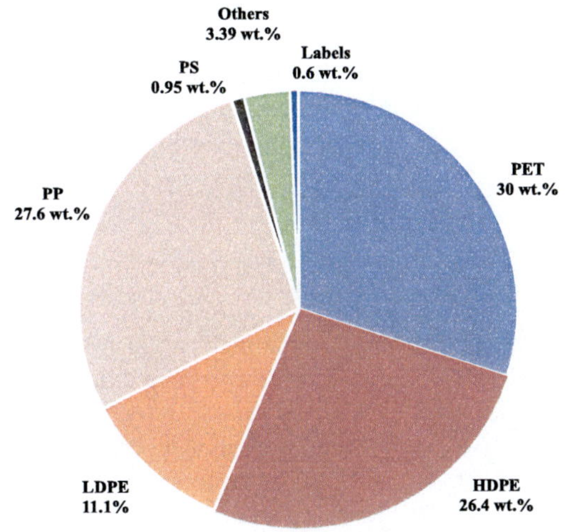

Fig. 33.1 Distribution of plastic types in HPW

Table 33.1 Equipment used for the physicochemical characterization of HPW samples

Analysis	Equipment
CHN/S	Leco TruSpec CHN Analyzer
Heating value	Parr 6300 Oxygen Bomb Calorimeter
Metal content	Agilent 7500c Octopole Reaction (ICP-MS) CEM Mars 6 (Microwave Digestion System)
TGA	Perkin Elmer Diamond TG/DTA
DSC	Perkin Elmer Diamond TG/DTA
SEM-EDX	FEI QUANTA 250 FEG

Plastic samples were characterized via the ultimate and proximate analyses, ICP-MS, TGA, DSC, and SEM-EDX. The brands of equipment used for the physicochemical characterization of HPW samples were summarized in Table 33.1. Proximate analyses were performed according to the related ASTM standards. ASTM D3173, ASTM D3174, and ASTM D3175 were implemented for the determination of moisture, ash, and volatile matter contents of HPW specimens, respectively. Fixed carbon was determined by difference. Ultimate analysis was used to determine the CHN/S contents of plastic samples. Oxygen content was also calculated by difference.

TGA and derivative thermogravimetric (DTG) curves were used to evaluate the thermal stability of HPW samples and obtain their behaviour under a certain temperature regime. For the TGA tests, the sample temperature was increased from 25 to 800 °C by applying a heating rate of 10 °C/min with an N_2 flow rate of 40 mL/min. DSC was performed to determine the characteristic endothermic phase transition temperatures that are commonly used for the characterization of polymer mixtures. During the DSC analysis, the sample was heated from 25 to 350 °C with a heating rate of 10 °C/min with an N_2 flow rate of 40 mL/min. Metal content was investigated by ICP-MS. As a pretreatment step prior to the ICP-MS analyses, microwave digestion was performed at 210 °C with the addition of 10 mL HNO_3 to the 0.25 g of sample for transferring metal content from the solid phase into the liquid phase. Another characterization technique that is applied to HPW is SEM-EDX used for determining any trace elements which were not detected by ICP-MS since their quantity is below the limit.

33.3 Results and Discussion

Table 33.2 shows the proximate, ultimate analysis results and higher heating values of waste plastic types. Based on the proximate analysis results, the carbon content of plastic samples is higher than 80 wt.% except for PET. Also, PET has the highest oxygen content, resulting in oxygen content liquid fuel obtained by pyrolysis. The carbon, hydrogen, and oxygen content of plastic samples are in line with other studies in the literature (Sharuddin et al. 2017; Chhabra et al. 2019). In this study, nitrogen and sulphur contents were not determined in the ultimate analysis.

Table 33.2 Results of ultimate analysis (wt.%), proximate analysis (wt.%), and heating values (cal/g and MJ/kg) of individual HPW samples

	PET	HDPE	LDPE	PP	PS	OTHER
Ultimate analysis						
C	60.0	85.1	81.6	80.9	90.4	79.9
H	4.4	14.4	13.2	13.5	8.3	12.6
O[a]	35.5	0.5	5.2	5.6	1.3	7.4
N	0	0	0	0	0	0
S	0	0	0	0	0	0
Proximate analysis						
Ash	0.08	0.96	1.74	2.30	1.17	1.16
Moisture content	0.3	0.3	1.3	0.4	0.5	0.4
Volatile matter	89	97.1	93.3	94.8	96.3	96.1
Fixed carbon[a]	10.6	1.6	3.7	2.6	2.1	2.4
Heating values						
HHV (cal/g)	5463	10,839	10,332	10,279	9885	9881
HHV (MJ/kg)	22.9	45.4	43.3	43.0	41.4	41.3

[a]Calculated by the difference

Table 33.3 Results of proximate analysis (wt.%) of HPW mix samples

	HPW MIX 1	HPW MIX 2
Proximate analysis		
Ash	1.01	0.64
Moisture content	0.29	0.22
Volatile matter	96.06	97.01
Fixed carbon*	2.64	2.12

*By difference

Table 33.3 shows the proximate analysis results of HPW mix samples. HPW MIX 1 which consists of PET has higher ash and moisture content compared to HPW MIX 2 which results in low liquid and high solid product yield at the end of the pyrolysis process.

Figure 33.2 shows the TGA curves of different types of plastics. It can be seen from the graph that the thermal degradation starts between 250 and 320 °C for individual plastic types. PP degraded thermally at the lowest temperature (250 °C), while the "Labels" required the highest temperature (320 °C). Sharuddin et al. (2017) have studied TGA analyses for plastic types, and the results were in the same line as these results.

Figure 33.3 shows the DTG curves of plastic types to investigate the temperature for maximum mass loss. The maximum mass losses have occurred between 425 and 480 °C for all plastic types. In another study, the temperatures for maximum mass losses for all individual plastic types were reported between 450 and 500 °C (Manivannan and Seehra 1997).

Figure 33.4 shows the melting points of plastic types which were obtained by DSC. In this study, melting points of LDPE, HDPE, PP, and PET were observed as

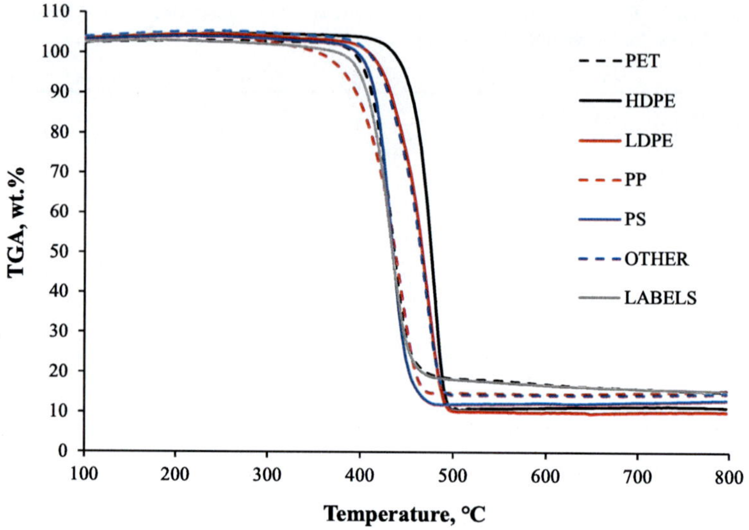

Fig. 33.2 TGA curves of individual plastic types

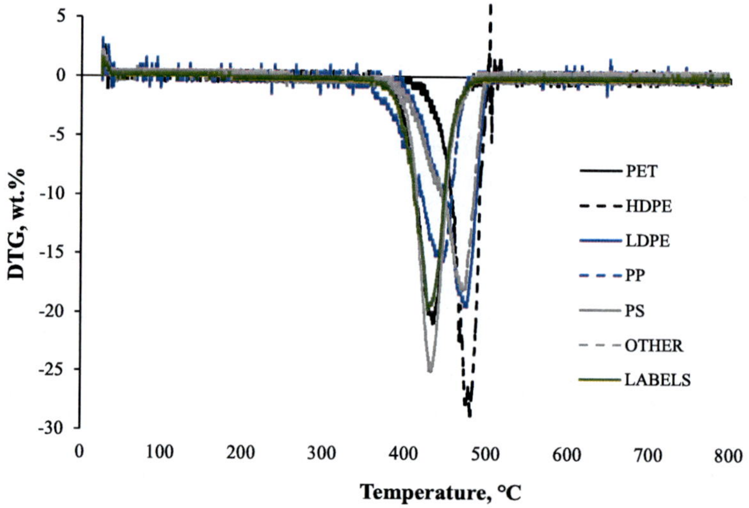

Fig. 33.3 DTG curves of individual plastic types

110 °C, 120 °C, 160 °C, and 250 °C, respectively. Determined melting points are in line with other studies in the literature. The research of Majewsky et al. (2016) indicated the melting temperatures of both individual and mixed plastic samples; a melting temperature of PE in mixed plastic and individually is observed as 100 °C and PP of 165 °C, where PET showed the highest melting temperature of 250 °C.

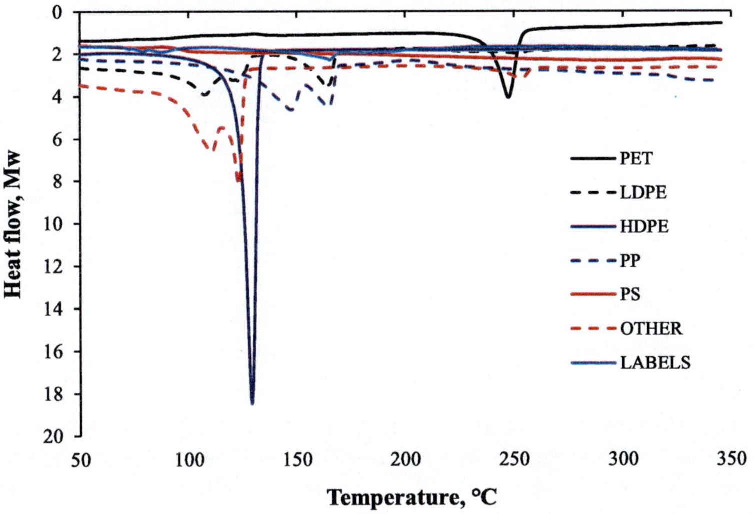

Fig. 33.4 DSC curves of individual plastic types

Moreover, Calero et al. (2018) studied the melting temperature of HDPE, LDPE, and PP samples, and the results showed that PE samples melted at 120–130 °C, whereas PP samples melted at 160 °C. A small amount of deviation was observed, and this was because of the variety of use of plastics.

While TGA and DSC analyses were conducted for assessing the thermal behaviour, ICP-MS and SEM-EDX analyses were conducted to obtain information about the elemental composition of HPW samples. Calcium, aluminium, magnesium, and sodium were indicated as dominant elements by ICP-MS results due to the utilization of several additives in plastic production. Calcium is mainly used as a filler and a photo-stabilizer in the forms of calcium carbonate, calcium silicate, and calcium zinc, and it can be used as an antacid and lubricant in the forms of calcium sulphate, etc. (Gala et al. 2020; Canopoli et al. 2020). Aluminium and magnesium compounds are widely used as flame retardants and as additives for providing high chemical resistance (Turku et al. 2017). Moreover, iron may come from inorganic pigments, while the reason for high concentrations of potassium and sodium may be explained due to K and Na being used as thermal stabilizers (Gala et al. 2020; Klöckner et al. 2021).

Table 33.4 shows the elemental compositions of scanned areas of individual HPW samples by weight percentages that were obtained by SEM-EDX. SEM-EDX results prove the presence of some metals in the sample which were not detected by ICP-MS since their quantity is below the limit. Titanium, silicon, and chlorine were the main elements detected by SEM-EDX. In general, titanium, in the form of TiO_2, is used as a pigment, while calcium and silicon oxides may be used as a reinforcing filler that provides tensile strength (Turku et al. 2017). Moreover, halogenated flame-retardant additives may contain bromine and chlorine that cause environmental concerns about hazardous gas emissions (Zhang et al. 2016).

Table 33.4 Elemental compositions of individual plastic types used in this study

	Element, wt.%									
Plastic type	C	O	Na	Si	Ti	Cl	N	Cu	Al	Ca
PET	64.9	35.0	0.08	0	0	0	0	0	0	0
HDPE	98.8	1.25	0	0	0	0	0	0	0	0
LDPE	99.4	0.55	0	0.04	0	0	0	0	0	0
PP	75.2	16.2	0	0.95	2.90	1.73	2.60	0.30	0.06	0
PS	97.5	1.98	0	0	0.40	0	0	0	0	0.35
OTHER	76.9	22.5	0	0	0.46	0	0	0	0	0.00

33.4 Conclusions

The detailed characterization of different plastic types of HPW collected in Izmir provided information for the evaluation of their suitability as raw feedstock to achieve high conversion to the liquid product by pyrolysis process. In this study, it was concluded that the elemental composition of plastic waste can vary since several additives are used in the plastic production industry to improve the plastic properties for utilization in different sectors. According to TGA-DSC curves, all plastic types complete their thermal degradation at 500 °C, and melting points vary depending on the plastic type. A pretreatment unit/reactor can be added to the system to either provide HCl removal in the presence of PVC in the HPW mixture or provide feeding of the molten plastic to the pyrolysis reactor that will increase the homogeneity of HPW. The determined melting temperatures can be used for process optimization if pretreatment will be used for the pyrolysis system. The temperatures obtained with the DTG curves can be evaluated in optimizing the pyrolysis process to achieve maximum mass conversion. Consequently, through a proper and efficient sorting process, a liquid product with high yields could be obtained with polyolefins, since they have high volatile and low ash and moisture contents. For future studies, a detailed characterization of HPW mix samples and the liquid product yields of plastic types and HPW mix samples in the pyrolysis process can be assessed.

Acknowledgements This work is supported and funded by the UK Department for Business, Energy and Industrial Strategy together with the Scientific and Technological Research Council of Turkey (TÜBİTAK; Project No. 119N302) and delivered by the British Council.

References

Areeprasert C, Asingsamanunt J, Srisawat S, Kaharn J, Inseemeesak B, Phasee P, and Chiemchaisri C (2017) Municipal plastic waste composition study at transfer station of Bangkok and possibility of its energy recovery by pyrolysis. Energy Procedia 107:222–226. https://doi.org/10.1016/j.egypro.2016.12.132

Calero M, Martín-Lara MA, Godoy V, Quesada L, Martínez D, Peula F, Soto J M (2018) Characterization of plastic materials present in municipal solid waste: Preliminary study for their mechanical recycling. https://doi.org/10.31025/2611-4135/2018.13732

Canopoli L, Coulon F, Wagland ST (2020) Degradation of excavated polyethylene and polypropylene waste from landfill. Science of the Total Environment 698:134125. https://doi.org/10.1016/j.scitotenv.2019.134125

Chhabra V, Bhattacharya S, Shastri Y (2019) Pyrolysis of mixed municipal solid waste: Characterisation, interaction effect and kinetic modelling using the thermogravimetric approach. Waste Management 90:152–167. https://doi.org/10.1016/j.wasman.2019.03.048

Del Rey Castillo E, Almesfer N, Saggi O, Ingham JM (2020) Light-weight concrete with artificial aggregate manufactured from plastic waste. Construction and Building Materials 265:120199. https://doi.org/10.1016/j.conbuildmat.2020.120199

Elliott T, Gillie H, Thomson A (2020) European Union's plastic strategy and an impact assessment of the proposed directive on tackling single-use plastics items. In Plastic waste and recycling, Academic Press, pp. 601–633 https://doi.org/10.1016/B978-0-12-817880-5.00024-4

Eriksen MK, Damgaard A, Boldrin A, Astrup TF (2019) Quality Assessment and Circularity Potential of Recovery Systems for Household Plastic Waste. Journal of Industrial Ecology 23: 156–168. https://doi.org/10.1111/jiec.12822

Fortelný I, Michálková D, Kruliš Z (2004) An efficient method of material recycling of municipal plastic waste. Polymer degradation and stability 85(3):975–979. https://doi.org/10.1016/j.polymdegradstab.2004.01.024

Gala A, Guerrero M, Serra JM (2020) Characterization of post-consumer plastic film waste from mixed MSW in Spain: A key point for the successful implementation of sustainable plastic waste management strategies. Waste Management 111:22–33. https://doi.org/10.1016/j.wasman.2020.05.019

Klöckner P, Reemtsma T, Wagner S (2021) The diverse metal composition of plastic items and its implications. Science of the Total Environment 764:142870. https://doi.org/10.1016/j.scitotenv.2020.142870

Kumagai S, Yoshioka T (2020) Latest trends in pyrolysis gas chromatography for analytical and applied pyrolysis of plastics, Journal of the Japan Petroleum Institute 63:345–364. https://doi.org/10.1627/jpi.63.345

Manivannan A, Seehra MS (1997) Identification and quantification of polymers in waste plastics using differential scanning calorimetry. Preprints of Symposia-Division of Fuel Chemistry American Chemical Society 42:1028–1030.

Majewsky M, Bitter H, Eiche E, Horn H (2016) Determination of microplastic polyethylene (PE) and polypropylene (PP) in environmental samples using thermal analysis (TGA-DSC). Science of the Total Environment 568:507–511. https://doi.org/10.1016/j.scitotenv.2016.06.017

Sharuddin SDA, Abnisa F, Daud WMAW, Aroua MK (2016) A review on pyrolysis of plastic wastes. Energy conversion and management 115:308–326. https://doi.org/10.1016/j.enconman.2016.02.037

Sharuddin SDA, Abnisa F, Daud WMAW, Aroua MK (2017) Energy recovery from pyrolysis of plastic waste: Study on non-recycled plastics (NRP) data as the real measure of plastic waste. Energy Conversion and Management 148:925–934. https://doi.org/10.1016/j.enconman.2017.06.046

Turku I, Kärki T, Rinne K, Puurtinen A (2017) Characterization of plastic blends made from mixed plastics waste of different sources. Waste Management & Research 35(2):200–206. https://doi.org/10.1177/0734242X16678066

Zhang M, Buekens A, Li X (2016) Brominated flame retardants and the formation of dioxins and furans in fires and combustion. Journal of hazardous materials 304:26–39. https://doi.org/10.1016/j.jhazmat.2015.10.014

Chapter 34
Analysis of Pyrolysis Process Parameters for the Maximized Production of Gasoline-Range Renewable Fuels from High-Density Polyethylene

Ecrin Ekici and Güray Yildiz

Nomenclature

HDPE	High-density polyethylene
WtE	Waste to energy
NC	Non-catalytically
CSTR	Continuously stirred reactor
CSBR	Continuously spouted bed reactor
BFBR	Bubbling fluidized bed reactor
TGA	Thermal gravimetric analysis
DTG	Derivative thermogravimetry
CAPEX	Capital expenditures
OPEX	Operational expenses

34.1 Introduction

Owing to their durability, being cheap, and easy to process, the demand for the production of plastics has been growing since the middle of the twentieth century (Armenise et al. 2021). This demand, which is mainly driven by packaging, building and construction, automotive, and electrical equipment industries, resulted in the production of 55 Mt and 367 Mt of plastics in EU27+3 countries and the world,

E. Ekici · G. Yildiz (✉)
Department of Energy Systems Engineering, Faculty of Engineering, Izmir Institute of Technology, Izmir, Turkey
e-mail: ecrinekici@iyte.edu.tr; gurayyildiz@iyte.edu.tr

M. Z. Sogut et al. (eds.), *Proceedings of the 2022 International Symposium on Energy Management and Sustainability*, Springer Proceedings in Energy,
https://doi.org/10.1007/978-3-031-30171-1_34

respectively (PlasticsEurope 2022). There is no doubt that plastic waste generation, linked directly to the rate of plastic production, is threatening the environment. One of the approaches to deal with this problem is the waste-to-energy (WtE) approach aiming to recover energy, monomer, and fuel from waste plastics by incineration, mechanical recycling, and chemical recycling, respectively. Incineration is meant to generate heat and power, while mechanical recycling mainly involves extruding and reshaping used plastics. Pyrolysis, on the other hand, is a thermochemical conversion technology leading to the production of fuel-like products from plastics.

In the perspective of life-cycle assessments, Jeswani et al. (2021) stated that incineration releases 50% more carbon emissions compared to pyrolysis. Mechanical recycling is the least harmful to the environment; however, the requirement of utilizing properly cleaned and classified plastics generate additional OPEX. Pyrolysis can deal with individual or mixed plastics regardless of their physical condition. The primary product of plastic pyrolysis is a liquid that contains hydrocarbons with varying carbon numbers, generally starting from C5 up to C40 (e.g., wax). Gasoline (C5–C12) and diesel (C13–C20) range hydrocarbons (Elordi et al. 2011) show similar physicochemical properties with petroleum-based conventional fuels. Wax, non-condensable gases, and char are the by-products of plastic pyrolysis, and their combined energy contents are more than enough to satisfy the energy requirements of the process (Jeswani et al. 2021). Incinerating the plastic feedstock equivalent to 10% of what is fed to the pyrolysis reactor is enough to sustain the energy requirements of a pyrolysis process (Dispons 2006). The utilization of by-products as an energy supply decreases the carbon footprint of the process and reduces external electricity consumption for the operation of the system, making it an economically viable process (Fivga and Dimitriou 2018).

HDPE is reported to be the third most abundant type of plastic in waste plastic streams (PlasticsEurope 2022); this makes it a suitable and considerable feedstock for pyrolysis for the production of fuel-like products. Virgin HDPE has a high C/H ratio (ca. 6.6) and contains high volatile matter (ca. 99.2 wt.%) and trace amounts of moisture and ash (Kumagai et al. 2020). However, the high viscosity and low heat conductivity of HDPE limit the heat transfer during its pyrolysis. Thus, the type of the pyrolysis reactor, among other process parameters such as the reaction temperature, residence time, etc., directly affects the quantity and quality of the products of HDPE pyrolysis (Czajczyńska et al. 2017). The main product of non-catalytically pyrolyzed HDPE is wax. Although wax is a valuable product, it cannot be utilized as a fuel for conventional internal combustion engines without being upgraded. Upgrading wax requires additional process units (e.g., feeders, pressurized reactors) and promoters (e.g., steam, catalysts), creating additional CAPEX and OPEX. Limiting wax formation and enhancing fuel production is possible by setting up optimum operating parameters and thermal (non-catalytic) pyrolysis units. In the pyrolysis of HDPE, increasing the pyrolysis temperature favors the cracking of waxes into smaller hydrocarbons (e.g., gasoline range) and even to gases (Elordi et al. 2011). Another option is increasing the vapor residence time; it is stated in many studies that longer vapor residence times result in hydrocarbons with lower molecular weights (Berrueco et al. 2002; Mastral et al. 2006). To set an optimum pyrolysis temperature and a vapor residence time, the pyrolysis reactor should be adjustable. Batch reactors allow easier operations, but their non-flexible structure restricts controlling and tailoring the operational parameters. On the other hand, continuously operated reactors such as bubbling fluidized bed (BFBR) and continuously stirred tank reactors (CSTR) overcome mass and heat transfer limitations (Miandad et al. 2016), and, hence, prone to be scaled and used in large-scale industrial processes.

This work compiles the literature reports addressing the results of HDPE pyrolysis obtained from non-catalytic and continuously operated pyrolysis systems. The ultimate goal of this review article is to perceive the optimum process parameters for the maximized production of liquid products within the gasoline range (C5–C12). The review will summarize the current status and the achievements of research and technology development, based on the scientific literature, the available reports on commercial/industrial attempts, and also the scientific expertise in plastic pyrolysis obtained in our research group. The main goal is to come up with some recommendations and suggestions regarding the design of a plastic pyrolysis process on a commercial/industrial scale.

34.2 Methodology

In this work, Web of Science and Scopus were used for the collection of literature data from the articles published between 1984 and 2021. Keywords of "pyrolysis" and "high-density polyethylene" were utilized for article searching. Obtained articles were categorized into three groups by the operation mode of pyrolysis: *batch*, *semi-batch*, and *continuous*. The papers reporting the results of *continuous* operation were further split up into two sets considering the catalyst usage: *non-catalytic* and *catalytic*. In total, 18 articles were collected that fall into the category of *non-catalytic*; they were precisely examined, and the reported information, such as the operational parameters (i.e., temperature and residence time), process units (i.e., reactors), and product characteristics (i.e., yields), was recorded in a Microsoft Excel file. The analysis was used to determine the optimum conditions for maximum production of C5–C11 hydrocarbons by thermal and continuous pyrolysis of HDPE.

Based on the reported results of TGA analyses, HDPE starts to crack at temperatures between 413 and 565 °C under a nitrogen atmosphere. The highest degradation rate was obtained around 480 °C via DTG (Diaz-Silvarrey et al. 2018; Saad et al. 2021). Several studies report a 99 wt.% of conversion of the initial mass of HDPE in TGA (Kumagai et al. 2020). In light of these data, 600 °C and 70 wt.% were set as the highest allowable pyrolysis temperature and the minimum yield for pyrolysis oil production from HDPE (shown as dashed lines in Fig. 34.1), respectively.

Gasoline-range hydrocarbon production and wax cracking as a function of residence time were also investigated in CSBRs and BFBRs (see Fig. 34.2). Since no data were reported about residence time of hot vapors in CSTRs in the non-catalytic set, CSTRs were omitted. For CSBRs, only the data falling into the

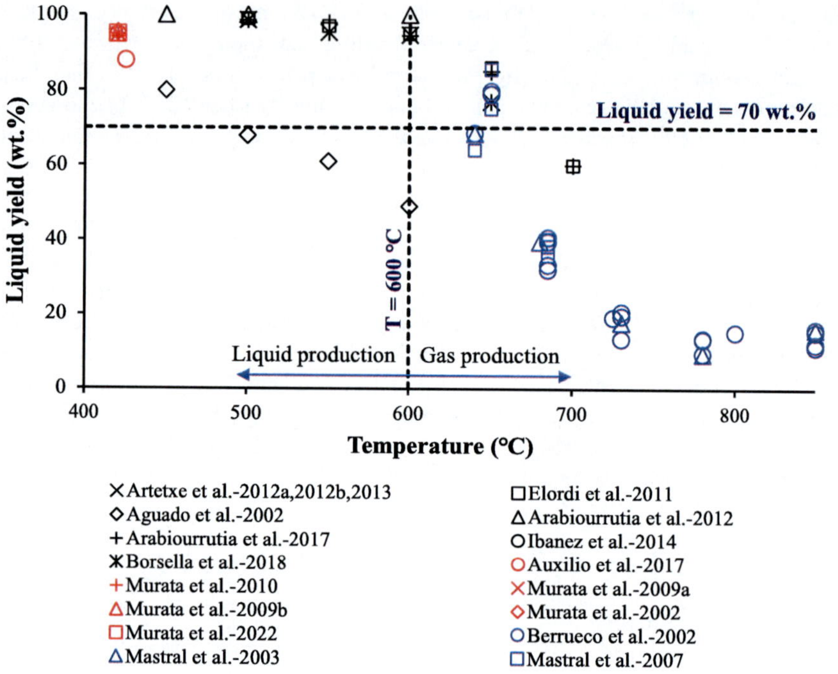

Fig. 34.1 Liquid production from pyrolysis of HDPE in CSBR (black), CSTR (red), and BFBR (blue)

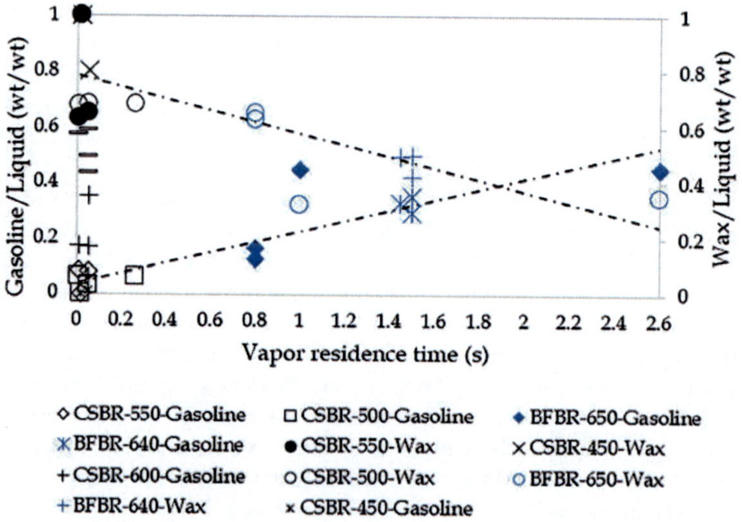

Fig. 34.2 Effect of vapor residence time on liquid gasoline production in CSBRs (black) and BFBRs (blue)

pyrolysis oil production range were selected. Since all the reported data for BFBRs were out of the temperature limit, only the yield limitation mentioned above was applied to give a point of view for the vapor residence time. Based on the reported results, primary pyrolysis vapors were not allowed to stay longer than 0.27 and 1.5 s in CSBRs and BFBRs, respectively.

34.3 Results and Discussion

Figure 34.1 shows that the pyrolysis temperatures higher than 600 °C resulted in an overall decrease in the liquid yield. At the same time, gas production was enhanced due to secondary cracking, independent of the reactor type (Lopez et al. 2017). Due to their allowance for elevated HDPE residence times, and, hence, their efficient heat transfer capabilities during pyrolysis, CSTRs are found to be the best-performing reactors for gasoline-range fuel production among the examined reactors for non-catalytic and continuous pyrolysis of HDPE. CSBRs were operated at higher temperatures compared to CSTRs (Murata et al. 2002, 2009, 2010, 2022; Auxilio et al. 2017). The studies reported that wax is the main product. The trend can be explained by the heat and mass transfer rates in the reactors. As seen in Fig. 34.1, BFBRs were operated out of the liquid product range compared to CSBRs (Lopez et al. 2017). Thus, gas production was the target for these experiments.

Figure 34.2 shows the effects of vapor residence time on gasoline-range hydrocarbon and wax production in CSBRs and BFBRs. According to Fig. 34.2, wax formation in pyrolysis oil decreases. In contrast, gasoline formation is enhanced by increased vapor residence time. Thermal pyrolysis of HDPE at 450 °C produces almost just wax in CSBRs since the temperature is lower than the maximum cracking temperature stated above. At this temperature, instant removal of hot vapors in less than 0.1 s from a CSBR is suitable for wax formation rather than gasoline production (Aguado et al. 2002; Elordi et al. 2011; Arabiourrutia et al. 2012, 2017; Artetxe et al. 2012, 2013; Ibáñez et al. 2014). Although increasing the temperature of pyrolysis from 450 to 500 °C or 550 °C converts 20 wt.% of wax to diesel range hydrocarbons, both increasing temperature and vapor residence time have an insignificant effect on the formation of C5–C12 hydrocarbons. Pyrolysis of HDPE at 600 °C results in the highest yield of gasoline-range hydrocarbons in CSBRs, but it is still not the optimum condition. CSBRs are known as suitable reactors for flash pyrolysis and provide flexibility for tailoring residence time. Thus, increasing residence time at lower temperatures provides a more economical process for gasoline production in CSBRs since vapor residence time is insufficient to further crack at this temperature. As the target product is generally gas for BFBRs, the reported pyrolysis temperatures and vapor residence times for BFBRs are higher compared to CSBRs. However, pyrolysis of HDPE in BFBRs at 650 °C leads to more than 70 wt.% liquid yields and a significant amount of C5–C12, so BFBRs are also examined in this case concept.

The optimum temperature and residence time for non-catalytic pyrolysis of HDPE for maximizing the yield of C5–C12 hydrocarbons are 650 °C and 1 s, respectively. Residence time longer than a second in BFBRs causes excessive cracking of vapors so that the gas production is enhanced (Berrueco et al. 2002; Mastral et al. 2003, 2007). For instance, there is no significant difference in the gasoline yield at 650 °C for vapor residence times of 1 s and 2.6 s. However, a longer vapor residence time in the BFBR should not be allowed if a maximum liquid yield is aimed at. On the other hand, no fair comparison can be developed for the reactors regarding vapor residence time due to the lack of information for CSTRs. But it should be known that the gasoline-range hydrocarbon yield for CSTRs is reported between 36 wt.% and 60 wt.%, which is higher than the yields obtained in the other reactors (i.e., BFBRs and CSBRs).

34.4 Conclusion

Published articles reporting the results of pyrolysis of HDPE performed in non-catalytically and continuously operated systems were collected and further examined with an aim of determining the best-performing reactor type, optimum pyrolysis temperature, and vapor residence time that yields a maximization in gasoline-range (C5–C12) hydrocarbons. To give a detailed explanation for the optimum conditions for gasoline-like fuel production, vapor residence times are also investigated for the reactors except for CSTRs since no data is available in the data set.

- CSTRs are found to be the best-performing reactors for both liquid and gasoline production at lower temperatures (ca. 420 °C) compared to the operating temperatures CSBRs and BFBRs.
- The maximum residency of vapors of HDPE in CSBRs is reported to be 0.05 s.
- CSBRs are suitable reactors for wax formation rather than lighter fuel production for all the operating conditions.
- The data obtained in BFBRs are in the gas formation range.
- A temperature of 650 °C and vapor residence time of 1, which lead to the maximum gasoline yield (35 wt.%) in BFBRs, are the optimum conditions for BFBRs.

CSTRs are highly recommended to produce hydrocarbons in the gasoline range (C5–C12) from HDPE because less wax forms and the C5–C12 yield is higher under mild conditions.

Acknowledgments This work is supported and funded by the UK Department for Business, Energy and Industrial Strategy together with the Scientific and Technological Research Council of Turkey (TÜBİTAK; Project No. 119N302) and delivered by the British Council.

References

Aguado R, Olazar M, San José MJ (2002) Wax formation in the pyrolysis of polyolefins in a conical spouted bed reactor. Energy and Fuels. 16:1429–1437. https://doi.org/10.1021/ef020043w

Arabiourrutia M, Elordi G, Lopez G (2012) Characterization of the waxes obtained by the pyrolysis polyolefin plastics in a conical spouted bed reactor. J Anal Appl Pyrolysis. 94:230–237. https://doi.org/10.1016/j.jaap.2011.12.012

Arabiourrutia, M, Elordi, G, Olazar, M, Bilbao, J (2017). Pyrolysis of Polyolefins in a Conical Spouted Bed Reactor: A Way to Obtain Valuable Products. In: Samer M (ed) Pyrolysis. IntechOpen, London, pp. 285–304. https://doi.org/10.5772/67706

Armenise S, SyieLuing W, Ramírez-Velásquez JM (2021) Plastic waste recycling via pyrolysis: A bibliometric survey and literature review. J Anal Appl Pyrolysis. 158:. https://doi.org/10.1016/j.jaap.2021.105265

Artetxe M, Lopez G, Amutio M (2012) Light olefins from HDPE cracking in a two-step thermal and catalytic process. Chem Eng J. 207–208:27–34. https://doi.org/10.1016/j.cej.2012.06.105

Artetxe M, Lopez G, Amutio M (2013) Cracking of high density polyethylene pyrolysis waxes on HZSM-5 catalysts of different acidity. Ind Eng Chem Res. 52:10637–10645. https://doi.org/10.1021/ie4014869

Auxilio AR, Choo WL, Kohli I (2017) An experimental study on thermo-catalytic pyrolysis of plastic waste using a continuous pyrolyser. Waste Manag. 67:143–154. https://doi.org/10.1016/j.wasman.2017.05.011

Berrueco C, Mastral EJ, Esperanza E, Ceamanos J (2002) Production of waxes and tars from the continuous pyrolysis of high density polyethylene. Influence of operation variables. Energy and Fuels. 16:1148–1153. https://doi.org/10.1021/ef020008p

Czajczyńska D, Anguilano L, Ghazal H (2017) Potential of pyrolysis processes in the waste management sector. Therm Sci Eng Prog. 3:171–197. https://doi.org/10.1016/j.tsep.2017.06.003

Diaz-Silvarrey LS, McMahon A, Phan AN (2018) Benzoic acid recovery via waste poly(ethylene terephthalate) (PET) catalytic pyrolysis using sulphated zirconia catalyst. J Anal Appl Pyrolysis. https://doi.org/10.1016/j.jaap.2018.08.014

Dispons J (2006) Continuous Thermal Process for Cracking Polyolefin Wastes to Produce Hydrocarbons. In: John S, Kaminsky W (eds) Feedstock Recycling and Pyrolysis of Waste Plastics: Converting Waste Plastics into Diesel and Other Fuels, 1st edn. Wiley, pp 595–604.

Elordi G, Olazar M, Lopez G (2011) Product yields and compositions in the continuous pyrolysis of high-density polyethylene in a conical spouted bed reactor. Ind Eng Chem Res. 50:6650–6659. https://doi.org/10.1021/ie200186m

Fivga A, Dimitriou I (2018) Pyrolysis of plastic waste for production of heavy fuel substitute: A techno-economic assessment. Energy. 149:865–874. https://doi.org/10.1016/j.energy.2018.02.094

Ibáñez M, Artetxe M, Lopez G (2014) Identification of the coke deposited on an HZSM-5 zeolite catalyst during the sequenced pyrolysis-cracking of HDPE. Appl Catal B Environ. 148–149:436–445. https://doi.org/10.1016/j.apcatb.2013.11.023

Jeswani H, Krüger C, Russ M (2021) Life cycle environmental impacts of chemical recycling via pyrolysis of mixed plastic waste in comparison with mechanical recycling and energy recovery. Sci Total Environ. 769:. https://doi.org/10.1016/j.scitotenv.2020.144483

Kumagai S, Nakatani J, Saito Y (2020) Latest trends and challenges in feedstock recycling of polyolefinic plastics. J Japan Pet Inst. 63:345–364. https://doi.org/10.1627/JPI.63.345

Lopez G, Artetxe M, Amutio M (2017) Thermochemical routes for the valorization of waste polyolefinic plastics to produce fuels and chemicals. A review. Renew Sustain Energy Rev. 73:346–368. https://doi.org/10.1016/j.rser.2017.01.142

Mastral FJ, Esperanza E, Berrueco C (2003) Fluidized bed thermal degradation products of HDPE in an inert atmosphere and in air-nitrogen mixtures. J Anal Appl Pyrolysis. 70:1–17. https://doi.org/10.1016/S0165-2370(02)00068-2

Mastral JF, Berrueco C, Ceamanos J (2007) Modelling of the pyrolysis of high density polyethylene. Product distribution in a fluidized bed reactor. J Anal Appl Pyrolysis. 79:313–322. https://doi.org/10.1016/j.jaap.2006.10.018

Mastral JF, Berrueco C, Gea M, Ceamanos J (2006) Catalytic degradation of high density polyethylene over nanocrystalline HZSM-5 zeolite. Polym Degrad Stab. 91:3330–3338. https://doi.org/10.1016/j.polymdegradstab.2006.06.009

Miandad R, Barakat MA, Aburiazaiza AS (2016) Catalytic pyrolysis of plastic waste: A review. Process Saf Environ Prot. 102:822–838. https://doi.org/10.1016/j.psep.2016.06.022

Murata K, Brebu M, Sakata Y (2009) Thermal degradation of polyethylene into fuel oil over silica-alumina by a continuous flow reactor. J Anal Appl Pyrolysis. 86:354–359. https://doi.org/10.1016/j.jaap.2009.08.009

Murata K, Brebu M, Sakata Y (2010) The effect of silica-alumina catalysts on degradation of polyolefins by a continuous flow reactor. J Anal Appl Pyrolysis. 89:30–38. https://doi.org/10.1016/j.jaap.2010.05.002

Murata K, Hirano Y, Sakata Y, Uddin MA (2002) Basic study on a continuous flow reactor for thermal degradation of polymers. J Anal Appl Pyrolysis. 65:71–90. https://doi.org/10.1016/S0165-2370(01)00181-4

Murata K, Sakata Y, Brebu M (2022) Thermal degradation of polyethylene in the presence of a non-acidic porous solid by a continuous flow reactor. J Anal Appl Pyrolysis. 161:105395. https://doi.org/10.1016/j.jaap.2021.105395

PlasticsEurope (2022) Plastics – the Facts2021. https://plasticseurope.org/knowledge-hub/plastics-the-facts-2021/

Saad JM, Williams PT, Zhang YS (2021) Comparison of waste plastics pyrolysis under nitrogen and carbon dioxide atmospheres: A thermogravimetric and kinetic study. J Anal Appl Pyrolysis. 156:105135. https://doi.org/10.1016/j.jaap.2021.105135

Chapter 35
Green Smart Home Model with Integrated Home Energy Management System Optimization

Ugurcan Uzunkaya, Irem Top, Simal Uzgur, Zehra Kamisli Ozturk, and Gurkan Ozturk

35.1 Introduction

In recent years, energy efficiency and sustainability have gained importance due to global climate change, nonrecyclable energy, increasing waste consumption, and the danger of depletion of natural resources. The fact that the resources are not endless has led to new searches in the society, resulting in the need to design self-sufficient models.

The increase in energy demand due to population growth has revealed the problem of ineffective management of energy. In the future, more efficient energy saving and decision support systems can be established with advanced and effective artificial intelligence systems by using cloud-based systems. Smart home usage is expected to increase from 12% to 22% (IPCC 2014). In order to eliminate similar concerns, this study adopts the "zero energy" philosophy and gives importance to user comfort while optimizing energy consumption. It is foreseen that this study, which is planned to contribute to the literature in terms of bringing together the purposes that were studied separately in previous studies, will contribute economically and socially in terms of the commercialization of the green smart house, which is planned to be produced as a final product, and its use in areas that will benefit the public. It is planned to present customer-specific solution with the help of the decision support system by evaluating the concern through different alternatives. Optimizing energy costs through the use of renewable energy sources and storage

U. Uzunkaya (✉) · I. Top · S. Uzgur · Z. Kamisli Ozturk · G. Ozturk
Faculty of Engineering, Eskisehir Technical University, Eskisehir, Turkey
e-mail: irem.top@vitra.com.tr; simal.uzgur@boschrexroth.com.tr; zkamisli@eskisehir.edu.tr; gurkan.o@eskisehir.edu.tr

M. Z. Sogut et al. (eds.), *Proceedings of the 2022 International Symposium on Energy Management and Sustainability*, Springer Proceedings in Energy,
https://doi.org/10.1007/978-3-031-30171-1_35

systems in smart homes will become an important part of the transition to smart cities. It is thought that positioning the green smart home model, which can be used for different purposes, with clustering algorithms will also contribute to urban planning (Dameri 2013).

35.2 Method

The idea of mobile/tiny houses, known as "tiny houses," has become widespread in recent years with the increasing human population, the cities' crowdedness, and the decrease in living spaces. Based on this idea, the green smart home model was created. Green smart houses will be suitable for use in the desired sector, large-scale efficiency will be achieved thanks to these houses, and it will support the creation of self-sufficient house models using clean energy resources, going beyond the definition of a constantly consuming society. Another aim of the study is to increase the awareness that resources are not endless, reduce carbon footprint, provide social and individual economic benefits, and create sustainable environmental awareness. This study consists of six stages:

- Building a mathematical model for smart home energy optimization
- Creating a decision support system for the user
- Making a prototype of the cube home with a decision support system with a 3D printer
- Inclusion of renewable energy sources with metaheuristic algorithm
- Creating the digital twin of the prototyped cube home
- Positioning the product in the city using artificial intelligence

In the first stage, the proposed multi-objective mixed integer nonlinear mathematical model has two objectives. Firstly, we focus on cost and user comfort optimization. Then, renewable energy sources are included in the optimization model. This basic situation can be seen in Fig. 35.1.

When weighted sum scalarization method was applied, different Pareto solutions are obtained in the multi-objective optimization model by multiplying the objective functions with different weight values. These Pareto solutions are obtained using GAMS-appropriate solver.

In the study, exact and heuristic algorithms are used simultaneously. A solution is also obtained with heuristic algorithms using the Python language. In the first stage

Fig. 35.1 Base case Including renewable energy sources in the model using GA

Fig. 35.2 Green Smart home model

of the study, the devices were classified, and the base case formed with six device types according to their categories. By increasing the number of devices, alternative model cases are being studied. The study's contribution in the first stage is the creation of a multi-objective energy optimization model. In the second stage, the design of the green smart home model and the creation of its prototype, the clustering algorithms used in artificial intelligence studies, and the positioning of the green smart homes according to the needs were made. The sample drawing of the green smart home model designed based on the cube shape is shown in Fig. 35.2 (Khan et al. 2017).

35.2.1 Classification of Appliances

Devices used in the energy management systems in smart homes are classified as shown in Fig. 35.3. Devices that fall into the inflexible category vary mostly depending on consumer preferences and lifestyles. Devices such as irons, toasters, televisions, and ovens can be given as examples of this category. The amount and timing of power usage of a flexible device can be controlled. For this reason, it is examined in two categories as power and time flexibility. Operation times of power flexible devices are limited, which does not allow delays and interruptions. Instant power-based devices, one of the subcategories of power flexible devices, are devices whose operating time cannot be shifted to different time zones and whose power can be changed instantly. Examples of these devices are adjustable lamps and thermostats with adjustable temperature. Temperature power-based devices, another subcategory of power flexible devices, operate within predefined time periods, and they are generally used for cooling and heating systems. Energy consumption can be adjusted in these devices, where the air conditioner can be shown as an example. Devices with energy storage capability are considered in the energy power-based category. Time flexible devices are also divided into two subcategories. Preemptive

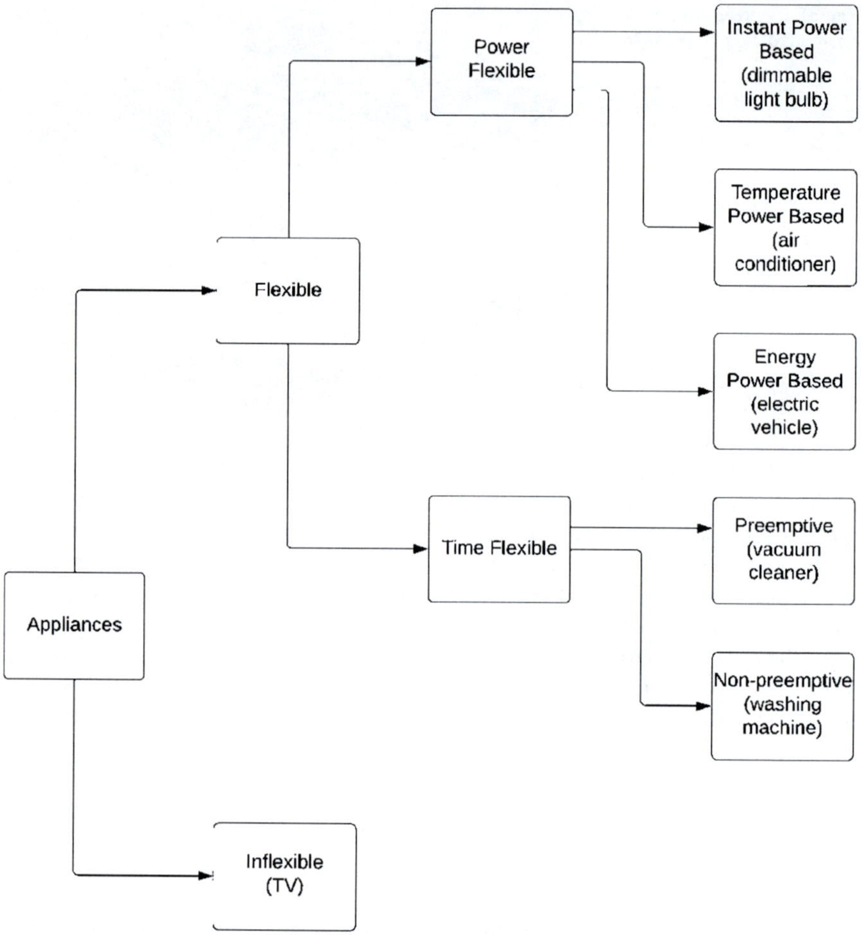

Fig. 35.3 Classification of appliances (Su et al. 2020)

devices consume a constant amount of energy when turned on. A delay in the start time is allowed. They are devices whose operation can be interrupted without adversely affecting their performance. Non-preemptive devices, on the other hand, consume a constant amount of energy when turned on; their difference from preemptive devices is that they cannot be interrupted. They must be run sequentially until the task period is complete (Sintov and Schulz 2017).

35.2.2 *Mathematical Model*

As a result of the literature research, the most suitable mathematical model for the study was determined as the model suggested by Alıç and Başaran Filik (2021).

Sets

Appliances' set:	$I = \{i\mid i = 1,\ldots,m\}$
Inflexible appliances' set:	$I_{\text{inf}} = \{i\mid i = 1,\ldots,b\} \subset I$
Flexible appliances' set:	$I_{\text{flx}} = \{i\mid i = 1,\ldots,c\} \subset I$
Power flexible appliances' set:	$I_{\text{pf}} = \{i\mid i = 1,\ldots,d\} \subset I$
Time flexible appliances' set:	$I_{\text{tf}} = \{i\mid i = 1,\ldots,e\} \subset I$
Instant power-based appliances' set:	$I_{\text{ipf}} = \{i\mid i = 1,\ldots,f\} \subset I_{pf}$
Temperature power-based appliances' set:	$I_{\text{tpf}} = \{i\mid i = 1,\ldots,h\} \subset I_{pf}$
Interior-exterior temperature set:	$O = \{o\mid o = 1,2\}$
Energy power-based appliances' set:	$I_{\text{epf}} = \{i\mid i = 1,\ldots,k\} \subset I_{pf}$
Preemptive appliances' set:	$I_{\text{pt}} = \{i\mid i = 1,\ldots,l\} \subset I_{tf}$
Non-preemptive appliances' set:	$I_{\text{npt}} = \{i\mid i = 1,\ldots,z\} \subset I_{tf}$
Type set:	$G = \{g\mid g = 1,2,3\}$
Illumination level set:	$A = \{a\mid a = 1,2\}$
Time horizon set:	$T = \{t\mid t = 1,\ldots,96\}$

Parameters

α_i : Earliest starting time of ith appliance

β_i : Latest finishing time of ith appliance

$p_{i,\ g} : g -$ type power of ith appliance

$$p_{i,g} = \begin{cases} \text{Minimum power of } i\text{th appliance}, & g=1 \\ \text{Maximum power of } i\text{th appliance}, & g=2 \\ \text{Desired power of } i\text{th appliance}, & g=3 \end{cases}$$

n : Number of dimmable light bulb

θ_i : Preference parameter of ith appliance

Q_a : Preference parameter for illumination level (a = 1: day, a = 2: night)

$$Q_a = \begin{cases} 2, & a=1 \\ 10, & a=2 \end{cases}$$

$E_{i,\ b}$: Starting energy level of ith appliance

$E_{i,\ g} : g -$ type energy of ith appliance

$$E_{i,g} = \begin{cases} \text{Minimum energy level of device } i, & g=1 \\ \text{Maximum energy level of device } i, & g=2 \end{cases}$$

$\Omega_{1,\ i}$: Temperature parameter for device i

$\Omega_{1,\ i} = 0,\ 18,\ \ \exists\, i \mid i \in I_{tpf}$

$\Omega_{2,\ i}$: Functional parameter device i

$$\Omega_{2,i} = \begin{cases} 0,9, & \text{Device } i \text{ is working in cooling} \\ -0,9, & \text{Device } i \text{ is working in heating} \end{cases} \exists i \mid i \in I_{tpf}$$

μ_i : Coefficient of performance of device i

R_i : Temperature resistance for device i

$T_{i,\ o,\ g} : g -$ type temperature in that environment of device i

$$T_{i,o,g} = \begin{cases} \text{Minimum indoor temperature}, & o=1, g=1 \\ \text{Maximum indoor temperature}, & o=1, g=2 \end{cases}$$

γ_i : Runtime of device i

Δh : One $-$ hour time interval

$T_{2,\ t}$: When $o = 2$, outdoor temperature in time interval t

$TT_{3,\,t}$: When $g=3$, the desired temperature in time interval t
$T_{i,\,1,\,0}$: Indoor initial temperature when $o=1$ and $t=0$
ρ_t : Electricity price in t time interval
ρ_{ort} : Daily average electricity unit price
ψ_{taban} : Base price for ψ_t
k_i : Preset coefficient of device i
ω_1 : Weighted factor
ω_2 : Weighted factor
N : Division factor

Decision Variables

$p_{i,\,t}$ = Power amount for device i at time interval t
$E_{i,\,s}$ = Final energy level for device i
$T_{i,\,1,\,t}$ = Indoor temperature in the time interval
$t_{i,\,b}$ = Actual start time for device i
$t_{i,\,s}$ = Actual end time for device i

$$y_{i,t}=\begin{cases}1, & \text{if device } i \text{ is using in } t \text{ time interval}\\ 0, & ow\end{cases}$$

Functions

U_i : Utilitiy of device i
$F_{1,\,i}$: Cost of device i
$\overline{F}_{1,i}$: Cost of normalized device i
$F_{2,\,i}$: Discomfort due to device i
$\overline{F}_{2,i}$: Normalized device i – discomfort
ψ_t : Deviation suppression function
d_i : Delay rate.

The equation suppressing extreme deviations in power depending on the unit price of electricity:

$$\psi_t=\psi_{taban}+\epsilon|\rho_t-\rho_{ort}|,\quad 0\le\psi_{taban,}\le 1, 0\le\epsilon\le 1,\quad \forall t\in T$$

Utility function for instant power flexible devices:

$$U_{i,g}=\frac{\left(n_i p_{i,g}\right)^{1-\theta_i}}{1-\theta_i}, \forall i\in I_{\text{ıpf}}$$

Utility function for energy power-based devices:

$$U_{i,1}=\frac{\left(\sum_{t\in T_i} p_{i,1}15+E_{i,b}\right)^{1-\theta_i}}{1-\theta_i}, \forall i\in I_{\text{epf}}$$

$$U_{i,2} = \frac{(E_{i,2})^{1-\theta_i}}{1-\theta_i}, \forall i \in I_{\text{epf}}$$

Utility function for temperature power-based flexible devices:

$$U_{i,1} = \frac{(T_{i,1,2})^{1-\theta_i}}{1-\theta_i}, \forall i \in I_{\text{tpf}}$$

$$U_{i,2} = \frac{(TT_{3,t})^{1-\theta_i}}{1-\theta_i}, \forall i \in I_{\text{tpf}}$$

when the relevant mathematical model can be given as follows:

Constraints	
The actual start and end time for each *i* device should be between the earliest start and the latest finish time:	
$\alpha_i \leq t_{i,\,b}, \forall\, i \in I$	(35.1)
$t_{i,\,b} \leq t_{i,\,s}, \forall\, i \in I$	(35.2)
$t_{i,\,s} \leq \beta_i, \forall\, i \in I$	(35.3)
The power of each device *i* at time *t* must be between the minimum and maximum power limits for that device:	
$p_{i,\,1} \leq p_{i,\,t}, \forall\, t, \forall\, i \in I$	(35.4)
$p_{i,\,t} \leq p_{i,\,2}, \forall\, t, \forall\, i \in I$	(35.5)
For inflexible devices, which is highly dependent on consumer preferences, the power consumption should equal the desired amount:	
$p_{i,\,t} = p_{i,\,3}, \forall\, t, \forall\, i \in I_{\text{inf}}$	(35.6)
No delays or interruptions are allowed with power flexible devices:	
$t_{i,\,s} - t_{i,\,b} = \gamma_i, \forall\, i \in I_{pf}$	(35.7)
The power of *n* instant power flexible devices at time *t* must be between the minimum and maximum power limits of these devices:	
$n_i p_{i,\,1} \leq p_{i,\,t}, \forall\, t, \forall i \in I_{\text{ipf}}$	(35.8)
$p_{i,\,t} \leq n_i p_{i,\,2}, \forall\, t \in, \forall\, i \in I_{\text{ipf}}$	(35.9)
The power to be consumed for the total illumination should be at least as much as the predetermined illumination power in a certain time interval:	
$\sum_{t \in T_i} p_{i,t} \geq \frac{n_i p_{i,2}}{Q_a}(\beta_i - \alpha_i), \forall a, \forall i \in I_{\text{ipf}}$	(35.10)
The total energy consumed should be at least half of the final maximum energy level by the device at the end of charging the energy power-based device:	
$\sum_{t \in T_i} p_{i,t} 15 \geq \frac{E_{i,2}}{2} - E_{i,b}, \forall i \in I_{\text{epf}}$	(35.11)
Total energy and initial energy drawn from the grid *i* is equal to the final energy of the device.	
$E_{i,s} = E_{i,b} + \sum_{t \in T_i} p_{i,t} 15, \forall i \in I_{\text{epf}}$	(35.12)
For a temperature power-based device, the internal temperature at time *t* must be between the minimum and maximum internal temperature limits for that device:	
$T_{i,\,1,\,1} \leq T_{i,\,1,\,t}, \quad \forall\, t, \quad \forall\, i \in I_{\text{tpf}}$	(35.13)
$T_{i,\,1,\,t} \leq T_{i,\,1,\,2}, \quad \forall\, t, \quad \forall\, i \in I_{\text{tpf}}$	(35.14)
$T_{i,\,1,\,t} = T_{i,\,1,\,t-1} + \Omega_{1,\,i}[T_{i,\,2,\,t} - T_{i,\,1,\,t-1}] - \Omega_{2,\,i} p_{i,\,t}, \forall\, t, \forall\, i \in I_{\text{tpf}}$	(35.15)

(continued)

For a temperature power-based device, the internal temperature at time t must be between the minimum and maximum internal temperature limits for that device:	
$p_{i,t} = y_{i,t} p_{i,2} + [1 - y_{i,t}] p_{i,1}, \forall t, \forall i \in I_{tf}$	(35.16)
For a preemptive device, the sum of the time periods the device has been running is equal to the operating time of that device:	
$\sum_{t=\alpha_i}^{\beta_i} y_{i,t} = \gamma_i, \forall i \in I_{pt}$	(35.17)
For a non-preemptive device, the time periods it works must be consecutive:	
$\sum_{t=t+1}^{t+\gamma_i} y_{i,t} \geq \gamma_i \left[y_{i,t+1} - y_{i,t}\right], \forall i \in I_{npt}$	(35.18)
i monetary cost of the device:	
$F_{1,i} = \sum_{t \in T_i} \rho_t p_{i,t} 15, \quad \forall i \in I$	(35.19)
Discomfort function for instant power flexible devices:	
$F_{2,i,t} = \psi_t (U_i[p_{i,t}] - U_{i,2})^2, \quad \forall t, \quad \forall i \in I_{ipf}$	(35.20)
Discomfort function for energy power-based flexible devices:	
$F_{2,i} = (U_i E_i - U_{i,2})^2, \quad \forall i \in I_{epf}$	(35.21)
Discomfort function for temperature power-based flexible devices:	
$F_{2,i,t} = \psi_t (U_i[T_{i,1,t}] - U_{i,2})^2, \forall t, \forall i \in I_{tpf}$	(35.22)
Discomfort function for preemptive time flexible devices:	
$F_{2,i} = \theta_i (t_{i,s} - (\alpha_i + \gamma_i - 1))^{k_i}, \quad \forall t_{i,s} \in [\alpha_i + \gamma_i - 1, \beta_i], \forall i \in I_{pt}$	(35.23)
Discomfort function for non-preemptive time flexible devices:	
$F_{2,i} = \theta_i^{d_i}, \qquad \forall t_{i,s} \in [\alpha_i, \beta_i - \gamma_i + 1], \quad \forall i \in I_{npt}$	(35.24)
Total discomfort for instant power flexible devices:	
$F_3 = \sum_i \sum_{t \in T_i} F_{2,i,t}, \forall i \in I_{ipf}$	(35.25)
Total discomfort for energy power based:	
$F_4 = \sum_{i=1} F_{2,i}, \forall i \in I_{epf}$	(35.26)
Total discomfort for temperature power-based flexible devices:	
$F_5 = \sum_i \sum_{t \in T_i} F_{2,i,t}, \forall i \in I_{tpf}$	(35.27)
Total discomfort for preemptive devices:	
$F_6 = \sum_i \sum_{t \in T_i} F_{2,i,t}, \forall i \in I_{pt}$	(35.28)
Total discomfort for non-preemptive devices:	
$F_7 = \sum_i \sum_{t \in T_i} F_{2,i,t}, \forall i \in I_{npt}$	(35.29)
Total monetary cost of all devices used:	
$F_1 = \sum_i F_{1,i}, \quad \forall i \in I$	(35.30)
Total discomfort of all devices used:	
$F_2 = F_3 + F_4 + F_5 + F_6 + F_7$	(35.31)
$p_{i,t}, E_i, \quad T_{i,1,t}, t_{i,b}, t_{i,s} \geq 0, \quad y_{i,t} \in \{0, 1\} \quad \forall i \in I$	(35.32)
$\min Z = \omega_1 \overline{F}_1 + \omega_2$	(35.33)

Table 35.1 Dimensions of alternative cases

	Base case	Case 1	Case 2	Case 3
Number of appliances	6	12	6	6
Time slot	96	96	288	24
Number of energy sources	1	1	1	1
Value set	3	3	3	3
Illumination level	2	2	2	2
Indoor-outdoor temperature	2	2	2	2
Total number of variables	2322	4644	6930	594
Total number of constraints	6420	12,840	19,092	1668

35.3 Application

35.3.1 *Dimension Analysis*

The dimensional analysis of the model is given in Table 35.1 to show the complexity of the multi-objective mathematical model established for the energy optimization study of the green smart house model. In the basic case, the total number of constraints of the model was determined as 6420, and the total number of variables was determined as 2322. As can be seen, the smallest sample operated with six devices among the alternatives.

The model size is quite large even for major time slots. It is seen that the size will increase as the number of variables and constraints increases.

35.3.2 *Python-Gurobi Optimization*

For reaching optimal solution with the exact algorithms the fastest, Python-Gurobi library was used. Gurobi is used for multi-objective mathematical models to find optimal solution(s) faster than GAMS. It is used by both academic and industrial sectors.

35.3.3 *Decision Support System*

Web-based decision support system was created with Python Django framework. It allows users to create a green smart home or other use cases. It has 60 different appliances with 3 different energy resources. There are five different use cases as office, hotel room, café, home, and working space. Decision support system can be seen in Figs. 35.4 and 35.5.

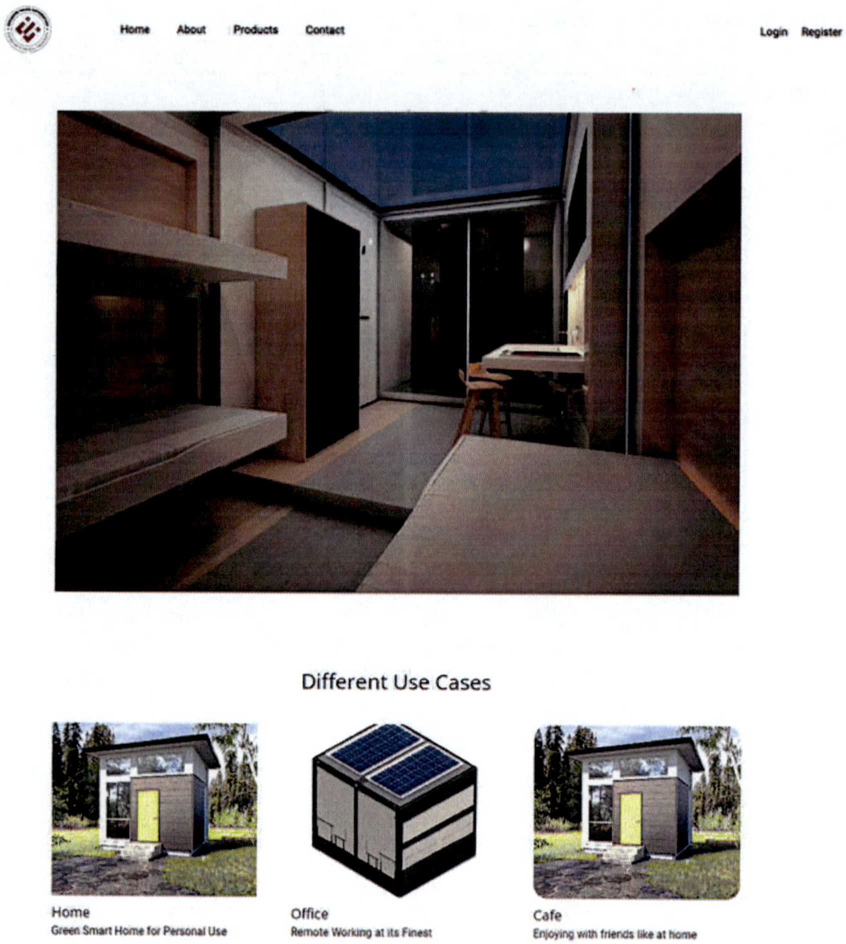

Fig. 35.4 Home page of decision support system

35.4 Conclusion

As a conclusion, this chapter aimed to contribute to the understanding of smart urbanism by commercializing the prototypes of the green smart home model, which is designed with the support of the optimization model created for the most effective use of energy resources. With the digitalization of homes, new business areas can be created, and professionals can be trained in these areas. More advanced systems can

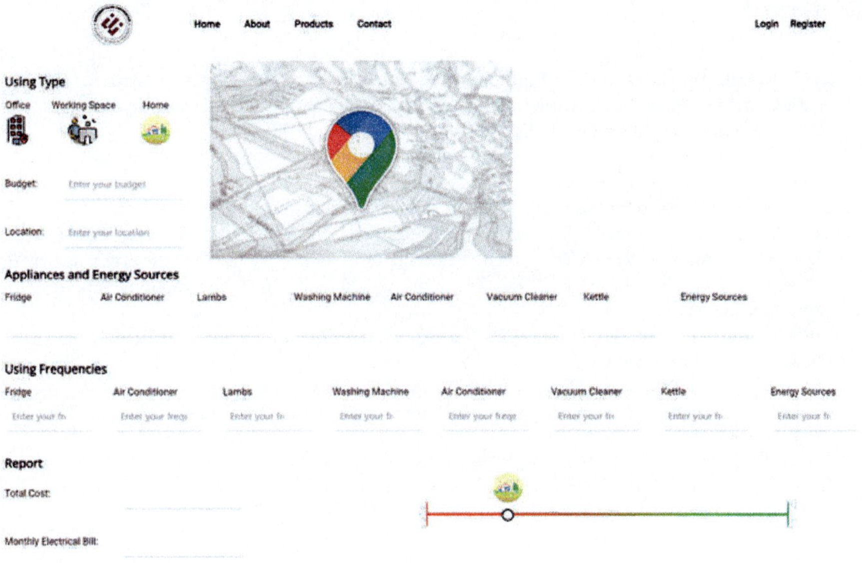

Fig. 35.5 Products page of decision support system

be designed and produced through collaborations between different sectors. This study also serves the purpose of Responsible Consumption and Production within the scope of the UN Sustainable Development Goals.

The weather, consumption, and green energy production data of the system in Remourban in Eskişehir were collected, and the linear regression method was used for the annual production/consumption amounts (kW) on a monthly basis. As a result of the analysis, the statistical distribution of the data was obtained. Increases/decreases were observed in the amount of energy produced in the same month of different years. After the regression analysis, estimation studies for the future periods were started by using the time series analysis method. Since the prototype of the green smart home model to be built will be made in Eskişehir, statistical analyses of sunshine durations based on weather data are being studied. Energy usage, type, capacity, and relevant data of smart home appliances are to be used for the user interface of the web-based decision support system, which is created with predetermined green values of every appliance to see how greener your home model is.

This project was developed on Python-Gurobi to get optimal solutions with exact algorithms.

References

Alıç, O., & Filik, B. Ü. (2021). A Multi-Objective Home Energy Management System for Explicit Cost-Comfort Analysis Considering Appliance Category-Based Discomfort Models and Demand Response Programs. Energy and Buildings, 240, 110868

Dameri, R. P. (2013). Searching For Smart City Definition: A Comprehensive Proposal. International Journal of Computers & Technology, 11(5), 2544–2551.

IPCC, (t.y). IPCC AR5 Synthesis Report, 2014. https://www.ipcc.ch/report/ar5/syr/

Khan, M., Khalid, R., Zaheer, B., Tariq, M., ul Abideen, Z., Malik, H., & Javaid, N. (2017). Residential Demand Side Management in Smart Grid Using Meta-Heuristic Techniques. In International Conference on P2P, Parallel, Grid, Cloud, and Internet Computing (pp. 76–88). Springer, Cham.

Nicole D. Sintov & P. Wesley Schultz (2017). "Adjustable Green Defaults Can Help Make Smart Homes More Sustainable"

Su, Y., Zhou, Y., & Tan, M. (2020). An Interval Optimization Strategy of Household Multi-Energy System Considering Tolerance Degree and Integrated Demand Response. Applied Energy, 260, 114144.

Chapter 36
Renewable Energy Usage in Wastewater Treatment Plants: A Case Study

Alper Alp, Ümmühan Başaran Filik, and Emine Esra Gerek

36.1 Introduction

Domestic and industrial wastewater treatment plants are of great importance for the removal and post-use of polluted water (Prasad et al. 2019). In many cases, the treatment strategy must handle some classic and well-known hazardous materials and pollutants. However, for the case of specific industrial production plants, the corresponding treatment must also be specific. In that case, the treatment plant must be designed or decided according to the materials of interest and the regulatory requirements. A good example of such is the case of industrial zones of major cities. The industrial zones are packed with factories and manufacturing plants that are usually specialized according to a major concept determined by a few locomotive industries. For example, Eskişehir Industrial Zone contains mostly metal manufacturing and ceramic industries accompanied by various other industries ranging from food manufacturing to plastic formwork. In the following sections, the wastewater profile of the zone is briefly explained, and its current treatment activities are presented. The current treatment plant is capable of supporting the waste input in total; however, the energy cost is steep. Besides, although the regulatory conditions are met, the treatment process itself has a high carbon footprint. In order to improve the efficiency of the system, the incorporation of renewable energy systems seems essential. This study starts by measuring the renewable energy potential of the particular site and continues with a proposal of a matched renewable energy harvesting plant sizing.

A. Alp · Ü. B. Filik · E. E. Gerek (✉)
Faculty of Enginering, Eskişehir Technical University, Eskişehir, Turkey
e-mail: eegerek@eskisehir.edu.tr

M. Z. Sogut et al. (eds.), *Proceedings of the 2022 International Symposium on Energy Management and Sustainability*, Springer Proceedings in Energy,
https://doi.org/10.1007/978-3-031-30171-1_36

36.2 The Industrial Zone of Eskişehir

The current industrial zone of Eskişehir was established in 1969, and its first plant and factory formation was activated in 1973. As the site is relatively new, it remains to be suitable for further extension with new plants. As a result of previous extensions, its current size reaches up to over 32,406,000 m^2.

36.2.1 Wastewater Treatment Plant

The wastewater treatment plant of the industrial zone was activated in 2008 and was extended to an area of 137,000 m^2 with a daily treatment capacity of 24,000 m^3 in the year 2018 (Fig. 36.1).

On par with classical treatment plants with similar capacities, the plant consists of five parts: physical treatment, chemical treatment, biological treatment, dehydration, and solar sludge drying, precisely with this particular order.

The feasibility of the treatment plant was calculated according to a population equivalent of 200,000 people yielding 20,000 COD/day, COD meaning the pollution unit of chemical oxygen demand.

Fig. 36.1 Eskişehir Organized Industrial Zone (EOSB) waste treatment plant

36.2.2 Electricity Consumption of the Wastewater Plant

The energy consumed by the wastewater treatment plants and the disposal of the treated sludge as a by-product are among the problems of these plants (Gu et al. 2017). The treatment plant of consideration has a detailed list of electric-consuming items with corresponding electric power ratings, ranging from relatively small consumptions of below 1 kW (such as various dosing units) to quite powerful units of above 150 kW (such as various blowers). The rated power of the total of 293 electric-consuming units inside the plants reaches up to a sum of over 3400 kW. Obviously, the total electric consumption depends on how long each unit is activated. Unfortunately, the instantaneous consumption is not monitored, and only weekly total consumptions are recorded and charged by the electric company. Within the specific collaboration dates, starting from December 2020 going up to May 2021, the weekly total consumptions gradually increased from 620,000 to 700,000 kWh. It is argued that, without considering the possibility of new manufacturing plants, economic situations, or area extensions, the production increase might reach to a wastewater treatment plant electrical consumption of well over 1000,000 kWh every month.

36.3 Renewable Energy Potential of the Site

Possible renewable energy sources of the region comprise solar energy, wind energy, and biomass.

According to the GIS simulation reference from the same site (39.76 lat; 30.63 lon), a fixed 1 m^2 PV system with a slope angle of 35° (southbound) is expected to generate a yearly total energy of 1433 kWh (measured irradiation power is 1835 kWh/m^2) (Mohammedreza et al. 2020; Ağçay and Attar 2007; European Commission 2019). The monthly variation of the produced energy is given in Fig. 36.2.

In order to maintain the wastewater treatment plant, the minimum monthly generation should be able to provide uninterrupted operation for the considered month. As an example, in January 2021, the electric consumption of the treatment plant was 623,122 kWh. Since the solar potential provides about 55 kW per month for a 1 m^2 panel, the necessary number of panels could be calculated as 11,262 panels. This figure is obtained if one considers only the solar potential. It must be noted that, with 11,262 panels, during July, the monthly total electric generation capacity becomes 169.97 kWh × 11,262 = 1,919,202 kWh, whereas the requirement is hardly anything above 700,000 kWh. This calculation shows that the sustainable system for uninterruptible operation gives a significant surplus in the summer season, when solar energy is considered. Furthermore, the actual implementation of this many solar PV panels is calculated to occupy an area of about 140,000 m^2 and require well over 120 million TL. If the surplus of 1000,000 kWh electric energy is

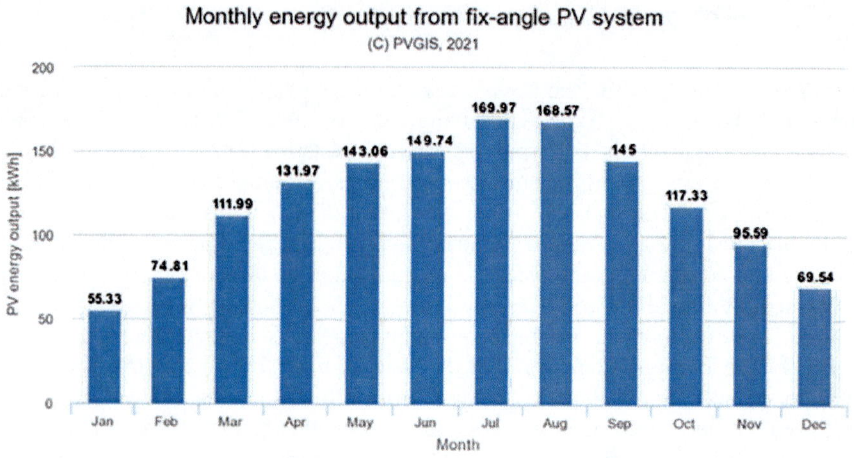

Fig. 36.2 Monthly energy output of sample PV at EOSB

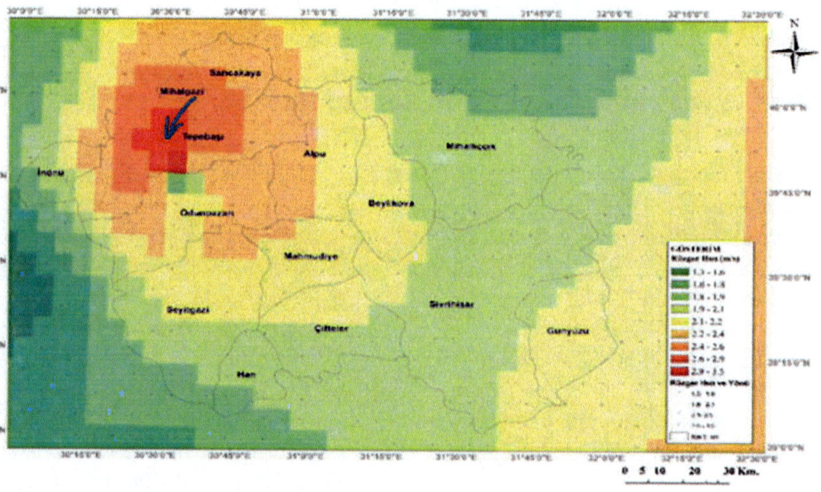

Fig. 36.3 Wind speed and direction distribution around Eskişehir city center. The location of the industrial zone of Eskişehir is indicated with an arrow

sold back to the grid during the summer times, that would recover about 3 million TL back (except the yearly 5–6 million TL electric bill saving). With the inflation rates, the financial turnover could be calculated as 10–12 years.

The wind speed profile according to the same GIS tool with the WindPRO software is observed to be as shown in Fig. 36.3.

The map indicates that the wind turbine electric generation potential around the city center seems feasible as compared to more rural areas. Furthermore, wind speed is a more uniform and stable energy supply as it does not vary as highly as the solar energy variations from winter to summer. Consequently, the industrial zone was found to be first grade suitable for a turbine installation with 1 MW power capacity at 60–80 m tower height and a sweeping radius of 54 m. Unfortunately, such systems typically cost about 1000 USD per kW capacity, making the installation cost around 1 million USD. On the other hand, it must be noted that the energy generation capacity of such a system would be well beyond the treatment plant requirements. If the excess generated energy is sold back to the grid, the economic turnover is calculated to be around 4–5 years.

Since the site naturally generates organic wastes, their processed and recycled outcomes automatically provide a third renewable energy source alternative: biogas. Unfortunately, information regarding biogas generation and potential calculations are not available for the industrial zone of Eskişehir. The size-wise analogous system is the Seyhan wastewater treatment plant (Adana, Turkey), which accepts an inflow of over 15,000 m^3/day, causing a sludge process of 1000 m^3/day and a treatment sludge of 350 m^3/day. These numbers show that the case of Eskişehir requires about the half of whatever is obtained for Seyhan. In Seyhan treatment plant, the decayed sludge gas contains about 60–70% methane. Considering the ideal caloric value of 35,800 kJ/m^3 for methane, the decay gas has a potential of about 22,400 kJ/m^3. The actual production in Seyhan with 350 m^3 of daily sludge was 3831 m^3/day biogas, providing a daily energy of 21,000 kWh (which is about the half of their energy requirement). In Eskişehir, the sludge amount and the process energy requirements are about half of Seyhan, giving us a hint that such a process would readily supply half of the electric consumption requirement by simple biogas recycling.

36.4 Discussions and Suggestions

Wastewater treatment, its capacity sizing, and its energy requirement planning are multiparameter optimization problems that require careful planning for efficient and clean operation. Depending on the site, the environment, produced waste characteristics, and renewable energy potential, an optimal treatment plant is supposed to run without interruption or overflow for the required treatment process time. The local renewable energy sources are expected to be:

- Low cost
- Sustainable
- Reliable
- Sufficient

with a short turnover time.

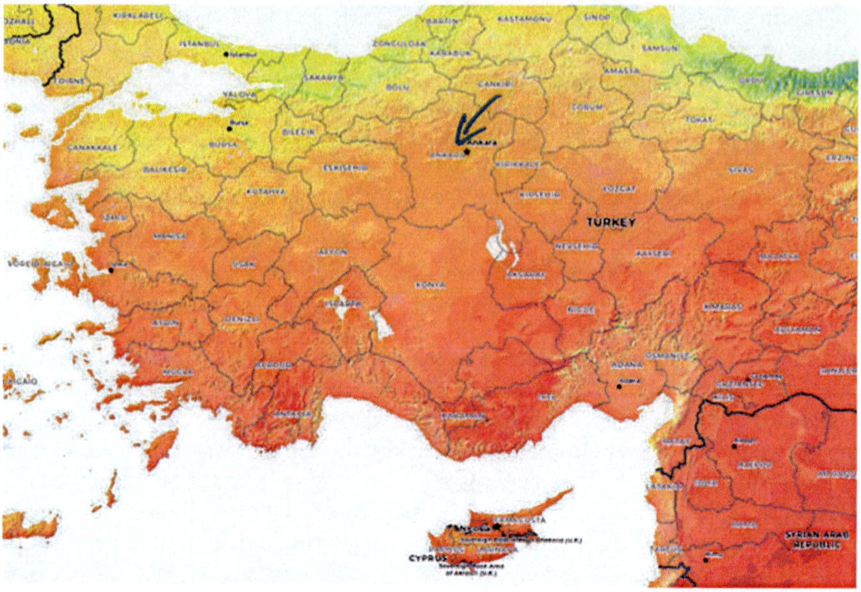

Fig. 36.4 Solar energy potentials in various regions of Turkey (darker tones mean higher solar potential). Eskişehir city center is shown with an arrow

The example case of Eskişehir Industrial Zone wastewater treatment plant shows that the particular choice of the geographical location is very suitable for all of the three renewable energy source candidates.

As we can see from Fig. 36.4, Eskişehir is not among the particularly advantageous places for solar energy potential. Yet, the monthly values given in Fig. 36.2 indicate that solar PV panels are still strong candidates for harvesting the necessary energy for the electrical requirements of the wastewater treatment plant.

Considering the total area of over 30 million m^2 for the industrial zone, this much efficiency necessitates a solar PV panel farm area that is about the same as the area of the treatment plant itself, which is manageable. Besides, most of the area of the treatment plant consists of either rooftop buildings or completely empty areas, which shows that about half of the required PV area can be readily supplied from within the treatment plant itself.

The calculations in Sect. 36.3 show that the solar-only solution has a financial turnover of over 10 years. However, it must be noted that this solution also significantly reduces the total carbon footprint. The world (and Turkish) average of CO_2 emission per kWh of grid electric energy is still 0.534 kg (in the USA, this value is even higher, 0.600 kg). Although over half of the energy is supplied from hydroelectric and renewable sources, the proportion of natural gas and coal still sums up to a significant CO_2 consumption (see Fig. 36.5) (Filik et al. 2015). This means that, by consuming 1 kWh of electricity, for whatever purpose, using grid-connected

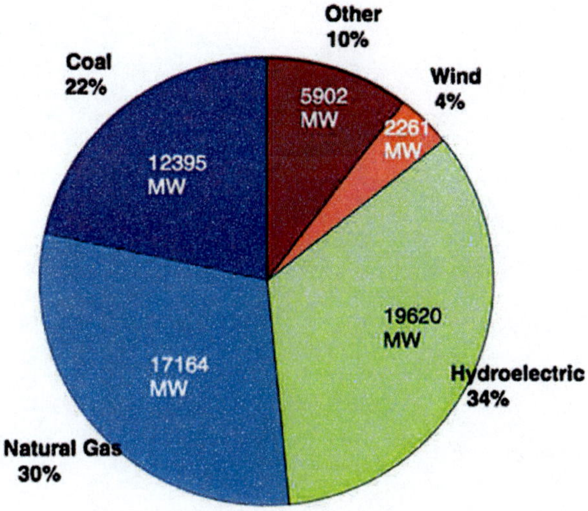

Fig. 36.5 Distribution of Turkey's electric energy generation in 2012

electric mains, 0.534 × 32/44 = 388 grams of chemical oxygen demand (i.e., pollution) is generated and 146 grams of carbon is injected to the atmosphere (Gerek et al. 2019).

The incorporation of solar energy is only a part of the solution candidates. Another strong renewable source is the wind power. Although Eskişehir region is not particularly rich in wind potential (Fig. 36.6), the analysis in Sect. 36.3 shows that the considered region is moderately suitable for wind turbines. With an investment of 1 million USD, a 1 MW power wind turbine can be achieved which can supply a large portion of the required electric demand with a turnover time of 4–5 years. Since this renewable source outputs can be shared with the solar panels, it may reduce the panel area necessity to below half of the required 140,000 m^2 of panel farm, further reducing the investment cost by 60 million TL (which is significantly more than the investment cost of the wind turbine itself).

A further renewable energy source option is biogas. Although the wastewater treatment plant in the industrial zone of Eskişehir has sludge drying facilities, the methane production and its recycling were not considered before. As a result, comparisons with existing plants are made in Turkey. It was found that a similar (but with double capacity and size) plant in Adana region was successfully generating electricity which supplies about the half of the energy requirements using the methane as a result of sludge decay (Türkmenler 2019). On the other hand, the industrial zone of Eskişehir is clearly producing less organic waste, and the share of methane-producing compounds is estimated to be half of the rate for the Seyhan treatment plant case. Since the compared plant can supply exactly half of the consumed energy, a candidate biogas recycling system is expected to recover a quarter of the electric consumption in the wastewater treatment plant of Eskişehir Industrial Zone. Therefore, excluding the investment cost of such a system, this means that the necessary electric consumption could be further halved with this kind of an operation, which would reduce the PV panel investment cost by 30 million TL, leaving a necessity of 35,000 m^2 panel field area and a cost of 30 million TL.

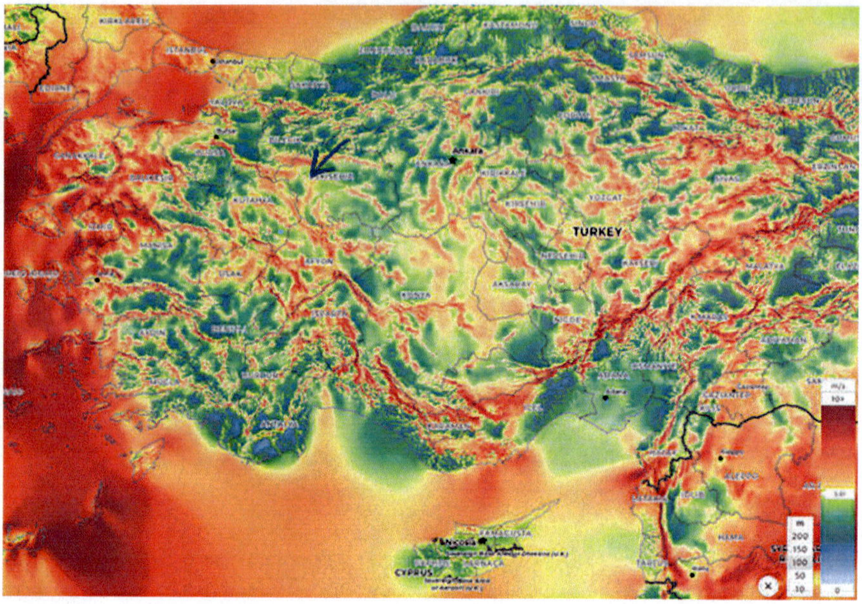

Fig. 36.6 Wind energy potentials in various regions of Turkey (red tones mean high, blue-green tones mean low potential). Eskişehir city center is indicated with an arrow

The proposed overall renewable energy installment in the wastewater treatment plant consists of the following:

- 35,000 m^2 of PV panel installment area (half of which is the actual PV area). Cost, 30 million TL; partial turnover time, 10 years
- 1 MW power wind turbine. Cost, 13 million TL; partial turnover time, 4 years
- Biogas generation and combustion unit. Cost: unknown.

With the incorporation of the renewable energy system, further electric bill benefits of about half a million TL worth of saving **was achieved** every month, With the incorporation of the renewable energy system, further electric bill benefits of about half a million TL worth of saving every month, reducing the turnover time to well below 4 years in total. Of course, the plant further requires extra investment for AC invertors for conversion to useful 220 V 50 Hz AC line voltage generation and possible sell-back operations to the national grid system. Besides, certain circuitry, such as solar charge control units, and other grid controllers are necessary. However, these electronic equipment are usually included in the costs of PV panel and wind turbine installations. It is noteworthy that the resulting need for the solar area is reduced to a quarter of the initial total estimate, so it perfectly fits into the actual treatment plant (Fig. 36.1) over the rooftops and on empty fields. Besides, the proposal does not need installment of any backup batteries, as the combined system already provides uninterruptable energy delivery, thanks to a variation of three distinct sources. If an extreme case "no-sun" + "no-wind" + "no-biogas" ever happens, since the treatment is not necessarily an online process, a daily delay can be perfectly applied.

Finally, it is concluded that the plant, its buildings, and its location are capable of delivering well over its own energy requirements. It is argued that the described renewable installments not only yield a green approach to the overall system but also provide a direct financial profit by means of selling the excess electrical energy back to the national grid.

Acknowledgments This work was supported by ESTU research fund under project no: 21GAP071.

References

Gu, Y., Li, Y., Li, X., Luo, P., Wang, H., Wang, X, Wu, J., Li, F. (2017) Energy self-sufficient wastewater treatment plants: feasibilities and challenges. Energy Procedia, 105:3741–3751.

Prasad, MNV., Favas, PJdC., Vithanage, M., Mohan, SV. (2019) Industrial and Municipal Sludge. Elsevier Publishing.

Ağçay, M. and Attar, F. (2007) Türkiye'nin Elektrik Enerjisi Arz Talep Dengesinin Tespiti (Tur). EMO Project competition report.

Mohammadreza, A., Nallapaneni, MK., Aref, E., Hamsa, A., Aline, K., Shauhrat, CS. (2020) Solar PV systems design and monitoring. Photovoltaic Solar Energy Conversion, 117–145.

European Commission (2019) Photovoltaic Geographical Information System. Online: https://re.jrc.ec.europa.eu/pvg_tools/en/#PVP

Türkmenler, H. (2019) Seyhan Atıksu Arıtma Tesisi'nde Biyogaz Üretim Verimliliğinin Araştırılması. Adıyaman Üniversitesi Mühendislik Bilimleri Dergisi, 95–101.

Filik Ü.B., Filik, T., Gerek, O.N. (2015) "New Electric Transmission Systems: Experiences from Turkey", book chapter in Wiley: Handbook of Clean Energy Systems. John Wiley & Sons.

Gerek, E., Afşar, S.Y., Koparal, A.S., Gerek, O.N. (2019) Combined energy and removal efficiency of electrochemical wastewater treatment for leather industry. Journal of Water Process Engineering, Vol. 30.

Chapter 37
Planning Electric Energy Consumption for Individuals

Sebnem Demirkol Akyol

Nomenclature

GWh	Gigawatt hours
TL	Turkish lira

37.1 Introduction

In the twenty-first century, most of the energy we use is electrical energy. Electricity, which is indispensable in our era, has a direct impact on the daily life of almost everyone in the world. It has been one of the most basic needs of people in every area such as our homes, workplaces, streets, etc.

Electrical energy is mainly produced from thermal, hydro and nuclear sources. In addition to these resources, renewable energy sources have also taken a significant share in electricity production with increasing technology in recent years. Table 37.1 shows the distribution of electric production in Turkey in 2019 and 2020 according to resources. Data is given in gigawatt hours (GWh) (TEDAŞ 2020). It is observed from Table 37.1 that a huge amount of electricity is gathered from natural gas in Turkey.

Since Turkey imports natural gas, it is important to decrease electricity consumption as much as possible. The aim of this study is to make consumers conscious of electricity consumption planning. For this purpose, a comprehensive survey is implemented and analysed in this study.

S. D. Akyol (✉)
Department of Industrial Engineering, Dokuz Eylül University, Izmir, Turkey
e-mail: sebnem.demirkol@deu.edu.tr

M. Z. Sogut et al. (eds.), *Proceedings of the 2022 International Symposium on Energy Management and Sustainability*, Springer Proceedings in Energy,
https://doi.org/10.1007/978-3-031-30171-1_37

Table 37.1 The ratio of licensed electricity generation by resources in 2020–2021

Source	2019 (GWh)	Percentage (%)	2020 (GWh)	Percent (%)	Difference (%)
Natural gas	108837.19	37.2	91227.14	30.88	−16.18
Imported coal	51172.22	17.49	62949.64	21.31	23.02
Dam	41269.59	14.1	40961.45	13.86	−0.75
Lignite	40581.02	13.87	45055.29	15.25	11.03
Wind	17859.86	6.1	19891.37	6.73	11.37
Stream	17124.40	5.85	18975.98	6.42	10.81
Geothermal	5969.48	2.04	7611.58	2.58	27.51
Stone coal	3453.87	1.18	3005.55	1.02	−12.98
Asphaltite coal	2394.64	0.82	2328.50	0.79	−2.76
Biomass	1939.72	0.66	2410.00	0.82	24.24
Diesel	1008.83	0.34	0.98	0	−99.9
Fuel oil	957.86	0.33	957.98	0.32	0.01
Sun	24.56	0.01	65.56	0.02	166.97
LNG	2.2	00	1.12	0	−48.83

37.2 Problem Definition

Nowadays, electric energy consumption planning and optimization is a hot topic for researchers due to several reasons such as the increase in electricity unit prices, the recent popularity of the concept of sustainable energy, the developments in the field of technology and the fact that income per capita is inadequate (see Yan et al. 2012; Silva et al. 2018; Yang et al. 2018; Zhu et al. 2019; Nel and de Kock 2021).

Many countries offer various electricity suppliers to the consumers as a result of free market economy, so that a consumer could evaluate the alternatives and select the most suitable one for her/his lifestyle. There are various suppliers in Turkey, too. However, consumers are not aware of these alternative suppliers and their tariffs. In this study, a comprehensive survey is implemented to express this gap.

37.3 Survey Study

This survey study is applied on individuals over the age of 18 who are subject to electricity bills. The age range is depicted in Fig. 37.1. The education levels of the participants are summarized in Fig. 37.2.

Lastly, the number of individuals living in the household is asked to participants and the results are given in Fig. 37.3.

After the demographic features, questions about consumption habits are asked to participants. Figures 37.4 and 37.5 show the time intervals that consumers are at home on weekdays and weekends, respectively.

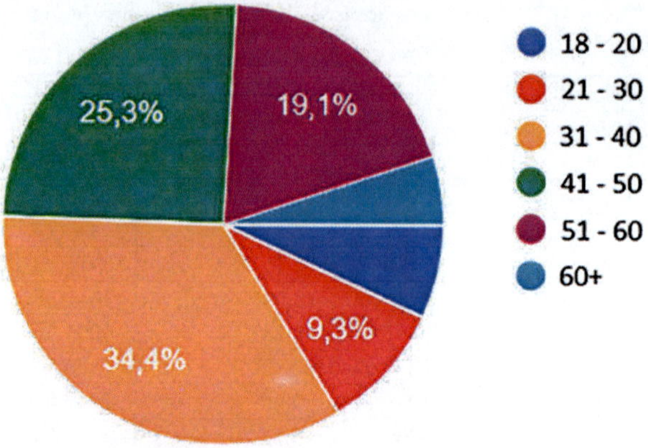

Fig. 37.1 Age range

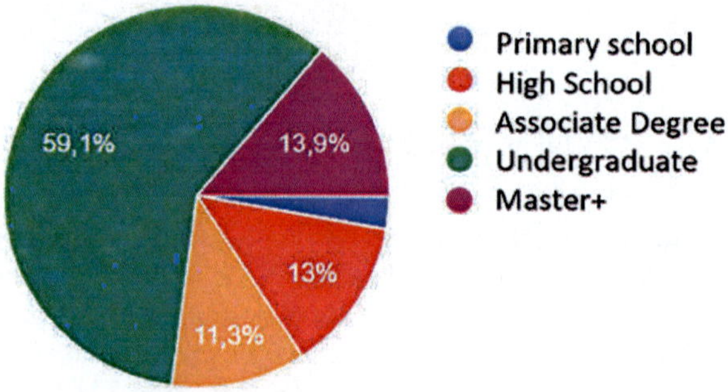

Fig. 37.2 Education level

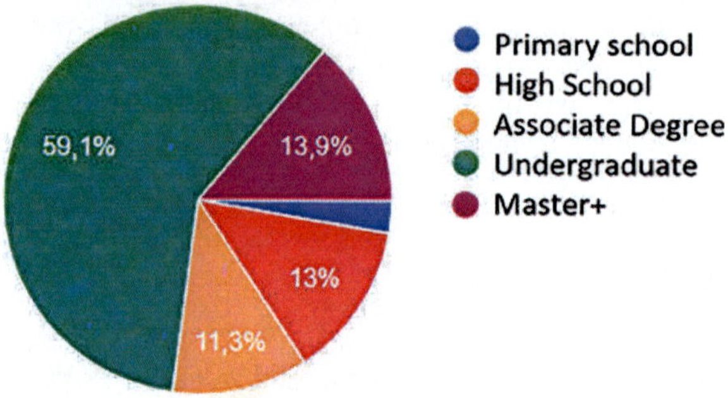

Fig. 37.3 Number of individuals in households

It can be concluded from Figs. 37.4 and 37.5 that individuals spend most of the time before 07:30 AM and after 07:00 PM on weekdays. Moreover, it is seen that electricity consumption increases between 06:00 AM and 12:00 PM on weekends.

Average monthly electricity bills of participants are stated in Fig. 37.6.

Consumer satisfaction is asked by "Are you satisfied with your electric bill?" in the questionnaire, and the results are given in Fig. 37.7.

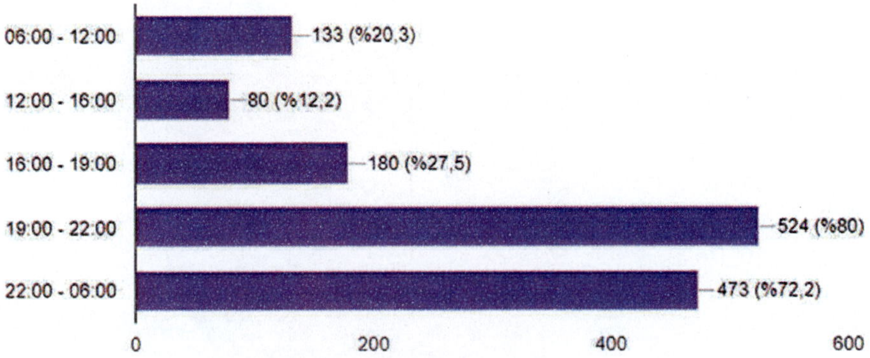

Fig. 37.4 Time period of electricity consumption on weekdays

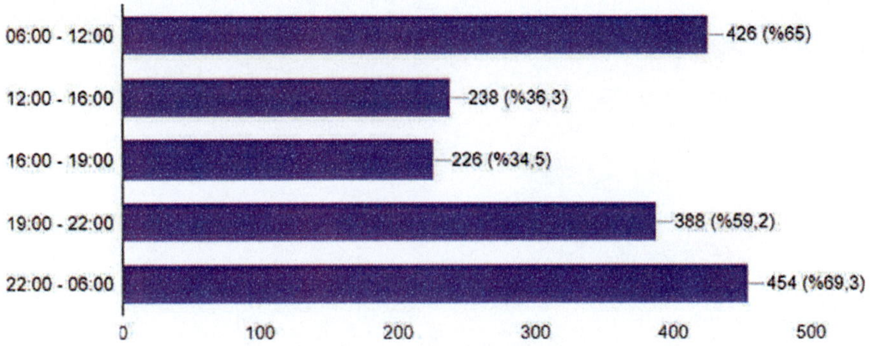

Fig. 37.5 Time period of electricity consumption on weekends

Fig. 37.6 Average monthly electricity bills

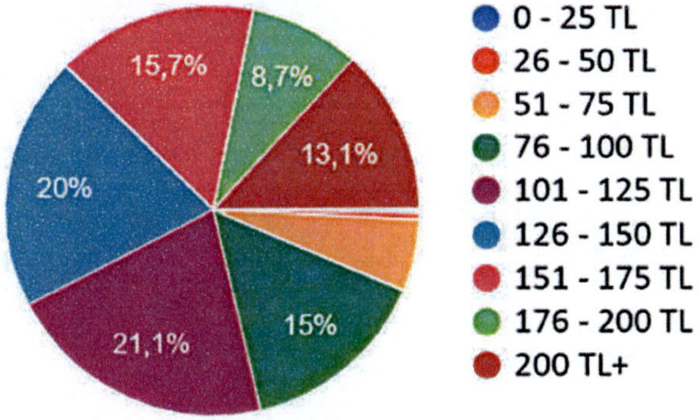

The subscribers are asked if they are willing to change their daily electricity consumption routine in order to reduce the bill. About 69% of them are willing to change, as demonstrated in Fig. 37.8.

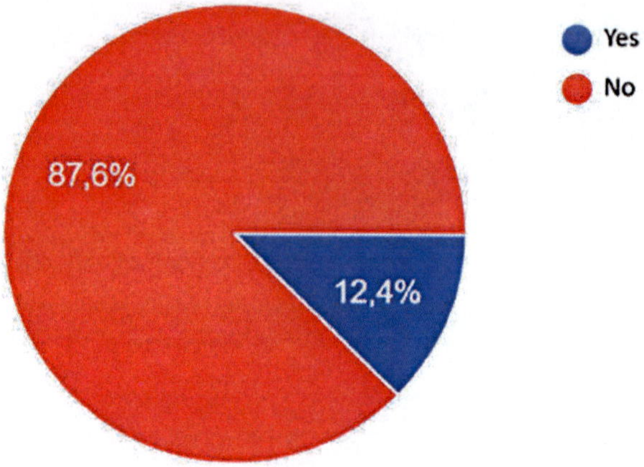

Fig. 37.7 Invoice satisfaction rate

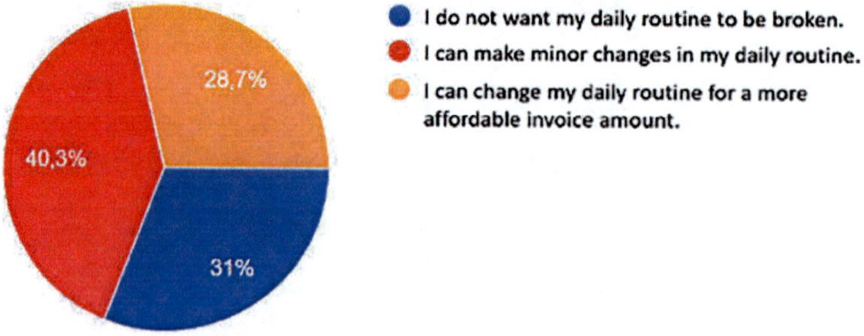

Fig. 37.8 Consumer's eagerness to change

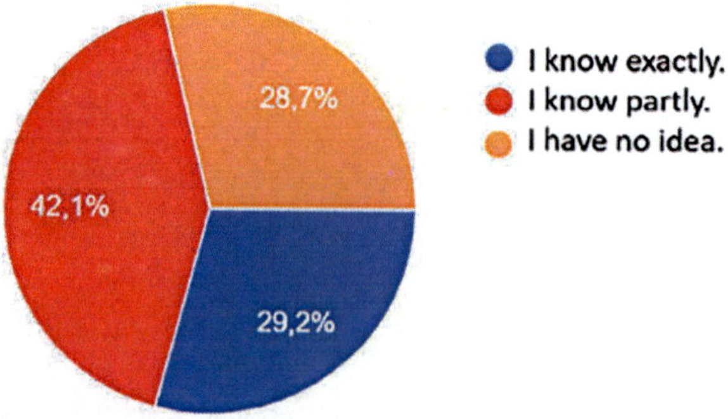

Fig. 37.9 Consumer's information level about tariffs

This survey points out that consumers should be informed about the various electricity suppliers and tariffs. By selecting the most appropriate tariff, a subscriber could reduce the average monthly electricity bill (Fig. 37.9).

37.4 Components of an Electricity Bill

The electricity bill is the information text sent by the distribution company, which provides service to the subscriber, covering a period of 1 month, showing the consumption amounts, prices and other related invoice items between the first day of the meter reading and the last day in detail. The explanations of some invoice items are as follows:

- First Index: It is the index value of the customer's electricity meter at the end of the previous period.
- Last Index: It is the index value of the day when the customer's electricity meter was last read.
- Consumption Price: It is the amount that corresponds to the product of the consumption amount in kilowatts and the unit price of electricity in Turkish liras (TL).
- Unit Price (TL): It is the TL value of the consumption in terms of kWh of the tariff group to which the customer belongs. The relevant unit price is determined by government authorities and notified to the supply companies.

  There are two types of tariffs in Turkey:

- Single-Time Tariff: A single price is applied to the electrical energy consumed at all times of the day. If you are a single-time tariff user, you will see the term "TTTZ Dwelling," which is the abbreviation of the term "Single-Term Single-Time Dwelling" in the "Subscriber No" section. For example, the unit price to be seen in the June 2020 invoice of a house in Izmir using the one-time tariff is 0.574033 TL. In other words, regardless of the time of day the electricity is used, the consumption price will always be multiplied by the unit price, 0.574033 TL.
- Multi-Time Tariff: It is a type of tariff applied to reduce electricity consumption from peak hours to lower hours. This tariff type is also called the three-time tariff. It is called a three-time tariff because the unit pricing is made by dividing the day into three parts. These three time zones are as follows:
  - Daytime Period: It is the priced period between 06:00 and 17:00.
  - Puant Period: It is the priced period between 17:00 and 22:00.
  - Night Period: It is the priced period between 22:00 and 06:00.

  For example, in the December 2019 bill of a house in Izmir using multi-time tariff, three different unit prices are seen for daytime, puant and night periods. These prices are 0.581299, 0.845897 and 0.369701 TL, respectively.

- Active Consumption Amount: It is the TL value obtained by multiplying the total consumption of the customer in kWh by the unit price.
- Energy Fund: It is the amount collected at the rate of 1% over the Active Energy Consumption Amount and transferred to the Energy Fund before the Ministry of Energy and Natural Resources in order to provide financial support to the facilities to be established in the energy sector and the works to be carried out.

- TRT Share: It is the fee collected at the rate of 2% over the Active Energy Consumption Amount and transferred to TRT in accordance with the TRT Income Law.
- ETV: It is the tax fee charged according to the tariff group calculated over the Active Energy Cost Consumption Amount for the municipality serving within the subscription zone of the customer.
- KDV: It is the amount of tax (value-added tax) calculated on the energy cost of the customer, other miscellaneous money additions and municipal tax.

37.5 Conclusion

Studies on electricity consumption have become popular recently due to excessive electricity bills, lack of electricity consumer profile, uncertainty in customer comfort and increasing technological developments and scheduling in smart homes. In this study a comprehensive questionnaire has been conducted in order to point out the gap between participants' rights and awareness. The questionnaire shows that the majority of the subscribers do not have adequate information about the supplier company and its tariffs. It is important because a subscriber could reduce the average monthly invoice by simply switching to another tariff.

References

Nel, M., and de Kock, I.H. 2021. An investigation into the applicability of super grids in a sub-saharan african context. South African Journal of Industrial Engineering, 32(3), 162–172.

Silva, F.L., Souza, R.C., Oliveira, F.L.C., Lourenco, P.M., and Calili, R.F. 2018 A bottom-up methodology for long term electricity consumption forecasting of an industrial sector-Application to pulp and paper sector in Brazil. Energy, 144, 1107–1118. https://doi.org/10.1016/j.energy.2017.12.078

Yan, Y., Qian, Y., Sharif, H., and Tipper, D. 2012. A survey on smart grid communication infrastructures: Motivations, requirements and challenges. IEEE communications surveys & tutorials, 15(1), 5–20.

Yang, T., Ren, M., and Zhou, K. 2018. Identifying household electricity consumption patterns: A case study of Kunshan, China. Renewable and Sustainable Energy Reviews, 91, 861–868. https://doi.org/10.1016/j.rser.2018.04.037

Zhu, J., Lin, Y., Lei, W., Liu, Y., and Tao, M. 2019. Optimal household appliances scheduling of multiple smart homes using an improved cooperative algorithm. Energy, 171, 944–955. https://doi.org/10.1016/j.energy.2019.01.025

http://www.tedas.gov.tr/#!tedas_hakkimizda accessed on June 1, 2021.

Chapter 38
With the Adoption of the Paris Climate Agreement, Turkey's Decarbonization Roadmap and Its Position in the 26th Conference of the Parties (COP26)

Hatice Merve Başar, Zafer Yalçınpınar, and Ahmet Feyzioğlu

Nomenclature

IPCC	Intergovernmental Panel on Climate Change
AR6	Sixth Assessment Report
UN	United Nations
COP	Conference of the Parties
GHG	Greenhouse gas
TUIK	Turkish Statistical Institute
NDC	Nationally determined contribution
BAU	Business as usual
Mt CO_2-eq.	Million tonnes of carbon dioxide equivalent
Gt	Giga tonnes (billion tonnes)
ETS	Emissions Trading System
CBAM	Carbon Border Adjustment Mechanism

38.1 Introduction

Climate change is a crucial global challenge on the nations' sustainable development. The long-term goal of the Paris Climate Agreement, which entered into force on 04 November 2016, was to keep the global average temperature increase well

H. M. Başar (✉) · Z. Yalçınpınar (✉) · A. Feyzioğlu (✉)
InnoEM Innovation & Environmental Management, Istanbul, Turkey

Faculty of Technology, Marmara University, Istanbul, Turkey
e-mail: merve.basar@innoem.com; zafer@innoem.com; ahmet.feyzioglu@marmara.edu.tr

M. Z. Sogut et al. (eds.), *Proceedings of the 2022 International Symposium on Energy Management and Sustainability*, Springer Proceedings in Energy,
https://doi.org/10.1007/978-3-031-30171-1_38

below 2 °C and, additionally, limit it to 1.5 °C, if possible, compared to the pre-industrial period. In the "Global Warming of 1.5 °C-Special Report" of the Intergovernmental Panel on Climate Change (IPCC) published in 2018, the importance to limit the temperature rise to 1.5 °C and the requirement to reduce global emissions to 45% below 2010 level by 2030 and net-zero by 2050 to achieve this target were mentioned. In the Sixth Assessment Report (AR6) of the IPCC in 2021, it has been revealed that if the current global emission level is maintained, the 1.5 °C limit will be exceeded in the global average temperatures in the 2030s. Under the guidance of the IPCC, the Parties have started to declare that they have accepted the Paris Agreement's target of limiting the global average temperature rise to 1.5 °C and net-zero by 2050 (Şahin et al. 2021).

The Paris Agreement was also ratified by the Grand National Assembly of Turkey on 06 October 2021 and entered into force on 10 November 2021 in Turkey. With the Presidential Decree No. 31643 (29 October 2021), the name of the Ministry of Environment and Urbanization was changed to the Ministry of Environment, Urbanization and Climate Change. The 26th UN Climate Change Conference of the Parties (COP26/Glasgow), which took place between 31 October and 13 November 2021, had a critical priority due to global evaluation of the progress achieved by the countries to set out increasingly ambitious climate action (every 5 years) since the signing of the Paris Agreement (2015) for the first time. While COP26 was a remarkable summit in terms of demonstrating the resolution of the countries to build a common mind, Turkey has also taken its place clearly and decisively by committing to 2053 net-zero emissions and green development. Declarations to which Turkey is a party at COP26 are listed as follows: *Declaration on Forests and Land Use, Statement on the Breakthrough Agenda, International Aviation Climate Ambition Coalition Declaration and Zero-Emission Vehicles Pledge,* respectively.

The intersection of the sustainability and energy management is climate change-based carbon management and decarbonization strategies. Decarbonization plays a crucial role for governments, companies and society in a global manner to limit the worst impacts of climate change. In line with Turkey's 2053 net-zero target, the purpose of this study is to assess overall decarbonization strategies being developed on a national scale with a focus on international trade, business, urban life/industry, finance, environment and the action plans.

38.2 Methodology

In this study, national-basis decarbonization strategies addressed by international authorities, governmental authorities, trade/finance authorities, industrial authorities and academia were discussed step by step for Turkey's 2053 net-zero target. Initially, six key action steps that can be taken into account by Turkey to achieve 2053 net-zero target were discussed. Then, other actual and potential decarbonization strategies were evaluated on key focus areas enriched by Turkey's greenhouse gas (GHG) emissions statistics.

38.3 Results and Discussion

By the end of 2021, 80+ countries (comprising approximately 75% of global emissions) declared their commitments to achieve net-zero emissions by 2050 in order to keep the world's temperature from rising above 1.5 °C. But just setting a target is not enough. Net-zero targets need to be tied to real policymaking. Thus, six key steps can be addressed for net-zero target-based actions by countries (Fig. 38.1) (WRI 2022):

1. *Define Net-Zero:* The crucial points are: the year countries intend to reach net-zero, the GHGs and sectors covered under their target and the scope of international offset purchases to be relied on.
2. *Model Pathways to Achieve Net-Zero:* Net-zero target defined clearly must be followed by a comprehensive monitoring to demonstrate how decarbonization can be achieved. Sectoral emission reduction pathways can be graphed by economic models, while these sector-specific pathways, orderly, can underline key decision points and priority actions in the near term to divert locking in future emissionsEmissions.

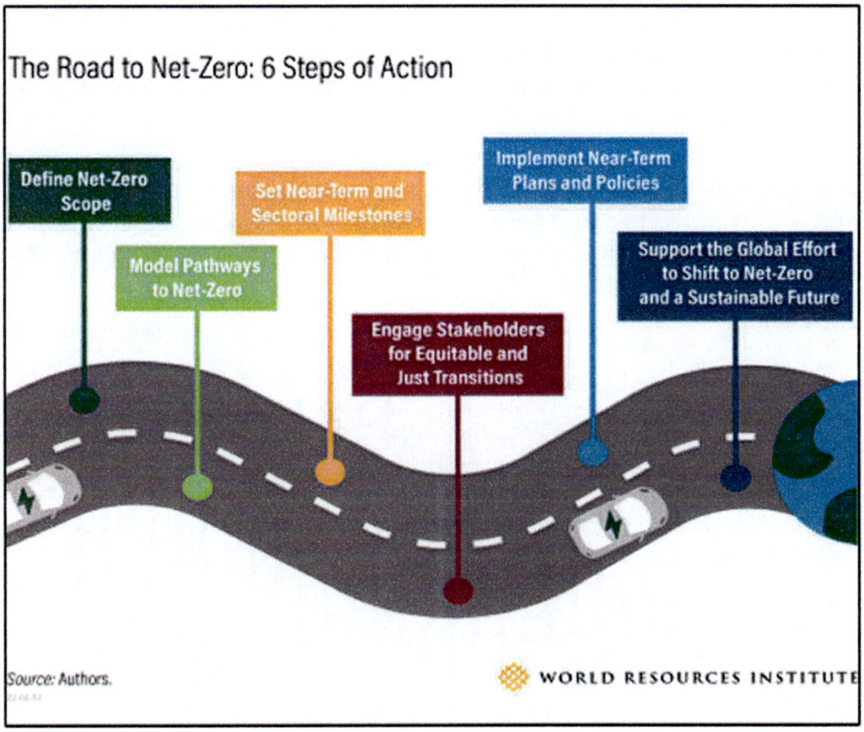

Fig. 38.1 The six-step road to net-zero. (WRI 2022)

3. *Setting Near-Term and Sectoral Goals and Milestones:* While reaching net-zero target will basically depend on economic transformation (from improving energy efficiency to reducing industrial emissions, deforestation and more), multiple policies will also be implemented in distinct sectoral areas. Countries will set specific interim and sectoral goals and targets, many of which may be aimed for before mid-century. These goals/targets can contribute to tracking the progress towards the net-zero target; thereby, accountability will be promoted.
4. *Engaging Stakeholders for Equitable and Just Transitions:* To provide decarbonization benefits to the entire society, negatively affected communities should be part of the solution of critical transformations and that impacts are equitably distributed. Some examples are collaborative public-private local investment schemes, reskilling the labour force and job for new fields, shifting locations of economic opportunity, etc.
5. *Implement Near-Term Plans and Policies Aligned with Net-Zero:* Tangible steps in order to implement near-term economic and fiscal policy interlinked with net-zero targets are taken such as to set legally binding carbon budgets that restrict the total amount of GHGs, to gather independent policy expert advice and/or to enable supportive policy and financing mechanisms in terms of emission taxes or emissions trading systems. Carbon pricing is a notably frequently used tool for promoting emission reductions in multiple sectors.
6. *Support the Global Effort to Shift to Net-Zero and a Sustainable Future:* It isn't enough to focus only on domestic emissions' mitigation if the efforts simply affect other countries negatively. The global efforts should also be taken by countries (WRI 2022).

According to the Turkish Statistical Institute (TUIK), the total GHG emission of Turkey in 2019 was calculated as 506.1 Mt CO_2-eq. (Fig. 38.2). The rate of GHG emissions by sectors was recorded as 72% energy, 13.4% agriculture, 11.2% industrial processes and product use and 3.4% waste, respectively. In terms of gases, CO_2

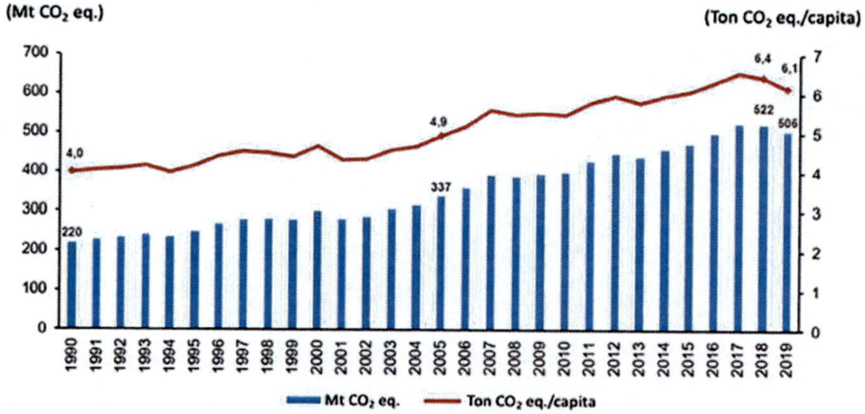

Fig. 38.2 GHG emissions of Turkey till 2019. (TUIK 2021)

had the biggest share in GHG emissions with 78.9% (TUIK 2021). That would mean roughly 5.6% annual GHG reductions to reach the 2053 net-zero target for Turkey. Thus, the overall decarbonization strategies being developed on a national scale were assessed with a focus on international trade, urban life/industry, business, finance, environment and the national action plans, respectively.

International Trade: With the publication of Turkey's "Green Deal Action Plan" on 16 July 2021, which is a roadmap that aims to ensure our country's adaptation to the policies of tackling with climate change, gained momentum in international trade in recent years, and that will strengthen our competitiveness in exports, issues such as Emissions Trading System (ETS) and Carbon Border Adjustment Mechanism (CBAM) come forward. As it is known, one of the tools used to reduce GHG emissions of countries is the implementation of an effective carbon pricing mechanism. In line with the suppressive targets in the fight against climate change worldwide, an increasing number of countries are implementing national carbon pricing mechanisms (Ministry of Trade 2021).

Urban Life/Industry: In a recent national decarbonization study, CO_2 emissions originating from the electricity sector, transport, buildings and energy/industrial process were discussed based on Turkey's 2018 economic indicators and emissions. Turkey's 2018–2050 CO_2 emissions were compared under two scenarios:

- Base scenario (without any policy to reduce emissions).
- Net-zero scenario (with essential policies to reduce emissions by preserving macroeconomic assumptions such as Turkey's current economic structure, population growth and economic growth).

Base Scenario: The increase in CO_2 emissions from 2018 to 2050 and 2070 (Mt CO_2) is as follows: Turkey's total CO_2 emissions will increase by 66% (700 Mt) in 2050 compared to 2018 levels, and in 2070, increasing by 120% (920 Mt). Total GHG emissions, on the other hand, will increase by 70% (890 Mt) compared to 2018 level in 2050, and in 2070, increasing by 125% (1170 Mt). Cumulative CO_2 emissions from energy are 18 Gt CO_2 between 2018 and 2050 and 34 Gt CO_2 between 2018 and 2070. The share of the electricity sector and industrial processes is increasing over the years, while the share of electricity consumption in transport, buildings and industry is decreasing (Table 38.1).

Net-Zero Scenario: Sectoral assumptions and decarbonization strategies that enable the reduction of emissions were added on the basis of the electricity sector, transport, buildings, industry and other production sectors together with the assumptions related to the basic macroeconomic indicators in the base scenario. CO_2 emissions from energy consumption in all sectors decrease by 37% (to 225 Mt) in 2030 compared to 2018 levels, and decrease by 80% (to 74 Mt) in 2050 (Şahin et al. 2021).

(a) *Electricity sector:* In order to increase the share of renewable energy in the electricity sector and to reduce the share of fossil fuels, storage systems consisting of battery-powered/pumped dams and market consolidation

Table 38.1 CO_2 emissions by sectors and years in two scenarios (2018–2050) (Mt CO_2) (Şahin et al. 2021)

	Base scenario				Net-zero scenario			
Sectors	2018	2020	2030	2050	2018	2020	2030	2050
Electricity generation from fossil fuels	149.0	133.9	184.0	281.9	149.0	133.9	72.7	15.0
Transport	82.8	80.7	96.5	119.5	82.8	80.7	65.3	28.9
Buildings	50.9	59.3	69.3	73.8	50.9	58.0	27.5	0.0
Energy in industry and other production sectors	77.2	68.7	99.4	108.5	77.2	69.3	60.0	30.2
CO_2 emissions from energy consumption	359.9	342.6	449.3	583.6	359.9	341.9	225.5	74.1
Process emissions	59.8	55.3	73.9	106.9	59.8	55.3	61.8	57.6
Total CO_2 emissions	419.7	397.9	523.2	690.5	419.7	397.2	287.3	131.6

mechanisms by increasing international interconnection capacity come to the fore, providing grid flexibility in the electricity system. Renewable energy potential is used at the highest level especially for wind and solar. Thus, coal is largely removed from electricity generation by 2035, while natural gas is reduced to very low capacity in 2050, thus decarbonizing the electricity sector to a large extent by 2050 (Şahin et al. 2021).

(b) *Transport:* The transition from road to railway in individual and public transportation, increase in efficiency in fossil fuel transportation vehicles and transition from fossil fuel vehicles to electric vehicles and another emission-free fuel type (i.e. green hydrogen) in individual, public and freight transportation are the main assumptions affecting emissions. In addition, the fact that transportation vehicles that cause emissions are not preferred have been added to the assumptions at low rates as a change in travel behavior (Şahin et al. 2021).

(c) *Buildings:* To reduce emissions in buildings, issues such as the rate of building renovation in residential and commercial/corporate buildings and new building construction and demolition rates have been addressed: energy performance improvement in electrical appliances, transition from coal, liquid fuels and natural gas to electricity and low rates of green hydrogen for heating in old and new buildings, heat pump use and improvement in heat pump performance, and behaviour change (Şahin et al. 2021).

(d) *Industry and other production sectors:* In reducing emissions caused by energy consumption, demand change in high-energy density industrial sectors in line with global demand projections, energy efficiency and electrification; direct use of renewable energy in low-energy density industrial sectors, agriculture and services sectors; and low rates of green hydrogen and carbon capture and storage technologies after 2040 in high-energy density sectors were added to the assumptions. Process emissions originating from industries, on the other hand, have been studied very limitedly in the macroeconomic model, as there are not enough assumptions in the literature to reduce them (Şahin et al. 2021).

Business: In the "SOS 1.5: The road to a resilient, net-zero carbon future" Roadmap (WBCSD 2020), which is a science-based climate change framework for companies to reach net-zero emissions and drive the transition to 1.5 °C economy, three key messages were highlighted:

(a) *Responsibility:* Companies are responsible for taking action to keep global warming at 1.5 °C for a sustainable world. The main framework is creating green jobs, ensuring economic growth and developing a more climate-resilient society.
(b) *Action Plans:* All companies need to implement climate action plans based on scientific data to achieve the net-zero carbon target by 2050. SOS 1.5 provides a framework for all companies to develop and implement climate action plans in order to achieve their net-zero carbon target, and within this scope, it supports all the commitments and initiatives of all companies towards the 1.5 °C target.
(c) *Collaboration:* No company can achieve net-zero carbon target alone. Companies need to collaborate with their stakeholders and value chains to accelerate the transition to a net-zero carbon economy. In the SOS 1.5 project, climate action projects were aligned with energy, transportation, construction, agriculture, environment and industry.

Finance: Within the scope of supporting sustainable finance practices as a long-term policy, both the Central Bank and the Ministry of Treasury and Finance offer a framework for green, social and sustainable finance. In terms of developing financial decarbonization strategies, the Central Bank decided to establish a new unit called the "Green Economy and Climate Change Directorate" in order to identify the vulnerabilities and opportunities that climate change may create in the financial system and reduce the related risks without changing the main objectives of the monetary policy (TCMB 2021). Besides, the "Sustainable Finance Framework Document" and the "Second Party Opinion", which will form the basis for borrowing transactions that can be carried out in the international "Environmental, Social and Governance" bond market, were published by the Ministry of Treasury and Finance in November 2021. The document comprises of various topics from Turkey's climate change commitments to the Green Deal Action Plan and a framework for green, social and sustainable finance (HMB 2021).

Combating Environmental Pollution: The Ministry of Environment, Urbanization and Climate Change will also create an electronic database for "The Registration of Emission/Transport of Pollutants" to protect the environment and reduce environmental pollution originating from diffuse sources and industry which will be open-public. All the relevant procedures and principles were regulated with the "Pollutant Release and Transport Registration Regulation" dated 04 December 2021 and numbered Off.Gaz.31679.

Environmental Inspection Activities: In the "Air Pollution Prevention Circular (10.12.2021–2021/22)" published by the Ministry of Environment, Urbanization and Climate Change and sent to the Governorships of 81 Provinces, it was stated that the environmental inspection activities will be increased by taking into account (a) the facilities' operations according to the provisions and principles of the "Regulation on Control of Industrial Air Pollution (20.12.2014; Off. Gaz.29211)" and (b) the facilities' Continuous Emission Measurement Systems and Air Quality Monitoring Data in order to control the emissions that may arise from those, respectively.

National Actions: Turkey has two crucial national actions before COP27, which was held in Sharm El-Sheikh, Egypt in November 2022: setting out/updating its new science-based Nationally Determined Contribution (NDC) and preparing long-term climate change strategy with a vision of 2053 net-zero target and green development. Turkey's previous NDC announced in 2015, before the Paris Agreement entered into force, is not compatible with 2053 net-zero target. It was foreseen that GHG emissions will be reduced by 21% compared to the reference scenario (BAU) in 2030 and must be updated. Besides, Turkey's Climate Council will also determine the medium- and long-term strategic targets and policies regarding the sectors with a common mind in 21–25 February 2022 in Konya, Turkey, with the contribution of public institutions, private sector, civil society and academia, respectively. The Council will be focused on GHG reduction, green finance and carbon pricing, adaptation to the climate change, local governments, science and technology, migration, just transition and other social policies, respectively. It is obvious that these national actions will also provide a technical framework for the Turkish Parliament's "Climate Law," which will be prepared within 6 months (Önsal 2021).

38.4 Conclusion

The following conclusions are stated from this study:

- Total GHG emissions of Turkey in 2019 were calculated as 506.1 Mt CO_2-eq.
- While 2021 was the year of net-zero commitments in the light of Paris Agreement and Glasgow COP26 Conference, 2022 must be the year of actions to identify Turkey's national decarbonization roadmap in order to achieve net-zero targets by 2053.
- Actual and potential national decarbonization strategies for Turkey with a focus on international trade, business, urban life/industry, finance, environment and the national actions were assessed under two different scenarios (base and net-zero), respectively.
- Turkey will have two crucial national actions before COP27: setting out/updating its new science-based NDC and preparing long-term climate change strategy to reach net-zero target and green development by 2053.

As a consequence, it is seen that the decarbonization strategies that our country will follow for transition to the low-carbon economy until 2053 gain special importance in the light of the adoption of Paris Agreement by the end of 2021.

References

HMB (2021) Sustainable Finance. Ministry of Treasury and Finance, Ankara. https://www.hmb.gov.tr/surdurulebilir-finansman

Ministry of Trade (2021) Green Deal Action Plan, 60 p., Ankara (in Turkish).

Önsal A (2021) Webinar on New Strategies of Turkey and Business World After COP26, November 22, 2021, Istanbul.

Şahin Ü, Tör OB, Kat B, Teimourzadeh S, Demirkol K, Künar A, Voyvoda E, Yeldan E (2021) Turkey's Decarbonization Roadmap-Net Zero in 2050, Executive Summary. Sabancı University Istanbul Policy Center, Istanbul. https://ipc.sabanciuniv.edu/Content/Images/CKeditorImages/20211103-19115588.pdf (in Turkish).

TCMB (2021) Monetary Policy Committee Meeting Summary 2021-52. Central Bank of the Republic of Turkey, Ankara. https://www.tcmb.gov.tr/wps/wcm/connect/tr/tcmb+tr/main+menu/duyurular/basin/2021/duy2021-52

TUIK (2021) Greenhouse Gas Emission Statistics, 1990-2019, Bulletin No. 37196, March 30, 2021, Turkish Statistical Institute, Ankara.

WBCSD (2020) SOS 1.5: The road to a resilient, net-zero carbon future, World Business Council for Sustainable Development, Geneva, Switzerland.

WRI (2022) Your Country Set a Net-Zero Target: What's Next? Explainer By Cynthia Elliott, Clea Schumer, World Resources Institute. https://www.wri.org/insights/net-zero-target-whats-next

Chapter 39
Short-Term Prediction for Wind Energy Systems Using Atmospheric Models

Irem Selen Yoldas and Ferhat Bingol

Nomenclature

NWP	Numerical weather prediction
MOS	Model output statistics
AR	Auto-regressive Model
MA	Moving Average Model
ARMA	Auto-regressive Moving Average Model
ARIMA	Auto-regressive Integrated Moving Average Model
SARIMA	Seasonal Auto-regressive Integrated Moving Average Model
SARIMAX	Seasonal Auto-regressive Integrated Moving Average with Exogenous Factors
SC	Spatial Correlation
LSTM	Long short-term memory
ANN	Long short-term memory with wavelet kernels
LSSVM	Least squares support vector machine
GP	Gaussian process
WN-LSTM	Long short-term memory with wavelet kernels
GRU	Gated recurrent unit
RNN	Recurrent neural network
MSE	Mean square error
RMSE	Root-mean-square error
MAE	Mean absolute error
R^2	R square

I. S. Yoldas · F. Bingol (✉)
Department of Energy Systems Engineering, Faculty of Engineering, Izmir Institute of Technology, Izmir, Turkey
e-mail: iremyoldas@iyte.edu.tr; ferhatbingol@iyte.edu.tr

M. Z. Sogut et al. (eds.), *Proceedings of the 2022 International Symposium on Energy Management and Sustainability*, Springer Proceedings in Energy,
https://doi.org/10.1007/978-3-031-30171-1_39

39.1 Introduction

Wind energy has been one of the most rapidly growing renewable energy sources recently (IEA 2021). As is well known, it is an environment-friendly and cost-effective energy source that contributes to pollution reduction and economic development. Because of the randomness and intermittency of wind characteristics, increasing wind power penetration into modern grids can impact power system operation safety and power quality. Energy storage and accurate wind speed forecasts can help to solve these issues.

Limitations of excess energy storage lead to wind speed forecasting. To use wind energy safely and efficiently, it is necessary to improve the accuracy of wind speed prediction; since wind speed affects the wind energy produced, accurate wind speed prediction models can enhance the safety of energy systems. However, compared to other traditional power plants, wind speeds are highly dependent on a variety of meteorological factors (e.g., temperature, relative humidity, barometric pressure, and wind direction); wind power is not easily predictable due to its highly probabilistic and fluctuating properties.

Research and contributions are currently being made on wind speed prediction. Many methods have been proposed in the literature to improve the accuracy of predictions. These methods can be divided into four groups: physical, statistical, artificial intelligence, and hybrid methods. In addition, forecasting algorithms can be categorized based on very short-term (several seconds and 30 min ahead), short-term (30 min and 72 h ahead), medium-term (72 h and 1 week ahead), and long-term (more than a week and a year) forecast periods (Meka et al. 2021). Recent studies have focused primarily on short-term wind forecasts due to the importance of wind forecasts for energy systems. In particular, the day before market, regulation, disposal, planning, load tracking, and other system operations occur during these periods. Therefore, this study will examine the subgroup of short-term prediction methods.

Each model has its natural strengths and weaknesses. For example, statistical models cannot correctly handle nonlinear problems, artificial intelligence methods do not simply choose some critical parameters, and physical processes give rough predictions in the short term, even outperforming others in the medium and long term. To overcome the weaknesses of the models, hybrid methods have been developed that combine the qualities of several forecasting methods (Tascikaraoglu and Uzunoglu 2014). This study aims to develop a new hybrid model to improve the accuracy of forecasting and address the shortcomings of individual models.

The paper is organized as follows: an overview of wind forecasting methods, the used data sets, the employed forecasting model comparison, and the main conclusion of the paper.

39.2 Overview of Wind Forecasting Methods

The first group considers physical factors such as terrain, obstacles, temperature, and pressure to predict wind speed. Another usage of these methods is an auxiliary input for the first step of other forecasting methods. Numerical weather prediction (NWP) is one of the most used physical methods meteorologists develop. Generally, it is utilized for large-scale weather prediction. Physical methods are mainly based on NWP, and in case of missing historical data, the manufacturer's power curves are used. These methods may not provide accurate results for short-term wind prediction. To increase accuracy, NWP numerically solves the conservation equations at the specified region. Simultaneously, digital elevation models to describe the topography should be employed in NWP to obtain better results. Model output statistics (MOS) can be applied to minimize the residual error (Tascikaraoglu and Uzunoglu 2014; Lei et al. 2008; Foley et al. 2011; Hu et al. 2021).

Bessac et al. (2018) developed a forecasting model that combines the sources of multiple physical model outputs that give better results. Louka et al. (2008) indicated that the Kalman filter could eliminate systematic errors in the forecasting results of the NWP model. Jung et al. stated that physical methods outperform statistical methods in long-term wind prediction. As the NWP model slowly updates and lags behind the historical data, it might cause significant errors in the forecasting results. Due to the computational complexity, their applicability in short-term wind prediction is limited.

The forecasting issue is approached by the second group using traditional statistical techniques. Based on a historical data set, statistical approaches use pattern discovery, parameter estimates, and model checking to create a mathematical model of a problem. An autoregressive model (AR), a moving average model (MA), an autoregressive moving average model (ARMA), and an autoregressive integrated moving average model (ARIMA) can be used to categorize Jenkins' approaches. (Guoyang et al. 2005). One of these approaches is the Kalman filter. It evaluates the predictions and observations obtained from the NWP model with statistical methods and establishes a regression between the predictions and observations. It corrects erroneous predictions from the model based on the observations (Cassola and Burlando 2012). Another method is the spatial correlation (SC). It considers the interaction between nearby wind farms based on temporal correlations that help increase prediction accuracy compared to the traditional statistical methods. SC models utilize the data from nearby wind farms to develop a wind resource model for the identified wind farm (Ezzat et al. 2019; Hu et al. 2021).

Li et al. (2015) defined a dynamic SC model with a tracking framework created by the Kalman filter between the geographically distributed wind farms for short-term wind predictions. Zhu et al. (2019) studied methods that could predict wind speed in multiple regions by adding the SC model. Spatial characteristics were obtained by long short-term memory (LSTM) and a convolutional neural network. The forecasting limitations caused by a rapid change in wind speed were overcome by considering the geographical characteristics of the wind farms in the dynamic SC model.

The third group contains the prediction models established with the development of artificial intelligence techniques and other forecasting methods. Some of these methods are artificial neural networks, fuzzy logic methods, and support vector machines. ANN is a commonly used method that consists of many layers. It predicts the wind speed by learning from a data set with input-output mapping. Input and output data are required to train and test these networks. The features of ANN, such as being fault tolerant, fast, and straightforward, able to learn and generalize, and adaptable to different situations, are significant. Another method is the fuzzy logic model, which is used when it is difficult to model a system (Lei et al. 2008).

Filik and Filik (2017) proposed ANN-based models that differentially combine multiple local meteorological measurements such as wind speed, temperature, and pressure values. This method could improve wind speed forecasting for various situations. Wang et al. (2004) described a mathematically nonlinear neural network model. In this algorithm, ANNs capture the short-term pattern in wind speed data, and the long-term pattern is categorized as increasing, decreasing, and almost stable. The process is divided into short-term forecasting and adapting results to long-term forecasting. The results were compared to other linear regression approaches. This model improves the forecasting accuracy in short-term and long-term predictions. Damousis et al. (2004) developed a fuzzy logic model based on the spatial correlation model for wind speed and power generation forecasting. In contrast, it performs good results on flat terrain, but its performance declines in a complex landscape.

The fourth group is hybrid methods which combine the final prediction performance of individual forecasting models and ensure significant advantages compared to the unique models. Hybrid models usually consist of a linear and a nonlinear model. Combining a linear and nonlinear model to forecast the hidden components embedded in the wind speed could show better performance to improve prediction accuracy (Tascikaraoglu and Uzunoglu 2014).

Zhang (2003) combined a seasonal ARIMA (SARIMA) and least squares support vector machine (LSSVM) model. LSSVM improved results by forecasting the residuals of SARIMA outputs. Hu et al. (2021) proposed a hybrid short-term forecasting method that integrates the corrected NWP and SC models into a Gaussian process (GP). Compared with the primary GP, the forecasting accuracy in different seasons is developed at 7.02–29.7% using the corrected NWP, 0.65–10.23% after integrating SC, and 10.88–37.49% using the proposed hybrid model. Shahid et al. (2020) developed a hybrid prediction model for nonlinear mapping using long short-term memory with wavelet kernels (WN-LSTM), which enriches deep learning for vanishing gradient and wavelet transformations. A percentage improvement of up to 30% was obtained compared with the well-known existing models. Liu et al. (2020a, b) proposed a model consisting of three stages: The empirical wavelet transform decreases the nonstationary of the wind speed data in the initial phase by subdividing the data into subarrays. The forecasting model is built and all subseries' outcomes are calculated in the second step, which employs three different types of deep neural networks. The three deep networks are combined in the last step of the reinforcement learning process. The ultimate forecast results are obtained by combining the outcomes of each series. Compared with 19 alternative models, it provides the best accuracy.

39.3 Result and Discussion

39.3.1 Data Sets

In the current study, the data used are collected from 100 m IZTECH met. mast. The proposed model is aimed to be validated using the production data set from a nearby wind farm. The collected wind time series are between May 2019 and June 2020 with 10 min temporal resolution from 30, 76, and 101 m.

Figure 39.1 shows the wind speed time series. The time series is divided into the training and test data sets to verify the forecasting results.

On the other hand, the determination of prevailing wind directions plays an essential role in wind energy studies. Figure 39.2 shows the wind roses obtained based on the blow frequency of data taken from 28 to 74 m heights of met. mast. Although most of the winds come from the north (0°), intensity is observed in the range of 330–30°. Slightly, winds are also seen between 180° and 210°. For this reason, the prevailing winds are from the north and south directions.

In addition to wind speed and direction, topography and meteorological variables also mainly affect forecasting accuracy. Factors affecting wind speed such as temperature, relative humidity, and barometric pressure should also be considered (Liu et al. 2020a, b). Temperature (°C), relative humidity (%), and barometric pressure (hPa) data are shown in Fig. 39.3.

39.3.2 Performance Evaluation Metrics

To quantitatively investigate the prediction performance of forecasting models, the mean square error (MSE), root-mean-square error (RMSE), mean absolute error (MAE), and R square (R^2) are used as evaluation metrics which are expressed as follows:

$$\mathrm{MSE} = \frac{1}{N}\sum\nolimits_{t=1}^{N}\left(w(i) - \overline{w}(i)\right)^2 \tag{39.1}$$

$$\mathrm{RMSE} = \sqrt{\frac{1}{N}\sum\nolimits_{t=1}^{N}\left(w(i) - \overline{w}(i)\right)^2} \tag{39.2}$$

$$\mathrm{MAE} = \frac{1}{N}\sum\nolimits_{t=1}^{N}\left|w(i) - \overline{w}(i)\right| \tag{39.3}$$

$$R^2 = 1 - \frac{\sum_{t=1}^{N}\left(w(i) - \overline{w}(i)\right)^2}{\sum_{t=1}^{N}\left(w(i) - w\_m\right)^2} \tag{39.4}$$

where N is the number of samples, $w(i)$ is the actual wind speed value, $\overline{w}(i)$ is the forecast value of wind speed, and w_m is the average wind speed value (Tian et al. 2019).

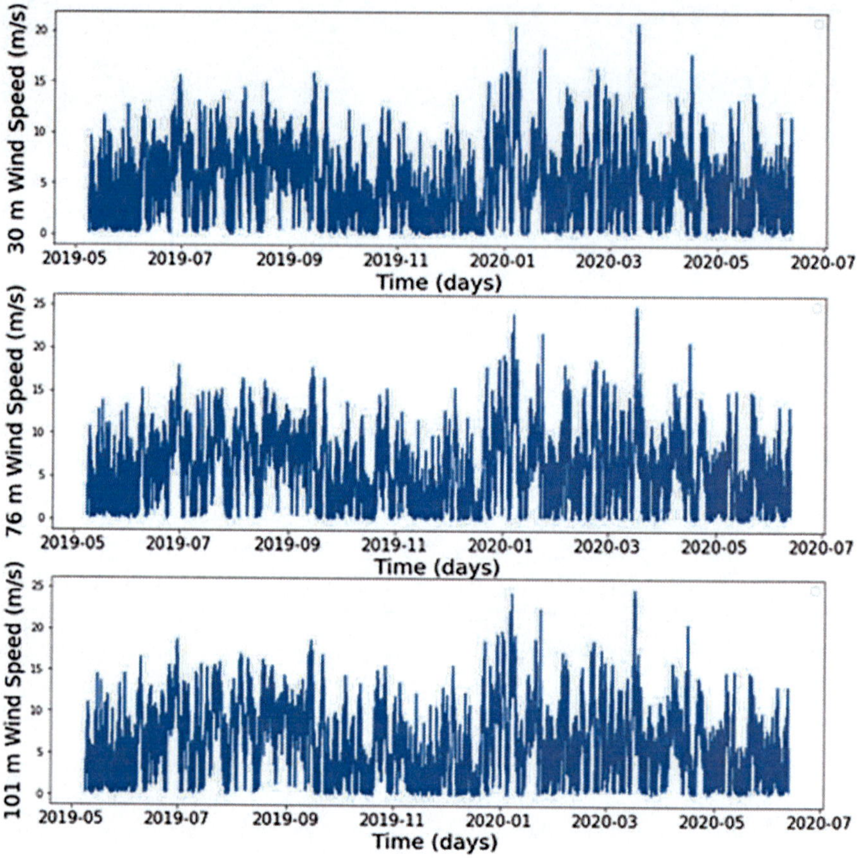

Fig. 39.1 Wind speed data sets at 30, 76, and 101 m

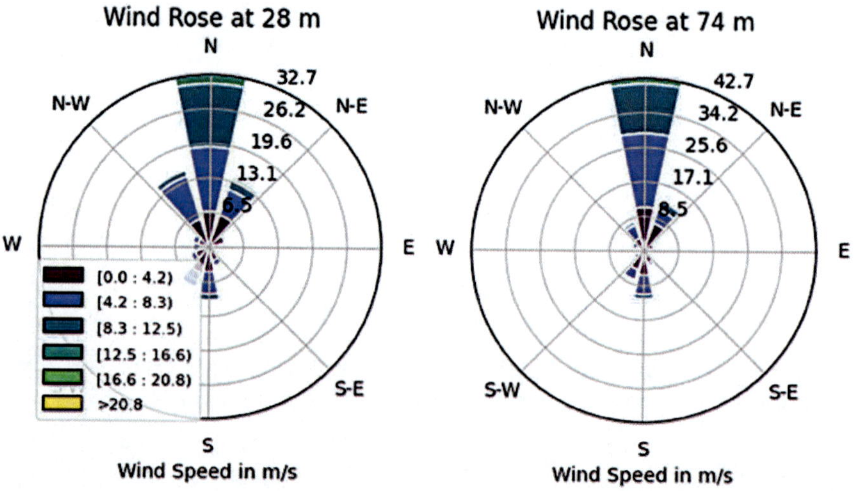

Fig. 39.2 Wind roses at 28 m and 74 m

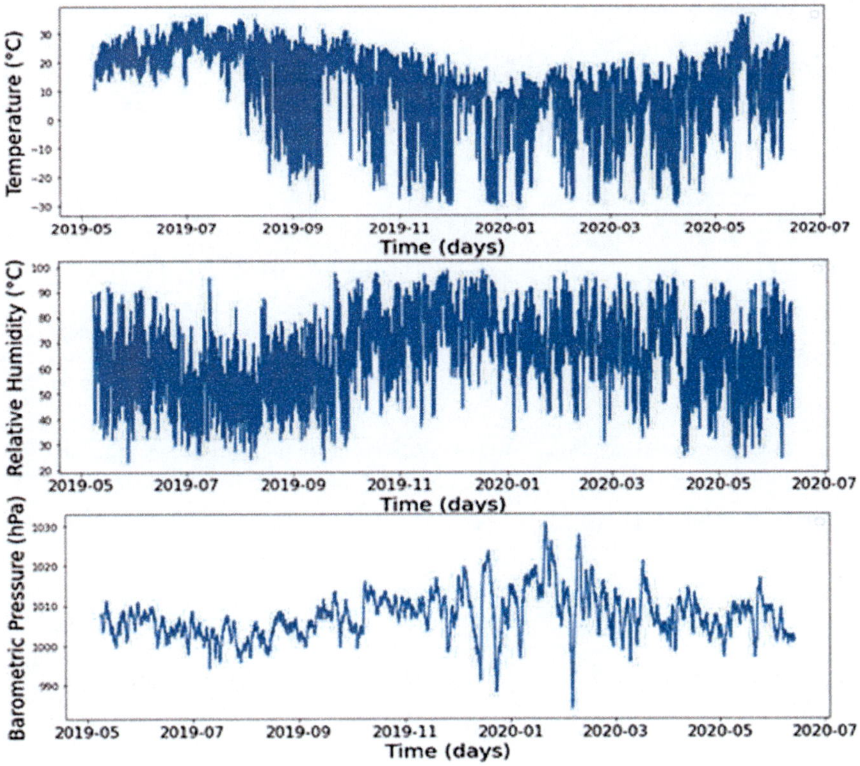

Fig. 39.3 Temperature, relative humidity, and barometric pressure data sets

39.3.3 Case Study

In this study, an analysis of the forecasting performance of the five approaches based on the results is performed. Besides these methods, the selected models, according to the literature review, will be investigated to obtain the best approaches for the hybrid model. After choosing the models, the remaining methods will be used as benchmarks. A brief of forecasting results is provided to analyze the observations further. After obtaining the model parameters, temperature, relative humidity, barometric pressure, and wind speed are forecasted. Figure 39.4 provides the absolute difference between the real-time data and predictions for the wind speeds based on the multivariate Facebook Prophet, SARIMA, SARIMAX, GRU, and LSTM models. Facebook Prophet is a method that utilizes the general additive model to fit the nonlinear trends for time series with daily, weekly, and yearly seasonality (Asha et al. 2020). SARIMA is an ARIMA model including seasonal effects, and SARIMAX has an additional exogenous factor (Mangayarkarasi et al. 2021). They are the conventional statistical methods and are generally used as benchmarks. A

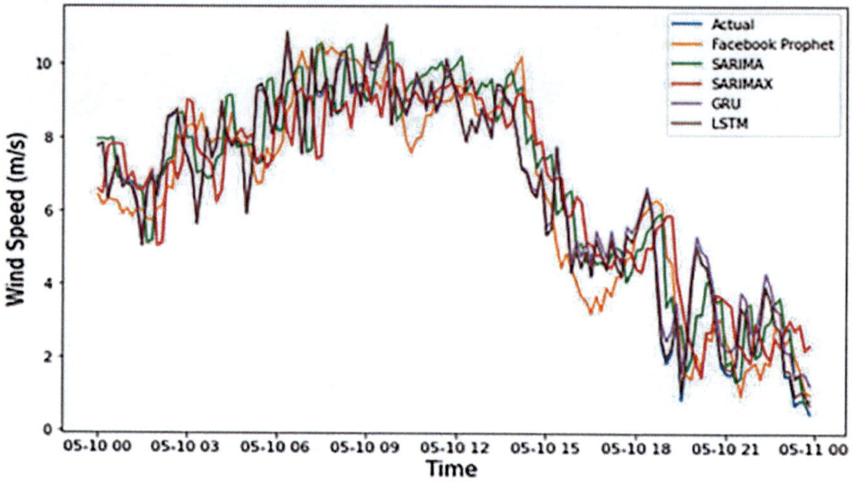

Fig. 39.4 Daily wind speed forecasting results

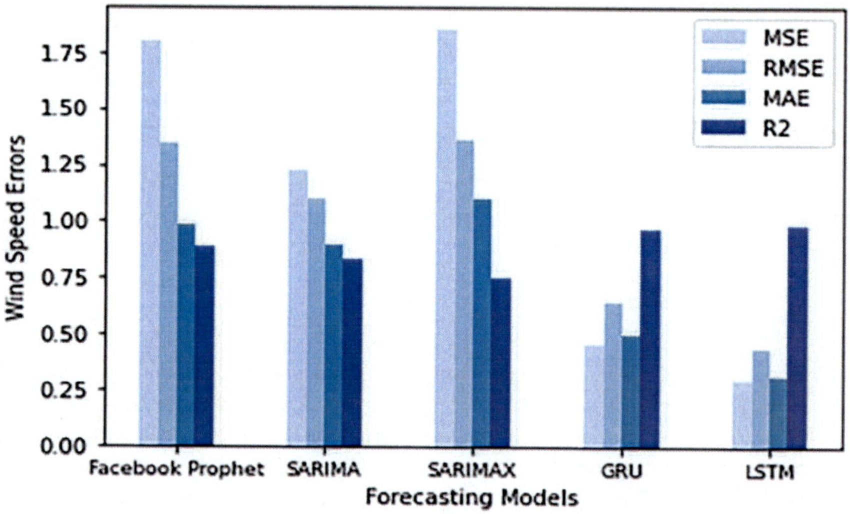

Fig. 39.5 Wind speed forecasting performance metrics

deep neural network is a promising approach based on machine learning, and GRU and LSTM are the types of recurrent neural networks (RNN) (Hossain et al. 2021).

The forecasting results are presented daily in Fig. 39.4. Figure 39.5 illustrates the comparison of forecasting performance metrics. As can be seen from Fig. 39.4, the prediction results of LSTM almost coincide with the real-time wind speed, which is also supported by the performance metrics. The performance metrics shown in

Fig. 39.5 represent that the MSE, RMSE, and MAE of LSTM are smaller than the other methods. Meanwhile, the R^2 value of LSTM is closer to 1 than the other methods. These results indicate that LSTM outperforms the different methods for the utilized real-time data sets.

39.4 Conclusion

The research aims to create a novel hybrid model that can improve prediction accuracy and overcome the shortcomings of individual models by embedding atmospheric models into them. An overview of forecasting methods is firstly presented. Then, a case study is performed using the real-time data from IZTECH met. mast of 10 min sampling period according to the overview. Five methods are investigated: Facebook Prophet, SARIMA, SARIMAX, GRU, and LSTM. To compare the methods, the performance indicators are MSE, RMSE, MAE, and R^2. According to the results, the LSTM model outperforms the other models for the utilized real-time data sets. Besides these methods, the selected models based on the literature review will be investigated to obtain the best models for the aimed hybrid method. Finally, the proposed model will be validated using the real-time met. mast data and production data set from a nearby wind farm.

Acknowledgment This work is supported and funded by the ADHERE project, a part of the ERA-Net SES call, and the Technological Research Council of Turkey (TÜBİTAK; Project No. 120N498).

References

Asha J, Rishidas S, SanthoshKumar S, Reena P (2020) Analysis of Temperature Prediction Using Random Forests and Facebook Prophet Algorithms. Innovative Data Communication Technologies and Application 46: 432–439. https://doi.org/10.1007/978-3-030-38040-3_49

Bessac J, Constantinescu E, Anitescu M (2018) Stochastic Simulation of Predictive Space–time Scenarios of Wind Speed Using Observations and Physical Model Outputs. The Annals of Applied Statistics 12(1):432–458. https://doi.org/10.1214/17-AOAS1099

Cassola F, Burlando M (2012) Wind speed and wind energy forecast through Kalman filtering of Numerical Weather Prediction model output. Applied Energy 99: 154–166. https://doi.org/10.1016/j.apenergy.2012.03.054

Damousis IG, Alexiadis MC, Theocharis JB, Dokopoulos PS (2004) A fuzzy model for wind speed prediction and power generation in wind parks using spatial correlation. IEEE Transactions on Sustainable Energy 19(2):352–361. https://doi.org/10.1109/TEC.2003.821865

Ezzat, A., Jun, M., Yu, D., 2019, Spatio-temporal Short-term Wind Forecast: A Calibrated Regime-switching Method. The Annals of Applied Statistics 13(3):1484–1510. https://doi.org/10.1214/19-AOAS1243

Filik ÜB, Filik T (2017) Wind Speed Prediction Using Artificial Neural Networks Based on Multiple Local Measurements in Eskisehir. Energy Procedia 107: 254–269. https://doi.org/10.1016/j.egypro.2016.12.147

Foley AM, Leahy PG, Marvuglia A, McKeogh EJ (2011) Current methods and Advances in Forecasting of Wind Power Generation. Renewable Energy 37: 1–8. https://doi.org/10.1016/j.renene.2011.05.033

Guoyang W, Yang X, Shasha W (2005) Discussion about Short-term Forecast of Wind Speed on Wind Farm. Jilin Electric Power 181(5): 21–24.

Hossain MA, Chakrabortty RK, Elsawah S, Gray EM, Ryan MJ (2021) Predicting Wind Power Generation Using Hybrid Deep Learning with Optimization. IEEE Transactions on Applied Superconductivity 31(8): 1–5. https://doi.org/10.1109/TASC.2021.3091116.

Hu S, Xiang Y, Zhang H, Xie S, Li J, Gu C, Sun W, Liu J (2021) Hybrid Forecasting Method for Wind Power Integrating Spatial Correlation and Corrected Numerical Weather Prediction. Applied Energy 293: 116951. https://doi.org/10.1016/j.apenergy.2021.116951

IEA (2021) Renewables 2021: Analysis and forecast to 2026. OECD Publishing. https://doi.org/10.1787/6dcd2e15-en

Lei M, Shiyan L, Chuanwen J, Hongling L, Yan Z (2008) A Review on the Forecasting of Wind Speed and Generated Power. Renewable & Sustainable Energy Reviews 13(4): 915–920. https://doi.org/10.1016/j.rser.2008.02.002

Li P, Guan X, Wu J, Zhou X (2015) Modeling Dynamic Spatial Correlations of Geographically Distributed Wind Farms and Constructing Ellipsoidal Uncertainty Sets for Optimization-based Generation Scheduling. IEEE Transactions on Sustainable Energy 6(4):1594–1605. https://doi.org/10.1109/TSTE.2015.2457917

Liu H, Yu C, Wu H, Duan Z, Yan G (2020a) A New Hybrid Ensemble Deep Reinforcement Learning Model for Wind Speed Short-term Forecasting. Energy 202: 117794. https://doi.org/10.1016/j.energy.2020.117794

Liu X, Zhang H, Kong X, Lee KY (2020b) Wind Speed Forecasting using Deep Neural Network with Feature Selection. Neurocomputing 397: 393–403. https://doi.org/10.1016/j.neucom.2019.08.108

Louka P, Galanis G, Siebert N, Kariniotakis G, Katsafados P, Kallos G (2008) Improvements in Wind Speed Forecasts for Wind Power Prediction Purposes using Kalman Filtering. Journal of Wind Engineering and Industrial Aerodynamics 96(12):2348–2362. https://doi.org/10.1016/j.jweia.2008.03.013

Mangayarkarasi R, Vanmathi C, Khan MZ, Noorwali A (2021) COVID19: Forecasting Air Quality Index and Particulate Matter (PM2.5). Computers, Materials and Continua 67(3): 3363–3380. https://doi.org/10.32604/cmc.2021.014991

Meka R, Alaeddini A, Bhaganagar K (2021) A robust deep learning framework for short-term wind power forecast of a full-scale wind farm using atmospheric variables. Energy 221: 119759. https://doi.org/10.1016/j.energy.2021.119759

Shahid F, Zameer A, Mehmood A, Raja MAZ (2020) A novel wavenets long short-term memory paradigm for wind power prediction. Applied Energy 269: 115098. https://doi.org/10.1016/j.apenergy.2020.115098

Tascikaraoglu A, Uzunoglu M (2014) A Review of Combined Approaches for Prediction of Short-term Wind Speed and Power. Renewable & Sustainable Energy Reviews 34: 243–254. https://doi.org/10.1016/j.rser.2014.03.033

Tian Z, Ren Y, Wang G (2019) Short-term wind speed prediction based on improved PSO algorithm optimized EM-ELM. Energy Sources, Part A: Recovery, Utilization, and Environmental Effects, 41(1): 26–46. https://doi.org/10.1080/15567036.2018.1495782

Wang X, Sideratos N, Hatziargyriou L, Tsoukalas LH (2004) Wind Speed Forecasting for Power System Operational Planning. International Conference on Probabilistic Methods Applied to Power Systems.

Zhang GP (2003) Time series forecasting using a hybrid ARIMA and neural network model. Neurocomputing 50: 159–175. https://doi.org/10.1016/S0925-2312(01)00702-0

Zhu Q, Chen J, Shi D, Zhu L, Bai X, Duan X (2019) Learning Temporal and Spatial Correlations Jointly: A Unified Framework for Wind Speed Prediction. IEEE Transactions on Sustainable Energy 11(1):509–523. https://doi.org/10.1109/TSTE.2019.2897136

Chapter 40
Energy and Exergy Analysis of Organic Rankine Cycle Driven by the Low-Temperature Geothermal Energy Sources

Hatice Narin Ucan, Merve Senturk Acar, and M. Ziya Sogut

Nomenclature

ORC Organic Rankine cycle
EES Engineering equation solver

40.1 Introduction

Geothermal energy is the thermal energy of hot water, steam, gas, or hot dry rocks under pressure accumulated at various depths of the earth's crust. In other words, geothermal energy can also be defined as earth heat energy. Since the center of the earth is very hot, the heat flows towards the surface and therefore the temperature increases as one goes deeper from the surface. The average temperature increase towards the center of the earth is 30 °C/km. The temperature increase is 0–40 °C/km in non-thermal regions, 70 °C/km in semithermal regions, and more than 70 °C/km in hypothermal regions (Lund et al. 1998).

Geothermal energy can be used in many applications where heat energy can be used as input. The main application areas where geothermal energy is used are:

H. N. Ucan (✉) · M. S. Acar
Faculty of Engineering, Bilecik Şeyh Edebali University, Bilecik, Turkey
e-mail: merve.senturkacar@bilecik.edu.tr

M. Z. Sogut
Maritime Faculty, Piri Reis University, İstanbul, Türkiye
e-mail: mzsogut@pirireis.edu.tr

M. Z. Sogut et al. (eds.), *Proceedings of the 2022 International Symposium on Energy Management and Sustainability*, Springer Proceedings in Energy,
https://doi.org/10.1007/978-3-031-30171-1_40

1. Conversion of heat energy into electrical energy.
2. Utilizing direct heat energy in industrial heating and drying processes (in products such as sugar, textile, paper, medicine, canning, etc.)
3. Its use in central heating and cooling (heating or cooling of greenhouses, housing estates, campuses, etc.)
4. Chemical production (fresh water, mineral and chemical salt production, etc.)
5. Use in touristic and therapeutic spas, swimming pools, and touristic facilities (Can 1994; Lund et al. 1998).

In Turkey, many geothermal regions ranging between 40 and 232 °C have been discovered. There are more than 1000 hot water sources up to 100 and 140 °C. As of 2008, the direct use in our country is 795 MWt, while the installed geothermal power-generation capacity is approximately 32.65 MWe (Serpen et al. 2009).

The Clausius-Rankine cycle is a thermodynamic cycle that converts heat energy into work. In the classical Rankine cycle, water is used as the working fluid. The system in which organic working fluid is used instead of water in the Rankine cycle is called the organic Rankine cycle (ORC). Since organic working fluids have lower critical temperature and pressure than water, they can operate at lower temperatures. ORC systems are used for waste heat, geothermal energy, solar energy, biomass energy, etc. They can be widely used in many fields.

When the slope of the saturated vapor curve of the fluid used in the Rankine cycle is defined as ($\xi = dT/ds$), fluids with $\xi 0$ are seen as dry type (Chen et al. 2010). Fluid type affects the degree of dryness of the steam at the turbine exit. T-s diagrams of different types of fluids are shown in Fig. 40.1: (a) wet type fluid (R718, R717, etc.), (b) isentropic type fluid (R123, R142b, etc.), and (c) dry type fluid (R600, R601, etc.)

Kaynaklı et al. (2017) analyzed a system that provides 1 MW turbine power in an ORC sourced from geothermal energy. In this study, the effects of changing geothermal energy source temperature (90–140 °C) and organic working fluid on system performance were analyzed. As a result of this analysis, they suggested R245fa as the organic working fluid.

Kavasoğulları and Cihan (2015) conducted energy and exergy analysis of a system in which waste heat is used as a heat source and where the organic Rankine cycle and classical cooling cycle work together. Five different fluids, R123, R600, R245fa, R141b, and R600a, were used in the system. As a result of their analysis, they suggested the R141b fluid for this system.

The thermodynamic properties of some fluids used in ORC systems are given in Table 40.1 (Chen et al. 2010; Calm and Hourahan 2007). It has been stated that among the organic fluids with hydrocarbon components used in ORC systems, those with high molecular weight, low critical temperature and pressure, and dry-isentropic type are more suitable (Cihan 2014).

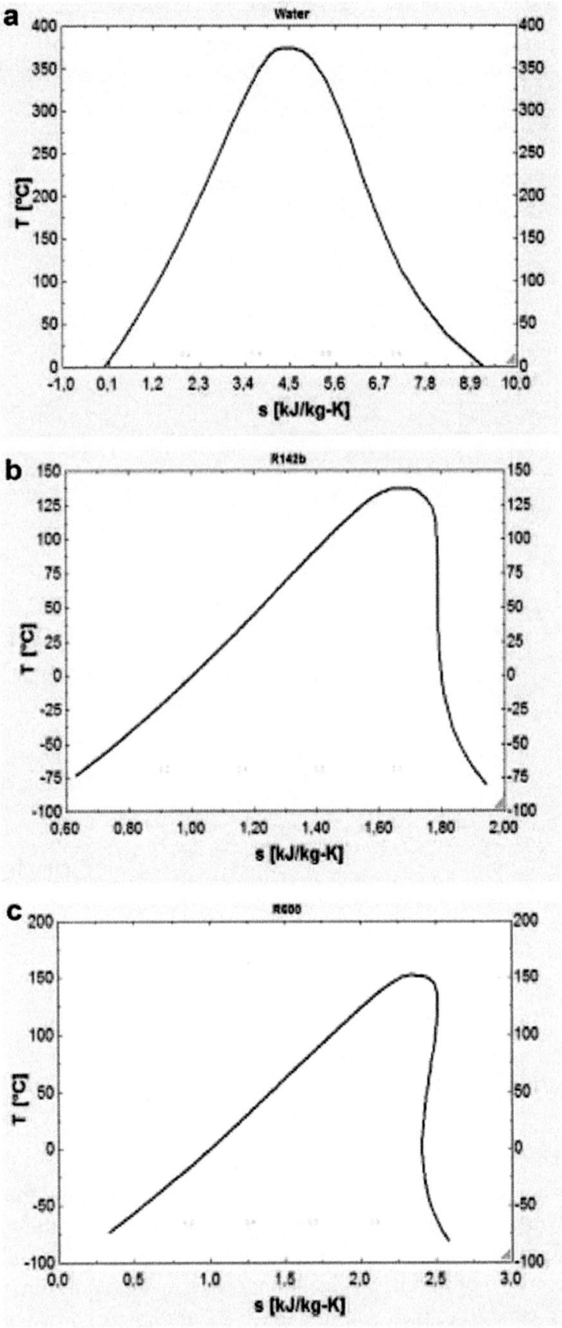

Fig. 40.1 T-s diagrams of different types of fluids in the organic Rankine cycle: (**a**) wet type fluid (R718, R717, etc.), (**b**) isentropic type fluid (R123, R142b, etc.), and (**c**) dry type fluid (R600, R601, etc.)

Table 40.1 Thermodynamic properties of some working fluids

Organic fluid	dT/ds	Fluid type	Molecular weight (g/mol)	Critical temperature (°C)	Critical pressure (Mpa)	ASHRAE 34 safety group
R134a	–0,39	Wet	102,3	101,1	4,06	A1
R141b	0	Isentropic	116,95	204,4	4,21	
R152a	–1,14	Wet	66,05	113,3	4,52	A2
R236fa		Isentropic	153,04	124,9	3,2	A1
R245fa	0,19	Isentropic	134,05	154	3,65	B1
R600	1,03	Dry	58,12	152	3,8	A3
R600a	1,03	Dry	58,12	134,7	3,63	A3
R601	1,51	Dry	72,15	196,6	3,37	
R601a	1,51	Dry	72,15	187,2	3,38	A3
R717	–10,48	Wet	17,03	132,3	11,3	B2
R718	–17,78	Wet	18,02	373,9	22	A1

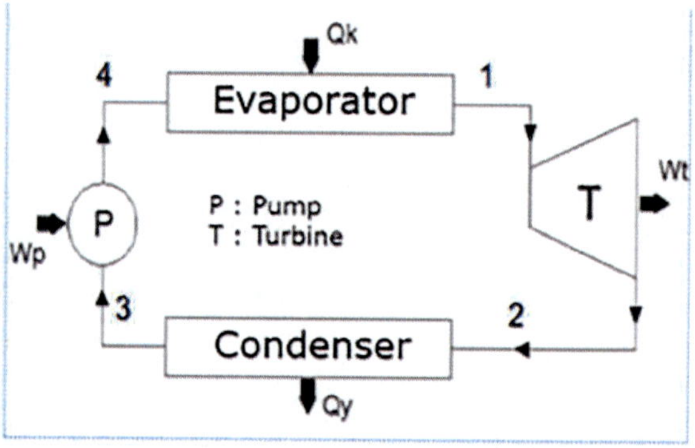

Fig. 40.2 Schematic representation of the organic Rankine cycle

40.2 Material and Method

40.2.1 Organic Rankine Cycle

The working principle of the organic Rankine cycle is the same as the Rankine cycle. The only difference between the Rankine cycle and the organic Rankine cycle is the working fluid. While water is used as the working fluid in the Rankine cycle, organic fluids are used as the working fluid in the organic Rankine cycle. The ORC system consists of pump, turbine, condenser, and evaporator. The schematic diagram of the organic Rankine cycle and the T-s diagram are shown, respectively, in Figs. 40.2 and 40.3.

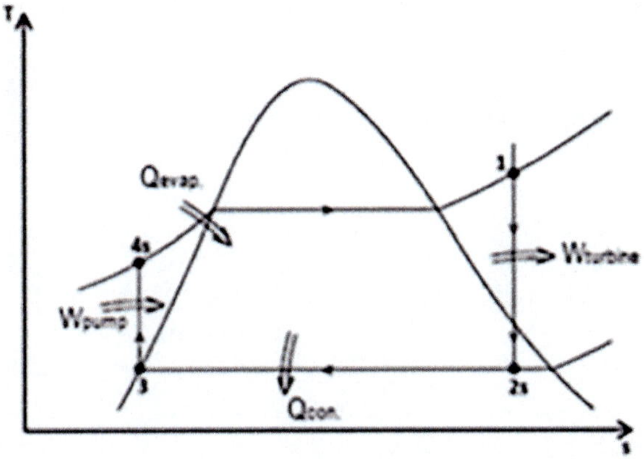

Fig. 40.3 Organic Rankine cycle T-s diagram

At the inlet of the evaporator, the organic working fluid at the evaporation pressure coming from the pump is in liquid form. In the evaporator, the heat from the geothermal energy source is transferred to the organic working fluid 5 and the organic working fluid evaporates with the heat it receives from the geothermal energy. Steam at high pressure and temperature expands in the turbine and exits the turbine as steam at low pressure. As the steam expands in the turbine, it turns the turbine. The generator, which is connected to the turbine shaft, rotates with the turbine and generates electricity. The steam entering the condenser at the condenser pressure condenses by throwing its heat to the atmosphere in the condenser and leaves the condenser as a saturated liquid. The pressure of the fluid leaving the condenser is brought from the condenser pressure to the evaporator pressure with the help of a pump. The fluid leaving the pump enters the evaporator and the cycle is complete.

40.2.2 *Thermodynamic Analysis of Organic Rankine Cycle*

In the analysis of the organic Rankine cycle, the first and second laws of thermodynamics are used to determine the performance of the system elements and the system efficiency (Cengel and Boles 2011).

40.2.2.1 Energy Calculations

The equations obtained according to T1K for the evaporator, turbine, condenser, and pump used in the organic Rankine cycle are given below.

Equations used for turbine:

$$W_T = m(h_1 - h_2) \tag{40.1}$$

$$\eta_T = \frac{(h_1 - h_2)}{(h_1 - h_{2s})} \tag{40.2}$$

The equations used for the condenser:

$$Q_y = m(h_3 - \mathrm{h}_2) \tag{40.3}$$

Equations used for the pump:

$$W_P = m(h_3 - h_4) \tag{40.4}$$

$$\eta_T = \frac{(h_{4s} - h_3)}{(h_4 - h_3)} \tag{40.5}$$

The equations used for the evaporator:

$$Q_b = m(h_1 - h_4) \tag{40.6}$$

The network of the system:

$$W_{\mathrm{Net}} = W_T - W_P \tag{40.7}$$

T1K efficiency of the system:

$$\eta = \frac{W_T - W_P}{Q_b} \tag{40.8}$$

40.2.2.2 Exergy Calculations

The equations of irreversibility (I), exergy (E), specific exergy (e), and T2K efficiency ($\eta 2$) for the evaporator, turbine, condenser, and pump used in the organic Rankine cycle are respectively given below.

Specific exergy current for all points in the system:

$$e_i = h_i - h_0 - [T_0 * (s_i - s_0)] \tag{40.9}$$

Exergy current for all points in the system:

$$E_i = m_i * e_i \tag{40.10}$$

Turbine:

$$I_T = E_1 - E_2 - W_T \tag{40.11}$$

$$\eta_{T,2} = \frac{W_T}{E_1 - E_2} \tag{40.12}$$

Condenser:

$$I_y = m * T_0 \left(s_3 - s_2 + \frac{q_y}{T_y} \right) \tag{40.13}$$

Pump:

$$I_P = E_4 - E_3 + W_P \tag{40.14}$$

Evaporizer:

$$I_b = m * T_0 \left(s_1 - s_4 - \frac{q_b}{T_b} \right) \tag{40.15}$$

Carnot yield:

$$\eta_c = 1 - \left[\frac{T_0 + 273,15}{T_b + 273,15} \right] \tag{40.16}$$

T1K efficiency of the system:

$$\eta_2 = \frac{\eta}{\eta_c} \tag{40.17}$$

The total irreversibility value of the system:

$$I_{\text{top}} = I_T + I_y + I_P + I_b \tag{40.18}$$

40.3 Calculation Results

The design parameters and operating temperatures used in the analysis of the organic Rankine cycle are given in Table 40.2. In this thermodynamic analysis, the design conditions given in Table 40.2 were used. The evaporator saturation temperature was chosen 5 °C lower than the geothermal source temperature. The turbine inlet temperature was chosen 2 °C lower than the geothermal source temperature. The heat taken from geothermal energy was kept constant as 10 MW, the condensation temperature was kept constant at 30 °C, the evaporator temperature was increased gradually from 85 to 110 °C by 5°, and the changes in other parameters were analyzed in response to the increase in the evaporator temperature.

Table 40.2 Design parameters and operating temperatures

Parameters	Values	Parameters	Values
Geothermal source temperature	90–115 °C	Turbine isentropic efficiency	75%
Turbine inlet temperature	88–113 °C	Pump isentropic efficiency	80%
Evaporation temperature	85–110 °C	Ambient temperature	20 °C
Condensing temperature	30 °C	Evaporator heat power	10 MW

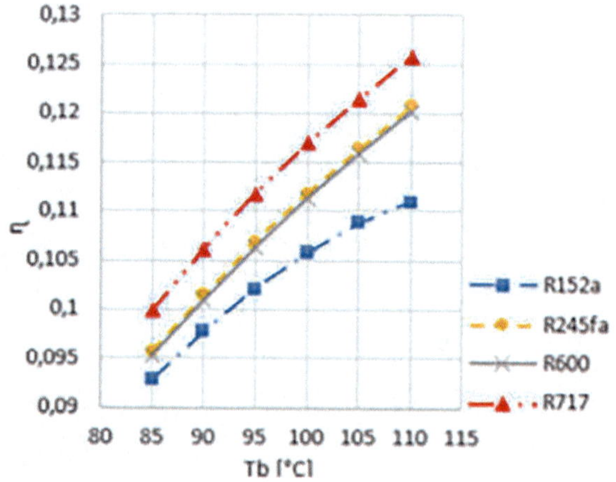

Fig. 40.4 Variation of the first law efficiency of the system against 10 MW constant heat input and varying evaporation temperatures

Figure 40.4 shows the change in the first law efficiency of the system in response to the changing evaporator temperatures. As can be seen from the figure, it is observed that the first law efficiency of the system increases with increasing evaporator temperature for all fluids. The highest value was observed in R717 and the lowest value in R152a. It has been observed that the first law efficiencies of R245fa and R600 fluids are very close to each other.

Figure 40.5 shows the change of the second law efficiency of the system in response to the changing evaporator temperatures. As can be seen from the figure, it is observed that the second law efficiency of the system decreases with increasing evaporator temperature for all fluids. As with the first law efficiency, the highest value was observed in R717 and the lowest value in R152a, and it was observed that the second law efficiencies of R245fa and R600 fluids were very close to each other.

Figure 40.6 shows the variation of turbine power in response to varying evaporator temperatures. As seen in the figure, it is observed that turbine power increases with increasing evaporator temperature for all fluids. The highest turbine power was observed in R717, and the lowest turbine power was observed in R245fa.

Figure 40.7 shows the variation of pump power in response to varying evaporator temperatures. As seen in the figure, it is observed that the pump power increases with increasing evaporator temperature for all fluids. The highest pump power was observed in R152a, and the lowest pump power was observed in R245fa.

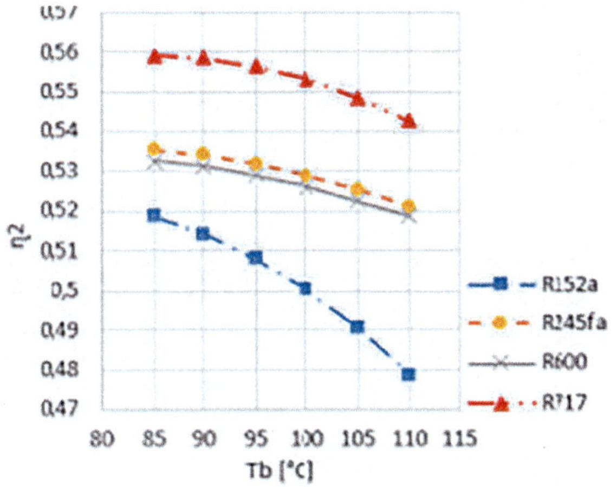

Fig. 40.5 Variation of the second law efficiency of the system against 10 MW constant heat input and changing evaporation temperatures

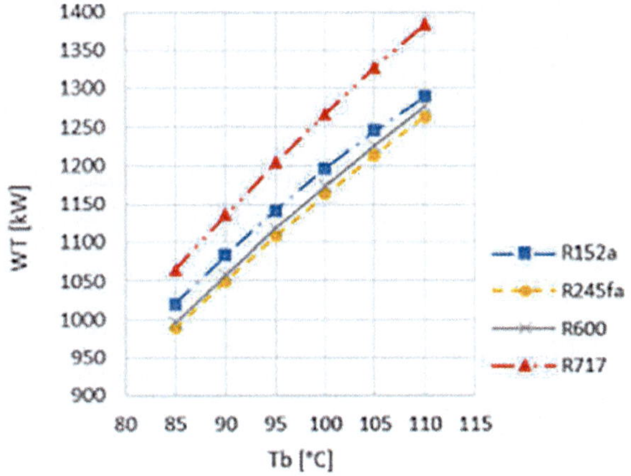

Fig. 40.6 Variation of turbine power with 10 MW constant heat input and changing evaporation temperatures

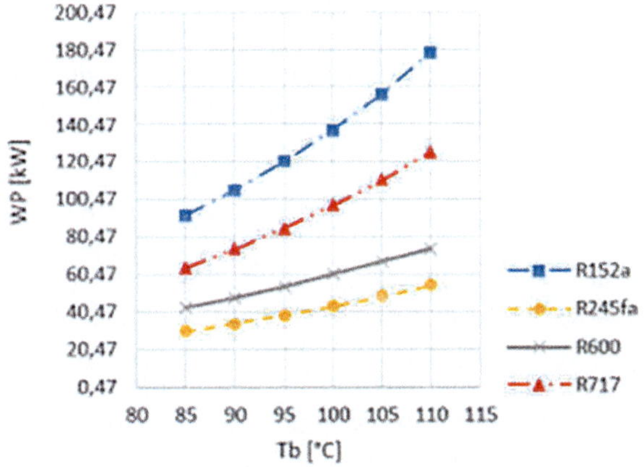

Fig. 40.7 Variation of pump work against 10 MW constant heat input and varying evaporation temperatures

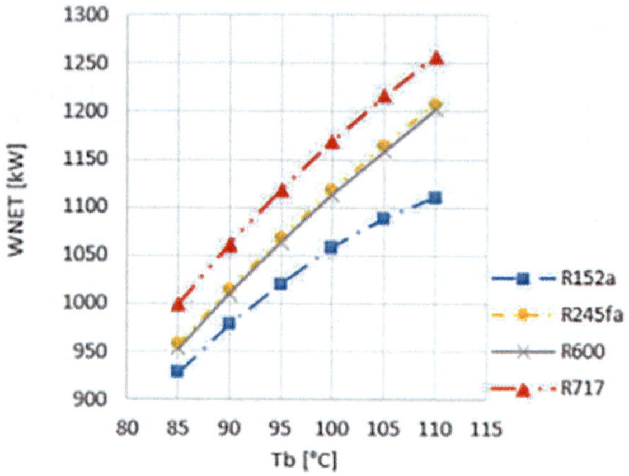

Fig. 40.8 Variation of network of the system against 10 MW constant heat input and varying evaporation temperatures

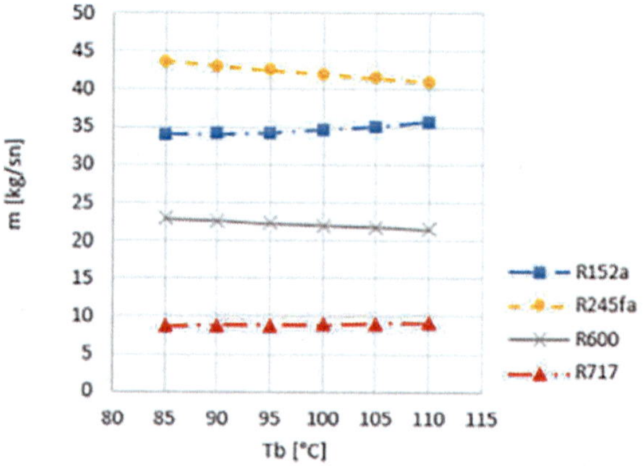

Fig. 40.9 Variation of total exergy destruction versus 10 MW constant heat input and varying evaporation temperatures

Figure 40.8 shows the change in the network of the system in response to changing evaporator temperatures. As can be seen from the figure, it is observed that the network of the system increases with increasing evaporator temperature for all fluids. The network value of the system is highest at R717 and lowest at R152a. It is observed that the network values of the systems in which R245fa and R600 fluids are used are very close to each other.

Figure 40.9 shows the variation of exergy destruction in response to the changing evaporator temperatures. As can be seen from the figure, it is observed that exergy destruction decreases with increasing evaporator temperature for all fluids. The highest exergy destruction was observed in R152a, and the lowest exergy destruction was observed in R717.

Figure 40.10 shows the change in the flow rate of the organic working fluid in response to the changing evaporator temperatures. As can be seen from the figure, the highest organic working fluid flow rate was observed in R245fa and the lowest organic working fluid flow rate was observed in R717. It was observed that R245fa and R600 fluid flow rates decreased with increasing evaporator temperature, and R717 and R152a fluid flow rates increased with increasing evaporator temperature.

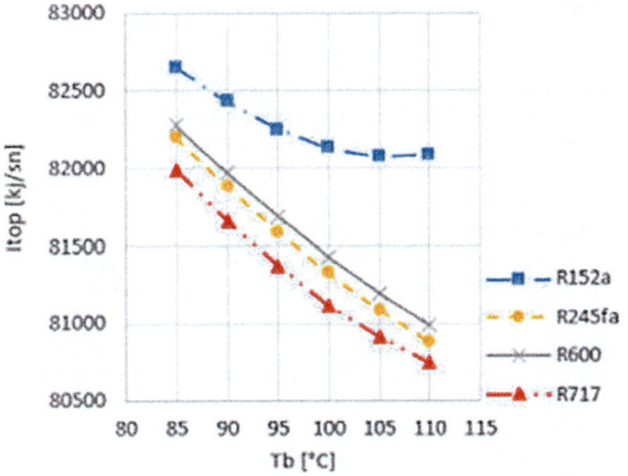

Fig. 40.10 Variation of organic working fluid flow rate versus 10 MW constant heat input and varying evaporation temperatures

40.4 Conclusion

In this study, energy and exergy analysis was performed according to the change of evaporator temperature of a system that produces electricity from low-temperature geothermal energy with the help of organic Rankine cycle. The first law efficiency, second law efficiency, turbine power, pump power, net power of the system, exergy destruction, and working fluid flow rate of the system were examined separately. When the fluids were examined in terms of the first law efficiency, second law efficiency, network, and total exergy destruction, R152a showed the worst performance, R245fa and R600 fluids performed close to each other, and the R717 fluid showed the best performance.

It is not possible to interfere with the source temperature and the amount of heat provided from the source in energy sources such as geothermal energy, waste heat, solar energy, etc. In energy production facilities where such resources are used, it is important how much we can benefit from the existing resource rather than the system efficiency. From this point of view, the network of the system is the most important performance parameter in fluid selection. When we examine the graph showing the network of the system, the R717 fluid was the fluid that produced the most power by producing approximately 1000 kW of power and the R152a fluid was the fluid that produced the least power by producing approximately 930 kW of power, in response to 85 °C evaporator temperature and 10 MW heat power input. In addition, R600 and R245fa fluids produce approximately 950 kW of power under the specified conditions. The R717 fluid is recommended as the most suitable fluid for the analyzed system.

References

Batık, H., Koçak, A., Akkuş, I., Şimşek, S., Mertoğlu, O., Dokuz, İ., Bakır, N. (2000) Geothermal energy utilisation development in Turkey-present geothermal situations and projections, Proc. World Geothermal Congress, Japan, 85–91

Calm J. M., Hourahan G. C., (2007) "Refrigerant Data Update", Hpac Engineering, 79, 50–64

Can, M. (1994) Bursa'da jeotermal enerjinin merkezi ısıtma sistemlerinde kullanılabilirliğinin incelenmesi, Ekoloji, 13, 44–49

Chen H., Goswami D.Y., Stefanakos E.K., (2010) A review of thermodynamic cycles and working fluids for conversion of low-grade heat, Renewable and Sustainable Energy Reviews, 14, 3059–3067

Cihan E., (2014) Organik Rankine Çevrimi İle Çalışan Atık Isı Kaynaklı Bir Soğutma Sisteminin performansının Araştırılması, Isı Bilimi Ve Tekniği Dergisi, 34, 1, 101–109

Çengel, Y. ve Boles, M.A. (2011) Thermodynamics: An Engineering Approach. McGraw Hill Book Co. Seventh Edition, [7] Kavasoğulları B. ve Cihan E. (2015) Organik Rankine Çevrimi (ORÇ) ile Birlikte Çalışan Sıkıştırmalı Bir Soğutma Çevriminin Ekserji Analizi, Tesisat Mühendisliği Dergisi, 150, 74–85

Kaynaklı Ö., Bademlioğlu A.H., Nurettin Y., Recep Y., (2017) Thermodynamic analysis of the organic rankine cycle and the effect of refrigerant selection on cycle performance, International Journal of Energy Applications and Technologies, 4(3), 101–108

Lund, J.W., Lienau, P.J. and Lunis, B.C. (1998) Geothermal Direct-Use Engineering and Design Guidebook, United States Department of Energy, Idaho

Lund, J.W., Lienau, P.J. and Lunis, B.C. (1998) Geothermal Direct-Use Engineering and Design Guidebook, United States Department of Energy, Idaho

SERPEN, U., AKSOY, N., ÖNGÜR, T., KORKMAZ, E.D., (2009) Geothermal Energy in Turkey: 2008 update, Geothermics, 38 (2), 227–237

Chapter 41
Waste Heat Recovery from Cooling Systems of Data Centers

Ulaş Ülkü, Ziya Haktan Karadeniz, and Gülden Gökçen Akkurt

41.1 Introduction

Data centers are the environments where computer systems produce, store, analyze, and use data in any way. They are installed as a corridor in a built environment. There is a lot of equipment in the cabinets installed along the corridors. The computer hardware arranged as a rack arrangement in these cabinets enables the data center to fulfill its task. In today's conditions, the need for data centers has increased. Moreover, cloud systems and many new technologies evolve around blockchain networks and data analytics centers.

The workload and capacity of data centers are also increasing day by day. The energy consumption in the data centers cannot be underestimated due to the increasing need and usage rates and high-capacity information systems. Information technology (IT) equipment accounts for most of the electricity consumption, while cooling systems also consume a lot of electricity. For this reason, the design of the cooling systems of data centers is crucial. Figure 41.1 shows the energy consumption distribution in a data center. As can be seen from Fig. 41.1, IT equipment consumes 52% of electricity, cooling system 38%, and the remaining equipment (electricity distribution, UPS, etc.) 10% (Cho et al. 2012).

On the other hand, the temperature of the equipment operating in these environments is constantly increasing and continuous cooling need arises. In other words, these systems produce waste heat, and to remove this waste heat, they consume electricity. Therefore, electricity consumption and the efficiency of this consumption have been the subject of special attention. Because fossil resources are decreasing in

U. Ülkü (✉) · Z. H. Karadeniz · G. G. Akkurt
Department of Energy Engineering, Izmir Institute of Technology, Izmir, Turkey
e-mail: ulasulku@iyte.edu.tr; haktankaradeniz@iyte.edu.tr; guldengokcen@iyte.edu.tr

M. Z. Sogut et al. (eds.), *Proceedings of the 2022 International Symposium on Energy Management and Sustainability*, Springer Proceedings in Energy,
https://doi.org/10.1007/978-3-031-30171-1_41

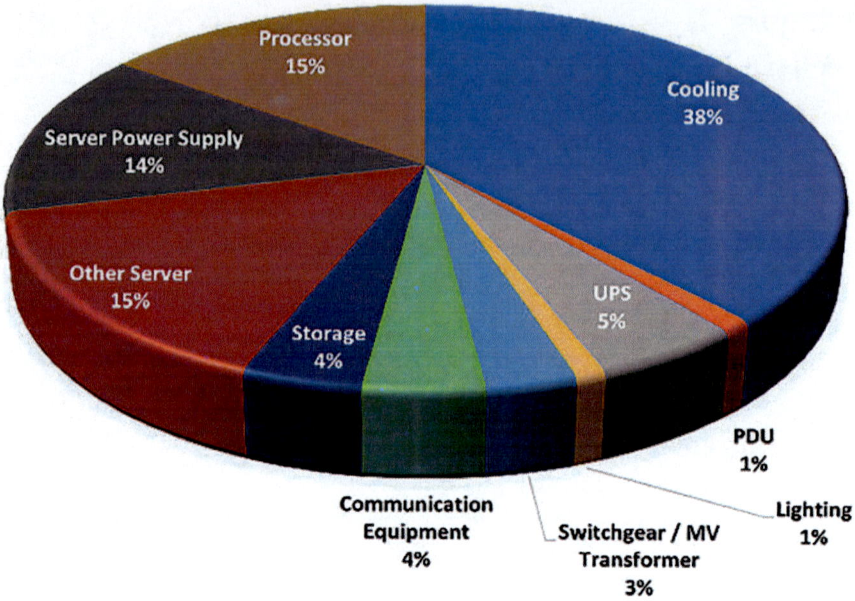

Fig. 41.1 Energy consumption of data centers

the world, the importance of efficient consumption is increasing. This situation has drawn attention to renewable energy and energy efficiency.

Today, cooling systems applied in data centers systems are developing very rapidly, and interesting solutions are being researched to reduce electricity consumption. This study aims to review the literature on the importance of designing data center cooling systems and establishing a waste heat recovery system.

41.2 Cooling Systems of Data Centers

Cooling systems can mainly be classified as active and passive systems, further classified as follows:

- Active cooling
- Passive cooling
- Free cooling
  - Direct air source free cooling
  - Indirect air source free cooling
- Liquid cooling systems
  - Direct liquid cooling systems
  - Rack level liquid cooling systems
  - Immersion type liquid cooling systems

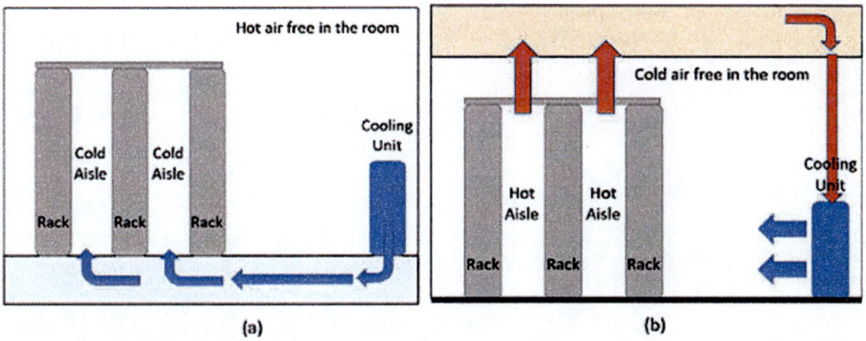

Fig. 41.2 (**a**) Cold aisle principle (**b**) Hot aisle principle

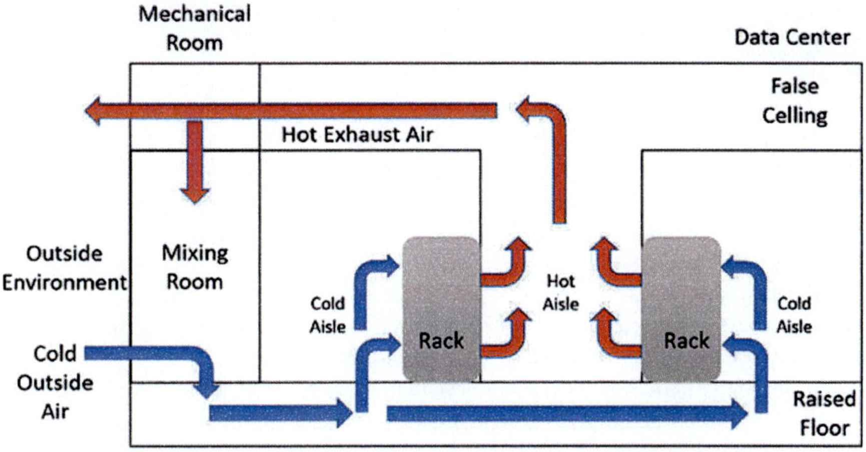

Fig. 41.3 Simplified scheme of direct air source free cooling with economizer system

The most common technique used in active cooling is to keep the cold air produced by the computer room air conditioning unit (CRAC) on a raised floor (Nadjahi et al. 2018). This design, shown in Fig. 41.2, is also called the hot-cold aisle principle. On the other hand, free cooling, which is a type of passive cooling, is the principle of cooling system design using natural ventilation (Oró et al. 2015). In direct air source free cooling, given in Fig. 41.3, the process takes place by fans that draw the cold outside air into the data center. In indirect air source free cooling, shown in Fig. 41.4, heat transfer is provided by an air-to-air cross-flow heat exchanger.

On the other hand, liquid cooling systems are among the alternatives in cases where the power density of data centers is high. In addition, liquid waste can be used for heat recycling. Water and other heat transfer fluids can transfer about 4.2 times more heat by weight than air (Pricing 2016).

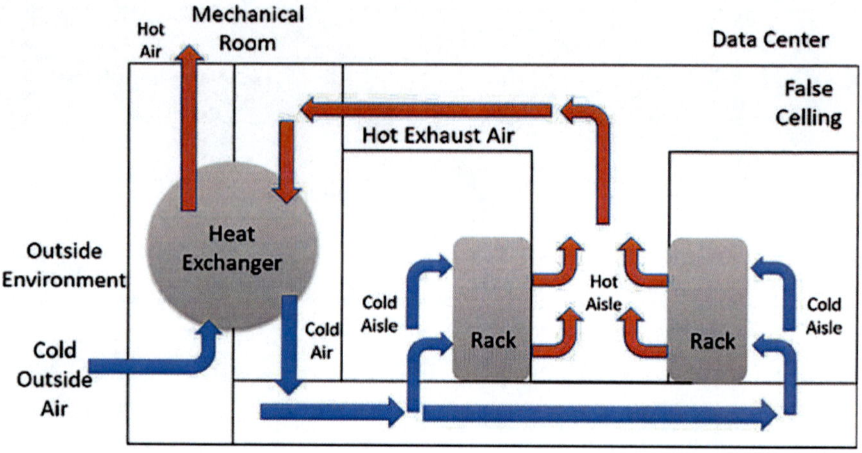

Fig. 41.4 Simplified scheme of cooling system using air-to-air heat exchanger

Another system is the direct liquid cooling system. In this system, the CPU is connected by a heat sink and other components are cooled by airflow. Copper pipes in the system carry the refrigerant. In this way, the processor is cooled by heat transfer from the plate (Nadjahi et al. 2018; Zimmermann et al. 2012). Also, Microsoft has an experimental cooling system study. This liquid cooling system is the immersion type. Servers are usually immersed in mineral oils. With this system, heat can be transferred directly to an external cycle. The transferred heat can be reused or released (Pricing 2016).

Additionally, a rack-level cooling system can be created. This system features an air-to-liquid finned heat exchanger behind the rack. An example of this system is given in Fig. 41.5. This technique can be very useful in a traditional hot/cold aisle arrangement. It will also help prevent mixing of air of different temperatures. In addition, different cooling systems can be used with this system. Moreover, waste heat recovery can be achieved with different integrations (He et al. 2018; Zimmermann et al. 2012).

41.3 Waste Heat Recovery from Data Centers

Table 41.1 shows the roles and temperatures of the components in total heat in the data centers (Huang et al. 2020). 63 percent of microprocessors have a temperature rating of 85 °C. Also, microprocessors need constant cooling to work efficiently. Additionally, IT equipment in typical data centers can convert more than 99% of its power into heat. Considering these situations, the recovery of the waste heat generated is an important study subject. There are many studies on waste heat recovery. Systems created for the recovery of waste heat will reduce the working

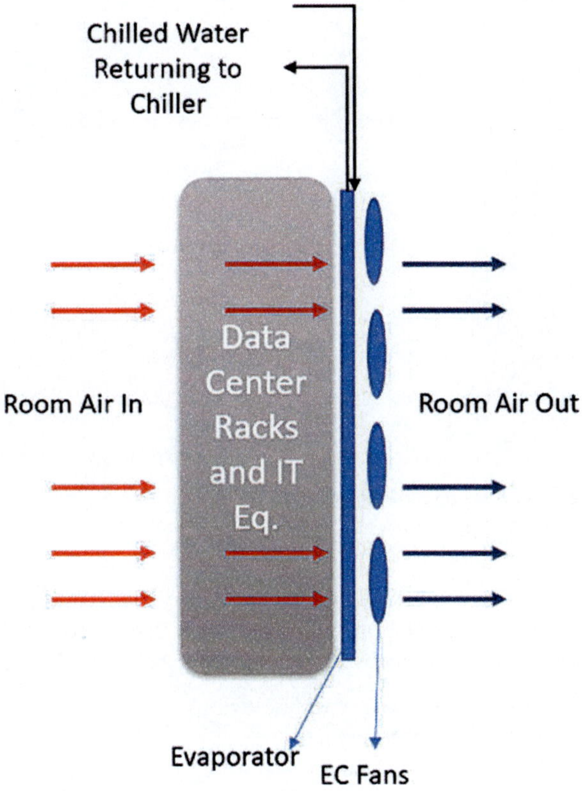

Fig. 41.5 Simplified scheme of a rack-level cooling system

Table 41.1 Heat dissipation rates of component

	Component	Proportion of total heat (%)	Temperature (°C)
For standard server	Microprocessors	30	85
	DC/DC conversion	10	50
	I/O processor	3	40
	AC/DC conversion	25	55
	Memory chips	11	70
	Fans	9	30
	Disk drivers	6	45
	Motherboard	3	40
For high-performance cluster (HPC)	Microprocessors	63	85
	DC/DC conversion	13	115
	I/O processor	10	100
	Memory chips	14	40

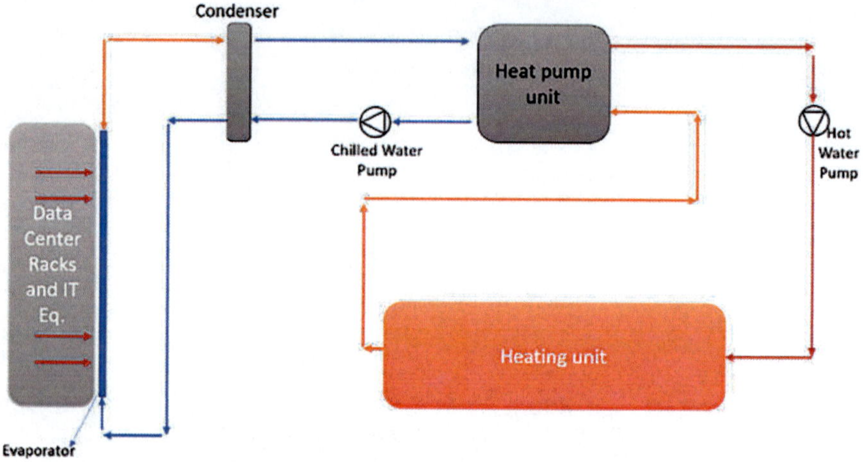

Fig. 41.6 Simplified scheme of the waste heat recovery system in the data center

load of air conditioning units. In this way, both energy efficiencies will increase, and electricity production will decrease in data centers. There are some examples of systems used to recover waste heat, further classified as follows:

- Waste heat recovery technologies
- Plant or district heating
- Power plant co-location and using thermodynamic cycle
- Absorption cooling
- Organic Rankine cycle

District heating is a common low-quality waste heat recovery method that is both cost-effective and environmentally friendly. As a result, this system can improve efficiency while also generating an income stream for the data centers (Ebrahimi et al. 2014). Figure 41.6 shows a district heating system that uses a heat pump to increase waste heat in a data center.

Another common waste heat recovery technique is shown in Fig. 41.7. This system is to use waste heat to heat water in the thermal Rankine cycle of a power plant. Data center heat can be used to preheat the boiler feedwater. In this way, fossil fuel consumption is generally reduced (Ebrahimi et al. 2014).

Also, the use of waste heat in the absorption refrigeration cycle is an important source of efficiency, for systems that require significant cooling, such as data centers. Figure 41.8 shows a schematic of a simple absorption cooling system powered by data centers. Liquids can also work more efficiently for this cycle because their specific volume is much lower than vapors. That is, replacing the vapor compression system with a liquid absorption cooling system can save a lot of energy (Ebrahimi et al. 2014).

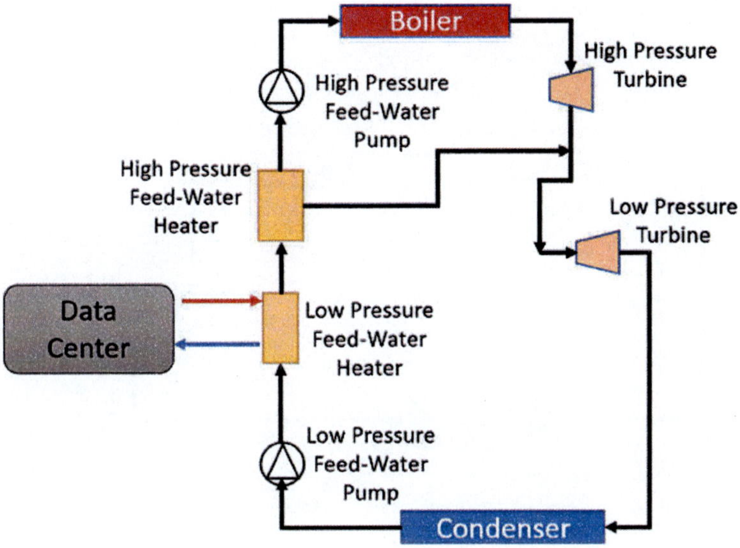

Fig. 41.7 Simplified scheme of waste heat recovery with Rankine cycle in data centers

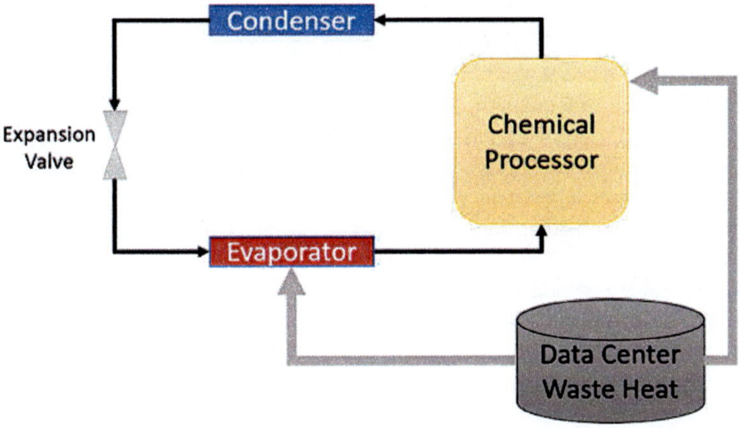

Fig. 41.8 Simplified scheme of data center waste heat recovery system using absorption cooling system

Additionally, an organic Rankine cycle (ORC) can be used to generate electricity directly from waste heat generated in data centers. An example schematic for this application is given in Fig. 41.9. On the other hand, the working fluid is an organic liquid with significantly lower boiling points. Lower boiling points allow data center waste heat to be used as a heat source. It is a suitable system for using low-temperature waste heat.

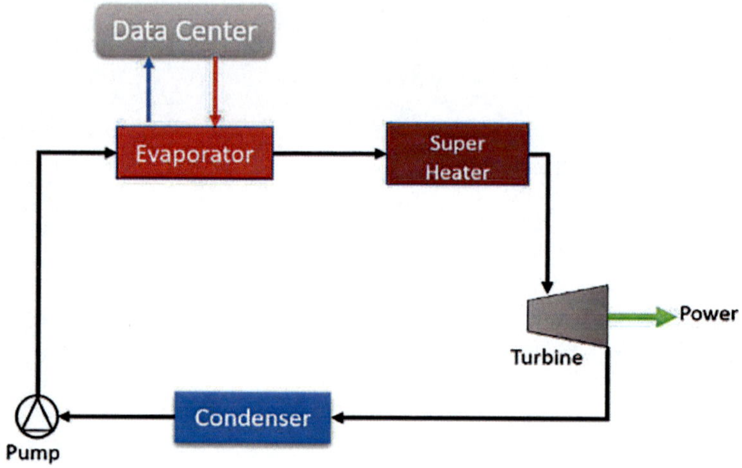

Fig. 41.9 Simplified scheme of data center waste heat recovery system using absorption cooling system

41.3.1 Cabinet Door-Type Heat Exchangers for Waste Heat Recovery

As it is known, data centers consist of shelves. There is airflow over the cabinet doors to cool the components inside. The refrigerated air circulating inside can be directed to the IT equipment and then exhausted out of the cabinet. Considering this cycle, waste heat can be captured from the cabinet doors. This is possible with the cabinet door-type heat exchanger (HEX) system. This system captures heat with the finned tube heat exchanger design on the cover. Thus, a simple and low-cost waste heat capture system for cooling systems is installed. In this way, the load on the cooling system is reduced and efficiency is increased (Fig. 41.10).

In addition to all this, capturing heat using a heat exchanger mounted on the cabinet door has not been studied enough yet. This presents a different research topic to increase efficiency.

41.4 Energy Efficiency Measures for Data Centers

Efficiency in data centers can be expressed as power usage efficiency (PUE) and data center infrastructure efficiency (DCIE). Equations 41.1 and 41.2 show how these expressions are calculated. The PUE can be anything between 1.0 and infinite. A PUE score of 1.0 represents 100 percent efficiency, indicating that all energy is used by IT equipment. PUE levels of 1.2 or less are found in state-of-the-art installations (Dreibholz et al. 2007). Efficiency values are presented in Table 41.2.

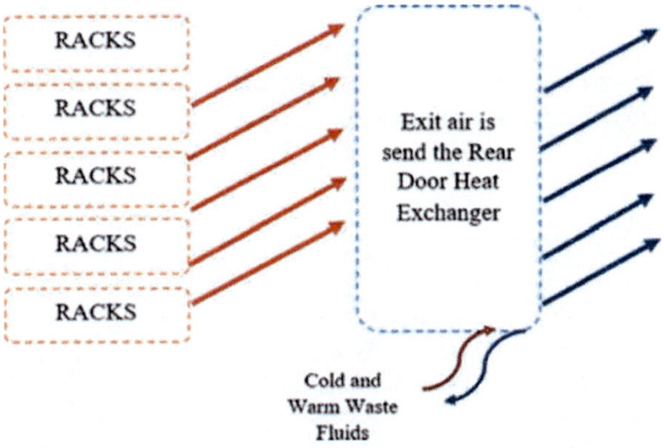

Fig. 41.10 Simplified scheme of the cabinet door-type HEX system

Table 41.2 Performance values of data centers

PUE	Level of efficiency	DCiE (%)
3.0	Very inefficient	33
2.5	Inefficient	40
2.0	Average	50
1.5	Efficient	67
1.2	Very efficient	83

$$\mathrm{PUE} = \frac{\text{Total Facility Power}}{\text{IT equipment power}} \tag{41.1}$$

$$\mathrm{DCiE}(\%) = \frac{1}{\mathrm{PUE}} \times 100 \tag{41.2}$$

41.5 Conclusions

The number of data centers is increasing day by day. The increased use of technological tools with Covid-19 has increased the need for data centers. However, this also indicates an increase in energy consumption. In this study, the literature of data center cooling systems and waste heat recovery systems is reviewed. Important systems in the literature are briefly introduced. As a result of the studies, it has been found that the energy consumption of the cooling systems is quite high. In addition, it has been observed that the waste heat is at a considerable level. The equipment used needs constant cooling, which indicates that there is a constantly generated waste heat. It is also estimated that 68% of this waste heat can be recovered. Different cooling systems can be selected according to the characteristics of the data center, as well as different waste heat recovery systems. Due to its advantages such as ease of installation and use, the use of cabinet type heat exchanger is also emphasized as an alternative to recovery. On the other hand, data centers usually have low waste heat due to the low cooling temperature. And this is a big hurdle for large-scale applications. For this purpose, the use of waste heat from data centers can be facilitated by the development of many low-temperature heat recovery technologies.

References

Cho, J., Lim, T., & Kim, B. S. (2012). Viability of datacenter cooling systems for energy efficiency in temperate or subtropical regions: Case study. Energy and Buildings, 55, 189–197. https://doi.org/10.1016/J.ENBUILD.2012.08.012

Dreibholz, T., Becke, M., & Adhari, H. (2007). Report to Congress on Server and Data Center Energy Efficiency Public Law 109–431. Tdr.Wiwi.Uni-Due.De, 109(August), 431. http://www.tdr.wiwi.uni-due.de/fileadmin/fileupload/I-TDR/SCTP/Paper/ConTEL2011.pdf

Ebrahimi, K., Jones, G. F., & Fleischer, A. S. (2014). A review of data center cooling technology , operating conditions and the corresponding low-grade waste heat recovery opportunities. Renewable and Sustainable Energy Reviews, 31, 622–638. https://doi.org/10.1016/j.rser.2013.12.007

He, Z., Ding, T., Liu, Y., & Li, Z. (2018). Analysis of a district heating system using waste heat in a distributed cooling data center. Applied Thermal Engineering, 141(June), 1131–1140. https://doi.org/10.1016/j.applthermaleng.2018.06.036

Huang, P., Copertaro, B., Zhang, X., Shen, J., Löfgren, I., Rönnelid, M., Fahlen, J., Andersson, D., & Svanfeldt, M. (2020). A review of data centers as prosumers in district energy systems: Renewable energy integration and waste heat reuse for district heating. Applied Energy, 258 (October 2019), 114109. https://doi.org/10.1016/j.apenergy.2019.114109

Nadjahi, C., Louahlia, H., & Lemasson, S. (2018). A review of thermal management and innovative cooling strategies for data center. Sustainable Computing: Informatics and Systems, 19(October 2017), 14–28. https://doi.org/10.1016/j.suscom.2018.05.002

Oró, E., Depoorter, V., Garcia, A., & Salom, J. (2015). Energy efficiency and renewable energy integration in data centres. Strategies and modelling review. Renewable and Sustainable Energy Reviews, 42, 429–445. https://doi.org/10.1016/j.rser.2014.10.035

Pricing, N. (2016). Building Automation Systems and Cybersecurity.

Zimmermann, S., Meijer, I., Tiwari, M. K., Paredes, S., Michel, B., & Poulikakos, D. (2012). Aquasar: A hot water cooled data center with direct energy reuse. Energy, 43(1), 237–245. https://doi.org/10.1016/j.energy.2012.04.037

Chapter 42
Wind Turbine Condition Monitoring Using Failure Analysis

Betül Sena Çağlar, Hasan Burak Ketmen, and Barış Bulut

42.1 Introduction

Wind turbines are machines that convert the kinetic energy of the air in motion, first into mechanical energy and then into electrical energy (Department of Energy 2014). Every wind turbine has subsystems that enable it to perform this transformation. These subsystems consist of parts that are expensive, and some take a long time to supply in case of failure. It is beneficial to predict a future problem in a certain subsystem before it occurs, both in terms of not negatively affecting the remaining life of other parts and in terms of improved availability.

Condition monitoring is important to improve the availability and stability of running for wind turbine, which is of great significance to utilize the wind power efficiently and reliably (Yang et al. 2010). The vital components of a wind turbine, such as gearbox, bearing, CCU (current converter unit), generator and blades, are the most important concerns in the study of fault diagnosis.

In this study, the goal is to detect fault symptoms in advance and to inform the maintenance personnel by creating an alarm at the earlier stages of a potential fault. This shall allow sufficient time to set the replacement schedule of the component. Indirectly, operating and maintenance costs will also be reduced.

Fault analysis methods could usually be divided into two categories: data-driven approaches (Yin et al. 2016; Zhang et al. 2018) and model-based methods (Dey et al. 2015; Beretta et al. 2020). Model-based methods rely on accurate system modelling, which can perform efficient and accurate fault detection if the system dynamics can be well described but model-based methods have difficulty in practice if the application cannot fit sufficiently well with the assumptions about the models and

B. S. Çağlar (✉) · H. B. Ketmen · B. Bulut
Enforma Bilisim A.S., Istanbul, Turkey
e-mail: sena.caglar@enforma-tr.com; burak@enforma-tr.com; baris@enforma-tr.com

M. Z. Sogut et al. (eds.), *Proceedings of the 2022 International Symposium on Energy Management and Sustainability*, Springer Proceedings in Energy,
https://doi.org/10.1007/978-3-031-30171-1_42

objectives. On the other hand, system modelling is not needed for data-driven methods, only requiring data that can be stored easily with the condition monitoring process using appropriate sensors. This chapter offers both model-based and data-driven methods.

Based on this data exploration, new features were extracted, and they were used in the LSTM-AE (long short-term memory-autoencoder) model. They are also used in pre-failure detection with a simple data-driven condition monitoring and a well-defined threshold. Since it is important to know in which component potential failures will occur, component-specific LSTM-AE models have been developed for important components that want to reduce the failure rate.

42.2 Materials and Methods

The study was carried out with the CRISP-DM (cross-industry standard process for data mining) methodology, which has a systematic and iterative approach. This methodology provides a structured approach to planning a data science project.

42.2.1 Data Description

All data were collected from wind turbines at the wind farm located in the south of Turkey. The SCADA dataset contains more than 130 features as the main systems of the turbine are all monitored (pitch, shaft bearing, gearbox, generator, etc.). Raw data's sampling frequency is 1 s, but the data was collected as 10-min averages where each feature is a signal from a sensor.

42.2.2 Data Pre-processing

Due to the miscommunications or defects of the sensors, real-life data is affected by outliers. The filter of absurd data readings is necessary to reduce the chances of generating false alarms. In this study manual threshold values based on the technical knowledge of turbine behaviours are used to filter nonsense data. Duplicate recordings were deleted, and the original data was resampled at 10 min due to missing recordings. Also the values were missing in some timestamps in the dataset, which may have been because of errors in the sensors or the recording system. These missing values are filled with the interpolation method considering the wind speed distribution.

Table 42.1 Event data

System	Start time	End time	Event ID	Category	Description
T02	08.01.2020 05:56:50	08.01.2020 10:08:08	422	Converter	CCU reports: Line CB close fail
T22	07.02.2020 09:08:53	07.02.2020 21:27:34	358	Unused	Malfunction triple-pitch system
T02	08.02.2020 04:54:35	08.02.2020 07:43:58	052	Gearbox	Gearbox oil pressure too low
T27	11.02.2020 00:54:55	11.02.2020 09:50:10	141	Frequency Converter	Gen. side CCU collective faults
T51	03.03.2020 22:16:35	04.03.2020 15:37:44	051	Gearbox	Gearbox oil level too low

After cleaning the operational data, the event data, that is, the fault records of the turbines, and the operational data were merged. Table 42.1 shows the event logs that took place in the turbines. Fault features, including event logs, have been used to assess the veracity of the predictions and to label pre-failure time steps.

42.2.3 Feature Engineering and Feature Selection

Feature engineering refers to the process of using domain knowledge to select and transform the most relevant variables from raw data when creating a predictive model using machine learning. Feature selection is the process of selecting a subset of relevant parameters that contribute the most to the predicted output. This is a vital step in the model design, as irrelevant or redundant parameters can negatively affect the model's performance. For this study, EBM (explainable boosting machine) was used, and it shows how each feature contributes to the model's prediction of the problem (Microsoft 2020).

First, operational data was labelled as pre-event 5 days before the failure occurred (1). Then, non-pre-event records were accepted as normal (0) operational data. The EBM model tries to classify whether records are pre-event data points or normal data points. During classification, overall model behaviour is explored and top features affecting model predictions using feature importance are found.

The importance of the features used in the classification for T27 is seen in Fig. 42.1. It is seen that the correlation between shaft bearing temperature and nacelle revolution is the most important feature for prediction.

When we examine the contribution of the correlation, the prediction score is stronger when the shaft bearing temperature is between 38 and 50 °C and the nacelle revolution is less than -1 rad. Similarly, Fig. 42.2 shows that the prediction score increases when the nacelle revolution is above 0.5 rad and the shaft bearing temperature is below 25 °C (Fig. 42.3).

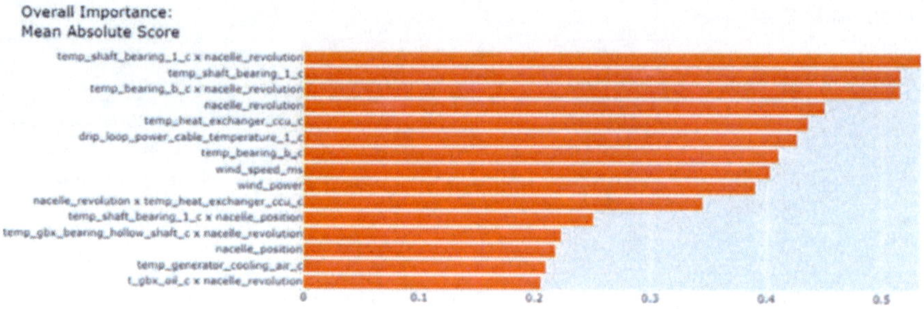

Fig. 42.1 Overall importance of features

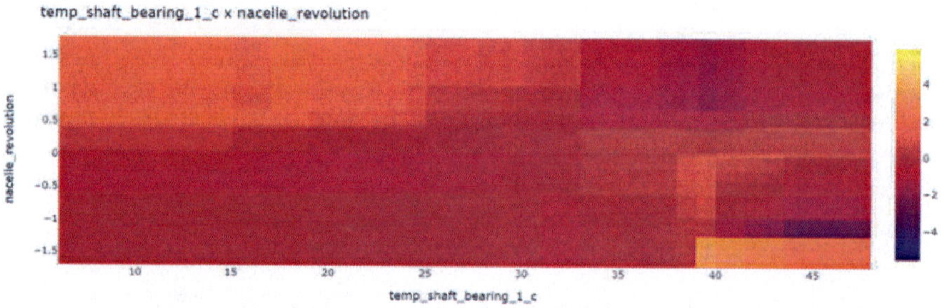

Fig. 42.2 Shaft bearing temperature and nacelle revolution heatmap

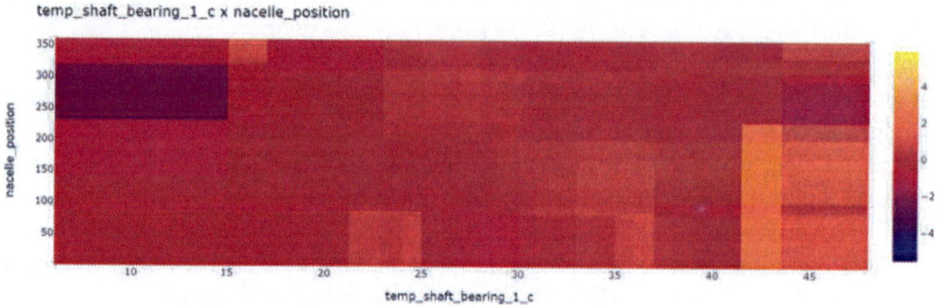

Fig. 42.3 Shaft bearing temperature × nacelle position heatmap

There are similar significant relationships between different features. When the nacelle position is between 0° and 220° and the shaft bearing temperature is higher than 40 °C, the prediction score increases. However, when the nacelle position is between 220° and 320° and the shaft bearing temperature is lower than 15 °C, the prediction score decreases. Feature importance relations in the pre-event classification were examined for four turbines with 17 months of operational data. In the pre-event classification of other turbines, feature and data combinations were analysed.

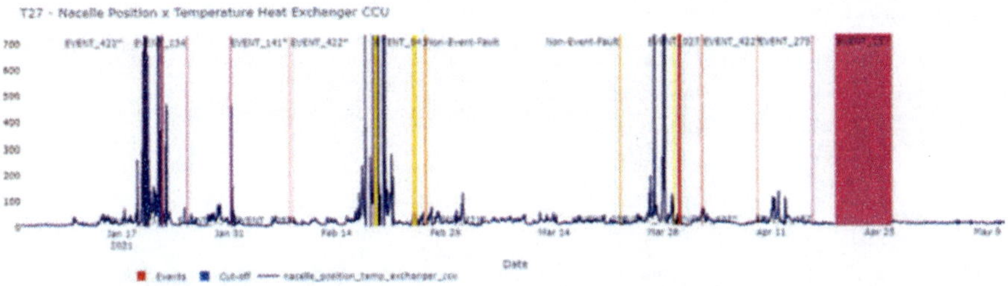

Fig. 42.4 Nacelle position × temp. Heat exchanger CCU

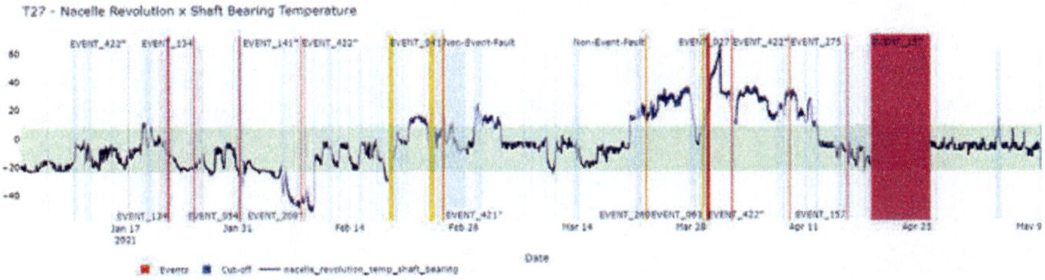

Fig. 42.5 Nacelle revolution × shaft bearing temperature

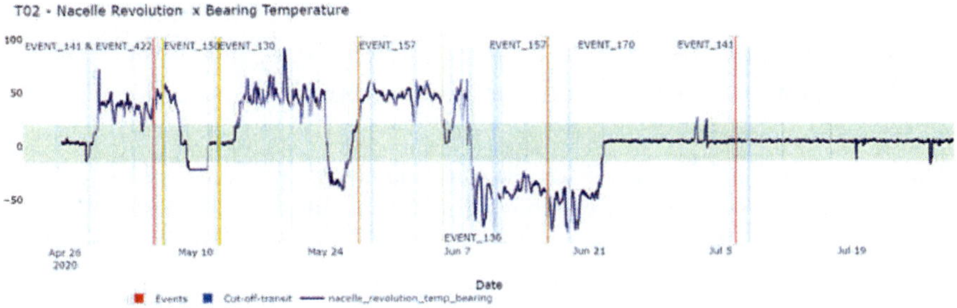

Fig. 42.6 Nacelle revolution × bearing temperature

The progression of the nacelle position × temp. Heat exchanger CCU temperature feature over time is shown in Fig. 42.4. This feature appears to be low under normal conditions, but extremely high a few days before most events occur. In Fig. 42.4, yellow-marked events belong to "Vibration detector" component, red-marked events "Converter" and pink-marked events "Brake" and "Pitch" components.

Normality analysis was performed by using nacelle revolution × shaft bearing temperature. The event occurred a maximum of 5 days after encountering values exceeding the normality range, shown in Fig. 42.5. Faults have occurred when the values of the nacelle revolution × bearing temperature feature exceed the normality range, shown in Fig. 42.6.

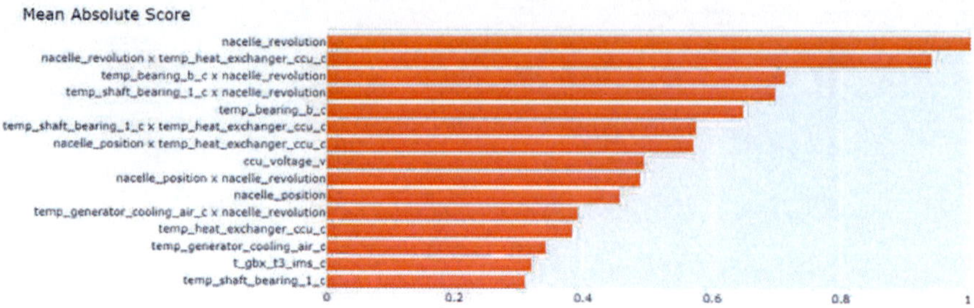

Fig. 42.7 Overall importance of features for CCU classification model

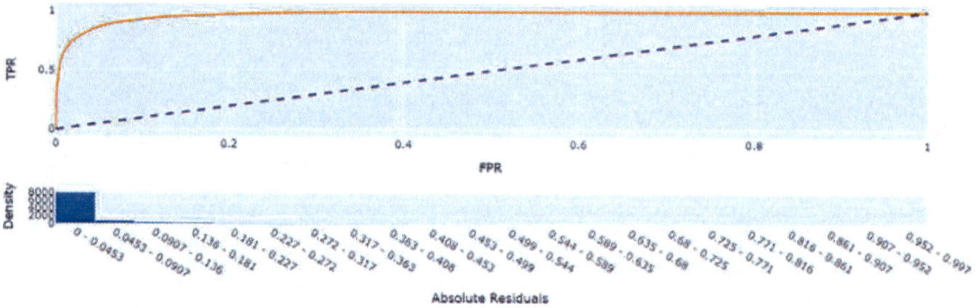

Fig. 42.8 ROC curve

42.2.4 Converter and Electrical Subsystem

The features that are important in the classification model, in which only the faults of the CCU subsystem are predicted 5 days in advance, are shown in Fig. 42.7.

The success of the classification model on the test data is 97.6% (AUC). The features that are important in the classification of the model formed the input data of the LSTM-AE model developed for condition monitoring (Fig. 42.8).

42.2.5 LSTM Autoencoder Model

The autoencoder neural network architecture is used for our condition monitoring model. Also, LSTM neural network cells were used in the autoencoder model. The reason for using LSTM is that it is useful in learning order dependence in sequence prediction problems. Autoencoder architecture is a neural network with the same number of neurons in the input and output layers. This type of architecture learns to construct an identity function (Larzalere 2019). It will take input data, build a compressed representation of the primary properties of that data and then learn to reconstruct it. The resulting reconstruction error can be considered as the anomaly score.

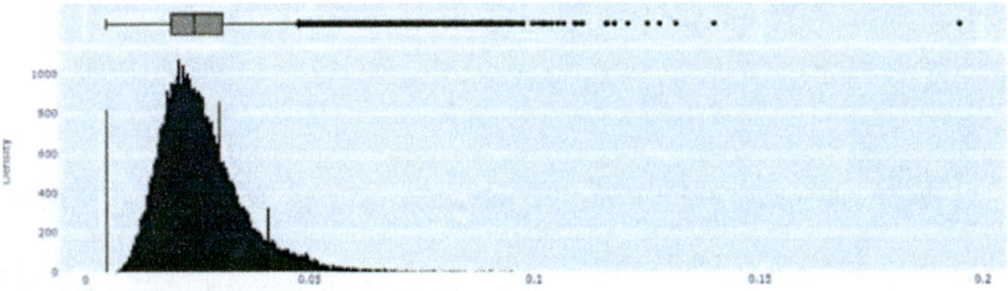

Fig. 42.9 Loss distribution for T27 normal data

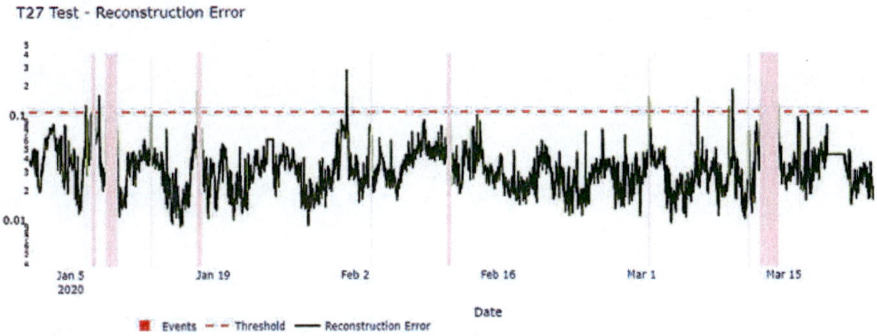

Fig. 42.10 Reconstruction error for T27 test data

42.2.5.1 General Turbine Faults-Based Model and Results

The training dataset is separate from the events/faults and the 5-day data points before the faults. There are 36 neurons in the input and output layers of the neural network, which was created by selecting 36 important features. ReLu is used as activation function in LSTM layers. While the mean absolute error (MAE) is preferred as loss function, Adam is used as optimizer, and 32 batch sizes and 100 epochs are used for the training. After the model reconstructed the normal data, the distribution of the calculated loss values was plotted. In order not to create a false alarm, a threshold value was determined by examining the loss distribution (Fig. 42.9).

The first 3 months of 2020 is reserved for the test data. Normal data for the remaining 14 months were used in the training. A separate model was developed for each wind turbine. This is because the normal data range in each turbine differs.

Figure 42.10 shows the reconstruction error calculated for the test data in the T27 model. The red areas show the times when the fault occurred. The abnormality score exceeds the threshold value 1–5 days before the fault occurs (Fig. 42.11).

The threshold value was determined as 0.11 according to the reconstruction error distribution of the T52 model. In the T52 model, it is seen that the anomaly score increases considerably before major faults.

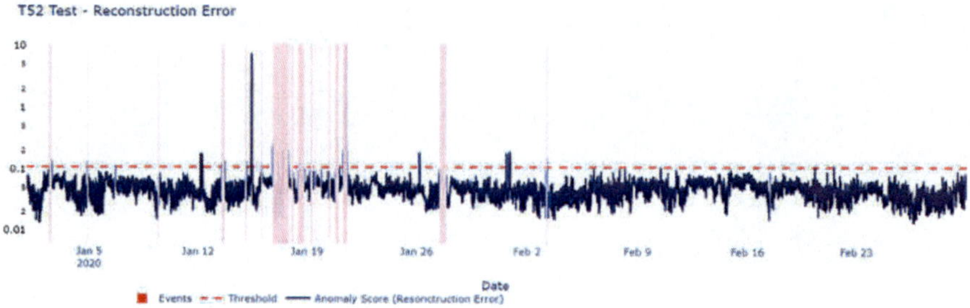

Fig. 42.11 Reconstruction error for T52 test data

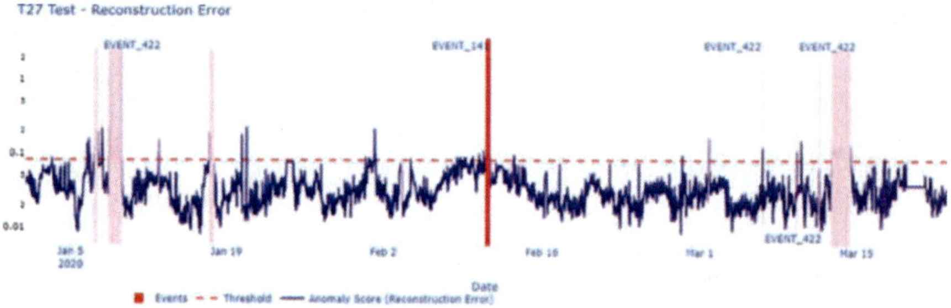

Fig. 42.12 Reconstruction error for T27 test data and CCU events

42.2.5.2 CCU Subsystem Faults-Based Model and Results

In the model developed specifically for the CCU subsystem in T27, unlike the general T27 model, it is seen that the threshold value is exceeded before CCU failures. It can be known that a fault that causes significant energy losses, such as Event 141 fault, will occur 3 days in advance. CCU models developed specifically for other turbines perform slightly better in predicting CCU faults than general models (Fig. 42.12).

42.3 Conclusions

In this chapter, we illustrated the use of LSTM-AE models to model wind turbine subsystems. The LSTM autoencoders were successful and effective in creating pre-fault alarms. The faulty subsystem was also identified by monitoring the new features representing the relationship between two important features. As expected, the model developed specifically for the CCU subsystem delivered more accurate predictions for CCU failures than the more general model. As part of future work, we plan to modify the model parameters and develop CM models for other subsystems such as pitch and gearbox.

Acknowledgement Part of the underlying work that led to this chapter has been funded by the Scientific and Technological Research Council of Turkey (TÜBİTAK) under the EUREKA programme.

References

Beretta M, Cárdenas JJ, Koch C, Cusidó J (2020) Wind Fleet Generator Fault Detection via SCADA Alarms and Autoencoders. Applied Sciences 10:. https://doi.org/10.3390/app10238649

Department of Energy (2014) How a Wind Turbine Works | Department of Energy. https://www.energy.gov/articles/how-wind-turbine-works . Accessed 31 Mar 2022

Dey S, Pisu P, Ayalew B (2015) A Comparative Study of Three Fault Diagnosis Schemes for Wind Turbines. IEEE Transactions on Control Systems Technology 23:1853–1868. https://doi.org/10.1109/TCST.2015.2389713

Larzalere B (2019) LSTM Autoencoder for Anomaly Detection | by Brent Larzalere | Towards Data Science. In: Towards Data Science. https://towardsdatascience.com/lstm-autoencoder-for-anomaly-detection-e1f4f2ee7ccf .

Microsoft (2020) InterpretML. https://interpret.ml/. Accessed 31 Mar 2022

Yang W, Tavner PJ, Crabtree CJ, Wilkinson M (2010) Cost-Effective Condition Monitoring for Wind Turbines. IEEE Transactions on Industrial Electronics 57:263–271. https://doi.org/10.1109/TIE.2009.2032202

Yin S, Wang G, Gao H (2016) Data-Driven Process Monitoring Based on Modified Orthogonal Projections to Latent Structures. IEEE Transactions on Control Systems Technology 24:1480–1487. https://doi.org/10.1109/TCST.2015.2481318

Zhang D, Qian L, Mao B, et al (2018) A Data-Driven Design for Fault Detection of Wind Turbines Using Random Forests and XGboost. IEEE Access 6:21020–21031. https://doi.org/10.1109/ACCESS.2018.2818678

Chapter 43
The Selection of a Renewable Energy System in Kayseri with Multi-criteria Decision-Making Method

İhsan Kılcı, Teyfik Şahnaz, İsmet Söylemez, and Muhammed Sütçü

Nomenclature

AHP	Analytic hierarchy process
ANP	Analytic network process

43.1 Introduction

There are many problems related to energy and its need today; it constitutes the basis of energy discussions especially in three topics. Firstly, fossil fuels are a limited resource. In spite of the increasing demand, although the energy supply is limited and there are large amounts of coal, oil and natural gas resources, this situation is the result of the signals that these supplies are likely to be exhausted in the coming years and the sources start to run out. Secondly, it can be defined as the issue of climate change. It is seen that this situation draws attention to a new level in the relationship between economic growth, energy consumption and environmental pollution. Lastly, it is the energy supply security problem which expresses the problems that may arise as a result of the difficulties in accessing the current energy sources and the increase in the demand for energy. The supply security problem, which will arise in case of a problem that may occur in the energy source of economies which are dependent on foreign sources and which have to import energy, is of particular importance.

İ. Kılcı · T. Şahnaz · İ. Söylemez (✉) · M. Sütçü
Department of Industrial Engineering, Abdullah Gul University, Kayseri, Turkey
e-mail: ismet.soylemez@agu.edu.tr; muhammed.sutcu@agu.edu.tr

M. Z. Sogut et al. (eds.), *Proceedings of the 2022 International Symposium on Energy Management and Sustainability*, Springer Proceedings in Energy,
https://doi.org/10.1007/978-3-031-30171-1_43

What makes us look at the concept of energy from a different point in the recent period is the increase in greenhouse gas use and the impact of the effects on our world. It has been observed that environmental degradation has caused global warming and climate change in the last two decades. In the energy sector, carbon dioxide (CO_2) resulting from the oxidation of carbon during combustion dominates the total greenhouse gas emissions. The increase in energy demand is due to worldwide economic growth and development. This results in more greenhouse gas emissions.

Given the fact that carbon dioxide is the main component of greenhouse gases, which are at the heart of the above-mentioned problems in general, there is a global concern about reducing carbon emissions. All countries in the world are concerned with the problem of energy security and global warming, and the increasing use of renewable energy is a way of responding to both of these problems. In this context, different policies that reduce carbon emissions such as increasing the use of renewable energy and encouraging technological innovations have become necessary.

43.2 Literature Review

Energy investment decisions are inherently multi-criteria. Multi-criteria decision-making methods help governments to evaluate plans and policies in the energy sector. Today, many studies on multi-criteria decision-making methods have focused on energy problems. The studies on this subject are summarized.

In the literature, there are plenty of articles about multi-criteria decision-making techniques. Analytic hierarchy process (AHP) and Vikor techniques for the selection of a suitable solar power plant location are applied (Özdemir et al. 2017). They considered the potential electric capacity, square meter unit prices, earthquake potential, distance to solar power plant and potential of terrorism as some criteria. TOPSIS method is used for the course selection (Kecek and Soylemez 2016). AHP and TOPSIS methods are applied as hybrid methods for the selection of the appropriate programming language for the graduate students (Soylemez and Soylemez 2017).

More than 90 publications for energy planning decisions identified the most popular decision-making methods as analytic network process (ANP), Promethee and Electre (Pohekar and Ramachandran 2004). Haralambopoulos and Polatidis (2003) evaluated renewable energy projects and Georgopoulou et al. (1998) evaluated the renewable energy planning process using Promethee II, a multi-criteria decision-making method. Ulutaş (2005) evaluated alternative energy for Turkey's energy resources. ANP has analysed the problem using the method of energy policy. Afgan and Carvalho (2002) presented the selection of options and criteria for evaluating new and renewable energy technologies on the basis of synthesis and analysis of parameters to identify energy indicators used to assess the energy system that meets the sustainability requirement.

In recent years, it has been observed that decision-making methods related to energy problems have been enriched with fuzzy clusters. Wang et al. (2009) evaluated multi-criteria decisions on sustainable energy as objective, subjective and mixed on the basis of criteria. Energy supply system criteria are summarized in technical, economic, environmental and social aspects. Beccali (1998) aimed at promoting energy planning by comparing the fuzzy set method to assist decision makers in understanding and choosing complex problems in a multi-criteria decision-making approach. Borges and Antunes (2003) proposed an interactive approach for fuzzy multi-criteria linear programming problems based on parsing analysis of parametric (weight) schemes with indifferent regions corresponding to simple basic problems. The model defined to deal with certain and uncertain coefficients of the input-output energy-economic planning model is developed to provide decision support to decision makers in the analysis of the interaction between the economies.

Çapik et al. (2012) investigated the present energy situation, the renewable energy potential and the energy politics of Turkey. Turkey's potential renewable energy sources as hydropower, geothermal, wind, biomass and solar are given in the study. Simsek and Simsek (2013) evaluated the availability and potential of renewable energy sources in Turkey and also gave the information about the government policies and economic aspects. They mentioned the laws about renewable energy. The authors investigated that relationship between renewable energy consumption and gross domestic product (GDP) in Turkey. They used data between 1995 and 2015 and reached that renewable energy consumption and GDP have no causality (Bulut and Muratoglu 2018).

Sahin (2021) proposed the new model considering the gross final energy consumption, energy consumption of renewable energy sources and its share in France, Germany, Italy, Spain, Turkey and the United Kingdom by 2030. The author improved the fractional nonlinear grey Bernoulli model for forecasting.

Tekin et al. (2021) investigated the potential best renewable energy site alternatives using the maximum entropy model and geographical information systems at the Eastern Mediterranean Region in Turkey. They considered the receiver operating characteristic curve for evaluation performance of the model.

43.3 Problem Definition

Energy is the basic requirement for human life. Before the industrial age, energy needs were met by the basic sources such as wood, wind, water and muscle power of humans and animals, while the discovery of steam engines completely changed the energy sources used. Coal, oil, natural gas, hydroelectric power plants and nuclear power plants are the most common energy sources used now. Today, fossil fuels used to generate energy cause emissions of certain substances (CO_2, SO_x, NO_x) harmful to the environment and human health. The release of these substances into the environment not only results in regional hazards such as water and air pollution but also in climate change with global impact. Therefore, some restrictions have been imposed on states through global agreements and efforts have been made to reduce emissions. For example, countries that signed the Kyoto Protocol, which entered into force on 11 December 1997, pledged to reduce emissions in 2012 to their values below the 1990s.

Renewable energy sources with a net emission value have gained importance not only for environmental reasons but also because of the search for sustainable energy sources instead of rapidly decreasing fossil resources. Turkey has not signed the Kyoto Protocol, but renewable energy sources are expected to meet a large part of our energy needs in the future. Therefore, it is necessary to develop effective strategies for planning the transition from fossil fuels to renewable energy sources both for our country and for all countries in the world. Developed countries have accelerated the transition to sustainable energy. However, when we look at the developing countries, they have experienced some problems during the transition period. In our country, the government supports investments in sustainable energy. Unfortunately, in some cases renewable energy systems are installed only to benefit from this support. Some criteria should be considered when installing these systems. In this study, we made the selection of the most suitable source considering the power, investment rate, application time, operating hours, useful life and operating and maintenance costs and avoided CO_2 emissions of renewable energy systems that can be installed in Kayseri region. The AHP method is considered to be used by taking into account seven criteria and the best sustainable energy alternatives were determined.

A decision model was created for the 11 alternatives. In general, we have determined the energy systems that have the potential to be installed in Kayseri and have classified these systems among themselves according to their power.

43.4 Methodology

When Turkey's solar energy potential is analysed, Southeastern Anatolia, Mediterranean and Eastern Anatolia regions are seen to be more efficient. Kayseri has a very high potential when sunbathing times and global radiation values are taken into consideration. The southern parts of Kayseri Province, Pınarbaşı, Hacılar, Yahyalı, Sarız, Develi and some parts of Tomarza, have high potential in terms of solar energy. Since Kayseri receives more sunlight especially in the summer months, electricity production from solar energy is mostly in these months. The highest sunshine duration of Kayseri is 12.03 h in July. The highest global radiation value is in July (6.86 kWh/m^2 day).

When the wind atlas of Turkey is examined, the wind energy potential in the Aegean Region and southern Marmara region is efficient. In addition, the wind speed must be at least 7 m/h at a height of 50 m and the capacity factor should be at least 35% for the wind power plant to be economical. Wind turbines do not harm the land where they are installed. In addition, it does not constitute an obstacle for agricultural activities in the field. It is advantageous that the land used as wind farm can be easily

Table 43.1 Alternatives for electric generation

Type of alternatives		Power
A1	Wind power	$P \leq 5$ MW
A2	Wind power	$5 \leq P \leq 10$ MW
A3	Wind power	$10 \leq P \leq 50$ MW
A4	Hydroelectric	$P \leq 10$ MW
A5	Hydroelectric	$10 \leq P \leq 50$ MW
A6	Hydroelectric	$50 \leq P \leq 100$ MW
A7	Solar power	$1 \leq P \leq 5$ MW
A8	Solar power	$6 \leq P \leq 10$ MW
A9	Biomass (forest and agricultural wastes)	$P \leq 5$ MW
A10	Biomass (farming industrial wastes)	$P \leq 5$ MW
A11	Biomass (co-combustion in conventional central)	$P \leq 50$ MW

restored and the turbines can be easily removed. It is stated that the total wind power of Kayseri province is 1,885.28 MW.

Kızılırmak and Yeşilırmak pass through Kayseri. Kızılırmak is born within the borders of Sivas Province, and after passing Sivas and Kayseri provinces, it is directed to the Black Sea from the west of Yozgat. In this context, it can be said that Kayseri is also effective in hydroelectric potential. According to the 2012 energy report prepared by the Turkish National Committee of the World Energy Council, Kızılırmak's electricity generation technical potential is 19.55 GWh and its economic potential is 6.32 GWh. Similarly, the technical potential of Yeşilırmak electricity generation is 18.68 GWh and the economic potential is 5.2 GWh.

In Kayseri, biofuel raw materials such as wheat, sugar beet and safflower are suitable for regional climate, and wheat and sugar beet are widely grown in the region. Energy crops can make a significant contribution to rural development if appropriate programmes are developed. In addition, the potential of converting wastes from common livestock activities into energy can be evaluated. Biogas energy in Kayseri is provided in Kayseri landfill biogas power plant and its installed capacity is 5.78 MW.

The designed model will be evaluated according to the criteria shown in Table 43.1. The attributes considered are: power (P), investment ratio (IR), implementation period (IP), operating hours (OH), useful life (UL), operation and maintenance costs (O&M) and tons of emissions of CO_2 avoided per year (tCO_2/y). And 11 alternatives are developed according to their power (Table 43.2). This study aims to decide which renewable energy resources should be invested in Kayseri by using multi-criteria decision-making method, especially analytical hierarchy process (AHP), taking into consideration the criteria that we determined.

In the objective tree, there are seven measures that affect the selection of renewable energy system at Kayseri. The study was made using Logical Decisions software. Pairwise comparison matrix is created and filled with values between 1 and 9 for each measure. These values were determined by the decision maker. The least

Table 43.2 Evaluation criteria

Definition	Unit
Power (P)	KW
Investment ratio (IR)	€/KW
Implementation period (IP)	Years
Operating hours (OH)	Hours/year
Useful life (UL)	Years
Operation and maintenance costs (O&M)	€*103/KWh
Tons of CO_2 avoided (tCO_2/y)	tCO_2/y

preferred and most preferred values were also determined for each type of sustainable energy system. After data collection, Logical Decisions software was run to choose the best alternative (Fig. 43.1).

Our main objective is to determine the best renewable energy system (RES) at Kayseri. There are seven measures which are CO_2 avoided, implementation period, investment ratio, operating hours, operating and maintenance cost, power and useful life. We examined our alternatives under these seven measures. Measures and measure units are constant. But the least and most preferred values can change for each type of renewable energy system individually. Of the attributes considered, power, operating hours, useful life and avoided CO_2 emissions are beneficial attributes and so higher values are desirable. Investment ratio, implementation period and operating and maintenance costs are non-beneficial attributes and so lower values are desirable.

There are seven measures which are CO_2 avoided, implementation period, investment ratio, operating hours, operating and maintenance cost, power and useful life (Table 43.2).

43.5 Results

Alternatives for all measures were evaluated among themselves and a pairwise matrix was created for each one. Consistency ratio was taken into consideration while creating matrices. It is desirable that the consistency ratio is less than 0.1. In the sensitivity analysis, it is clearly observed how the alternatives are in relation to each other in different measures. Sensitivity analysis was done through Logical Decisions software outputs. The ranking results compare the alternatives and show which alternative is the best for each goal and overall goal. It can be figured out from the ranking results whether any alternative is dominant or not.

The model we created was run for the selection of the best alternative which has the potential for installation. Seven measures were determined in the model and all pairwise matrices are created to each measure. The results are as follows.

When we consider the criteria according to the model we created, it is seen that the best renewable energy in Kayseri is biomass (co-combustion in conventional

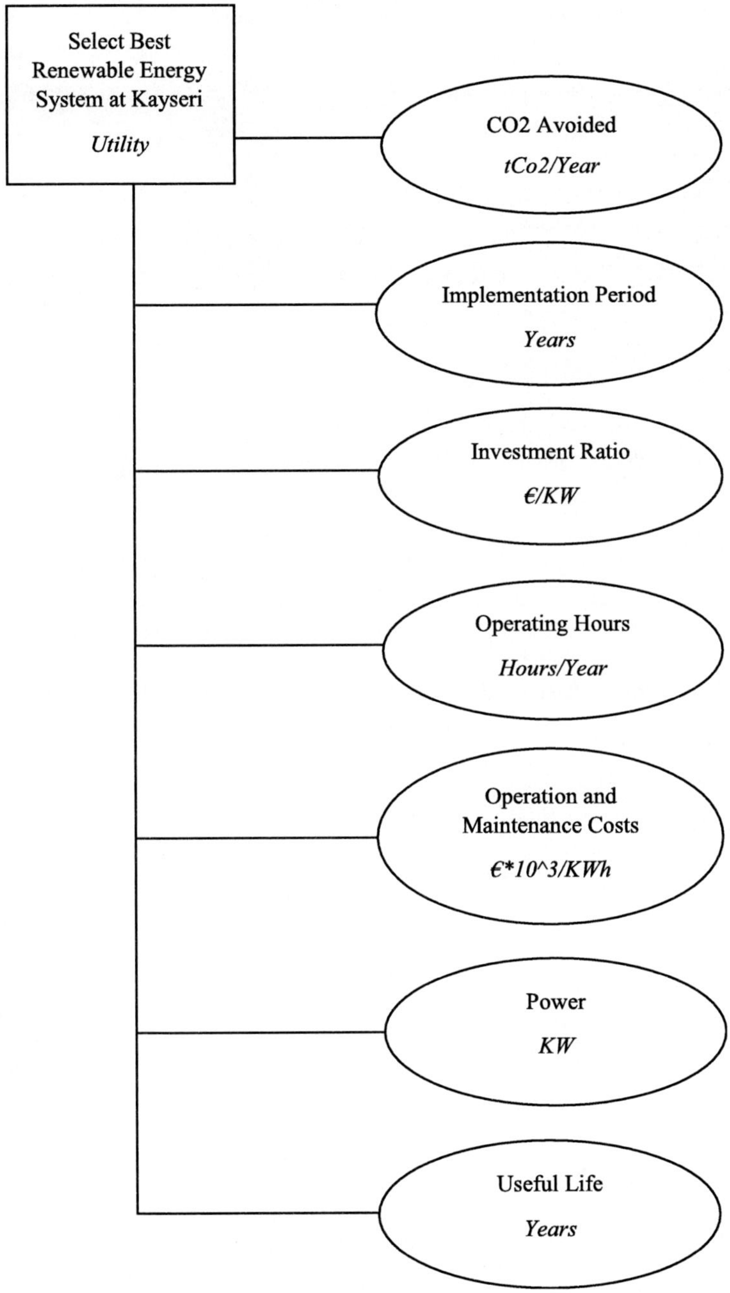

Fig. 43.1 Objective tree of the decision model

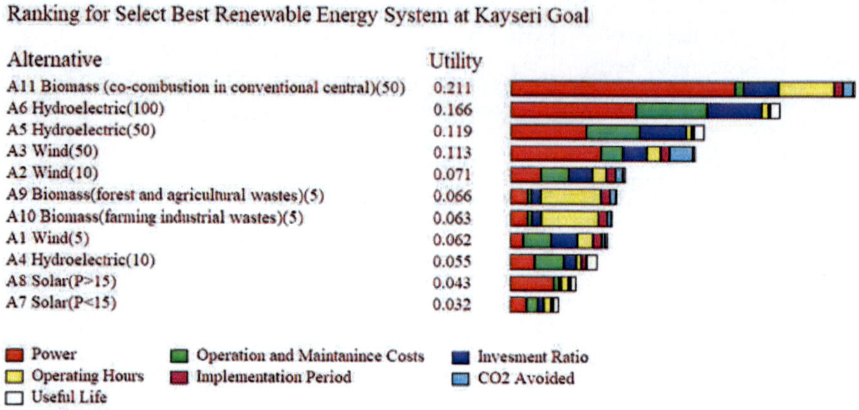

Fig. 43.2 Ranking results of alternatives

central). For a 50 MW plant, the Kayseri region is a very reasonable region in terms of cost, power and operating hours. There is currently no biomass plant in Kayseri. The second alternative is the 100 MW hydroelectric power plant. In fact, there are 19 dams in Kayseri and 4 of them are still under construction. However, only 7 of these dams produce electricity. Others are used for irrigation purposes. The utility values of the hydroelectric (50) and wind (50) alternatives are very close to the second alternative. In other words, the installation of these systems in some regions may be logical (Fig. 43.2).

43.6 Conclusion

As a result, renewable energy systems have gained importance due to the limited energy resources and damages to the environment. However, it is very important to choose the places to be installed in order to make maximum use of these systems. In this model we made for Kayseri, we determined which renewable energy systems are suitable for this region. This study provides maximum benefit to the investors.

References

Afgan N H, Carvalho M G (2002) Multi-criteria assessment of new and renewable energy power plants. Energy 27, 739–55. https://doi.org/10.1016/S0360-5442(02)00019-1

Beccali M, Cellura M, Ardente D (1998) Decision making in energy planning: the ELECTRE multicriteria analysis approach compared to a Fuzzy-Sets methodology. Energy Conversion and Management 39(16): 1869–1881. https://doi.org/10.1016/S0196-8904(98)00053-3

Borges AR, Antunes C H (2003) A fuzzy multiple objective decision support model for energy-economy planning. European Journal of Operation Research 145(2):304–31616 https://doi.org/10.1016/S0377-2217(02)00536-2

Bulut U, Muratoglu G (2018) Renewable energy in Turkey: Great potential, low but increasing utilization, and an empirical analysis on renewable energy-growth nexus. Energy Policy 123, 240–250. https://doi.org/10.1016/j.enpol.2018.08.057

Çapik M, Yılmaz A O, Çavuşoğlu İ (2012) Present situation and potential role of renewable energy in Turkey. Renewable Energy 46, 1–13. https://doi.org/10.1016/j.renene.2012.02.031

Georgopoulou, E., Sarafidis, Y., & Diakoulaki, D. (1998). Design and implementation of a group DSS for sustaining renewable energies exploitation. European Journal of Operational Research, 109(2), 483–500.

Haralambopoulos, D. A., & Polatidis, H. (2003). Renewable energy projects: structuring a multi-criteria group decision-making framework. Renewable energy, 28(6), 961–973.

Kecek G, Soylemez C (2016) Course Selection in Postgraduate Studies through Analytic Hierarchy Process and Topsis Methods. British Journal of Economics, Finance and Management Sciences 11 (1), 142–157.

Özdemir B, Özcan B, Aladağ Z (2017) Güneş enerjisi santrali kuruluş yerinin AHS ve VIKOR yöntemlerine dayalı bütünleşik yaklaşım ile değerlendirilmesi. Erciyes Üniversitesi Fen Bilimleri Enstitüsü Fen Bilimleri Dergisi 33(2), 16–34

Pohekar, S. and Ramachandran, M. (2004) Application of Multi-Criteria Decision Making to Sustainable Energy Planning:A Review. Renewable and Sustainable Energy Reviews, 8, 365–381.

Sahin U (2021) Future of renewable energy consumption in France, Germany, Italy, Spain, Turkey and UK by 2030 using optimized fractional nonlinear grey Bernoulli model. Sustainable production and consumption 25, 1–14. https://doi.org/10.1016/j.spc.2020.07.009

Simsek H A, Simsek N (2013) Recent incentives for renewable energy in Turkey. Energy Policy 63, 521–530. https://doi.org/10.1016/j.enpol.2013.08.036

Soylemez C, Soylemez I (2017) The Selection of Appropriate Programming Language for Graduate Students: A Case Study. Innovation and Global Issues in Social Sciences Extended Abstracts, 82–84. https://doi.org/10.6084/m9.figshare.13634555

Tekin S, Guner E D, Cilek A, Cilek M U (2021) Selection of renewable energy systems sites using the MaxEnt model in the Eastern Mediterranean region in Turkey. Environmental Science and Pollution Research 28:51405–51424. https://doi.org/10.1007/s11356-021-13760-6

Ulutaş, B. H. (2005). Determination of the appropriate energy policy for Turkey. Energy, 30(7), 1146–1161.

Wang, J. J., Jing, Y. Y., Zhang, C. F., & Zhao, J. H. (2009). Review on multi-criteria decision analysis aid in sustainable energy decision-making. Renewable and sustainable energy reviews, 13(9), 2263–2278.

Chapter 44
The Performance Assessment of TiO_2/ITO-PET TENG Device

Gizem Durak Yüzüak and Ercüment Yüzüak

Nomenclature

TENG	Triboelectric nanogenerator
TiO_2	Titanium dioxide
ITO	Indium tin oxide
PET	Polyethylene terephthalate
Si	Silicone
XRD	X-ray diffraction
XRR	X-ray reflectometry
SEM	Scanning electron microscopy
AFM	Atomic force microscopy
I–V	Current-voltage
I_{SC}	Short circuit current
V_{OC}	Open circuit voltage

44.1 Introduction

The expression of "energy", which is defined as the work capacity of a system, has changed from ancient times to the present, according to the need created by technological developments. Since fossil fuels are becoming exhausted, green energy, which is plentiful and renewable, is becoming more and more prevalent in

G. D. Yüzüak (✉) · E. Yüzüak
Functional Materials Research Laboratory, Faculty of Engineering and Architecture, Recep Tayyip Erdoğan University, Rize, Turkey
e-mail: gizemyuzuak@erdogan.edu.tr

M. Z. Sogut et al. (eds.), *Proceedings of the 2022 International Symposium on Energy Management and Sustainability*, Springer Proceedings in Energy,
https://doi.org/10.1007/978-3-031-30171-1_44

our lives. With the fast development of smart devices, attention has been paid to discovering reusable, modular power for the energy needs of the contemporary day. The efforts to evaluate different types of energy and to develop methods of harvesting in order to provide continuous power to low-energy portable electronic devices attract considerable attention. In terms of efficiency, decentralized power generation by energy harvesters offers a significant advantage in that this approach can minimize energy transmission losses and balance energy supply costs. The rapid increase in the number of portable electronics has led to the development of technologies related to energy storage of these battery-operated systems.

However, it is not possible to ensure the continuity of this much system based on battery power. For this reason, it is necessary to collect energy by utilizing the environment, to recycle it efficiently in line with the needs and to obtain the desired power. In the energy collection method, generally waste mechanical energy is converted into electrical energy. Various methods such as electromagnetic (Williams et al. 2001), electrostatic (Torres and Rincon-Mora 2009) and piezoelectric (Wang and Song 2006; Seol et al. 2013) are used to convert waste mechanical energy into useful electrical energy. As an alternative to all these, new, environmentally friendly, low-cost, highly reliable triboelectric nanogenerators (TENGs) emerge as a new field of study (Hinchet et al. 2019).

Triboelectric effect and electrostatic induction are the bases of TENGs' operation. During contact, when the tribo-pairs are separated, the triboelectric effect causes charges to develop on both dielectric surfaces that have the opposite sign. Charges are induced on the electrodes at the same time that the activity of electrostatic induction takes place on them. The charges on the metal electrodes may be transferred to the other electrodes through the external circuit in order to restore equilibrium to the potential difference. TENG's output current, which is also known as Maxwell's displacement current, is effectively created as a result of this (Xia et al. 2019; Zi et al. 2015).

TiO_2 is one of the suitable semi-conductor materials compared to using weak polymers as a friction layer in TENGs. With good mechanical and chemical stability and high breakdown electric field and electron saturation velocity, TiO_2 films are the most widely used oxide for transparent electronic applications owing to low cost and high activity (Durak Yüzüak et al. 2022). To replace inappropriate polymers for long-term usage in hostile settings as the friction layer in TENGs, one of the conceivable options is to employ durable thin films made of semiconductor materials, which are resistant to damage. Since TiO_2 films have such high activity and cheap cost, they are the most regularly used oxide for electronic applications. They also have excellent chemical and mechanical stability, making them the most widely used oxide for electronic applications.

44.2 Experimental

100 nm TiO_2 thin film grown on Si substrate using magnetron co-sputtering system in a high vacuum chamber is located in TENG with Cu-electrode and the TENG secondary electrode part consists of ITO-PET thin film structure. The structural properties of the triboelectric friction layers were investigated using X-ray diffraction (XRD), X-ray reflectometry (XRR), scanning electron microscopy (SEM) and atomic force microscopy (AFM), and its electrical properties were investigated using the current-voltage (I–V) measurements and frequency-dependent capacitance measurement system. TENG is formed by combining these two different parts of electrodes with electrical connections (Fig. 44.1). The potential changes in the contact density of the semiconductor thin film and the electrical output power of TENG will be examined due to the different physical properties and different surface roughness of the TiO_2 thin films used in the produced TENGs; for better performance the TiO_2 thin films are heat-treated at 900 °C.

44.3 Results and Discussion

The XRD patterns and AFM and SEM images of the TiO_2 and PET/ITO thin films are shown in Fig. 44.2a–f. The rutile phase and cubic structure can be seen in the XRD reflections for TiO_2 and PET/ITO thin films, respectively. In accordance with the XRD peaks, the TiO_2 crystal structure indexed by reference is associated with ICSD # #16636.

The surface morphologies of each thin film were obtained from SEM and AFM studies. The granular structure was obtained for TiO_2 directly deposited on Si (100) substrates and PET/ITO thin films. Achieving uniformity in contact area and total number of transferred charges is critical for device performance. This homogeneous structure was clearly determined from the obtained SEM and AFM images. A charge density difference is believed to occur in semiconductor surfaces, causing charges to be transferred from one surface to another. In the meantime, triboelectrification is thought to occur in semiconductors as a result of this transfer.

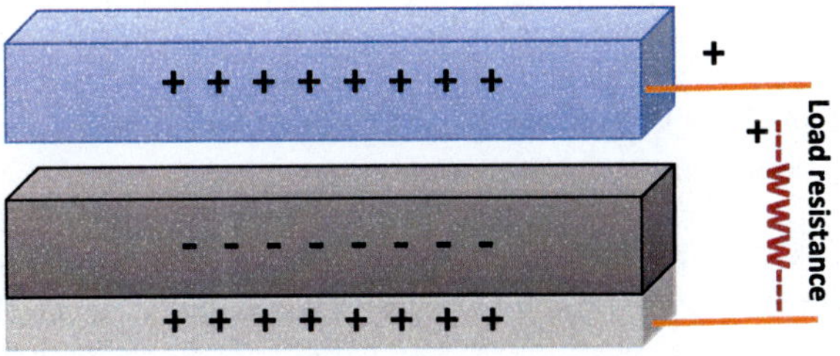

Fig. 44.1 Working principle of TENG

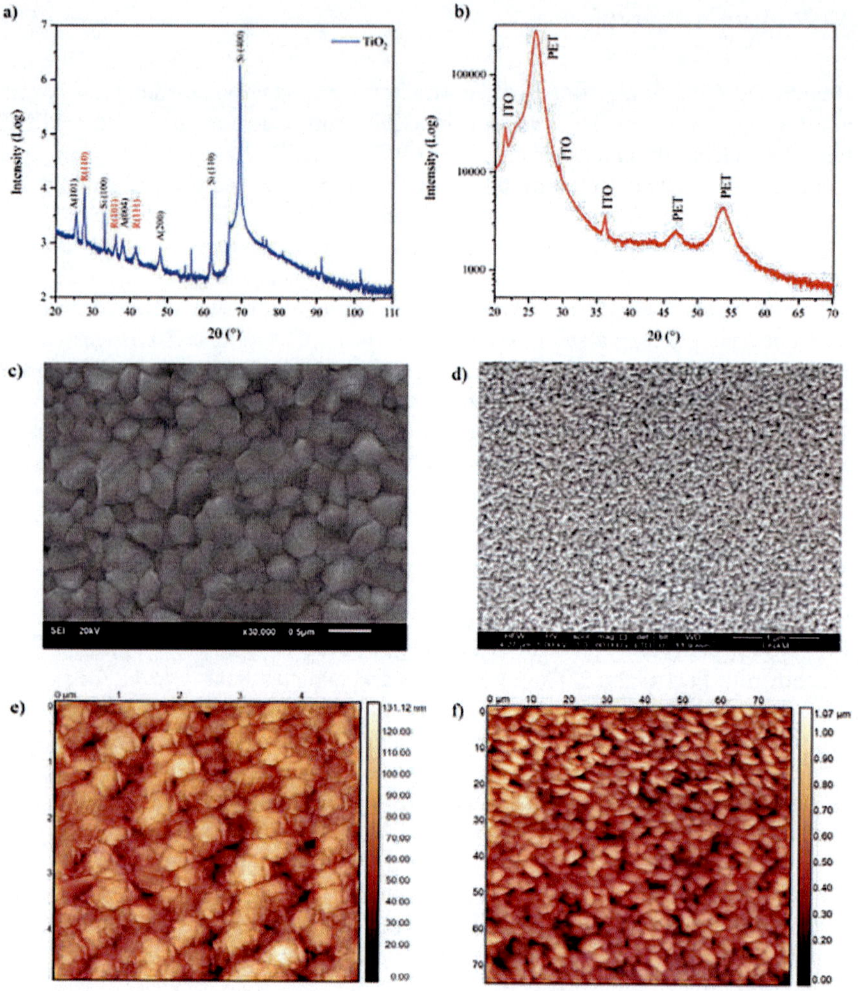

Fig. 44.2 The XRD patterns (**a, b**) and SEM (**c, d**) and AFM (**e, f**) images of the TiO_2 and PET/ITO thin films, respectively

The current-voltage (I–V) curve and frequency-dependent capacitance measurement of TiO_2 thin film are seen in Fig. 44.3 to determine the charge transfer resistance. It can be seen from the obtained I–V curve that the thin film exhibits a typical diode property. It has a peak current value of about 0.45 μA under a varying potential of + −2 V.

According to the capacitance curve of TiO_2 thin film (inset of Fig. 44.3), it can be observed that the low- and high-frequency zones are separated into two groups: one for low frequency and one for high frequency. It can be observed that the capacitance values increase rapidly in line with the rapid decrease in the frequency value when

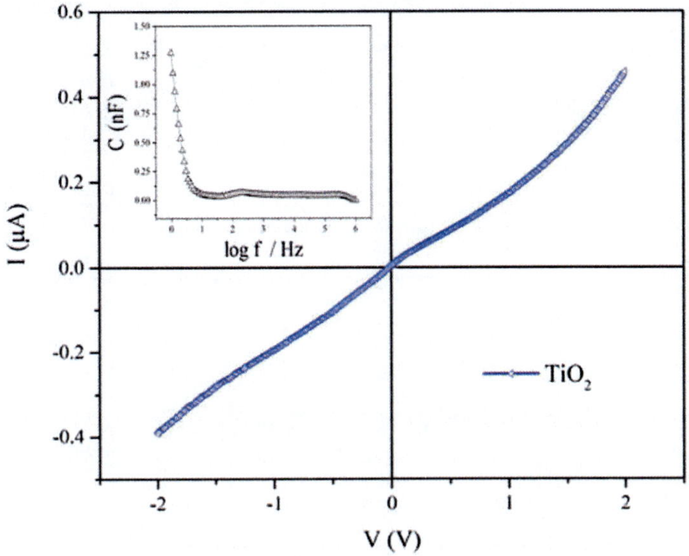

Fig. 44.3 The I–V curve and frequency-dependent capacitance measurement (inset) of the TiO_2 thin film

the frequency value is less than 10 Hz. Triboelectric devices, particularly those with high-frequency values of 10 Hz and lower, must match the efficiency expectations expected in the low-frequency zone if they are to perform as expected. So the maximum capacitance value at very low frequencies is regarded as a further contribution for the triboelectric, which is quite close to the natural oscillation frequency due to the coupling of triboelectrification and electrostatic induction.

The electric output of the TENG device was measured using the triggering frequencies and regulated amplitude under mechanical deformation (Fig. 44.4). In the first stage, the triboelectric charge accumulation increases and reaches equilibrium after numerous cycles, resulting in an increase in output. As a result, an oscilloscope with an infinite input resistance will show a stable value for the open-circuit voltage (V_{OC}). Short-circuit current (I_{SC}) achieves its maximum peak value in Fig. 44.4b, which corresponds to the half-cycle of pressing that is occurring at a faster straining rate than the release rate. The sum of the charges transferred in a half-cycle of deformation may be calculated by integrating the current peaks.

There are several factors to consider while designing a triboelectric nanogenerator (TENG), including triboelectric charge density and plate separation distance. In order to understand how TENG's output is affected by frequency, it is important to look at the mechanical energy from the environment. As a result, the triggering motor's amplitude was held constant while we evaluated the TENG device at three distinct frequencies ranging from 1 to 3 Hz. Figure 44.4a shows that the V_{OC} virtually stays the same at various frequencies. The dynamic process of charge

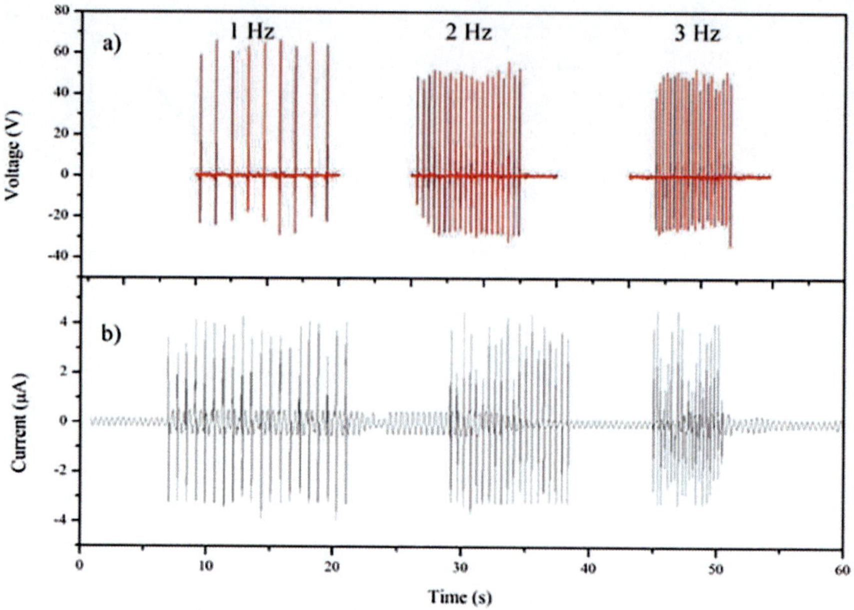

Fig. 44.4 The output performance of the TiO_2-ITO/PET-based TENG

transfer is absent in the open-circuit situation, which is most likely the cause. A plate's spacing and charge density are the sole factors affecting voltage at any given moment. Because of this, the short-circuit current is seen in Fig. 44.4b to rise somewhat with increasing frequency, from 3 μA at 1 Hz up to 4 μA at 3 Hz, as the deformation rate increases.

44.4 Conclusion

The purpose of this research is to use small-scale mechanical energy to power electrical components so that a self-powered system may be achieved. Energy conversion devices like TENGs, on the other hand, may open up a slew of new possibilities. The results show an environmentally friendly, low-cost and increased output performance triboelectric nanogenerator based on semiconductor-based layers. In this framework, the structural, electrical and device properties of TiO_2/ITO-PET structure were systematically attained. The maximum output power of 20 W/m^2 was reached without any sublayer or electrical poling of the TENG surfaces.

Acknowledgement This work was supported by the Scientific and Technological Research Council of Turkey (TÜBİTAK) under grant number 119M972.

References

Hinchet R, Yoon HJ, Ryu H, Kim MK, Choi EK, Kim DS, Kim SW (2019) Transcutaneous ultrasound energy harvesting using capacitive triboelectric technology, Science 365: 491–494. https://doi.org/10.1126/science.aan3997

Seol ML, Choi JM, Kim J.Y, Ahn JH, Moon D II, Choi YK (2013) Piezoelectric nanogenerator with a nanoforest structure, Nano Energy, 2:1142. https://doi.org/10.1016/J.NANOEN.2013.04.006

Torres, EO, Rincon-Mora EA (2009) Electrostatic Energy-Harvesting and Battery-Charging CMOS System Prototype, IEEE Transactions on Circuits and Systems I: Regular Papers, 56:1938. https://doi.org/10.1109/TCSI.2008.2011578

Wang ZL, Song J (2006) Piezoelectric Nanogenerators Based on Zinc Oxide Nanowire Arrays, Science, 312: 242. https://doi.org/10.1126/science.1124005

Williams CB, Shearwood C, Harradine MA, Mellor PH, Birch TS, Yates RB (2001) Development of an electromagnetic micro-generator, IEE Proc.-Circuits Devices Syst., 148(6):337. https://doi.org/10.1049/ip-cds:20010525

Xia X, Fu J, Zi Y (2019) A universal standardized method for output capability assessment of nanogenerators, Nature Communications, 10:4428. https://doi.org/10.1038/s41467-019-12465-2

Durak Yüzüak G, Karagöz C, Yüzüak E (2022) Exploring the Sputtering Conditions in ZnO Thin Film for Triboelectric Nanogenerator Electrode, Int. Energy Research. 1–7. https://doi.org/10.1002/er.7777

Zi Y, Niu S, Wang J, When Z, Tang W, Wang ZL (2015) Standards and figure-of-merits for quantifying the performance of triboelectric nanogenerators, Nature Communications, 6:8376. https://doi.org/10.1038/ncomms9376

Chapter 45
How a Good Lightning Protection Program Contributes to Energy Management and Sustainability

Shadreck Mpanga, Ackim Zulu, Mabvuto Mwanza, and Koray Ulgen

Nomenclature

LPP	Lightning protection program
ETAP	Electrical transient analyzer program
IEC	International Electrotechnical Commission
IEEE	Institute of Electrical and Electronics Engineers
DEHN	Germany Lightning Protection Guide
SPD	Surge protection device
LPS	Lightning protection system
ESS	Energy supply system
RES	Renewable energy sources
CHP	Combined heat and power
PPE	Personal protective equipment
AFB	Arc-flash boundary, cm
PV	Photovoltaic
AF	Arc flash, cal/cm^2
WD	Working distance, cm
V_s	Sending end voltage, V
I_s	Sending end current, A
V_r	Receiving end voltage, V

S. Mpanga (✉) · A. Zulu · M. Mwanza
School of Engineering, University of Zambia, Lusaka, Zambia
e-mail: shadreck.mpanga@unza.zm; ackim.zulu@unza.zm; mabvuto.mwanza@unza.zm

K. Ulgen
Solar Energy Institute, Ege University, Izmir, Turkey
e-mail: koray.ulgen@ege.edu.tr

M. Z. Sogut et al. (eds.), *Proceedings of the 2022 International Symposium on Energy Management and Sustainability*, Springer Proceedings in Energy,
https://doi.org/10.1007/978-3-031-30171-1_45

I_r Receiving end current, A
V_p Protective voltage level, V

45.1 Introduction

Lightning occurrences on earth are a necessity as they fix nitrates into the soil, and the importance of nitrates to plants that are a source of food and energy to the animal kingdom cannot be overemphasized. For instance, bioenergy comes from plants and is one of the dependable renewable energy sources (RES). However, lightning is a major source of danger and damage to energy systems around the world (Mpanga et al. 2021; DEHN 2014). The damages caused result in huge costs to energy utilities and consumers, thereby posing a challenge to energy management. Energy management and sustainability involve every aspect of power supply such as power generation, transmission, and distribution and utilization sections. The increased integration of RES into the power system and its corresponding research around the world (Mwanza et al. 2017a, b; Mwanza and Ulgen 2021; Kaoma et al. 2017) are all an effort to ensure energy management and sustainability. Energy management can also involve paying attention to the dynamic features of a transmission line like the generated electric and magnetic fields as these affect the lives near the line, clearance requirements, and reliability (Mpanga et al. 2013, 2014). These fields can be several magnitudes higher when lightning strikes an energy line. Energy management also concerns low-energy appliances like video surveillance systems (Lubobya et al. 2018) and home and office electronic gadgets. The management and sustainability aspects are even more critical now because of climate change effects on our environment (Mpanga et al. 2016; Lungomesha and Zulu 2018). However, safety of both energy equipment and operation personnel is priority number one for every energy distribution company.

A typical lightning stroke can induce up to 30×10^8 kW of power at about 125 MV and an average current of more than 20 kA into a grounded installation (Das 2010). The effects of such a lightning stroke can be summarized as follows: (i) the lightning stroke crest current is responsible for ohmic voltage drops, especially in the ground resistance; (ii) the steepness of the lightning current accounts for the inductive voltage drop, induced voltage, and magnetic couplings; (iii) the electric charge of the lightning current is a measure of the destructive energy transferred by the lightning arc to every object in the current path; (iv) the integral of the lightning current squared and then multiplied by time is the basis for the mechanical effects and impulse heating up of objects struck by lightning; (v) lightning parameters vary over wide ranges and the current and wave shape data are determined statistically. The lightning stroke frequency is an important parameter for determining the surge arrester duty (Das 2010; DEHN 2014).

Thus, energy management and sustainability strategies cannot be complete without including the lightning protection aspect in their planning. Most infrastructures that bring electricity into our homes, schools, commercial buildings, or factories are exposed to the open environment where lightning can easily occur due to static charge buildup. The lightning currents can easily reach the electronic gadgets in the buildings, and if the lightning protection system is not good enough, the damage to the installations can be huge. A good lightning protection program (LPP) must take care of the lightning transients that can cause havoc on the energy equipment such as generators, motors, computers, and so on. Some specific energy systems requiring a good LPP are photovoltaics (PV), wind turbines, combined heat and power (CHP), diesel, battery, and electric vehicles (Lee et al. 2016).

Most published works on energy management (Lee et al. 2016; Aziz et al. 2019; Etzion 2018; Kumar et al. 2021; Lewis et al. 2011; Singh and Kumar 2018a, b; Su and Ramírez-Gómez 2018; Pietrosemoli and Monroy 2013) have tackled the design, construction, operation, forecast, optimization, data analysis, human machine interface, and maintenance of the energy systems. They show that there is an interdependence among energy management, maintenance, and sustainability of critical energy assets. Integration of RES is supported by all as it helps to improve system stability and smoothen out voltage fluctuations in conventional systems. However, the lightning protection aspect doesn't come out explicitly in most papers on energy management. Only one (Aziz et al. 2019) discusses a good energy management strategy that improves stability and reliability and protects assets against damage caused by overloads. In fact, most of the maintenance works done on the energy equipment are due to lightning-induced faults in summer of many countries around the world. If maintenance is linked to energy management and sustainability, as evidenced in published literature, then a good lightning protection program (LPP) is as well.

Hence, this chapter shows that in addition to all the strategies of energy management mentioned in published literature, there is a need for a good LPP too. It is a case study of a 640-kW mini-hydropower system in the north-western part of Zambia that articulates how a good LPP protects critical energy assets against lightning transients that can cause their destruction or electrical power blackouts, thereby leading to poor reliability of energy delivery. This is done using an electrical transient analyzer program (ETAP) to show the importance of devices like Faraday cage, lightning arresters, circuit breakers, earth conductors, and types of surge protection devices (SPD) appropriate for RES systems like hydro and photovoltaics (PV). The importance of maintenance is also emphasized and shown how it is linked to the lightning phenomenon. Switching transients are qualitatively discussed to explain how they were found to affect the operation of a 33-kV transmission line in rural areas during the case study.

45.2 Methodology

The methodology involved two parts: tour of the mini-hydropower system to collect the necessary parameters and the simulation of the 640-kW mini-grid in ETAP software.

45.2.1 *Tour of the Mini-Grid*

A hydro mini-grid with a generation capacity of 640 kW was chosen for a case study to demonstrate the importance of a good lightning protection program (LPP). Some data collected were generator information as shown in Fig. 45.1, transformer name plate data, conductor spacing, and so on.

Figure 45.2 shows the lightning arresters on the high voltage side of the 33/0.4-kV step-down transformer at the receiving end of the mini-hydropower system. There is another set of lightning arresters on the pole where the drop-down cable picks up power from the incoming 33-kV overhead conductors. This is in addition to the drop-out fuses connected there too to enhance the protection of the transformer and thereby ensure the reliability of power supply to the communities around the area.

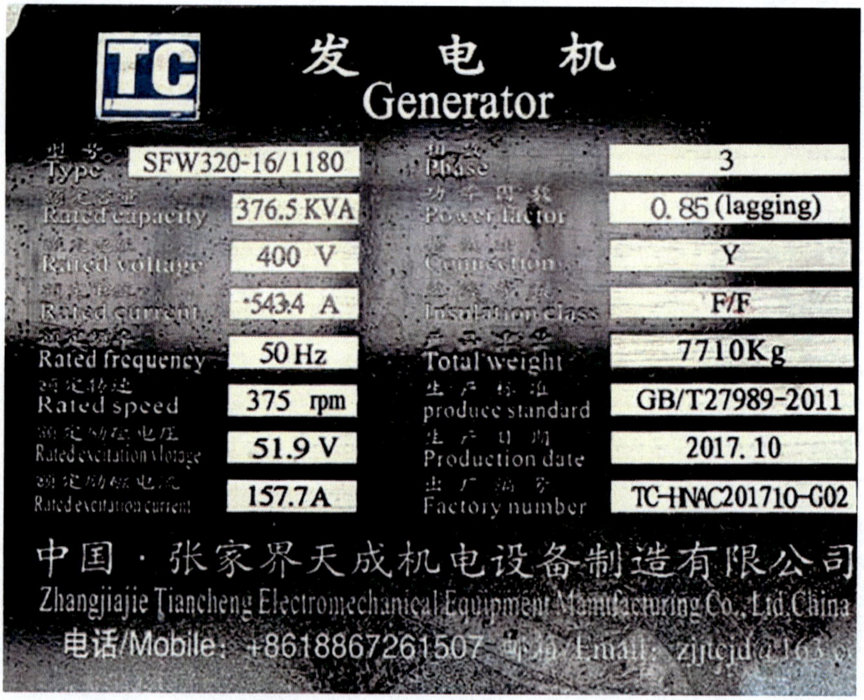

Fig. 45.1 Generator parameters used for the mini-grid

Fig. 45.2 Lightning arresters on the receiving end of the step-down transformer

45.2.2 Simulation of the Mini-Grid Using ETAP

Figure 45.3 shows the model of a 30-km long, 33-kV transmission line that was used for simulations in ETAP software. It also shows two hydro-generators (Gen1 and Gen2) each rated at 400 V, 320 kW, 375 rpm, and 50 Hz, feeding a common bus bar (Bus 1) where a cable that feeds the low voltage side of the 0.4/33-kV step-up transformer T_1 is connected. A good LPP divides such a power system into zones of protection as shown. Zone 1 is for the generators housed inside a building, zone 2 an outdoor generator step-up transformer and the associated auxiliaries, zone 3 the 33-kV overhead transmission line, zone 4 the 33/0.4-kV step-down transformer T_2 (Fig. 45.2) at the customer premises, and zone 5 the utilization centers (Lump 1). A load flow simulation was done first to ensure the voltage profiles were okay from the sending end to the receiving end of the mini-grid. Arc-flash analysis based on IEEE 1584-2018 standard was then done to determine the safe working distances and incident energy levels. An electromagnetic transient simulation results could not be presented in this chapter because a distributed parameter model could not be used for a short line to get more accurate results. However, an attempt to simulate transients for a short line has been done by Garbelim Pascoalato et al. (2021) using Carson's approach. This chapter leans towards renewable energy sources (RES) because RES is a buzzword worldwide currently and mini-grids based on them are helping communities far away from the national utility grids to access power at affordable prices.

45.3 Results and Discussion

Direct or indirect lightning strikes to energy systems can cause voltage rises that lead to equipment insulation stresses and these can affect their reliability. The line in Fig. 45.3 is a short line whose ABCD parameters are $A = D = 1$, $B = Z$, and $C = 0$ as depicted in Eq. 45.1 and the wave equation may not be used for analysis. Only its fault performance is analyzed herein.

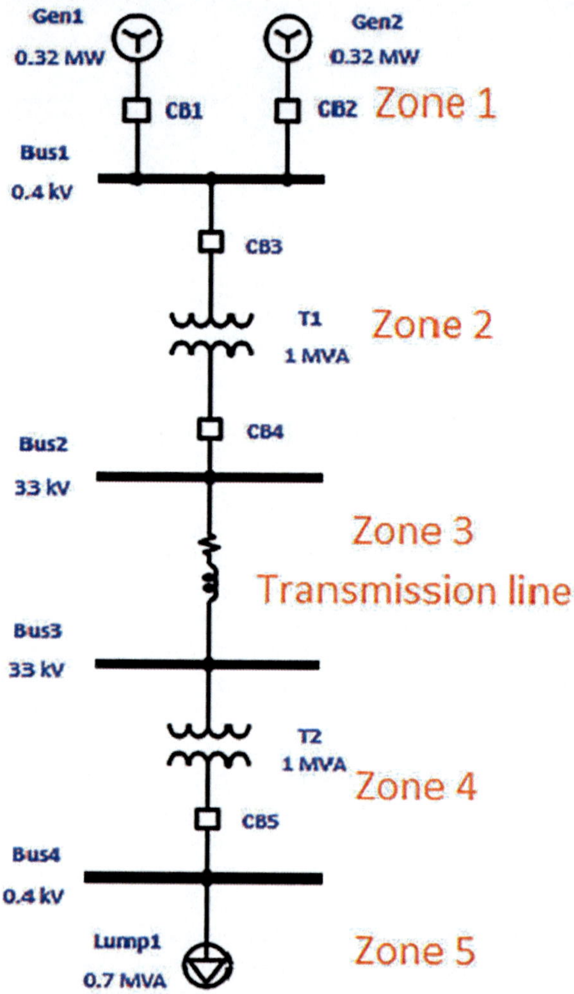

Fig. 45.3 A 640-kW mini-hydropower transmission system

$$\begin{bmatrix} V_s \\ I_s \end{bmatrix} = \begin{bmatrix} A & B \\ C & D \end{bmatrix} \begin{bmatrix} V_r \\ I_r \end{bmatrix} \tag{45.1}$$

Electromagnetic transients due to lightning strikes are so fast that the breakers numbered CB1 to CB5 that depend on relays to detect the faults in 20–30 ms and then trip in 40–60 ms cannot be effectively used to protect the expensive equipment like the generators, transformers, and control equipment. The lightning energy must be restricted to areas where damage cannot be done to equipment. The lightning arresters are used for this purpose to restrict the overvoltages to the protective voltage level V_p that cannot harm the equipment when impressed on them. Thus, the lightning arrester in a LPP is the foundation of protection even in mini-grids.

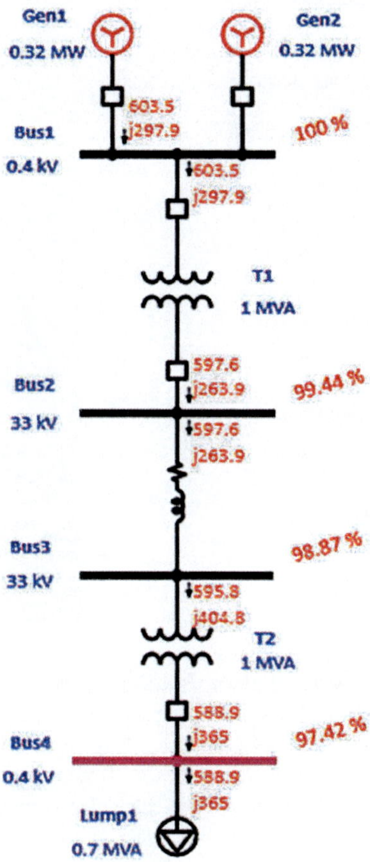

Fig. 45.4 Load flow analysis results of the mini-hydropower system

Figure 45.4 is the simulation result for the load flow analysis to ensure the voltage profiles are okay from the sending end to the receiving end as it can be observed. The voltage drops are within the standard stipulated margins of ±5%.

In Fig. 45.3 the equipment in zone 1 are housed in a building which must be protected by the air termination on top of the building. The air terminations are connected to the down conductors that take away the incident lightning energy on the air terminals to the earth electrodes in the ground (DEHN 2014). This ensures that most of the dangerous energy goes into the ground. The Faraday cage arrangement in this case ensures both equipment and operation personnel inside are well protected. For any internal faults, the surge protection devices (SPDs) connected to Bus 1 do protect the generators by diverting all the energy to the earth grid immediately there is an electromagnetic transient. Zone 2 contains a step-up power transformer and other sensitive equipment in an outdoor substation. The electrical equipment in this zone are protected by the SPDs at Bus 1 and the lightning arresters

Fig. 45.5 Lightning mast providing the angle of protection to the 640-kW power substation equipment

in each phase at Bus 2. However, the primary protection in this zone is offered by a 21-m-high lightning mast seen in Fig. 45.5. Both the mast and the SPDs are connected to the same earth grid for quick absorption of energy into the ground.

Zone 3 comprises a 30-km long, 33-kV overhead transmission line that delivers the much-needed electrical power to a mission hospital and the surrounding rural community. It has an aerial earth conductor connected to ground at regular intervals to ensure any transients are diverted to ground. It has an auto-recloser that tries to close the line when struck by lightning. The protection is also provided by lightning arresters at Buses 2 and 3. The surge arresters at Buses 3 and 4 do provide protection to the distribution transformer in zone 4. The lumped load is protected by the surge arresters at Bus 4 and other type 1, type 2, and type 3 SPDs within the premises of the customer. Type 1 SPDs are used at the point of entry of a distribution line into the building. Then just inside the building type 2 SPDs are used and then type 3 SPDs are dedicated to each apparatus being used in an establishment (DEHN 2014). The reason for this arrangement is that type 1 SPDs remove from the electrical circuit the surge energy that type 2 SPDs cannot handle, whereas type 2 SPDs remove from the circuit the energy levels that type 3 SPDs cannot handle.

Figure 45.6 shows the arc-flash results that outline a safe working distance (WD) of 45.72 cm from the switchgear and how the heat flux varies with distance from the live electrical equipment. WD is more than the arc-flash boundary (AFB) of 43 cm. The incident energy of 1.09 cal/cm^2 is lower than the standard value of 1.2 cal/cm^2 known to be the onset of second-degree burns. WD and AFB values were based on a fault current of 4.091 kA and an arcing current of 2.646 kA. If there

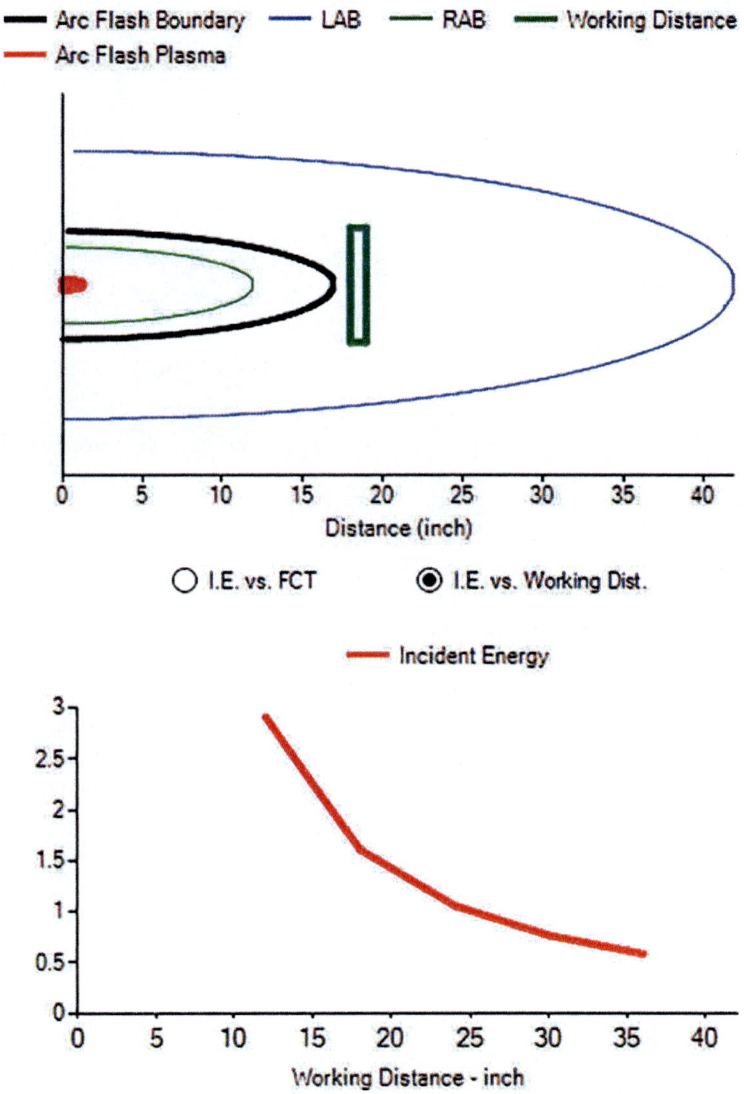

Fig. 45.6 Arc-flash results at the receiving end of the mini-hydropower system

happens to be no protection against lightning, these clearances can be encroached upon if lightning strikes the installation. That poses a danger to both equipment and operating personnel. In turn, that compromises the reliability of power supply that can affect energy management and sustainability. Thus, a good LPP would need to take care of all these aspects.

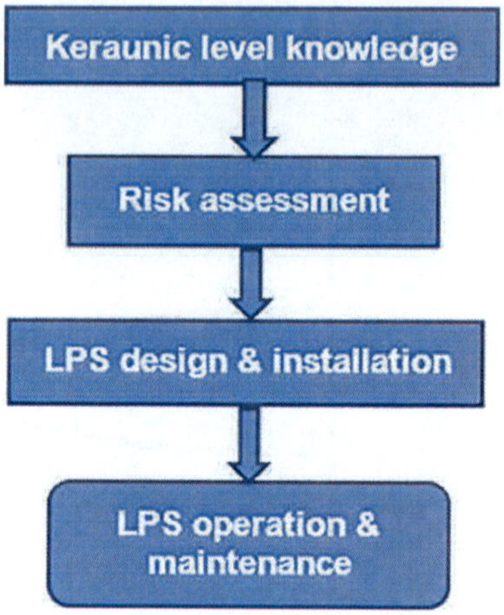

Fig. 45.7 LPP algorithm

Arc flash (AF) is the unwanted electrical discharge through the air from one conductor to another or to the ground facilitated by a low-impedance connection. This causes a rapid increase in temperature and pressure of the air between the electrical conductors that can result in fires, pressure waves, and flying shrapnel that can pose a danger to electrical equipment and lives of personnel (IEEE Std. 1584-2018). The explosions occur without any warning and can increase when lightning strikes or during switching operations and hence the need for a good LPP. AF analysis helps eliminate potential hazards and helps operation personnel to choose the right personal protective equipment (PPE) for certain areas. It helps the energy company to be aware of lightning or switching-related risks. It was learned during the case study that the 33-kV overhead line is also sensitive to overgrown vegetation in the transmission corridor. The line would trip during switching or while in operation whenever the vegetation encroached on the clearance to ground. When the vegetation was cleared, the tripping stopped. Thus, a good maintenance program must ensure this aspect of the transmission line is paid attention to as it affects the reliability of the power supply. The other maintenance tasks are monitoring of the earthing system and ensuring that the resistance of the installation is below 10 Ω, physical inspection of all the equipment to ensure there are no strange smells due to burning of components, and so on.

Figure 45.7 outlines the importance of keraunic knowledge about an area before carrying out any risk assessment and lightning protection system (LPS) design. The installation of the surge arresters all the way from the power station to the load centers of the mini-grid in Fig. 45.3 ensures a reliable supply of power to the surrounding areas. Reliability of power supply is very important for energy management and sustainability. A good LPP thus adds resilience to a power system and helps sustain the electricity supply.

45.4 Conclusion

This article has shown that for an energy system to be reliable, it must be sustainable. If the pillars of sustainability such as safety and a good LPP are not in place, the operation of an energy supply system (ESS) cannot be reliable. The planning, construction, operation, and maintenance of a reliable ESS must therefore have an effective LPS installed. The lightning arresters are quick in diverting all the dangerous lightning strike energy to the ground before it reaches the vulnerable protected electrical equipment. The most effective arresters for the renewable energy systems such as PV and hydro installations are the metal-oxide varistors as opposed to silicon carbide arresters (DEHN 2014). By restricting all the lightning transients to the ground, a good LPP ensures that all the aspects of energy management and sustainability, such as optimization, economic dispatch, etc., are maintained.

References

Aziz AS, Tajuddin MFN, Adzman MR, Ramli MAM, Mekhilef S (2019) Energy Management and Optimization of a PV/Diesel/Battery Hybrid Energy System Using a Combined Dispatch Strategy, Int. J. Sustainability, 11, 683. https://doi.org/10.3390/su11030683.

Das JC (eds) (2010) Transients in electrical systems analysis, recognition and mitigation. McGraw-Hill, New York.

DEHN (2014) Lightning protection guide, 3rd edition, Germany.

Etzion D (2018) Management for sustainability. Int. J. Nat Sustain, (1):744–749. https://doi.org/10.1038/s41893-018-0184-z.

Garbelim Pascoalato TF, Justo de Araújo AR, Caballero PT, Leon Colqui JS, Kurokawa S (2021) Transient analysis of multiphase transmission lines located above frequency-dependent soils, Int. J. Energies, (14):5252. https://doi.org/ 10.3390/en14175252.

IEEE std 1584-2018: IEEE guide for performing arc-flash hazard calculations.

Kaoma M, Mwanza M, Mpanga S (2017) Biomass resource potential and enabling environment for bioenergy production in Zambia, in: Proceedings of the Engineering Institution of Zambia Symposium, April 7–8, 2017, Livingstone, Zambia.

Kumar M, Shenbagaraman VM, Shaw RN, Ghosh A (2021) Predictive Data Analysis for Energy Management of a Smart Factory Leading to Sustainability, in: Favorskaya MN, Mekhilef S, Pandey RK, Singh N (eds) Innovations in Electrical and Electronic Engineering. Lecture Notes in Electrical Engineering, Vol. 661. Springer, Singapore. https://doi.org/10.1007/978-981-15-4692-1_58.

Lee EK, Shi W, Gadh R, Kim W (2016) Design and Implementation of a Microgrid Energy Management System, Int. J. Sustainability, 8(11):1143. https://doi.org/10.3390/su8111143.

Lewis A, Elmualim A, Riley D (2011) Linking energy and maintenance management for sustainability through three American case studies, Int. J. Facilities, 29(5/6):243–254. https://doi.org/10.1108/02632771111120547.

Lubobya SC, Dlodlo ME, De Jager G, Zulu A (2018) Mesh IP video surveillance systems model design and performance evaluation. Journal of Wireless Personal Communication, 100(2): 227–240.

Lungomesha E, Zulu A (2018) Environmental and economic benefits of railway electrification of Southern African countries. Journal of Transport and the City, WIT Press.

Mpanga S, Feng W, Chun C (2013) Electromagnetic field evaluation of a 500 kV high voltage overhead line, TELKOMNIKA, 11(2):789–796.

Mpanga S, Kaoma M, Zimba K, Zulu A (2016) Mitigating the effects of adverse climatic conditions in Zambia, in: International Workshop on Mitigation of disasters due to Severe Natural Events: From Policy to Practice, March 10–13, 2016, Colombo, Sri Lanka.

Mpanga S, Zulu A, Mwanza M, Holle RL (2021) Articulating the threat of the lightning phenomenon in Zambia, in: Proceedings of the International Conference on Electrical, Computer and Energy Technologies, Dec. 8–10, 2021, Cape Town, South Africa. https://ieeexplore.ieee.org/document/9698551.

Mpanga S, Zulu A, Ngoyi L (2014) Study of key theoretical and technical features of dynamic loading of high voltage overhead transmission lines. International Journal of Engineering Innovation and Research, 3(1):67–73.

Mwanza M, Chachak J, Cetin NS, Ulgen K (2017b) Assessment of solar energy source distribution and potential in Zambia, Periodicals in Engineering and Natural Sciences, 5(2):103–116.

Mwanza M, Kaoma M, Bowa CK, Cetin NS, Ulgen K (2017a) The potential of solar energy for sustainable water resource development and averting national social burden in rural areas of Zambia, Periodicals of Engineering and Natural Sciences, 5(1).

Mwanza M, Ulgen K (2021) GIS-based assessment of solar energy harvesting sites and electricity generation potential in Zambia, African Handbook of Climate Change Adaptation.

Pietrosemoli L, Monroy CR (2013) The impact of sustainable construction and knowledge management on sustainability goals. A review of the Venezuelan renewable energy sector, in: Renewable and Sustainable Energy Reviews, (27):683–691.

Singh AG, Kumar N (2018a) MEnSuS: An efficient scheme for energy management with sustainability of cloud data centers in edge–cloud environment. International Journal of Future Generation Computer Systems, (86):1279–1300.

Singh AG, Kumar N (2018b) SDN-based energy management scheme for sustainability of data centers: An analysis on renewable energy sources and electric vehicles participation. Journal of Parallel and Distributed Computing, (117):228–245.

Su JJ, Ramírez-Gómez Á (2018) Optimizing the location of a biomass plant with a fuzzy-decision-making Trial and Evaluation Laboratory (F-DEMATEL) and multi-criteria spatial decision assessment for renewable energy management and long-term sustainability. Journal of Cleaner Production, (182):509–520.

Chapter 46
Techno-economic Analysis of Wind/PV Hybrid System for Sustainable and Clean Energy Production for Shang'ombo District of Zambia

M. Mwanza, K. Mwansa, C. K. Bowa, M. Sumbwanyambe, J. H. Pretorius, and K. Ulgen

Nomenclature

CO_2	Carbon dioxide
GHG	Greenhouse gas
IEC	International electrotechnical commission
kW	Kilowatt
KWh	Kilowatt-hour
MW	Megawatt
MWh	Megawatt-hour
NASA	National Aeronautics and Space Administration

M. Mwanza (✉) · K. Mwansa
School of Engineering, University of Zambia, Lusaka, Zambia
e-mail: mabvuto.mwanza@unza.zm

C. K. Bowa
School of Mechanical Engineering, Copperbelt University, Kitwe, Zambia

M. Sumbwanyambe
Department of Electrical and Mining Engineering, University of South Africa, Pretoria, South Africa

J. H. Pretorius
Faculty of Engineering and Built Environment, University of Johannesburg, Johannesburg, South Africa

K. Ulgen
Solar Energy Institute, Ege University, Bornova/Izmir, Turkey
e-mail: koray.ulgen@ege.edu.tr

M. Z. Sogut et al. (eds.), *Proceedings of the 2022 International Symposium on Energy Management and Sustainability*, Springer Proceedings in Energy,
https://doi.org/10.1007/978-3-031-30171-1_46

NPV	Net present value
NOCT	Nominal operating cell temperature
PV	Photovoltaic

46.1 Introduction

Shang'ombo district is located in the Western Province of Zambia with its headquarters at Shang'ombo town. As of 2017, the district had a population of 114,506 inhabitants with about 95.2% of the population living in rural areas and 4.8% in urban (CSO 2013). It has a total surface area of 14,3669 km^2. At present, Shang'ombo is one of the districts in Zambia with lower electricity access levels estimated at less than 7%. This is being met through the use of mini off-grid system using an installed diesel generator with a capacity of 1000 kW and producing 1030.70 MWh per year (ZICTA 2015; ERB 2017). This lack of access to electricity is negatively affecting the development of the district and it is ranked amongst the poorest districts in Zambia (Masumbu and Mahrt 2014). However, according to the Government of Zambia Rural Electrification Master Plan, the government intends to increase the electricity access levels from the current less than 7% to 51% by 2030 (EAA 2017; Bowa et al. 2017; Mwanza et al. 2017; EPRI 2015). In essence, this means that the electricity consumption will rise from the current 1030.70 MWh per year to approximately 7.51 GWh per year by 2030 (USAID 2005). It is to this end that, ZESCO, the utility company, intends to replace the existing generation technology with renewable energy technologies mainly because of high operation and maintenance cost of diesel generator. The replacement of the diesel generator with renewable resources is in line with the government goals of achieving increased electricity access levels, reducing energy production cost, increasing economic activities in the district and also reducing greenhouse gas emission. The motivation behind the replacement of the diesel generator by renewable sources is based purely on the potential of the district in renewable energies and to less extent in its potential for economic activities as can be seen in solar energy atlas (Fig. 46.2) (Mwanza and Ulgen 2020a) and wind energy atlas (Fig. 46.1).

Currently, there exist no studies on techno-economic analysis of hybrid solar and wind systems in Zambia that are considered for replacing the currently installed mini off-grid diesel generator-based systems across the country. Therefore, this chapter is aimed at determining the wind and solar PV hybrid system based on technical and economic analysis, capable of meeting the current energy consumption of 1030.70 MWh per year for Shang'ombo district, which would replace the existing 1000 kW diesel generator at a lowest cost of energy and land use (see Table 8) (Musonda 2019). The decision variables adopted in sizing the system include yearly system energy yield, system energy yield per land used, system capacity factor, unit cost of energy generation, system initial investment cost, payback period, net present value of the system, land use and amount of CO_2 emission reduction. The system

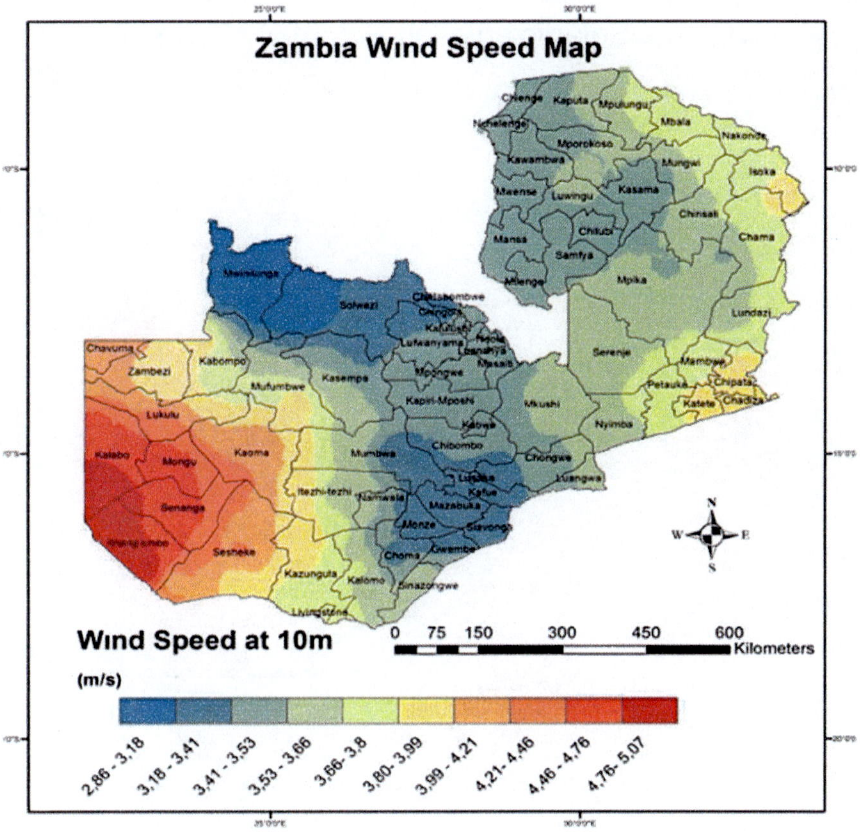

Fig. 46.1 Wind speed distribution map

design was based on the mean monthly variation of the renewable energies obtained from NASA (wind data) and PVGIS (solar data) (Suprava and Pradip 2015) and current monthly energy consumption of the district obtained from the utility company (ZESCO Ltd.).

The essential benefits of this method are matching the energy consumption with the typical seasonal variation of wind and solar energy at the selected sites (see Table 46.1) (Markvart 1996). Hence, this chapter provides preliminary studies for hybrid renewable energy system (HRES) studies in Zambia. It also provides the reader with complete analysis approach for research in the area of HRES.

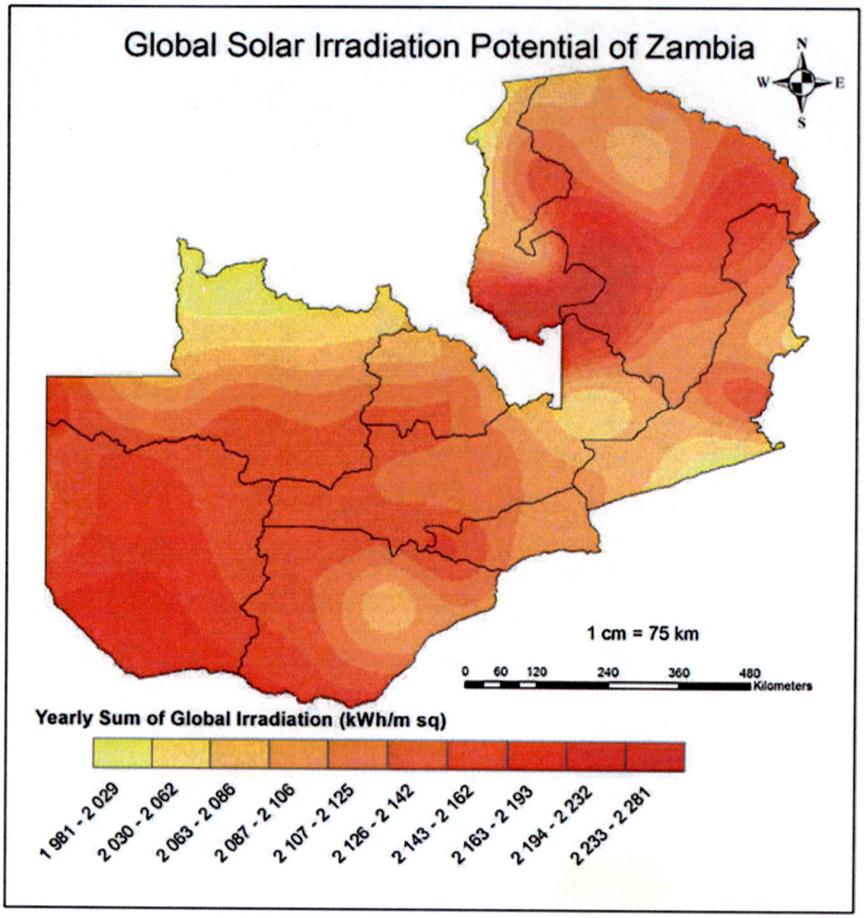

Fig. 46.2 Solar energy distribution map. (Mwanza and Ulgen 2020a)

Table 46.1 Comparison of the technical and environmental performance of hybrid system options

Hybrid options	Technologies		Technical factors			Environmental factors	
	Wind	Solar	AEY	EYLU	SCF	SLU	Avoid CO_2
(kWp)	(kWp)	(kWp)	(MWh/a)	(kWh/m^2)	(%)	(Ha)	(tons/year)
612	500	112	2686.2	227.80	50.11	1.18	3584
593	250	343	1814.6	173.02	34.93	1.05	2486
750	750	0	3753.3	250.22	57.13	1.50	4767
792	0	792	1301.3	102.69	18.76	1.27	1398

AEY annual energy yield, *EYLU* energy yield per land use, *SCF* system capacity factor, *SLU* system land use; avoided annual CO_2 reduction

46.2 Material and Methods

46.2.1 Hybrid System Energy

A typical hybrid system usually comprises of more than one energy generation system which is either supplying energy direct to the load or integrated to the utility grid (Mahesh and Sandhu 2015). In this chapter, the primary energy sources considered are wind and solar without battery system since the system will be connected to the mini off-grid system currently being supplied by diesel generator. Hence, in cases where the hybrid system fails to meet the energy demand, the generator will work as a backup supply to cover the excess energy needs. For the technical evaluation of the total monthly hybrid system energy generation, the main criteria adopted to be fulfilled for sizing the hybrid system were maximum power reliability and minimum cost (Sinha and Chandel 2015). The energy generation by the hybrid system EHE has been estimated using Eq. 46.1 as given below (Hongxing et al. 2009).

$$E_{\mathrm{HE}} = E_{\mathrm{WTE}} + E_{\mathrm{PVE}} \tag{46.1}$$

where E_{WTE} is the total energy generated by the wind energy system and E_{PVE} is the total energy generated by the solar PV energy system. The energy generated by the hybrid system may or may not meet the energy demand; hence, three different situations may occur depending on the energy consumption and the total energy generation by the hybrid system as follows (Markvart 1996; Eke et al. 2005):

- The total energy demand is less than the total energy generation by the hybrid system, so there is excess energy, hence the need to send the excess energy to the utility grid.
- The total energy demand is equal to the total energy generation from the hybrid system, so that there is no deficiency in the energy or excess energy.
- The total energy demand is greater than the total energy generated by the hybrid system; hence, there is energy deficiency and the system may not meet all the energy needs and the diesel generator will have to supply the excess energy demand.

The sizing of the system was formalized using the monthly energy consumption data that matched the supply-demand conditions for each month, with the lowest hybrid system energy generation month being considered for the design. Hence, for each month the system has been designed to meet the following supply-demand condition (Markvart 1996; Eke et al. 2005):

$$D_{\mathrm{m}} \leq E_{\mathrm{HE}} \tag{46.2}$$

where D_{m} is the monthly energy demand of Shang'ombo and system energy demand varies throughout the year.

46.2.2 Hybrid System Economical Analysis

In the hybrid system economic analysis, the initial capital cost, unit cost of energy and payback period are considered as the main criteria for selecting the optimal hybrid alternative suitable for Shang'ombo district. The hybrid system initial cost depends mainly on three parameters, namely, the geographic location, i.e. country, the size of the system and the technology used in the power plant (Mukund 1999). The unit cost of energy (COEPV) of electricity delivered to the grid is one of the most important decision-making financial parameters for an electrical system project (Mukund 1999; Ayompe and Duffy 2014; Binayak et al. 2015). It can be used to compare with feed-in tariff to check if the project is viable. The payback period is defined as the length of time that is required for the project to recover the project initial investment costs (IRENA 2012). In this chapter, the project has been considered as generating inconsistent or an uneven cash inflow due to system degradation of 0.5% per year. Hence, the payback period is computed based on the cumulative cash inflow.

46.2.3 Hybrid System Environmental Analysis

In the hybrid system environmental analysis, two criteria were considered: CO_2 emission avoidance and land requirement for each proposed hybrid alternative since these have direct impact on the environment.

46.3 Shang'ombo RES Potential and Electricity Production

46.3.1 RES Potential of Shang'ombo

The selection process of potential sites took into consideration a number of factors that influence the technical and economic performance of the system. The selection process also considered the accessibility of the system and its proximity to the local load centre, roads and transmission lines (Azadeh et al. 2011; Mohammadi et al. 2014). Also taken into consideration were factors such as legal framework, distance from wildlife areas, bare lands free of trees and land availability, as suggested by Ahmed et al. (2013), Amir et al. (2016), and Mwanza and Ulgen (2020b).

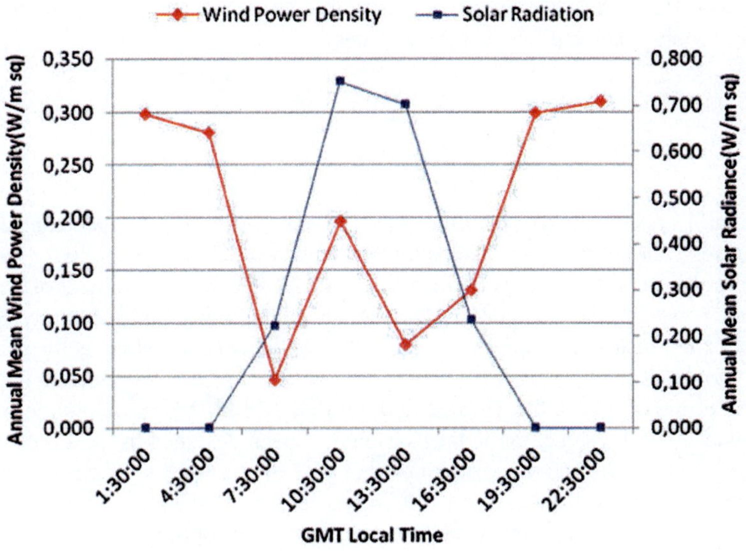

Fig. 46.3 Annual solar radiation and wind density

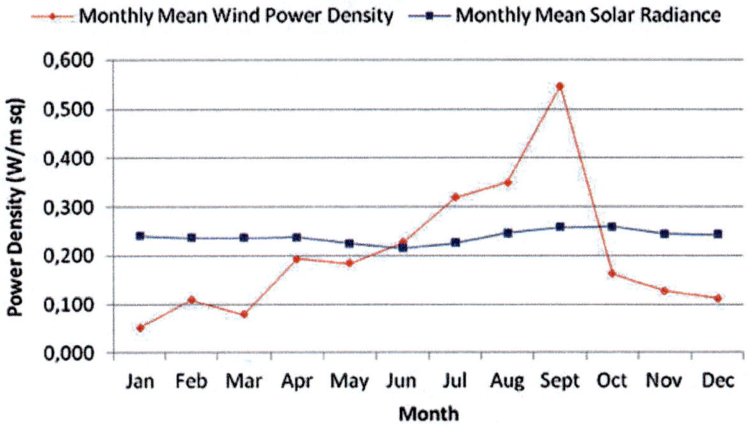

Fig. 46.4 Monthly solar radiation and wind density

46.3.2 Wind/Solar Characteristics at Selected Sites

Figures 46.3 and 46.4 present the annual hourly average solar radiation and wind speed variation of the selected sites. It can be noted that the hourly variation of wind speed and solar radiation complements very well for hybrid system implementation. The wind speed is always low during the day when the solar radiation is higher and vice versa. Hence, the two energy sources can complement to meet the hourly and monthly energy demand for the selected sites.

46.3.3 Hybrid System Energy Balance

The hybrid system was designed based on the monthly energy balance considering the current monthly energy consumption and availability of primary energy resources for the selected sites (British Standard 1998). The Government of Zambia with the national utility company, ZESCO, has plans to increase electricity access levels for the district through the use of a renewable energy source as the primary energy for electricity generation and also grid extension, that is, connecting the mini off-grid of the district to the national grid through a 33 kV transmission line (www.lusakavoice.com). Hence, the hybrid system has been designed to meet the current energy consumption with the assumption that the excess energy will be injected back into the grid once the mini off-grid system is connected to the main national grid. In the analysis several options were assessed and four optimal options selected. Table 46.1 presents the summarized results of the technical and environmental performance of the selected hybrid system options considering the technology characteristics and local site weather conditions (Paul, 2004; Pham and Thananchai 2015; Hocaoglu 2012). Option 3 had the highest performance followed by options 1 and 4 being the least of all the indices considered in the analysis as presented in Table 46.1.

46.3.4 Hybrid System Economics

The economic analysis of the hybrid system was conducted based on the capital cost break-down for wind and solar power plant below considering the current system component prices as obtained from various sources in the sector (IRENA 2012). In the technical and economic analysis of the system, the system degradation was assumed at 0.5% considering the panel degradation as provided by manufacturers, while the annual system operation and maintenance was assumed at 3% of capital cost with inflation rate of 7% and hybrid system economical life span estimated at 30 years (Farivar et al. 2015; Satyanarayana and Shiva 2016). The hybrid system operation expenditure (OpEx), feed-in tariff (FiT) and electricity price (EP) have been considered constant for the entire economic life span of the hybrid system (Satyanarayana and Shiva 2016; EPRI 2015; IRENA 2012).

Figure 46.5 and Table 46.2 provide the summary of the results of the hybrid system options analysed for the selected sites. The initial capital cost and cost of energy for options 1, 2, 3 and 4 were found to be US$63, US$87, US$56 and $158 per MWh, respectively. Hence, developing options 1 and 3 of hybrid systems for the district provides more advantages for generating energy at affordable prices than

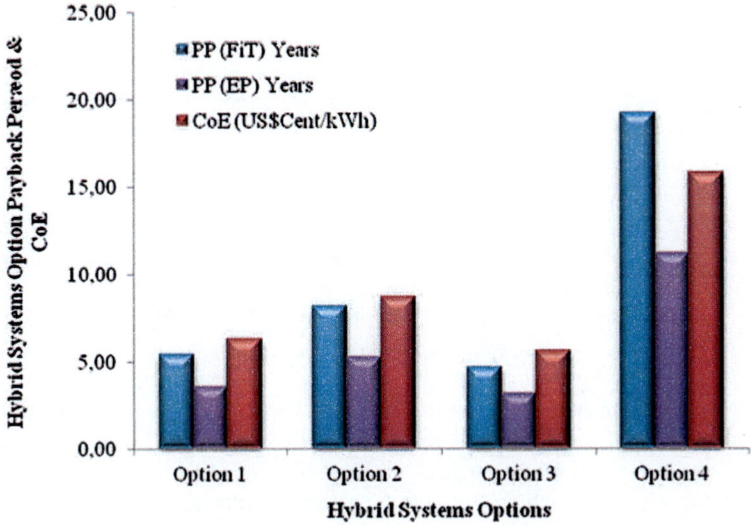

Fig. 46.5 Comparison of hybrid system option PP at the proposed FiT and current electricity price

Table 46.2 Comparison of the financial performance of hybrid system options

	ICC	CoE	CEP (US$ cent/MWh)		PP (years)	
Hybrid options	(US$)	(US$/MWh)	FiT	EP	FiT	EP
612k Wp	865,080.00	63	70	100	5.41	3.59
593 kWp	817,120.00	87	70	100	8.15	5.27
750 kWp	1,072,500.00	56	70	100	4.70	3.14
792 kWp	1,061,280.00	158	70	100	19.22	11.17

ICC initial capital cost, *CoE* cost of energy, *REFiT* feed-in tariff, *EP* electricity price, *CEP* unit electricity prices, *PP* payback period

options 2 and 4 when the solar energy resource for the system is the majority resource for the hybrid system. This is due to the lower energy generation and higher initial capital cost. As seen in Table 46.2 the present proposed REFiT, i.e. 70 $ MWh.1, is almost the same as the cost of energy for the hybrid system options 1 (63$ MWh-1) and 3 (56$ MWh-1) and less for options 2 (87$ MWh-1) and 4 (158 $ MWh.1). Therefore, this will be a major barrier against investment in wind/PV hybrid energy harvesting systems in Zambia at the present proposed REFiT.

When the four hybrid energy system options are compared while considering their simple payback periods (PP) at the current electricity prices and proposed REFiT, it is seen that option 3 has the shortest PP of 3.14 and 4.70 years, followed by option 1 with values of 5.41 and 3.59 years. The longest is seen for option 4 with values of 19.22 and 11.17 years as presented in Table 46.2 and Fig. 46.5. Therefore, according to the indices considered in the economic analysis for the hybrid options, it is seen that hybrid energy system can greatly benefit the district by replacing the diesel generator or reducing the expensive primary energy importation (fuel), increasing access to energy and achieving sustainable development goals. However, to encourage or attract investment in hybrid energy projects in the district, there is a need to adjust the present proposed REFiT to cost reflective levels and also to extend the electrical grid of these regions.

The analysis shows that hybrid system option 1 has the second highest capacity factor, final energy yield, CO_2 emission reduction, energy yield per land use and initial capital cost as presented in Tables 46.1 and 46.2. This hybrid option is also the second highest in CoE and PP based on both (70$ MWh^{-1}) FiT for microgenerations proposed by the Energy Regulation Board of Zambia (ERB) and the current EP (100 $ MWh^{-1}) offered by the utility company to energy consumers in the district. Therefore, considering the long-term benefit of new electrification and grid connection, the first alternative hybrid system stands to be the best option for the selected sites. At the same time, the first alternative has the best cost of energy generation which is good for society in terms of access to affordable energy, but it also has shorter payback period which presents good opportunity for investments. On the other hand, the option of using only one of the sources either wind system or solar system would present a challenge of failing to meet the energy needs in some cases whenever there is no wind or solar radiation during the day or at night, respectively, which will require frequent use of diesel generator for backup or use of battery bank as compared to when both resources are employed. In addition, the use of only one resource and battery bank would result in huge initial investment and operation costs. This in turn will result in unaffordable electricity prices with reduced CoE and PP. Hence, due to first alternative outstanding in the majority of the criteria as compared to the second one, it can be concluded that in the present situation and with the plans of new electrifications and connection to the national grid, the first alternative presents the best option for the selected sites to replace the current diesel generator system and reduce the operation cost and greenhouse gas emissions.

46.3.5 Conclusion

This chapter has presented the methodology for sizing of a hybrid system capable of meeting the current energy needs of Shang'ombo district and replacing the existing 1000 kW diesel generator designed based on the technical and economic analysis of the system. The technical and economic analysis shows that the system has provision for the excess energy to be fed into the grid once it is connected to the national grid. However, in case the system is not connected to the grid immediately, the excess energy provides an opportunity for adding new electrification to the system without any need to expand the system. Adopting option 1 of the hybrid system provides the best cost of energy generation which is suitable for Shang'ombo as it does not only provide access to affordable energy but also has higher net present value and shorter payback period which present good opportunity for investments. The analysis provides a solution to the current energy problem being faced in the area with a clear indication of how the hybrid system would best benefit the area and ZESCO in facilitating their decisions. Hence, the presented system provides the best solution in reducing greenhouse gas emission and reducing the current high operation and maintenance costs faced by ZESCO. In addition, based on the availability of renewable energy resource in the selected sites, the proposed system provides the best solution in that the solar radiation is high in the day at the selected sites and these sites have relatively good wind speed in the night, thus complementing each other well. Therefore, it can be concluded that the wind/solar PV hybrid system can be adopted to meet the current and future energy needs in Shang'ombo district as well as help to increase access to electricity at an affordable price while reducing greenhouse gas emissions in the district.

Acknowledgement The authors gratefully acknowledge the support provided by the Turkish Government, University of Zambia and Ege University.

Declaration of Interest The authors declared no potential conflicts of interest with respect to the research, authorship and/or publication of this article.

References

Ahmed B, Muhamad M, and Mahmoud AY (2013) Assessment of Wind and Solar Energy Potentials in Malaysia, 2013 IEEE Conference on Clean Energy and Technology (CEAT).

Amir AM, Mezhab AB, and Mezhab A.(2016) CSP site suitability analysis in the Eastern Region of Morocco, 49; 2270–2279

Asif I, Mohammad SI, and Mohammad ZI (2012) Monthly and Seasonal Assessment of Wind Energy Potential in Coastal Area of Bangladesh,

Ayompe LM, and Duffy A (2014) An assessment of the energy generation potential of photovoltaic systems in Cameroon using satellite-derived solar radiation datasets, Sustainable Energy Technologies and Assessments 7: 257–264.

Azadeh K, Alireza M, Ghanim P, et al (2011) Optimal Sizing of Grid Connected Hybrid Wind-PV systems with Battery bank Storage, Northumbria University, Newcastle, UK.

Binayak B, Kyung-Tae L, Gil-Yong L, et al (2015) Optimization of Hybrid Renewable Energy Power Systems: A Review, International Journal of Precision Engineering and Manufacturing-Green technology 2(1): 99–112.

Bowa CK, Mwanza M, Sumbwanyambe, and Pretorius JH, (2017), SAUPEC 2017, 25th Southern African Universities Power Engineering Conference

British Standard (1998) Photovoltaic System Performance Monitoring, Guidelines for measurement, data exchange and analysis; BS EN 61724: 1998, IEC 61724: 1998", BSI 05.1999.British Standards (BS).

CSO (2013) 2010 Census of Population and Housing: Population and Demographic Projections 2011–2035, Report, July 2013. Zambia: Central Statistical Office of Zambia (CSO).

EAA (2017) Electricity Access in Zambia . Available at: https://energyaccess-africa.com/2017/08/11/electricity-access-in-zambia/ (Accessed January 2018). Energy Access-Africa (EAA).

Eke R, Kara O, and Ulgen K (2005) Optimization of a wind/PV Hybrid Power Generation System, International Journal of Green Energy 2(1):57–63. https://doi.org/10.1081/GE-200051304.

EPRI (2015) Budgeting for solar PV plant Operations and Maintenance: Practices and Pricing, SAND2016-0649R, Sandia Laboratories, USA: Electric Power Research institute (EPRI).

ERB (2017) Energy Sector Report 2016, Zambia: Energy Regulation Board(ERB).

Farivar F, Nima S, Sina S, et al. (2015) Assessment of wind energy potential and economics in the north-western Iranian Cities of Tabriz and Ardabil, Renewable and Sustainable Energy Reviews 45: 87–99.

Hocaoglu FO (2012) Review of Wind-PV Sizing Algorithms and a Case Study, Journal of Engineering Science and Technology Review 5(4): 24–29.

Hongxing Y, Wei Z, and Chengzhi L (2009) Optimal design and techno-economic analysis of a hybrid solar-wind power generation system, Applied Energy 86: 163–169.

IRENA (2012) Renewable Energy Technologies: Cost Analysis Series.1(4/5). International Renewable Energy Agency(IRENA).

Mahesh A, and Sandhu KS (2015) Hybrid wind/photovoltaic energy system developments: Critical Review and Findings, Renewable and Sustainable Energy Reviews 52: 1135–1147.

Markvart T (1996) Sizing of Hybrid Photovoltaic-Wind Energy Systems, Solar Energy, 57(4): 277–281.

Masumbu G, and Mahrt K (2014) Comparison of Welfare Status of Districts in Zambia. Working Paper 17. Lusaka, Zambia, Zambia Institute for Policy Analysis and Research. Available at : https://opendocs.ids.ac.uk/opendocs/handle/123456789/4148 (Accessed October 2017)

Mohammadi K, Mostafaeipour A, and Sabzpooshani M (2014) Assessment of solar and wind energy potentials for three free economic and industrial zones of Iran, Energy, 67:117–129.

Mukund RP (1999) Wind and Solar Power Systems; Design, Analysis, and Operation, Taylor & Francis Group CRC Press LLC, p 283–313.

Musonda G (2019) Shangombo Monthly Electricity Consumption for period 2017–2018, Zambia: Zesco Ltd.

Mwanza M, and Ulgen K (2020a), Sustainable electricity generation fuel mix analysis using an integration of multicriteria decision making and system dynamic approach, International Journal of Energy research, https://doi.org/10.1002/er.5216

Mwanza M, Chakchak J, Cetin NS et al (2017), Assessment of solar energy source distribution and potential in Zambia, Periodicals of Engineering and Natural Sciences, 2(5), 2017.

Mwanza M, and Ulgen K, (2020b), GIS-Based Assessment of Solar Energy Harvesting Sites and Electricity Generation Potential in Zambia, African Handbook of Climate Change Adaptation, 1–48

Paul G (2004) Wind Power: Renewable Energy for Home, Farm, and Business, 2nd Edition 2004, White River Junction, VT: Chelsea Green Pub. Co., 1993, p 36–38.

Pham Q, and Thananchai L (2015) Assessment of wind energy potential for selecting wind turbines: An application to Thailand, Sustainable Energy Technologies and Assessments 11:17–25.

Satyanarayana G, and Shiva PKK (2016) Wind energy potential and cost estimation of wind energy conversion (WECSs) for electricity generation the selected locations of Tigray region Ethiopia, 10.1186/s 40807-016-0030-8, Gaddada and Kodicherla Renewables (2016):10, Renewable: Wind, Water, and Solar, Springer Open.

Sinha S, and Chandel SS (2015) Review of recent trends in optimization techniques for solar photovoltaic-wind based hybrid energy systems, Renewable and Sustainable Energy Reviews 50: 755–769

Suprava C, and Pradip KS (2015) Technical mapping of solar photovoltaic for the coal city of India, Renewables: Wind, Water, and Solar 11(2), https://doi.org/10.1186/s40807-015-0013-1

USAID (2005) Zambia Rural Electrification Master Plan: Phase 1: Rapid Resource Assessment, Final Report December 30, 2005. Zambia: United States Agency for International Development (USAID)

ZICTA (2015) ICT Survey Report-Households and Individuals: Survey on Access and Usage of Information and Communication Technology by Households and Individuals in Zambia, Zambia: Zambia Information and Communication Technology Authority (ZICTA).

Chapter 47
Numerical Analysis of Tank Coating Selection in Chemical Tanker Ships

Murat Mert Tekeli, Emre Akyuz, Muhammed Fatih Gulen, and Omer Berkehan Inal

Nomenclature

MARPOL	International Convention for the Prevention of Pollution from Ships
IBC Code	International Code for the Construction and Equipment of Ships Carrying Dangerous Chemicals in Bulk
TOPSIS	Technique for Order of Preference by Similarity to Ideal Solution
AHP	Analytical Hierarchy Process

47.1 Introduction

Today, following the enormous growth of the chemical industry, the chemical shipping industry transports a large number of chemicals, and the world petrochemical is increasing day by day. As a result of this increase, it is clear that the demand for more useful and efficient chemical tankers will increase in the coming days.

A chemical tanker is a type of tanker vessel designed to transport chemicals in bulk. As defined in MARPOL Annex II, chemical tanker means a ship constructed or adapted to carry in bulk any liquid product listed in section 17 of the IBC Code. The range of cargoes carried by these vessels is very wide and includes not only chemical products but also vegetable oils, animal fats, molasses, wine, solvents, and some clean petroleum products. Also, a chemical tanker can transport inorganic substances such as sulfuric acid, phosphoric acid, and caustic soda.

M. M. Tekeli (✉) · E. Akyuz · M. F. Gulen · O. B. Inal
Maritime Faculty, Maritime Transportation Management Engineering Department, Istanbul Technical University, Tuzla/Istanbul, Turkey
e-mail: tekelimu@itu.edu.tr

M. Z. Sogut et al. (eds.), *Proceedings of the 2022 International Symposium on Energy Management and Sustainability*, Springer Proceedings in Energy,
https://doi.org/10.1007/978-3-031-30171-1_47

The chemical tanker is a very special type of vessel due to the complexity and nature of the cargo. This is because the cargoes it can be transported to are corrosive, such as methanol, sulfuric acid, caustic soda, acetic acid, and pure naphtha. For this reason, much attention is often paid to cargo tanks and their ability to preserve the integrity and purity of the cargo. Often chemical tankers are built with stainless steel cargo tanks. Generally, stainless steel is considered the ideal building material, which is non-corrosive and easy to clean. However, not all cargoes can be transported in stainless steel tanks. Many ships carrying chemicals have cargo tanks made of mild steel lined with special coating systems that prevent corrosion of the steel and protect the cargo from contamination by contacting the steel.

Cargo tank coatings, which are widely used today, are divided into four types: epoxy, zinc silicate, stainless steel, and marine line. Due to the variety of cargoes in chemical tanker transportation, tank lining is very important for the cargo to be transported. The type of coating to be applied in chemical tankers, the mechanical properties of the pavement, the maintenance and repair actions of the pavement, the method of application of the pavement, the role of the pavement in ventilation and moisture removal, the absorption and efflux of the pavement, cargo cleanliness, and cost should be evaluated. In this study, a decision mechanism by consulting the expert opinions about the tank coatings used in chemical tankers, the most important criterion required for the type of coating to be applied, and the most appropriate tank coating selection to be applied within the framework of these criteria has been tried to be established.

In a systematic laboratory study on the control and energy loss of an anticorrosive coating, the composition and manufacturing process of a three-layer anticorrosive coating for oil storage tanks is introduced. What is meant to be explained in the study is that solar radiation with an intensity of 1373 kW/m^2 on the earth's atmospheric surface causes the surface temperature of oil or petroleum product storage tanks to rise. As a result, the light hydrocarbon components of the oil evaporate in the space between the oil and the tank roof, known as evaporation loss. The loss in the evaporation of the oil reduces the quality of the oil and causes environmental pollution (Zhang et al. 2013).

Ozsever and Solmaz evaluated the coating types over seven criteria and used the fuzzy TOPSIS method in their study. As criteria, the first application of the coating, freight income, coating life, durability of the coating, and cargo compatibility were considered. In this study, which was carried out using the fuzzy TOPSIS method, the tanker companies were asked about their preferred coating type, and it was concluded that stainless steel tank coating was used predominantly (Solmaz et al. 2020).

47.2 Methodology

Analytical Hierarchy Process (AHP), according to Saaty's definition, is a multi-criteria decision-making mechanism. In this context, the method exhibits an eigen-value approach in pairwise comparisons. It also provides a method for measuring

Table 47.1 Scale table

Significance	Scale definition	Comparison
1	Equally important	Two options are equally important
3	Moderately important	Experience and judgment slightly favor one criterion over another
5	Strongly important	Experience and judgment strongly favor one criterion over another
7	Very strongly important	One criterion is deemed superior to another
9	Absolute importance	Evidence showing that one criterion is superior to another is of the greatest credibility
2,4,6,8	Intermediate values	The value between two consecutive judgments to be used when compromise is required

Table 47.2 Implementation steps of TOPSIS technique

Step number	Explanation
1	In order to evaluate the scales, it is necessary to determine an N decision matrix.
2	To determine the weighted out-of-scale matrix, the V matrix is determined. In this step, the v matrix that we created without scale is multiplied by the diagonal weight matrix (Wn * n).
3	In this step, the positive ideal solution and the negative ideal solution are determined. As the positive ideal solution, the vector of the best values of each feature of the matrix V is taken (Vj+), and as the negative ideal solution, the vector of the worst values of each feature of the matrix V is taken (Vj-).
4	In this step, the distance of each option from the positive and negative ideals is determined.
5	In this step, the relative closeness of an option to the ideal solution is determined.
6	In this step, the grade values of the options are determined.

quantitative and qualitative performances. This scale covers all values of the comparisons, ranging from 1/9 for least valuable, 1 for equal, and 9 for most important, shown in Table 47.1 (Saaty 1994).

The TOPSIS technique can identify solutions after a limited set of alternatives has been established. According to Hwang and Young, the logic of fuzzy TOPSIS is to define the positive ideal solution and the negative ideal solution (Hwang et al. 1993). The positive ideal solution is the one that maximizes the benefit measures and minimizes the cost measures, while the negative ideal solution is the one that maximizes the cost measures and minimizes the benefit measures (Chen 2000). The best alternative is the one that is the shortest from the positive ideal solution and the farthest from the negative ideal solution. However, it is often difficult for a decision maker to assign a precise performance rating to an alternative for those qualifications (Hwang and Yoon 1981). The benefit of using a fuzzy TOPSIS approach, then, is to assign different criteria values using fuzzy numbers. The implementation steps of TOPSIS method are shown in Table 47.2.

In this study, experts on tank coatings in chemical tankers and the criteria determined as a result of the literature review were evaluated by the AHP method. As a result of the calculations, the weight found for each criterion was integrated with the TOPSIS method, and the linguistic expression data were digitized with the fuzzy method.

47.3 Results and Discussion

In this study, the measurable criteria that can affect the process of choosing four different tank coatings used today in chemical tankers have been determined, and the degree of impact and priority order of these criteria on the coating preference have been determined with the opinion of field experts. These criteria were determined by consulting the field experts and by conducting a detailed literature review, taking into account the most probable problems that may arise in tank linings and the problems that may occur as a result of these problems. The said criteria are the mechanical properties of the coating, the maintenance and repair actions of the coating, the difficulty of application of the coating, the role of the coating in ventilation and dehumidification, the absorption and expansion properties of the coating, cargo cleaning, and the cost of the coating (Appleman 2000). Among these, the most effective one on the tank lining was selected, and support was received from a total of 13 field experts, consisting of the captain, the chief navigator, the ship operator, and the tank coating specialist.

During the interviews with the experts, they were asked to fill in the questionnaires prepared within the scope of the thesis. Obtained expert opinions were evaluated within the scope of Analytical Hierarchy Process, one of the criteria decision-making methods. Each of the seven criteria determined was compared with each other, and a scaling between 9 and 1 was made. As a result of these evaluations, it has been determined that the most important criterion for tank coatings is cargo cleanliness with a weight of 0.25, with the sum of all criteria weights being 1.00. The cargo cleaning criteria were determined by the mechanical properties of coatings with a weight of 0.24, maintenance and repair actions in the coatings with a weight of 0.23, absorption and exhaustion properties of the coatings with a weight of 0.14, the role of the coating in ventilation and dehumidification with a weight of 0.08, and the coating with a weight of 0.05. Difficulties in application tank coating with a weight of 0.05 and the cost of coating with a weight of 0.02 with the lowest impacts. The obtained values are shown graphically in Fig. 47.1.

After the criterion weights were determined, each of the four coating types used in chemical tankers today, consisting of epoxy, zinc silicate, stainless steel, and APC marine line, was evaluated separately within the framework of these criteria, and the result found in the last step was normalized with the criteria weights. In the selection of the determined coating type alternatives, the TOPSIS method, which is also a multi-criteria decision-making mechanism, was used since the determined criteria were not based on numerical data. Since it is more suitable for the evaluation and

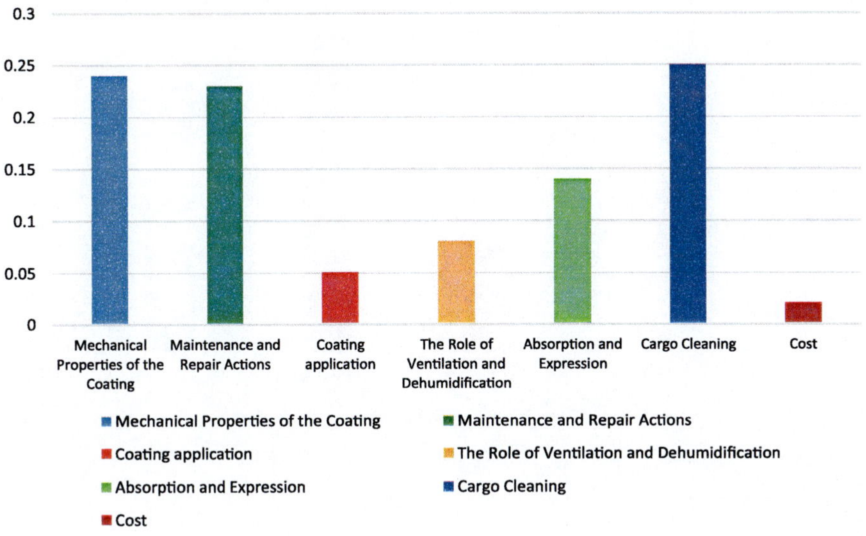

Fig. 47.1 Criterion weights from AHP method

decision-making mechanisms in the minds of the experts, it was preferred to present the evaluations with linguistic expressions rather than numerical comparisons. In other words, fuzzy logic approach was used in order to obtain the results more objectively and accurately.

In the fuzzy method, the highest linguistic expression determined for each criterion, very high (VH), corresponds to 0.9-1.0-1.0 and the lowest score is very low (VL), which corresponds to its value 0.0-0.1-0.1 (Ganga et al. 2011). Within the framework of these numerical data, all criteria for each alternative were evaluated by experts through a survey, and the alternative with the highest weight was stainless steel with 0.423. Following the stainless steel, marine line with a weight of 0.357, epoxy with a weight of 0.247, and zinc silicate with a weight of 0.153 as the lowest were found. The obtained values are shown graphically in Fig. 47.2.

47.4 Conclusion

Chemical tankers have a significant share in the number of ships in the world. This share is growing due to the developing technologies of transportation and their increasing need. The most important factors for chemical tanker businesses are to ensure the safe navigation of the ship and to deliver the cargo on the ship by taking all precautions against any deterioration. In this context, the tank lining used in chemical tankers is very important for the cargo to be transported safely. Tank coatings used today are epoxy, zinc silicate, stainless steel, and marine line coating.

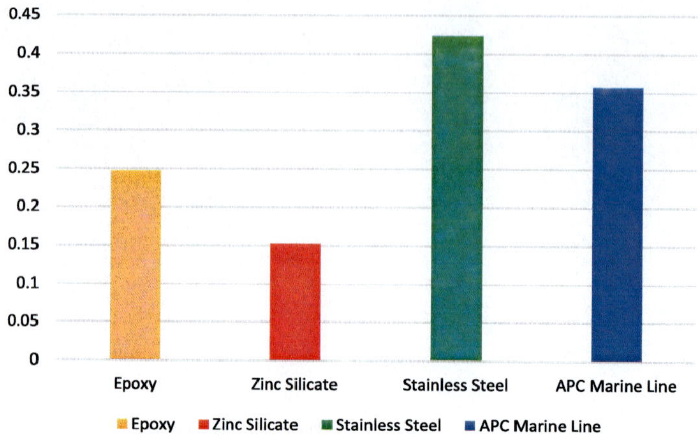

Fig. 47.2 Alternative weights from fuzzy TOPSIS

Each of these coatings has advantages and disadvantages in terms of their own characteristics. When a tank coating is to be made on chemical tankers, many criteria should be evaluated together by the decision makers of this coating (Tekeli 2022).

In this study, the order of importance of tank coatings in chemical tankers, within the framework of the criteria determined as a result of experts and research, and the problem of choosing the most suitable one for ship enterprises among the tank coating types used today are included. While seeking solutions to these problems, AHP and TOPSIS methods, which are multi-criteria decision-making methods, were used. While determining the order of importance of the criteria, the questionnaires filled in by the experts were formulated with the content of the AHP method and a ranking was made by weight ratios. According to this result, the most effective criterion determined for tank linings is the effect of the coating on cargo cleanliness. Among the tank coatings used today, the TOPSIS method was used when choosing the most effective one in the context of the determined criteria. In this method, linguistic expressions were taken as a result of interviews with experts, and fuzzy logic was used to transform these linguistic expressions into mathematical expressions. According to the fuzzy TOPSIS method applied on the selection of tank coatings, the most suitable tank coating type is the stainless steel within the framework of the determined criteria. The coating type with the least value in terms of suitability to use was determined as zinc silicate.

Acknowledgment This conference paper is derived from one of the authors' MSc thesis.

References

Appleman BR (2000). Ventilation and dehumidification of ship ballast tanks for blasting and coating work. Journal of Protective Coatings & Linings 17(4):43–52.

Chen CT (2000). Extensions of the TOPSIS for group decision-making under fuzzy environment. Fuzzy sets and systems 114(1):1–9. https://doi.org/10.1016/S0165-0114(97)00377-1

Ganga GMD, Carpinetti LCR, Politano PR (2011). A fuzzy logic approach to supply chain performance management. Gestão & produção 18(4):755–774. https://doi.org/10.1016/j.ijpe.2011.06.011

Hwang CL, Yoon K (1981). Methods for multiple attribute decision making. In Multiple attribute decision making (pp. 58–191): Springer.

Hwang CL, Lai YJ, Liu TY (1993). A new approach for multiple objective decision making. Computers & operations research 20(8):889–899. https://doi.org/10.1016/0305-0548(93)90109-V

Saaty TL (1994). How to make a decision: the analytic hierarchy process. interfaces 24(6):19–43. https://doi.org/10.1016/0377-2217(94)90222-4

Solmaz MS, Eyupoglu A, Karabulut N (2020). Decısıon Makıng in Cargo Tank Coatıngs for Chemıcal Tanker Companıes. 18th Annual General Assembly of the International Association of Maritime Universities, January 2020, Varna, Bulgaria.

Tekeli MM, 2022, Numerical Analysis on Selection of Tank Coatings in Chemical Tankers. MSc Thesis. Istanbul Technical University.

Zhang W, Song Z, Song J, Shi Y, Qu J, Qin J, Zhang R (2013). A systematic laboratory study on an anticorrosive cool coating of oil storage tanks for evaporation loss control and energy conservation. Energy 58:617–627. https://doi.org/10.1016/j.energy.2013.06.044

Chapter 48
A System Dynamics Analysis of Impact of Feed in Tariff Policy on Renewable Energies in Zambia

C. K. Bowa, M. Mwanza, M. Sumbwanyambe, J. H. Pretorius, and K. Ulgen

Nomenclature

REFiT	Renewable energy feed-in tariff
FiT	Feed-in tariff
ERB	Energy Regulation Board
RE	Renewable energy
PV	Photovoltaic
ROI	Return on investment
SADC	Southern African Development Community
MWh	Megawatt-hour
MW	Megawatt
CO_2	Carbon dioxide

C. K. Bowa (✉)
Solar Energy Institute, Ege University, Izmir, Turkey

M. Mwanza
School of Mechanical Engineering, Copperbelt University, Kitwe, Zambia
e-mail: mabvuto.mwanza@unza.zm

M. Sumbwanyambe
Department of Electrical and Mining Engineering, University of South Africa, Pretoria, South Africa

J. H. Pretorius
School of Engineering, University of Zambia, Lusaka, Zambia

K. Ulgen
Faculty of Engineering and Built Environment, University of Johannesburg, Johannesburg, South Africa
e-mail: koray.ulgen@ege.edu.tr

M. Z. Sogut et al. (eds.), *Proceedings of the 2022 International Symposium on Energy Management and Sustainability*, Springer Proceedings in Energy,
https://doi.org/10.1007/978-3-031-30171-1_48

VAT	Value-added tax
ZRA	Zambia Revenue Authority
CEC	Copperbelt Energy Corporation

48.1 Introduction

As the economy of the country grows, the energy demand of the country also grows, resulting in a need for increased energy supply systems. In Zambia the only method currently being used is the increase in hydropower station or an upliftment of the old hydropower stations. The options regardless of how good that might be have their pitfalls. Generally, the conventional hydropower generation takes about 15–30 years lead time to construct which in certain instance may not correspond with the rapid demands in electrical energy. Thus, to rapidly address the energy challenges, the deployment of alterative renewable energies provides the best option. To do this, there is a need for favourable renewable energy policies to be in place. In Zambia the plan is in achieving these options is sustainable policy implementation so as to speed up the increase of renewable energies in the country. To that effect several policy plans have been brought unto the fore. The recent one being that of renewable energy feed-in tariffs (REFiTs) (Bowa et al. 2017). Within the ambit of this policy the government envisaged that the renewable energy development and deployment will only be achieved by adopting REFiTs and, by association, continued zero taxes on all the imported renewable energy products to the country. This will attract the public as well as private companies to invest in and use renewable energy systems and encourage more participation of independent power producers in the sector. REFiT is defined as a mechanism to promote the deployment of renewable energies that places an obligation on specific entities (such as ZESCO) to purchase the output from qualifying renewable energy generators at predetermined prices. The pricing varies depending on the country penetration level of renewable energy and the ability of the government to subsidize the whole process of deployment to implementation (IRENA, OECD/IEA and REN21 2018; Mwanza and Ulgen 2020).

48.1.1 Renewable Energy Perspectives

The REFiT pricing factors from the perspective of project developers and the utilities are underwritten by the cost- and value-based approaches (Mwila 2015). These are the two author citations in the text. These are more than two author citations in the text. This is only one author citation in the text. These are two separate citations in the text. These are the same two author citations in the text.

Despite a standardized process and documents for development of projects eligible for REFiT programmes, they are sometimes differentiated by the technology and the project size. Several projects in Zambia, we must make mention, and within the context of renewable energy, are eligible for REFiT. The projects eligible for REFiT include solar PV, hydropower and biomass with the project limit size of about 20 MW for individual projects. This paper provides a detailed analysis of a system dynamics (SD) model based on subsidy and REFiT policy to simulate the impacts of renewable energy deployment in Zambia. This model is country specific and is based on a complexity of the energy sector, power generation dependency as well as set targets of the country. As a background, SD has over the years proved to be a good method to analyse a long-term change of variables. This makes it suitable for energy policy investigation (Mwanza and Ulgen 2020).

Literature according to Ford (1999) points out that SD is a valuable tool for policy analysis in both government and power companies. Similarly, Hsu (2012) assessed the effect of capital subsidies and FiT on solar PV installation using SD model to simulate the economic benefit of policy combinations (Hsu 2012). Aslani et al. (2014) developed an SD model to simulate renewable energy generation based on different scenario policies in Finland (Aslani et al. 2014). Sheikhifini et al. (2014) applied SD method to model the impact of policies on the expansion of wind resources, combined heat and power and photovoltaic resources.

Hou (2015) designed a GPU-based system and developed a combinational fast algorithm, which was used to enhance the simulation speed and improve the simulation efficiency of complex system (Hou 2015). On the one hand, scholars have proposed that it is necessary for REFiT to be high enough so that the recovery of cost can be achieved. For instance, Dusonchet and Telaretti (2010) state that the REFiT must be high enough to recover the investment cost within a reasonable timeframe (Dusonchet and Telaretti 2010). On the other hand, Rühter and Zilles (2011) argue, to the contrary, that REFiT should be small enough to avoid enforcing a big financial burden on the states (Rühter and Zilles 2011).

Shahmohammadi et al. (2014) analysed the long-term FiT policy effect on renewable energy funding in Malaysia. The model showed by simulation that funding a stock for renewable energy sources shrinks exponentially. In the same model, the authors conducted a sensitivity analysis for the recommended reduction in FiT rate or increasing surcharges on electricity bills. The authors considered the electricity demand as endogenous and investors' trust effects on "willingness of investment" were not mentioned (Shahmohammadi et al. 2015).

Hejazi et al. (2016) developed a model as a virtual laboratory in which policy makers could assess the effects of different policies to assist them in implementing more efficient budget systems towards sustainability in electricity generation. In the model, the author concluded that the policy makers could carry out analyses to forecast the future conditions of renewable energies in Iran under different circumstances created by different periods. All these studies analysed the influence of policies on energy development and provided suggestions for policy making (Hejazi et al. 2016).

48.2 Methodology

48.2.1 Model Validation

Model validation is vital for the precision of the model and credibility to depict the real scenario of the existing polices in the country. According to Barlas, the model credibility has two aspects: structural and behavioural, which involve the examination of a meaningful description if the real relations of the stock flows as well as the dynamic patterns generated by the model to be close enough to the real dynamic patterns of concern. The CLDs were examined using available knowledge from reviews and previous studies by the authors with respect to findings on renewable energy in Zambia (Mwanza and Ulgen 2020; Bowa et al. 2017).

48.2.2 Model Structure and Development

This paper analyses the effectiveness and efficiency of the RE policies in the promotion of solar PVC. A solar PV technology customized model was used in analysing the impact of REFiT and subsidy policy on the renewable energy deployment in Zambia. The historical data used in this paper is limited to 2 years. That is the period the REFiT programme was in existence in Zambia. A system dynamic model was applied to simulate the impact of capital subsidy and REFiT policies on the solar PV deployment and their implementation in Zambia. This was in line with the fact that both initial REFiT prices and capital subsidies offered by government generally had a direct influence on public tendency to invest in renewable energy systems.

Hence, to better illustrate the impact of these policies on solar PV implementation and government policy promotion cost, the relationship between solar PV accumulation and cost of policy promotion by government was analysed for the period 2018–2035. The simulations have been carried out in several scenario settings using capital subsidy proportion and REFiT price as the main essential variables for solar PV system implementation.

48.2.3 Scenario Setting

Generally, capital subsidies and the initial REFiT price offered by governments as energy off-taker to independent power producers (IPPs) usually have direct impact to public tendency to invest in solar PV systems. Within such locality there is a need to analyse the impact of both initial REFiT and capital subsidies including the combination of initial REFiT and capital subsidies on solar implementation in Zambia. A total of 49 scenarios were analysed: 9 on initial REFiT price, 8 on capital subsidies and 32 on the combination of the initial REFiT and capital subsidies.

Table 48.1 Parameters for analysing solar PV implementation in Zambia 2019–2035

Parameter	Value	Sources
Installed solar PV systems	1000 kW (1.0 MWp)	CEC
Initial solar PV system	1000 kW (1.0 MWp)	CEC
Average cost of solar PV system in Zambia	US$ 1550/kWp	Calculated
Solar PV system capacity factor	5.54 kWh/d/kWp	Meteorological Department/ PVGIS
Electricity emission factor	0.619 kgCO_2/kWh	Emissionfactor.com
Electricity price	US$ 100/MWh	Zesco
Feed-in tariff	US$ 70/MWh	ERB
Subsidy: 16% VAT, 25% CD and 20% AIT/SG	61%	ZRA

In the analysis, the initial FiT price was reduced annually to avoid a long-term impact on government expenditure, while the initial capital subsidies were kept constant during the analysis. Essentially, the annual FiT price reduction depends on technology progress in terms of capacity factor, cost of PV installation, annual return on investment and the fixed upper limit of ROI by the government.

48.2.4 *Effect of Initial FiT on Solar PV Implementation*

Factors such as the proposed REFiT, current electricity price (CEP) and REFiT price adopted by other African countries are adopted. Each parameter in analysing solar PV implementation in Zambia as shown in Table 48.1 was assigned an initial parameter in the simulation. In the scenarios of FiT70 to FiT200, 9 different initial REFiT price values in the range of 70–200 $/MWh were adopted to check the impact of different initial REFiT price on achieving the solar PV accumulation set target of 600 MW between the period 2018 and 2035 with fixed upper limit of ROI (7.5%) and capital subsidy (61%).

Explicitly, the investor has to pay 0% (16% VAT, 25% duty clearance and 20% AIT/SG) of the solar PV system initial investment capital cost to the government. The government of Zambia's proposed REFiT price of 70 $/MWh was used as the lowest value, while the current electricity price of about 100 $/MWh was used as the average value and the REFiT price of 200 $/MWh, based on other African countries, was used as the highest value. Figure 48.1 shows the CLD for implementation of solar PV in Zambia.

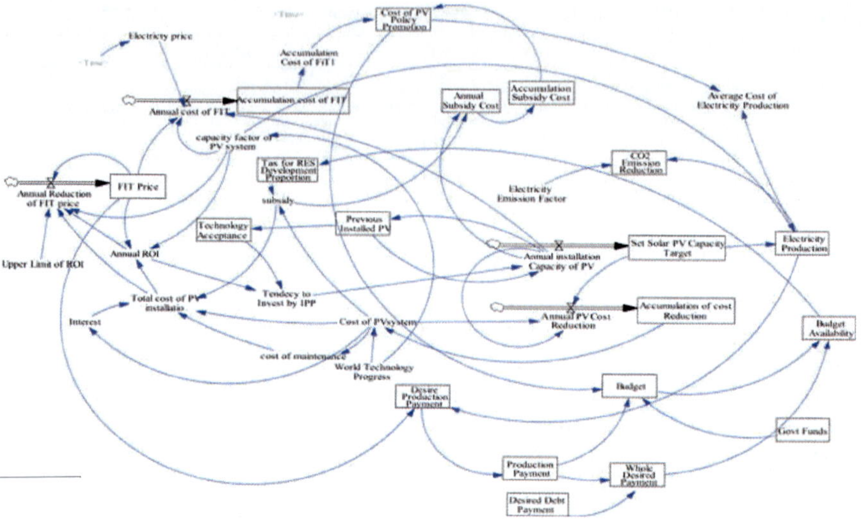

Fig. 48.1 System dynamic model for simulating cost of solar PV policy promotion and solar PV implementation

48.3 Results and Discussion

In this section we present some of the result that was obtained using the model that we developed. Figures 48.3 and 48.4 show the simulation results for fixed capital subsidy at 61% and variable initial REFiT. As seen in Fig. 48.2, the highest accumulative solar PV capacity is achieved at higher initial REFiT as compared to lower REFiT as expected from the previous research in other countries. At proposed REFiT price of 70 \$/MWh, the accumulative solar PV capacity in 2035 grows from current 1 MW to 339 MW, whereas, in a case where the government adopts the initial REFiT price as other African countries (OAC), such as Uganda and Kenya, i.e. at a value of 200 \$/MWh, the accumulative solar PV capacity grows from 1 MW current installed capacity to 484 MW. However, with REFiT between the proposed and current electricity price, it is observed that almost the same solar PV accumulative capacity could be achieved, with the difference only in the initial years of policy implementation. The result as presented in Fig. 48.2 shows that with an increase in initial REFiT by 10 \$/MWh, the increase in the annual average accumulative solar PV capacity of 16,15 MW in the initial year of policy implementation can be achieved which reduces every year due to REFiT reduction in 2035.

In general, as seen in Fig. 48.2, increasing the initial REFiT price at fixed capital subsidy results in an increased government budget expenditure on policy promotion. As observed in Figs. 48.3 and 48.4, a higher initial REFiT on solar energy system implementation leads to increased investments in solar energy systems and higher solar energy systems implementation in the first years of policy implementation.

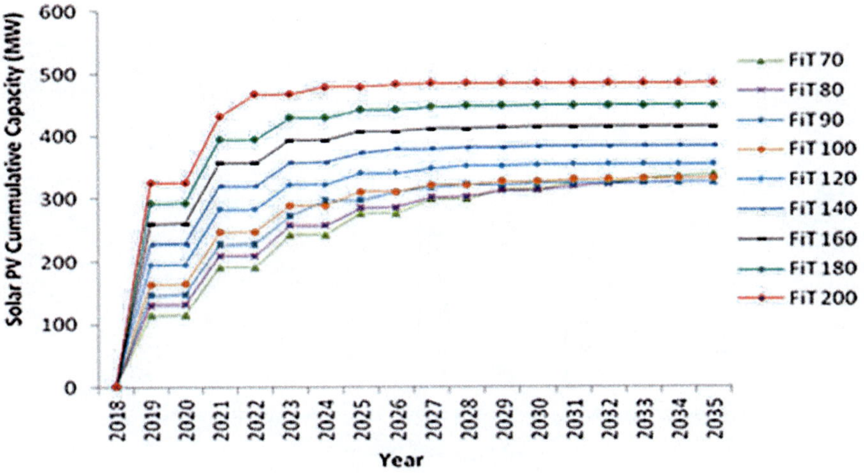

Fig. 48.2 Solar PV cumulative capacity at different FiT prices

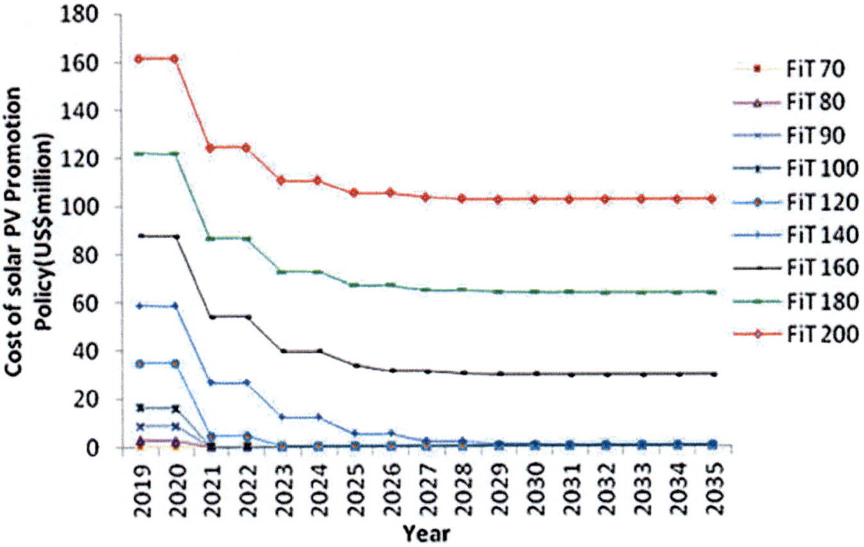

Fig. 48.3 Cost of solar PV Policy promotion at different FiT prices

However, it can be seen that due to continuous reduction in the REFiT annually with increased accumulation of solar PV capacity, the public investment in solar PV eventually reduces, decreasing government expenditure on policy promotion. In this study, considering the initial REFiT price in the range of 70–100 $/MWh, the highest accumulative solar PV capacity is given by the lowest initial REFiT price despite giving a lower initial accumulative capacity in the first year of policy

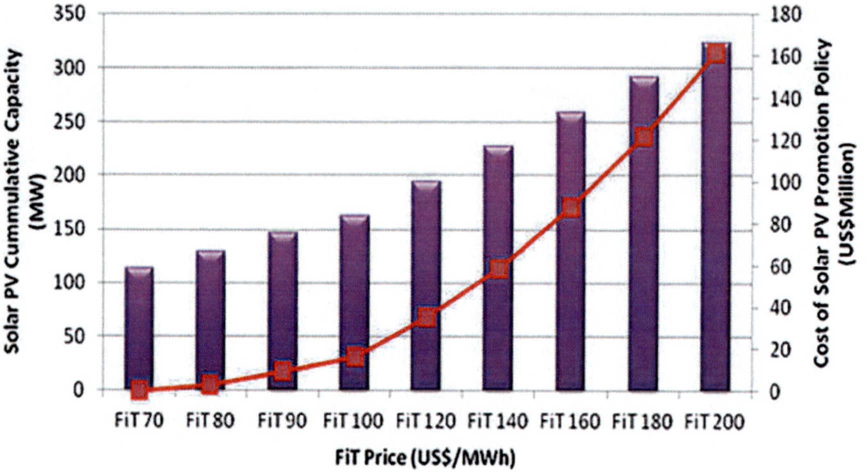

Fig. 48.4 Solar PV cumulative capacity and policy promotion cost in the first year of policy implementation

implementation. Considering the cost of policy promotion on the government and long-term accumulation of solar PV capacity, the REA proposed a REFiT of 70 $/MWh stands as the best alternative for a long-term solar energy system deployment. Further analysis suggests that an adjustment of the proposed REFiT price would attract more investments in the solar energy sector which would be essential to Zambia and by extension to the SADC region.

For instance adjusting the REFiT price to 90 $/MWh could lead to deployment of more than 200 MW before 2022 with less cost of policy promotion since the REFiT price would still be below current electricity price of 100 $/MWh. This would benefit the country's economy by reducing dependency on imported expensive energy (120 $/MWh) from other countries.

48.3.1 Effect of Subsidy on Solar PV Implementation

In these scenarios, eight subsidy proportions were used based on the current government charges (16% VAT, 20% AIT/SG and 25% duty clearances) on the initial investment capital cost to all imported equipment for power plants considered in the simulation. These subsidy proportions are taken into account based on the fact that all the solar PV system equipment are imported from foreign countries. Eight different capital subsidy proportion values in the range of 0–61% were applied in the simulation in analysing the impact of different capital subsidy on achieving the government of Zambia solar PV accumulation set target of 600 MW between the period 2018 and 2035. The initial REFiT price and upper limit of ROI were kept constant at the proposed REFiT of 70 $/MWh and 7,5%, respectively.

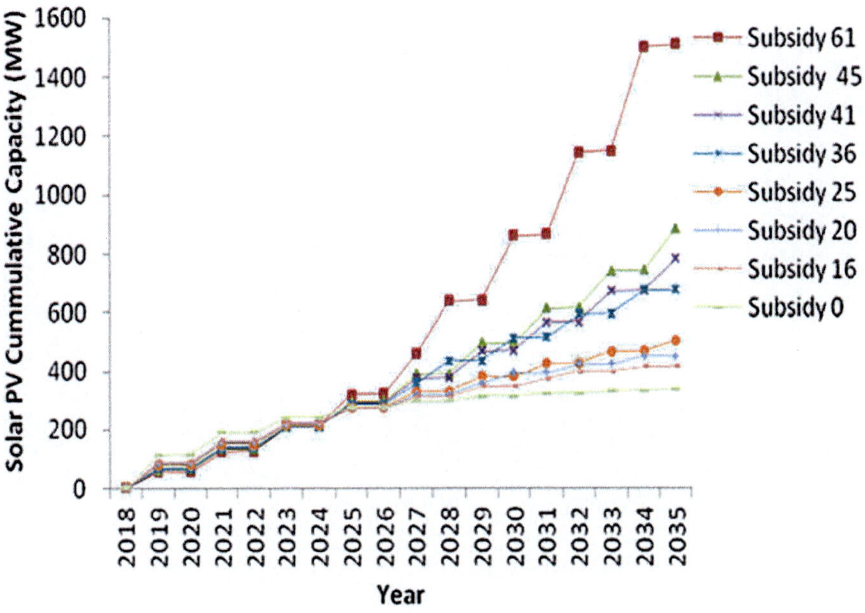

Fig. 48.5 Solar PV cumulative capacity at different subsidies

Figure 48.5 shows the simulation results; it is observed that an increase in subsidy proportion results in an increase in accumulative solar PV capacity. The solar PV accumulative capacity trend shows that a linear correlation does not exist. However, the trend is different to that seen for the different initial REFiT prices. It is observed that when the capital subsidy is 61%, the predicted accumulative solar PV capacity grows from 1 to 1508 MW similar to the findings by Mwanza and Ulgen (2020), whereas, if the subsidy proportion is reduced to 36%, the accumulative solar PV capacity in 2035 decreases to 503 MW.

Also, with zero subsidy proportion, i.e. decreasing capital subsidy from 36% to 0%, the accumulative solar PV capacity decreases to 337 MW. The results show that for every 1% increase in the capital subsidy proportion for the period of 16 years, approximately an increase of 19.20 MW in solar PV system could be achieved. Analysis of the accumulative solar PV capacity in 2035 based on the initial REFiT and capital subsidy shows that capital subsidies have greater influence in the long term as compared to the initial REFiT price. Hence, the results indicate that increasing subsidy, i.e. zero cost on imported solar PV system equipment, has a greater influence on the tendency of the public to invest in solar PV project in the long term.

In terms of the cost of solar PV policy promotion at various capital subsidies in 2035, Fig. 48.6 illustrates the values obtained at subsidy range of 0–61%. The results show the cost in the range of 0–0.17 million dollars for promoting solar PV policy at various subsidy proportions. This is similar to the initial REFiT, which also increases

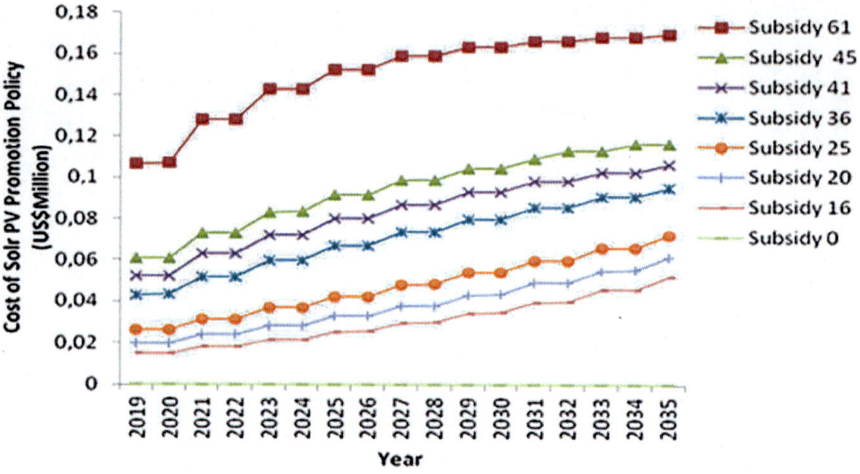

Fig. 48.6 Cost of solar PV policy promotion at different subsidies

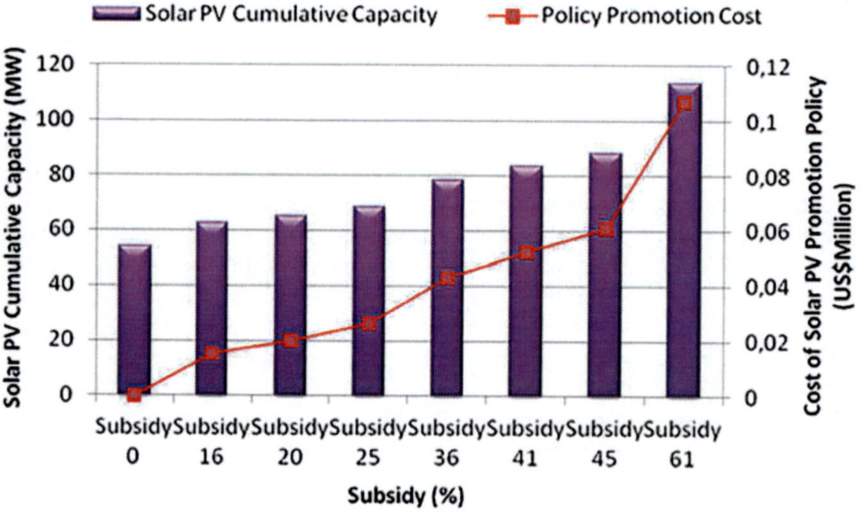

Fig. 48.7 Solar PV cumulative capacity and policy promotion cost in the first year of policy implementation

the government expenditure on the policy promotion. However, as compared to increasing initial REFiT, increasing subsidy proportionately leads to less expenditure on cost policy promotion while achieving high long-term accumulative solar PV capacity. Figure 48.6 shows the cost of policy promotion and Fig. 48.7 shows the accumulative solar PV capacity. These summarized simulation results would be achieved at different initial REFiT prices and capital subsidy proportions.

48.4 Conclusion

This chapter has presented a comprehensive method for analysing renewable energy policy for the promotion of solar PV system implementation in Zambia. It presented the evaluation of the expected solar PV cumulative capacity and cost of policy promotion for the long-term solar investment at different REFiT and subsidy rates:

- The results not only indicate the impact of initial REFiT and capital subsidy on accumulative solar PV capacity but also show the potential cost of solar PV policy promotion, accumulative CO_2 emission reduction and average cost of CO_2 emission reduction.
- The results also show that increasing either REFiT or capital subsidy is a better approach to attracting investment in renewable energy sector.

It can be concluded that increasing both REFiT and subsidies has greater influence in increasing public participation willingness in the RET implementation but at government expense. Hence, a balance of both is a must. In the context of Zambia, a higher subsidy, such as complete import duty exemptions and tax holiday for the first few years of project operation, is necessary. This must be coupled with the REFiT price close to current electricity price to avoid huge policy promotion by the government. In this chapter it was noted that increasing REFiT results in initial sharp and rapid investment in solar PV accompanied by higher cost of policy promotion. However, for a gradual long-term accumulation of solar PV capacity, increasing capital subsidies is the best option. The combination of REFiT and capital subsidies analysis indicates that the lower cost of policy promotion coupled with gradual and higher accumulation of solar PV capacity can be achieved at lower REFiT and higher capital subsidy. But to rapidly address the current electricity deficits in Zambia, increasing the initial REFiT to about 90 $/MWh, at zero capital subsidies, can help in quickly attaining the set target (600 MW). In summary, this study provides a method for policy implementation and an analysis on the impact of policy options while balancing competing government policy budget. The results will directly help the government and electric utility to recognize the policy combination potential in achieving government set targets and in the setting of REFiT for solar PV systems

Acknowledgement The authors gratefully acknowledge the support provided by the Turkish Government, University of Zambia and Ege University.

References

Aslani, A., Helo, P., Naaranoja, M., (2014), Role of Renewable Energy Policies in Energy Depencdency in Finland: System Dynamics Approach, Applied Energy 113: 758–765

Bowa, CK., Mwanza, M., Sumbwanyambe M., (2017), Solar Photovoltaic Energy Progress in Zambia: A Review, SAUPEC, South Africa.

Dusonchet, L., Telaretti, E., (2010), Economic Analysis of Different Supporting Policies for the Production of Electrical Energy by Solar Photovoltaics in Western Union Countries, Energy Policy 38(7): 3297–3308.

Ford, A., (1999), Modelling the Environment, Island Press.[An introductory textbook on System Dynamics, as well as environmental and energy applications, 2nd Edition, Island Press, ISBN-13: 978-1597264730

Hejazi MM., Shakouri, HG:, Sedaghat, S., Mashayekhi, AN., (2016), Evaluation of feed in tariff policy effects on sustainable development of renewable energies in Iran: A System Dynamics Approach, Proceedings of the 34th International Conference of the System Dynamics Society.

Hou, L., (2015), System Dynamic Simulation of Large-Scale Generation System for Designing Wind Power Policy in China, Discrete Dynamics in Nature and Society, 1–11.

Hsu, CW., (2012), Using a system dynamics model to assess the effects of capital subsidies and feed in tariffs on solar PV installations, Applied Energy, 100: 981–990

IRENA, (2018), Renewable Energy Policies in a Time of Transition, International Renewable Energy Agency (IRENA).

Mwanza M., Ulgen, K., (2020), Sustainable electricity generation fuel mix analysis using an integration of multicriteria decision-making and system dynamic approach, International Journal of Energy Research, 44(12): 9560–9585.

Mwila, AM., (2015), Renewable Energy Feed in Tariff Stakeholder Workshop Objectives, Lusaka Energy Regulation Board (ERB).

Rühter R., Zilles, R., (2011), Making the case for off grid – connected photovoltaic in Brazil, Energy Policy 39(3): 1027–1030

Shahmohammadi, MS., Yusuff, RM. Keyhanian, S., Shakouri, HG. (2015), A Decision Support System for Evaluating Effects of Feed in Tariffs Mechanism: Dynamic Modelling of Malaysia's Electricity Generation Applied Energy 146: 217–229.

Chapter 49
Energy-Efficient Yacht Design: An Investigation on the Environmental Impacts of Engine Selection for Bodrum Gulets

Mehmet Akman and Bülent İbrahim Turan

Nomenclature

B	Breadth, m
E	Annual amount of CO_2 emission (ton/year)
FC	Fuel consumption (ton/year)
k	Form factor
L	Length, m
LNG	Liquefied natural gas
R	Resistance, kN
SFOC	Specific fuel oil consumption
T	Draught, m
V	Speed, kt
C_P	Prismatic coefficient

Greek Letters

ρ	Density, kg/m^3
g	Gravitational acceleration, m/s^2

M. Akman (✉) · B. İ. Turan
Department of Motor Vehicles and Transportation Technologies, Muğla Sıtkı Koçman University, Muğla, Turkey
e-mail: mehmetakman@mu.edu.tr; bulentturan@mu.edu.tr

M. Z. Sogut et al. (eds.), *Proceedings of the 2022 International Symposium on Energy Management and Sustainability*, Springer Proceedings in Energy,
https://doi.org/10.1007/978-3-031-30171-1_49

∇ Displacement volume, m^3
γ Emission factor (kg/kg)
η Efficiency

Subscripts

b	Brake
F	Frictional
e	Effective
OA	Overall
T	Total
V	Viscous
W	Wave

49.1 Introduction

Ship design is a multidisciplinary and complex process including a wide range of analysis steps. The design process is completed in three phases – concept, contract and detail design where the ship characteristics, such as stability, manoeuvrability, strength and motions, are determined (Turan 2021). The design process of a ship is defined by design spirals that consist of a set of iterative processes (Hamlin 1996; Larsson and Eliasson 2007). Determining the performance in terms of power requirements is one of the primary keystones of these processes. The speed, weight and power are three elements that form a triangle, which has to be balanced with the design process of a ship (Thomas 2015). In the speed-power calculations after the determination of the main dimensions and form of the hull, it is aimed to select the appropriate machine by using the weight and usage conditions of the boat (Turan 2021).

The powering of a design, on the other hand, is directly related to energy efficiency and emissions which have been the major concerns of the International Maritime Organization (IMO), recently. The IMO put strict measures into effect with the International Convention for the Prevention of Pollution from Ships (MARPOL Annex VI) to minimize the shipping greenhouse gas emissions (GHGs) and aims to reduce the emissions by 50% as of 2050 by baselining the emission level of 2008 (IMO 2021). However, the regulations are applied to all merchant ships of 400 GT and above, and there is no obligation for below 400 GT group where 14000 ships were reported to be included (Wu and Bucknall 2016).

The literature studies and applications on machinery systems mainly focus on alternative fuels and prime movers (Inal et al. 2021), carbon capture and storage (Güler and Ergin 2021), waste heat recovery (Akman and Ergin 2019) and emission control systems. Moreover, hull form optimization, trim and ballast optimization, slow steaming and planned and timely maintenance are the available design and operational measures for decarbonization and energy-efficient shipping, respectively (Akman and Ergin 2021). In this study, the main and auxiliary engines of in-service 20 Bodrum Gulets with different lengths are analysed and the hull forms of these yachts are modelled for resistance calculations at service and design speeds using the Holtrop-Mennen method (Holtrop and Mennen 1982). The design power requirements (MAN 2013) are determined and the engine load factors are calculated to compare the actual data for the environmental impact evaluation.

49.2 Bodrum Gulets

Bodrum Gulet, a type of sailing boat, is among the boats specific to the Bodrum region in Turkey (Gammon et al. 2005; Turan et al. 2021). The origin of the term "gulet" refers to a specific sail and rigging configuration (Köyağasıoğlu 2014). Although the word gulet defines a sail and rigging type rather than a hull type, it is now mostly used to describe a special hull type (Turan and Akman 2021; Turan et al. 2021). Gulets are the ancestors of Bodrum Gulets and have been used for different purposes such as fishing, sponge-fishing and transportation of commercial goods and in the military (Turan and Akman 2021). With the use of gulets for marine tourism on the South Aegean coast of Turkey, the hull form of these boats has changed and the types of boats known today as Bodrum Gulets have emerged (Büyükkeçeci and Turan 2018). Round stern form and violin-shaped stempost are among the characteristic features of Bodrum Gulets (Köyağasıoğlu 2014). Figure 49.1 illustrates a profile drawing of a Bodrum Gulet with two masts. Generally, the range of L_{OA} of the Bodrum Gulets is between 16 and 50 m (Turan et al. 2021). Even though these yachts are equipped with qualified sails that enable a sailing cruise, Bodrum Gulets mainly use diesel inboard engines for cruising.

49.3 Methodology

The analyses were conducted in six steps as shown in Fig. 49.2. The target group was determined considering the typical lengths (15–30 m) (Turan et al. 2021) of Bodrum Gulets designed and built in the Bodrum region. Then the data related to the hulls and machinery systems of active 20 Bodrum Gulets which are in the range of 16–28.1 m in length overall are collected. According to the collected data, the gross tonnage and displacement tonnage of the yachts range between 40–152 GT and 18.83–76.58 t, respectively. The L_{OA}/displacement tonnage and GT of the analysed gulets are shown in Fig. 49.3.

Hull forms of Bodrum Gulets were modelled by using Rhinoceros software and the geometric data were obtained. The resistance and power calculations were conducted using Maxsurf software. The Holtrop-Mennen method, which is derived from the regression analysis of a set of various scale model tests and sea trial data and

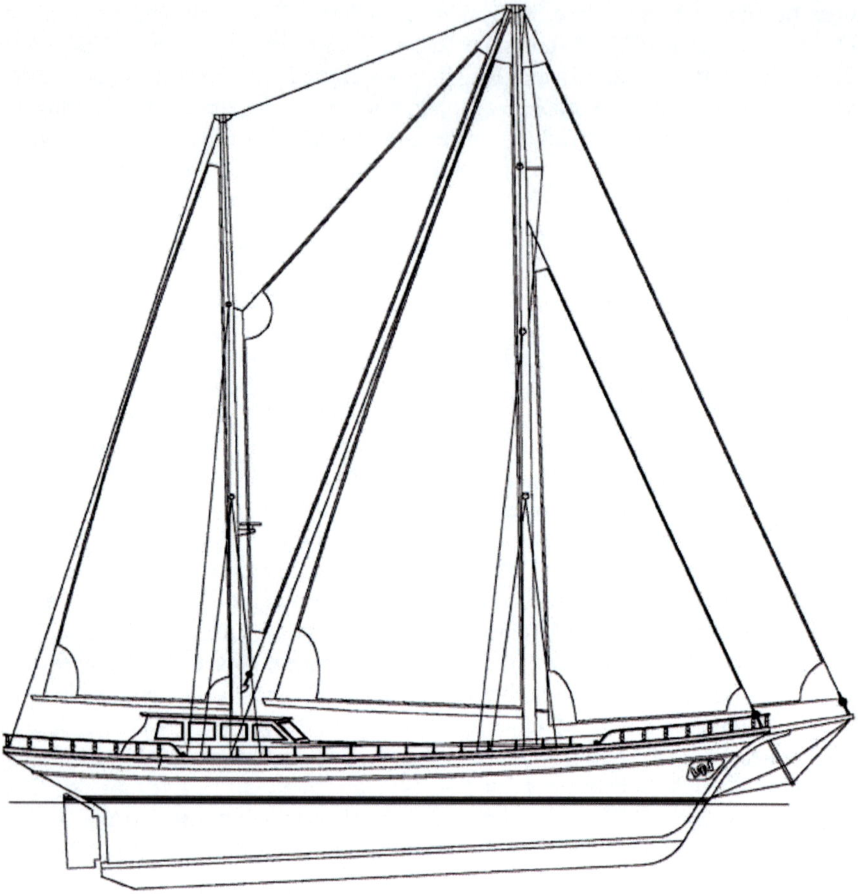

Fig. 49.1 Profile of a Bodrum Gulet with two masts

is a convenient tool for displacement-type hulls (Turan 2009; Birk 2019), was used for resistance estimations of Bodrum Gulets at 8 kts, 10 kts and 12 kts.

In the mathematical model (Holtrop and Mennen 1982; Elkafas et al. 2019) of the Holtrop-Mennen method, the total resistance can be calculated as

$$R_T = R_V + R_W \tag{49.1}$$

where R_V is the viscous resistance and R_W is the wave-making resistance of the hull. The total resistance consists of friction and residual parts and the viscous part of the total resistance is calculated as follows:

$$R_V = (1 + k)R_F \tag{49.2}$$

where R_F is the frictional resistance formulated in ITTC-1957.

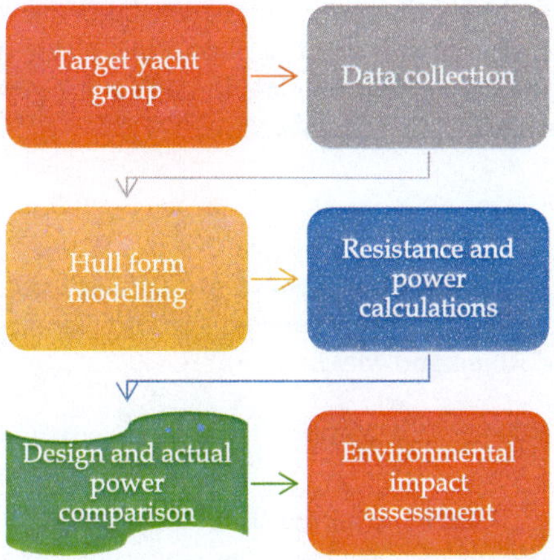

Fig. 49.2 Flow chart of the analysis and evaluation steps

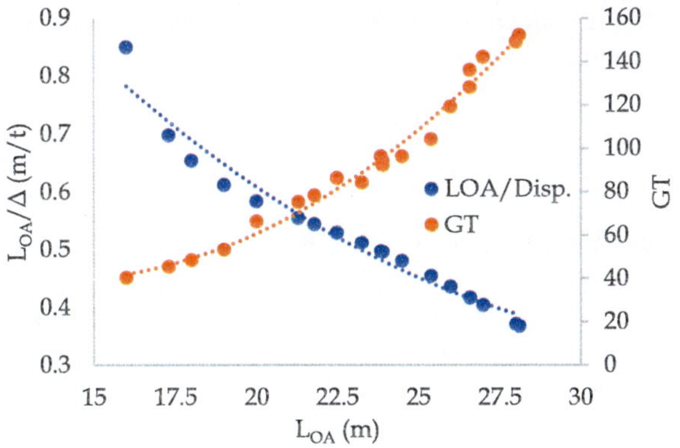

Fig. 49.3 Length and the gross tonnage of the target yachts

The form factor k is a function of the main characteristics and can be written as follows (Elkafas et al. 2019):

$$k = f\left(\frac{B}{L}, \frac{T}{L}, \frac{L}{L_R}, \frac{L^3}{\nabla}, C_P, c\right) \tag{49.3}$$

where L_R is the after-body length and c is the coefficient based on the after-body shape. Then the wave resistance is calculated as

$$R_W = c_1 c_2 c_3 \nabla \rho g e^{\left(m_1 F_n^d + m_2 \cos\left(\lambda F_n^{-2}\right)\right)} \tag{49.4}$$

where c_1, c_2, c_3, m_1, m_2 and λ are the coefficients related to the form of the hull (Holtrop and Mennen 1982). F_n is the Froude number depending on the velocity and length of the yacht. After calculating the total resistance, the effective power of the yachts at different speeds can be calculated as given in Eq. 49.5. The effective power is the starting power for determining the brake engine power, which depends on the propeller and shaft performances along with the engine and sea margins (Tupper 2004). In Bodrum Gulets, 10 kt is the general service speed and 12 kt is the design speed (Turan and Akman 2021).

$$P_e = R_T V \tag{49.5}$$

Then the brake power is calculated using the total efficiency (η_T), which consists of the hull, propeller and shaft efficiencies (MAN 2013) and is assumed to be 0.55 (Blount and McGrath 2009).

$$P_b = P_e / \eta_T \tag{49.6}$$

After the determination of the service and design brake powers, the environmental analysis is conducted by calculating the amount of excess power and the corresponding excess fuel consumption. During analysis, the excess amount of CO_2 is calculated by using the emission factor (γ) as 3.206 kg-CO_2/kg-fuel (Kristensen 2015):

$$E = \left(P_{b,\text{application}} - P_{b,\text{design}}\right).SFOC.\gamma \tag{49.7}$$

where *SFOC* is the specific fuel oil consumption at service and design speeds determined by the engine manufacturers and taken as 220 g/kWh and 224 g/kWh (Kristensen 2015) for high-speed marine diesel engines, respectively. According to the engine collected data, it was observed that the yachts generally have two high-speed diesel engines used as main prime movers and two high-speed diesel engines used as service generators.

49.4 Results

The total resistance of the modelled yachts is plotted at different speeds and overall lengths as shown in Fig. 49.4. According to Fig. 49.4, the increase in the length and speed of the hulls increases the total resistance exponentially. The calculated maximum resistance is 31.5 kN at 12 kt and the calculated effective powers range between 22.3 kW and 281.6 kW at 8 kt–12 kt, respectively.

The brake power ranges of the main and auxiliary engines are 167.7–529.3 kW and 13.4–65.6 kW, respectively. According to the operational engine data, the engine load factors change between 26.8% and 69.7% at service speed and the average load factor is calculated as 40%. The results show that there are incompatibilities between the installed engine and calculated service design powers. Figure 49.5 shows the power differences between the actual data marked with application and the calculated design data with respect to the overall length of the yachts. It is hard to find the exact engine fulfilling the designated powers available in

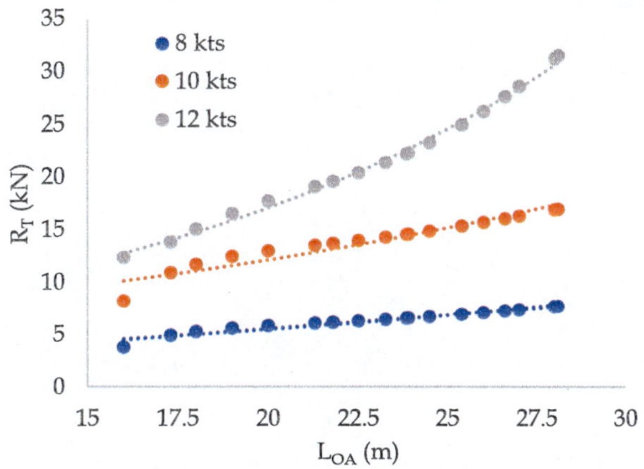

Fig. 49.4 Change of the total resistance with respect to overall length

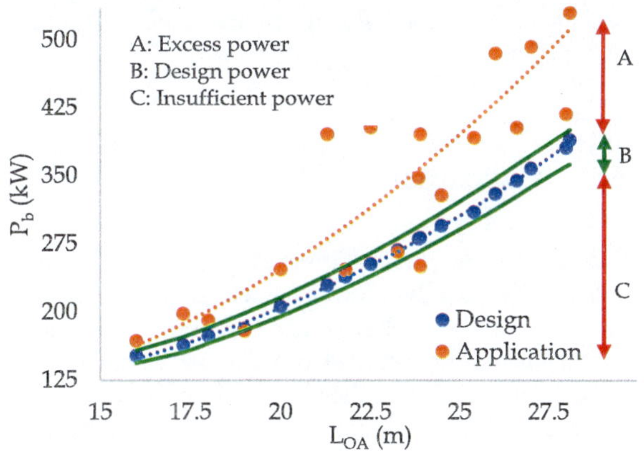

Fig. 49.5 Comparison of brake powers

the market; therefore, 10% market margin is considered and marked with the design power region – B in Fig. 49.5. Despite this flexibility, most of the engines are above the design power curves and have very low load factors at service speed. According to Fig. 49.5, the longest yacht in the target group with 28.1 m has a total of 529.3 kW from main engines and the calculated design power is 389.6 kW at 12 kt. On the other hand, two yachts in the group have main engines below the design power region and have higher fuel consumption based on the high-power factors at service speed.

The power gap between the design and application causes higher fuel consumption (FC), as shown in Fig. 49.6, where the actual and design fuel consumptions, as well as excess CO_2 emissions, are plotted.

According to the results, using overcapacity engines increases fuel consumption by more than 5% which corresponds to approximately 10 tons CO_2 annually under 1500 h of operation. The calculated annual CO_2 emitted by the target yacht group for design and application cases are approximately 1167.2 tons and 1227.1 tons,

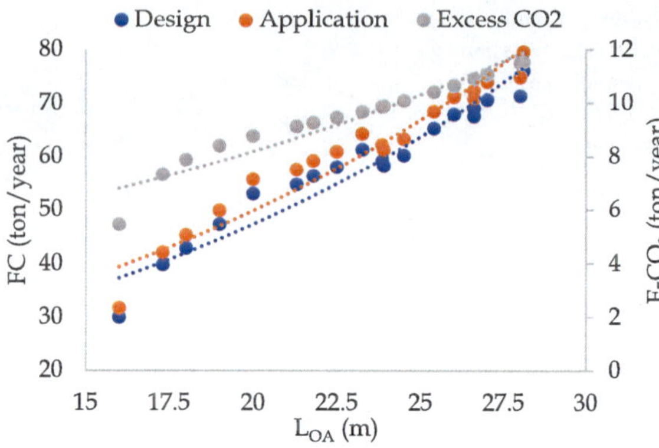

Fig. 49.6 Comparison of fuel consumption and amount of excess CO_2 emission

respectively. The total amount of the excess CO_2 of the target yacht group is about 192 tons/year.

49.5 Conclusions

This study underlines the importance of suitable engine selection for the designed power ranges. The hull form characteristics and machinery data of 20 Bodrum Gulet types of yachts are collected and analysed to evaluate the suitability of the installed engine powers. The resistance and power calculations are conducted to compare the design and actual data then the fuel consumption of the engines is calculated to evaluate the environmental impacts of incompatible engine powers. According to the obtained data and calculations, the following conclusions are drawn:

- Considering the efficient engine operating conditions defined in engine manuals, the load factors of main engines at some yachts are very low or very high causing up to 6% higher fuel consumption.
- The investigated yacht group has conventional diesel engines. Even though they have no obligation of complying with emission regulations for yachts below 400 GT, the engines which are capable of burning alternative and environmentally friendly fuels such as LNG and even ammonia or hydrogen can be used as prime movers.
- The properties of engines available in the market can be taken as inputs at the design and form optimization phases.
- The yachts' machinery systems can be supported with smart energy management modules capable of doing load and power optimizations, active engine, rudder and speed control.

In future studies, the applicability of alternative propulsion systems integrated with the modules to increase the energy efficiency of yachts can be investigated from the design perspective.

References

Akman, M. and Ergin, S. (2019) 'An investigation of marine waste heat recovery system based on organic Rankine cycle under various engine operating conditions', *Proceedings of the Institution of Mechanical Engineers Part M: Journal of Engineering for the Maritime Environment*, 233(2), pp. 586–601. DOI: https://doi.org/10.1177/1475090218770947.

Akman, M. and Ergin, S. (2021) 'Energy-Efficient Shipping : Thermo-Environmental Analysis of an Organic Rankine Cycle Waste Heat Recovery System Utilizing Exhaust Gas From a Dual-Fuel Engine', in *TEAM 2020/21, Dec. 6–8, Istanbul, Turkey*, pp. 329–335.

Birk, L. (2019) Fundamentals of Ship Hydrodynamics: Fluid Mechanics, Ship Resistance and Propulsion. John Wiley & Sons, Ltd.

Blount, D. L. and McGrath, J. A. (2009) 'Resistance characteristics of semi-displacement mega yacht hull forms', *Transactions of the Royal Institution of Naval Architects Part B: International Journal of Small Craft Technology*, 151(2), pp. 19–30. DOI: https://doi.org/10.3940/rina.ijsct.2009.b2.95.

Büyükkeçeci, E. and Turan, B. I. (2018) 'Türkiye'de Tekne Tasarımında Tasarımcının Rolünün Araştırılması: Gulet ve Motor Yat Karşılaştırması', in *Türkiye'de Tekne Tasarımında Tasarımcının Rolünün Araştırılması: Gulet ve Motor Yat Karşılaştırması*. Ankara: ODTU Faculty of Architecture, pp. 159–171.

Elkafas, A. G., Elgohary, M. M. and Zeid, A. E. (2019) 'Numerical study on the hydrodynamic drag force of a container ship model', *Alexandria Engineering Journal*, 58(3), pp. 849–859. DOI: https://doi.org/10.1016/j.aej.2019.07.004.

Gammon, M., Kükner, A. and Alkan, A. (2005) 'Hull Form Optimisation Of Performance Characteristics of Turkish Gulets For Charter', in *Hull Form Optimisation Of Performance Characteristics of Turkish Gulets For Charter*. Annapolis, pp. 79–90.

Güler, E. and Ergin, S. (2021) 'An investigation on the solvent based carbon capture and storage system by process modeling and comparisons with another carbon control methods for different ships', *International Journal of Greenhouse Gas Control*, 110(October 2020), p. 103438. DOI: https://doi.org/10.1016/j.ijggc.2021.103438.

Hamlin, C. (1996) *Preliminary Design of Boats and Ships*. Centreville: Cornell Maritime Press.

Holtrop, J. and Mennen, G. G. J. (1982) 'An approximate power prediction method', pp. 166–170.

IMO (2021) Fourth IMO GHG Study.

Inal, O. B., Zincir, B. and Deniz, C. (2021) 'Hydrogen and Ammonia for the Decarbonization of Shipping', *5th International Hydrogen Technologies Congress*, (May).

Köyağasıoğlu, Y. (2014) *Denizin Kanatlı Perileri Yelkenliler*. Istanbul: Naviga Publishings.

Kristensen, H. O. (2015) 'Energy demand and exhaust gas emissions of marine engines', *Clean Shipping Currents*, 1, pp. 18–26.

Larsson, L. and Eliasson, R. E. (2007) *Principles of Yacht Design*. Third Edit. Camden, Maine: International Marine/McGraw-Hill.

MAN (2013) Basic Principles of Ship Propulsion.

Thomas, T. (2015) 'The owner's guide to superyacht naval architecture part 1 – learning your lines', *Boat International*. Available at: https://www.boatinternational.com/yachts/yacht-design/the-owner-s-guide-to-superyacht-naval-architecture-part-1-learning-your-lines%2D%2D723 (Accessed: 25 August 2021).

Tupper, E. C. (2004) *Introduction to Naval Architecture, Fourth Edition*. Oxford: Elsevier Butterworth-Heinemann.

Turan, A. E. (2009) *Türk Tipi Gulet Yatlarının Formunun Prizmatik Katsayıya Göre Belirlenmesi.* Istanbul Technical University. Available at: https://polen.itu.edu.tr/bitstream/11527/4819/1/9185.pdf.

Turan, B. İ. (2021) 'Design-engineering relation in yacht design process', *IDA: International Design and Art Journal*, 3(2).

Turan, B. İ. and Akman, M. (2021) 'Modeling and Comparison of Bodrum Gulets' Hull Forms with Round and Transom Sterns', *Journal of ETA Maritime Science*, 9(2), pp. 118–127.

Turan, B. İ, Akman, M. and Özbey, T. (2021) 'Design Comparison of Bodrum Gulets and Tırhandils', in *2nd International Congress on Ship and Marine Technology*, pp. 491–497.

Wu, P. and Bucknall, R. (2016) 'Marine propulsion using battery power', *Shipping in Changing Climates Conference 2016*, pp. 1–10.

Chapter 50
The Effect of Acid Pretreatments on Biomass Pyrolysis

E. Pehlivan and E. Fatullayev

50.1 Introduction

The conversion of biomass into valuable materials plays an important role for a renewable and sustainable environment. Converting lignocellulosic biomass into usable energy and chemicals is one way of promoting society to achieve a sustainable economy.7

Biomass pyrolysis has been considered as one of the most popular and promising approaches for creating liquid fuels and even high-value compounds during the previous three decades. As one of the primary pyrolysis products, bio-oil can be used as fuel directly, further transformed into fuels or fuel additives, or as a commodity chemical source. Nevertheless, the final utilization of bio-oil is limited due to its high contents of water and oxygen, instability, corrosiveness and other undesirable characteristics. Therefore upgrading the biomass raw materials through pretreatment before improves the quality of bio-oil. Acid washing is a promising pretreatment means that was particularly effective for lignocellulose [LC] biomass (Chen et al. 2021; Kuglarz et al. 2018). The common goal of the pretreatment methods is to make carbohydrates of complex LC biomass available for further biofuel production. Without pretreatment the presence of indigestible lignin and crystalline cellulose makes the biomass almost unavailable for hydrolytic enzymes to generate simple sugars. During pretreatment crystalline cellulose is converted to an amorphous form which is easier to be degraded, while lignin is removed by delignification, and hemicellulose is totally or partially hydrolysed (Rodionova et al. 2022).

E. Pehlivan (✉) · E. Fatullayev
Faculty of Engineering, Chemical Engineering Department, Bilecik Seyh Edebali University, Bilecik, Turkey
e-mail: eylem.onal@bilecik.edu.tr

M. Z. Sogut et al. (eds.), *Proceedings of the 2022 International Symposium on Energy Management and Sustainability*, Springer Proceedings in Energy,
https://doi.org/10.1007/978-3-031-30171-1_50

Agricultural residue is one of the main renewable energy sources and the typical indicative of biomass in Turkey, which is obtained in large quantities when crops are harvested and processed every year. Due to the clover food quality, high yield at large areas and adaptability to different environmental conditions, it is an important fodder plant grown in many parts of the world. Clover is rich in minerals and vitamins and is very important in terms of animal feed as roughage, for use in medicine and as a source of pollen. The main producers of 255 million tons of clover in the 35 million hectares in the world are the USA and Western Europe. Clover yield in Turkey is 2675 kg, which is approximately quadruple of the world's clover yield (729 kg/ha). In 2018, 17,544,948 tons of alfalfa was produced on an area of 635,000 hectares in Turkey. During the harvesting, baling and silage of the plant, losses may occur due to transportation and fragmentation. These losses correspond to approximately 25% of the total production (Karadaş 2019; Acar et al. 2020; Yozgatlı et al. 2019). Therefore, the evaluation of these wastes is of economic importance.

In this study, clover wastes (*Medicago sativa* L.), which were field wastes, are selected as biomass samples. In order to improve the yield of bio-oil products, the effects of acid concentrations changed in acid washing and pyrolysis temperatures on the yield and quality of products were investigated.

50.2 Materials and Methods

Clover wastes (CWs) as biomass samples were selected. The residues have been taken from Bilecik, Turkey. The samples were dried at room temperature and then ground mechanically. A mean particle size range (1.094 mm) was selected. Chemical pretreatment of clover wastes by washing three different concentrations of phosphoric acid (H_3PO_4) (Merck, 85%) with 0.5 wt.%, 1 wt.% and 2 wt.% acid was carried out in a water bath at 30 °C under stirring conditions (300 r/min) for 3 h and in an autoclave heat of 180 °C with a residence time of 30 min. (solid/liquid ratio 1:15) After the pretreated process, the solid sample was isolated by filtration, washed with deionized water until the filtrate reached neutral pH and then dried at 100 °C for 24 h. In the second step, the pyrolysis of the raw and pretreated biomass samples was carried out at 400, 500, 550 and 600 °C in the fixed bed pyrolysis reactor.

50.3 Result and Discussion

Proximate, ultimate and structure analysis results of clover wastes are given in Table 50.1. Lower ash and moisture content of CW shows that the biomass materials are suitable to use in pyrolysis since higher moisture and ash content of biomass may cause slagging, corrosion that lowers the heating value and fouling. TG curve reveals weight loss as a function of temperature arising from phase transitions during

Table 50.1 Properties of raw materials

Proximate analysis	CW (%)
Moisture	6.5
Ash	12.87
Volatile	67,46
Fixed carbon[a]	13.17
Bulk density(kg/m^3)	223.4
Component analysis	
Hemicellulose	22.09
Lignin	25.09
Extractive material	11.99
Ultimate analysis	
C	42.36
H	5.45
N	1.96
S	0.14
O[a]	50.09
H/C	1.53
O/C	0.89
Calorific value (Mj/kg)	13.16

[a]Calculated by difference

pyrolysis reactions. Hence, TG and dTG curves of the samples and their blends are given in Fig. 50.1. Thermal decomposition of the biomass material generally constitutes three steps: (i) decomposition of unbound water, (ii) decomposition of volatile substances, and (iii) decomposition of cellulose, hemicellulose and lignin. Thermal decomposition of CW mainly consisted of three weight loss stages: loss of moisture bound from ambient temperature up to approximately 131 °C. Active pyrolysis region is between approximately between 200 °C and 391 °C because the most weight loss occur at this region. At this last region passive pyrolysis due to secondary decomposition at high temperatures over 392 °C. In this region weight remain stable due to char and ash.

Figures 50.2, 50.3, 50.4 and 50.5 show the yields of bio-oil, biochar and gas produced from CW. The effect of temperature and acid pretreatment on product yields is seen in the figure. It is known that temperature is a very effective parameter that directly affects both product yield and product quality in pyrolysis processes. The effect of temperature is further increased by the pretreatment applied to the raw material. The highest liquid product yields were determined as 25.16%, 40.39%, 44.04% and 42.27% for raw materials and 0.5%,1% and 2% for H3PO4 pretreatment applications at 600 °C, respectively. In acidic pretreatment applications, the liquid product yield increased as the temperature increased. The ratio of levoglucosan and other sugars increases, and hemicellulose is largely removed from the structure with acidic pretreatment. In this case, the structure of the biomass changes to a large extent, which causes an increase in the bio-oil yield with an increase in temperature (Şenol et al. 2021). Furthermore, acid washing caused the metals to be greatly

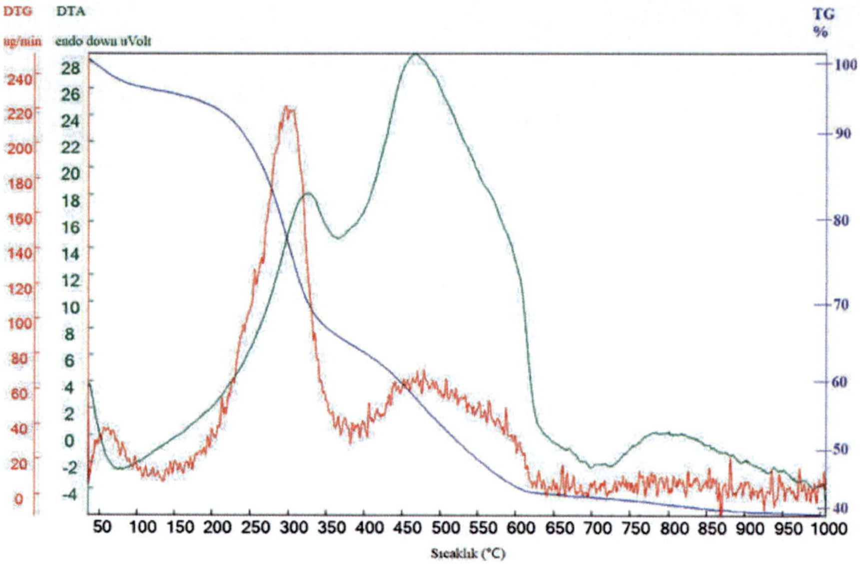

Fig. 50.1 TG curves of CW

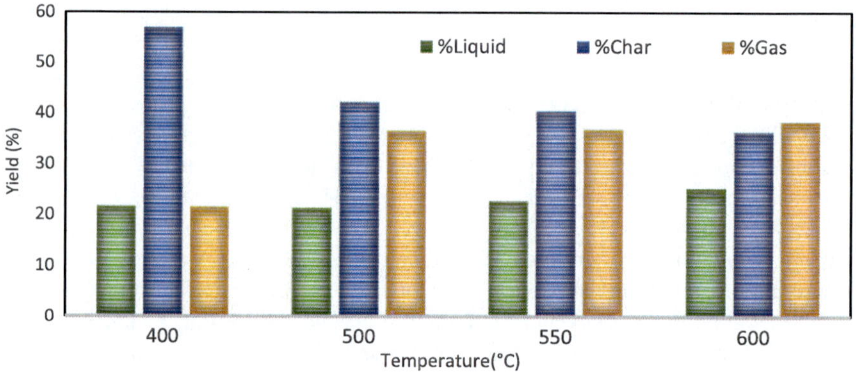

Fig. 50.2 Yields of pyrolysis products (untreated)

removed, especially the alkali and alkaline earth metals. This process changed the pyrolysis pathway to some extent, decreased the secondary decomposition of volatiles, increased the production of bio-oil and reduced the yield of biochar (Dengyu et al. 2017).

In the temperature biochar formation graph, as a result of the pyrolysis experiments, the amount of solid product decreased with the increase in temperature, as expected. This is because decomposition reactions dominate with increasing temperature. In the gas product formation graph, it is seen that all trials exhibited parallel

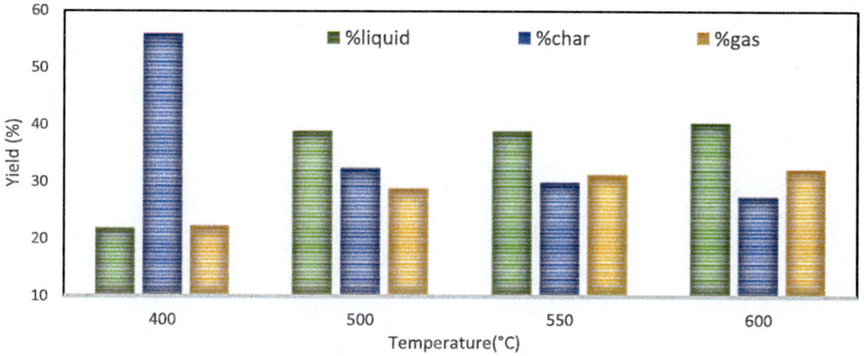

Fig. 50.3 Yields of pyrolysis products (0.5% H3PO4)

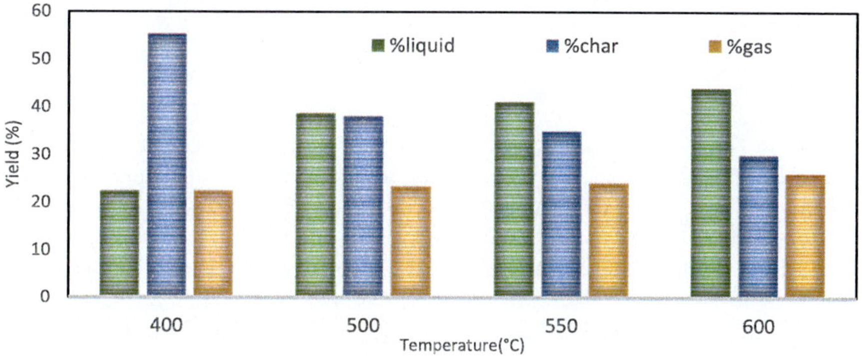

Fig. 50.4 Yields of pyrolysis products (1% H3PO4)

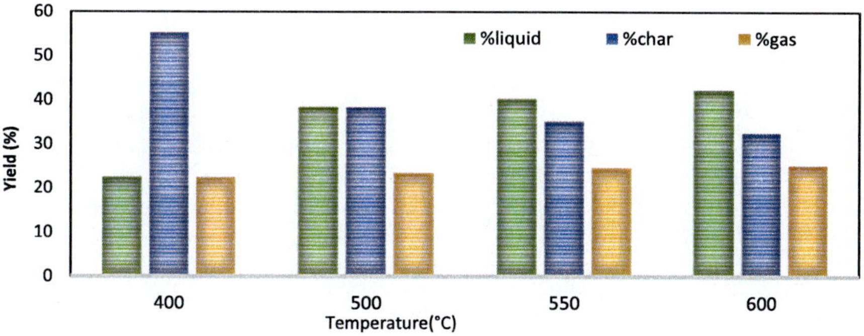

Fig. 50.5 Yields of pyrolysis products (2% H3PO4)

Table 50.2 Codes of biochars

Biochar samples	Acid concentration	Pyrolysis temperature (°C)
CW1	–	400
CW2	0.5	400
CW3	1	400
CW4	2	400
CW5	–	500
CW6	0.5	500
CW7	1	500
CW8	2	500
CW9	–	550
CW10	0.5	550
CW11	1	550
CW12	2	550
CW13	–	600
CW14	0.5	600
CW15	1	600
CW16	2	600

behaviour. In all experiments, gas product formation increased with increasing temperature. This shows that gasification reactions with temperature are more effective in both liquid products and solid products. The highest gas product yield was obtained at 600 °C as 38.45% (Table 50.2).

FT-IR spectrums of CW and the raw and pretreated biochars are given in Fig. 50.6. The FTIR analyses also indicated that the functional groups of biochars obtained at under different conditions were similar and that aromatic and aliphatic groups were predominant. It referred to the peak hydroxyl group observed around 3000–3500 cm^{-1}. These peaks indicate the presence of hydrogen bonds within the structure. Broad band for the OH in-plane bend decreased with increase in pyrolysis temperature. The region in 2887 cm^{-1} belonged to the C-H tension vibrations in the methyl and methylene groups. Carbonyl groups (C=O) were located in the region of 1860 cm^{-1}. 1534 cm^{-1} indicates the variance of lignin's aromatic skeleton. Severe peaks observed in the range of 1000–1200 cm^{-1} indicate the presence of O-H and the 1037 cm^{-1} region belonged to the C-O tension vibration of the phenol, ether and ester groups. They were the out-of-plane bending vibrations of the C-H bond in the peak benzene derivatives observed after the 800 cm^{-1} region. C-C stress vibration peaks between 2200 and 2100 cm^{-1} were observed. Peaks in the range of 1300–1600 cm^{-1} and 1150–1085 cm^{-1} represent functional groups of aromatic C=C and to C-O aliphatic ether tension vibrations respectively.

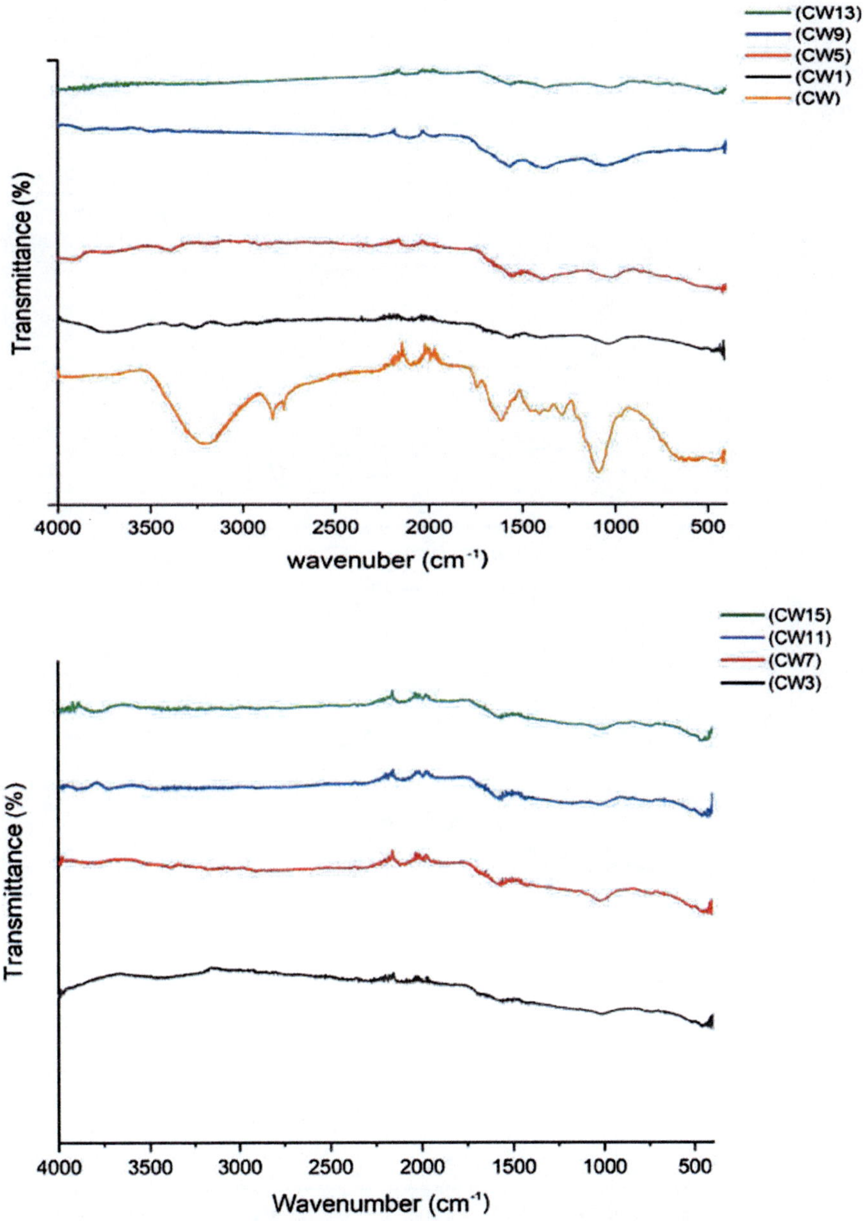

Fig. 50.6 FT-IR spectrums of biomass and untreated and pretreated biochars

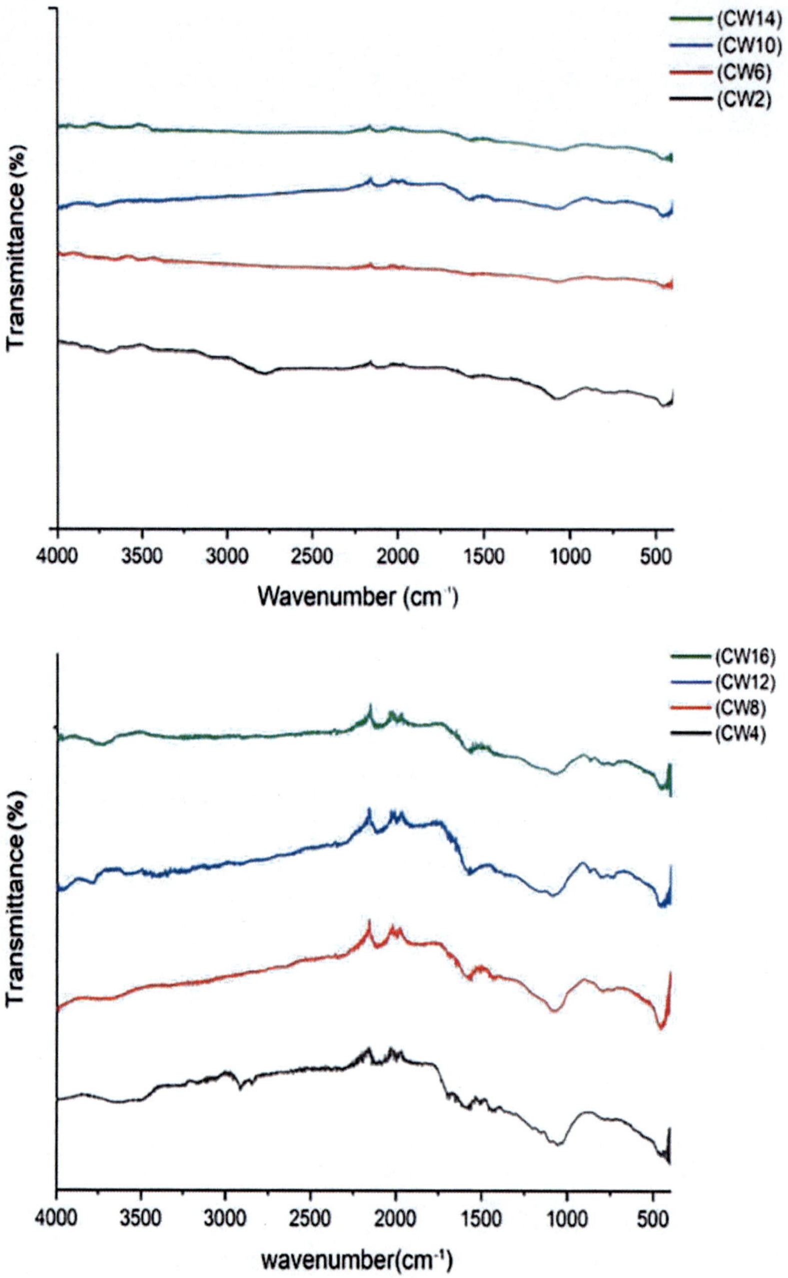

Fig. 50.6 (continued)

50.4 Conclusion

Acid-based pretreatment has an important role in the conversion of lignocellulosic biomass into sugar for further processing into biofuels and other industry-grade bio-products. The operating process parameters, such as acid types, acid concentration and reaction temperature, for the biomass pretreatment process directly influence the yield of post-pretreated products (Hoang et al. 2021). The effect of acid pretreatment on product yields resulted in higher liquid product yield and lower solid product yield. According to the results of the optimum pretreatment condition assessment, it was determined that the highest tar yield was reached as 44.04% with the CW15 sample at 600 °C pyrolysis temperature.

References

Acar Z, Tan M, Ayan İ, Önal Aşçı Ö, , Mut H, Başaran U, Gülümser E, Can M, Kaymak G (2020) Forage Crops in Turkey Status of Agriculture and Development Opportunities. Turkish Agricultural Engineering IX. Technical Congress, January 13–17, Ankara Turkey 529–551.

Chen D, Dongxiao G, Shunchao H, Capareda S, Xinyue L, Ying W, Ting Z, Yueyang Y, Weisheng Y (2021) Influence of acid-washed pretreatment on the pyrolysis of corn straw: A study on characteristics, kinetics and bio-oil composition. Journal of Analytic and Applied Pyrolysis 155: 1–7. https://www.sciencedirect.com/science/article/pii/S0165237021000139

Dengyu C, Mei J, Li H (2017) Combined pretreatment with torrefaction and washing using torrefaction liquid products to yield upgraded biomass and pyrolysis products. Bioresource Technology 228: 62–8. https://www.sciencedirect.com/science/article/pii/S0960852416317667

Hoang AT, Nizetic S, Ong HC, Chong CT, Atabani AE, Pham VV (2021) Acid-based lignocellulosic biomass biorefinery for bioenergy production: Advantages, application constraints, and perspectives. Journal of Environmental Management 296: 113194. https://www.sciencedirect.com/science/article/pii/S0301479721012561

Karadaş K (2019) Alfalfa Production And Economic Importance in Iğdır Province. Zeugma II. International Multidisciplinary Studies Conference January 18–20, Gaziantep Turkey, 334–343.

Kuglarz M, Alvarado-Morales M, Dabkowska K, Angelidaki I (2018) Integrated production of cellulosic bioethanol and succinic acid from rapeseed straw after dilute-acid pretreatment. Bioresource Technology 265: 191–199. https://www.sciencedirect.com/science/article/pii/S0960852418307673

Rodionova MV, Bozieva AM, Zharmukhamedov SK, Leong YK, Lan JC, Veziroglu A, Veziroglu TN, Tomo T, Changan J, Allakhverdiev SI (2022) A comprehensive review on lignocellulosic biomass biorefinery for sustainable biofuel Production. International Journal of Hydrogen Energy 47: 1481–1498. https://www.sciencedirect.com/science/article/pii/S0360319921041549

Şenol A, Didem BK and Halil D, (2021) The role of acidic, alkaline and hydrothermal pretreatment on pyrolysis of wild mustard (Sinapis arvensis) on the properties of bio-oil and bio-char. Bioresource Technology Reports. 17: 1–9. https://www.sciencedirect.com/science/article/pii/S2589014X22000378

Yozgatlı O, Başaran U, Gülümser E, Mut H, Çopur Doğruöz M (2019) Morphological Traits, Yield and Silage Qualities of Some Corn Varieties Under Yozgat Ecological Condition. KSU Journal of Agriculture and Nature. 22 (2): 170–177. 10.18016/ksutarimdoga.vi.450938

Chapter 51
Using an E-fuel Method to Meet the 2030 Decarbonization Target: A Case Study

Bugra Arda Zincir, Burak Zincir, Hasan Bora Usluer, and Yasin Arslanoglu

Nomenclature

CAPEX	Capital expenditure
CII	Carbon Intensity Indicator
DAC	Direct air capture
DCS	Data collection system
DWT	Deadweight tonnage
EEDI	Energy Efficiency Design Index
EEXI	Energy Efficiency Existing Ship Index
GHG	Greenhouse gases
IMO	International Maritime Organization
LHV	Lower heating value
LNG	Liquefied natural gas
MDO	Marine diesel oil
OPEX	Operational expenditure
SEEMP	Ship Energy Efficiency Management Plan
WTW	Well to wake

B. A. Zincir (✉)
Maritime Faculty, Istanbul Technical University, Istanbul, Tuzla, Turkey

Maritime Vocational School, Galatasaray University, Istanbul, Besiktas, Turkey
e-mail: bazincir@gsu.edu.tr

B. Zincir · Y. Arslanoglu
Maritime Faculty, Istanbul Technical University, Istanbul, Tuzla, Turkey
e-mail: bzincir@itu.edu.tr; arslanoglu@itu.edu.tr

H. B. Usluer
Maritime Vocational School, Galatasaray University, Istanbul, Besiktas, Turkey
e-mail: hbusluer@gsu.edu.tr

M. Z. Sogut et al. (eds.), *Proceedings of the 2022 International Symposium on Energy Management and Sustainability*, Springer Proceedings in Energy,
https://doi.org/10.1007/978-3-031-30171-1_51

51.1 Introduction

Climate change has many effects, one of which is on the environment. A major reason for climate change is fossil fuels, and day by day fossil fuel consumption is increasing due to the energy demand. Maritime transportation is the major transportation mode that forms 90% of global trade (Deniz and Zincir 2016). Therefore, the energy demand of maritime transportation is high, and 72% of the total fuel consumption is poor quality heavy fuel oil, 26% is MDO, and 2% is liquefied natural gas (LNG) (Gray et al. 2021). The GHG emissions which are related to fossil fuel use at maritime transportation are increased from 977 million tons to 1076 million tons from 2012 to 2018 (IMO 2020). However, recent rules and regulations in maritime transportation encourage the use of alternative energy sources to reduce GHG emissions. Although alternative systems are not mature enough compared to conventional ones, recent studies raised attention to them. Moreover, some measures need to be taken to limit the global temperature rise agreed in the Paris Agreement below 2.5 °C. In addition to the Paris Agreement, IMO has brought attention to the topic and announced its GHG strategy in 2018. The strategy aims to reduce CO_2 emission by 40% until 2030 and 70% until 2050 compared to the 2008 level. Furthermore, Energy Efficiency Design Index (EEDI), Ship Energy Efficiency Management Plan (SEEMP), and data collection system (DCS) are the rules already in effect to mitigate global carbon emissions, and in 2023, it is expected that Energy Efficiency Existing Ship Index (EEXI) and Carbon Intensity Indicator (CII) rules will enter into force for the same purpose.

After the announcement of the IMO Initial GHG Strategy, various systems are proposed to meet the limits, such as renewable energy sources, alternative fuels produced from fossil fuels, and E-fuels. Nowadays, studies on E-fuels have increased since carbon atoms of the fuel are obtained from CO_2 or direct air capture (DAC), and hydrogen is derived from electrolysis while the processes are powered by renewable electricity. The last step is the synthesis of carbon and hydrogen atoms to produce the fuel (Lindstad et al. 2021a). During the combustion of E-fuels, no harmful emissions are produced.

The carbon freeness of E-fuels has raised attention, and recently, some studies were made for maritime transportation. In a study conducted by Carvalho et al. (2021), 14 carbon-neutral fuels are investigated in terms of economic, environmental, and technical criteria for the maritime sector. In a study by Linstad et al. (2021a), alternative fuels, including E-fuel, are investigated according to GHG reduction, energy utilization, feasibility, and cost. Another study is made by Linstad et al. (2021b) about GHG emission reduction by utilizing E-fuels. Authors have evaluated options with respect to emissions, energy, and the cost of E-fuels. Moreover, Lester et al. (2020) analyzed the role of E-fuels for the IMO's 2050 targets.

In this paper, some of the alternative fuels are evaluated in a general cargo ship concerned with carbon footprint and cost criteria. The evaluation is made by a case study, where the data are gathered from an actual ship sailing from Turkey to Israel.

Table 51.1 Ship specifications

Ship type	General cargo
Deadweight	10,300 mtons
Gross tonnage	6177 GRT
Net tonnage	3680 NRT
Length overall	128 m
Beam	18 m
Draught	7.6 m
Keel laid date	2004
Main engine	S.X.D. – Daihatsu 8DKM–282,500 kW at 750 rpm
Speed	12.3 knot
Fuel type	HFO/MDO
Specific fuel consumption	200 g/kWh at 75% engine load

51.2 Methodology

In this study, a case study is made using a 10,300 deadweight tonnage (DWT) general cargo ship. Details of the reference ship can be seen in Table 51.1. The vessel does liner shipping between Nemrut/Turkey and Haifa/Israel, where the distance is 648 nm between two ports. Data used in the study is gathered from an actual ship using MDO. The main engine load of the ship is assumed as 75%, and the actual ship speed is calculated by Eq. (51.1) (Dere and Deniz 2019), and the actual ship speed is found as 11 knots.

$$P_{actual}/P_{design} = \left(V_{actual}/V_{design}\right)^{\alpha} \tag{51.1}$$

where P_{actual} is the actual main engine power during the voyage; P_{design} is the design power of the main engine, which is 2500 kW for the case ship; V_{actual} is the actual ship speed during the voyage; V_{design} is the design speed of the ship, which is 12.3 knots for the case ship; and α is the speed coefficient between 2.5 and 3 (Moreno-Gutierrez et al. 2015). α is taken as 2.5 in this study.

The voyage-based MDO consumption is calculated by using Eq. (51.2).

$$FC_{voy} = SFC \times P_{actual} \times duration_{voy} \tag{51.2}$$

In the equation, FC_{voy} is the voyage-based MDO consumption and $duration_{voy}$ is the voyage duration in hours which equals 58.8 h at 11 knots ship speed.

In the case study, MDO, E-ammonia, E-LNG, E-methanol, and E-diesel are investigated. The result is then compared according to the IMO's initial GHG strategy limits through 2030, assuming the ship uses alternative fuels and E-fuels.

MDO is a distillate diesel oil that is a mixture of heavy fuel oil and gas oil. MDO has a wide usage area in maritime transportation, especially for medium- to high-speed engines. E-ammonia consists of one nitrogen and three hydrogen atoms

Table 51.2 Fuel input data

Fuel types	WTW (gCO_2/MJ)	LHV (MJ/kg)	Fuel cost (USD/ton)	CAPEX (USD)
MDO	87.8	43.5	895	400
E-ammonia	5.3	18.6	940–1750[a]	1400
E-LNG	1.7	49.2	1350–3000[a]	1400
E-methanol	0.9	19.9	1360–3235[a]	800
E-diesel	1.3	42.7	1530–3575[a]	400
USD	E-ammonia: 1345 USD E-LNG: 2175 USD E-methanol: 2297.5 USD E-diesel: 2552.5 USD			

Ship and Bunker (2022), Al-Aboosi et al. (2021), Lindstad et al. (2021a, b)
[a]The mean value of the fuel costs is used in the calculations

produced by air separation. It is obtained from the synthesis of nitrogen and hydrogen, which occurs in a Haber-Bosch reactor. In addition, E-LNG is a green fuel generated from CO_2 and hydrogen by the Sabatier reaction (VoltaChem 2020). Methanol is obtained from a mix of carbon dioxide, carbon monoxide, and hydrogen, also known as syngas. However, while producing E-methanol, captured CO_2 is hydrogenated; thus, green fuel is gathered. Furthermore, E-diesel is a combination of green carbon and hydrogen atoms, and it is produced after the Fischer-Tropsch processes (VoltaChem 2020).

Except for E-diesel, remaining E-fuels require pilot MDO fuel to combust. Therefore, it is assumed that 5% pilot fuel and 95% E-fuel energy fraction is burned at the main engine with the same engine efficiency. To calculate the voyage-based fuel consumption of E-diesel, Eq. (51.2) is used and for other E-fuels, Eq. (51.3) is used.

$$FR_{e-fuel} = \frac{m_{e-fuel} \times LHV_{e-fuel}}{m_{e-fuel} \times LHV_{e-fuel} + m_{MDO} \times LHV_{MDO}} \tag{51.3}$$

where $FR_{e\text{-}fuel}$ is the energy fraction of E-fuel, which is 0.95 for 95% E-fuel energy fraction, $m_{e\text{-}fuel}$ is E-fuel consumption mass, $LHV_{e\text{-}fuel}$ is E-fuel lower heating value, m_{MDO} is MDO consumption mass, and LHV_{MDO} is MDO lower heating value.

In Table 51.2, input data of the fuels are shown, and the evaluation of the fuels are made according to well to wake (WTW) CO_2 emissions, lower heating value (LHV), fuel costs, capital expenditure (CAPEX), and operational expenditure (OPEX).

Equation (51.4) is used for the calculations of WTW emissions of the fuels.

$$V_{CO2_i} = \frac{CF_{CO2_i} \times LHV_i \times FC_{voy}}{1000} \tag{51.4}$$

where V_{CO2_i} is the voyage-based CO_{2e} emission of i type of fuel, CF_{CO2_i} is well-to-propeller CO_{2e} emissions of i type of fuel, LHV_i is lower heating value of i type of fuel, and FC_{voy} is voyage-based fuel consumption.

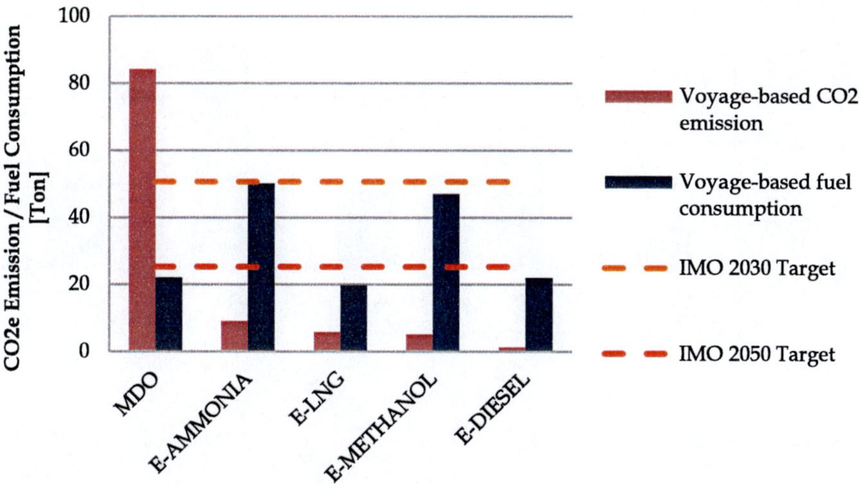

Fig. 51.1 Voyage-based CO_{2e} emission and fuel consumption

51.3 Results and Discussion

The case study results include voyage-based CO_{2e} emission, fuel consumption, fuel expense, and CAPEX and OPEX costs of the fuel systems.

Figure 51.1 shows the voyage-based CO_{2e} emission and fuel consumption in tons. MDO is the baseline scenario for the case study with the voyage-based fuel consumption of 22.1 tons and voyage-based CO_{2e} emission of 84.2 tons. It can be seen that the current CO_{2e} emission is higher than the IMO 2030 target of 40% reduction, which equals 50.5 tons for the case study voyage. To achieve this target, E-fuels are important options. The CO_{2e} emissions are 9.0, 5.8, 5.0, and 1.2 tons for E-ammonia, E-LNG, E-methanol, and E-diesel, respectively. The voyage-based emission reduction levels are 89.3%, 93.2%, 94%, and 98.5% for the respective E-fuels. The emission amounts are far below the IMO 2030 target and also the IMO 2050 target of 25.3 tons for the respective voyage. The E-fuels in this study, except E-diesel, are combusted by a 5% energy fraction of pilot MDO since the auto-ignition temperature of these fuels is high and is not suitable for diesel engines. The pilot MDO amount is calculated as 1.1 tons for each E-fuel for the voyage. The voyage-based fuel consumption of E-ammonia is higher than MDO due to its low LHV, and 49 tons of E-ammonia is required for the same voyage. E-methanol has also had a lower LHV than MDO but higher than E-ammonia. 45.8 tons of e-methanol is needed for the same distance. E-LNG has a higher LHV than MDO. Therefore, the required fuel amount is 19.6 tons for the case study voyage. From the perspective of CO_{2e} emission, E-diesel is the leading E-fuel, and from the fuel consumption perspective, E-LNG is the prominent E-fuel with lower voyage-based fuel consumption than the baseline scenario.

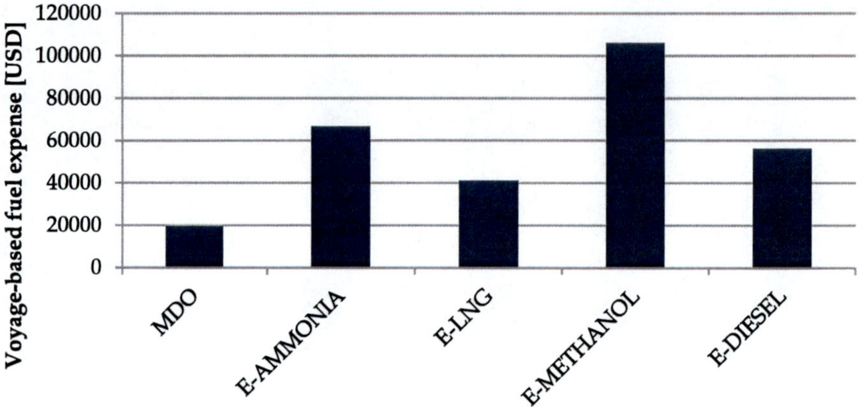

Fig. 51.2 Voyage-based fuel expense

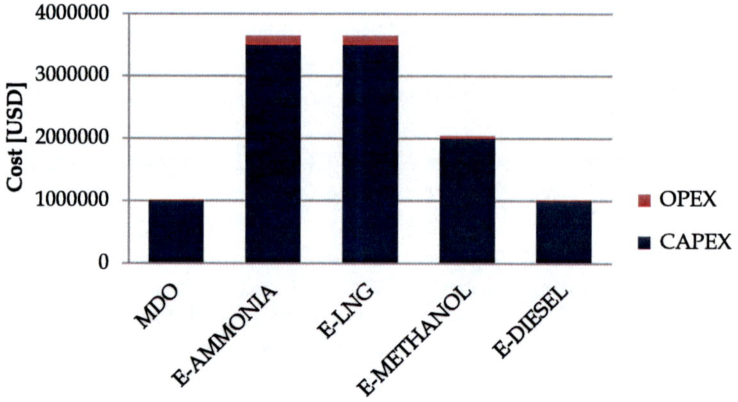

Fig. 51.3 CAPEX and OPEX costs of fuel systems

Voyage-based fuel expense depending on the fuel type is shown in Fig. 51.2. MDO and E-diesel include expenses of their selves, while remaining E-fuels contain both E-fuel and pilot fuel, MDO, and expenses. The baseline scenario, MDO, has a voyage-based fuel expense of 19,735 USD. After then E-LNG, E-diesel, E-ammonia, and E-methanol follow with 41,238, 56,283, 66,879, and 106,190 USD, respectively. It can be seen that all E-fuels are more expensive than conventional MDO fuel recently. Moreover, low LHV of E-fuels, except E-diesel, results in higher voyage-based fuel expenses.

Figure 51.3 shows the CAPEX and OPEX costs of the fuel systems. OPEX costs are calculated by assuming that it is 2.5% for diesel and methanol engines 4.5% for LNG and ammonia (Korberg et al. 2021). And OPEX cost for E-diesel is the same as MDO-fuelled engine. The CAPEX of the baseline scenario, MDO, is 1,000,000 USD, and the OPEX is 25,000 USD. E-diesel has the same CAPEX and OPEX costs as MDO since E-diesel can be used at the same engine and fuel system without any modification requirement. The E-fuel that has the lowest CAPEX and OPEX cost after MDO and E-diesel is E-methanol with 2,000,000 USD and 50,000 USD, respectively. Methanol is in the liquid state at room temperature and can be stored in similar fuel tanks with MDO. Therefore, fuel system requirements are lower than gaseous fuels. On the other hand, methanol-compliant engine parts and designs are needed to use E-methanol effectively (Zincir and Deniz 2021). E-ammonia and E-LNG are gaseous E-fuels and have the same CAPEX and OPEX costs of 3,500,000 USD and 157,500 USD, respectively. The storage and fuel delivery systems and special main engines for gaseous fuel are required. Besides, the operational expenses of gaseous fuels are higher than liquid fuels (Zincir 2022). E-diesel is the leading E-fuel with the same CAPEX and OPEX costs with MDO and E-methanol follows it.

Results of the case study in Figs. 51.1, 51.2, and 51.3 demonstrate that some E-fuels stand out more than others. Considering the voyage-based CO_2 emissions, all meet the IMO 2030 and 2050 targets; thus, they are valid options. In terms of fuel consumption, E-LNG performed the best even compared to the baseline scenario, and its fuel expense is the lowest after the MDO. However, its CAPEX and OPEX results are highest with the same value as E-ammonia. The second least fuel consumption is observed by using E-diesel. Even though its fuel expense is higher than the E-LNG, both CAPEX and OPEX are the same with the baseline scenario and almost one-fourth of the E-LNGs. Another option is E-methanol; in terms of fuel consumption, it performed better than E-ammonia and has the lowest CAPEX and OPEX after the E-diesel, but its fuel expense is higher compared to others. The last one is E-ammonia; it performed poorly both in fuel consumption, CAPEX, and OPEX and in the middle ground considering fuel expense. Thus, it is the least desirable choice between them. With only E-LNG and E-diesel left, all the expenses have to be evaluated. In Fig. 51.3, it can be seen that the CAPEX of E-LNG is higher than the E-diesel with a quite big margin. Fuel expense of E-LNG and E-diesel is 41,238 and 56,283 USD, while OPEX is 25,000 and 157,500 USD, respectively. Even though the E-LNG fuel expense is 15,045 USD lower, its OPEX is 132,500 USD higher. Consequently, E-diesel is the most cost-efficient option, despite E-LNG having lower fuel consumption.

51.4 Conclusion

This study shows the environmental benefit of E-fuels and the expenses due to these fuels on ships. A case study is made from an actual ship sailing Nemrut/Turkey to Haifa/Israel. All the data used in this paper is gathered from the ship and evaluated. As it can be seen from the results, to meet the IMO 2030 and 2050 targets, MDO cannot be used as a standalone energy source. Thus, E-ammonia, E-LNG, E-methanol, and E-diesel are investigated for CO_2 reduction and expenses. Case study results indicate that E-diesel is the best solution, but it has to be reminded that the current fuel costs are much higher than the MDO, yet further studies aiming to decrease it should be made.

References

Al-Aboosi, F. Y., El-Halwagi, M. M., Moore, M. and Nielsen, R. B., 2021, Renewable ammonia as an alternative fuel for the shipping industry. Current Opinion in Chemical Engineering. 31: 100670. https://doi.org/10.1016/j.coche.2021.100670

Carvalho, F., Müller-Casseres, E., Poggio, M., Nogueira, T., Fonte, C., Wei, H. K., Portugal-Pereira, J., Rochelo, P. R. R., Szklo, A., Schaeffer, R., 2021, Prospects for carbon-neutral maritime fuels production in Brazil. Journal of Cleaner Production 326:129385. https://doi.org/10.1016/j.jclepro.2021.129385

Deniz, C. and Zincir, B., 2016, Environmental and economical assessment of alternative marine fuels. Journal of Cleaner Production 296:438–449. https://doi.org/10.1016/j.jclepro.2015.11.089

Dere, C. and Deniz, C., 2019, Load optimization of central cooling system pumps of a container ship for the slow steaming conditions to enhance the energy efficiency. Journal of Cleaner Production 222, 206–217. https://doi.org/10.1016/j.jclepro.2019.03.030

Gray, N., McDonagh, S., O'Shea, R., Smyth, B. and Murphy J. D., 2021, Decabonising ship, planes and trucks: an analysis of suitable low-carbon fuels for the maritime, aviation and haulage sectors. Adv. Appl. Energy 1:100008. https://doi.org/10.1016/j.adapen.2021.100008

International Maritime Organization (IMO), 2020, Fourth Greenhouse Gas Study 2020. https://wwwcdn.imo.org/localresources/en/OurWork/Environment/Documents/Fourth%20IMO%20GHG%20Study%202020%20-%20Full%20report%20and%20annexes.pdf

Korberg, A. D., Brynolf, S., Grahn, M. and Skov, I. R., 2021, Techno-economic assessment of advanced fuels and propulsion systems in future fossil-free ships. Renewable and Sustainable Energy Reviews 142, 110861. https://doi.org/10.1016/j.rser.2021.110861

Lester, M. S., Bramstoft, R., Münster, M., 2020, Analysis on Electrofuels in Future Energy Systems: A 2050 Case Study. Energy 199:117408. https://doi.org/10.1016/j.energy.2020.117408

Lindstad, E., Gamlem, G. M., Rialland, A. and Valland, A., 2021a, Assessment of alternative fuels and engine technologies to reduce GHG. SNAME Maritime Convention 2021, October 27–29, 2021, Providence, Rhode Island. https://doi.org/10.5957/SMC-2021-099

Lindstad, E., Lagemann, B., Rialland, A., Gamlem, G. M., Valland, A., 2021b, Reduction of maritime GHG emissions and the potential role of E-fuels. Transportation Research Part D 101:103075. https://doi.org/10.1016/j.trd.2021.103075

Moreno-Gutierrez, J., Calderay, F., Saborido, N., Boile, M., Rodriguez Valero, R. and Duran-Grados, V., 2015, Methodologies for estimating shipping emissions and energy consumption: a comparative analysis of current methods. Energy 86, 603-616. https://doi.org/10.1016/j.energy.2015.04.083

Ship & Bunker, 2022, Top Bunker Prices. https://shipandbunker.com/

VoltaChem, 2020, E-fuels: Towards a More Sustainable Future For Truck Transport, Shipping and Aviation. https://smartport.nl/wp-content/uploads/2020/09/20-11482-whitepaper-Voltachem-10.pdf

Zincir, B. and Deniz, C., 2021, Methanol as a fuel for marine diesel engines. In book: Alcohol as an Alternative Fuel for Internal Combustion Engines. Springer, Singapore.

Zincir, B., 2022, Ammonia for decarbonized maritime transportation. In book: Clean Fuels for Mobility. Springer, Singapore.

Chapter 52
Heat Transfer Enhancement of Biomass-Based Stirling Engine

Nik Kechik Mujahidah and Syamimi Saadon

Nomenclature

W_{out}	Power output (W)
p	Pressure (Pa or bar)
V	Volume (m^3)
l_d	Height of displacer (m)
W_{out}	Power output (W)
l_1, l_2, l_3, l_4	Lengths of the linkages of the rhombic drive mechanism (m)
R_d	Offset distance from the crank to the center gear (m)
L_{pt}	Height from the linkage l_1 to the top surface of the piston (m)
L_{dt}	Length from the linkage l_4 to the top surface of the displacer (m)
R_d	Offset distance from the crank to the center of gear (m)
T, T_H, T_C	Temperature, hot-end temperature, cold-end temperature (K)

52.1 Introduction

The usage of biomass in industrial boilers for electricity generation has received significant attention in recent years. The significant attention is due to Malaysia has an advantage over other types of renewable energy as the largest palm oil producer in the world. This helps in reducing carbon dioxide by replacing coal or natural gas (Syakirah 2019). However, during the process, a significant fraction of thermal energy is often lost to the environment by the flue gas. The heat loss with exhaust

N. K. Mujahidah (✉) · S. Saadon
Department of Aerospace Engineering, Universiti Putra Malaysia, Selangor, Malaysia
e-mail: mimisaadon@upm.edu.my

M. Z. Sogut et al. (eds.), *Proceedings of the 2022 International Symposium on Energy Management and Sustainability*, Springer Proceedings in Energy,
https://doi.org/10.1007/978-3-031-30171-1_52

flue gas which ranges from 150 to 180 °C (B.D.B.L.Wang 2020) has resulted in the importance of recovering the waste heat of the flue gas to overcome the reliance on fossil fuel. The exhaust heat recovery as much as possible will also result in environmental protection and energy conservation (Li et al. 2016).

Stirling engine as an external combustion engine with high efficiencies that is able to use any of heat source is the best candidate to recover waste heat of the exhaust gas and heat by converting it into power (Song et al. 2015). However, when connected to lower heat sources, Stirling engines' performance becomes disrupted (Bianchi and De Pascale 2011). For this reason, simulation test was performed on Stirling engine model used to recover low-temperature waste heat from flue gas in biomass combustion to evaluate methods to improve the heat transfer method in the Stirling engine.

Computational fluid dynamics (CFD) approach is one of the numerical models that can simulate the multidimensional components and complicate processes in Stirling engine, hence giving an accurate prediction on the overall engine performance (Chen et al. 2014). Therefore, CFD is used to perform this study. ANSYS fluent 20.1 is used to solve the continuity, momentum, and energy. The wall features and thickness that were detailed in the wall boundary condition to include the conduction heat transfer through the external wall were solved by the software. There are very few computational works in the literature that fully provide a complete information about geometry and boundary conditions that are needed to design geometry for the CFD simulations. The computational analysis from Ben-Mansour et al. provides a concise information needed for the CFD simulation (B.D.B.L.Wang 2020). Hence, his paper will be referred for the initial geometry of the model of Stirling engine before adding the flue gas and heater chamber at the tubular heater of the engine. To ensure that the model is viable, further analysis on the work done by the engine is performed and compared with the experimental data of the work done reported by Aksoy (Aksoy 2013).

This paper presents the initial model for further improvisation for low-temperature heat transfer process from the heat source to the external part of the tubular heater of the Stirling engine. This paper shows the validation of the primary model of Stirling engine according to the previous researcher by applying different engine speed and comparing the trendlines of the power output of every engine speed. This work is useful for further research in heat transfer enhancement method of Stirling engine for low-temperature heat source.

52.2 Modeling and Geometry

Computational fluid dynamics (CFD) modeling is used in this study. ANSYS fluent 20.1 is used to solve the continuity, momentum, and energy. The wall features and thickness that were detailed in the wall boundary condition to include the conduction heat transfer through the external wall were solved by the software. There are very few computational works in the literature that fully provide a complete information

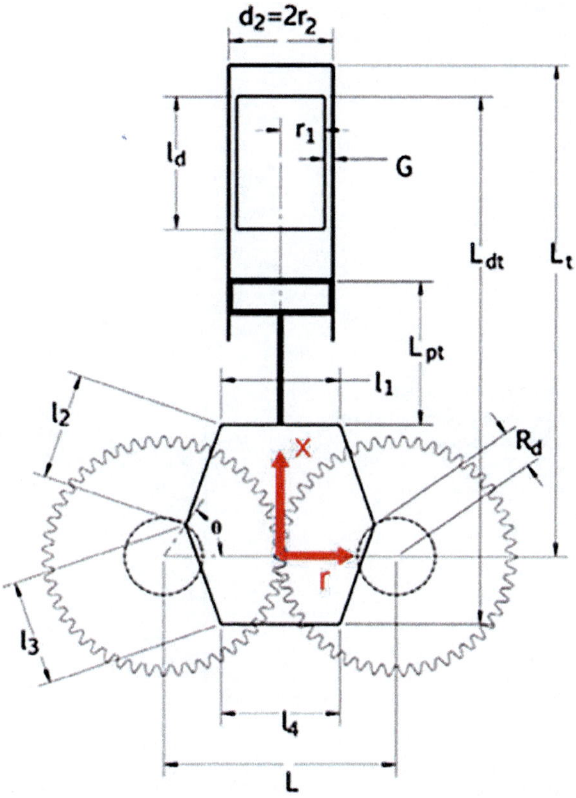

Fig. 52.1 Engine configuration. (Abuelyamen et al. 2017)

about geometry and boundary conditions that are needed to design geometry for the CFD simulations. The computational analysis from R.Ben-Mansour et al. (Abuelyamen et al. 2017) provides a concise information needed for the CFD simulation. Hence, his paper will be referred for the initial geometry of the model of Stirling engine before adding the flue gas and heater chamber at the tubular heater of the engine.

In this study, a rhombic drive mechanism β-type Stirling engine which consists of three zones, i.e., compression zone, expansion zone, and narrow zone, as shown in Fig. 52.1 is taken as an example. The narrow zone which connects the expansion and compression zone is initially assumed to have no regenerator materials. The dimensions of the geometries are shown in Table 52.1.

As shown in Fig. 52.2, the simulation will be solved as 3D geometry. Air is utilized and treated as an ideal gas in this study. Air is assumed to be dependent on the gas temperature. The dimension of simulated geometry of the Stirling engine is shown in Table 52.1. The cylinder wall features and thickness are detailed in the wall boundary condition as shown in Table 52.2. The thermal boundary conditions are limitedly between 775 K for hot temperature (T_H) and 300 K for cold temperature (T_C).

Table 52.1 Dimension of simulated geometry of the Stirling engine in (mm)

Geometry	Dimensions
r_1	43
r_2	42.25
l_d	155
$l_1 = l_2 = l_3 = l_4$	66
R_d(mm)	l_1/2.6666768
L_{pt}	50.93
L_{dt}	163.74
R_d	3.5
Engine speed	360 ~ 575 rpm

R. Ben-Mansour (2017)

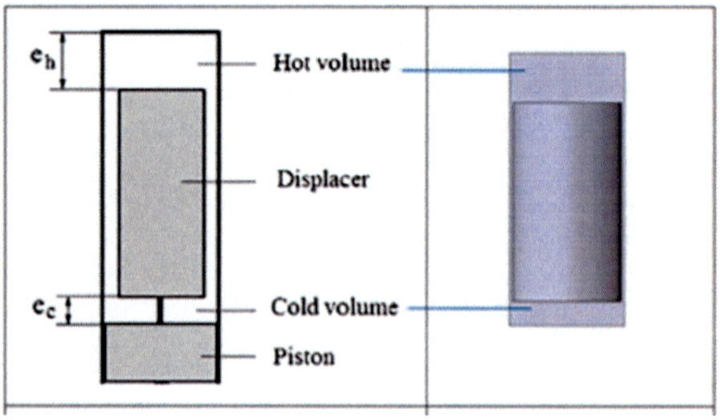

Fig. 52.2 Schematic diagram and CFD domain. (Abuelyamen et al. 2017)

Table 52.2 Cylinder wall dimensions and properties

Cylinder wall dimensions	Properties
Working fluid	Air
Thickness (mm)	1
Material	Steel
Density (kg/ m^3)	7840
Thermal conductivity (W m^{-1}. k^{-1})	43
Specific heat (J. kg^{-1}. K^{-1})	450

R. Ben-Mansour (2017)

To ensure that the model is viable, further analysis on the work done by the engine is performed and compared with the experimental data of work done (Fig. 52.3) reported by Aksoy (Aksoy and Cinar 2013). The model then will be used for further evaluation with lower temperature of heat sources.

Geometry of Stirling engine used in CFD simulation is generated in SolidWorks 2019. Figure 52.2 shows the physical domain of Stirling engine with an outer diameter of 90 mm and a height of 213.12 mm. The displacer is located 13.98 mm from the bottom surface and has a diameter of 132.73 mm and a height of 155 mm (Rahman et al. 2022).

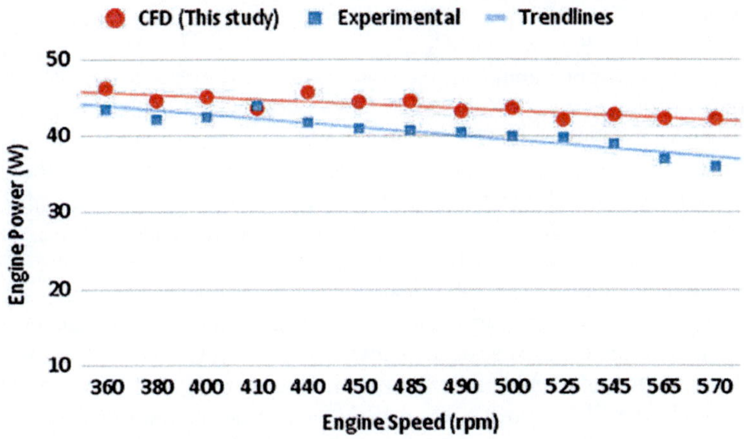

Fig. 52.3 Engine speed vs power

52.3 Equations

The work done of the system can be derived by integrating the area of P-V diagram (power vs volume) according to the Eq. (52.1) below.

$$W_{\text{out}} = \frac{\omega}{2\pi} \sum_{i=1}^{N} 0.5(p_{i-1} + p_i) \cdot (V_i - V_{i-1}) \qquad (52.1)$$

The model is validated by comparing the simulation results obtained from this study with the experimental results that were obtained in F. Aksoy (2013). Figure 52.3 shows the comparison of engine power-engine speed relation of experimental result and CFD result obtained from this study. The engine power is calculated by multiplying the engine work done with engine speed. The work done by the system can be derived by integrating the area of this P-V diagram according to the Eq. (52.1) stated (Abuelyamen 2017). The pressure value is taken from the pressure in the cold volume every time step of the engine cycle, while the volume is the area of the displacer surface and the height of cold volume in every time step. From this result, the trendlines is sketched in Fig. 52.3.

52.4 Results and Discussion

This section discusses the results of Stirling engine validation. The model is validated by comparing the simulation results obtained from this study with the experimental results. From the trendlines of CFD data and experimental data in Fig. 52.3, engine speed of 360–565 rpm, cold source temperature of 300 K, and hot source temperature of 775 K were determined.

For CFD results, maximum engine power was found to be 46.186 (W) 5.54 J at 360 rpm of engine speed which is 43.5(W), while the minimum engine power obtained was 42.07(W) which is 5.048 J at 525 rpm engine speed. The trendlines of experimental result and CFD result shows similar trendlines. i.e., the engine power declines when the engine speed increases. The trendlines of both results also show the average deviation of 6.93% for every engine speed applied.

The difference that occurs between this study and experimental results is due to the high engine speeds where the heat-exchange time decreases with increasing engine speed. Therefore, the different values of heat transfer coefficients used are crucial. In addition, the inconsistency between both models is due to the simplifications made in the modeling for the simulation process.

Above all, the CFD model of the Stirling engine designed in this study can be used for further studies.

52.5 Conclusion

A beta-type Stirling engine with a three-dimensional cylinder model is investigated in this study to present a baseline CFD model of Stirling engine. The validation of the baseline model of Stirling engine according to the previous researcher is presented in this study. The validation was done with different engine speeds and by comparing the trendlines of the power output based on every engine speed. It shows that the CFD results obtained are deviated averagely 6.93% from the actual experimental results.

- From the validated model shown, this study can be further improved to determine the best method for the heat transfer in Stirling engine as the waste heat recovery system in low-temperature heat source applications.
- The development of the Stirling engine in the waste heat recovery technology will give advantages in reducing the usage of the fossil fuel to generate power consumption which will lead to lower fuel use and ultimately reduced CO_2 emissions. It can also reduce the operational cost if the exhaust heat from the industrial boilers can be recovered.

Acknowledgments This research was funded by a grant from the Ministry of Higher Education of Malaysia (FRGS Grant: FRGS/1/2018/TK07/UPM/02/2).

References

A. Abuelyamen and E. M. A. Mokheimer, "CFD analysis of radiation impact on Stirling engine performance," Energy Convers. Manag., vol. 152, no. September, pp. 354–365, 2017. https://doi.org/10.1016/j.enconman.2017.09.056

B. D. B, L. Wang, M. Motta, and N. Karimi, "Modelling of waste heat recovery of a biomass combustion plant through ground source heat pumps- development of an efficient numerical framework," Appl. Therm. Eng., vol. 166, no. November 2019, p. 114625, 2020. https://doi.org/10.1016/j.applthermaleng.2019.114625

F. Aksoy and C. Cinar, "Thermodynamic analysis of a beta-type Stirling engine with rhombic drive mechanism," Energy Convers. Manag., vol. 75, pp. 319–324, 2013. https://doi.org/10.1016/j.enconman.2013.06.043

F. Li, L. Duanmu, L. Fu, and X. L. Zhao, "Research and Application of Flue Gas Waste Heat Recovery in Co-generation Based on Absorption Heat-exchange," Procedia Eng., vol. 146, pp. 594–603, 2016. https://doi.org/10.1016/j.proeng.2016.06.407

M. Bianchi and A. De Pascale, "Bottoming cycles for electric energy generation: Parametric investigation of available and innovative solutions for the exploitation of low and medium temperature heat sources," Appl. Energy, vol. 88, no. 5, pp. 1500–1509, 2011. https://doi.org/10.1016/j.apenergy.2010.11.013

N. K. M. N. A. Rahman, S. Saadon, and M. H. C. Man, "Waste Heat Recovery of Biomass Based Industrial Boilers by Using Stirling Engine," J. Adv. Res. Fluid Mech. Therm. Sci., vol. 89, no. 1, pp. 1–12, 2022. https://doi.org/10.37934/aram.81.1.110

W. Syakirah, W. Abdullah, M. Osman, M. Zainal, and A. Ab, "The Potential and Status of Renewable Energy," 2019. https://doi.org/10.3390/en12122437

W. L. Chen, K. L. Wong, and Y. F. Chang, "A computational fluid dynamics study on the heat transfer characteristics of the working cycle of a low-temperature-differential γ-type Stirling engine," Int. J. Heat Mass Transf., vol. 75, pp. 145–155, 2014. https://doi.org/10.1016/j.ijheatmasstransfer.2014.03.055

Z. Song, J. Chen, and L. Yang, "Heat transfer enhancement in tubular heater of Stirling engine for waste heat recovery from flue gas using steel wool," Appl. Therm. Eng., vol. 87, pp. 499–504, 2015. https://doi.org/10.1016/j.applthermaleng.2015.05.028

Chapter 53
A Feasibility Study of GCPV Solar Panels for Commercial Buildings

Parsa Kaviani and Zeynep Gergin

Nomenclature

C_t	Net cash flow in the time of t, US$
i	Real interest rate, %
I_{cc}	Initial capital cost, US$
t	Time, year
T	Project lifetime, years

53.1 Introduction

The consequences and impacts of climate change become more and more vivid every year, and countries are starting to adopt sustainable approaches in generating and utilizing energy through replacement of the use of fossil fuels with benefiting from renewable energies such as wind and solar natural resources. The European Union (EU) announced its vision in 2018, to introduce policies with the objective of preventing global temperature to rise 2 °C. The European Green Deal sets the targets for 2050 and Europe strives to be the first climate-neutral continent in the globe (2050 Long-Term Strategy 2017). According to the 2021 Global Status Report regarding Renewables in Cities published by REN21, 51% of the global energy consumption is related to heating and cooling and only 10.1% of this share is renewable energy (REN21 2021).

P. Kaviani (✉) · Z. Gergin
Department of Industrial Engineering, İstanbul Kültür University, Istanbul, Turkey
e-mail: z.gergin@iku.edu.tr

M. Z. Sogut et al. (eds.), *Proceedings of the 2022 International Symposium on Energy Management and Sustainability*, Springer Proceedings in Energy,
https://doi.org/10.1007/978-3-031-30171-1_53

Implementation of grid-connected photovoltaic (GCPV) solar panels is an option to support the use of renewable energy. One of the determinative elements in the feasibility assessment of photovoltaic panel application is the solar radiation rate and intensity of the region. Turkey's annual average solar radiation is 1525 kWh/m^2 year with an average daily sunshine duration of 7.5 h (Ekmekci and Şen 2016). The country is possessing 12.5% of the total geothermal capacity of the world and has considerable potential in fulfilling energy demand with benefiting from solar power, and this capacity can be utilized not only to reduce the consumption of coal, natural gas, and oil for energy production but also to decrease the country's dependency on other countries for energy supply (Simsek and Simsek 2013).

The variety of findings in the literature in the assessment of photovoltaic solar panels' efficiency and applicability can generally be divided into two main categories: the ones who found this application cost-effective and economically justifiable and those who hold the opposite view. Various measures were adopted to gauge the level of viability of the project such as discounted payback period (DPBP), profitability index (PI), internal rate of return (IRR), net present value (NPV), and levelized cost of energy (LCOE) via utilizing diverse methods to measure, such as HOMER Grid software and PVsyst software. Moreover, different billing schemes are also considered in such feasibility studies. Watts et al. (2015) compared the cost of energy with levelized cost of energy (LCOE) among residential households' photovoltaic panel systems across ten cities in Chile. Two billing schemes, namely, "net metering" and "net billing" were evaluated which differ in the rates that electric utility companies buy back the excess energy produced. This study showed that "net billing" is the superior approach as Chile benefits from high solar radiation and this can cover the cost of utility of a household. Bakhshi and Sadeh (2018) conducted an economic evaluation of 5 kW rooftop photovoltaic solar panels among 14 different cities in Iran. PVsyst software is utilized to apply solar radiations and calculate the efficiency of solar panels. The assessment showed that installing these panels would not only be beneficial in reducing the air pollution in major cities in Iran but also carries a noticeable profitability especially among commercial sector, and the payback period time (PBT) of the system would be 3.5 years. A comparative analysis between Germany and Spain's household photovoltaic panel installations from 2004 to 2014 indicated that comprehensive and long-term plan is the decisive element in prosperity or failure of these systems in a country (López Prol 2018).

In spite of the importance of this subject, the number of researches being conducted in this field is scarce, especially in Turkey and for the commercial sector. Duman and Güler (2020) modeled 5 kW rooftop photovoltaic panels on HOMER Grid software in order to assess the cost efficiency of these solar panels on residential buildings, considering current feed-in tariff (FiT) billing scheme. They selected nine different regions across Turkey to expand the comparison with regard to different solar radiations. It was concluded that based on current FiT and the initial costs of photovoltaic panels, only Antalya province in the southern part of Turkey benefits from the ideal conditions regarding solar radiation.

This study examines the feasibility of GCPV solar panels in commercial buildings in Istanbul with consideration of current capacities and regulations. The geographical characteristics and governmental supportive policies can be named as two main elements in determining the feasibility of a project in a country. However, the effect of these factors on commercial buildings in Turkey has not yet been studied, and the focus has mainly been on residential buildings. Evaluating the practicality of installing GCPV setups in commercial projects is especially valuable because of their energy utilization and carbon footprint which are considerably higher than residential buildings. Moreover, this consideration provides an opportunity to review and assess the present regulations with regard to renewable resources in the country and assist the corporations in their decision-making in the future.

53.2 Methodology

The methodology is comprised of the following order. The first step is the collection of solar radiations' data, and the location selection. The next step is introducing current FiTs as well as other applicable incentives following by current electricity rates and usage in commercial projects. Then all the attained data and information are entered to HOMER Grid software for simulation and calculation of the most efficient solution. Finally, the calculated results are assessed regarding three economic indices: discounted payback period (DPBP), simple payback period (SPBP), and internal rate of return (IRR). DPBP below 11 years, SPBP below 8 years, and IRR above 12% are considered as economically justifiable.

The following is the calculation of the economic determinants. The equations used in the calculation of DPBP, IRR, and SPBP are mentioned below as Eqs. 53.1, 53.2, and 53.3, respectively.

$$\sum_{t=1}^{\text{DBPB}} \frac{C_t}{(1+i)^t} = I_{cc} \tag{53.1}$$

$$\sum_{t=1}^{T} \frac{C_t}{(1+\text{IRR})^t} - I_{cc} = 0 \tag{53.2}$$

$$\text{SPBP} = \frac{I_{cc}}{\text{Annual Savings}} \tag{53.3}$$

Both DPBP and SPBP indicate the number of years required for the initial investment to be paid back by the gradual savings during the life cycle of the project. However, they do not represent the profitability of the project after the initial cost is met. In the calculation of DPBP, compound inflation is also being considered. This means that this indicator is more accurate than SPBP as it measures the time value of the money. IRR is an economical tool to gauge the profitability of a project. It can be explained as the annual discount rate at which the sum of initial capital of the project and the positive cash flow of profits equal to zero.

53.3 Implementation

53.3.1 Location and Solar Radiation Data

The province of Istanbul is selected as the target of this study due to its geographical significance and large population. Also, solar radiation in Istanbul is on the lower level in comparison with southern provinces, and this implies that Istanbul can be an appropriate base line for future decision in expanding renewable solutions' adaptation. The months of December and January experience the least amount of solar radiation with an average of 1.23 kWh/m2d. On the other hand, maximum radiation can be observed in the month of July with a daily average of 6.55 kWh/m2d. The annual daily average of solar radiation in Istanbul is 3.81 kWh/m2d (Topçu et al. 1995).

53.3.2 Current Rate of FiT and Electricity Rate in Turkey

The FiT for electricity generated from solar panels was set to be 13.3 USD cent/kWh (Kiliç 2011). It is better to mention "sunny days" as that is the sentence's emphasis on differentiating between sunny and cloudy daysIt should be mentioned that additional incentives would be applicable to the aforementioned FiT, and it can raise up to 20 USD cent/kWh if the components are manufactured in Turkey. However, for the simplicity of future calculations, it is assumed that the solar panel modules are imported. Also, electricity rate for businesses is 0.9 Turkish Lira/kWh, 11 USD cent/kWh with conversion rate of 7.8 TL for 1 USD (Turkey Electricity Prices 2020).

53.3.3 Electricity Usage Data and Optimization

Regarding electricity usage, a medium size shopping mall, with the surface area of 20,000 m^2, is selected. The annual consumption is selected as 430 kWh/m^2 referring to the work of Lam and Li (2003), and the daily and hourly electrical load are tailored according to climatic characteristics of Istanbul. July is the month with the highest average daily electrical usage of 31.6 MWh, and December has the lowest energy consumption of 15 MWh. The next step is defining the components of the system. In this study, Schneider Conext Core XC 630 kW is selected as the inverter with generic photovoltaic solar panel modules. The cost of the system is estimated to be one million US dollars. This includes the cost of the inverter, photovoltaic modules, taxes, labor, and logistics which are assigned as USD1.6/W (Cox 2020). Additionally, half-cut modules with 144 cells and the estimated power output of 620 W and the size of 1.3 m by 2.4 m are selected (Svarc 2021). To convert the DC power generated by the solar panels to AC, Leonics GTP-518HET(P) 680 kW

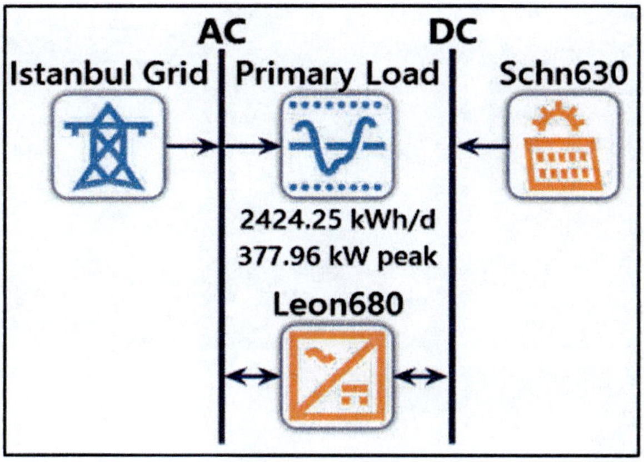

Fig. 53.1 Schematic presentation of the system

Table 53.1 PV module dimensions and required surface area

Module	Dimensions	Surface area of each module	Power output per module	Total power output required	Total number of modules require	Total surface area required
Half-cut 144 cell	1.3 m × 2.4 m	3.12 m^2	620 kW	630 kW	1016	3167 m^2

converter is selected with the capacity of 680 kW. The diagram of the modeled system is shown in Fig. 53.1.

53.3.4 Surface Area Required for PV Modules

The size characteristics of the modules are listed in Table 53.1. Accordingly, 3167 m^2 of space is required to get an output of 630 kW with the selected modules. To put 3167 m^2 of rooftop space in perspective, Fig. 53.2 shows the top view of Marmara Park Shopping Mall in Esenyurt suburb in Istanbul. The specified area in red is the C-shape front section of the shopping center. The surface area of this section is 3742 m^2 which is less than 20% of the total surface of the roof and 575 m^2 more than the required space for the model. Therefore, it can be stated that the necessary space for the model in this research is reasonable.

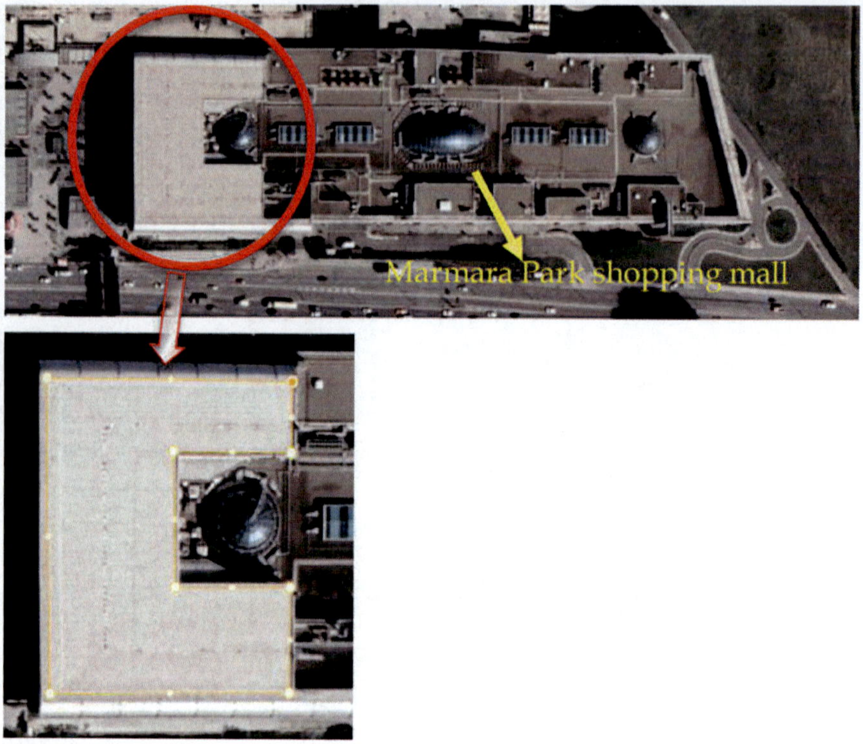

Fig. 53.2 Marmara Park rooftop space. (Google Earth, 2021)

53.4 Results and Discussion

HOMER software simulated 85 possible scenarios considering all the possible combinations with or without solar panels and suggested the most cost-effective option. It should be noted that solar panels can only generate energy during the sunny days and their electricity production is reduced on cloudy and rainy days. Therefore, grid connectivity and purchase form the grid is inevitable.

DPBP is calculated as 14.6 years which is relatively high and indicates with current FiT and cost of equipment; thus, it is economically hard to justify investing in solar panels. IRR for this project is 9.28% which is higher than the real interest rate of 8%. This indicates that the project is profitable and generates cash. However, the difference is not large enough for the project to be easily justifiable. The value of SPBP for this project is 9.6 years which is acceptable; however, the combination of various indices should be evaluated to decide on the feasibility of a project.

Carbon dioxide emissions' reduction can also be an influential factor that demonstrates another valuable side of integrating sustainable approaches in energy consumption. In the case of absence of solar panels (i.e., using grid line only), the annual CO_2 emission production is 559.2 metric tons. After the installation of panels, this number reduces to 217.5 tons per year, that is, a total of 341.7 tons per annum savings in CO_2 emissions which is a noticeable amount.

53.4.1 Sensitivity Analysis

A sensitivity analysis is being conducted to consider four possible impactful changes to feed-in tariff rate as well as capital cost of GCPV systems. The reduction in GCPV systems can be accommodated by applying incentives, tax exemptions, and low interest loans as well as promoting local production of modules and inverters. The alternative scenarios are listed below:

Scenario 1. Considering nationally manufactured equipment with no increase in the cost of the system. This is the most reasonable assumption as the manufacturing of equipment for PV systems is growing in Turkey and also the FiT rate for domestically manufactured modules, inverters, cells, and structural elements required for the assembly of the system is 20 USD cent/kWh which can provide an advantage for both the manufacturers and the enterprises willing to invest in renewable resources. The result of this change gives the DPBP, SPBP, and IRR values of 10.9, 7.9, and 11.96% respectively, which all are at the limit set in this research.

Scenario 2. Considering nationally manufactured equipment with 10% reduction in the initial cost of the system. The possibility of this situation is also predictable, because of locally produced materials and equipment that generally cost less than the imported systems. The results showed that the values of the indices can further drop to 9.8, 7.3, and 13.13% for DPBP, SPBP, and IRR respectively.

Scenario 3. 20% reduction in the prices of solar panels and inverters combined with 20% increase of the current FiT rate. Assuming this will positively affect the viability of the project, the value of FiT will raise to 16 USD cent/kWh and the cost of the system reduces to US$800,000 with DPBP of 10.1 years, SPBP of 7.5 years, and IRR value of 12.73%. Hence, installation of GCPV panels is being considered as economically practical for this scenario. The values of aforementioned indices indicate that the system can be announced as economically justifiable. This change also increases the annual saving of energy bills and lifetime saving of the project to $204,133 and $5,103,326, respectively.

Scenario 4. 30% reduction in the prices of solar panels and inverters combined with 30% increase of the current FiT rate. This amendment further supports the viability of the system. The cost of the PV system is cut to US$700,000 and the rate of FiT increases to 17.29 USD cent/kWh in this scenario. The result of this combination is evidently superior to the previous case with DPBP of 8.4 years, SPBP of 6.5 years, and IRR of 15%.

53.5 Conclusion

It is evident that multiple criteria should be monitored in the process of applying sustainable methods in commercial buildings. The results of feasibility assessment in this study indicate that, in spite having a savings of $3,802,651 during the lifetime of the project, with DPBP value of 14.6 years, SPBP of 9.6 years, and IRR of 9.28%, installation of GCPV system is assessed as infeasible. The sensitivity analysis is also conducted among four different possible situations to investigate the effects of changes in current FiT rate as well as capital cost of PV systems. It is concluded that 30% reduction in equipment cost and 30% increase in the FiT rate is the most cost-effective scenario with DPBP of 8.4, SPBP of 6.5, and IRR value of 15%. DPBP below 11 years, SPBP below 8 years, and IRR above 12% are considered as economically justifiable.

Considerable energy savings and reduction in harmful gas emissions are the two major values of this application. This is especially applicable in commercial buildings as their environmental footprint is noticeably larger than residential buildings and the use of sustainable resources in their energy production can have a positive impact on the planet. On the other hand, some barriers may complicate the process and make the project infeasible such as lack of infrastructure or plan for a national adaptation, cost of equipment, lack of suitable natural potential, and lack of available space.

Taking everything into consideration, commercial buildings that are located in sunny and open areas can benefit the most from this environmentally friendly tactic not only to save on their energy bills but also to contribute to minimizing their carbon emissions and preserve the planet for future generations.

Lastly, it is highly recommended for future studies to closely observe governmental legislations as they are dynamic, and it is likely for new and supportive regulations to be introduced, especially in a country like Turkey with a great potential with regard to solar resources and assist the companies that are willing to take the sustainable way in their projects.

References

2050 long-term strategy. 2017, February 16, Climate Action – European Commission. Retrieved at 01.03.2021 from https://ec.europa.eu/clima/policies/strategies/2050_en

Bakhshi, R., & Sadeh, J., 2018, Economic evaluation of grid–connected photovoltaic systems viability under a new dynamic feed–in tariff scheme: A case study in Iran. Renewable Energy, 119: 354–364.

Cox, M., 2020, December 17, Key 2020 US Solar PV Cost Trends and a Look Ahead. Wood Mackenzie. Retrieved at 01.03.2021 from https://www.greentechmedia.com/articles/read/key-2020-us-solar-pv-cost-trends-and-a-look-ahead#:%7E:text=In%20the%20residential%20segment%2C%20average,monofacial%2C%20monocrystalline%20PERC%20solar%20modules.&text=However%2C%20without%20the%20bifacial%20exemption,than%20monofacial%2C%20monocrystalline%20PERC%20prices

Duman, A. C., & Güler, N., 2020, Economic analysis of grid-connected residential rooftop PV systems in Turkey. Renewable Energy, 148: 697–711.

Ekmekci, İ., & Şen, Y. 2016, A Study on Solar Radiation Calculation for Istanbul with Measured Data. International Journal of Solar Energy Research, 1(2):1–6.

Google Earth. (2021, April 3). Istanbul, Turkey. https://earth.google.com/web/@41.00970992,28.65976763,135.88524039a,348.95913147d,35y,24.45902676h,0t,0r

Kiliç, F., 2011, Recent renewable energy developments, studies, incentives in Turkey. Energy Education Science and Technology Part A-Energy Science and Research, 28: 37–54.

Lam, J. C., & Li, D. H., 2003, Electricity consumption characteristics in shopping malls in subtropical climates. Energy Conversion and Management, 44(9): 1391–1398.

López Prol, J., 2018, Regulation, profitability and diffusion of photovoltaic grid-connected systems: A comparative analysis of Germany and Spain. Renewable and Sustainable Energy Reviews, 91: 1170–1181.

REN21. 2021, Renewables In Cities 2021 Global Status Report. Retrieved at 01.03.2021 from https://www.ren21.net/wp-content/uploads/2019/05/REC_2021_full-report_en.pdf

Svarc, J., 2021, April 21, Most efficient solar panels 2021 — Clean Energy Reviews. Clean Energy Reviews. Retrieved at 01.02.2021 from https://www.cleanenergyreviews.info/blog/most-efficient-solar-panels

Simsek, H. A., & Simsek, N. 2013, Recent incentives for renewable energy in Turkey. Energy Policy, 63: 521–530.

Topçu, S., Dİlmaç, S., & Aslan, Z., 1995, Study of hourly solar radiation data in Istanbul. Renewable Energy, 6(2): 171–174.

Turkey electricity prices_September 2020, 2020, October 1, Retrieved at 03.01.2021 from https://www.globalpetrolprices.com/Turkey/electricity_prices/#hl226

Watts, D., Valdés, M. F., Jara, D., & Watson, A., 2015, Potential residential PV development in Chile: The effect of Net Metering and Net Billing schemes for grid-connected PV systems. Renewable and Sustainable Energy Reviews, 41: 1037–1051.

Chapter 54
Going on Energy Control Management Framework Based on Trigeneration Systems: A Case Study

Ozay Kas and M. Ziya Sogut

Nomenclature

CHP	Combined heat and power
CCHP	Combined cooling heat and power
CHP	Combined heat and power
ASHRAE	American Society of Heating Refrigerating and Air-Conditioning Engineers
HVAC	Heating, ventilation, and air-conditioning
LiBr	Lithium bromide
USD	United States dollar
PES	Primary energy saving

54.1 Introduction

The building sector is a prominent player in total energy consumption with different building usage characteristics. In this building formation, integrated structures use multiple energy resources depending on their multipurpose utilization. In this context, due to the increasing awareness in recent years, the tendency to low carbon technologies and integrated power generation technologies stands out. These technologies are mainly used as cogeneration (CHP) systems where heat energy is evaluated together with electrical energy and trigeneration (CCHP) systems where cooling power is added to these outputs (Nesheim and Ertesvag 2007). CCHP

O. Kas (✉) · M. Z. Sogut
Maritime Faculty, Piri Reis University, İstanbul, Türkiye
e-mail: ozaykas@alumni.pru.edu.tr; mzsogut@pirireis.edu.tr

M. Z. Sogut et al. (eds.), *Proceedings of the 2022 International Symposium on Energy Management and Sustainability*, Springer Proceedings in Energy,
https://doi.org/10.1007/978-3-031-30171-1_54

systems can be prioritized within the scope of efficient energy management by considering multiple energy demands in integrated structures. However, energy consumption control, efficient planning of load distributions, and technology management of system components are prominent issues in such system preferences for such evaluation processes. Within this structure, it can be seen in studies that especially CCHP systems have the potential to increase operational efficiency (Cho et al. 2014).

Today, the economic and environmental efficiency of CCHP technologies is evaluated by many criteria (Wang et al. 2018). During the design phase, structural features are primarily considered, and energy demand in buildings generally depends mainly on the building components and the utilization characteristics of the areas. However, environmental parameters such as climate conditions are still included as the main criteria in system capacity calculations. It has been observed in field studies that especially cold climate conditions directly affect the performance of CCHP systems (Boschiero 2014). The performance analysis of such systems generally depends on thermodynamic principles. In addition to electricity demand for many integrated building examples, heating and cooling requirements are concurrent demands. Single-effect LiBr-water type absorption cooling technology is mostly preferred in such applications because of its ability to serve heating and cooling systems simultaneously (Ghafurian and Niazmand 2018). Thus, the heat energy obtained from the cogeneration unit can be served to both the cooling and the heating system simultaneously with the flowchart formed in serial connection. While heating efficiency is related to the efficient use of the heat source in thermal processes, the coefficient of performance (COP) value of an absorption cooling unit depends on the temperature of the heat energy supplied to the generator circuit.

In integrated structures, parts of the building or areas are subject to dynamic variations in energy demands due to different utilization purposes and the purpose of use of the building or its related part, climatic conditions, human density, equipment, and equipment infrastructure that are the primary inputs of this change (Martínez-Lera and Ballester 2010). However, determining the design capacity and operating strategy of a CCHP system for the air-conditioning requirements of such structures is a complicated process due to the complex interactions between efficiency and economic parameters (Wu et al. 2019).

In CCHP systems and applications, the heat energy obtained from the system is prioritized for utilization in the heating system in some cases, and it provides higher economic savings depending on the fuel prices (Agarwal et al. 2020). However, prioritizing the heating system in this way in the flow diagram creates an inevitable drop in the coefficient of performance of the absorption cooling unit (Manu and Chandrashekar 2016). In this study, the prioritization of the waste heat obtained from the CCHP unit depending on the energy demand is examined in terms of thermo-economic analysis, and the decision processes are determined. In addition, a flowchart has been proposed to ensure that cooling performance does not degrade under heating priority operating conditions. Thus, this study aims to contribute to the efficient and effective operation of CCHP systems.

54.2 Trigeneration Systems

CCHP technologies are advanced systems designed to meet different energy demands in the building directly and industrial sectors. However, such technologies directly involve effective design and engineering studies. In this respect, while CCHP control strategy and optimization algorithms to be developed accordingly provide investment advantages for operations, they also provide economic sustainability (Mavromatidis et al. 2018). In absorption technologies, their effectiveness is defined by COP, as in vapor compression refrigeration applications. In these systems, based on waste heat characteristics, as the heat and temperature entering the generator increase, the COP value of the system will increase (ASHRAE 2015).

Although the absorption chillers utilizing high-temperature steam and exhaust media have higher COP values than hot water types, they cannot serve the systems in integrated structures where hot water is used for heating and domestic hot water consumption and chilled water for heating and domestic hot water consumption HVAC systems concurrently. A conventional single-stage LiBr-water CCHP flow diagram where the heat supplied from CHP unit(s) is utilized simultaneously at the cooling and heating system allows the building of the system cooling system primarily, because otherwise, it will be an inevitable decrease in the COP value of the absorption chiller. A modified single-stage LiBr-water CCHP flow diagram has been developed to overcome this disadvantage, as a part of this study, where it is possible to build the heating system primarily without any decrease in COP value of absorption chiller as shown in Fig. 54.1.

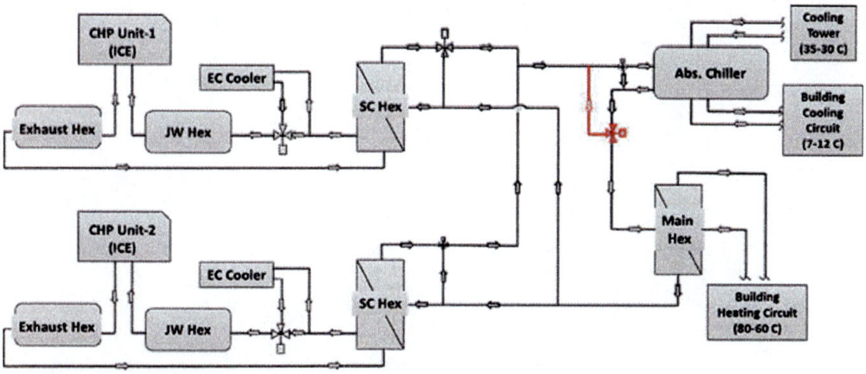

Fig. 54.1 Modified single-stage LiBr-water CCHP flow diagram

54.3 Thermo-economic Analysis

CCHP systems are complex structures that operate with multiple thermodynamic cycles. As with all thermodynamic systems, CCHP processes are also affected by environmental conditions. In these systems operating under continuous flow conditions, each cycle works based on the conservation of mass principle and heat demand management. In this study, the thermodynamic analysis of the CCHP system application for an integrated structure determined in reference conditions was made. The system diagram used for the proposed system model is presented in Fig. 54.2.

The total electricity demand of the integrated structure is the sum of the facility itself and the conventional chiller where it is supplied from CHP and grid network, which is expressed as follows:

$$\dot{W}_f + \dot{W}_{ch} = \dot{W}_{chp} + \dot{W}_g \tag{54.1}$$

where $\dot{W}_f$ is the total electrical demand of integrated structure (kW), $\dot{W}_{ch}$ is the electrical input of a conventional chiller (kW), $\dot{W}_{chp}$ is the electrical output of the cogeneration unit (kW), and $\dot{W}_g$ is the electricity from the grid to the integrated structure (kW). The total heating demand of the integrated structure is supplied from cogeneration and hot water boiler and expressed as follows:

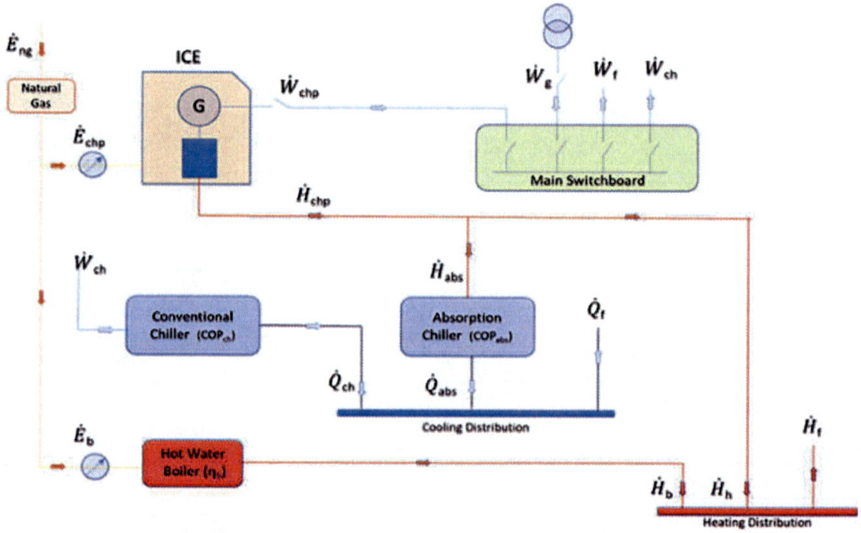

Fig. 54.2 Simplified diagram of an integrated structure with trigeneration system

$$\dot{H}_f = \dot{H}_h + \dot{E}_b.\eta_b \tag{54.2}$$

where $\dot{H}_f$ is the total heating demand of integrated structure (kW), $\dot{H}_h$ is the heat input of the heating system from the CHP unit (kW), $\dot{E}_b$ is the fuel input of the hot water boiler (kW), and η_b is the efficiency of the hot water boiler. Heat supplied from the CHP unit can be utilized in the cooling or heating system, or both at the same time:

$$\dot{H}_{chp} = \dot{H}_{abs} + \dot{H}_h \tag{54.3}$$

where $\dot{H}_{chp}$ is the heat output of the cogeneration unit (kW) and $\dot{H}_{abs}$ is the heat input of the absorption chiller from the CHP unit (kW). The total cooling demand of the integrated structure is supplied from the absorption chiller and conventional hot water boiler and expressed as follows:

$$\dot{Q}_f = \left(\dot{H}_{abs}.COP_{abs}\right) + \left(\dot{W}_{ch}.COP_{ch}\right) \tag{54.4}$$

where $\dot{Q}_f$ is the total cooling demand of integrated structure (kW), COP_{abs} is the absorption chiller's coefficient of performance, and COP_{ch} is the conventional chiller's coefficient of performance. If the heat supplied from the CHP unit is utilized in the *cooling system primarily* as shown in Eq. (54.4), then the heating demand of the integrated structure is supplied from the CHP unit, and the hot water boiler is expressed as follows:

$$\dot{H}_f = \left(\dot{H}_{chp} - \dot{H}_{abs}\right) + \dot{E}_b.\eta_b \tag{54.5}$$

If $\dot{Q}_f > \dot{H}_{abs}.COP_{abs}$ which means the cooling supplied by the CHP unit for the cooling system is not enough for the integrated structure's cooling demand, then there will be no remaining heat for the heating system by the CHP unit; $\left(\dot{H}_{chp} - \dot{H}_{abs}\right) = 0$. All heating demand should be supplied by a hot water boiler: $\dot{H}_f = \dot{E}_b.\eta_{b.}$

If the heat supplied from the CHP unit is utilized in the *heating system primarily* as shown in Eq. (54.2), then the cooling demand of the integrated structure is supplied by the absorption chiller, and the conventional chiller is expressed as follows:

$$\dot{Q}_f = \dot{Q}_{abs} + \dot{Q}_{ch} = \left(\dot{H}_{abs}.COP_{abs}\right) + \left(\dot{W}_{ch}.COP_{ch}\right) \tag{54.6}$$

where $\dot{Q}_{abs}$ is the cooling output of the absorption chiller (kW). If $\dot{H}_f > \dot{H}_h$ which means the heat supplied by the CHP unit for the heating system is not enough for the integrated structure's heating demand, then there will be no remaining heat for the cooling system by CHP unit; $\left(\dot{H}_{chp} - \dot{H}_h\right) = 0$. All cooling demand should be supplied by a conventional chiller: $\dot{Q}_f = \dot{W}_{ch}.COP_{ch}$ Economic saving of the heat supplied by the CHP unit for the heating system is expressed as follows:

$$C_{\text{save}_h} = \sum_{t=1}^{n} \frac{\dot{H}_{chp(t)} \cdot C_f}{\eta_b} \tag{54.7}$$

where C_{save_h} is the economic saving of the heat supplied by CHP unit for the heating system and C_f is the unit cost of natural gas. Similar to Eq. (54.7), the economic saving of the heat supplied by the CHP unit for the cooling system is expressed as follows:

$$C_{\text{save}_q} = \sum_{t=1}^{n} \frac{\dot{H}_{chp(t)} \cdot COP_{abs} \cdot C_e}{COP_{ch}} \tag{54.8}$$

where C_{save_q} is the economic saving of the heat supplied by the CHP unit for the cooling system and C_e is the unit cost of electricity. If $C_{\text{save}_h} > C_{\text{save}_q}$, which means the economic saving of the heat supplied by the CHP unit for the heating system is more than for the cooling system, then the heat shall primarily be utilized in the heating system.

Thus, if

$$\frac{C_f}{\eta_b} > \frac{COP_{abs} \cdot C_e}{COP_{ch}} \tag{54.9}$$

Heating System Primarily.

Thus, if

$$\frac{C_f}{\eta_b} < \frac{COP_{abs} \cdot C_e}{COP_{ch}} \tag{54.10}$$

Cooling System Primarily.

54.4 Results and Discussion

Real field data belong to an integrated building with a two-block hospital structure with a 72,000 m^2 indoor area. The integrated structure serves seven days and 24 h without any interruption. The structure is located in Sariyer province of Istanbul. All climate conditions and parameters of Istanbul are applicable to this complex. The energy demand distribution of the facility is shown in Fig. 54.3, where annual electricity demand (excluding chillers' consumption) is 11,444 MWh/year, cooling demand is 12,103 MWh/year, heating demand is 6958 MWh/year, and domestic hot water demand is 3140 MWh/year.

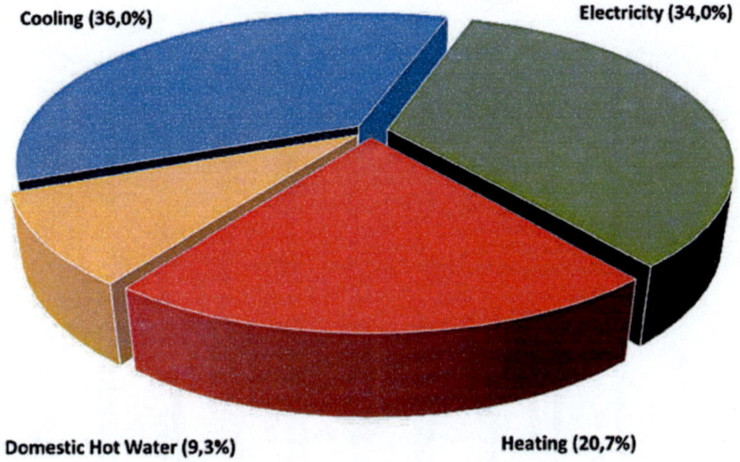

Fig. 54.3 Energy demand distribution

The COP value of the conventional chillers is 5.5, and the annual electricity demand of the chillers is 2200 MWh/year. Therefore, yearly electricity demand (including chillers' consumption) is 13,644 MWh/year.

The active use of air-conditioning processes in the integrated structure emphasizes electricity consumption. In addition, demand for heating and cooling in air-conditioning plants has a potential of approximately 21% and 36%, respectively, in direct conventional load. The share of consumption distribution according to resources in energy management is also examined. According to the review, monthly change in electricity, cooling, heating, and domestic hot water demand distribution is shown in Fig. 54.4. Seasonal effects are remarkable in energy consumption. 17% of the total energy heat demand stands out, especially for the winter months. Controlling seasonal transitions in heat demand management, thermal management of air-conditioning processes should be evaluated in a structure that needs attention. The cooling demand required depending on the monthly consumption in the enterprise is 12,103 MWh/year, and it has a potential of 36% of total energy demand.

The trigeneration processes designed for the enterprise have been developed as a model based on primary energy consumption, primarily electricity. The facility has a trigeneration plant composed of two internal combustion engines (ICE), two exhaust heat exchangers, and one piece of single-effect LiBr-water absorption chiller. Basic parameters of the trigeneration plant are shown in Table 54.1 according to technical specifications by manufacturers.

The effectiveness of trigeneration processes is primarily aimed at optimum covering energy demand. Resource priority and coverage rates are valuable in system preferences in this context. The CCHP plant realized 10,196 MWh of electricity (74.7% of total electricity demand), 6527 MWh of heating (64.6% of total heating demand), and 4190 MWh of cooling energy (34.6% of total cooling

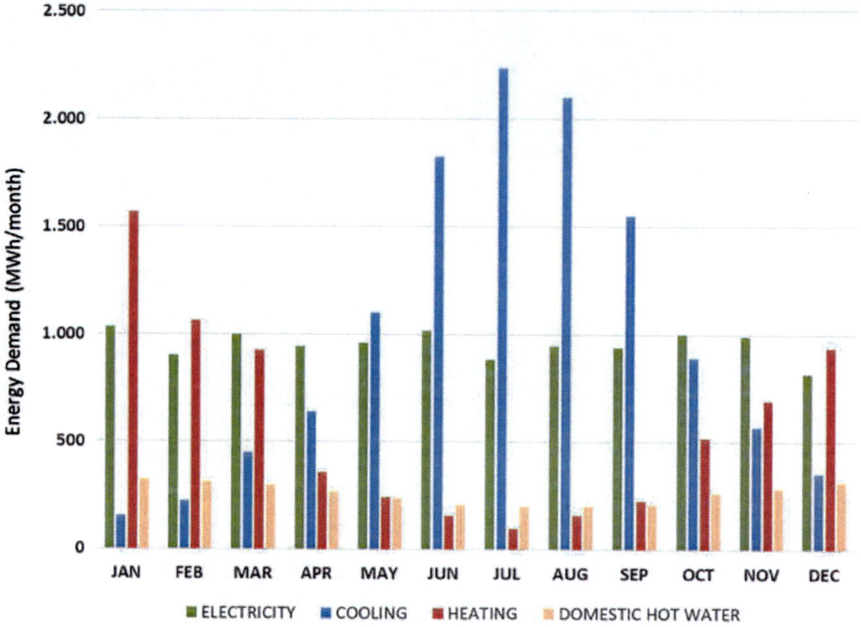

Fig. 54.4 Energy demand distribution monthly change

Table 54.1 Basic parameters of the CCHP plant

Description	Unit	Data
ICE electrical output	kW	800
ICE heat output (95–75 °C)	kW	826
ICE fuel input	kW	1845
ICE electrical efficiency	%	43.3
ICE heat efficiency	%	44.8
ICE total efficiency	%	88.1
Exh. T at ICE outlet	°C	425
Exh. T at Exh. Exch. outlet	°C	120
Exh. gas amount	kg/h	4348
Exh. heat Exch. capacity	kW	404
Abs. chiller cooling capacity	kW	1100
Abs. chiller COP value	–	0.7

demand) for the facility, while the system is operated in heating priority mode as shown in Fig. 54.1. CCHP plant's entire annual operation period is 8.140 h, where the total electricity generation capacity utilization rate is 78.3%. During the same period, total generated waste heat is 13,045 MWh/year, and 96% of it is utilized by heating and/or cooling system.

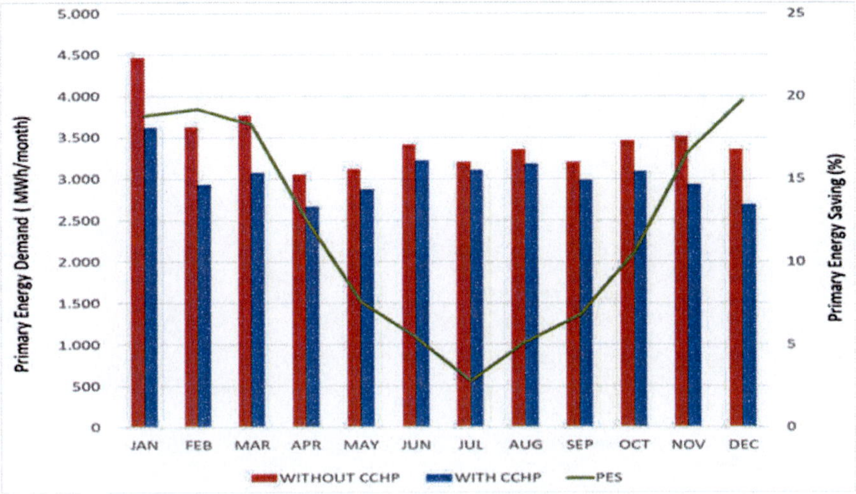

Fig. 54.5 Primary energy distribution and savings monthly change

The primary energy demand of the facility was 41,442 MWh/year regarding 0.525 thermal power plant's reference electricity efficiency, 14% transmission and distribution loss, and 0.90 hot water boiler efficiency before the CCHP plant's installation. Primary energy demand is reduced to 36,330 MWh after the CCHP plant's installation, which provides 12.33% primary energy saving (PES). The primary energy distribution of monthly change and primary energy saving (PES) is shown in Fig. 54.5.

Trigeneration plant provides 421,596 USD/year cost saving with utilized heat for heating and cooling system in heating priority mode, where the share for heating is 66.6% and 33.4% for cooling. If the plant were operated in cooling priority mode, then cost saving would be 327,833 USD/year, where the share for heating is 44.2% and 55.8% for cooling. Thus, the modified system developed depending on demand management provides approximately 28.6% extra savings from the utilized heat for heating and cooling.

A cost-saving comparison of heating and cooling priority modes is shown in Fig. 54.6. The cost-saving difference is significant in the spring and autumn seasons, where heating and cooling loads are relatively near to each other.

54.5 Conclusions

This study has investigated efficiency analysis to be provided in the system according to the preference of heating and cooling demands in a trigeneration system according to thermo-economic principles. The study has shown a direct relation between thermo-economic parameters, energy cost, and efficiencies.

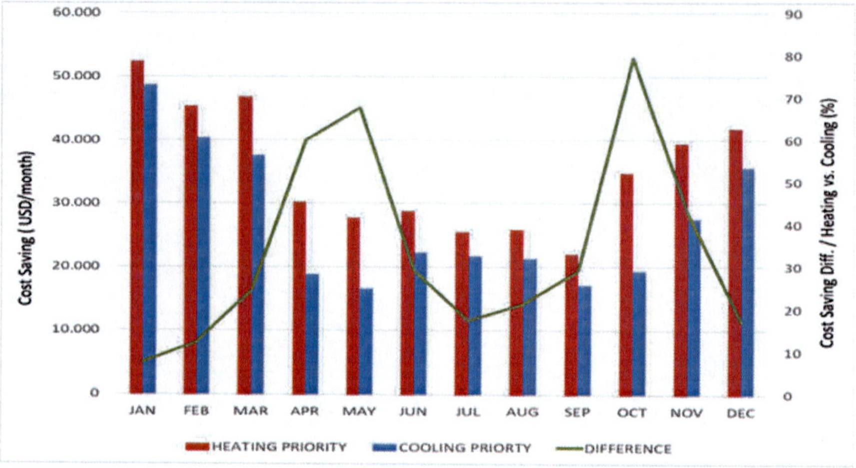

Fig. 54.6 Cost-saving comparison of heating and cooling priority modes

Conventional trigeneration and modified systems were examined comparatively based on demand management optimization on real field study data. In the analysis, it was seen that the modified system developed depending on demand management provides approximately 28.6% extra savings from the utilized heat for heating and cooling.

References

Agarwal S, Arora A, Arora B.B (2020) Energy and exergy analysis of vapor compression–triple effect absorption cascade refrigeration system. Engineering Science and Technology and International Journal 23: 625–641. https://doi.org/10.1016/j.jestch.2019.08.001

ASHRAE (2015) Combined Heat and Power Design Guide, RP-1592, ISBN 978-1-936504-87-9

Boschiero D (2014) An energy and exergy analysis of a high-efficiency engine trigeneration system for a hospital: A case study methodology based on annual energy demand profiles. Energy and Buildings, 76: 185–198. https://doi.org/10.1016/j.enbuild.2014.02.014

Cho H, Smith A.D, Mago P (2014) Combined cooling, heating and power: A review of performance improvement and optimization. Applied Energy 136: 168–185. https://doi.org/10.1016/j.apenergy.2014.08.107

Ghafurian M, Niazmand H (2018) New approach for estimating the cooling capacity of the absorption and compression chillers in a trigeneration system. International Journal of Refrigeration, 86: 89–106. https://doi.org/10.1016/j.ijrefrig.2017.11.026

Manu S, Chandrashekar T.K (2016) A simulation study on performance evaluation of single-stage LiBr–H_2O vapor absorption heat pump for chip cooling. International Journal of Sustainable Built Environment 5: 370–386. https://doi.org/10.1016/j.ijsbe.2016.08.002

Martínez-Lera S, Ballester J (2010) A novel method for the design of CHCP systems for buildings. Energy 35: 2972–2984. https://doi.org/10.1016/j.energy.2010.03.032

Mavromatidis G, Orehounig K, Carmeliet J (2018) Design of distributed energy systems under uncertainty: A two-stage stochastic programming approach. Applied Energy 222: 932–950. https://doi.org/10.1016/j.apenergy.2018.04.019

Nesheim S.J, Ertesvag I.S (2007) Efficiencies and indicators defined to promote combined heat and power. Energy Conversion and Management 48:1004–1015. https://doi.org/10.1016/j.enconman.2006.08.001.

Wang X, Yang C, Huang M, Ma X (2018) Off-design performances of gas turbine-based CCHP combined with solar and compressed air energy storage with organic Rankine cycle. Energy Conversion and Management 156: 626–638. https://doi.org/10.1016/j.enconman.2017.11.082

Wu D., Han Z, Liu Z, Zhang H (2019) Study on configuration optimization and economic feasibility analysis for combined cooling, heating and power system. Energy Conversion and Management 190: 91–104. https://doi.org/10.1016/j.enconman.2019.04.004

Chapter 55
Comparison of Biofuels for Decarbonized Maritime Transportation

Cagatayhan Sevim and Burak Zincir

Nomenclature

IMO	International Maritime Organization
LCA	Life cycle assessment
ILUC	Indirect land-use change
UCO	Used cooking oil
FAME	Fatty acid methyl ester
HVO	Hydrotreated vegetable oil
FT	Fischer-Tropsch
DME	Dimethyl ether
WTW	Well-to-wake

55.1 Introduction

Maritime transportation is the most efficient type of transportation and logistics, thanks to the internal combustion diesel engines used. The United Nations Conference on Trade and Development (UNCTAD) announced that the commercial shipping fleet has grown by 3% between 1 January 2020 and 2021, and the number of ships of 100 gross tons and above has reached 99,800 (UNCTAD 2021). Maritime

C. Sevim (✉)
Naval Architecture and Maritime Faculty, Yildiz Technical University, Istanbul, Turkey
e-mail: sevimca@yildiz.edu.tr

B. Zincir
Maritime Faculty, Istanbul Technical University, Istanbul, Turkey
e-mail: bzincir@itu.edu.tr

M. Z. Sogut et al. (eds.), *Proceedings of the 2022 International Symposium on Energy Management and Sustainability*, Springer Proceedings in Energy,
https://doi.org/10.1007/978-3-031-30171-1_55

transportation emits a substantial amount of greenhouse gas as a result of the vast majority of these ships consuming fossil fuels. According to the European Environment Agency (EEA), ships are responsible for 0.61% of global CO emissions, 9.84% of SO_x emissions, 14.74% of NOx emissions, and 6.75% and 3.56% of PM2,5 and PM10 emissions, respectively (EEA 2021). In addition, in its fourth greenhouse gas study, the International Maritime Organization (IMO) stated that 2.89% of global CO_2 emissions originate from shipping (IMO 2020). The IMO, the maritime transportation policymaker, is tightening its emission policies day by day to diminish shipboard GHG emissions. IMO adopted the first greenhouse gas strategy plan in 2018 to reduce greenhouse gases and has set a goal within this framework. The main goals of IMO's initial strategy plan are to reduce global ship-source emissions by 50% and mitigate CO_2 emissions per unit transport work to at least 70% by 2050, compared to that of 2008 (Joung et al. 2020). The strategy plan also includes short-, mid-, and long-term candidate measures to achieve these basic objectives. Short-term measures cover 2018–2023, mid-term measures cover 2023 and 2030, and long-term measures cover 2030 and beyond. Among these measures, alternative marine fuels have an important position, especially for the mid- and long term (Rutherford and Comer 2018). New rules and regulations regarding emissions are pushing researchers to seek alternative marine fuels from clean and renewable sources. There are several alternative fuel pathways to reach decarbonized maritime transportation such as liquefied natural gas (LNG), liquefied petroleum gas (LPG), hydrogen, ammonia, fully electric, methanol, and biofuels (Ryste et al. 2019). Each fuel pathway has some advantages and disadvantages in terms of its use in ships, due to its characteristic features. In this study, biofuels were examined because biofuels have a significant advantage in the transition to alternative fuels for decarbonizing maritime transportation. They can be blended with petroleum-derived conventional fuels or used directly as a drop-in in the existing ship infrastructure (Ryste et al. 2019; Sevim and Zincir 2022).

Biofuel encompasses a variety of liquids or gases that can be produced from many different types of biomass or biological wastes. For example, vegetable oils such as palm, soybean, and rapeseed; inedible vegetable oils such as jatropha and pongamia; lignocellulosic biomass such as miscanthus, corn stover; and used cooking oils (UCO), tallow, and microalgae biomass (Lin and Lu 2021; Mohd Noor et al. 2018). Biofuels are expressed as three generations according to the raw material used and the production process. The first generation mainly consists of edible food crops. Second-generation biofuels consist of nonedible plant oils and lignocellulosic raw materials. The third-generation biofuel is obtained from microalgae biomass and studies are still ongoing (Sevim and Zincir 2022).

This study aims to determine the most suitable biofuel to decarbonize shipping. In this context, selected biofuels were grouped according to raw material and production process. To compare the net carbon impact of selected biofuels, the LCA approach is explained. The indirect land-use change (ILUC) effect of selected biofuels was also examined in the LCA approach adopted. After this preliminary evaluation, the feedstock availability, cost, technological maturity, safety, and compatibility of the selected biofuels with existing ships (required modifications and challenges) were examined, and the most suitable one for maritime transportation was tried to be determined.

55.2 Preliminary Grouping of Biofuels

Biofuels can be produced using different raw materials and different processes. The biofuels selected in accordance with the purpose and scope of this study are shown in Fig. 55.1 with the raw materials and the generation they belong to.

55.3 Life Cycle Assessment Approach

The current monitoring regulatory infrastructure in maritime transportation focuses more on the exhaust emissions from combustion on ships. Today, there are two regulations for monitoring, reporting, and verification of emissions from maritime transport. These are the European Union MRV (EU MRV) regulation and the IMO

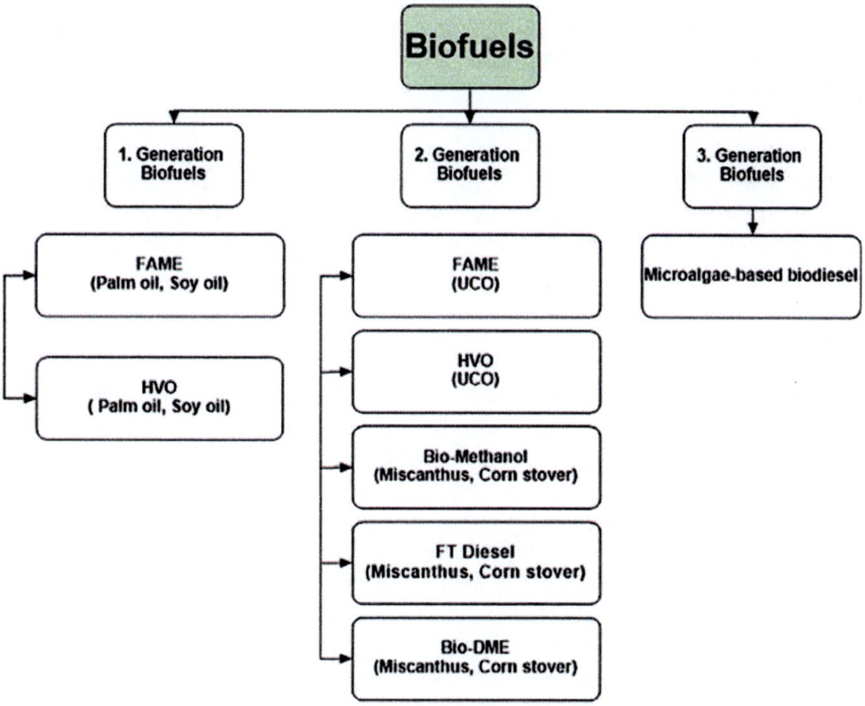

Fig. 55.1 Selected biofuels

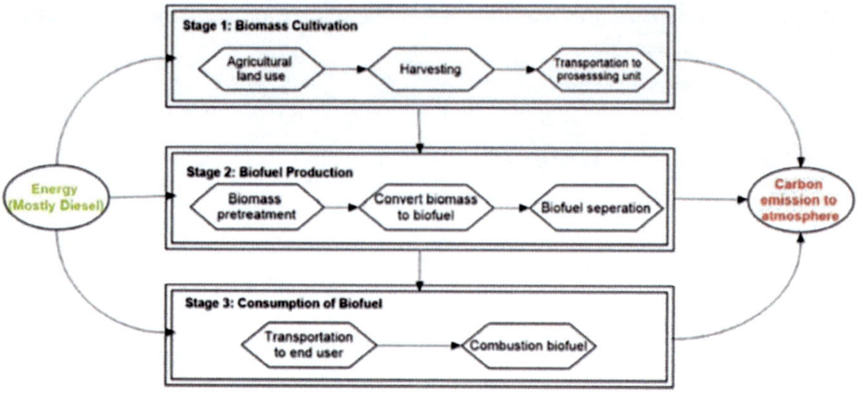

Fig. 55.2 LCA approach

global Data Collection System (IMO DCS), both of which calculate emission data on the total fuel consumption on board (Boviatsis and Tselentis 2019). Even though there is no official directive, it would be more reasonable to evaluate the net carbon emissions of fuels with an LCA approach. The LCA approach evaluates not only the exhaust emissions resulting from combustion but also the production process of the raw material of the fuel, the production of the fuel, the logistics, and the emissions as a result of combustion (Osman et al. 2021). The stages of the LCA approach are shown in Fig. 55.2.

The LCA approach basically consists of three main stages. Stage 1 biomass cultivation, stage 2 biofuel production process, and stage 3 transportation and consumption of biofuel. Energy is required for each stage, and this energy is mostly derived from diesel. For example, agricultural machinery and tractors are used to increase biomass, machinery is used in plants for the production of biofuel from biomass, and heavy-duty vehicles are used for the logistics of the biofuel produced. In addition, carbon is released into the atmosphere at each stage. Emissions from the biofuel produced from scratch until it reaches the final consumer are called well-to-tank (WTT) emissions, and emissions resulting from combustion are called tank-to-wake (TTW) emissions. The total value of these emissions is called the well-to-wake (WTW) emission, which expresses the net carbon effect of the fuel (ABS 2021). Biofuels have an important advantage in terms of well-to-tank emissions because the raw materials to be used in production have the effect of reducing the amount of carbon in the atmosphere since they carry out photosynthesis until the harvest time. This means that it is negative carbon emissions.

Another point to consider when evaluating biofuels is the ILUC. ILUC arises when farmland that was previously devoted for food and feed is used to harvest biomass. Since there will be no decrease in the food demand of people and animals, new agricultural lands will be needed to meet this demand. The need for new agricultural land will result in the conversion of forests and wetlands to agricultural land (Finkbeiner 2014; Juncker 2019). However, forests play a critical role in nature's carbon cycle with biogenic carbon uptake ability, thanks to photosynthesis. Therefore the conversion of forests to farmland is undesirable and the ILUC factor has a negative impact on a fuel LCA.

55.4 Comparison of Selected Biofuels

Fatty acid methyl ester (FAME), hydrotreated vegetable oil (HVO), bio-methanol, Fischer-Tropsch (FT) diesel, bio-dimethyl ether (bio-DME), and microalgae-based biodiesel have the potential to decarbonize maritime transportation with their different properties. A comparison was made to find the most suitable of these fuels for maritime transportation and the results are shown in Table 55.1. The comparison was made with seven criteria for each fuel and shown in three scales: good (***), average (**), and poor (*).

55.4.1 Technological Maturity

The level of technological maturity is an important factor that affects the fuel supply chain and price. FAME and HVO performed best on this criterion, while biodiesel derived from microalgae performed the worst. FAME and HVO have started to be used in certain blending ratios or purely in ships as in road transport (Nayyar 2010; Ushakov and Lefebvre 2019). Continuing studies on second and third generation biofuels are increasing their technological level over time.

55.4.2 Cost

FAME with the most mature technology performed well with the lowest fees. Since the hydrotreating process in HVO's production process requires more energy, current market prices are slightly higher than FAME (Sevim and Zincir 2022) and show average performance. Other second- and third-generation biofuels performed poorly in terms of cost due to the excess energy and equipment costs required by the production processes (Brown et al. 2020; Swanson et al. 2010).

55.4.3 Feedstock Availability

The second-generation FAME and HVO raw materials are waste cooking oils and performed the worst performance in this study. These waste oils require the collection of the food industry and households. For this reason, there is a limited supply of

Table 55.1 Comparison results of selected biofuels with evaluation criteria

Fuel	Technological maturity	Cost	Feedstock availability	Compatibility	Safety	ILUC impact	WTW emission performance
FAME 1.Gen	***	***	**	**	***	*	*
FAME 2.Gen	***	***	*	**	***	***	**
HVO 1. Gen	***	**	**	***	***	*	*
HVO 2. Gen	***	**	*	***	***	***	***
Bio-methanol	**	*	**	*	*	***	***
Bio-DME	**	*	**	*	*	***	***
FT diesel	**	*	**	***	***	***	***
Microalgae-based biodiesel	*	*	***	**	***	***	*

waste oil and it is not sustainable for maritime transport. The raw material of the third-generation biofuel is microalgae, and it showed the best performance in terms of raw material availability because microalgae have the capacity to multiply rapidly in the aquatic ecosystem. For this reason, the interest of researchers on the subject has increased in recent years. Since it is a photosynthetic creature, it provides its nutritional needs by photosynthesis with the help of carbon dioxide and light (Francisco et al. 2009). Second-generation biofuels are mostly derived from lignocellulosic biomass and showed an average performance. Lignocellulosic raw materials can be composed of agricultural residues such as corn stover, wheat straw, inedible plant oils, plants that are grown for energy purposes such as miscanthus, and forest residues (Brown et al. 2020).

55.4.4 Compatibility

HVO and FT diesel can be used as a drop-in fuel in ship diesel engines in a blend or pure form. These fuels, which do not need to make any modifications to the ship's engines and fuel systems (Ushakov and Lefebvre 2019; Zhou et al. 2020), showed the best performance in terms of compatibility. Bio-DME and bio-methanol performed the worst in the compatibility section because it is not possible to burn these fuels on diesel engine-powered ships. They require retrofit modifications on the engine and fuel system or require a new dedicated engine (Yuanrong Zhou 2020; Zincir and Deniz 2021). FAME, on the other hand, can be used directly in diesel-powered engines up to certain mixing ratios, but if it is desired to be used in pure form, it requires minor changes in the fuel system (Mohd Noor et al. 2018).

55.4.5 Safety

Bio-methanol and bio-DME performed poorly in safety due to their low flash point (Zhou et al. 2020). Bio-methanol is also toxic. FAME, HVO, FT diesel, and microalgae-based biodiesel performed well in safety.

55.4.6 ILUC Impact

First-generation biofuels had a high ILUC effect and performed the worst. This is because their raw materials require a lot of agricultural land use and conflict with the food industry. In this paper, since second-generation biofuels are obtained from UCO, agricultural residue biomass, and high agricultural field-efficient energy crops, there is no ILUC effect and they show good performance. Likewise, since the third-generation biofuel raw material will be installed in the aquatic ecosystem, there is no ILUC effect (Searle and Jacopo 2018).

55.4.7 WTW Emission Performance

Determining the net carbon emissions of biofuels is a difficult issue, because there are many different raw materials, production techniques, and processes. There are various studies in the literature to determine the net carbon emission of biofuels with the LCA approach. In this study, including the ILUC effect of biofuels as well, an evaluation was made by considering the raw materials used and the production process by also utilizing the literature. The WTW performance of microalgae-based biodiesel is currently poor (Carneiro et al. 2017) due to the steps in the production process and the energy requirement. However, it still has a long-term potential with the development of production technology. Similarly, first-generation FAME and HVO performed poorly for a completely different reason. This is due to the very high ILUC effect. Biofuels from second-generation waste oils and ligno-cellulosic biomass showed the best WTW performance. The raw materials for the second-generation biofuels selected in this paper are UCO, miscanthus, and corn stover. These three raw materials have no ILUC effect. In addition, miscanthus can be grown in marginal areas that cannot be used as farmland. It also has a high carbon capture ability, such as forests. Corn stover is an agricultural waste and provides biogenic carbon uptake until harvest time (Pavlenko and Searle 2018).

55.5 Conclusion

In this study, biofuels, which is one of the promising alternative fuel pathways within the scope of decarbonization of maritime transportation, were discussed. When examining the carbon effect of a fuel, the LCA approach was considered to be more logical. For this reason, the LCA approach and the ILUC impact are explained. Different generation biofuels selected for this study were compared with determined evaluation criteria. The main findings of the study are as follows:

- Despite the mature technology and low cost, first-generation biofuels do not seem to be attractive due to their high ILUC effect and WTW emissions.
- The performance of the second-generation FAME and HVO obtained from waste oil is average and good in all criteria except feedstock availability. However, a biofuel without feedstock adequacy cannot be sustainable.
- Microalgae-based biodiesel has the best performance in feedstock availability but is not a good option today due to its technological maturity and cost. It is promising for the future with research studies and developing technology.

- Bio-methanol and bio-DME are not compatible with existing ship infrastructure and are less secure. Although it is not very preferred today, engine manufacturers have studies on engines compatible with these fuels.
- FT diesel showed good and average performance in all criteria except cost. In case its cost decreases to reasonable levels, it is the best option, thanks to its compatibility with existing ships.

References

ABS. (2021). Sustainability whitepaper – Biofuels as marine fuel. May, 28.

Boviatsis, M., & Tselentis, B. (2019). A comparative analysis between EU MRV and IMO DCS – the need to adopt a harmonised regulatory system. 16th International Conference on Environmental Science and Technology, September, 2018–2019.

Brown, A., Waldheim, L., Landälv, I., Saddler, J., Mahmood, E., McMillan, J. D., Bonomi, A., & Klein, B. (2020). Advanced Biofuels - Potential for Cost Reduction, IEA Bioenergy: Task 41: 2020:01. IEA Bioenergy, 1–88.

Carneiro, M. L. N. M., Pradelle, F., Braga, S. L., Gomes, M. S. P., Martins, A. R. F. A., Turkovics, F., & Pradelle, R. N. C. (2017). Potential of biofuels from algae: Comparison with fossil fuels, ethanol and biodiesel in Europe and Brazil through life cycle assessment (LCA). Renewable and Sustainable Energy Reviews, 73(January), 632–653. https://doi.org/10.1016/j.rser.2017.01.152

European Environment Agency (EEA) (2021) Emissions of air pollutants from transport https://www.eea.europa.eu/data-and-maps/indicators/transport-emissions-of-air-pollutants-8/transport-emissions-of-air-pollutants

Finkbeiner, M. (2014). Indirect land use change - Help beyond the hype? Biomass and Bioenergy, 62, 218–221. https://doi.org/10.1016/j.biombioe.2014.01.024

Francisco, Ë. C., Jacob-Lopes, E., Neves, D. B., & Franco, T. T. (2009). Microalgae as feedstock for biodiesel production: carbon dioxide sequestration, lipid production and biofuel quality. New Biotechnology, 25, S278–S279. https://doi.org/10.1016/j.nbt.2009.06.630

IMO. (2020). Fourth IMO Greenhouse Gas Study: Executive Summary. IMO Greenhouse Gas Study, 4(1), 46.

Joung, T.-H., Kang, S.-G., Lee, J.-K., & Ahn, J. (2020). The IMO initial strategy for reducing Greenhouse Gas (GHG) emissions, and its follow-up actions towards 2050. Journal of International Maritime Safety, Environmental Affairs, and Shipping, 4(1), 1–7. https://doi.org/10.1080/25725084.2019.1707938

Juncker, J.-C. (2019). ILUC. European Comission Delegated Regulation, C (2019) 2055 Final, 2013–2015.

Lin, C.-Y., & Lu, C. (2021). Development perspectives of promising lignocellulose feedstocks for production of advanced generation biofuels: A review. Renewable and Sustainable Energy Reviews, 136, 110445. https://doi.org/10.1016/j.rser.2020.110445

Mohd Noor, C. W., Noor, M. M., & Mamat, R. (2018). Biodiesel as alternative fuel for marine diesel engine applications: A review. Renewable and Sustainable Energy Reviews, 94(February 2017), 127–142. https://doi.org/10.1016/j.rser.2018.05.031

Nayyar, M. (2010). The use of biodiesel fuels in the US marine industry. Bulletin, April, 1–88. http://origin-www.marad.dot.gov/documents/The_Use_of_Biodiesel_Fuels_in_the_US_Marine_Industry.pdf

Osman, A. I., Mehta, N., Elgarahy, A. M., Al-Hinai, A., Al-Muhtaseb, A. H., & Rooney, D. W. (2021). Conversion of biomass to biofuels and life cycle assessment: a review. In Environmental Chemistry Letters (Vol. 19, Issue 6). Springer International Publishing. https://doi.org/10.1007/s10311-021-01273-0

Pavlenko, N., & Searle, S. (2018). A Comparison of Induced Land- Use Change Emissions Estimates From Energy Crops. February.

Rutherford, D., & Comer, B. (2018). International Maritime Organization's Initial Greenhouse Gas Strategy. International Council on Clean Transportation (ICCT), April 2018, 8. https://theicct.org/publications/IMO-initial-GHG-strategy

Ryste, J. A., Wold, M., & Sverud, T. (2019). Comparison of Alternative Marine Fuels. 1–65.

Searle, S., & Jacopo, G. (2018). Analysis of high and low indirect land-use change definitions in European Union renewable fuel policy. Icct, 26.

Sevim, C., & Zincir, B. (2022). Biodiesel and Renewable Diesel as a Drop-in Fuel for Decarbonized Maritime Transportation. In A. K. Agarwal & H. Valera (Eds.), Potential and Challenges of Low Carbon Fuels for Sustainable Transport (pp. 319–345). Springer Singapore. https://doi.org/10.1007/978-981-16-8414-2_10

Swanson, R. M., Platon, A., Satrio, J. A., & Brown, R. C. (2010). Techno-economic analysis of biomass-to-liquids production based on gasification. Fuel, 89(SUPPL. 1), S11–S19. https://doi.org/10.1016/j.fuel.2010.07.027

United Nations Conference on Trade and Development (UNCTAD). (2021). Review of Maritime Report 2021. In United Nations Publications. http://unctad.org/en/PublicationsLibrary/rmt2015_en.pdf

Ushakov, S., & Lefebvre, N. (2019). Assessment of Hydrotreated Vegetable Oil (HVO) Applicability as an Alternative Marine Fuel Based on Its Performance and Emissions Characteristics. SAE International Journal of Fuels and Lubricants, 12(2), 109–120. https://doi.org/10.4271/04-12-02-0007

Yuanrong Zhou, N. P. (2020). The potential of liquid biofuels in reducing ship emissions | International Council on Clean Transportation. International Council on Clean Transportation, September. https://theicct.org/publications/marine-biofuels-sept2020

Zhou, Y., Pavlenko, N., Rutherford, D., Osipova, L., & Comer, B. (2020). The potential of liquid biofuels in reducing ship emissions. International Council on Clean Transportation, 1 (September), 31. https://theicct.org/publications/marine-biofuels-sept2020

Zincir, B., & Deniz, C. (2021). Methanol as a Fuel for Marine Diesel Engines. In P. C. Shukla, G. Belgiorno, G. Di Blasio, & A. K. Agarwal (Eds.), Alcohol as an Alternative Fuel for Internal Combustion Engines (pp. 45–85). Springer Singapore. https://doi.org/10.1007/978-981-16-0931-2_4

Chapter 56
Investigating the Effects of Design Parameters on the Performance of an Ejector–Expansion Refrigeration Cycle for Different Refrigerants

Ibrahim Karacayli, Lutfiye Altay, and Arif Hepbasli

Nomenclature

$\dot{E}$	Rate of energy, W or kW
$\dot{m}$	Mass flow rate, kg/s
h	Specific enthalpy, kJ/kg
P	Pressure, kPa
$\dot{Q}$	Heat transfer rate, W or kW
s	Specific entropy, kJ/kgK
T	Temperature, °C or K
$\dot{W}$	Power, W or kW
$\dot{X}$	Rate of energy, W or kW
η	Efficiency, %
ψ	Rational exergy efficiency, %
ε	Specific flow exergy, kJ/kg

56.1 Introduction

In recent years, more efficient use of limited energy resources has become very important. One of the studies carried out for this purpose is the development of energy saving methods due to the increasing energy consumption in air conditioning

I. Karacayli · L. Altay (✉) · A. Hepbasli
Department of Mechanical Engineering, Faculty of Engineering, Ege University, Izmir, Turkey
e-mail: ibrahim.karacayli@ege.edu.tr; lutfiye.altay@ege.edu.tr; arif.hepbasli@yasar.edu.tr

M. Z. Sogut et al. (eds.), *Proceedings of the 2022 International Symposium on Energy Management and Sustainability*, Springer Proceedings in Energy,
https://doi.org/10.1007/978-3-031-30171-1_56

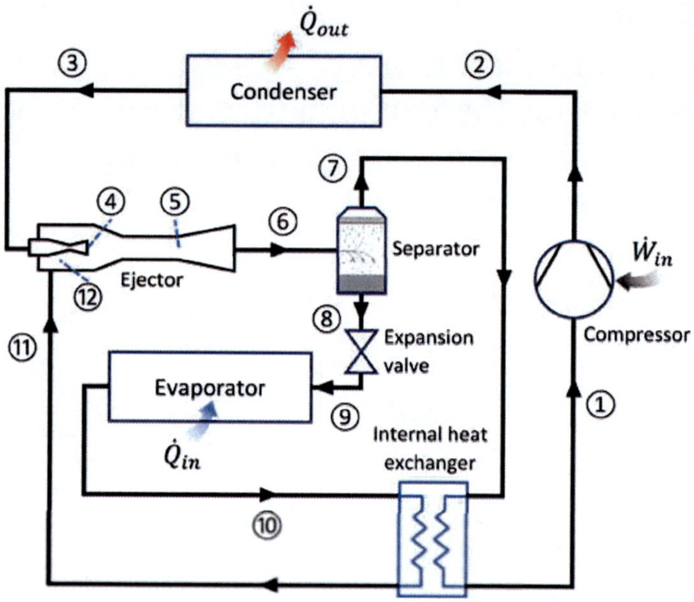

Fig. 56.1 The EERHX cycle

and refrigeration systems. For this reason, alternative solutions, such as the use of renewable energy sources, the selection of new refrigerants, or the development of cycles, are emphasized.

The EER cycles were improved from the conventional refrigeration cycles to decrease the compressor capacity as well as to reduce the irreversibility in the throttle valve (Rostamnejad and Zare 2019; Zhang et al. 2020). The EER cycles can be subcritical or supercritical like conventional refrigeration cycles.

The main difference between the conventional refrigeration cycle and the EER system is that an ejector is used instead of an expansion valve as a pressure reducing device. The ejector includes a motive nozzle, a mixing chamber, and a diffusion nozzle. The EERHX system consisting of a compressor, a condenser, an ejector, a liquid separator, an internal heat exchanger (IHX), an expansion valve, and an evaporator is illustrated in Fig. 56.1.

Bai et al. (2022) reported that the position of the IHX significantly affected the system performance of the EER systems. In the present study, IHX was placed between the liquid separator and the evaporator.

Liu et al. (2021) investigated the exergetic performance of the transcritical CO_2 EER cycle with thermoelectric subcooler. Iskan and Direk (2021) experimentally examined the EER system using R134a and R456a with double evaporator for two different configurations.

The main objectives of this study are (i) to develop an EERHX system, (ii) to perform parametric studies at different superheating and subcooling degrees, and (iii) to evaluate the effects of design parameters on system performance energetically and exergetically.

56.2 Thermodynamics Modeling

The mathematical model of the EERC system can be obtained by the assumptions given below. Then, the formulations for the rate of exergy transfer, rate of entropy generation, and exergy efficiency of all components and whole system are established in the exergy analysis. In conservation of energy, the net rate of energy transfer through heat, work, and mass must be equal to the net rate of energy transfer by heat, work, and mass, as shown in Eq. 56.1:

$$\dot{E}_{\text{in}} = \dot{E}_{\text{out}} \tag{56.1}$$

The coefficient of performance (COP) of the EERC is calculated from Eq. 56.2:

$$COP = \dot{Q}_{\text{in}} / \dot{W}_{\text{in}} \tag{56.2}$$

The exergy balance is defined as the difference between inlet and outlet exergy transfer rates by heat, work, and mass equals to rate of the exergy destruction and can be written as follows:

$$\dot{X}_{\text{heat,in}} + \dot{X}_{\text{mass,in}} - \dot{X}_{\text{work,out}} = \dot{X}_{\text{dest}} \tag{56.3}$$

The exergy destruction rate can be evaluated by

$$\dot{X}_{\text{dest}} = (1 - T_o/T)\dot{Q}_{\text{in}} + \dot{W}_{\text{in}} + \dot{m}(\varepsilon_{\text{in}} - \varepsilon_{\text{out}}) \tag{56.4}$$

The rational exergy is defined as

$$\psi = \dot{X}_{\text{dest,out}} / \dot{X}_{\text{cons,in}} \tag{56.5}$$

Calculations were performed for different superheating and subcooling degrees and various refrigerants, such as R134a, R410A, and R32. The dead state conditions were assumed to be same as the environmental conditions. The dead state properties for the refrigerant are shown in Table 56.1.

The operating conditions for performing energy and exergy analyses in the EERHX cycle are given in Table 56.2. Isentropic efficiencies of the motive nozzle, suction nozzle, and diffuser were taken as 0.85. The temperature rise at the compressor inlet after IHX was selected as 3 °C.

Table 56.1 Dead state properties

Properties	R134a	R32	R410A
Pressure (kPa)	101.3	101.3	101.3
Temperature (°C)	35 °C	35 °C	35 °C
Specific enthalpy (kJ/kg)	284.97	568.68	468.15
Specific entropy (kJ/kgK)	1.13331	2.65562	2.1966

Table 56.2 Operating conditions

Evaporation temperature		12 °C	
Condensation temperature		45 °C	
Environmental temperature		35 °C	
Refrigerated ambient temperature		22 °C	
Superheating value		+7 °C	
Subcooling value		– 3 °C	
Optimal pressure drop	R134a	R32	R410A
	15 kPa	40 kPa	40 kPa

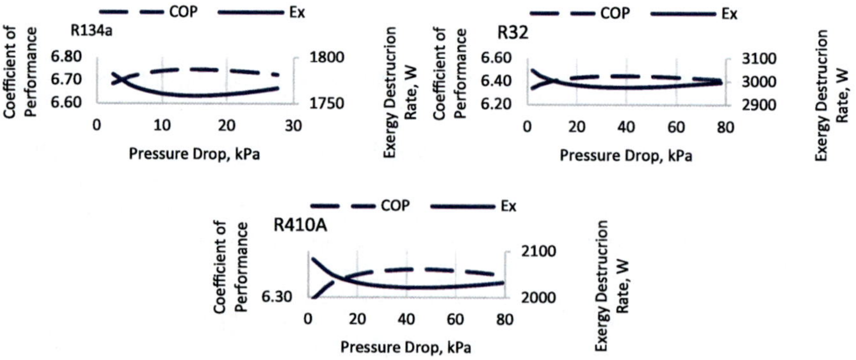

Fig. 56.2 COP and total exergy destruction rate variations with pressure drop

56.3 Results and Discussion

The optimal pressure drop in the suction nozzle of the ejector was determined according to the maximum COP of cooling and minimum total exergy destruction rate of EERHX (Fig. 56.2).

The calculated results for the superheating and subcooling degrees of 7 °C and 3 °C, respectively, for R32 are summarized in Table 56.3.

In the energy and exergy analyses, the energy transfer rate, exergy destruction rate, and functional exergy efficiency of the EERHX were determined. The results of the energy and exergy analyses for R32 are given in Table 56.4.

Figure 56.3 shows the exergy destruction rate and rational exergy efficiency of components and the EERHX system using R134a, R23, and R410A as the refrigerants. Total exergy destruction rates of R134a, R32, and R410A were 1758.2 W, 2974.0 W, and 2021.3 W, respectively. Although the exergy destruction rate in the

Table 56.3 Results calculated for R32

Point	P (kPa)	T (°C)	h (kJ/kg)	s (kJ/kgK)	$\dot{m}_r$ (kg/s)	ε (kJ/kg)	$\dot{E}x$ (W)
1	1257.7	17.4	521.09	2.1174	0.111	118.3	13.11
2	2794.6	74.0	558.55	2.1344	0.111	150.5	16.69
3	2794.6	42.0	279.61	1.2636	0.111	139.9	15.51
4	1134.3	10.8	275.51	1.2661	0.111	135.0	14.97
5	1134.3	10.8	384.60	1.6503	0.200	125.7	25.14
6	1257.7	14.4	387.13	1.6516	0.200	127.8	25.57
7	1257.7	14.4	516.91	2.1030	0.111	118.5	13.14
8	1257.7	14.4	225.67	1.0900	0.089	139.4	12.43
9	1174.3	12.0	225.67	1.0904	0.089	139.3	12.42
10	1174.3	19.0	526.09	2.1435	0.089	115.2	10.27
11	1174.3	15.0	520.89	2.1256	0.089	115.5	10.30
12	1134.3	13.1	519.79	2.1263	0.089	114.2	10.18

Table 56.4 Results of the energy and exergy analyses

Component	$\dot{E}$ (kW)	$\dot{X}_{\text{dest}}$ (W)	ψ
Compressor	4.15	1175.46	93.0%
Condenser	30.93	578.55	86.1%
Ejector	–	969.37	54.9%
Expansion valve	–	239.02	99.1%
Evaporator	26.77	9.66	99.9%
IHX	–	1.93	28.4%
EERHX (total)	–	2973.99	93.0%

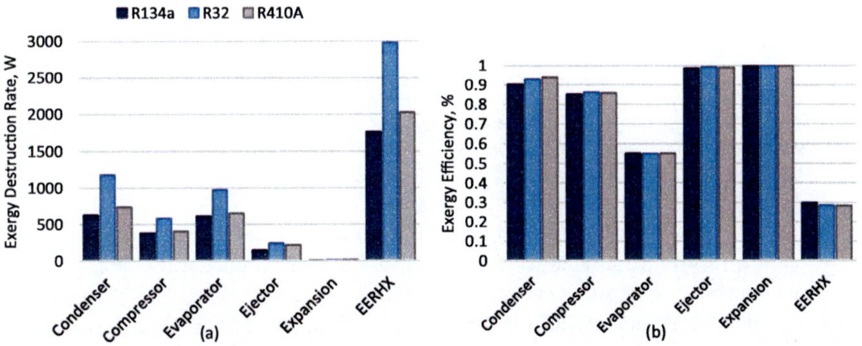

Fig. 56.3 (**a**) Exergy destruction rates and (**b**) rational exergy efficiencies of components of EERHX system for different refrigerants

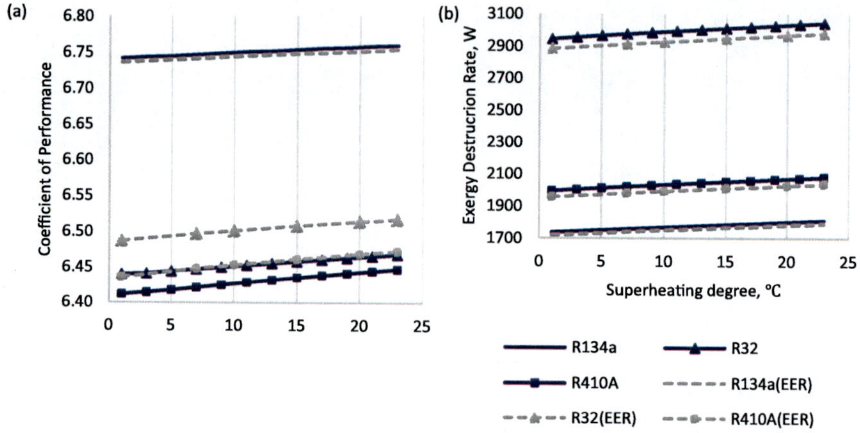

Fig. 56.4 Variation of (**a**) COP and (**b**) exergy destruction rate with superheating degrees for different refrigerants

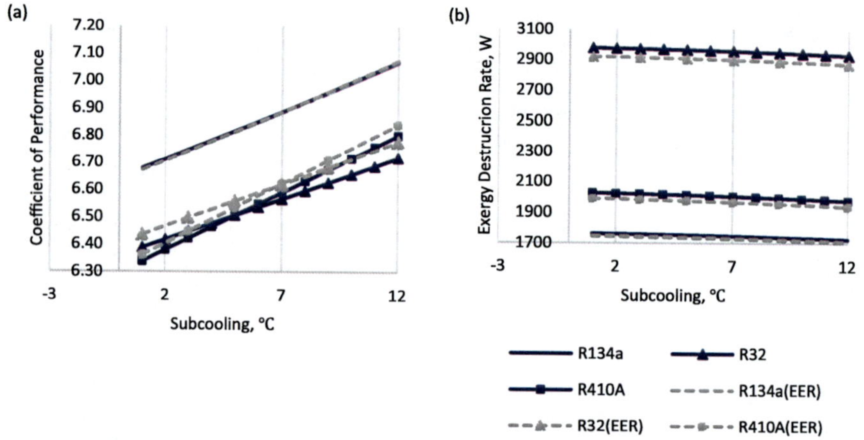

Fig. 56.5 Variation of (**a**) COP and (**b**) exergy destruction rate with subcooling degrees for different refrigerants

EERHX using R32 is significantly higher than other refrigerants, the rational exergy efficiencies for all refrigerants were obtained to be approximately 28–29%.

The superheating degree was gradually increased from 1 °C to 23 °C, while the superheating degree increased from 1 °C to 12 °C. The effect of the superheating and subcooling degrees on COP and the exergy destruction rate of EERHX and EER were presented in Figs. 56.4 and 56.5, respectively. It is seen from Fig. 56.4 that there was an increase in both COP and the exergy destruction rate with superheating degree. However, the increase in the exergy destruction rate is dominant than the increase in COP. The highest COP and the lowest exergy destruction rate were obtained in the EERHX system using R134a. On the other hand, the lowest COP and the highest exergy destruction rate were obtained for EERHX system using R32.

20 °C of superheating resulted in an increase of 3.9% in the exergy destruction rate while improving the COP and the rational exergy efficiency by 0.25% in the EERHX system working with R134a. The increases in exergy destruction rate were 3.1% and 3.9% for the R32 and R410A, respectively. On the other hand, the use of a HX in the EER did not cause a significant change (Fig. 56.5). For instance, the EERHX using R134a, R32, and R410A system instead of EER system without HX caused 1.0%, 2.1%, and 2.0% increases, respectively, in the total exergy destruction rate.

As can be seen in Fig. 56.5, the performance of the EERHX in terms of COP and exergy destruction rate was positively affected by subcooling. The COP of the EERHX for R134a, R32, and R410A improved by 5.84%, 5.16%, and 7.30%, respectively, when the subcooling degree increased from 1 to 12 °C. Besides, the total exergy destruction rates for R134a, R32, and R410A decreased by 2.1%, 1.7%, and 2.5%, respectively.

56.4 Conclusion

The effect of subcooling and superheating on the COP and exergy destruction rate of the EERHX system was examined in the present study. The following concluding remarks may be obtained from the results of this study:

- In the EERHX system, the highest exergy destruction rate was obtained in the condenser unit, and the lowest exergy destruction rate was obtained in the expansion element, which is considered as adiabatic.
- The COP and the rational exergy efficiency for R134a, R32, and R410A increased by 5.84%, 5.16%, and 7.30%, respectively, when the subcooling increased from 1 °C to 12 °C. Further, the total exergy destruction rates for R134a, R32, and R410A decreased by 2.1%, 1.7%, and 2.5%, respectively.
- The highest COP with the lowest exergy destruction rate was obtained for R134a. On the other hand, the highest exergy destruction rate and the lowest COP were obtained for R32.
- Operation of the system with lower superheating degrees and with higher subcooling degrees will increase the COP and the exergy efficiency.
- The EERHX using R134a, R32, and R410A system instead of EER system without HX caused 1.0%, 2.1%, and 2.0% increment in total exergy destruction rate, respectively.
- For a further study, performing (advanced) exergy analyses for different HX positions to achieve optimum working conditions is recommended.

References

Bai T, Yan G, Yu J (2022) Influence of internal heat exchanger position on the performance of ejector-enhanced auto-cascade refrigeration cycle for the low temperature freezer. Energy, 238: 1–13. https://doi.org/10.1016/j.energy.2021.121803

Liu X, Yu K, Wan X, Zheng M, Li X (2021) Conventional and advanced exergy analyses of transcritical CO_2 ejector refrigeration system equipped with thermoelectric subcooler. Energy Reports, 7: 1765–1779. https://doi.org/10.1016/j.egyr.2021.03.023

Iskan U, Direk M (2021) Experimental investigation on the effect of expansion valves in a dual evaporator ejector refrigeration system using R134a and R456. Energy Sources, Part A: Recovery, Utilization, and Environmental Effects. https://doi.org/10.1080/15567036.2021.1982076

Rostamnejad H, Zare V (2019) Performance improvement of ejector expansion refrigeration cycles employing a booster compressor using different refrigerants: Thermodynamic analysis and optimization. International Journal of Refrigeration, 101:56–70. https://doi.org/10.1016/j.ijrefrig.2019.02.031

Zhang Z, Feng X, Tian D, Yang J, Chang L (2020) Progress in ejector-expansion vapor compression refrigeration and heat pump systems. Energy Conversion and Management, 207. https://doi.org/10.1016/j.enconman.2020.112529

Chapter 57
Analysis of Sustainable Development Goals in Airports Using Stepwise Weight Assessment Ratio Analysis (SWARA)

Beste Pelin Çelem, Vildan Durmaz, and Ebru Yazgan

57.1 Introduction

In recent years, there has been an increasing interest in air transportation, and this rapid development in the sector reveals the concept of "airport cities." Aerotropolises need to maintain the balance between the city and the environment in sustainable ways. Airport sustainability is a holistic approach that deals with economic sustainability, operational efficiency, protection of natural resources, and social problems. In September 2015, the United Nations introduced the 2030 Agenda for Sustainable Development and 17 Sustainable Development Goals (SDGs) to achieve a more sustainable future for all industries (UN 2015). Airports are aware of the importance and benefits of putting the SDGs into action. Aviation authorities are also paying attention to this issue and supporting the implementation of SDGs at airports (ICAO 2015; ACI 2021).

Based on the expert interviews and literature review, (SDG7) Affordable and Clean Energy, (SDG8) Decent Work and Economic Growth, (SDG11) Sustainable Cities and Communities, (SDG12) Responsible Consumption and Production, and (SDG13) Climate Action have critical importance in airport sustainability (Sreenath et al. 2021; Dube and Nhamo 2021; Dube 2021; Di Vaio and Varriale 2020). Therefore, in this paper, we concentrated on these five criteria. This study is considered a multi-criteria decision-making problem. In the context of environmental challenges, MCDM techniques have been used to improve decision-making processes (Lu et al. 2018; Ahmad et al. 2021; Mukul et al. 2021; Lee et al. 2017).

B. P. Çelem (✉) · V. Durmaz · E. Yazgan
Faculty of Economics, Administrative and Social Sciences, Nisantasi University, Istanbul, Turkey

Faculty of Aeronautics and Astronautics, Eskisehir Technical University, Eskisehir, Turkey
e-mail: bestepelin.celem@nisantasi.edu.tr; vkorul@eskisehir.edu.tr; eyazgan@eskisehir.edu.tr

M. Z. Sogut et al. (eds.), *Proceedings of the 2022 International Symposium on Energy Management and Sustainability*, Springer Proceedings in Energy,
https://doi.org/10.1007/978-3-031-30171-1_57

In this study to evaluate and rank the criteria, the MCDM approach SWARA is used. Stepwise weight assessment ratio analysis (SWARA) is a method that considers the experts' priorities based on their experience and knowledge.

57.2 Methodology

The SWARA (stepwise weight assessment ratio analysis) approach was used to calculate the criteria weights in this study. In the first section, the SWARA method and the steps are explained. Following that, the approach for weighing the selected SDGs based on the sustainability of the airports is presented.

57.2.1 SWARA Method

The SWARA (stepwise weight assessment ratio analysis) method was developed by Kersuliene et al. (2010) for evaluating and weighing criteria. The approach is a multi-criteria decision-making process that has a wide range of applications in domains like accounting, management, business, product design, architecture, policy, and environmental sustainability (Ghenai 2020). SWARA gives decision-makers a key role in evaluating criteria and weights (Zolfani et al. 2018). SWARA steps are as follows (Kersuliene et al. 2010):

Step 1. Identification of prioritization of criteria

Criteria are prioritized based on their importance and the opinions of experts.

Step 2. Evaluation of the relative importance

Decision-makers determine the relative importance of each criterion j in relation to the previous ($j - 1$) criterion. This ratio called as s_j^k for each decision-maker (DM) and ($0 \leq s_j^k \leq 1$).

Step 3. Determination of the coefficient (k_j^k)

k_j^k is determined for each decision-maker with Eq. (57.1).

$$k_j^k = \left\{ 1 j = 1 \mid s_j^k + 1 j > 1 \right\} \tag{57.1}$$

Step 4. Determination of recalculated weight (q_j^k)

q_j^k is determined for each decision-maker with Eq. (57.2).

$$q_j^k = \left\{ 1j=1 \mid \frac{k_{j-1}^k}{k_j^k} j > 1 \right\} \quad (57.2)$$

Step 5. Calculation of the relative weights of the criteria

w_j^k is determined for each decision-maker with Eq. (57.3).

$$w_j^k = \frac{q_j^k}{\sum_{j=1}^{n} q_j^k} \quad (57.3)$$

Final weights are generated by taking the arithmetic average of the relative weights calculated from the evaluation of each decision-maker.

57.2.2 Application

As a result of the literature review and expert opinions, SDG 7, SDG 8, SDG 11, SDG 12, and SDG 13, which are considered the most important criteria in terms of the sustainability of airports, were selected (Table 57.1).

In this study, the SWARA method was used to evaluate the weights of the criterion. The five selected criteria were ranked and evaluated by three airport experts. Then, the criteria were weighted by applying the SWARA method and the following results were obtained:

Step 1. The SDGs were examined in terms of airport sustainability parties and scored accordingly by three aviation specialists. The rankings of the decision-makers are shown in Table 57.2.

Step 2. The decision-makers determine each criterion's relative importance (s_j^k) compared to the previous one starting with the second. The relative importance of the criteria are shown in Table 57.3.

Step 3. The coefficient (k_j^k) for each criterion was calculated using Eq. (57.1). The results are shown in Table 57.4.

Table 57.1 Sustainable Development Goals evaluation criteria

Criteria(C)	Sustainable development goals
C_1	(SDG7) affordable and clean energy
C_2	(SDG 8) decent work and economic growth
C_3	(SDG 11) sustainable cities and communities
C_4	(SDG 12) responsible consumption and production
C_5	(SDG 13) climate action

Table 57.2 Decision-makers' ranking

Criteria (C)	D1	D2	D3
C_1	3	3	4
C_2	2	1	1
C_3	1	2	2
C_4	4	5	3
C_5	5	4	5

Table 57.3 Relative importance of criteria

Criteria (C)	s_j^1	s_j^2	s_j^3
C_1	0.10	0.05	0.10
C_2	0.15		
C_3		0.20	0.10
C_4	0.20	0.05	0.05
C_5	0.10	0.10	0.10

Table 57.4 The coefficient of criteria

Criteria (C)	k_j^1	k_j^2	k_j^3
C_1	1.10	1.05	1.10
C_2	1.15	1.00	1.00
C_3	1.20	1.10	
C_4	1.20	1.05	1.05
C_5	1.10	1.10	1.15

Table 57.5 Recalculated weights of criteria

Criteria (C)	q_j^1	q_j^2	q_j^3
C_1	0.7905	0.8333	0.7870
C_2	0.8695	1	1
C_3	1	0.8333	0.9090
C_4	0.6587	0.7215	0.8658
C_5	0.5988	0.7575	0.6844

Step 4. The recalculated weights (q_j^k) for all criteria were calculated with Eq. (57.2). The recalculated weights are shown in Table 57.5.

Step 5. The relative weights (w_j^k) of the criteria were calculated with Eq. (57.3) for each decision-maker and results are shown in Table 57.6.

The final weights were determined using the arithmetic average of the relative weights, as shown in Table 57.7.

Table 57.6 Relative weights of criteria

Criteria (C)	w_j^1	w_j^2	w_j^3
C_1	0.2017	0.2010	0.1853
C_2	0.2219	0.2412	0.2354
C_3	0.2552	0.2012	0.2140
C_4	0.1681	0.1740	0.2038
C_5	0.1528	0.1827	0.1611

Table 57.7 Final weights of criteria

Criteria (C)	Computed weights
C_1	0.1960
C_2	0.2328
C_3	0.2234
C_4	0.1820
C_5	0.1656

Table 57.8 Final ranks of criteria

Criteria (C)	Rank order
C_1	3
C_2	1
C_3	2
C_4	4
C_5	5

57.3 Results and Discussion

When the SWARA method's results were evaluated, it was determined that the criteria were ranked in order of importance as follows: $C_2 > C_3 > C_1 > C_4 > C_5$. The final ranks of the criteria are shown in Table 57.8.

As a result of the evaluations of the decision-makers, it was concluded that the most important criterion was the "C2" criterion with a value of 0.2328. The study showed that (SDG 8) Decent Work and Economic Growth, which is among the Sustainable Development Goals published by the UN, is seen to be the most important goal for airports compared to other criteria. The aviation industry's ability to create employment opportunities by supporting several key areas is highly effective for the economic development of the countries. Also, findings confirm that the other goals support airports to achieve (SDG11) Sustainable Cities and Communities in their sustainability practices.

57.4 Conclusion

The environmental impact and importance of the aviation industry are increasing more than ever before. In terms of social and economic aspects, the global synergy created by airports leads to regional and international development. On the other hand, airports are expanding in response to rising demand, and facilities are becoming increasingly urbanized with the concept of "aerotropolises." As a result of this situation, airports must improve the sustainability of their operations and structures. This rapid increase in airport activities also brings many environmental problems such as air pollution, noise distress, natural resource, and energy consumption, degradation of the local ecosystem, carbon, and greenhouse gas emissions.

The United Nations' 2030 Sustainable Development Goals are essential in identifying environmental challenges and drawing lessons for the aviation industry. In this case, the SDGs become a roadmap for airports to achieve sustainability and environmental goals.

This study focused on the importance and priority of the United Nations' 2030 Agenda for Sustainable Development in terms of airports. The SWARA approach, which is an MCDM method, is used to analyze five SDGs in this research. In the following study, 17 SDGs and airports will be evaluated using a multi-criteria decision-making process.

References

ACI. (2021). Sustainability Report A Review of ACI'S Environmental, Social and Governance Efforts

Ahmad, S., Ouenniche, J., Kolosz, B. W., Greening, P., Andresen, J. M., Maroto-Valer, M. M., & Xu, B. (2021). A stakeholders' participatory approach to multi-criteria assessment of sustainable aviation fuels production pathways. International Journal of Production Economics, 238, 108156. https://doi.org/10.1016/j.ijpe.2021.108156

Di Vaio, A., & Varriale, L. (2020). SDGs and airport sustainable performance: Evidence from Italy on organizational, accounting and reporting practices through financial and non-financial disclosure. Journal of Cleaner Production, 249, 119431 https://doi.org/10.1016/j.jclepro.2019.119431

Dube, K. (2021). Climate Action at International Airports: An Analysis of the Airport Carbon Accreditation Programme. In Sustainable Development Goals for Society Vol. 2 (pp. 237–251). Springer, Cham.

Ghenai, C., Albawab, M., & Bettayeb, M. (2020). Sustainability indicators for renewable energy systems using multi-criteria decision-making model and extended SWARA/ARAS hybrid method. Renewable Energy, 146, 580–597 https://doi.org/10.1016/j.renene.2019.06.157

Hashemkhani Zolfani, S., Yazdani, M., & Zavadskas, E. K. (2018). An extended stepwise weight assessment ratio analysis (SWARA) method for improving criteria prioritization process. Soft Computing, 22(22), 7399–7405. https://doi.org/10.1007/s00500-018-3092-2

ICAO. (2015). UN Sustainable Development Goals (SDGs). https://www.icao.int/

Keršuliene, V., Zavadskas, E. K., & Turskis, Z. (2010). Selection of rational dispute resolution method by applying new step-wise weight assessment ratio analysis (SWARA). Journal of business economics and management, 11(2), 243–258

Lee, G., Choi, J., & Jun, K. S. (2017). MCDM approach for identifying urban flood vulnerability under social environment and climate change. Journal of Coastal Research, (79 (10079)), 209–213. https://doi.org/10.2112/SI79-043.1

Lu, M. T., Hsu, C. C., Liou, J. J., & Lo, H. W. (2018). A hybrid MCDM and sustainability-balanced scorecard model to establish sustainable performance evaluation for international airports. Journal of Air Transport Management, 71, 9–19. https://doi.org/10.1016/j.jairtraman.2018.05.008

Mukul, E., Güler, M., & Büyüközkan, G. (2021, August). Evaluation of Clean Energy Alternatives with Hesitant Fuzzy Linguistic MCDM Methods. In International Conference on Intelligent and Fuzzy Systems (pp. 325–332). Springer, Cham. https://doi.org/10.1007/978-3-030-85626-7_39

Sreenath, S., Sudhakar, K., & Yusop, A. F. (2021). Sustainability at airports: Technologies and best practices from ASEAN countries. Journal of environmental management, 299, 113639. https://doi.org/10.1016/j.jenvman.2021.113639

United Nations (2015) Transforming our world: the 2030 Agenda for Sustainable Development A/RES/70/1. The General Assembly, New York

Chapter 58
Hydrogen as a Transition Fuel in Marine Engines

Caglar Dere

58.1 Introduction

The IMO (International Maritime Organization) Initial Strategy aims to reduce GHG (greenhouse gas) emissions by 50% by 2050, compared to that of the emission levels in 2008. While the aim is well defined, the ways of fulfilling this strategy remain to be operators' decision.

In recent years, alternative fuels are proposed as contemporary solutions in marine engines to comply with the green shipping regulations. In particular, carbon dioxide emissions are on the primary scope of the regulations. Therefore, besides the practice in carbon dioxide abatement methods, non-carbon fuels are proposed as promising solutions for future shipping. As a prime mover for ships, the dominancy of large bore engines with their reliability, high-power output, and efficiency rates show that the engines will continue their dominancy in the market for the next years. Since the alternative power sources will take a process to be used widely for the propulsion of the ship, the internal combustion engines will remain as a significant player for the next few decades.

A carbon-free fuel can vitalize internal combustion engines to continue its use in transportation. Accordingly, hydrogen utilization in transportation is a prominent solution and gains an importance in the recent years. Together with its capability of reversal fuel operation and compatibility with existing marine engines, burning hydrogen in internal combustion engines seems to be an attractive solution.

Hydrogen combustion generates quite low emissions with the ignition assistance of pilot conventional fuel. In contrast to exhaust gases being near-zero carbon dioxide, nitrogen oxide emission is the determining factor for the combustion of

C. Dere (✉)
Faculty of Naval Architecture and Maritime, Izmir Katip Celebi University, Izmir, Turkiye
e-mail: caglar.dere@ikc.edu.tr

M. Z. Sogut et al. (eds.), *Proceedings of the 2022 International Symposium on Energy Management and Sustainability*, Springer Proceedings in Energy,
https://doi.org/10.1007/978-3-031-30171-1_58

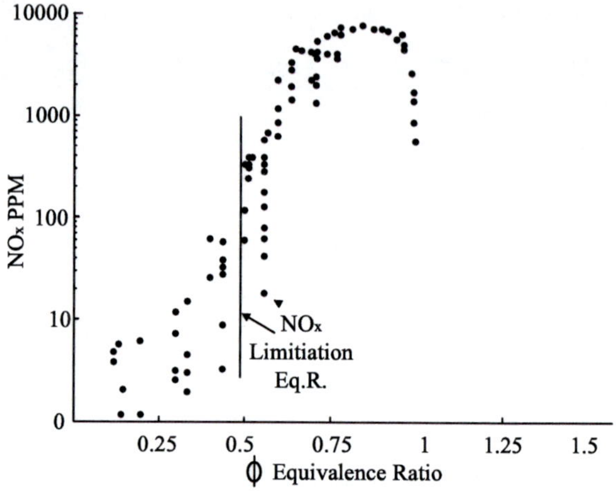

Fig. 58.1 NOx production with regard to equivalence ratio for H_2 combustion

hydrogen. Since the combustion temperature is higher with hydrogen, compared to diesel, the NO_x production is greater. Additionally, another factor for NO_x is fuel-air ratio that plays a crucial role at greater than 0.5 equivalence ratio as demonstrated in Fig. 58.1 (Inal et al. 2022). Therefore, lean combustion capability of hydrogen allows to operate engines at lower power outputs which decrease the power density of the power plant (White et al. 2006).

Hydrogen implementation alternatives to internal combustion engines such as port injection and direct injection (Serrano et al. 2021) have their own advantages and disadvantages. For instance, while pre-ignition and knocking can be overcame by direct injection, direct injection has a limited time to fuel-air mixing (Seddiek et al. 2015).

Hydrogen has different operational parameters in that the efficiency of the engine differs for the same output powers as higher combustion temperatures, lower mean effective pressures, and higher exhaust temperature values compared to that of the diesel engine (Seddiek et al. 2015).

There is another compelling alternative for the usage of hydrogen. It is fuel cell technology, but it seems will take a process to integrate into the shipping sector as a power source because of its cost disadvantage. Currently, the internal combustion engines are attractive solutions because of their compatibility with hydrogen. Hydrogen fuel applications are evaluated in the study with their advantages and disadvantages for internal combustion engines. The practical usage of hydrogen in current marine engines within operational perspectives, drawbacks, and findings is investigated.

58.2 Hydrogen as Fuel

Hydrogen has been used as a fuel since the 1970s (Escher and Ecklund 1976). Being flammable, hydrogen is recognized as a promising fuel for the internal combustion engines. Hydrogen is more flammable than other conventional fuels; moreover, its flammability concentration range is ultrawide as 4% to 75% by volume (Bechtold 1997). Furthermore, despite having a high auto-ignition temperature, it has a low minimum ignition energy and high flame speed.

Because hydrogen as a fuel has no carbon content, the challenge has been to generate high enough output power with reduced NOx levels in the hydrogen combustion processes. The research conducted in the past have been focused on how to comply with the NOx regulations. To comply with the regulations, besides the after-treatment systems, exhaust gas recirculation, lean combustion, and excessive air with supercharging are the methods to be used.

With increased boost pressure and exhaust gas recirculation (EGR), higher levels of power output and decreased NOx levels can be achieved. Thereby, the injection strategies together with EGR are used to decrease the NOx production by creating inhomogeneity in the combustion chamber with reduced injection pressures, which causes slightly lower efficiency as a result of improper combustion (Tsujimura and Suzuki 2019). Higher output power can be obtained by sacrificing energy efficiency and by slight increasing fuel consumption via after-treatment systems, de-NOx catalysts. On the other hand, within the higher fuel-air ratio operation, namely, over-stoichiometric, one more step improved operation with cooled EGR with the same boost and after-treatment systems can provide elevated torque-power output with the help of unburned hydrogen such as NOx reductant (Krishnan Unni et al. 2017).

Supercharging is needed to satisfy appropriate equivalence ratios so as to meet NOx limits. Unless boost pressure is increased, the hydrogen lean combustion cannot meet the required torque and power which can be provided by conventional fuels NOx remains as the only harmful combustion product of hydrogen operation.

Although the fuel is appealing in that it allows a carbon-neutral power generation, there are other hydrogen-assisted operations used in internal combustion engines. The adaptation of hydrogen into internal combustion engines is carried out by several methods called hydrogen-doped operations.

The study also analyzes the hydrogen usage in internal combustion engines in the light of the previous studies, carried out for LNG, gasoline, and diesel engines. Additionally, there are combinations of use of hydrogen with alternative fuels together with methanol and ethanol. There are wide range of studies with respect to the hydrogen usage as hydrogen-doped operations with different power outputs and different operating parameters. The physical and chemical properties of conventional fuels and alternative fuels are demonstrated in Table 58.1.

Having advantages in physical and chemical properties, hydrogen can also come up with disadvantages in internal combustion operations. Although hydrogen has a high auto-ignition temperature, which is 858 K, it has low ignition energy. Together

Table 58.1 The physical and chemical properties of hydrogen, methanol, ethanol, gasoline, methane, and diesel adapted from Yip et al. (2019) and Zhen et al. (2020)

Property	Hydrogen	Methanol	Ethanol
Carbon content (mass%)	0	37%	52%
Density (kg/m^3)	0.0899	795	790
Lower heat value (MJ/kg)	120	20.26	27
Auto-ignition temperature (K)	858	738	698
Stoichiometric air-fuel ratio	34.3	6.5	9
Laminar flame speed (m/s)	2.37	0.52	0.39
Adiabatic flame temperature (K)	2382	2143	2193
Minimum ignition energy in the air (mJ)	0.02	0.14	0.23
Quenching distance (mm)	0.6	1.85	1.65
	Gasoline	Methane	Diesel
Carbon content (mass %)	84%	75%	86%
Density (kg/m3)	730–780	0.83	830
Lower heat value (MJ/kg)	44.8	50.05	42.5
Auto-ignition temperature (K)	623	813	523
Stoichiometric air-fuel ratio	17.2	17.4	14.5
Laminar flame speed (m/s)	0.37–0.43	0.38	0.37–0.43
Adiabatic flame temperature (K)	2300	2225	2300
Minimum ignition energy in the air (mJ)	0.24	0.28	0.24
Quenching distance (mm)	2	2.03	–

with large air stoichiometry cause unfavorable hydrogen ignitions can occur when used with other fuels because of fuel residues in-cylinder, or hot spots near top surfaces. This phenomenon is called knocking and it is an undesirable operation which causes deterioration of proper combustion, efficiency degradation, and possible failures.

The injection strategy is another determining operating factor for the general engine design, namely, port fuel injection (PFI) and direct I=injection (DI). In the case of selecting port fuel injection, because of short quenching distance, high laminar burning velocity may cause back firing through back to intake manifold (Yip et al. 2019). Enhancing the engine design or engine operating conditions which is appropriate for hydrogen combustion can eliminate the risk. Besides, direct injection may completely eliminate the risk to its operation after the closure of the intake valve. Moreover, PFI causes reduced volumetric efficiency because of displaced volume of hydrogen instead of air, and the problem will also be eliminated by DI operation. The high unit volume brings drawback for its storage because hydrogen has a high-energy density per unit mass.

Additionally, higher value of adiabatic flame temperature of hydrogen also contributes to NOx formation. Thanks to flammability concentrations, higher EGR rates can be implemented to control the NOx emissions. Compared with natural gas, the diffusion rate of hydrogen is four to five times faster than the natural gas which allows reduced time for mixing in the combustion chamber.

For the marine engines, it is investigated that the required power for the propulsion of the ship is generated with neat or combination of fuels.

In the research of a RORO ship engine, the hydrogen operation efficiency could be achieved around 32% since the cooling loses in the hydrogen engine is higher (Seddiek et al. 2015). Additionally, it was reported that the operational load affects cooling loses significantly, while lean combustion operation has lower cooling loss than near-stoichiometric combustions (Shudo, et al. 2001). Bigger engine size is required to meet the power demand for a ship. Another study is conducted combining natural gas and hydrogen in marine engine. A turbocharged operation had been conducted for the lean-burn combustion. Then a modeling method had been used to illustrate the combustion mechanism in marine engines (Sapra et al. 2020). Another natural gas engine study is conducted by researchers (Leng et al. 2021). The hydrogen enrichment to the operation had been adopted by pre-chamber initiated jet ignition for a large bore engine. For the hydrogen-doped operation with a variable fraction, methane combustion had been modeled with the help of fluid dynamics software. The hydrogen fraction increases, the carbon emission decreases, and the NOx emission increases. The reformed exhaust gas recirculation system is adopted in another marine natural gas engine together with hydrogen enrichment (Li et al. 2019). Together with the proposed system with the specified air-fuel ratios, the elevated rates of reformed exhaust gases, the flame development, combustion duration, and the efficiency of the combustion could be improved.

In a two-stroke marine diesel engine, another hydrogen-doped operation had been conducted (Pan et al. 2014). Hydrogen enrichment had been implemented to the diesel engine with intake manifold integration, and a significant advantage had been seen at the idle speed when low-power requirement is needed that allows higher hydrogen ratio to be used in the combustion as compared to that of the diesel engine consumption. As an outcome of the study, significant reductions can be achieved when hydrogen contributes significantly in the inlet fuel energy portion. Another study looked into the combination of hydrogen and heavy fuel oil combustion (HFO) in a marine diesel engine (Serrano et al. 2021). In the study, water injection was recommended as a NOx emission reduction solution to meet the emission criteria. The research includes a modeling methodology for combustion prediction. With the dataset of HFO combustion for two-stroke low-speed (125 rpm) marine diesel engine with 16MW power rate, the modeling investigation and validation were performed.

58.3 Conclusion

The present paper evaluates the potential usage of hydrogen as a marine fuel in marine engines. It has been observed in the study that knocking problems, storage volumes, and lower thermal efficiency are the disadvantages of the hydrogen combustion engines compared to those of conventional fueled engines.

For the sake of brevity, the NOx emissions are the challenge for the neat hydrogen and hydrogen-doped operations. The lean combustion satisfies the NOx limitations without after-treatment systems. Near-stoichiometric conditions, EGR operations, and catalytic converters can help in the reduction of NOx emissions to near-zero values. Furthermore, lean combustion also helps to cope with pre-ignition characteristics of hydrogen. Other control strategies are intake charge cooling, valve timing, and effective scavenging

Additional efforts have been put by researchers with solution-oriented approaches to the problems. The storage, cost, incentive regulations, and ensuring safe operation will be key parameters for hydrogen utilization in marine engines.

References

Bechtold, R. L. (1997). Alternative Fuels Guidebook. Alternative Fuels Guidebook – properties storage, dispensing and vehicle facility modifications. https://doi.org/10.4271/r-180

Escher, W. J. D., and Ecklund, E. E. (1976). Recent progress in the hydrogen engine. In SAE Technical Papers. https://doi.org/10.4271/760571

Inal, O. B., Zincir, B., and Dere, C. (2022). HYDROGEN AS MARITIME TRANSPORTATION FUEL: A PATHWAY FOR DECARBONIZATION. In A. K. Agarwal and H. Valera (Eds.), Greener and Scalable E-fuels for Decarbonization of Transport (pp. 67–110). Springer US. https://doi.org/10.1007/978-981-16-8344-2

Krishnan Unni, J., Bhatia, D., Dutta, V., Das, L. M., Jilakara, S., and Subash, G. P. (2017). Development of Hydrogen Fuelled Low NOx Engine with Exhaust Gas Recirculation and Exhaust after Treatment. SAE International Journal of Engines. https://doi.org/10.4271/2017-26-0074

Leng, X., Huang, H., Ge, Q., He, Z., Zhang, Y., Wang, Q., … Long, W. (2021). Effects of hydrogen enrichment on the combustion and emission characteristics of a turbulent jet ignited medium speed natural gas engine: A numerical study. Fuel, 290(August 2020), 119966. https://doi.org/10.1016/j.fuel.2020.119966

Li, G., Long, Y., Zhang, Z., Liang, J., Zhang, X., Zhang, X., and Wang, Z. (2019). Performance and emissions characteristics of a lean-burn marine natural gas engine with the addition of hydrogen-rich reformate. International Journal of Hydrogen Energy, 44(59), 31544–31556. https://doi.org/10.1016/j.ijhydene.2019.10.007

Pan, H., Pournazeri, S., Princevac, M., Miller, J. W., Mahalingam, S., Khan, M. Y., … Welch, W. A. (2014). Effect of hydrogen addition on criteria and greenhouse gas emissions for a marine diesel engine. International Journal of Hydrogen Energy, 39(21), 11336–11345. https://doi.org/10.1016/j.ijhydene.2014.05.010

Sapra, H., Godjevac, M., De Vos, P., Van Sluijs, W., Linden, Y., and Visser, K. (2020). Hydrogen-natural gas combustion in a marine lean-burn SI engine: A comparative analysis of Seiliger and double Wiebe function-based zero–dimensional modelling. Energy Conversion and Management, 207(February), 112494. https://doi.org/10.1016/j.enconman.2020.112494

Seddiek, I. S., Elgohary, M. M., and Ammar, N. R. (2015). The hydrogen-fuelled internal combustion engines for marine applications with a case study. Brodogradnja, 66(1), 23–38.

Serrano, J., Jiménez-Espadafor, F. J., and López, A. (2021). Prediction of hydrogen-heavy fuel combustion process with water addition in an adapted low speed two stroke diesel engine: Performance improvement. Applied Thermal Engineering, 195, 117250. https://doi.org/10.1016/j.applthermaleng.2021.117250

Shudo, T., Nabetani, S., and Nakajima, Y. (2001). Analysis of the degree of constant volume and cooling loss in a spark ignition engine fuelled with hydrogen. International Journal of Engine Research. https://doi.org/10.1243/1468087011545361

Tsujimura, T., and Suzuki, Y. (2019). Development of a large-sized direct injection hydrogen engine for a stationary power generator. International Journal of Hydrogen Energy, 44(22), 11355–11369. https://doi.org/10.1016/j.ijhydene.2018.09.178

White, C. M., Steeper, R. R., and Lutz, A. E. (2006). The hydrogen-fueled internal combustion engine: a technical review. International Journal of Hydrogen Energy, 31(10), 1292–1305. https://doi.org/10.1016/j.ijhydene.2005.12.001

Yip, H. L., Srna, A., Yuen, A. C. Y., Kook, S., Taylor, R. A., Yeoh, G. H., . . . Chan, Q. N. (2019). A review of hydrogen direct injection for internal combustion engines: Towards carbon-free combustion. Applied Sciences (Switzerland). https://doi.org/10.3390/app9224842

Zhen, X., Li, X., Wang, Y., Liu, D., and Tian, Z. (2020). Comparative study on combustion and emission characteristics of methanol/hydrogen, ethanol/hydrogen and methane/hydrogen blends in high compression ratio SI engine. Fuel. https://doi.org/10.1016/j.fuel.2020.117193

Chapter 59
Performance Evaluation of R-290 as a Substitute for R-22 in a Domestic Refrigerator by Advanced Exergy Analysis Method

Alaattin Metin Kaya, Abid Ustaoglu, and Mustafa Alptekin

Nomenclature

$\dot{Q}$	Heat load (kW)
$\dot{W}$	Power (kW)
η_{ex}	Exergy efficiency
η_c	Compressor efficiency
dest	Destruction
amb	Ambient
AV	Avoidable
EN	Endogenous
EX	Exogenous
UN	Unavoidable
GWP	Global warming potential
ODP	Ozone depletion potential
CON	Condenser

A. M. Kaya (✉)
Department of Mechanical Engineering, Bursa Uludag University, Bursa, Turkey
e-mail: alaattinkaya@uludag.edu.tr

A. Ustaoglu
Department of Mechanical Engineering, Bartin University, Bartin, Turkey
e-mail: austaoglu@bartin.edu.tr

M. Alptekin
Graduate School, Bartin University, Bartin, Turkey
e-mail: mustafa.alptekin@ogrenci.bartin.edu.tr

M. Z. Sogut et al. (eds.), *Proceedings of the 2022 International Symposium on Energy Management and Sustainability*, Springer Proceedings in Energy,
https://doi.org/10.1007/978-3-031-30171-1_59

EVA	Evaporator
COM	Compressor
EXP.V.	Expansion valve

59.1 Introduction

Since refrigerating systems consume a weighty amount of energy, they have an enormous impact on climate change. For this purpose, environmentally harmful refrigerants used in refrigerating systems have been gradually restricted and removed from use, and more environmentally friendly ones have been used as a substitute.

After the restriction of the use of R-12 first in Europe and then in developing countries, R-22, which has a relatively lower ozone depletion potential (ODP), has been widely used. With increasing environmental concerns, R-22 is no longer considered environmentally friendly, and both ODP and global warming potential (GWP) values remain high. Therefore, the production and import of R-22 is prohibited in 2020. During this gradual period, the search for alternative refrigerants continued. Many refrigerants have been investigated as an alternative to R-22 and the search continues.

Kasera and Bhaduri compared R-22 with R-407C, and it was observed that although the ODP of R407C was zero, R-22 gave better results in terms of both COP and exergy efficiency (Kasera and Bhaduri 2017). On the other hand, it was stated that R-453A (Devecioğlu and Oruç 2016) and R-507 (Bolaji 2011) gave better results than R-22 in terms of energy consumption. One of the refrigerants that can be used as an alternative to R-22 is the natural R-290. Kaya stated that if R-22 is used in a multistage refrigeration system, it will have lower energy efficiency than R600 and higher than R227ea (Kaya 2020). Choudhari and Sapali stated that R-290 is a good alternative due to its environmental and thermophysical properties (Choudhari and Sapali 2017). Ustaoglu et al. investigated a multistage refrigeration system by using advanced exergy analysis method (Ustaoglu et al. 2022).

In this study, the energy and exergy analyses of the system are examined when R-22 and R-290 refrigerants are used in an ideal vapor compression refrigeration cycle. In addition, with advanced exergy analysis, it is possible to obtain information about the extent of exergy destruction in the components in the system, which is caused by the component itself, and to what extent it is caused by external factors. As a result, it is aimed to determine the effects of refrigerants on the exergy destruction occurring in the system and in each system component.

59.2 Materials and Method

59.2.1 *Refrigerants*

Chlorodifluoromethane ($CHClF_2$), also known as R-22 or HCFC-22, is a hydrochlorofluorocarbon (HCFC). ODP and GWP values of R-22 are 0.055 and 1.760, respectively. Although propane (R-290), a hydrocarbon (HC), is highly flammable, it has a low GWP value (3.3). Normal boiling points of R-22 and R-290 are −29.8 °C and −42.1 °C, respectively. Besides having a very low GWP compared to the R-22, propane has zero ODP. The critical temperatures of these refrigerants are quite close to each other, but the critical pressure of R-22 is higher.

59.2.2 *Advanced Exergy Analysis*

Exergy destruction was divided into avoidable-unavoidable and endogenous-exogenous parts by advanced exergy analysis (Morosuk and Tsatsaronis 2009; Tsatsaronis and Park 2002) (Eq. 59.1). Because of technological constraints such as material and manufacturing cost and availability, unavoidable exergy destruction cannot be avoided. The remaining represents avoidable exergy destruction. It is eliminated by improving the system and running it under optimal operating conditions.

$$\dot{E}_{D,k} = \dot{E}_{D,k}^{UN} + \dot{E}_{D,k}^{AV} \tag{59.1}$$

The other approach is dividing exergy destruction into endogenous and exogenous parts, giving information about whether the cause of exergy destruction is due to the component itself or to other components (Eq. 59.2).

$$\dot{E}_{D,k} = \dot{E}_{D,k}^{EN} + \dot{E}_{D,k}^{EX} \tag{59.2}$$

59.3 Results and Discussion

Quite a lot of conventional exergy analysis studies about refrigerating systems have been made and continue to be done. It gives information about irreversibilities in the system. Conventional exergy analysis, on the other hand, is incapable of explaining component interaction or estimating actual improvement potential (Petrakopoulou 2010; Tsatsaronis and Park 2002).

Table 59.1 Input parameters

	Parameter	Value	Unit
η_C	Compressor isentropic efficiency	0.80	–
T_{CON}	Condenser temperature	37	°C
T_{EVA}	Evaporator temperature	−5	°C
$T_{H,out}$	Outlet temperature of cooling water	32	°C
$T_{H,in}$	Inlet temperature of cooling water	27	°C
$T_{L,out}$	Evaporator outlet temperature of water	1	°C
$T_{L,in}$	Evaporator inlet temperature of water	6	°C
$\dot{Q}_L$	Cooling capacity of the evaporator	5	kW
P_{amb}	Ambient pressure	101	kPa
T_{amb}	Ambient temperature	25	°C

Table 59.2 COP, exergy efficiency, and energy input and output in the cycle

Parameter	Unit	R-22	R-290
$\dot{Q}_L$	kW	5	5
$\dot{Q}_H$	kW	6.207	6.241
$\dot{W}_C$	kW	1.207	1.241
COP	–	4.142	4.03
η_{ex}	–	41.46	40.37
$E_{x,dest}$	kW	0.7066	0.7398

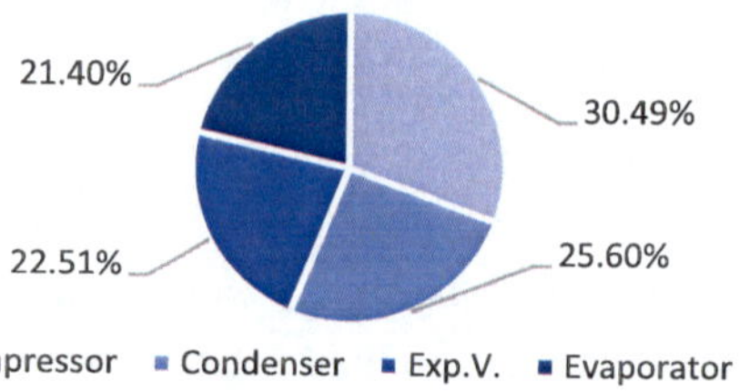

Fig. 59.1 Exergy destruction rates for R-22

In order to examine the energy and exergy efficiencies of the refrigerants on the system and system components, a 5 kW cooling capacity system was designed. System parameters and operating conditions are presented in Table 59.1.

As a result of energetic and exergetic analysis, it has been observed that in case of using R-290 refrigerant, the system needs more compressor work to provide the same amount of cooling load. It has a slightly lower COP (2.7%), and at the same time, the exergy efficiency of the system is lower and the exergy destruction is higher (less than 5%) (Table 59.2).

By examining the exergy destructions in the system, in both cases, it is seen that the highest exergy destruction in the system occurs in the compressors (Figs. 59.1 and 59.2).

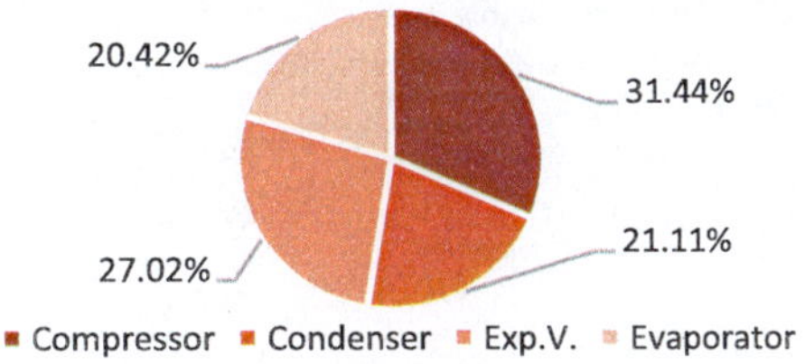

Fig. 59.2 Exergy destruction rates for R-290

Table 59.3 Components with their conditions

Component	Factor	Real	Ideal	Unavoidable
Condenser	ΔT_{CON}	+5 °C	0 °C	+0.5 °C
Evaporator	ΔT_{EVA}	−5 °C	0 °C	−0.5 °C
Compressor	η_C	80%	Isentropic	95%
Exp. valve	Isenthalpic		Isenthalpic	Isenthalpic

Comparing the exergy destruction rates of a refrigeration system working with R-290 and R-22, it is seen that while a higher rate of exergy destruction occurs in the compressor and expansion valve, less exergy destruction rates occur in the evaporator and condenser.

After the exergy destructions of the systems were determined, advanced exergy analyses were performed to determine how much of the exergy destruction in each component was avoidable and how much was unavoidable.

When Table 59.3, which shows the design and calculation values of the refrigeration system in real, ideal, and unavoidable conditions, is examined, it is seen that the throttling operation is isenthalpic in all cases, while the compression operation in the compressor is isentropic only in the ideal case. For the compression process in unavoidable conditions, the values of the compressor efficiency in the literature were taken (Li et al. 2012; Wang et al. 2016). Again, while determining the temperature differences to be used in the calculations, the studies in the literature were taken as reference (Bai et al. 2016; Ustaoglu 2020).

In the first stage of the advanced exergy analysis, the avoidable and unavoidable parts of the system components, which show how much exergy destruction can be reduced, are determined. In the analysis results, it is seen that the avoidable part of the compressor, where the highest exergy destruction occurs in the system, is the highest in both refrigerator cases (Figs. 59.3 and 59.4). Looking at this result, it gives the information that the most important contribution in reducing the exergy destruction in the system will be provided by the improvements to be made in the compressor operating conditions.

In the second stage of the advanced exergy analysis, the endogenous and exogenous parts of the system components, which are the indicators of the extent to which the exergy destruction occurred in the system components are caused by the external components and the component itself, were determined. In both cases, it has been observed that the exergy destruction of all components of the system is largely due to the component itself (Figs. 59.5 and 59.6). It is seen that the exergy destructions in the evaporator are completely caused by itself (Figs. 59.5 and 59.6).

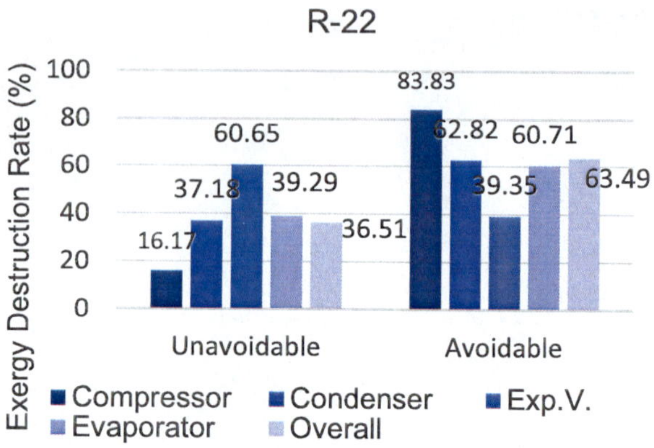

Fig. 59.3 Avoidable and unavoidable exergy destruction rates for R-22

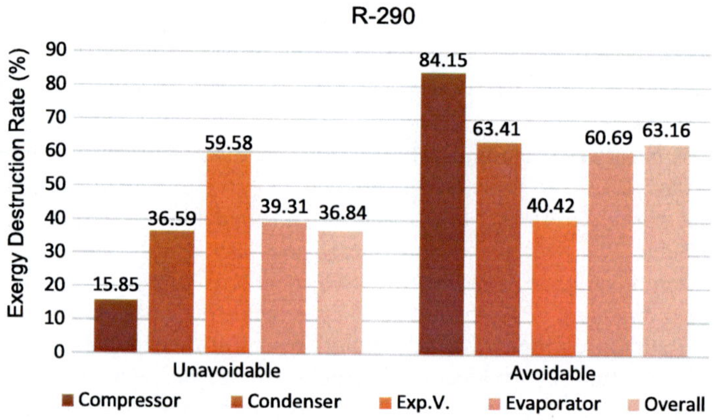

Fig. 59.4 Avoidable and unavoidable exergy destruction rates for R-290

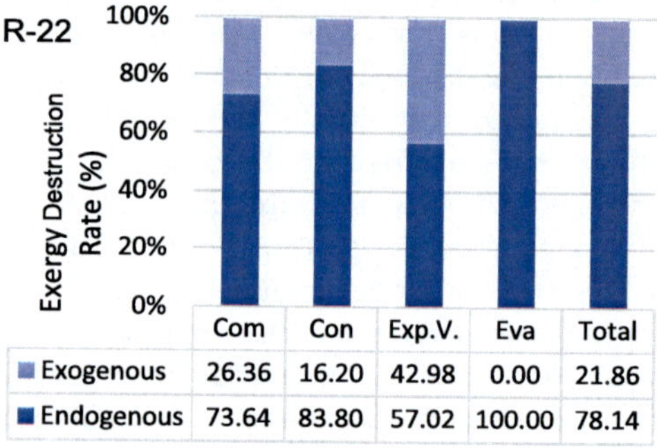

Fig. 59.5 Endogenous and exogenous exergy destruction rates for R-22

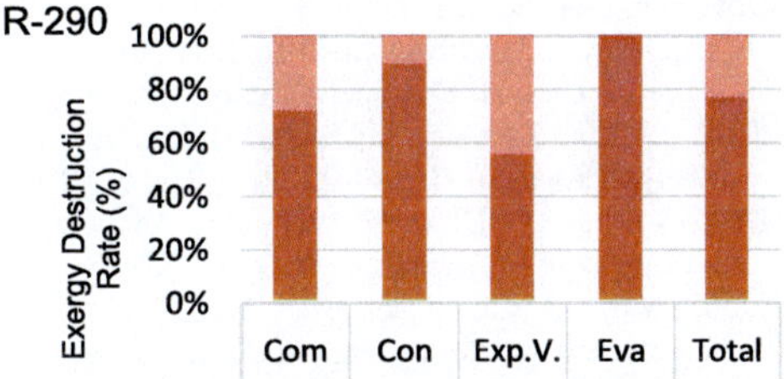

	Com	Con	Exp.V.	Eva	Total
Exogenous	27.82	10.24	44.12	0.00	22.83
Endogenous	72.18	89.76	55.88	100.00	77.17

Fig. 59.6 Endogenous and exogenous exergy destruction rates for R-290

In the given conditions, if R-290 is used instead of R-22 as a refrigerant in a refrigeration system, it is seen that the exogenous rate in the condenser decreases and the endogenous rate increases by approximately 7%.

59.4 Conclusion

In an ideal vapor compression refrigeration cycle, the use of R-290 instead of R-22 was investigated by energy, exergy, and advanced exergy analyses.

- As a result of the examinations, under the given conditions, R-22 showed a better performance by 2.7% in terms of energy performance.
- In the exergy analysis, it was calculated that the exergy destruction was lower by up to 5% when R-22 was used.
- In an ideal vapor compression refrigeration cycle working with R-290 instead of R-22, it is seen that the rate of exergy destructions in the condenser due to itself increases.
- The disadvantages in terms of COP and exergy are not significant, as well as zero ODP and relatively very low GWP values make R-290 a good alternative.

References

Bai, T., Yu, J., and Yan, G., (2016), Advanced exergy analyses of an ejector expansion transcritical CO_2 refrigeration system. Energy Conversion and Management, 126, 850–861. https://doi.org/10.1016/j.enconman.2016.08.057

Bolaji, B. O., (2011), Performance investigation of ozone-friendly R404A and R507 refrigerants as alternatives to R22 in a window air-conditioner. Energy and Buildings, 43(11), 3139–3143. https://doi.org/10.1016/j.enbuild.2011.08.011

Choudhari, C. S., and Sapali, S. N., (2017), Performance Investigation of Natural Refrigerant R290 as a Substitute to R22 in Refrigeration Systems. Energy Procedia, 109, 346–352. https://doi.org/10.1016/j.egypro.2017.03.084

Devecioğlu, A. G., and Oruç, V., (2016), The influence of plate-type heat exchanger on energy efficiency and environmental effects of the air-conditioners using R453A as a substitute for R22. Applied Thermal Engineering. https://doi.org/10.1016/j.applthermaleng.2016.10.180

Kasera, S., and Bhaduri, S. C., (2017), Performance of R407C as an Alternate to R22: A Review. Energy Procedia, 109, 4–10. https://doi.org/10.1016/j.egypro.2017.03.032

Kaya, A. M., (2020), Parametric Energy Analysis of the Performances of Various Re- frigerants on a Multistage Refrigeration System. ICSEEC: Sustainable Energy and Energy Calculations, 6.

Li, Y., Wang, X., Li, D., and Ding, Y., (2012), A trigeneration system based on compressed air and thermal energy storage. Applied Energy. https://doi.org/10.1016/j.apenergy.2012.04.048

Morosuk, T., and Tsatsaronis, G., (2009), Advanced exergetic evaluation of refrigeration machines using different working fluids. Energy, 34(12), 2248–2258. https://doi.org/10.1016/j.energy.2009.01.006

Petrakopoulou, F., (2010), Comparative Evaluation of Power Plants with CO_2 Capture: Thermodynamic, Economic and Environmental Performance. 230.

Tsatsaronis, G., and Park, M. H., (2002), On avoidable and unavoidable exergy destructions and investment costs in thermal systems. Energy Conversion and Management, 43(9–12), 1259–1270. https://doi.org/10.1016/S0196-8904(02)00012-2

Ustaoglu, A., (2020), Parametric study of absorption refrigeration with vapor compression refrigeration cycle using wet, isentropic and azeotropic working fluids: Conventional and advanced exergy approach. Energy, 117491. https://doi.org/10.1016/j.energy.2020.117491

Ustaoglu, A., Kursuncu, B., Metin Kaya, A., and Caliskan, H., (2022), Analysis of vapor compression refrigeration cycle using advanced exergetic approach with Taguchi and ANOVA optimization and refrigerant selection with enviroeconomic concerns by TOPSIS analysis. Sustainable Energy Technologies and Assessments, 52, 102182. https://doi.org/10.1016/j.seta.2022.102182

Wang, Z., Xiong, W., Ting, D. S. K., Carriveau, R., and Wang, Z., (2016), Conventional and advanced exergy analyses of an underwater compressed air energy storage system. Applied Energy, 180, 810–822. https://doi.org/10.1016/j.apenergy.2016.08.014

Chapter 60
Choosing the Best Solar Panel for Photovoltaic (Pv) System Analytical Hierarchy Process (AHP)

Abid Ustaoğlu and Samet Kuloğlu

Nomenclature

AHP	Analytical hierarchy process
P4	Number panel 4

60.1 Introduction

Energy is the most important factor today. It is an indispensable element for sustainable life. 80% of the energy need in the world is met from basic fossil fuels. The rapid depletion of fossil fuels has brought the use of renewable energy sources to the fore (Uyan 2013).

This study aims to make a binary comparison of any criteria of the system. It will optimize the system by comparing systems with multiple alternatives with multiple criteria. Various condenser types, geometries and photovoltaics, solar energy response intervals, yields, heat flux distributions and homogeneity, hot zone formations, sun exposure angles, weather conditions, polyts, coolants and geometries, material quantities and dimensions, thermal conductivity, radiation by evaluating absorbance, permeability and reflection, installation and maintenance costs in heat and electrical transformations of solar energy, and maintenance costs were tried to obtain maximum cost with minimum cost. For these systems, criteria were determined under electrical properties, mechanical properties, and environmental properties (Bruce 2011; Yılmaz et al. 2015) (Fig. 60.1).

A. Ustaoğlu (✉) · S. Kuloğlu
Faculty of Engineering, Architecture and Design, Bartın University, Bartın, Turkey
e-mail: austaoglu@bartin.edu.tr

M. Z. Sogut et al. (eds.), *Proceedings of the 2022 International Symposium on Energy Management and Sustainability*, Springer Proceedings in Energy,
https://doi.org/10.1007/978-3-031-30171-1_60

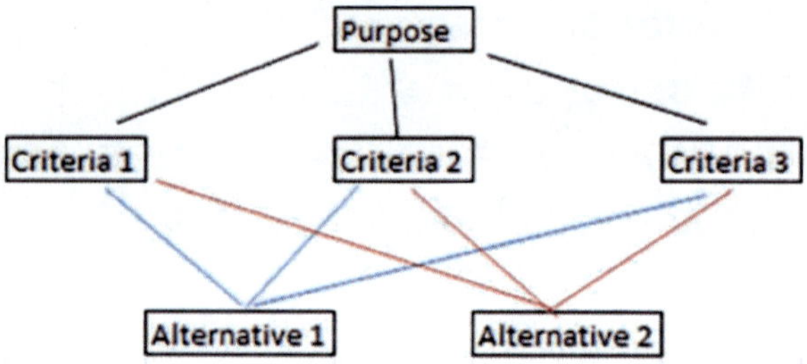

Fig. 60.1 AHP hierarchy model

In the literature, Balo and Şağbanşua (2016) worked on the selection of the best solar panel for photovoltaic system design using AHP. JianweiGao et al. (2021) tried a multi-criteria decision-making framework for the location of photovoltaic power coupling storage projects. Datta et al. (2014) tried are divided on the relevant choices from a perspective towards multi-criteria decision analysis for grid-connected photovoltaic system applications. Mukisa et al. (2021) completed a model on a multi-criteria analysis of countries importing solar photovoltaic modules (e.g., Uganda). Performance, parametric analysis, and multi-target optimization have been used by all NSGA-II users. Power efficiency is an improvement for global warming climate change's capacity to produce (Yan Cao et al. 2020).

In this study, photovoltaic system design and performance optimization was performed using multi-criteria decision-making methods. Thus, these systems offer work with less cost and higher efficiency. Concentrator photovoltaic design and performance optimization was also performed to make a multi-criteria decision-making system. File systems are systems that run less efficiently. Fundamental studies such as the selection to be made before have been made. This is not seen in the literature for the revision under review, and there are no similar publications when looking at the others. However, this study aims at all kinds of pairwise comparisons, i.e., to apply optimization by alternating multiple systems with multiple systems. No studies have been conducted on the educational arrangements of these elective alternatives.

In Table 60.1, various condenser types, geometries and photovoltaics, solar energy response intervals, yields, heat flux distributions and homogeneity, hot zone formations, sun exposure angles, weather conditions, polyts, coolants and geometries, material quantities and dimensions, thermal conductivity, radiation by evaluating absorbance, permeability and reflection, installation and maintenance costs in heat and electrical transformations of solar energy, and maintenance costs were tried to obtain maximum cost with minimum cost. For these systems, criteria were determined under electrical properties, mechanical properties, and environmental properties. Thus, low-cost high-performance systems will be obtained.

Table 60.1 Solar panel properties (First Solar Series Company 2016)

	P1	P2	P3	P4
Electrical characteristics				
Power rating(W)	170	175	172	183.4
Power/area (W/m^2)	138.2	145.6	131.7	166.5
Power tolerances (%)	(7.5/+7.5)	(0/+3)	(−2/+29)	(0/+12)
Mechanical characteristics				
Dimensions (mm)	1562.3*801.5*42	1460*755*32	1381*920*37	1210*852*42
Cost($)	200	200	385	500
Cost($)/watt	1.05	1.05	1.65	2.05
Environment				
Area(m^2)	1.3	1.18	1.33	1.08
Material	Policryst	Monocryst	Monocryst	Monocryst

60.2 Conclusion

AHP (analytical hierarchy process), one of the multi-criteria decision-making methods, and system alternatives were compared according to the determined criteria. The matrices formed as a result of the comparison were normalized and checked. Then, with the help of matrix algebra, the average score was created for each panel alternative. The most suitable system was created from the examined systems and the working conditions were optimized. After evaluating the electrical, mechanical, and environmental performance, it was concluded that each panel is the most suitable panel for P4 use. Thus, a low-cost high-performance system was obtained.

References

A. Datta, A. Ray, D. Mukherjee, H. Saha (2014) Selection of islanding detection methods based on multi-criteria decision analysis for grid-connected photovoltaic system applications. – Sustainable Energy Technologies and Assessments;Volume 7, Pages 111–122. https://doi.org/10.1016/j.seta.2014.04.003

Bruce T. 2011, Investigation of Cost and Performance Characteristic of Photovoltaic Panels. Research Dissertation, ENG 4111/4112 Project.

Figen Balo, Lütfü Şağbanşua (2016) The selection of the best solar panel for the photovoltaic system design by using AHP. – Kitakyushu, Japan. 3rd International Conference on Power and Energy Systems Engineering, CPESE 2016, 8–12. https://doi.org/10.1016/j.egypro.2016.10.151

First Solar Series Company (2016) The data sheets of solar panel brands.

Jianwei Gao, Huijuan Men, Fengjia Guo, Pengcheng Liang, Yuejin Fan (2021) A multi-criteria decision -making framework for the location of photovoltaic power coupling hydrogen storage projects. -Journal of Energy Storage. Volume 44, Part B. https://doi.org/10.1016/j.est.2021.103469

Nicholas Mukisa, Ramon Zamora, Tek Tjing Lie, Xudong Wu, Guoqian Chenb (2021) Multi criteria analysis ranking of solar photovoltaic modules manufacturing countries by an importing country: A case of Uganda. – Solar Energy; Volume 223, Pages 326–345. https://doi.org/10.1016/j.solener.2021.05.078

Uyan M. 2013, "GIS-Based solar farms site selection using analytic hierarchy process (AHP) in Karapinar region, Konya/Turkey". Renewable & Sustainable Energy Reviews, 28, 11–17. https://doi.org/10.1016/j.rser.2013.07.042

Yan Cao, Leonardus W.W. Mihardjo, Towhid Parikhani (2020) Thermal performance, parametric analysis, and multi-objective optimization of a direct-expansion solar-assisted heat pump water heater using NSGA-II and decision makings.– Applied Thermal Engineering Volume 181, 115892. https://doi.org/10.1016/j.applthermaleng.2020.115892

Yılmaz S, Ozcalik HR, Kesler S, 2015, The analysis of different PV power systems for the determination of optimal PV panel and system installation-A case study in Kahramanmaras. Renewable and Sustainable Energy Reviews; 52:1015–1024. https://doi.org/10.1016/j.rser.2015.07.146

Chapter 61
Wind Resource Assessment of the Selected Districts of Kütahya, Turkey

Murat Ertan and Onur Koşar

Nomenclature

U	Wind speed, m/s
P	Wind power, W/m^2
k	Dimensionless shape parameter
c	Scale parameter
f_w	Wind speed probability

61.1 Introduction

Interest in renewable energy sources will meet the sustainable energy demand of future generations. Renewable energy sources are unlimited, so they are sustainable based on economic and social needs (Verma et al. 2015).

It is estimated that about 2% of the total solar energy received by the world is converted into the kinetic energy of the wind. Considering that this amount is hundreds of times more than the total world energy consumption, the importance of wind energy will be better understood. Wind turbines can only produce energy within a certain wind speed range. Therefore, it is necessary to know the wind regime in the district where the wind farm will be established. The amount of wind

M. Ertan (✉)
Institute of Graduate Education, Kütahya Dumlupınar University, Kütahya, Turkey

O. Koşar
Simav Technology Faculty, Department of Mechanical Engineering, Kütahya Dumlupınar University, Kütahya, Turkey
e-mail: onur.kosar@dpu.edu.tr

M. Z. Sogut et al. (eds.), *Proceedings of the 2022 International Symposium on Energy Management and Sustainability*, Springer Proceedings in Energy,
https://doi.org/10.1007/978-3-031-30171-1_61

energy depends on the speed of the wind. The speed of the wind increases with height, and its power increases in proportion to the cube of its speed. The energy that the wind will provide depends on its power and blowing time (Lipman and Musgrove 1982).

Koşar and Özgür (2020) researched Kütahya's wind farm in Turkey. The median and quartiles (MQ) method is the best parameter estimation method to obtain the Weibull distribution parameters. Before the wind farm investment in the investigated area, a meteorology station should be established in the determined campus. With predicted error |ΔRIX| relationship must be determined. Then the WAsP results must be corrected to obtain definitive results. In addition, their work presents the feasibility of a wind farm in the province of Kütahya. It contributes to a better understanding of the relationships between key wind parameters such as wind speed, wind shear, turbulence intensity, and wind speed ramps.

Hulio and Jiang (2020) investigated the wind power potential of the site using wind speed, wind direction, and other meteorological data, including temperature and air density, collected over a one-year period. Data are calculated at 30 and 10 m altitude. The results of the wind resource assessment show that the site has the potential to install wind turbines for power generation.

Cakmakcı and Hüner (2020) conducted a study to evaluate the wind energy potential at Kırklareli University, Kayalı Campus. In the study, an altitude of 100 m was chosen to give a perspective on the wind power generation potential of the location. The results show that wind is a promising energy source for the studied area.

Our aim in the study is to determine the wind energy availability of five districts (Aslanapa, Altıntaş, Dumlupınar, Gediz, Simav) in the province of Kütahya. It will be concluded whether it is necessary for the installation of wind turbines in five different counties. In our study, hourly data averages of wind speed and directions of five different stations in Kütahya province were found. The data are shown in graphs. In addition, the monthly average wind speed distribution of five stations is shown. Bünelek Tepe, which is located in the Kütahya Dumlupınar University Central Campus area, was determined as a suitable land for wind measurement in Özgür's (2006) previous study. A wind measurement station has been established in the region. From the station established in the region, wind data for 36 months between July 2001 and June 2004 were obtained by the CallAlog 02 software. These data were evaluated by various statistical methods.

Kütahya may include candidate regions for possible wind farm installation sites due to its proximity to provinces with good wind energy potential such as Manisa and Balıkesir. Therefore, in this study, long-term recorded wind data for five different districts (Altıntaş, Aslanapa, Dumlupınar, Gediz, Simav) of Kütahya are analyzed. The average wind speed and the wind direction parameters that constitute the three-year wind data are analyzed in the statistical evaluation. Then, the daily, monthly, and seasonal variations of the wind speed are examined. As a result of the comparative analysis, wind energy calculations for the region with the best wind potential are made using the WAsP software.

61.2 Wind Status in Turkey

In Turkey, 9,84% of our total electricity needs are met with the electricity produced from wind power plants in 2022. With the installed mechanical power of 11.101 MWm, there was a growth of 19,31% as of the end of 2022. In 2022, there are 273 power plants in operation. The percentage of wind power plants was realized as 37,73 in the Aegean Region and 35,36% in the Marmara Region. Wind power plants in operation are 19,32% İzmir, 13,11% Balıkesir, and 8,07% Çanakkale (TWEA 2022) (Fig. 61.1).

61.3 Site Locations and Wind Data

Kütahya is located in the Aegean Region (Inner West Anatolia). This part of the Aegean Region is the threshold between the Central Anatolia Region and the original Aegean Region. The obvious character of the threshold is that it consists of plateaus with an average height of around 1.200 m. For this reason, it is called "Kütahya Highlands" in the language of geography.

Kütahya is located in the Aegean Region, Central West Anatolian part with coordinates 39°25′ latitude and 29°59′ longitude. It is surrounded by Eskişehir, Afyon from the east, Uşak from the south, Manisa and Balıkesir from the west, and Bursa and Bilecik from the north. Kütahya may include candidate regions for possible wind farm installation sites due to its proximity to provinces with good wind energy potential such as Manisa and Balıkesir. Therefore, in this study, long-term recorded wind data for the five different districts (Altıntaş, Aslanapa,

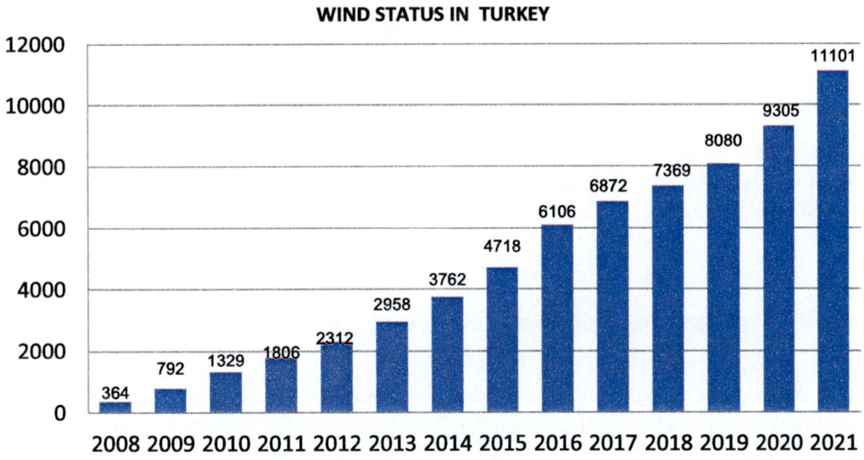

Fig. 61.1 Turkey's WPP status by years (MWm). (TWEA 2022)

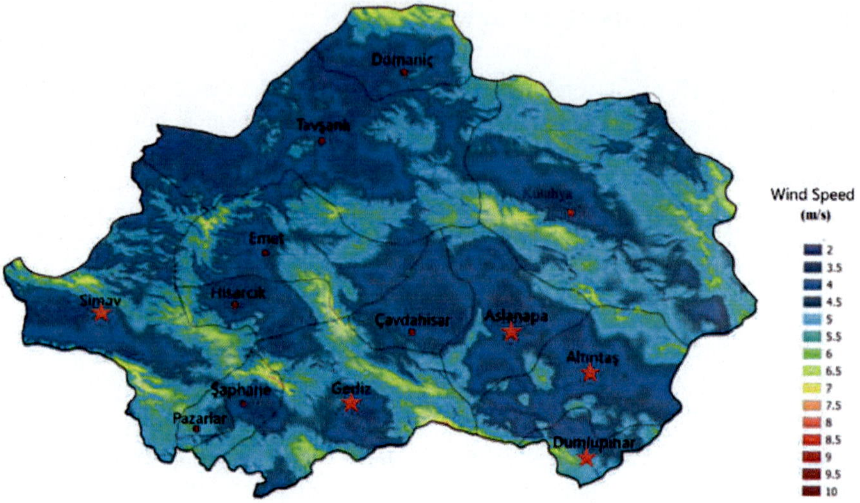

Fig. 61.2 Selected wind measurement stations on the global wind atlas. (TWEA 2022)

Table 61.1 Loss rates of the three-year wind data

District	Altitude (m)	Loss rate (%)
Simav	1021	2.21
Aslanapa	1041	6.85
Gediz	1250	4.16
Dumlupınar	736	5.74
Altıntaş	809	8.33

Dumlupınar, Gediz, Simav) of Kütahya are analyzed. Figure 61.2 shows the selected wind measurement stations on the global wind atlas.

Wind data includes hourly averages of wind speed and wind direction recorded from 10 m above ground by meteorology stations of the Turkish State Meteorological Service between 2016 and 2019. Loss rates of the wind data of each district are given in Table 61.1. According to the standards, the data loss rate should not exceed 10% (Bailey et al. 1997). The data loss rate in the wind dataset does not exceed 9%, and no value was found in the entire dataset beyond the measurement range of the speed and direction sensors.

61.4 Methodology of Wind Resource Assessment

The wind speed frequency distribution is required to estimate the wind energy potential of a location. Wind speed frequency can be described by various probability density functions such as Weibull, Beta, Rayleigh, Gamma, Gaussian, and lognormal distribution. The most widely used method is the Weibull distribution function because of its simplicity and accuracy. Thus, the average annual output of a given wind turbine can be quickly calculated. The Weibull probability density function is defined as (Teimourian et al. 2019)

$$f(v)_w = \frac{k}{A}\left(\frac{v}{A}\right)^{k-1} e^{-\left(\frac{v}{A}\right)k} \tag{61.1}$$

where v is the wind speed, $f(v)_w$ is the probability of wind speed, k is the shape parameter (dimensionless), and A is the scale parameter (m/s) (Fig. 61.3).

$$k = \left(\frac{\sigma}{v}\right)^{-1{,}086} \tag{61.2}$$

$$A = \frac{\overline{v}}{\Gamma\left(1 + \frac{1}{k}\right)} \tag{61.3}$$

61.5 Results and Discussion

One of the main parameters in the optimum positioning of a wind farm is the wind direction. The geographical features of the studied area and the atmosphere circulation significantly affect the wind direction, and as a result, wind direction fluctuations can be observed. A weather vane diagram is a useful diagram that shows wind speed frequency and wind direction together in one diagram. Thus, the respective wind speed and the corresponding wind direction can be represented in a weather vane diagram.

In the diagrams in the figure, our data rate at a height of 10 m is 97,79% for Altındaş, Aslanapa 93,15%, Dumlupinar 95,84%, Gediz 94,26%, and Simav 91,67%. The wind direction is intensely from the north and south in Altıntaş, in the northeast direction in Dumlupınar, in the north direction in Gediz, and in the east and west directions in Simav. Aslanapa in the northeast and west directions is blowing.

The Weibull parameters of the five counties are given in Table 61.2. According to the table, Aslanapa has the highest wind speed and wind power with a wind speed of 3,02 m/s and a wind power of 61 (W/m^2). Gediz, on the other hand, has the lowest wind speed and wind power values with 1.36 m/s wind speed and 4 W/m^2.

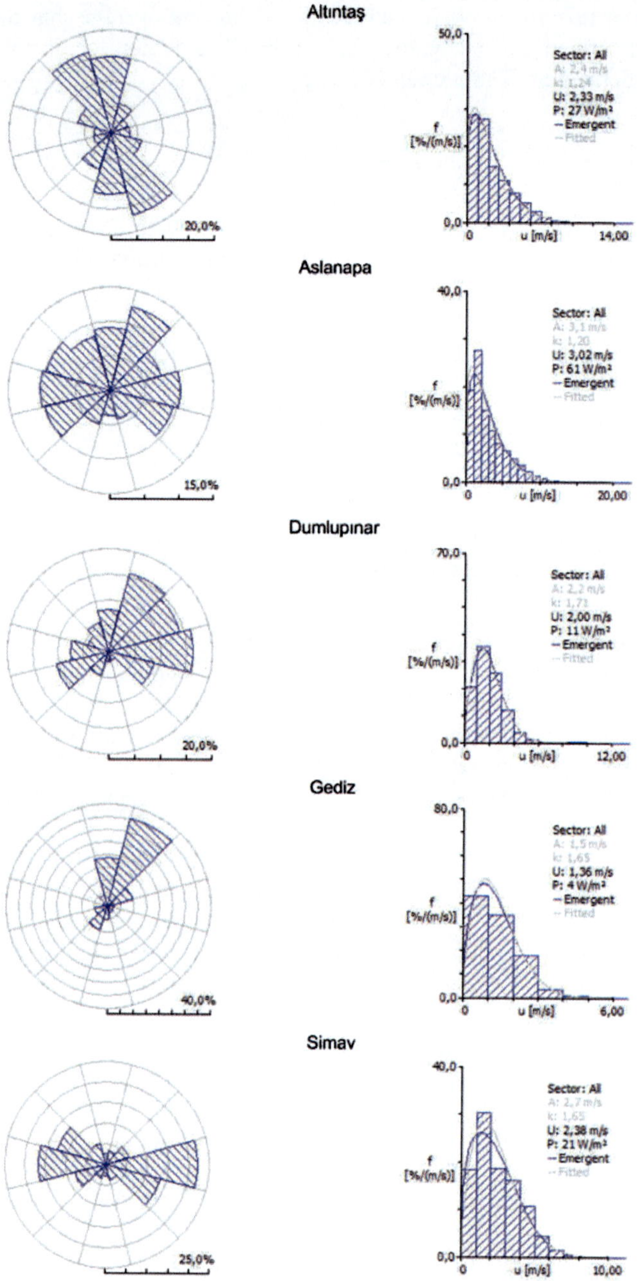

Fig. 61.3 Wind direction and velocity distributions for 10 m above ground level

Table 61.2 Weibull parameters and analysis for the years 2016–2019 based on meteorological data (10 m)

	Wind speed (m/s) (U)	Wind power (W/m^2) (P)	Scale parameter (m/s) (A)	Shape parameter (k)
Simav	2,38	21	2,7	1,65
Aslanapa	3,02	61	3,1	1,2
Gediz	1,36	4	1,5	1,65
Dumlupınar	2	11	2,2	1,71
Altıntaş	2,33	27	2,4	1,24

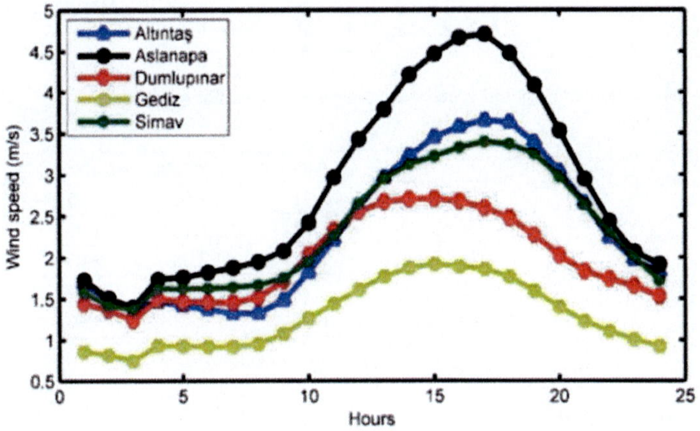

Fig. 61.4 Hourly wind speed averages of five stations

61.6 Wind Speed and Direction Measurements

The measurements of wind speed and direction were taken from the Ministry of Agriculture and Forestry Meteorology Station, starting from October 30, 2016, and giving 1 h until the end of July 2019. Wind data were taken hourly at a height of 10 m. The data used were collected from five different districts of Kütahya province. Hourly wind speeds of five different districts according to the months are given in the figure.

According to the data obtained, the highest wind speed was observed as 3,89 m/s in December at Aslanapa station, which is the average of the last 3 years. The average wind speed of five stations in December is 2,38 m/s. Aslanapa station wind speed average of December 2017 is 5,99 m/s. According to Fig. 61.4, wind speeds generally reach the highest level between 15:00 and 18:00.

According to the graph in Fig. 61.5, there is a general increase from November to December. A general increase in wind speed is observed in the summer months. As seen in Table 61.3, there is an increase in wind speeds at five stations from autumn to winter. The seasons with the highest wind speed potential are observed in winter and spring.

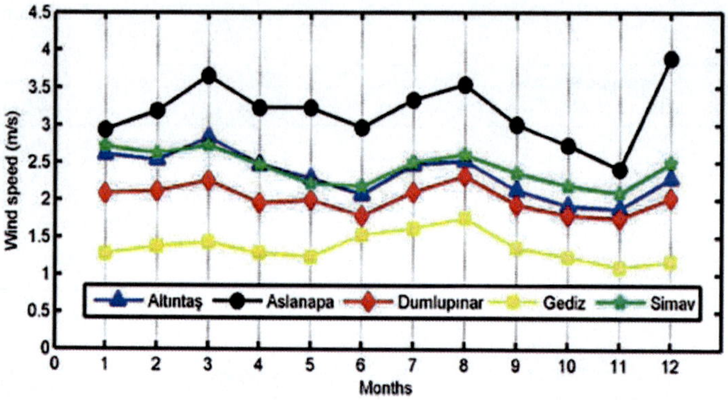

Fig. 61.5 Monthly average wind speed distribution of five different stations

Table 61.3 Seasonal average wind speeds of five districts (m/s)

	Autumn	Winter	Spring	Summer
Simav	2,21	2,61	2,47	2,42
Gediz	1,23	1,28	1,32	1,64
Dumlupınar	1,84	2,08	2,07	2,07
Aslanapa	2,72	3,34	3,37	3,28
Altıntaş	1,98	2,48	2,53	2,35

61.7 Conclusion

The result of the research is summarized as follows:

- Wind speeds are usually highest between 15:00 and 18:00. Aslanapa has the highest wind speed and wind power with a wind speed of 3,02 m/s and a wind power of 61 (W/m^2). Gediz, on the other hand, has the lowest wind speed and wind power values with 1.36 m/sec wind speed and 4 W/m^2.
- Aslanapa station wind speed average of December 2017 is 5,99 m/s. There is an increase in wind speeds at five stations from autumn to winter.
- The wind direction is intensely from the north and south in Altıntaş, in the northeast direction in Dumlupınar, in the north direction in Gediz, and in the east and west directions in Simav. Aslanapa in the northeast and west directions is blowing.

Acknowledgments The authors acknowledge the Turkish Republic General Directorate of Meteorology for providing wind data for the wind districts of Kütahya between 2016 and 2019.

References

Bailey B, McDonald S, Bernadett D (1997) Wind resource assessment handbook: Fundamentals for conducting a successful monitoring program. AWS Sci; 41: 65.

Cakmakcı BA and Hüner E (2020) Evaluation of wind energy potential: a case study. Energy Sources, Part A: Recovery, Utilization, And Environmental Effects. https://doi.org/10.1080/15567036.2020.1811810

Hulio, Z.H. and Jiang, W (2020) Wind energy potential assessment for KPT with a comparison of different methods of determining Weibull parameters, International Journal of Energy Sector Management, Vol. 14 No. 1, pp. 59-84. https://doi.org/10.1108/IJESM-09-2018-0007

Koşar, O. and Özgür M.A (2020) Wind Energy Resource Assessment of Kütahya, Turkey Using Wasp and Layout Optimization, Proc IMechE Part A: J Power and Energy 0(0) 1–12, https://doi.org/10.1177/0957650920936030

Lipman, N.H. and Musgrove P.J (1982) Wind Energy for the Eighties, England

Özgür, M, A (2006) Kütahya Statistical Analysis of Wind Characteristics and Applicability to Electricity Generation, Ph.D. Thesis, Osmangazi University, Department of Mechanical Engineering

Teimourian A, Bahrami A, Teimourian H, Vala M, Huseyniklioglu AO (2019) Assessment of wind energy potential in the southeastern province of Iran, Energy Sources, Part A: Recovery, Utilization, and Environmental Effects https://doi.org/10.1080/15567036.2019.1587079

TWEA, Wind Energy Association Of Turkey (2022)

Verma, M., Ahmed, S., Bhagoria, J.L (2015) Re-powering of wind farms: State of art. International Journal on Emerging Technologies 6 (2), 112.

Chapter 62
A Comparison of an Analytic Gaussian Wake Model with a Classical Model for Wind Farm Layout Optimization

Murat Ertan, Onur Koşar, and Mustafa Arif Özgür

Nomenclature

C_T	Thrust coefficient
d	Boeing Airplane Company
I_a	Lockheed California Company
k_{Jensen}	Wake expansion rate of the Jensen model
k_g	Growth rate of the Gaussian model

62.1 Introduction

Researchers need simple and high-accuracy analytical models to be used in processes involving a high number of iterations, such as the wind farm layout optimization problem. Pioneering Jensen wake model (Jensen 1983) assumes that the wake region behind the turbine expands linearly with downstream distance and the velocity deficit has at op-hat profile:

M. Ertan
Institute of Graduate Education, Kütahya Dumlupınar University, Kütahya, Türkiye

O. Koşar (✉)
Simav Technology Faculty, Department of Mechanical Engineering, Kütahya Dumlupınar University, Kütahya, Türkiye
e-mail: onur.kosar@dpu.edu.tr

M. A. Özgür
Faculty of Engineering, Department of Mechanical Engineering, Kütahya Dumlupınar University, Kütahya, Türkiye
e-mail: arif.ozgur@dpu.edu.tr

M. Z. Sogut et al. (eds.), *Proceedings of the 2022 International Symposium on Energy Management and Sustainability*, Springer Proceedings in Energy,
https://doi.org/10.1007/978-3-031-30171-1_62

$$\frac{\Delta U}{U_\infty} = \frac{1-\sqrt{1-C_T}}{\left(1+\frac{2k_{Jensen}x}{d}\right)^2} \tag{62.1}$$

Jensen used a value of $k_{jensen} = 0.1$ for the wake expansion rate. However, as a result of later research, the accepted value in the literature was 0.075 for onshore wind farms and 0.05 for offshore applications (Barthelmie et al. 2005). Researchers have conducted studies on more precise models because of the dramatic effect of the accurate definition of the velocity field in wind power plants on the energy production calculations.

Bastankhah and Porte-Agel (2014) proposed a new analytical model based on conservation of mass and momentum with ignoring the viscous and pressure terms from the momentum equation. Assuming that the velocity disturbance has a Gaussian shape and the wake region behind the turbine expands linearly, the normalized velocity deficit in the wake is obtained as a function of thrust coefficient, wake growth rate, and normalized coordinates:

$$\frac{\Delta U}{U_\infty} = \left(1-\sqrt{1-\frac{C_T}{8\left(\frac{k_g x}{d}+\varepsilon\right)^2}}\right) \times \exp\left(-\frac{1}{2\left(\frac{k_g x}{d}+\varepsilon\right)^2}\left\{\left(\frac{z-z_h}{d}\right)^2+\left(\frac{y}{d}\right)^2\right\}\right) \tag{62.2}$$

They stated that the proposed model is in good agreement with the experimental and LES data for five different case studies. They also showed that the top-hat models underestimate the velocity deficit in the center of the wake region and overestimate the velocity deficit at the edge. In this study, the effects of the two different wake models on the results of a wind farm layout problem are presented comparatively.

62.2 Comparison of the Wake Models

Before performing the optimization procedure, it will be useful to compare the results obtained by the models. For the Jensen wake model, the wake expansion rate was taken as 0.075. In the Gaussian wake model, the growth rate varies with certain site parameters such as surface roughness and turbulence intensity (Parada et al. 2017). Bastankhah and Porte-Agel (2014) did not make a specific definition for the growth rate in their related study. Ishihara and Qian (2018) expressed the kg and ε as a function of thrust coefficient and ambient turbulence intensity:

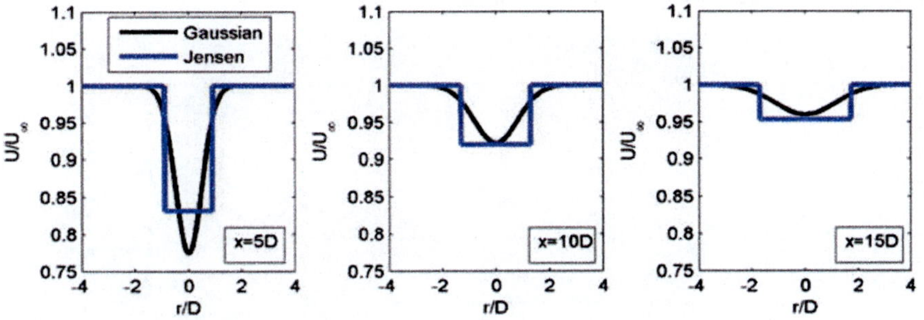

Fig. 62.1 Comparison of the velocity deficit profile at different downstream distances

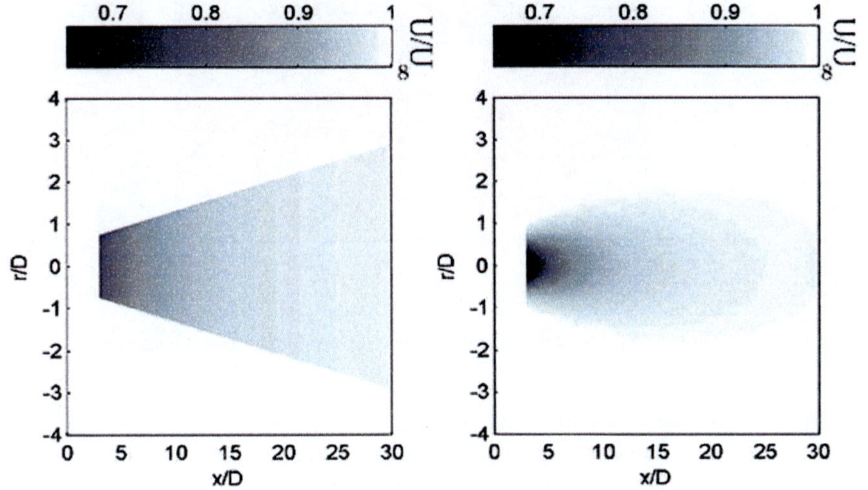

Fig. 62.2 Comparison of the (left) Jensen and (right) Gaussian models

$$k_g = 0.11 C_T^{1.07} I_a^{0.20}, \quad \varepsilon = 0.23 C_T^{-0.25} I_a^{0.17} \tag{62.3}$$

The results given in this section were obtained under the assumptions of $C_T = 0.8$, $I_a = 0.134$, $z_h = 70$ m, $d = 80$ m, and $U_{hub} = 10$ m/s. Figure 62.1 shows a comparison of the velocity deficit profile at different downstream distances. The Gaussian model predicts a larger velocity deficit in the region close to the turbine rotor than the Jensen model. As the distance from the turbine increases in the direction of the flow, the velocity deficit predicted by both models is structurally different, but generally in good agreement with each other.

Figure 62.2 presents the development of the wake region behind a wind turbine along the downstream from the perspective of both models at hub height. It is observed that there is no change in the radial direction in the Jensen model. The velocity deficit recovers faster in the radial direction with the Gaussian model.

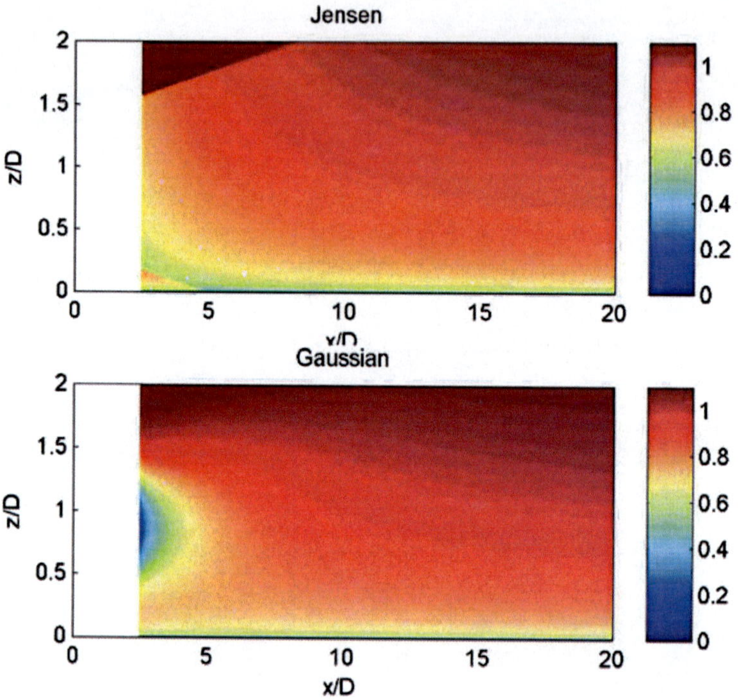

Fig. 62.3 Comparison of the velocity deficit profile at different downstream distances

Figure 62.3 shows contours of the streamwise velocity predicted by the analytical models on a vertical plane perpendicular to the wind turbine at $y = 0$. When the model results are compared with the wind tunnel measurements given in the related reference (Chamorro and Porté-Agel 2010), it is seen that the velocity field modeling ability of the Jensen model in the presence of turbulent boundary layer is less than the Gaussian model.

62.3 Optimization Approach

A single objective genetic algorithm was used in this study because of its effectiveness in discrete optimization problems. The structural details of the algorithm are given in detail in a previous study (Koşar and Özgür 2021). In the optimization study conducted for Enercon's 11 commercial wind turbines, not only the location of the wind turbines but also the hub height was optimized while the elevation change of the terrain was taken into account. Both wake models were tested for three desired nominal power values (40 MW, 80 MW, 120 MW) under the same design and constraint parameters. The wind farm layout optimization problem was conducted for a selected candidate region on the Aslanapa district of Kütahya, Turkey.

62.4 A Brief Summary of the Wind Potential of Aslanapa, Kütahya

Figure 62.4 shows the 3-year mean wind speed and wind direction data obtained from Aslanapa Meteorology Station at 39°13′17″ N and 29°52′.8″ E at an altitude of 10 m from the General Directorate of Meteorology of Turkey. This wind data was used for the wind resource assessment of Aslanapa as the loss rate (6.85%) is below 10%.

The Wind Atlas Analysis and Application Program (WAsP) is a software program at the Danish Meteorological Institute's Riso Meteorological Laboratory and the Technical University of Denmark to determine the wind energy potential (Mortensen et al. 2009). Wind speed and direction data, elevations, and surface roughness maps are required to set up a wind farm in the WAsP program. Figure 62.5 shows places with wind energy potential in Aslanapa. The area in the red rectangle is found suitable for wind farm installation due to its relatively high mean wind speed and low ΔRIX values.

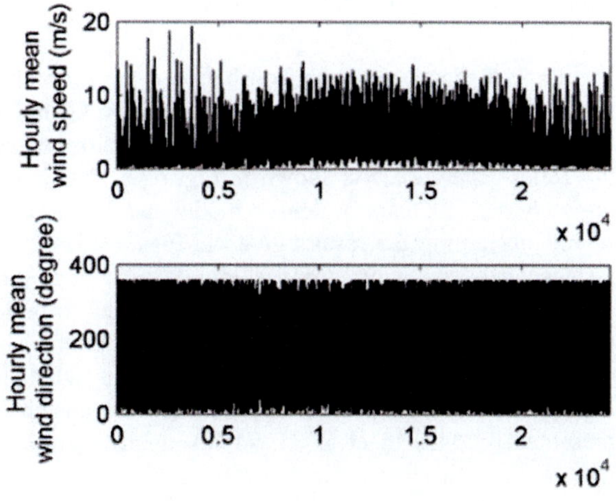

Fig. 62.4 Aslanapa's 3-year wind data (hourly – 10 m AGL)

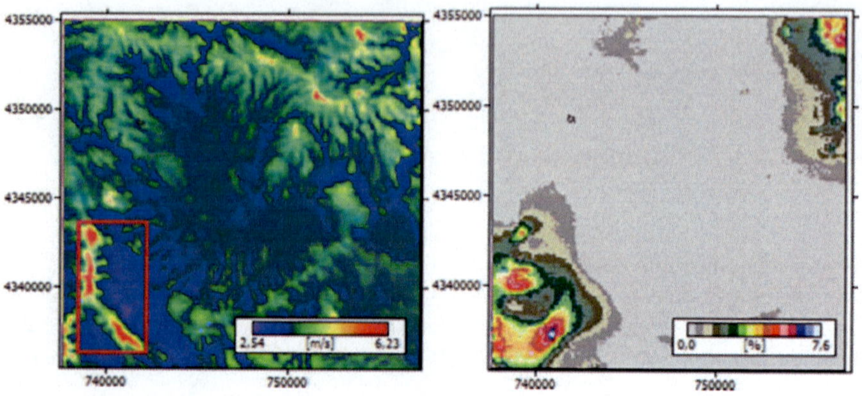

Fig. 62.5 WAsP results ((left) mean wind speed at 30 m AGL (right) ΔRIX) and selection of suitable location for a wind farm

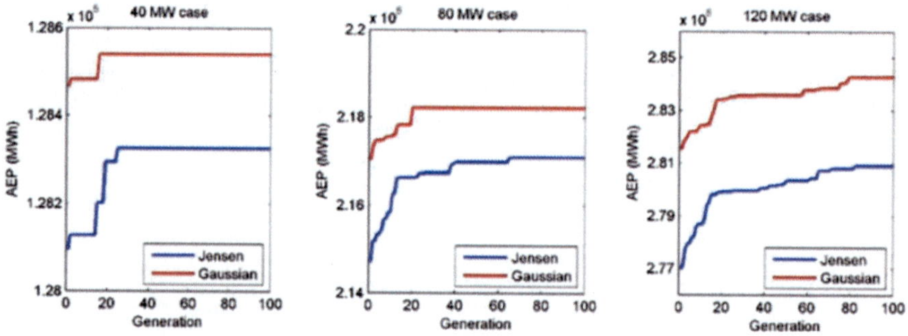

Fig. 62.6 Comparison of the (i) Jensen and (ii) Gaussian models

62.5 Results and Discussion

Figure 62.6 shows a comparison between the generation development of the best fitness (AEP) obtained with the Jensen and the Gaussian wake models. This figure demonstrates that the generation development of the genetic algorithm is limited at the 100th generation. As shown in Fig. 62.6, the Gaussian model calculates wake region losses less than the Jensen model and therefore AEP calculations are always high. Considering the results given in Figs. 62.1, 62.2, and 62.3, and limiting the distance between wind turbines to at least five rotor diameters, the wake loss calculated with the Gaussian model is expected to be less compared to the Jensen model, and therefore the AEP values are expected to be higher. Another proof of this expectation was observed in the results of the wind farm layout optimization by Parada et al. (2017). Parada et al. (2017) solved the optimization problem cases presented by Mosetti et al. (1994) using the Gaussian model and compared the results with the studies of Grady et al. (2005), who solved the same cases with the Jensen model. In the results given for all three cases, the wake losses obtained with the Gaussian model are less than the Jensen model.

Figure 62.7 shows the wind farm designs optimized for the three desired minimum nominal power values on the annual mean wind speed contour plot. The number on the turbine points indicates the rotor diameter of the turbine, and the number below indicates the hub height. The diameter of the red circles drawn centered on the turbine point is five rotor diameters of the relevant turbine and represents the prohibited areas due to the presence of the turbine. When the algorithm assigns a turbine to the area while generating possible solutions, it deletes the locations in the circle from the possible location pool. Looking at all the designs, it is seen that the E-142 model (4.2 MW), produced by Enercon for efficient operation at low wind speeds, is the dominant model. This trend seems reasonable given the fact that Aslanapa has low wind potential. The trend observed in this study, in which AEP maximization was applied, will continue until a different parameter such as cost is added to the objective function. In addition, it has been observed that the algorithm prefers the larger of the two different hub height options (129 m and 159 m) defined for this turbine. Considering the fact that the wind speed increases with height and the power obtained from the wind is proportional to the cube of the wind speed, it can be said that this choice is logical. The results produced for the 40 MW case appear to be the same for both wake models. The algorithm easily created a wind farm layout with the desired power value according to the given area. The algorithm was not challenged in terms of wake effects and placed the turbines at the points where the wind speed was highest. A similar solution is seen in the study of Parada et al. (2017). For the simplest case, the optimized layouts of Parada et al. (2017) and Grady et al. (2005) are the same. With the increase of the nominal power value, the results have slight differences, although they are generally similar to each other. However, when these minor changes are examined, it is seen that the turbines are located at points with higher mean wind speed in the layouts obtained with the Gaussian model.

62.6 Conclusions

In this study, two different analytical wake region models were used in wind farm layout optimization with real long-term wind data. The wake region estimates of the models are given comparatively. The results showed that the Gaussian model predicts higher velocity deficits near the rotor, and velocity disturbance recovery is faster in the radial direction. In addition, it was observed that it better modeled the flow behind the turbine in the presence of a boundary layer.

Table 62.1 presents a summary of the optimization results. According to the optimization results, when all conditions are equal, lower wake losses are calculated with the Gaussian model. It can be said that this trend will generally continue, since the distance between the turbines is a minimum of five rotor diameters. If the

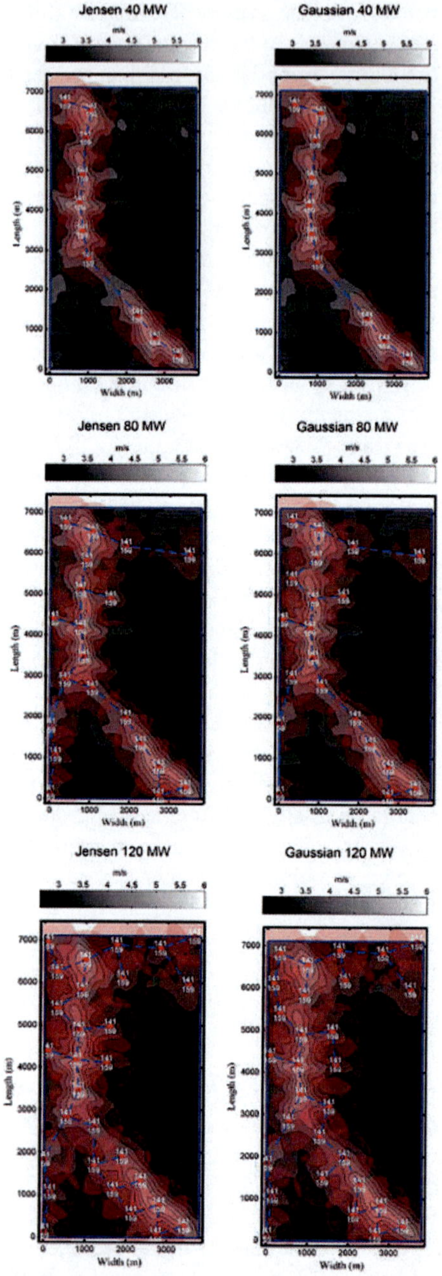

Fig. 62.7 Comparison of the optimized layouts

Table 62.1 Optimization result

Rated power of the optimized layout (MW)	Jensen		Gaussian	
	AEP (MWh)	Wake losses	AEP (MWh)	Wake losses
42–42 (40 MW case)	128,325	%0.89	128,540	%0.55
84–84 (80 MW case)	217,096	%2.16	218,221	%1.57
121.8–121.8 (120 MW case)	280,905	%4,05	284,285	%2.56

algorithm can efficiently distribute the turbines over the area, that is, if the wake region effects are not very dominant, it is seen that the layouts to be produced with both models will be similar. For this reason, the Jensen model is still seen as a useful tool in the design of relatively low-power capacity wind farms without area constraints, due to its simplicity.

Acknowledgments The authors acknowledge the Turkish Republic General Directorate of Meteorology for providing the wind data of Aslanapa, Kütahya.

References

Barthelmie RJ, Folkerts L, Larsen GC, Rados K, Pryor SC, Frandsen ST, et al. Comparison of wake model simulations with offshore wind turbine wake profiles measured by sodar. J Atmospheric Ocean Technol 2005;23:881–901.

Chamorro, L. P., Porté-Agel, F. 2010. Effects of thermal stability and incoming boundary-layer flow characteristics on wind- turbine wakes: a wind-tunnel study. Boundary-layer meteorology, 136(3), 515–533.

Grady, S. A., Hussaini, M. Y., Abdullah, M. M. 2005. Placement of wind turbines using genetic algorithms. Renewable energy, 30(2), 259–270.

Ishihara, T., Qian, G. W. "A new Gaussian-based analytical wake model for wind turbines considering ambient turbulence intensities and thrust coefficient effects." Journal of Wind Engineering and Industrial Aerodynamics 177 (2018): 275–292.

Jensen N. A note on wind turbine interaction. Technical report Ris-M-2411. Roskilde, Denmark: Risø National Laboratory; 1983.

Koşar, O., Özgür, M. A. 2021. Wind energy resource assessment of Kütahya, Turkey using WAsP and layout optimization. Proceedings of the Institution of Mechanical Engineers, Part A: Journal of Power and Energy, 235(3), 629- 640.

M. Bastankhah, F. Porte-Agel,. A new analytical model for wind- turbine wakes, Renew. Energy 70 (2014) 116–123.

Mosetti G, Poloni C and Diviacco B. Optimization of wind turbine positioning in large windfarms by means of a genetic algorithm. J Wind Eng Ind Aerodyn 1994; 51: 105–116.

N.G. Mortensen, D.N. Heathfield, L. Myllerup, L. Landberg, O. Rathmann, Getting Started with WAsP 9, Risø National Laboratory Technical University, 2009.

Parada, L., Herrera, C., Flores, P., Parada, V. "Wind farm layout optimization using a Gaussian-based wake model." Renewable energy 107 (2017): 531–541.

Chapter 63
Energy-Exergy Analysis of a Building Heated with Waste Heat Source District Heating Systems: Soma, Manisa, Case Study

Murat Ertan, Onur Koşar, and Selçuk Sarıkoç

Nomenclature

Q	Thermal power in the system
$\dot{m}$	Mass flow rate
S	Entropy
H	Enthalpy
C_p	Specific heat at constant pressure
I	Total irreversibility

63.1 Introduction

District heating is the distribution of energy generated from one or more centers to consumers in order to meet the heating and domestic hot water needs of residences, businesses, and commercial facilities in a residential area. The district heating system not only increases comfort indoors and reduces fuel costs for residents but

M. Ertan (✉)
Institute of Graduate Education, Kütahya Dumlupınar University, Kütahya, Türkiye

O. Koşar
Simav Technology Faculty, Department of Mechanical Engineering, Kütahya Dumlupınar University, Kütahya, Türkiye
e-mail: onur.kosar@dpu.edu.tr

S. Sarıkoç
Taşova Yüksel Akın Vocational School, Department of Motor Vehicles and Transportation Technologies, Amasya University, Amasya, Türkiye

M. Z. Sogut et al. (eds.), *Proceedings of the 2022 International Symposium on Energy Management and Sustainability*, Springer Proceedings in Energy,
https://doi.org/10.1007/978-3-031-30171-1_63

also reduces the amount of greenhouse gases and pollution in terms of dust and SO_2 (Lund et al. 2010; IDEA 2021).

Europe is promoting cogeneration (combined heat, electric power generation, and combined heat and power (CHP)) as a method to increase energy efficiency and reduce greenhouse gas emissions. CHP reduces primary energy consumption by about a third compared to heat and electrical power generation alone. Therefore, the heat recovered in a fossil fuel CHP plant results in about two-thirds reductions in carbon dioxide emissions. Its efficiency is doubled (Elliot and Kaasberg 2002).

In Turkey, the use of resources other than geothermal in district heating is limited, and it is provided only by waste heat obtained from the power plant in Esenyurt and Soma. In 2020, the Esenyurt Natural Gas Combined Cycle Power Plant was transferred to the Electricity Generation Corporation. Electricity production has stopped. As of 2020, the Soma district heating system is the only district heating system in Turkey that uses cogeneration method. In addition to geothermal and thermal power plants, there are applications such as the use of waste heat released in industrial facilities in administrative buildings and lodgings inside the facility. Generally, the heat released in sugar factories is used in these applications. Today, garbage and sludge incineration plants in Turkey are used only for electricity generation. The process of converting waste heat into economic gain continues throughout the country (Republic of Turkey Electricity Generation Company 2022). Waste energy has been identified by the International Energy Agency as a key technology enabling increased integration of renewable resources (Schmidt et al. 2017).

Exergy analysis is important for energy resource utilization, because exergy is part of energy analysis. The concepts of exergy, usable energy, and usability are basically similar. Exergy destruction and consumption are defined as lost work. Exergy is a measure of the maximum useful work that can be done by a system interacting with a medium at constant pressure and temperature (Rogers and Mayhew 1980; Rosen and Dincer 2001).

Sarıkoç et al.'s energy and exergy analyses were performed in a direct-injected (DI) diesel engine operating with various diesel-biodiesel butanol fuel blends (Sarıkoç et al. 2020). Yazıcı calculated the energy-exergy efficiency of the district heating system, whose source is geothermal energy, in Afyon province. As a result of the analysis, it was determined that the exergy destruction in the system was caused by the re-injected thermal water and the heat exchanger. The energy efficiency of the entire system was found to be 46,17% and the exergy efficiency 60,63% (Yazici 2016). Ozgener and Ozgener geothermal calculated the general energy, exergy, and technical availability analysis of geothermal district heating systems. Technical availability, actual availability, capacity factor, and energy and exergy efficiency values were analyzed (Ozgener and Ozgener 2009). Kanoğlu et al. offer suggestions regarding energy and exergy efficiencies. It provides important information about heat losses and exergy destructions in the plants considered as examples (Kanoğlu et al. 2007).

As seen in the studies, researchers emphasized the importance of energy and exergy analysis related to the development of a system. In this study, the energy-exergy analysis of a building heated by Soma district heating systems, which utilizes the waste heat of Turkey's third largest thermal power plant, has been studied. The heat losses of the building are calculated and the capacity of the underground power station is determined.

63.2 System Description

As of 2022, there are geothermal-sourced district heating systems (residences, workplaces, etc.) in 18 regions in Turkey. It has the equivalent of 130.000 houses with a total energy of 1205 MW. Geothermal fluid heating is available in many cities such as İzmir (Balçova-Bergama-Dikili-Narlıdere), Afyon (Ömer-Gecek-Sandıklı), Balıkesir (Gönen-Güre-Sındırgı-Edremit-Bigadiç), and Kütahya (Simav). The Balçova geothermal district heating system has been operating since 1996 and has reached 37.305 residences (Ministry of Energy and Natural Resources 2022).

Soma district is located in the northwest part of Manisa and it has a population of 110.935 and an area of 839 km^2 (Turkish Statistical Institute 2021). The district heating system in Soma is at 6000 (cubic meters/hour) and 99 °C. It has a heating potential of 40.000 houses equivalent. The planned maximum capacity of the district heating system is 270 MW of heat energy. Currently, only about 31% is used (Soma Dıstrıct Heatıng Systems 2022).

Operating in Soma district, each Soma B Thermal Power Plant has an installed power of 165 MW. It consists of six units. The total installed power of the power plant is 990 MW. Its share in the generation of electricity in the country is 2,21%. 8.000.000 tons of coal is burned in the plant on a yearly basis. There are four heat exchangers in the thermal power plant. It heats Soma district from the waste heat of four units. The temperature of the steam entering the heat exchangers in the thermal power plant is 120 °C. Soma district heating system comes to the pump station at 95 °C. The distance between the thermal power plant and the pumping station is 1,5 km. Three 45 MW heat exchanger groups are installed at the Waste Heat Transfer Station within the Soma Thermal Power Plant. As of 2022, the total installed capacity is 135 MW (Soma Thermal Power Plant 2022).

Soma district heating system distribution mountain consists of three closed circuits. Figure 63.1 shows the closed circuit rotating inside the switchboard, the closed circuit from the switchboard to the building, and the closed circuit inside the building. The waste steam inside the first closed-circuit units heats the Soma district heating system distribution line with heat exchangers. In the second closed circuit, the heated water in the 95 °C distribution line heats the underground power station in the independent sections and the network lines in the buildings. The third circuit is the distribution of the heated water in the building to the radiators with a circulation pump. The round-trip temperatures of the Soma district heating systems distribution line are 95/60 °C. The heated peak load is 80 MW/h.

The required amount of water is sent to each residential unit by using flow and temperature control valves at the sub-building power station. Thus, the heat balance of the system is ensured. The average round-trip temperatures sent to the buildings are 75/55 °C on average.

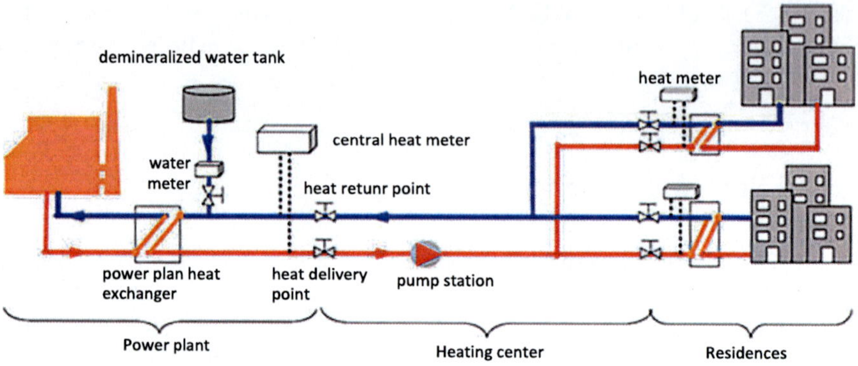

Fig. 63.1 Soma district heating flow schematic display

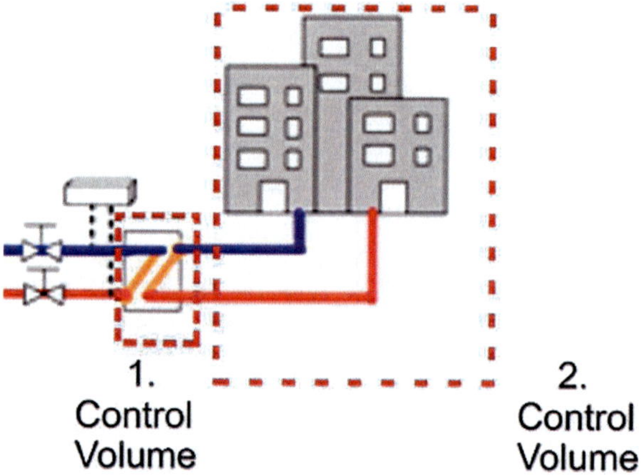

Fig. 63.2 Control volumes we examined in the third closed circuit

63.3 Analysis

The heat exchange of the buildings with the external environment is defined as heat loss and heat gain. The total heating load of the building is determined by summing the heat losses calculated for the spaces. The energy-exergy efficiencies of the system are calculated (Fig. 63.2).

The heat energy of the system:

$$Q = mc_p \Delta T \tag{63.1}$$

Continuity equation:

$$\sum \dot{m}_1 - \sum \dot{m}_e = 0 \tag{63.2}$$

First law of thermodynamics:

$$Q - W = \sum \dot{m}_e \left(h + \frac{V^2}{2} + gz\right)_e - \sum \dot{m}_1 \left(h + \frac{V^2}{2} + gz\right)_1 \tag{63.3}$$

Entropy generation:

$$\dot{S}_{prod.} = \sum \dot{m}_e s_e - \sum \dot{m}_l s_l - \sum \frac{\dot{Q}_k}{T_k} \tag{63.4}$$

Irreversibility:

$$I = T_o \Delta S_{prod} \tag{63.5}$$

Flow energy:

$$\psi_1 = (h_1 - h_O) - T_O(s_1 - s_O) \tag{63.6}$$

$$\psi_e = (h_e - h_o) - T_o(s_e - s_o) \tag{63.7}$$

Exergy destruction:

$$\dot{X}_{dest} = (\dot{m}_1 \psi_1 - \dot{m}_1 \psi_e) \tag{63.8}$$

II. Law efficiency:

$$\eta_{II} = 1 - \frac{\dot{X}_{exergy}\ dest}{\dot{m}(\psi_1 - \psi_e)} \tag{63.9}$$

63.4 Results and Discussion

The heat losses of the apartments in the second control volume examined were calculated. According to TS 825 (thermal insulation rules in buildings), the heat losses and gains of the apartment are calculated. Heat losses occur from the outer wall of the building, the ceiling, the floor, and the adjacent wall in the unheated indoor environment. The heat transfer coefficients used when calculating the heat loss are shown in Table 63.1.

Table 63.1 Heat conduction coefficients of structural components in the building

Building component	Heat transfer coefficient (W/m^2K)
Interior walls	1,10
Exterior walls	0,46
Floor	0,59
Ceiling	0,49
Glass window	2,40
Exterior door	3,50

Table 63.2 Properties of the studied regions in thermodynamic equilibrium positions

State no	Temperature (C°)	Pressure (kPa)	Specific enthalpy (kJ/kg)	Specific entropy (kJ/kgK)	Mass flow rate (kg/s)
1	70	737	293,07	0,9551	0,27
2	60	550	251,18	0.8313	0,27
3	54	250	230,26	0,7680	0,11
4	46	200	188,44	0,6386	0,11

The area we will examine in this study is the third closed circuit. The data in the building in January and February 2022 were examined. These are the months with the highest heat requirement in Soma. In Table 63.2, comfort temperatures, pressure values, and thermodynamic values of each area of the building we examined are given. With the values in the table, the total heat loss of the building is 69.165 W. Workplaces on the ground floor of the building are not included in the heat loss since they are not included in the district heating.

The average flow rate of the water coming from the district heating is 0,27 kg/s, and the average temperature is 70 °C. The return water temperature is 60 °C. The pressure of the flow line in the Soma district heating network is 737 kPa. The pressure of the return line in the Soma district heating network is 550 kPa.

The average temperature coming into the building is 54 °C, and the temperature of the water returning from the radiators is 46 °C. The mass flow rate of the water returning inside the building is 0,11 kg/s. The pressure of the flow line to the radiator inside the building is 250 kPa. The pressure of the return line from the radiator inside the building is 200 kPa.

There are adjacent buildings on the east and west facades of the building we examined. The outside temperature is calculated as 5 °C. The pressure of the air is 1 bar. Table 63.3 shows the thermodynamic results of the first control volume. The thermal efficiency of the building power station is 20% and the exergy efficiency is 63,19%.

Table 63.4 shows the thermodynamic results of the second control volume. The thermal efficiency of the apartment is 18,2% and the exergy efficiency is 65,4%.

Table 63.3 Results for the first control volume

Q (kW)	1883,74
Entropy generation (kW/K)	5618
İrreversibility (kW)	1562,6
Flow exergy	
Entering (kJ/kg)	31,69
Out (kJ/kg)	20,16
Irreversibility (kW)	349,58
Thermal efficiency	20%
Exergy efficiency	63,19%

Table 63.4 Results for the second control volume

Q(kW)	28,44
Entropy generation (kW/K)	0,087
İrreversibility (kW)	24,47
Flow exergy	
Entering (kJ/kg)	16,84
Out (kJ/kg)	11,02
Irreversibility (kW)	3,96
Thermal efficiency	18,2%
Exergy efficiency	65,4%

63.5 Conclusions

The usability of existing thermal power plants in Turkey for district heating should be examined. Thermal power plants where district heating can be made should be implemented with pilot projects (as in the example of the Soma Power Plant). For the thermal power plants to be built, a suitability analysis for district heating should be made. The aim of social benefit should be prioritized. Infrastructures in new residential areas should be compatible with district heating. Municipalities should be authorized to install district heating systems and deliver them to households. Financial support models should be created for municipalities. The use of domestic and renewable fuels in district heating should also be encouraged.

- In this study, energy-exergy analysis of a building heated by the waste heat sourced district heating system of a thermal power plant was carried out.
- The energy efficiency of the building was calculated as 18,2% and the exergy efficiency as 65,4%.
- The thermal efficiency of the building power station is 20% and the exergy efficiency is 63,19%.
- The exergy destruction of each component is calculated. It was found that the most exergy destruction occurred in the heat exchanger.
- The heat losses of the building are calculated. The capacity of the building power station is determined as 69.165 W.

- The average flow rate of the water coming from the district heating is 0,27 kg/s, and the average temperature is 70 °C. The return water temperature is 60 °C.
- The average temperature coming into the building is 54 °C, and the temperature of the water returning from the radiators is 46 °C. The mass flow rate of the water returning inside the building is 0,11 kg/s.
- According to the data of January 2022, the pressure data of both control volumes were taken. The pressure of the flow line in the Soma district heating network is 737 kPa. The pressure of the return line in the Soma district heating network is 550 kPa. The pressure of the flow line to the radiator inside the building is 250 kPa. The pressure of the return line from the radiator inside the building is 200 kPa.

Acknowledgments The authors acknowledge Kütahya Dumlupınar University, Amasya University, and Manisa Metropolitan Municipality Soma District Heating Systems for their contributions.

References

Elliot R, Kaasberg T (2002) Combined heat and power: saving energy and the environment. The Northeast-Midwest Institute.

International District Energy Association (2021) https://www.districtenergy.org/topics/district-heating

Kanoğlu M, Dincer I, Rosen M.A. (2007) Understanding energy and exergy efficiencies for improved energy management in power plants. Energy Policy 35 3967–3978. https://doi.org/10.1016/j.enpol.2007.01.015

Lund H, Möller B, Mathiesen BV, Dyrelund A. (2010) The role of district heating in future renewable energy systems. Energy.35:1381–90.

Ministry of Energy and Natural Resources of The Republic of Turkey (2022) https://enerji.gov.tr/bilgi-merkezi-enerji-jeotermal

Ozgener L, Ozgener O (2009) Monitoring of energy exergy efficiencies and exergoeconomic parameters of geothermal district heating systems (GDHSs). Applied Energy Volume 86, Issue 9, Pages 1704–1711

Republic of Turkey Electricity Generation Company (2022) https://www.euas.gov.tr/

Republic of Turkey Manisa Metropolitan Municipality Soma District Heating Systems Technical Department (2022)

Rogers G.F.C., Mayhew Y.R. (1980) Engineering Thermodynamics, Work and Heat Transfer. Longman Group Ltd., London and New York, pp. 150–156.

Rosen M.A., Dincer I (2001) Exergy as the confluence of energy, environment and sustainable development. Exergy Int. J. 1 (1), 3–13.

Sarıkoç S, Ors I, Unalan S (2020). An experimental study on energy-exergy analysis and sustainability index in a diesel engine with direct injection diesel-biodiesel-butanol fuel blends. Fuel, 268(117),321. https://doi.org/10.1016/j.fuel.2020.117321

Schmidt D, Kallert A, Blesl M, Li H, Svendsen S, Nord N (2017) IEA Annex TS1: Low Temperature District Heating for Future Energy Systems - Final report –Future low temperature district heating design guidebook. Frankfurt Am Main (Germany): AGFW-Project Company.

Soma Thermal Power Plant Company (2022) http://somatermik.com.tr/Tr/anasayfa

Turkish Statistical Institute (2021)

Yazici H (2016) Energy and exergy based evaluation of the renovated Afyon geothermal district heating system. Energy and Buildings 127 794–804 https://doi.org/10.1016/j.enbuild.

Chapter 64
The Effects of Climate Change on Water Resources in Turkey

Murat Pinarlik

Nomenclature

HadGEM2-ES	Hadley Centre Global Environment Model
MPI-ESM-MR	Max Planck Institute – Earth System Model
CNRM-CM5.1	Centre National de Recherches Météorologiques –Model
RCP	Representative Concentration Routes
IPCC	Intergovernmental Panel on Climate Change
EEA	European Environment Agency
EU	European Union

64.1 Introduction

About three-quarters of the world is covered by water. Freshwater constitutes only 2.5% of the total amount of water in the world. Considering that 90% of freshwater is located at the poles and underground, human beings can benefit from only 0.3% of the total water potential (Küçükklavuz 2009). While the percentage of water used in agriculture was 90.5% at the beginning of the twentieth century, it has decreased to 69% today and increased to 23% in the industry and energy sector and 8% in residences (Aytemiz and Kodaman 2006; Çetin et al. 2008). Ever-increasing water scarcity directly affected 26 countries with a total population of 300 million in the 1990s (Ateş 2008; Uzmen 2007; Filinte 2007). Also, it is known that the world's

M. Pinarlik (✉)
Faculty of Technology, Civil Engineering Department, Gazi University, Ankara, Turkey
e-mail: muratpinarlik@gazi.edu.tr

M. Z. Sogut et al. (eds.), *Proceedings of the 2022 International Symposium on Energy Management and Sustainability*, Springer Proceedings in Energy,
https://doi.org/10.1007/978-3-031-30171-1_64

population will be 9.3 billion in 2050 and water scarcity will be experienced in 60 countries due to climate change (Dündar 2007).

Depending on the increase in global temperatures, it is expected that changes such as changes in the hydrological cycle, a decrease in the volume and quality of water resources, a rise in sea level, drought, and flood may occur (Kibaroğlu 2008). In general, the significant effects of climate change on water resources can be summarized as depending on the regions where the basins are located, i.e., a decrease or an increase in surface water potentials, changes in the recharge and discharge of underground aquifers, changes in the frequency of extreme currents (floods and drought), changes in the seasons and sizes of their occurrence, and changes in precipitation regimes (Fıstıkoğlu and Biberoğlu 2008; Karaman and Gökalp 2010).

Although Turkey is seen as one of the trouble-free countries in terms of the use and evaluation of water resources, the situation is different when we look at the usable water potential per capita (Aküzüm et al. 2010). According to the water potential calculations of the State Hydraulic Works in Turkey, Turkey has an annual water potential of 1652 m^3 per capita. According to the estimates of the Turkish Statistical Institute, the population of the country will reach 100 million in 2030 and the water potential will decrease to 1120 m^3 per person per year (Yılmaz 2015). This situation shows that Turkey will be among the countries that will experience water shortages.

In this study, the current situation of water resources in the world and Turkey and the general effects that may be caused by climate change are evaluated and future water shortage scenarios that may occur with climate change are emphasized. In addition, what needs to be done about the sustainability of water resources was discussed and suggestions were made.

64.2 Effect of Climate Change on Water Resources

Global climate change will inevitably affect many sectors depending on the increase in temperature. It is expected that sectors such as health, forestry, biodiversity, and tourism, especially water resources and agricultural production, will be primarily affected. The important changes caused by climate change in the hydrological cycle can be listed as follows: changes in seasonal distribution and amount of precipitation, changes in average annual surface flow, changes in water quality, changes in groundwater, and its effects on floods and drought.

64.2.1 Climate Change Models and Scenarios

The climate models in this study were made by creating a regional climate model with the outputs of four global models selected from the CMIP5 archive. These models are HadGEM2-ES (Hadley Center Global Environment Model), MPI-ESM-

MR (Max Planck Institute – Earth System Model), and CNRM-CM5.1 (Centre National de Recherches Météorologiques – Model Version 5.1) and GFDL-ESM2M.

Scenarios play an important role in understanding and evaluating the possible future development of complex systems with high uncertainty such as climate (Turkish State Meteorological Service 2022). The new concentration scenarios are called Representative Concentration Routes (RCP). In the same meeting, the literature was reviewed in terms of the characteristics determined, and four RCP types were defined for radioactive forcing levels and routes.

These four scenarios describe the future climate by considering how much greenhouse gas will be emitted in 2100. The four RCP scenarios in question are RCP2.6, RCP4.5, RCP6, and RCP8.5 with references of +2.6, +4.5, +6.0, and +8.5 W/m^2 which take into account possible values of radiation stress. These scenarios are also the four greenhouse gas concentration routes adapted for the IPCC's Fifth Assessment Report.

64.2.2 Changes in Precipitation

The effects of climate change on precipitation are more complex and uncertain than the expected effects on temperature. Changes that may occur in the amount and distribution of precipitation show regional differences all over the world. As a result of temperature increases, there will be possible increases in global average precipitation and evaporation. Also, there will be more significant changes in the characteristics of local and regional precipitation due to global warming. For example, although the frequency of precipitation will generally decrease, the precipitation intensity will be greater. Since extreme weather events will increase the recurrence rate of floods and droughts, water storage will become more important.

According to another source (NOAA NCDC/CICS-NC), the optimistic scenario is RCP 2.6, which predicts a rapid decrease in greenhouse gas emissions, and the pessimistic scenario is RCP 8.5, which predicts an increase in emissions, over the 2071–2099 period (compared to the 1970–1999 period). The changes in annual precipitation amounts are given in Fig. 64.1. Investigating at the precipitation forecast map created, it is expected that the northern regions of the USA (especially the Northeast and Alaska) will receive more precipitation in general, while the southern regions (especially the Southwest) are expected to receive less precipitation.

64.2.3 Changes in Surface Flow

Changes in runoff are mainly dependent on changes in temperature and precipitation, among other variables. In the current studies, the increase in river flows was

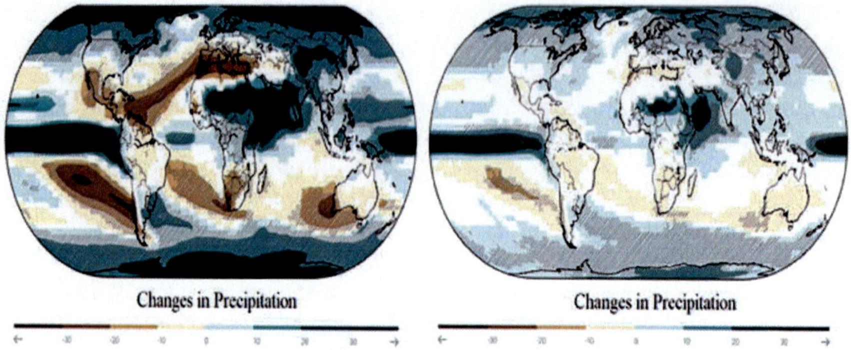

Fig. 64.1 Changes in average annual precipitation amounts during 2071–2099 periods based on 1970–1999 years according to RCP 2.6 (upper) and RCP 8.5 (lower) scenarios (IPCC 2014)

detected in the north and west of Europe during the 1963–2000 periods, especially in winter, while a decrease was detected in the south and east, especially in the summer months. According to the report titled "Climate Change, Impacts And Vulnerability in Europe 2016, An Indicator-Based Report" published by the European Environment Agency in 2017, significant seasonal changes are expected in river flows in Europe as a result of climate change (EEA 2017). It is expected that the dates when summer flows will decrease in many parts of Europe, snowfalls will be replaced by rain, and peak flow rates in spring and summer flows will shift forward.

64.2.4 Changes in Water Quality

Water quality is directly dependent on many factors such as the change in the amount of water, the change in water temperatures, and the increase in the pollution load due to migration. Therefore, it is expected that water quality will be adversely affected in many places as a result of climate change. The decrease in stream flows and the decrease in water levels in lakes cause deterioration of water quality due to the presence of nutrients and pollutants in less volume of water. The increase in water temperatures directly affects the water quality by reducing the amount of dissolved oxygen. Prolonged drought causes pollutants to accumulate on the soil surface, posing a risk to the quality of water resources when precipitation begins.

64.2.5 Changes in Groundwater

Groundwater is the main source of water for irrigation, drinking use, and industrial water supply in most regions of the world. As a result of climate change, in many aquifers around the world, spring discharges are being reduced more or less toward winter and summer discharges are decreasing dramatically. An increase in precipitation density may increase the flow, resulting in less discharge to the ground. Groundwater discharges are affected by the change in natural vegetation and crops with the reflection of climate change (SYGM 2020).

64.2.6 Impact of Climate Change on Floods

In the Fourth Evaluation Report of the IPCC, it is stated that excessive precipitation in the projected years will increase the frequency and severity of floods and will also increase the loss of life and property with effects such as deterioration in water quality and pollution of water resources. In the same report, it is stated that by 2080, 20% of the world's population will live in basins where flood damage has increased due to climate change, and it is reported that 100-year frequency floods will be experienced much more frequently (IPCC 4th Assessment Report 2007).

64.2.7 Effect of Climate Change on Drought

One of the most important effects of climate change is drought and water scarcity. The short-term effect of drought is a decrease in the amount of usable water. With the increase in water demand as a result of human needs, the decrease in the amount of water causes thirst pressure for humans and ecological systems. In the long run, this will cause underground water resources and reservoirs to become unable to meet excessive consumption, worsen the situation with the effect of new drought periods, and cause water scarcity. The frequency and severity of meteorological and hydrological droughts are increasing in Europe, particularly in Southwestern and Central Europe. Current projections indicate that these impacts will increase further in most of Europe in the twenty-first century, except Northern Europe regions.

64.3 The Effect of Climate Change on Water Resources of Turkey

Turkey is divided into 25 hydrological basins, and water, which forms the basis of these basins, is a vital and socially important resource. The precipitation regime of Turkey, which is located in a semiarid region, varies seasonally and regionally, and it is seen that some basins cannot meet the water needs. In a study conducted by the European Environment Agency (EEA), water stress levels were determined in Turkey and EU countries in 2000 and 2030. Accordingly, it is predicted that by

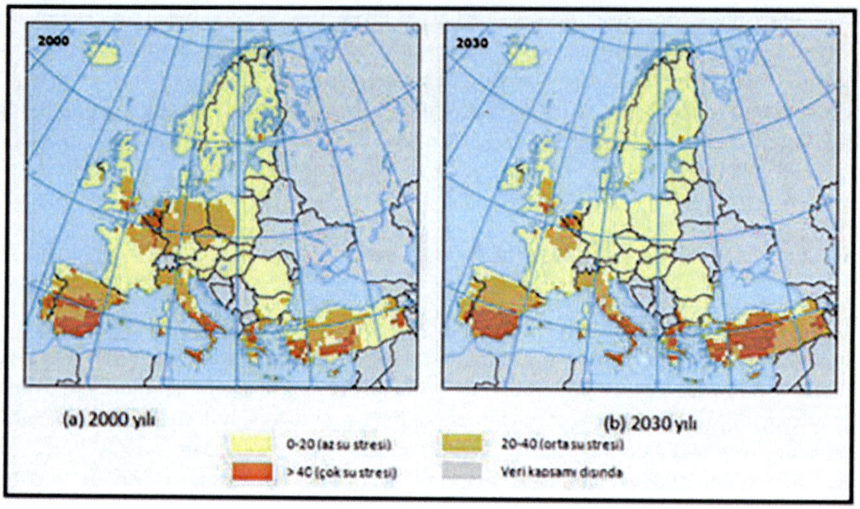

Fig. 64.2 Water stress levels in Turkey and EU countries (EEA 2017)

2030 in Turkey, water stress will exceed 40% in the Central and Western regions, and this rate will be between 20 and 40% in the Southeastern and Eastern regions. Drought is a recurring feature of the European climate. From 2006 to 2010, an average of 15% of the European region and an average of 17% of its population were affected by meteorological drought every year. In a study carried out by the European Environment Agency in 2009, water stress (the ratio of decrease in water amount to water availability) levels in Turkey and EU countries in 2000 and 2030 were determined. The change in water stress levels on Earth between the years 2000 and 2030 is shown in Fig. 64.2. As seen in Fig. 64.2, contrary to the improvement in Northern and Central Europe, it is understood that the situation will be negative for Turkey.

Possible changes in the water potential of the basins in the future have been determined using the different climate models. Also, the temperature and precipitation projections have been estimated for the 2015–2100 periods by RCP4.5 and RCP8.5 scenarios. According to the results of all climate models and two scenarios, total flow in Turkey is estimated to decrease when compared to the reference period. Hereunder, with the hydrological modeling based on the climate model outputs, it is estimated that the median gross water potentials for the three sub-periods in the 2015–2100 period will decrease by 40–45% compared to the reference period median value. It is estimated that the median gross water potential reduction rate obtained from the hydrological model projections performed with the other climate model outputs under the same conditions will remain in the 15–20% range. The net water deficit/surplus status of the river basins in Turkey for the period 2015–2100 was prepared separately for three models and two scenarios in thematic map format, for example, MPI-ESM-MR model RCP4.5 scenario Figs. 64.3, 64.4, and 64.5 is presented with.

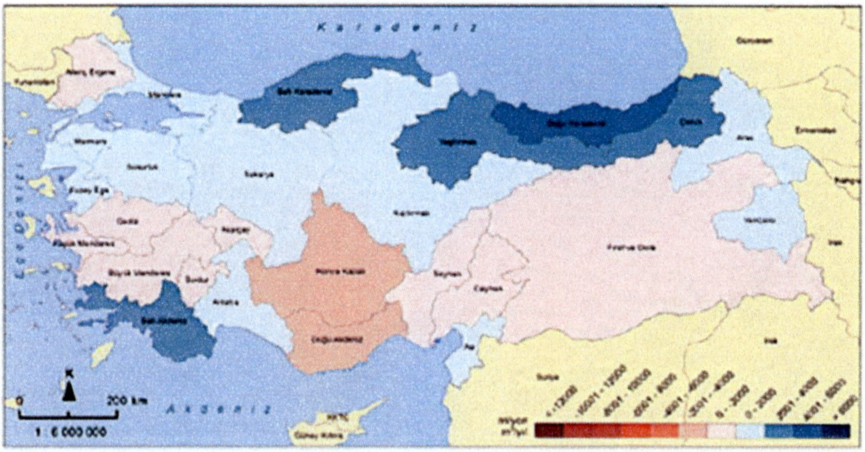

Fig. 64.3 Thematic map showing basin-based water excess/deficit for the period 2015–2040 (SYGM 2016)

Fig. 64.4 Thematic map showing basin-based water excess/deficit for the period 2041–2070 (SYGM 2016)

Thematic maps showing the water surplus/deficit can also be used to identify possible water transfer between neighboring basins in the future. The basins where the most significant water deficit is observed in all periods are the Euphrates-Tigris, Eastern Mediterranean, and Konya Closed Basins in general.

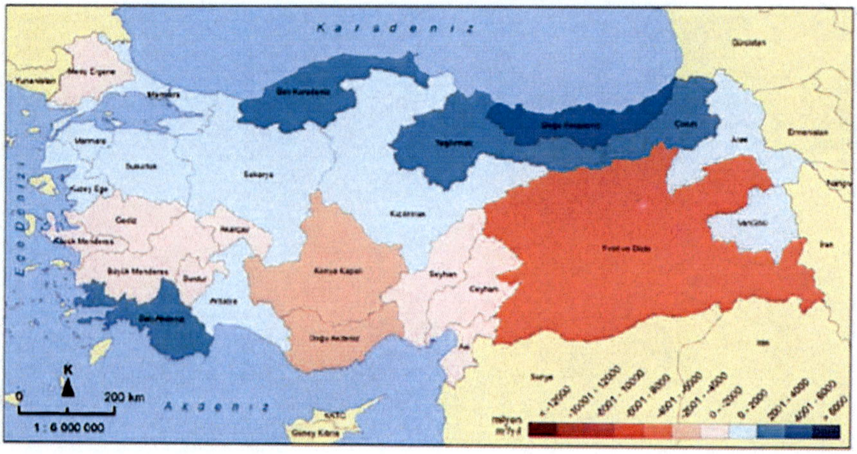

Fig. 64.5 Thematic map showing basin-based water excess/deficit for the period 2071–2100 (SYGM 2016)

64.4 Conclusion

It is a fact that our country and the whole world will experience water shortages in the future due to climate change. Therefore, it is necessary to take measures to eliminate water deficiencies that will occur due to climate change. Some works that can be done in this context are listed articled below:

- Dams should be built on the rivers so that the surface waters can be used and stored in a planned manner.
- To prevent evaporation losses due to increased temperature, underground dams should be given importance.
- Besides, losses and leaks in pipelines should be reduced to a minimum.
- With smart irrigation systems, the amount of water used in agriculture should be reduced and recyclable use of water should be ensured.
- Sustainable use of water resources should be ensured by considering climate change models.
- Water resources and dam operation policies should be developed against possible droughts.
- Industrial gray (waste) water must be recycled and reused.

It should be noted that if necessary precautions are not taken, great disasters, especially water scarcity, await humanity.

References

Aküzüm T, Çakmak B, Gökalp Z (2010) Evaluation of water resources management in Turkey, Research Journal of Agricultural Sciences, 3(1), 67–74.

Ateş, İ (2008) The politic and economic consequences of global warming and their effects on Turkey, Gebze High Technology Institute, Gebze, 76 p.

Aytemiz L, Kodaman T (2006) Sınıraşan sular kullanımı ve Türkiye-Suriye ilişkileri, TBMM Su Politikaları Kongresi Bildiriler Kitabı, Mart 2006, 527–536.

Çetin Ö, Eylen M, Üzen N (2008) İklim değişikliğine karsi GAP bölgesinde etkin sulama stratejileri, TMMOB İklim Değişimi Sempozyumu, 264–281, 13–14 Mart, Ankara.

Dündar, M (2007) International problems caused by water resources (Master Thesis), KTÜ, Trabzon, 157 p.

EEA (2017) European Environment Agency "Climate change, impacts and vulnerability in Europe 2016 Report, 2017.

Fıstıkoğlu O, Biberoğlu E (2008) Küresel iklim değişikliğinin su kaynaklarina etkisi ve uyum önlemleri, TMMOB İklim Değişimi Semp., 238–252, 13–14 Mart, Ankara.

Filinte, HM (2007) Yaklaşan küresel iklim krizi, Yeni İnsan Yayınevi, İstanbul, Eylül.

IPCC (2007) Intergovernmental Panel on Climate Change, Climate Change 2007: Synthesis Report. Contribution of Working Groups I, II and III to the Fourth Assessment Report of the Intergovernmental Panel on Climate Change, 2007.

IPCC (2014) Intergovernmental Panel on Climate Change (IPCC), 5th Assessment Report (AR5) Working Group III Contribution to the IPCC Fifth Assessment Report." Climate Change 2014: Mitigation of Climate Change.

Karaman S, Gökalp Z (2010) Impacts of global warming and climate change over water resources, Research Journal of Agricultural Sciences, 3 (1): 59–66.

Kibaroğlu, A (2008) Küresel iklim değişikliğinin sınıraşan su kaynakları politikasina etkileri, 20-22 Mart 2008, Ankara 347–355.

Küçükklavuz, E (2009) The effects of global warming about water resources: Turkey sample (Master Thesis), Harran University, 134 p. Şanlıurfa.

SYGM (2020) İklim Değişikliği ve Uyum, T.C. Mülga Orman ve Su İşleri Bakanlığı, Su Yönetimi Genel Müdürlüğü, Ankara.

SYGM (2016) İklim Değişikliğinin Su Kaynaklarına Etkisi Projesi Raporları, T.C. Mülga Orman ve Su İşleri Bakanlığı, Su Yönetimi Genel Müdürlüğü, Ankara.

Turkish State Meteorological Service, 2022, 11 March 2022 https://www.mgm.gov.tr/iklim/iklim-degisikligi.aspx?s=senaryolar

Uzmen, R (2007) Küresel ısınma ve iklim değişikliği: insanliği bekleyen büyük felaket mi?, Bilge Kültür Sanat Yayınları, İstanbul.

Yılmaz, A (2015) Küresel ısınmanın Dünya su rezervleri üzerindeki etkileri, Reviewed Journal of Urban Culture and Management, Urban Academy, Volume: 8 Issue: 2 Summer 2015.

Chapter 65
Recovery of Used and Aged Lithium-Ion Batteries by Impedance Analysis

Salim Erol and Selcuk Temiz

Nomenclature

EIS	Electrochemical impedance spectroscopy
OCP	Open circuit potential
CPE	Constant phase element
TLM	Transmission line model

65.1 Introduction

Li-ion batteries have increasing applications in a wide variety of fields, such as electric vehicles and portable electronic devices. Li-ion batteries have been preferred in many applications since the amount of energy that can be stored per unit volume and weight is the highest among rechargeable batteries (Yan 2015). In the selection of batteries to be used in the application, there are a wide range of battery features that are very important and must be considered. Size, weight, and shape are important physical properties. Commercial Li-ion batteries come in a variety of sizes and shapes (Muenzel et al. 2015).

A large number of Li-ion cells are usually placed in battery packs connected in series and in parallel especially for high-stored energy required applications. Some

S. Erol (✉)
Department of Chemical Engineering, Eskisehir Osmangazi University, Eskisehir, Turkey
e-mail: esalim@ogu.edu.tr

S. Temiz
Department of Physics, Eskisehir Osmangazi University, Eskisehir, Turkey
e-mail: stemiz@ogu.edu.tr

M. Z. Sogut et al. (eds.), *Proceedings of the 2022 International Symposium on Energy Management and Sustainability*, Springer Proceedings in Energy,
https://doi.org/10.1007/978-3-031-30171-1_65

cells inside the battery packs are occasionally subjected to capacity fade or degradation due to cycling effects and/or abuse conditions such as high temperature, overcharge, overdischarge, and thermal runaway. Even though the other cells in the battery packs are still in good condition, the whole battery pack or even the battery system might be discarded since it is thought to be useless. However, the cells that work properly can be detected and utilized for future use.

Battery characterization can be categorized as destructive and nondestructive methods. The destructive methods allow to investigate the local structural and chemical alteration of battery materials over aging and usage. The destructive methods require dismantling batteries to conduct characterization; hence, the obtained information is restricted to be offline. On the other hand, nondestructive methods provide information about performance of the system over usage, and this approach does not need disassembly of battery cell. EIS is able to provide both offline and online measurement data that allows monitoring the state of charge and state of health of battery system; hence, it can be considered as the principal component of battery management system (Wang et al. 2021).

In this study, using EIS the discarded batteries were aimed to be recovered for utilizing their second lives. In addition, the parameters of batteries were determined using equivalent circuit modeling.

65.2 Methodology

In this section, the experimental method and the mathematical model used in the current work are explained.

65.2.1 Experimental Methods

Several commercial rechargeable Li-ion battery cells in different shapes and chemistries were collected. Electrochemical impedance measurements at the open circuit potential (OCP) were performed using a Gamry PCI4/750 potentiostat connected to a desktop computer. The Gamry software packages – including Gamry's Virtual Front Panel (VFP600), Gamry Framework, and Gamry Echem Analyst – were used to run the experiments. The primary purpose of the potentiostat in these experiments was to maintain a constant cell potential while measuring the impedance. The impedances of each cell were measured using a 10 mV perturbation and 100 kHz–20 mHz frequency range.

All experiments were performed at room temperature (approximately 20 °C), and experiments with the same type of battery were repeated three times to ensure results were both consistent and reproducible. Consistent and precise procedures have been enabled to reduce errors between the software in the Gamry instrument and repeated experiments.

65.2.2 *Mathematical Modeling*

Equivalent circuit modeling is a broadly used method to analyze impedance results with physically meaningful parameters such as resistances (R), capacitances (C), and inductances (L). Along with these real circuit elements, there are hypothetical ones such as constant phase element (CPE), Warburg element (W), and Bisquert elements (Erol 2021). The proposed equivalent circuit for the collected commercial Li-ion batteries is shown in Fig. 65.1. The impedance responses and fits of two different types of Li-ion batteries are demonstrated in Figs. 65.2 and 65.3. The regressed parameters belonging to the fits are tabulated in Table 65.1.

Since cable, anodic, electrolyte, and cathodic impedances are connected in series as shown in Fig. 65.1, the overall impedance, Z, can be expressed as follows:

$$Z = Z_{cab} + Z_{ano} + Z_{ele} + Z_{cat} \tag{65.1}$$

where Z_{ele} is the electrolyte impedance that is just a resistance, i.e., $Z_{ele} = R_1$. In Eq. 65.1, Z_{cat} represents the cathodic impedance that can be defined as

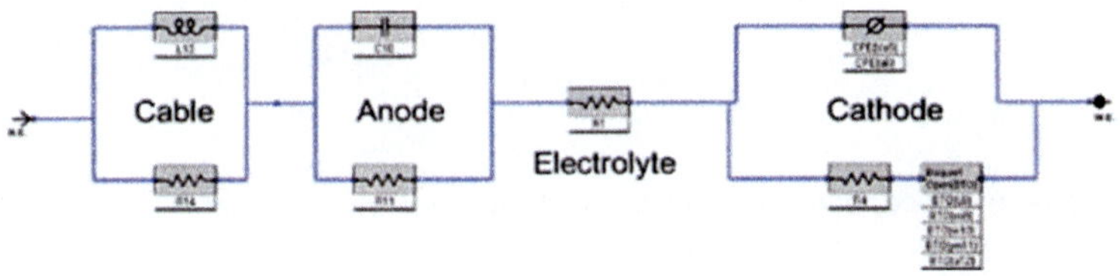

Fig. 65.1 Equivalent circuit representation of the model developed in the present work for the Li-ion battery system

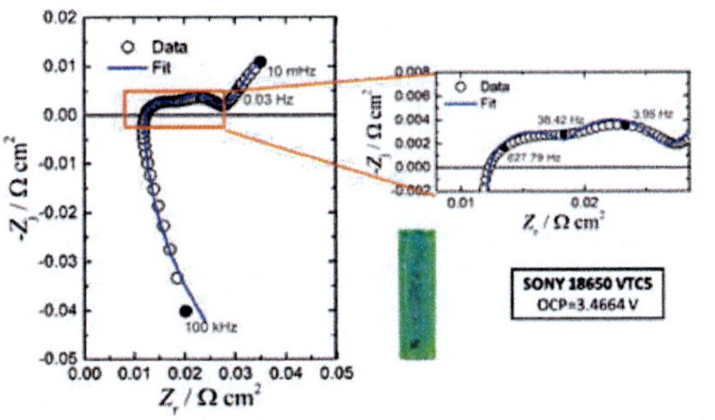

Parameter	Value	± Error	Units
R1	12.01e-3	193.2e-6	ohm
Yo2	145.6e-3	81.82e-3	S*s^a
a3	918.6e-3	74.34e-3	
R4	6.479e-3	591.1e-6	ohm
C10	2.963	533.9e-3	F
R11	6.640e-3	565.1e-6	ohm
L8	663.1e-3	630.7e-3	
rm9	22.25e-3	5.758e-3	ohm
rk10	28.33e-3	60.43e-3	ohm
ym11	1.352e3	160.8	S*s^a
a12	894.1e-3	31.85e-3	
L13	72.82e-9	1.419e-9	H
R14	160.3e-3	18.36e-3	ohm
Goodness of Fit	835.2e-6		

Fig. 65.2 Impedance response and the fit in Nyquist format for a Sony 18,650 VTC5 cell at a 3.4664 V cell potential and regression results and $\pm\sigma$ confidence intervals of a fit of the equivalent circuit

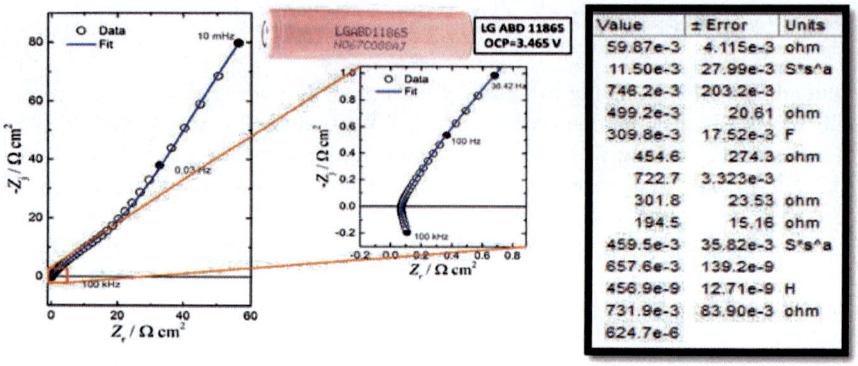

Fig. 65.3 Impedance response and the fit in Nyquist format for a LG ABD 11865 cell at a 3.4650 V cell potential and regression results and $\pm\sigma$ confidence intervals of a fit of the equivalent circuit

$$Z_{\text{cat}} = \frac{R_4 + Z_{\text{BTO}}}{1 + (\text{j}\omega)^{a_3}(R_4 + Z_{\text{BTO}})Y_{\text{o2}}} \tag{65.2}$$

where Z_{BTO} is the Bisquert-Open diffusion impedance at the cathode derived from the transmission line model (TLM) developed by Bisquert (Bisquert 2000). This mathematical model is derived for porous electrodes and expressed as

$$Z_{\text{BTO}} = \sqrt{\frac{r_{\text{m9}} r_{\text{k10}}}{1 + (\text{j}\omega)^{a_{12}} r_{\text{k10}} Y_{\text{m11}}}} \coth\left(L_8 \sqrt{\frac{r_{\text{m9}}\left[1 + (\text{j}\omega)^{a_{12}} r_{\text{k10}} Y_{\text{m11}}\right]}{r_{\text{k10}}}} \right) \tag{65.3}$$

The anodic impedance, Z_{ano}, in Eq. 65.1 can be expressed as

$$Z_{\text{ano}} = \frac{R_{11}}{1 + \text{j}\omega R_{11} C_{10}} \tag{65.4}$$

and Z_{cab} represents the impedance resulting from the cable effects in the experiments that can be defined as

$$Z_{\text{cab}} = \frac{\text{j}\omega L_{13} R_{14}}{\text{j}\omega L_{13} + R_{14}} \tag{65.5}$$

In Eqs. (65.2), (65.3), (65.4), and (65.5), j expresses the imaginary complex number, and ω is the angular frequency which is defined as $\text{j} = \sqrt{-1}$ and $\omega = 2\pi f$, respectively.

65.3 Results

The impedance responses at their received OCPs of different Li-ion battery cells and the regressed fitting parameter values with their confidence intervals are presented in Figs. 65.2, 65.3, 65.4, 65.5, 65.6, 65.6, and 65.7. The data are shown with symbols, and the fits are presented with lines in the figures.

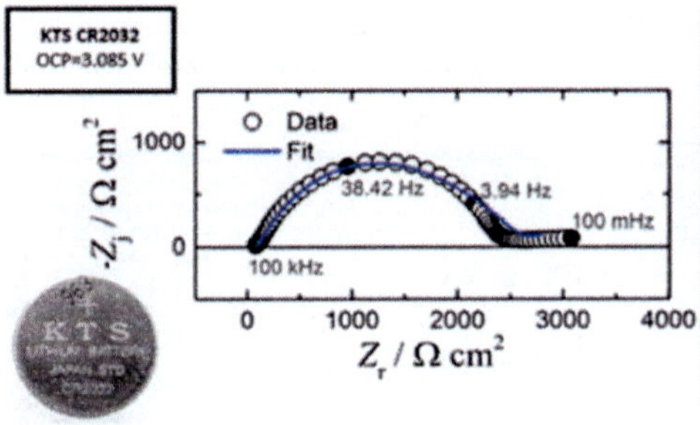

Parameter	Value	± Error	Units
R1	69.63	28.28	ohm
Yo2	21.33e-6	8.045e-6	S*s^a
a3	636.3e-3	33.44e-3	
R4	1.883e3	201.7	ohm
C10	7.283e-6	2.994e-6	F
R11	564.3	197.0	ohm
L8	6.855	36.86e-3	
rm9	12.56e3	4.137e3	ohm
rk10	33.18	10.95	ohm
ym11	566.4e-3	186.5e-3	S*s^a
a12	580.6e-3	64.71e-6	
L13	1.891e-3	8.989e-3	H
R14	9.049	28.60	ohm
Goodness of Fit	1.950e-3		

Fig. 65.4 Impedance response and the fit in Nyquist format for a KTS CR2032 cell at a 3.085 V cell potential and regression results and $\pm\ \sigma$ confidence intervals of a fit of the equivalent circuit

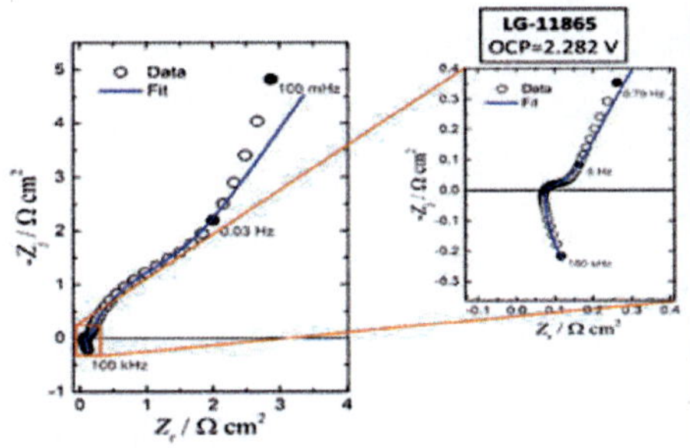

Parameter	Value	± Error	Units
R1	67.16e-3	4.510e3	ohm
Yo2	455.4e-9	1.378	S*s^a
a3	583.8e-3	857.3e3	
R4	694.0e-6	4.510e3	ohm
C10	1.608	399.3e-3	F
R11	819.8e-3	265.8e-3	ohm
L8	21.93e-3	1.366	
rm9	10.52	16.77	ohm
rk10	20.45	193.0e3	ohm
ym11	60.31	77.54	S*s^a
a12	688.2e-3	212.0e-3	
L13	386.5e-9	28.97e-9	H
R14	1.156	208.7e-3	ohm
Goodness of Fit	3.272e-3		

Fig. 65.5 Impedance response and the fit in Nyquist format for another LG ABD 11865 cell at a 2.282 V cell potential and regression results and $\pm\ \sigma$ confidence intervals of a fit of the equivalent circuit

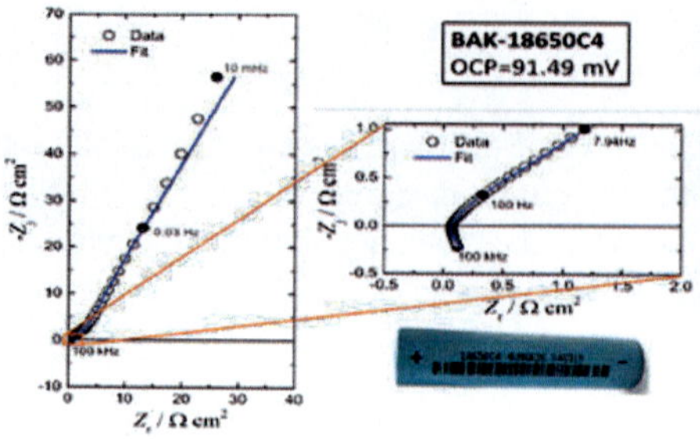

Parameter	Value	± Error	Units
R1	56.53e-3	6.452e-3	ohm
Yo2	34.71e-3	17.63e-3	S*s^a
a3	640.6e-3	151.9e-3	
R4	2.262	1.437e3	ohm
C10	88.18e-3	55.98e-3	F
R11	981.4e-3	602.4e-3	ohm
L8	257.0e-3	9.428e-3	
rm9	13.55e-3	20.19e3	ohm
rk10	629.7	1.981e15	ohm
ym11	338.3e-3	341.7e3	S*s^a
a12	779.3e-3	334.2	
L13	438.5e-9	9.800e-9	H
R14	727.2e-3	66.18e-3	ohm
Goodness of Fit	1.254e-3		

Fig. 65.6 Impedance response and the fit in Nyquist format for a BAK 18650C4 cell at a 91.49 V cell potential and regression results and ± σ confidence intervals of a fit of the equivalent circuit

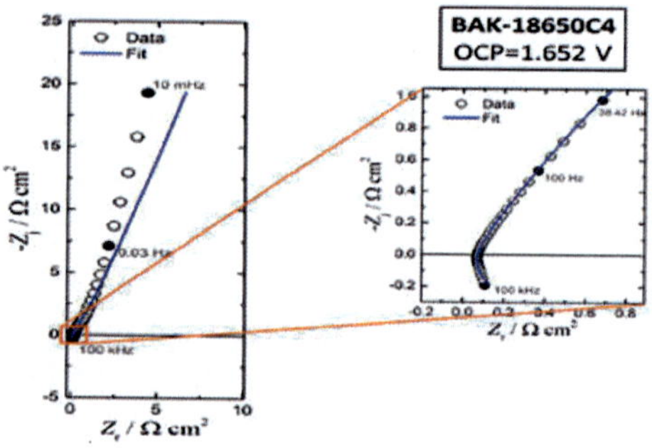

Parameter	Value	± Error	Units
R1	70.03e-3	2.111e-3	ohm
Yo2	81.50e-3	263.9e-3	S*s^a
a3	790.6e-3	177.0e-3	
R4	24.54e-3	278.6e-3	ohm
C10	8.996e-3	9.075e-3	F
R11	17.44e-3	13.43e-3	ohm
L8	101.9e-3	38.10e-3	
rm9	1.650	8.721	ohm
rk10	934.7	878.9e6	ohm
ym11	3.714	6.982	S*s^a
a12	774.5e-3	183.1e-3	
L13	413.1e-9	9.461e-9	H
R14	901.6e-3	109.3e-3	ohm
Goodness of Fit	2.200e-3		

Fig. 65.7 Impedance response and the fit in Nyquist format for another BAK 18650C4 cell at a 1.652 V cell potential and regression results and ±σ confidence intervals of a fit of the equivalent circuit

65.4 Conclusion

In this study, electrochemical impedance analysis at the open cell potential, constant temperature, and frequency range was performed for commercially purchased and discarded Li-ion batteries. The significant physical parameters for batteries were determined by the impedance responses and the developed equivalent circuit model of these batteries. The discarded Li-ion battery packs might have active cells. Some of these cells can be recovered by inspecting their impedance responses even though their open circuit potential values are in the normal range. In addition, the model developed with the electrochemical impedance spectroscopy technique has been shown to be effective and has a great potential for meeting the energy needs and design of future batteries.

References

Bisquert J (2000) Influence of the boundaries in the impedance of porous film electrodes. Physical Chemistry Chemical Physics 2(18): 4185-4192. https://doi.org/10.1039/B001708F

Erol S (2021) Comparative Study of Impedance Spectroscopy Between Nickel-Metal Hydride and Lithium-ion Batteries. European Journal of Science and Technology 28: 144-151. https://doi.org/10.31590/ejosat.993325

Muenzel V, Hollenkamp AF, Bhatt AI, de Hoog J, Brazil M, Thomas DA, Mareels I (2015) A comparative testing study of commercial 18650-format lithium-ion battery cells. Journal of the Electrochemical Society 162(8): A1592–A1600. https://doi.org/10.1149/2.0721508jes

Yan J. (2015) Rechargeable Battery Energy Storage System Design. Handbook of Clean Energy Systems. John Wiley & Sons. Hoboken, NJ, USA: 2801–2819.

Wang S, Zhang J, Gharbi O, Vivier V, Gao M, Orazem ME (2021) Electrochemical impedance spectroscopy. Nature Reviews Methods Primers 1(41): 1–21. https://doi.org/10.1038/s43586-021-00039-w

Chapter 66
Green Liner Ship Routing with Time Windows Considering Resistance Effects of Weather Conditions

Mesut Can Köseoğlu and Temel Öncan

Nomenclature

UNCTAD	United Nations Conference on Trade and Development
EEOI	The Energy Efficiency Operational Index
VRP	Vehicle routing problem
GSRP	Green ship routing and scheduling problem
V	Set of ports to be visited
K	Number of available ships
c_{ijk}	CO_2 emission generated by ship k sailing from port i to port j
$EEOI_{ijk}$	EEOI for ship k sailing from port i to port j
F_{ijk}	Fuel consumption of ship k sailing from port i to port j
EP_k	Engine power of ship k
t_{ijk}	Travel time of ship k sailing from port i to port j
v_{ijk}	Speed over ground of ship k sailing from port i to port j
d_{ij}	Distance from port i to port j
x_{ijk}	1 if and only if ship k visits port j immediately after port i
y_{ik}	1 if port i is served by ship k
Q_k	Capacity of each ship
q_i	Demand at port i
u_{ik}	The amount of load of ship k after visiting port i
w_{ik}	The leaving time of ship k from port i

M. C. Köseoğlu (✉)
Maritime Faculty, Piri Reis University, İstanbul, Türkiye
e-mail: mckoseoglu@pirireis.edu.tr

T. Öncan
Engineering Faculty, Galatasaray University, İstanbul, Türkiye
e-mail: ytoncan@gsu.edu.tr

M. Z. Sogut et al. (eds.), *Proceedings of the 2022 International Symposium on Energy Management and Sustainability*, Springer Proceedings in Energy,
https://doi.org/10.1007/978-3-031-30171-1_66

$[a_i, b_i]$	Time windows for port i
$[E, L]$	Early and late departure times for the depot
M	A sufficiently large number

66.1 Introduction

According to United Nations Conference on Trade and Development (UNCTAD 2021), approximately 83% of the global trade volume has been carried out by international shipping carriers. Due to the rise of containerization, economic globalization and the rapid development of foreign trade, container ships have become more and more common, and they have a widespread utilization for transportation of commodities between ports (Qi et al. 2021). Containerships are usually designed with capacities to achieve economies of scale, carrying thousands of containers, in an efficient manner (UNCTAD 2021). In addition, container ships are also designed for speed which conforms to the updated environmental regulations.

On the other hand, environmental concerns and regulations have been forcing shipping companies to take precautionary actions. The Energy Efficiency Operational Index (EEOI) is a benchmarking tool for monitoring ships or fleet's efficiency by evaluating CO_2 emissions. EEOI also enables measuring fuel efficiency of ships; hence, low EEOI values represent better performance.

In this study, a green vehicle routing problem has been introduced to minimize EEOI of a small-scale liner container fleet. The problem considers both wind and current effects between departure and destination ports on each voyage as well as time window restrictions. This approach provides flexibility for liner shipping companies to update their liner routes depending on the fuel consumption and produced CO_2 emission based on the dynamic effects with an official benchmarking with EEOI.

66.2 Green Ship Routing Problems

Broadly speaking, maritime routing problems are in strong relation with the vehicle routing problem (VRP) (Toth and Vigo 2002) which deals with the routing of several vehicles with fixed capacities, visiting all client destinations once and then returning to the initial departure point. Ever since the seminal work by Ronen (1983), there is a vast body of literature on the ship routing which is known to be NP hard (Dithmer et al. 2017; Brouer et al. 2014).

In most of the studies on the green routing problem, the objective function is to minimize the cost of transportation as well as consider environmental concerns. Christiansen et al. (2013) stated that increasing bunker costs has caused to divert the focus from ship routing and scheduling to the ship speed and emission. For a survey on the container ship routing and scheduling literature in the last three decades, we refer to Wang et al. (2013). Generally in the literature, studies addressing the liner shipping network optimization have been conducted addressing several objective functions such as the minimization of fuel consumption and running cost or the maximization of the profit from each voyage, which is subject to a variety of constraints such as capacity, minimum speed requirement, time windows and fleet size (Agarwal and Ergun 2008; Christiansen et al. 2020; Surury et al. 2021).

Kontovas (2014) has focused on the green ship routing and scheduling problem (GSRSP). The author states that the optimization of ship speeds benefits by reducing both the bunker operation costs and the environmental costs. Psaraftis and Kontovas (2016) and Bektaş et al. (2019) have separately carried out reviews of green freight transportation literature, with a classification of the studies on the liner shipping, the tramp shipping and the general shipping.

66.3 Problem Description

Ships are exposed to various forces when sailing; mainly they are affected by winds, waves and currents which are parts of added resistances. Ship speed and heading on water is different from its speed and course over ground due to the external effects (Bowditch 2018). Weather effects on ships' speed can be calculated by considering both the wind and wave resistance on ship surface (Kim et al. 2017; Prpić-Oršić et al. 2015). For winds to have a substantial effect on ships, they have to blow for a long time in a given area with high speed, making a current effect on the ship. The effect of winds on speed loss in a given time can be calculated by estimating the resistance caused by wind to the surface area of the ship affected by the wind (Kim et al. 2017; Prpić-Oršić et al. 2015). A thorough analysis has been carried out in Kim et al. (2017) for the speed loss of the container ships with added resistances from wind and waves. For the inclusion of the weather effects to the GSRPs, speed over ground of the ships should be taken into account regarding sailing time, while for emission and fuel consumption, propulsion speed (or speed through water) should be considered. Henceforth, benchmark values in Kim et al. (2017) have been utilized here to determine the ships' speed over ground values between ports.

Ships used in liner shipping have tighter time windows, causing liner ships to consume more fuel to reach their destinations in time, with high speed. In this study, we address a real-life problem of a shipping company operating in the Black Sea (arkasline.com.tr 2021). Accordingly, the ports in our sample problem consist of Odessa, Yuzhnyy, Novorossiysk (Novo), Constanta, Samsun, Poti, Varna and Burgas, while the northern exit of the İstanbul Strait is the depot. Strait passage and ports in Marmara Sea have not been included in the problem to reduce the scale of the problem for initial testing. A fleet consisting of three ships is assumed to be available at the depot.

Time windows for berthing, cargo operations and un-berthing in the ports have been considered, but for the sake of obtaining reliable emission values, fuel consumption in ports and manoeuvring have not been included into the model. Note that during manoeuvring, since a lot of engine commands are utilized, irregular burning affects reliable calculation and emission factors. Due to confidentiality reasons, time windows have been set according to the experts' opinions, and random demand values are generated for each port considering the real demand values.

The proposed model aims to enable preplanning of shipping routes taking into consideration seasons and hence updating the routes when necessary to optimize fleet EEOI in coastal areas. The following assumptions have been deemed for the mathematical model:

- All ships are well maintained and fully operational.
- Routes between ports are according to pilot-to-pilot routing.
- Ships burn very low sulphur fuel oil (VLSFO).
- Only emission and fuel consumed by the main engine have been considered.
- Ships' engine load is assumed to be always 85%.
- Specific fuel oil consumption (SFOC) is 0.305 kg/kwh.
- Each port is fully capable of hosting ships.
- Each ship is fully loaded during passages.

66.3.1 Weather Data

Wind and current effects have been incorporated in our model taking into consideration the seasonal averages. The data, which has been collected from the Black Sea Meteorological Atlas (SHOD 1991), was based on the monthly observations for different grids of Black Sea. Note that Black Sea has been separated to grids of 1° latitudes and 1° longitudes to apply the extracted data. Figs. 66.1, 66.2, 66.3 and 66.4 describe the wind direction and force in knots for winter, spring, summer and fall, respectively. In Fig. 66.5, dominant current sets and drifts are illustrated.

The calculation for the average ship speed on each route v_{ijk} is given in Eq. (66.1):

$$v_{ijk} = \sum_{n=1}^{N} v_{nk}^{sog} / N \quad \forall i, j \in V, i \neq j, k \in K \tag{66.1}$$

where N stands for the number of grids passed by ship k in her voyage from port i to port j and v_{nk}^{sog} denotes the speed over ground in each grid.

66.3.2 Mathematical Model

For each ship's fuel consumption, we have employed Eq. (66.2) which has also been used in Teodorovic and Janic (2016). Total CO_2 emission generated from the voyage

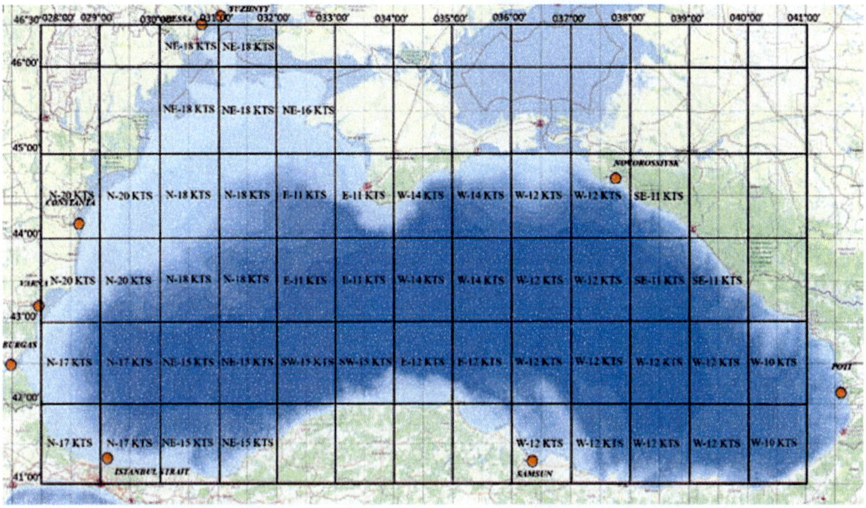

Fig. 66.1 Wind direction and force in knots for winter. (SHOD 1991)

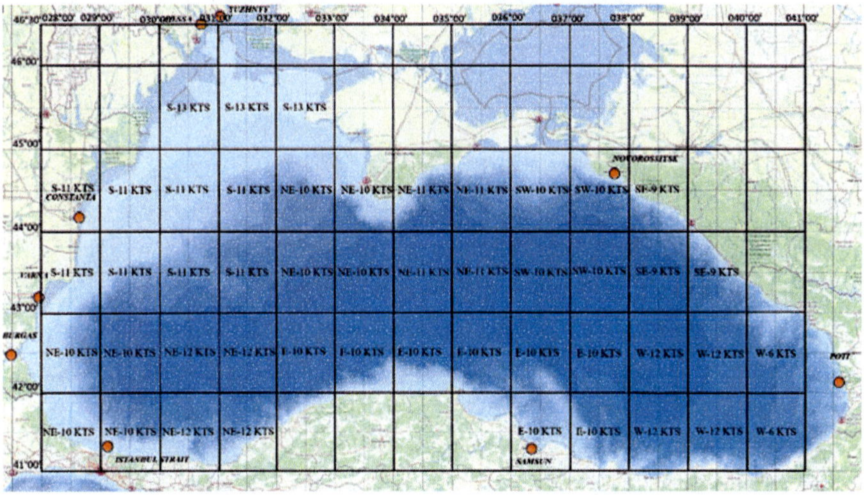

Fig. 66.2 Wind direction and force in knots for spring. (SHOD 1991)

is calculated with the product of fuel consumption value and the emission factor of CO_2 which is 3.1144 kg CO_2 per kg fuel oil consumed as given in Eq. (66.3) (Kontovas 2014). EEOI formula according to IMO (2009) is given in Eq. (66.4):

$$F_{ijk} = EP_k * 0.85 * \text{SFOC} * t_{ijk} \tag{66.2}$$

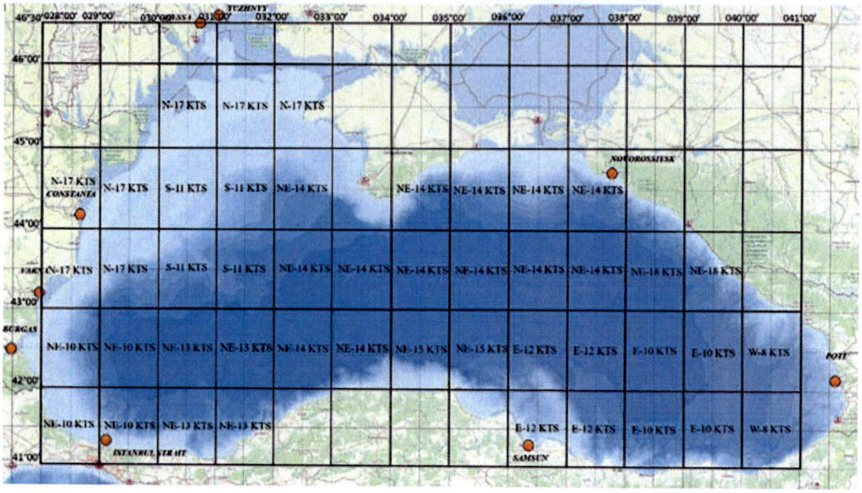

Fig. 66.3 Wind direction and force in knots for summer. (SHOD 1991)

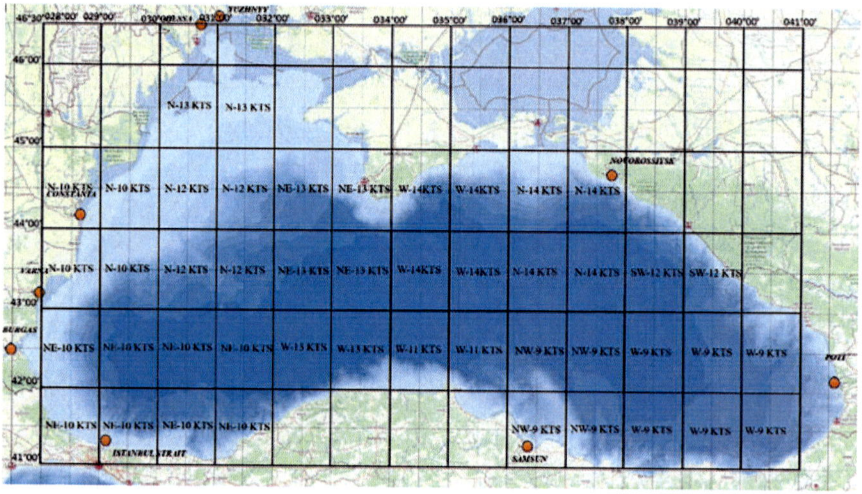

Fig. 66.4 Wind direction and force in knots for fall. (SHOD 1991)

$$c_{ijk} = F_{ijk} * 3.1144 \tag{66.3}$$

$$\mathrm{EEOI}_{ijk} = \frac{c_{ijk}}{Q_k * d_{ij}} \tag{66.4}$$

Traveling time is calculated in Eq. (66.5) with the average speed of traveling from port i to port j with vessel k, and hence, v_{ijk} represents the speed of the vessel that is affected by both the wind and the current:

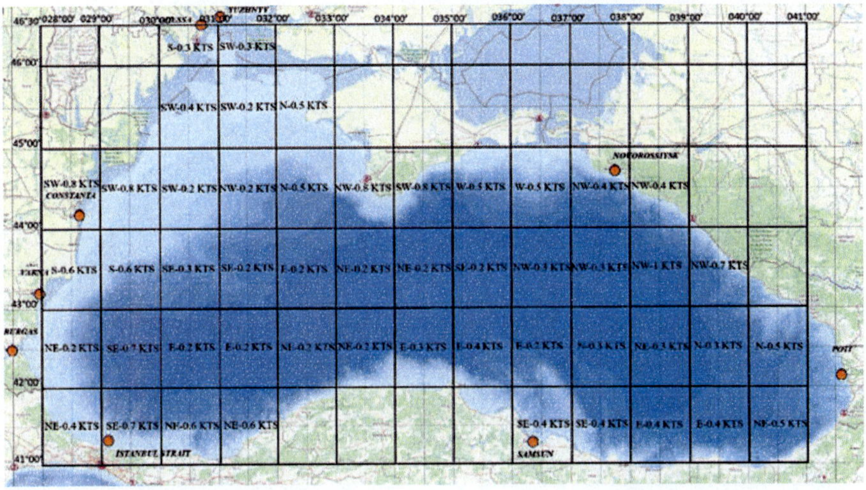

Fig. 66.5 Currents' sets and drifts. (SHOD 1991)

$$t_{ijk} = \frac{d_{ij}}{v_{ijk}} \tag{66.5}$$

Then we propose a mixed-integer linear programming (MILP) formulation as follows:

$$\text{minimize} \sum_{i,j \in V} \sum_{k \in K} \text{EEOI}_{ijk} * x_{ijk} \tag{66.6}$$

Subject to:

$$\sum_{k \in K} y_{ik} = 1, \quad \forall i \in V \backslash \{0\} \tag{66.7}$$

$$\sum_{k \in K} y_{0k} = K \tag{66.8}$$

$$\sum_{j \in V} x_{ijk} = \sum_{j \in V} x_{jik} = y_{ik}, \quad \forall i \in V \backslash \{0\}, \quad i \neq j, k \in K \tag{66.9}$$

$$q_i * y_{ik} \leq Q_k, \quad \forall i \in V \backslash \{0\}, k \in K \tag{66.10}$$

$$u_{ik} - u_{jk} + Q_k * x_{ijk} \leq Q_k - q_j, \quad \forall i,j \in V \backslash \{0\}, i \neq j, k \in K \tag{66.11}$$

$$q_i \leq u_{ik} \leq Q_k, \quad \forall i \in V \backslash \{0\}, k \in K \tag{66.12}$$

$$w_{ik} - w_{jk} + t_{ijk} \leq \left(1 - x_{jik}\right) * M, \quad \forall i, j \in V\backslash\{0\}, i \neq j, k \in K \tag{66.13}$$

$$a_i * y_{ik} \leq w_{ik} \leq b_i * y_{ik}, \quad \forall i \in V\backslash\{0\}, k \in K \tag{66.14}$$

$$E \leq w_{0k} \leq L, \quad k \in K \tag{66.15}$$

$$x_{ijk} \in \{0, 1\}, \quad \forall i, j \in V\backslash\{0\}, i \neq j, k \in K \tag{66.16}$$

$$y_{ik} \in \{0, 1\}. \quad \forall i \in V\backslash\{0\}, k \in K \tag{66.17}$$

Here the objective function (66.6) minimizes fleet EEOI. Constraints (66.7), (66.8), and (66.9) ensure that each port is visited exactly once, and K ships leave and return to İstanbul Strait. Moreover, these constraints make sure that the same ship enters and leaves a port, respectively. Constraints (66.10) impose the capacity restriction for ship k. Both constraints (66.11) and (66.13) obviate the subtours (Miller et al. 1960). Note that these constraints utilize continuous auxiliary variables u_{ik} and w_{ik}. In constraints (66.11), u_{ik} represents the load of each ship after departing from port i. When x_{ijk} takes the value of 1, $u_{ik} + q_j \leq u_{jk}$ holds, stating that the total demand on port j and the load of ship k after leaving port i must be less than or equal to the load of ship k after leaving port j. Constraint (66.13) ensures the feasibility of the voyage considering time windows in conjunction with constraint (66.14), which restricts w_{ik}, ensuring each port is visited within its time window. Constraints (66.14) guarantee that the time windows are imposed whenever port i is visited by ship k. Finally, constraints (66.15) apply a time window for the departure time of the ships from İstanbul Strait, representing early and late leaving options from İstanbul Strait.

66.4 Results and Discussion

The experiments are performed on a PC with 3.60 GHz Intel Core i-7 processor and 16 GB of RAM. To solve the MILP problems, CPLEX ver. 20.1.0 is called within Microsoft Visual Studio 2019 platform.

Even though a heterogenous model has been proposed, in order to comply with the benchmarks in Kim et al. (2017), we assume that a homogenous ship fleet comprising of three ships with 177.5 meters (m) length, 24.5 m breadth, 9 m draught, 29,783 metric tons displacement, 13,280 kWh engine power and 23 knots service speed is available.

For comparison, routes used in Black Sea (arkasline.com.tr 2021) are presented in Table 66.1 with fuel consumption, CO_2 emissions and fleet EEOI employing Eqs. (66.2, 66.3, and 66.4).

Numerical experiments have been conducted regarding three different factors: weather, time windows and demand. For weather factor, seasons were considered as scenarios. Time windows have been set as loose and tight and demands as high and low. Instead of following original routes, the results of the proposed model provided newly generated routes for each scenario which are given in Tables 66.2, 66.3, 66.4, and 66.5.

Table 66.1 Fuel consumption and CO_2 emissions for Black Sea shipping routes

Routes	Fuel cons. (kg)	CO_2 emissions (kg per kg fuel)
İstanbul-Poti-Constanta-İstanbul	197471.20	615004.30
İstanbul-Constanta-Odessa-İstanbul	101783.30	316993.90
İstanbul-novo-İstanbul	130094.10	405165.10
İstanbul-Burgas-Varna-İstanbul	43760.90	136288.90
İstanbul-Poti-Samsun-İstanbul	174397.00	543142.00
Yuzhnyy-Poti-Yuzhnyy	166349.00	518077.30
Fleet Total	813855.50	2534671.50
Fleet EEOI	0.093917	

Table 66.2 Optimal routes obtained from numerical experiments for summer

Weather period	Time windows	Demand	Routes
Summer	Loose	High	Ship 1: İstanbul-Samsun-Poti-novo-İstanbul Ship 2: İstanbul-Constanta-Varna-Burgas-İstanbul Ship 3: İstanbul-Yuzhnyy-Odessa-İstanbul
	Loose	Low	Ship 1: İstanbul-Constanta-Odessa-Yuzhnyy-novo-Poti-Samsun-İstanbul Ship 2: İstanbul-Varna-İstanbul Ship 3: İstanbul-Burgas-İstanbul
	Tight	High	Ship 1: İstanbul-Samsun-Poti-novo-İstanbul Ship 2: İstanbul-Constanta-Varna-Burgas-İstanbul Ship 3: İstanbul-Odessa-Yuzhnyy-İstanbul
	Tight	Low	Ship 1: İstanbul-Varna-Constanta-Odessa-Yuzhnyy-İstanbul Ship 2: İstanbul-Samsun-Poti-novo-İstanbul Ship 3: İstanbul-Burgas-İstanbul

Table 66.3 Optimal routes obtained from numerical experiments for spring

Weather period	Time windows	Demand	Routes
Spring	Loose	High	Ship 1: İstanbul-novo -Poti-Samsun-İstanbul Ship 2: İstanbul-Constanta-Varna-Burgas-İstanbul Ship 3: İstanbul-Yuzhnyy-Odessa-İstanbul
	Loose	Low	Ship 1: İstanbul-Constanta-Odessa-Yuzhnyy-novo-Poti-Samsun-İstanbul Ship 2: İstanbul-Varna-İstanbul Ship 3: İstanbul-Burgas-İstanbul
	Tight	High	Ship 1: İstanbul-Samsun-Poti- novo -İstanbul Ship 2: İstanbul-Yuzhnyy-Odessa-İstanbul Ship 3: İstanbul-Constanta-Varna-Burgas-İstanbul
	Tight	Low	Ship 1: İstanbul-Varna-Constanta-Yuzhnyy-Odessa-İstanbul Ship 2: İstanbul-Samsun-Poti-novo-İstanbul Ship 3: İstanbul-Burgas-İstanbul

Table 66.4 Optimal routes obtained from numerical experiments for winter

Weather period	Time windows	Demand	Routes
Winter	Loose	High	Ship 1: İstanbul-Odessa-Yuzhnyy-İstanbul Ship 2: İstanbul-novo -Poti-Samsun-İstanbul Ship 3: İstanbul-Constanta-Varna-Burgas-İstanbul
	Loose	Low	Ship 1: İstanbul-Varna-İstanbul Ship 2: İstanbul-Samsun-Poti-novo-Yuzhnyy-Odessa-Constanta-İstanbul Ship 3: İstanbul-Burgas-İstanbul
	Tight	High	Ship 1: İstanbul-Samsun-Poti-novo -İstanbul Ship 2: İstanbul-Constanta-Varna-Burgas-İstanbul Ship 3: İstanbul-Yuzhnyy-Odessa-İstanbul
	Tight	Low	Ship 1: İstanbul-Samsun-Poti-novo-İstanbul Ship 2: İstanbul-Burgas-İstanbul Ship 3: İstanbul-Varna-Constanta-Odessa-Yuzhnyy-İstanbul

Table 66.5 Optimal routes for numerical experiments for fall

Weather period	Time windows	Demand	Routes
Fall	Loose	High	Ship 1: İstanbul-Samsun-Poti-novo-İstanbul Ship 2: İstanbul-Burgas-Varna-Constanta-İstanbul Ship 3: İstanbul-Yuzhnyy-Odessa-İstanbul
	Loose	Low	Ship 1: İstanbul-Constanta-Odessa-Yuzhnyy-novo -Poti-Samsun-İstanbul Ship 2: İstanbul-Varna-İstanbul Ship 3: İstanbul-Burgas-İstanbul
	Tight	High	Ship 1: İstanbul-Yuzhnyy-Odessa-İstanbul Ship 2: İstanbul-Samsun-novo-Poti-İstanbul Ship 3: İstanbul-Burgas-Varna-Constanta-İstanbul
	Tight	Low	Ship 1: İstanbul-Varna-Constanta-Odessa-Yuzhnyy-novo-İstanbul Ship 2: İstanbul-Samsun-Poti-İstanbul Ship 3: İstanbul-Burgas-İstanbul

Spring and summer routes on each scenario have been observed to be similar to each other, considering weather effect is also similar on each route leg. Although the destination ports are similar, the scheduling of the ports is different on certain scenarios.

Compared to the original routes, different routes have been observed depending on the scenarios, but Istanbul-Poti-Samsun-Istanbul, one of the original routes, still holds on tight time window scenario with low demand during spring.

Winter routes differ significantly on each scenario from the original routes since weather effects are the highest. When the weather effect is high with strong winds and waves, the routes have become shorter or altered compared to original routes to take the wind from aft, by combining multiple route legs together to create a new line, increasing the travel speed, hence reducing emission and fuel consumption.

Table 66.6 Emission values for numerical experiments

Weather period	Time windows	Demand	Fleet fuel cons. (kg)	Fleet CO_2 emissions (kg per kg fuel)
Winter	Loose	High	357,372	1112999.3
	Loose	Low	312,935	974604.8
	Tight	High	360,814	1123719.1
	Tight	Low	342,381	1066311.4
Summer	Loose	High	357,128	1112239.4
	Loose	Low	304,692	948932.8
	Tight	High	364,561	1135389.8
	Tight	Low	343,998	1071347.4
Fall	Loose	High	355,836	1108215.6
	Loose	Low	307,712	958338.3
	Tight	High	374,846	1167420.4
	Tight	Low	397,902	1239226.0
Spring	Loose	High	357,128	1112239.4
	Loose	Low	304,692	948932.8
	Tight	High	364,561	1135388.8
	Tight	Low	343,998	1071347.3

For each scenario, fleet emission values (kg per kg fuel) and fleet fuel consumption values (kg) have been included in Table 66.6. Regarding the fleet EEOI values, the improvements in each season compared to fleet EEOI from the original routes are presented in Table 66.7.

According to the obtained routes, results show significant improvements in fuel consumption, CO_2 emission values and EEOI for the fleet compared to initial system for each season on each scenario.

66.5 Conclusion

In this study, a green vehicle routing approach has been introduced to optimize EEOI in small-scale liner container shipping. A mixed-integer linear programming model has been proposed for the green vehicle routing problem considering resistances sourced from winds and waves between departure and destination ports on each voyage as well as time window restrictions and cargo demand in each port. For each season, four different scenarios have been constructed based on time windows and demands in ports.

From the provided routes by the proposed model, EEOI, fuel consumption and CO_2 emission results of each scenario have been compared to the same values of the original routes. Results show that new routes provide a decrease in EEOI, fuel consumption and CO_2 emission varying between 56% and 60% depending on the season and scenario.

Table 66.7 Fleet EEOI compared to original routes

Weather period	Time windows	Demand	Fleet EEOI	Improvement (%)
Winter	Loose	High	0.053462	56.92
	Loose	Low	0.053148	56.59
	Tight	High	0.054369	57.89
	Tight	Low	0.054579	58.11
Summer	Loose	High	0.055552	59.15
	Loose	Low	0.055552	59.15
	Tight	High	0.056011	59.64
	Tight	Low	0.055761	59.37
Fall	Loose	High	0.054212	57.72
	Loose	Low	0.054212	57.72
	Tight	High	0.054802	58.35
	Tight	Low	0.055433	59.02
Spring	Loose	High	0.055114	58.68
	Loose	Low	0.055095	58.66
	Tight	High	0.056947	60.64
	Tight	Low	0.055552	59.15

References

Agarwal R and Ergun Ö (2008), Ship scheduling and network design for cargo routing in liner shipping, Transportation Science 42(2): 175–196. https://doi.org/10.1287/trsc.1070.0205

Arkasline.com.tr (2021) Routes and Schedules http://www.arkasline.com.tr/en/routes_and_schedules.html

Bektaş T, Ehmke JF, Psaraftis HN and Puchinger J (2019) The role of operational research in green freight transportation, European Journal of Operations Research 274(3): 807–823. https://doi.org/10.1016/j.ejor.2018.06.001

Bowditch N, (2018) The American practical navigator: an epitome of navigation, National Geospatial-Intelligence Agency, vol. 1.

Brouer BD, Alvarez JF, Plum CE, Pisinger D and Sigurd MM (2014) A base integer programming model and benchmark suite for liner-shipping network design, Transportation Science 48(2): 281–312. https://doi.org/10.1287/trsc.2013.0471

Christiansen M, Fagerholt K, Nygreen B and Ronen D (2013) Ship routing and scheduling in the new millennium, Eur. J. of Oper. Res 228 (3): 467–483. https://doi.org/10.1016/j.ejor.2012.12.002

Christiansen M, Hellsten E, Pisinger D, Sacramento D and Vilhelmsen C (2020) Liner shipping network design. Eur. J. of Oper. Res. 286(1): 1–20. https://doi.org/10.1016/j.ejor.2019.09.057

Dithmer P, Reinhardt L and Kontovas CA (2017) The Liner Shipping Routing And Scheduling Problem Under Environmental Considerations: The Case Of Emissions Control Areas, In: Bektaş, T., Coniglio, S., Martinez-Sykora, A., Voß, S. (eds) Computational Logistics. ICCL 2017. Lecture Notes in Computer Science(), vol 10,572. Springer, Cham. https://doi.org/10.1007/978-3-319-68496-3_23

International Maritime Organization (IMO) (2009) Guidelines for Voluntary Use of the Ship Energy Efficiency Operational Indicator; MEPC 59/Circ. 684: London, UK.

Kim M, Hizir O, Turan O, Day S and Incecik A (2017) Estimation of added resistance and ship speed loss in a seaway. Ocean Engineering 141: 465–476. https://doi.org/10.1016/j.oceaneng.2017.06.051

Kontovas CA (2014) The green ship routing and scheduling problem (GSRSP): A conceptual approach. Transp. Res. D: Transp. Environ.31: 61–69. https://doi.org/10.1016/j.trd.2014.05.014

Miller CE, Tucker AW and Zemlin RA (1960) Integer programming formulation of traveling salesman problems. Journal of the ACM 7(4): 326–329. https://doi.org/10.1145/321043.321046

Prpić-Oršić J, Vettor R, Faltinsen OM and Soares C (2015) Influence of ship routes on fuel consumption and CO2 emission. In Maritime Technology and Engineering: CRC Press: 857–864.

Psaraftis HN and Kontovas CA (2016) Green maritime transportation: speed and route optimization. In Green Transportation Logistics. International Series In Psaraftis H (eds) Operations Research & Management Science, vol 226, Springer, Cham.

Qi J, Zheng J, Yang L and Yao F (2021) Impact analysis of different container arrival patterns on ship scheduling in liner shipping. Maritime Policy Management 48 (3): 331–353. https://doi.org/10.1080/03088839.2020.1768316

Ronen D (1983) Cargo ships routing and scheduling: Survey of models and problems. Eur. J. of Oper. Res. 12(2): 119–126. https://doi.org/10.1016/0377-2217(83)90215-1.

Seyir, Hidrografi ve Oşinografi Dairesi (SHOD) (1991) Karadeniz meteorolojik atlası. İstanbul: Deniz Kuvvetleri Komutanlığı Hidrografi Yayınları.

Surury F, Syauqi A and Purwanto WW (2021) Multi-objective optimization of petroleum product logistics in Eastern Indonesia region. Asian Journal of Shipping and Logistics 37(3): 220–230. https://doi.org/10.1016/j.ajsl.2021.05.003

Teodorovic D and Janic M (2016) Transportation Engineering: Theory, Practice and Modeling. Oxford, UK: Butterworth-Heinemann.

Toth P and Vigo D (2002) The Vehicle routing problem. Society for Industrial and Applied Mathematics.

UNCTAD (2021) Review of Maritime Transport 2020. The United Nations Conference on Trade and Development.

Wang S, Meng Q and Liu Z (2013) Bunker consumption optimization methods in shipping: a critical review and extensions. Transp. Res. E: Log. and Transp. 53: 49–62. https://doi.org/10.1016/j.tre.2013.02.003

Chapter 67
Managerial Evaluations of Environmental and Energy Impact of Ship Cruising

Oktay Çetin

67.1 Introduction

The acceleration in maritime transport continues in parallel with the increase in the fuel consumption of the ships although a slight decrease was observed in the beginning even in 2020 and 2021, which was under the effect of the pandemic. In direct proportion to the increase in fossil fuel-based consumption, CO_2 emissions also increased, although not linearly. The International Maritime Organization (IMO) continues its studies aiming to provide efficiency in energy management processes utilizing technical, operational, and design measures that should be applied on ships in use today to reduce the emission impact. In this context, efforts to reduce greenhouse gas emissions are supported (Kirk and Ferreira 2012).

After 2000, studies have been increasingly carried out on the efficient use of energy onboard ships. The IMO continues to carry out numerous studies aimed at reducing the pollutants that originated from ships. Administrations have been invited to implement some of the amendments introduced by IMO to MARPOL Annex VI when developing and enforcing national laws. IMO developed some instruments like attained Energy Efficiency Design Index (EEDI) of MARPOL Annex VI and guidelines to calculate EEDI (IMO 2021). Mandatory EEDI for new ships and SEEMP for all ships in MEPC 62 (July 2011) can be considered the first legally binding climate change agreement since the Kyoto Protocol (IMO 2022). As it is known, there are some terms created by the IMO to describe necessary actions on energy-related issues onboard ships. For example, EEDI is in force in the new shipbuilding process. Ship Energy Efficiency Management Plan (SEEMP) is applied

---

O. Çetin (✉)
Faculty of Economics & Administrative Sciences, Piri Reis University, Istanbul, Turkey
e-mail: ocetin@pirireis.edu.tr

M. Z. Sogut et al. (eds.), *Proceedings of the 2022 International Symposium on Energy Management and Sustainability*, Springer Proceedings in Energy,
https://doi.org/10.1007/978-3-031-30171-1_67

during planning and improvement activities. Energy Efficiency Operational Indicator (EEOI) is the monitoring phase of the cycle (IMO 2015).

It is observed that maritime transportation companies are more sensitive to energy-saving measures compared to the past. However, when this issue is not handled with a holistic approach for ships, certain obstacles related to energy efficiency will inevitably arise. The issue of increasing energy efficiency onboard ships has a multifaceted structure. Therefore, the technical aspects of the subject, awareness, information/data management, and motivation of the ship's crew are also important.

In parallel with the increase in the awareness of maritime companies, with the impact of the determined efforts of IMO, a moderate recovery period has been observed in carbon intensity (EEOI) (CO_2e) at a slower rate than the growth in demand, starting from 2014. Emission projections were prepared in the Fourth Greenhouse Gas Study in the period up to 2050 (IMO 2020).

The greenhouse gas (GHG) emissions – including carbon dioxide (CO_2), methane (CH_4), and nitrous oxide (N_2O), expressed in CO_2e – of total shipping have steadily increased more than 9% beginning from 2012. Since 2015, the rate of reduction of carbon intensity varies between 1% and 2% and shows a slowing trend. Although it varies according to ship types, the annual carbon intensity performance of ships has fluctuated over the years. In terms of EEOI, fluctuation rates in oil tankers, bulk carriers, and container ships were generally above ±5% (IMO 2020).

This study has two aims: firstly, to emphasize the importance of economic sustainability by examining ships in terms of energy efficiency and considering environmental factors in the light of the IMO's studies, and secondly, to attract the attention of maritime business companies so that they are more directed toward energy-efficient practices in ship management.

67.2 Literature Review

The increasingly fierce competition conditions in the world's maritime transport are putting pressure on costs, leading businesses to search for additional measures in terms of efficiency. In cases where transportation costs can exceed 50%, reducing fuel costs is perhaps the first of the measures that come to mind. In this context, the development of energy efficiency onboard ships and the related measures will positively affect the energy consumption per transported cargo (Insel 2012).

There are numerous environmental effects of maritime transportation. The environmental impacts of ocean transport can be categorized as episodic or routine. Vessel-based pollution is an episodic environmental phenomenon. Engine emissions, on the other hand, cause oil spills because of routine environmental events. When it comes to air pollution caused by maritime transport, the first issues that come to mind can be listed such as emissions, including NOx, SOx, particulate matter (PM10), hydrocarbons and methane, black carbon and organic carbon, and refrigerants. Many emission control technologies have been developed and used in the field, especially for pollutants such as NOx and PM, within the scope of studies to reduce air pollution caused by ships, which are known to cause serious harm to human health. These controls are categorized as pre-combustion, in-engine, or post-combustion controls. Many of these technologies require an increase in CO_2 emissions due to increasing energy demand. In this case, it is important to study alternative energy sources and new technologies to solve environmental problems or to develop more sustainable solutions onboard ships (Corbett and Winebrake 2008).

The proportion of cargo ships in the world's maritime transport is more than 40%. For all ship types, fuel costs constitute the largest proportion of operating expenses. Cargo ships consume about 4% of the total fossil fuel. Considering this high rate, sustainable energy management processes gain special importance for ship operating companies. Power management, which is based on the fuel consumption of the ships, depends on the technical characteristics of the ship, the weather and sea conditions during the cruise, the effectiveness of the planned maintenance system processes, and the total efficiency. The most important process is the creation of a management system that will successfully integrate all the features and data of the ships in terms of operation.

67.3 Results and Discussions

This study aimed to present an energy efficiency strategy for ships by considering environmental factors. While doing this, the efficiency, emission effects, and economic performance processes of the ships during their voyages were examined and considered.

Energy management for ships is one of the topics that has been studied for a long time. The things to be done in terms of optimization of the energy consumed by the ships during the cruising and port periods and the harm caused to the environment by the gases that occur during the consumption of energy have been put forward by the relevant institutions (especially the IMO), and ship operating companies. In this context, companies operating ships also have to work on reducing energy costs per unit. Cetin and Sogut addressed the issue as shown in Fig. 67.1 in terms of energy efficiency, in a way to successfully support the applications of EEDI and SEEMP published by IMO for ships.

One of the important steps that should be considered on a preferential basis is "energy audit." Energy audit, which is an activity to determine energy efficiency, will reveal the losses in energy consumption in a building/workplace/ship and will enable accurate assessments of efficiency. The audit reports prepared as a result of these audits can make a significant contribution to the determination of energy use problems that have not been defined before, and to take corrective measures and thus reduce energy costs. There are three steps in an energy audit. These are evaluation, testing, and efficiency recommendations. Conducting energy audits at regular intervals helps to reduce the carbon footprint in the workplace/ships. It also contributes to

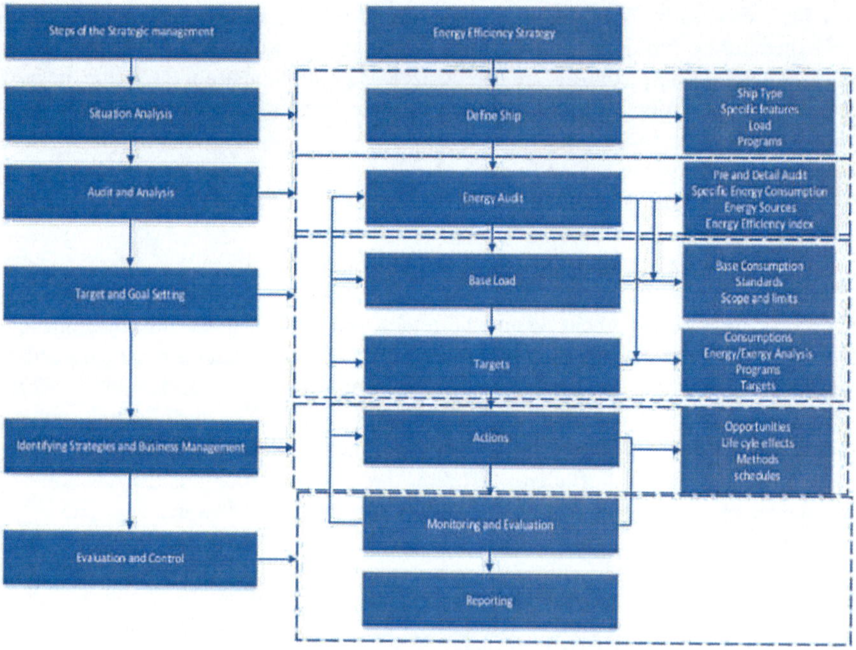

Fig. 67.1 Flow diagram of energy efficiency strategy. (Cetin and Sogut 2021)

the continuity of the energy-efficient structure by developing new improvement activities for energy saving and ensuring their continuity. According to a 2020 survey conducted at Stony Brook University in New York, only 9% of those survey respondents had passed an energy audit (Just Energy, 2022). It is not easy to express that the level of awareness about energy audits is high for maritime companies. Undoubtedly, an energy audit is not the only important factor for ships as shown in Fig. 67.1. However, it has special importance in terms of providing situational awareness.

If the strategic management process described in Fig. 67.1 is well understood and reinforced by regular training by shipping companies, it will be possible to establish and then develop a behavioral culture that will maintain energy efficiency onboard ships. In addition, it will provide a significant advantage in the management of parametric values, especially for EEDI.

Corbett and Winebrake predicted that the increase in maritime transportation will increase linearly every year and stated that the sectoral greenhouse gas potential will increase approximately three times if additional measures are not taken (Corbett and Winebrake 2008). This evaluation, made in the absence of pandemic conditions, shows that a realistic determination has been made in this period when an increase in maritime trade is observed despite the effects of the pandemic. Being aware of this situation, all relevant institutions, and organizations, especially IMO, obviously continue to carry out alternative studies to reduce total greenhouse gas emissions.

At the operational level, the Ship Energy Efficiency Management Plan (SEEMP) may easily support ships' crew to take necessary energy-saving actions. The plan determines difficulties on energy efficiency issues and deficiencies faced onboard ship operations, examining their complicated causes. The level of knowledge of crew on the energy-related issues onboard ships is not satisfactory enough, especially on the energy efficiency subjects. From the maritime companies' perspective, it is highly needed to have an energy management officer onboard ship. One of the educated and experienced officers can assume this job. Therefore, it is considered that it would be a correct approach to provide training on energy management for the ship's crew before they are assigned to the ship.

67.4 Conclusion

In this study, the importance of the energy efficiency strategy is emphasized by evaluating the parameters of a study in which the voyage-based values of the ships are examined. The factors that are effective in the energy efficiency strategy, which was developed considering environmental factors and sustainable economic values, were emphasized, and especially emphasis was placed on energy audit.

Maritime companies need to show more sensitivity to energy efficiency matters. Creating an energy efficiency strategy and sustainable energy management onboard ships will be able to reduce fuel-related costs.

In addition to the evaluation of environmental impacts, it is becoming increasingly important to look at the issue in terms of the economic empowerment of ship operating companies by increasing the energy efficiency of ships. In this context, it is considered important for companies to focus on training the crew working onboard the ships in a regular program on energy efficiency and developing an awareness of individual responsibility.

It is evaluated that it would be beneficial to include the subject of "energy efficiency onboard ships" in the curricula of all institutions providing training for officers and ship crews and to impose a certificate requirement. If this is done, besides raising awareness, a significant benefit will be provided in increasing energy efficiency onboard ships.

References

Corbett, JJ & Winebrake, J (2008) The Impact of Globalization on International Maritime Transport Activity: Past trends and future perspectives, Energy and Environmental Research Associates, the United States. Retrieved (26 February 2022) from https://www.researchgate.net/publication/255891211_The_impact_of_globalisation_on_international_maritime_transport_activity_Past_trends_and_future_perspectives

Çetin, O & Söğüt, MZ (2021) A new strategic approach of energy management onboard ships supported by exergy and economic criteria: A case study of a cargo ship, Ocean Engineering, 219 (2021) 108137. https://doi.org/10.1016/j.oceaneng.2020.108137

Ferreira JC & Kirk T (2012) Shipboard Energy Efficiency: Standards & Opportunities. Intertanko Latin America Panel, Rio de Janeiro. http://www. intertanko.com/upload/93963/LAPEnergyEfficiencyOct12.pdf. (Retrieved (08 March 2022) from Nizetic S & Papadopoulos. The Role of Exergy and the Environment, Green Energy and Technology. https://doi.org/10.1007/978-3-319-89845-2 (Springer)).

Insel, M (2012) Energy efficiency, Emissions and New Rules in the Marine Transportation. www.denizticaretodasi.org.tr/.../SektorelEgitim/EEDI_MINSE L2012.ppt.

IMO (2015) IMO Train the Trainer (TTT) Course on Energy-Efficient Ship Operation (Module 2: Ship Energy Efficiency Regulations and Related Guidelines) Retrieved (20 February 2022) from https://wwwcdn.imo.org/localresources/en/OurWork/Environment/Documents/Air%20pollution/M2%20Energy%20Efficiency%20Regulations%20-%20IMO%20TTT%20course%20presentation%20final1.pdf

IMO (2020) Fourth Greenhouse Gas Study 2020. Retrieved (19 February 2022) from https://www.imo.org/en/OurWork/Environment/Pages/Fourth-IMO-Greenhouse-Gas-Study-2020.aspx

IMO (2021) MEPC 76/15/Add.2, Annex 5, Resolution MEPC.332(76) (adopted on 17 June 2021), Amendments to the 2018 Guidelines on the Method of Calculation of the Attained Energy Efficiency Design Index (EEDI) For New Ships (Resolution MEPC.308(73), As Amended by Resolution MEPC.322(74)). Retrieved (19 February 2022) from https://www.imo.org/en/OurWork/Environment/Pages/Index-of-MEPC-Resolutions-and-Guidelines-related-to-MARPOL-Annex-VI.aspx

IMO (2022) Energy Efficiency Measures. Retrieved (04 March 2022) from https://www.imo.org/en/OurWork/Environment/Pages/Technical-and-Operational-Measures.aspx

Just Energy (2022) Understanding the Energy Audit: Why It's Worth Doing? Retrieved (04 March 2022) from https://justenergy.com/blog/understanding-energy-audit-what-why-important/

Chapter 68
Exergy Analysis of Cascade Refrigeration System for Different Refrigerant Couples

Hüsamettin Tan and Ali Erişen

Nomenclature

h	Enthalpy (kJ/kg)
s	Entropy (kJ/kg.K)
$\dot{W}$	Energy by work (kW)
$\dot{Q}$	Energy by heat (kW)
$\dot{m}$	Mass flow rate (kg/s)
$\dot{\psi}$	Stream exergy (kW)
$\dot{E}_D$	Exergy destruction (kW)

Subscripts

ltc	Low-temperature compressor
cascade	Cascade heat exchanger
gv	Expansion valve
evap	Evaporator
ihe	Internal heat exchanger
htc	High-temperature compressor
gc	Gas cooler
exp	Expander

H. Tan (✉) · A. Erişen
Faculty of Engineering and Architecture, Kırıkkale of University, Kırıkkale, Turkey
e-mail: Tan@univerist.edu; Erisen@uniersidad.edu

M. Z. Sogut et al. (eds.), *Proceedings of the 2022 International Symposium on Energy Management and Sustainability*, Springer Proceedings in Energy,
https://doi.org/10.1007/978-3-031-30171-1_68

68.1 Introduction

Nowadays, refrigeration systems have different designs according to the desired ambient temperature and refrigerant. The main factor in the invention of different designs is system performance. In addition, the effects of refrigerants on the environment are also an important issue.

In refrigeration systems, the single-stage vapor compression and absorption cycles are generally used. However, there are limitations in the use of these systems at low temperatures. These limitations are related to the system performance and the technical properties of the elements in the system. In this case, the use of cascade systems consisting of different refrigeration cycles is an effective solution.

In the studies on cascade refrigeration systems, there are two-stage vapor compression, two-stage absorption, and vapor compression-absorption and auto-cascade systems. In the literature, studies have been carried out in terms of design and refrigerant couples for these systems (Pan et al. 2020).

Cascade two-stage vapor compression refrigeration systems are widely used in application. Design improvements to increase system performance for low-temperature applications are investigated (Bhattacharyya et al. 2009; Petrenko et al. 2011; Ben et al. 2014; Dubey et al. 2014; Sun et al. 2019). As the studies were generally evaluated in themselves, there were positive and negative situations. In terms of refrigerants, thermodynamic, exergy, and economic analyses are generally made for the R717/R744 (Lee et al. 2006; Getu and Bansal 2008; Alberto et al. 2009; Bingming et al. 2009; Dopazo and Fernández-Seara 2011; Rezayan and Behbahaninia 2011; Alberto Dopazo and Fernández-Seara 2012; Aminyavari et al. 2014; Gholamian et al. 2018; Patel et al. 2019). The environmental friendliness and cost-effective contribution of R744 have been taken into account in other studies (Bhattacharyya et al. 2009; Di Nicola et al. 2011; Dubey et al. 2014; Sachdeva et al. 2014; Sanz et al. 2014; Massuchetto et al. 2019; Sánchez et al. 2019; Kauffeld et al. 2020). In addition, investigations were made in terms of system performance in different refrigerants (Di Nicola et al. 2005; Gong et al. 2009; Sarkar et al. 2013; Yilmaz et al. 2020).

Two-stage absorption cascade refrigerant systems can be used in applications where waste heat recovery is appropriate. However, there is a limitation for low-temperature applications with the used refrigerant couples (Pan et al. 2020). Therefore, there are not many studies in the literature on this design. In the studies, LiBr, LiCl, NH_3, and H_2O refrigerants were analyzed (Songara et al. 1998; She et al. 2015; Cui et al. 2019). In applications where electrical energy consumption is high, vapor compression-absorption refrigeration systems are used instead of two-stage cascade vapor compression systems. Here, vapor compression refrigeration cycles are used in the low-temperature cycle, and absorption refrigeration cycles are used in the high-temperature cycle. Comparisons of LiBr-H_2O and NH_3-H_2O fluids in the absorption refrigeration cycles of these systems were made (Cimsit and Ozturk 2012). Energy analyses were performed for R744, R717, R134a, R410A, R407c,

R1234yf, R1234ze, and R1233zd fluids in the low-temperature cycle (Han et al. 2013; Jain et al. 2013, 2015; Salhi et al. 2018).

Due to the high operating and maintenance costs of cascade systems, auto-cascade systems using mixed refrigerants with a single compressor have been widely used. Many studies have been carried out on increasing the system performance of auto-cascade systems, refrigerant selection, and optimization of system parameters (Kim and Kim 2002; Du et al. 2009; Tan et al. 2015; Chen et al. 2016; Sreenivas et al. 2017; Bai et al. 2018, 2019; Hao et al. 2018; Yan et al. 2018).

The scope of the studies carried out is generally in three different subjects: refrigerant couples, system parameters, and system designs. The aim of this study is to evaluate the efficiency of a cascade system consisting of two different refrigeration cycles operating at ultralow temperature in terms of II. law. Cascade system consists of gas and vapor compression cycles with a different novel design. Exergy analyses were made for different refrigerants with the same design assumptions. Comparisons were made within refrigerant couples with exergy analysis.

68.2 System Design and Exergy Analysis

The cascade system designed for the low-temperature applications is given in Fig. 68.1. The cascade system consists of mechanical vapor compression and gas refrigeration cycles.

In the LTC, the refrigerant (#5) that evaporates with the heat transferred from the environment increases the temperature at constant pressure without entering the LTC compressor with an intermediate heat exchanger (#6). The temperature and pressure of refrigerant (#1) increase with the LTC compressor and condense in the cascade heat exchanger as a saturated liquid (#2). The saturated liquid is cooled by an intermediate heat exchanger and enters the expansion valve (#3). The liquid-vapor mixture (#4) that is the outlet of the expansion valve enters into the evaporator and completes the cycle. In the cascade heat exchanger, heat is transferred from the LTC to the HTC at constant pressure, and the temperature and pressure of the gas refrigerant (#10) at the outlet of the cascade heat exchanger is increased by the HTC compressor (#7). The temperature of the refrigerant is reduced with a gas cooler at constant pressure (#8). The refrigerant enters the expander element to obtain the work output. The refrigerant (#9) whose pressure and temperature is decreased by the expansion element enters the cascade heat exchanger and completes the cycle.

The II. law analyses of the cascade system were made by the assumptions given below:

- All components are assumed to be in steady-state and steady-flow process.
- The potential and kinetic energy changes and fan power of gas cooler and evaporator are negligible.
- Pressure losses in fittings and heat exchangers are negligible.

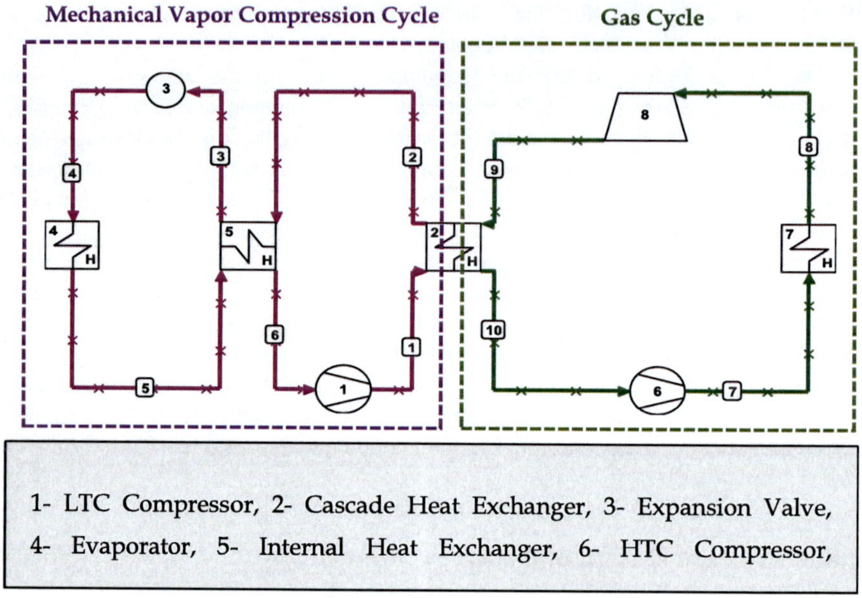

Fig. 68.1 The schematic of cascade refrigeration system

- In the LTC, the evaporator outlet is considered to be saturated vapor (# 4), and the condenser outlet is considered to be saturated liquid (# 2).
- The isentropic efficiency for both compressors and expander was fixed as 80%, based on recommended values by Oh et al. (2016).
- The expansion valves are isenthalpic devices.
- The evaporator and HTC cascade heat exchanger working pressure is higher than the atmospheric pressure.
- In the thermodynamic analysis, the refrigeration capacity of the system is 10 kW, the LTC evaporator temperature is −60 °C (T5), the HTC cascade heat exchanger outlet temperature is 25 °C (T10), the HTC expander outlet temperature is −10 °C (T9), and the medium temperature is −50 °C (Tmedium).
- Temperature and pressure at the environment for exergy analyses are considered to be 298 K and 101.325 kPa, respectively.

Exergy analysis is performed for each system element based on eqs. (68.1) and (68.2). The exergy equations of the system elements are given in Table 68.1.

$$\text{Stream exergy} \rightarrow \dot{\psi} = \dot{m}[(h - h_0) - T_0(s - s_0)] \quad (68.1)$$

$$\text{Exergy destroyed} \rightarrow \dot{E}_D = \sum \dot{\psi}_g - \sum \dot{\psi}_c + \sum \dot{Q}\left(1 - \frac{T_0}{T}\right) + \sum \dot{W} \quad (68.2)$$

Table 68.1 The exergy equations for system elements

No	System element	Exergy equation
1	LTC compressor	$\dot{E}_{D,\text{ltc}} = \dot{\psi}_6 - \dot{\psi}_1 + \dot{W}_{k,\text{ltc}}$
2	Cascade H.E.	$\dot{E}_{D,\text{cascade}} = (\dot{\psi}_1 - \dot{\psi}_2) + (\dot{\psi}_9 - \dot{\psi}_{10})$
3	Expansion valve	$\dot{E}_{D,\text{gv}} = \dot{\psi}_3 - \dot{\psi}_4$
4	Evaporator	$\dot{E}_{D,\text{evap}} = \left(1 - \frac{T_0}{T_{\text{medium}}}\right) * \dot{Q}_{\text{evap}} + (\dot{\psi}_4 - \dot{\psi}_5)$
5	Internal H.E.	$\dot{E}_{D,\text{ihe}} = (\dot{\psi}_5 - \dot{\psi}_6) + (\dot{\psi}_2 - \dot{\psi}_3)$
6	HTC compressor	$\dot{E}_{D,\text{htc}} = \dot{\psi}_{10} - \dot{\psi}_7 + \dot{W}_{k,\text{htc}}$
7	Gas cooler	$\dot{E}_{D,\text{gc}} = (\dot{\psi}_7 - \dot{\psi}_8) + (\dot{\psi}_{12} - \dot{\psi}_{13})$
8	Expander	$\dot{E}_{D,\exp} = \dot{\psi}_8 - \dot{\psi}_9 - \dot{W}_{\exp}$

Table 68.2 Refrigerant properties

Refrigerant	Critical temperature (°C)	Critical pressure (bar)	Boling temperature (°C)
R290	96.7	42.48	−42.1
R744	30.978	73.77	−78.914
R1234yf	94.7	33.81	−29.4
R170	32.172	48.72	−88.574
R1150	9.2	50.41	−103.77

The exergy equations for the system elements are solved by the EES package program. The selection of the refrigerants to be used in the system was determined by considering the boiling point temperature according to the design temperatures. R170 and R1150 in the low-temperature cycle and R290, R744, and R1234yf fluids in the high-temperature cycle are used. The properties of the selected fluids are given in Table 68.2.

68.3 Results and Discussion

Exergy analysis was performed for the cascade refrigeration system, and the II. law efficiencies and the exergy destruction rate for each system element were calculated. Compression ratios and flow rates of the cycles were investigated for different refrigeration couples. According to the results of the determined parameters, a comparison was made for the refrigeration couples.

According to the exergy analysis results given in Table 68.3, the highest II. law efficiency is in R1234yf/R170 and the lowest is in R744/R1150. The highest system performance value of 0.8229 under design conditions was calculated for R1234yf/R170. Looking at the compression ratios given in Table 68.4, it is the lowest for R1150 in the low-temperature cycle and the lowest for R744 in the high-temperature cycle.

Table 68.3 Exergy destruction rate and exergy efficiency

Refrigerant couples	Eevap	Eltc	Eev	Ecascade	Eihe	Egc	Ehtc	Eexp	COP	II
R290/R1150	0.6181	0.8660	0.2634	0.5744	0.2751	1.314	3.483	3.088	0.6943	0.2335
R744/R1150	0.6181	0.8660	0.2634	0.5781	0.2751	1.591	3.957	3.597	0.6379	0.2145
R1234yf/R1150	0.6181	0.8660	0.2634	0.5740	0.2751	1.142	3.037	2.630	0.7509	0.2526
R290/R170	0.6175	0.7745	0.0415	0.1611	0.2204	1.248	3.309	2.933	0.7585	0.2551
R744/R170	0.6175	0.7745	0.0415	0.1647	0.2204	1.512	3.759	3.417	0.6947	0.2337
R1234yf/R170	0.6175	0.7745	0.0415	0.1608	0.2204	1.085	2.885	2.498	0.8229	0.2768

Table 68.4 Compression ratio and mass flow ratio

Refrigerant couples	LTC compression ratio	HTC compression ratio	Mass flow ratio (HTC/LTC)
R290/R1150	4.84	2.81	8.13
R744/R1150	4.84	1.80	15.91
R1234yf/R1150	4.84	3.6	15.05
R290/R170	5.57	2.81	9.11
R744/R170	5.57	1.80	17.82
R1234yf/R170	5.57	3.6	16.86

68.4 Conclusions

In the study, a new cascade system design was made. Exergy analyses were carried out in line with the determined assumptions. In addition, the compression ratios in the system and the flow rates of the cycles were calculated. In the cascade system, analyses were made for the R290, R744, and R1234yf in the high-temperature cycle and for the R170 and R1150 in the low-temperature cycle. The main results of the study are given below:

- Among the different refrigerants analyzed, the R1234yf/R170 refrigerant couple has the best result in terms of the exergy efficiency, while R744/R1150 has the lowest one.
- Compression ratios are low for R1150 in LTC and for R744 in HTC.
- In terms of mass flow rates of the cycles, it is the lowest for R290/R1150 and the highest for R744/R170.
- The specific volumes of vapor of the fluids at the compressor inlet, the enthalpy difference in the evaporation temperature, and the condenser operating pressure are the main reasons for the different efficiencies.
- The mass flow rate difference between the cycles increases the amount of refrigerant in the system. This is an issue that needs to be considered in terms of both environmental impact and cost.

The results of the study contributed to the investigation of a different design in the cascade systems. Different refrigerants and system parameters need to be investigated in future studies.

References

Alberto Dopazo, J. et al. (2009) 'Theoretical analysis of a CO2-NH3 cascade refrigeration system for cooling applications at low temperatures', Applied Thermal Engineering, 29(8–9), pp. 1577–1583. doi: https://doi.org/10.1016/j.applthermaleng.2008.07.006.

Alberto Dopazo, J. and Fernández-Seara, J. (2012) 'Experimental evaluation of freezing processes in horizontal plate freezers using CO 2 as refrigerant', International Journal of Refrigeration, 35(8), pp. 2093–2101. doi: https://doi.org/10.1016/j.ijrefrig.2012.08.018.

Aminyavari, M. et al. (2014) 'Exergetic, economic and environmental (3E) analyses, and multi-objective optimization of a CO2/NH3 cascade refrigeration system', Applied Thermal Engineering, 65(1–2), pp. 42–50. doi: https://doi.org/10.1016/j.applthermaleng.2013.12.075.

Bai, T., Yan, G. and Yu, J. (2018) 'Experimental investigation of an ejector-enhanced auto-cascade refrigeration system', Applied Thermal Engineering, 129, pp. 792–801. doi: https://doi.org/10.1016/J.APPLTHERMALENG.2017.10.053.

Bai, T., Yan, G. and Yu, J. (2019) 'Experimental investigation on the concentration distribution behaviors of mixture in an ejector enhanced auto-cascade refrigeration system', International Journal of Refrigeration, 99, pp. 145–152. doi: https://doi.org/10.1016/J.IJREFRIG.2018.11.024.

Bhattacharyya, S., Garai, A. and Sarkar, J. (2009) 'Thermodynamic analysis and optimization of a novel N2O-CO2 cascade system for refrigeration and heating', International Journal of Refrigeration, 32(5), pp. 1077–1084. doi: https://doi.org/10.1016/j.ijrefrig.2008.09.008.

Bingming, W. et al. (2009) 'Experimental investigation on the performance of NH3/CO2 cascade refrigeration system with twin-screw compressor', International Journal of Refrigeration, 32(6), pp. 1358–1365. doi: https://doi.org/10.1016/j.ijrefrig.2009.03.008.

Chen, J., Yu, J. and Yan, G. (2016) 'Performance analysis of a modified autocascade refrigeration cycle with an additional evaporating subcooler', Applied Thermal Engineering, 103, pp. 1205–1212. doi: https://doi.org/10.1016/j.applthermaleng.2016.05.029.

Cimsit, C. and Ozturk, I. T. (2012) 'Analysis of compression-absorption cascade refrigeration cycles', Applied Thermal Engineering, 40, pp. 311–317. doi: https://doi.org/10.1016/j.applthermaleng.2012.02.035.

Cui, P. et al. (2019) 'Energy, exergy, and economic (3E) analyses and multi-objective optimization of a cascade absorption refrigeration system for low-grade waste heat recovery', Energy Conversion and Management, 184(January), pp. 249–261. doi: https://doi.org/10.1016/j.enconman.2019.01.047.

Dopazo, J. A. and Fernández-Seara, J. (2011) 'Experimental evaluation of a cascade refrigeration system prototype with CO2 and NH3 for freezing process applications', International Journal of Refrigeration, 34(1), pp. 257–267. doi: https://doi.org/10.1016/j.ijrefrig.2010.07.010.

Du, K. et al. (2009) 'A study on the cycle characteristics of an auto-cascade refrigeration system', Experimental Thermal and Fluid Science, 33(2), pp. 240–245. doi: https://doi.org/10.1016/j.expthermflusci.2008.08.006.

Dubey, A. M., Kumar, S. and Agrawal, G. Das (2014) 'Thermodynamic analysis of a transcritical CO2/propylene (R744-R1270) cascade system for cooling and heating applications', Energy Conversion and Management, 86, pp. 774–783. doi: https://doi.org/10.1016/j.enconman.2014.05.105.

Getu, H. M. and Bansal, P. K. (2008) 'Thermodynamic analysis of an R744-R717 cascade refrigeration system', International Journal of Refrigeration, 31(1), pp. 45–54. doi: https://doi.org/10.1016/j.ijrefrig.2007.06.014.

Gholamian, E., Hanafizadeh, P. and Ahmadi, P. (2018) 'Advanced exergy analysis of a carbon dioxide ammonia cascade refrigeration system', Applied Thermal Engineering, 137(October 2017), pp. 689–699. doi: https://doi.org/10.1016/j.applthermaleng.2018.03.055.

Gong, M. et al. (2009) 'Performance of R170 mixtures as refrigerants for refrigeration at -80 °C temperature range', International Journal of Refrigeration, 32(5), pp. 892–900. doi: https://doi.org/10.1016/j.ijrefrig.2008.11.007.

Han, W. et al. (2013) 'New hybrid absorption-compression refrigeration system based on cascade use of mid-temperature waste heat', Applied Energy, 106, pp. 383–390. doi: https://doi.org/10.1016/j.apenergy.2013.01.067.

Hao, X. et al. (2018) 'Hybrid auto-cascade refrigeration system coupled with a heat-driven ejector cooling cycle', Energy, 161, pp. 988–998. doi: https://doi.org/10.1016/J.ENERGY.2018.07.201.

Jain, V., Kachhwaha, S. S. and Sachdeva, G. (2013) 'Thermodynamic performance analysis of a vapor compression-absorption cascaded refrigeration system', Energy Conversion and Management, 75, pp. 685–700. doi: https://doi.org/10.1016/j.enconman.2013.08.024.

Jain, V., Sachdeva, G. and Kachhwaha, S. S. (2015) 'NLP model based thermoeconomic optimization of vapor compression-absorption cascaded refrigeration system', Energy Conversion and Management, 93, pp. 49–62. doi: https://doi.org/10.1016/j.enconman.2014.12.095.

Kauffeld, M. et al. (2020) 'N2O/CO2-Mixtures as Refrigerants for Temperatures below -50°C', International Journal of Refrigeration, 117, pp. 316–327. doi: https://doi.org/10.1016/j.ijrefrig.2020.04.026.

Kim, S. G. and Kim, M. S. (2002) 'Experiment and simulation on the performance of an autocascade refrigeration system using carbon dioxide as a refrigerant', International Journal of Refrigeration, 25(8), pp. 1093–1101. doi: https://doi.org/10.1016/S0140-7007(01)00110-4.

Lee, T. S., Liu, C. H. and Chen, T. W. (2006) 'Thermodynamic analysis of optimal condensing temperature of cascade-condenser in CO2/NH3 cascade refrigeration systems', International Journal of Refrigeration, 29(7), pp. 1100–1108. doi: https://doi.org/10.1016/j.ijrefrig.2006.03.003.

Ben Mansour, R., Ouzzane, M. and Aidoun, Z. (2014) 'Numerical evaluation of ejector-assisted mechanical compression systems for refrigeration applications', International Journal of Refrigeration, 43, pp. 36–49. doi: https://doi.org/10.1016/J.IJREFRIG.2014.04.010.

Massuchetto, L. H. P. et al. (2019) 'Thermodynamic performance evaluation of a cascade refrigeration system with mixed refrigerants: R744/R1270, R744/R717 and R744/RE170', International Journal of Refrigeration, 106, pp. 201–212. doi: https://doi.org/10.1016/J.IJREFRIG.2019.07.005.

Di Nicola, G. et al. (2005) 'Blends of carbon dioxide and HFCs as working fluids for the low-temperature circuit in cascade refrigerating systems', International Journal of Refrigeration, 28(2), pp. 130–140. doi: https://doi.org/10.1016/j.ijrefrig.2004.06.014.

Di Nicola, G. et al. (2011) 'Performance of cascade cycles working with blends of CO2 + natural refrigerants', International Journal of Refrigeration, 34(6), pp. 1436–1445. doi: https://doi.org/10.1016/J.IJREFRIG.2011.05.004.

Oh, J. S. et al. (2016) 'Improving the energy efficiency of industrial refrigeration systems', Energy, 112, pp. 826–835. doi: https://doi.org/10.1016/j.energy.2016.06.119.

Pan, M. et al. (2020) 'A review of the cascade refrigeration system', Energies, 13(9). doi: https://doi.org/10.3390/en13092254.

Patel, V. et al. (2019) 'An efficient optimization and comparative analysis of cascade refrigeration system using NH3/CO2 and C3H8/CO2 refrigerant pairs', International Journal of Refrigeration, 102, pp. 62–76. doi: https://doi.org/10.1016/j.ijrefrig.2019.03.001.

Petrenko, V. O., Huang, B. J. and Ierin, V. O. (2011) 'Design-theoretical study of cascade CO2 sub-critical mechanical compression/butane ejector cooling cycle', International Journal of Refrigeration, 34(7), pp. 1649–1656. doi: https://doi.org/10.1016/J.IJREFRIG.2010.11.012.

Rezayan, O. and Behbahaninia, A. (2011) 'Thermoeconomic optimization and exergy analysis of CO2/NH3 cascade refrigeration systems', Energy, 36(2), pp. 888–895. doi: https://doi.org/10.1016/j.energy.2010.12.022.

Sachdeva, G., Jain, V. and Kachhwaha, S. S. (2014) 'Performance Study of Cascade Refrigeration System Using Alternative Refrigerants', International Journal of Mechanical, Aerospace, Industrial, Mechatronic and Manufacturing Engineering, 8(3), p. 7. Available at: http://waset.org/publications/9997619/performance-study-of-cascade-refrigeration-system-using-alternative-refrigerants.

Salhi, K., Korichi, M. and Ramadan, K. M. (2018) 'Thermodynamic and thermo-economic analysis of compression–absorption cascade refrigeration system using low-GWP HFO refrigerant powered by geothermal energy', International Journal of Refrigeration, 94, pp. 214–229. doi: https://doi.org/10.1016/j.ijrefrig.2018.03.017.

Sánchez, D. et al. (2019) 'Energy assessment and environmental impact analysis of an R134a/R744 cascade refrigeration plant upgraded with the low-GWP refrigerants R152a, R1234ze(E), propane (R290) and propylene (R1270)', International Journal of Refrigeration, 104(2019), pp. 321–334. doi: https://doi.org/10.1016/j.ijrefrig.2019.05.028.

Sanz-Kock, C. et al. (2014) 'Experimental evaluation of a R134a/CO2 cascade refrigeration plant', Applied Thermal Engineering, 73(1), pp. 41–50. doi: https://doi.org/10.1016/j.applthermaleng.2014.07.041.

Sarkar, J., Bhattacharyya, S. and Lal, A. (2013) 'Performance comparison of natural refrigerants based cascade systems for ultra-low-temperature applications', International Journal of Sustainable Energy, 32(5), pp. 406–420. doi: https://doi.org/10.1080/14786451.2013.765426.

She, X. et al. (2015) 'A novel low-grade heat-driven absorption refrigeration system with LiCl-H2O and LiBr-H2O working pairs', International Journal of Refrigeration, 58, pp. 219–234. doi: https://doi.org/10.1016/j.ijrefrig.2015.06.016.

Songara, A. K., Fatouh, M. and Srinivasa Murthy, S. (1998) 'Thermodynamic studies on HFC134a-DMA double effect and cascaded absorption refrigeration systems', International Journal of Energy Research, 22(7), pp. 603–614. doi: https://doi.org/10.1002/(SICI)1099-114X(19980610)22:7<603::AID-ER379>3.0.CO;2-9.

Sreenivas, B., Nayak, H. G. and Venkatarathnam, G. (2017) 'Relationship between composition of mixture charged and that in circulation in an auto refrigerant cascade and a J-T refrigerator operating in liquid refrigerant supply mode', Cryogenics, 81, pp. 42–46. doi: https://doi.org/10.1016/j.cryogenics.2016.11.007.

Sun, Z. et al. (2019) 'Comparative analysis of thermodynamic performance of CO2 cascade refrigeration system assisted with expander and mechanical subcooling', International Journal of Energy Research, 43(14), pp. 7891–7907. doi: https://doi.org/10.1002/er.4719.

Tan, Y., Wang, L. and Liang, K. (2015) 'Thermodynamic performance of an auto-cascade ejector refrigeration cycle with mixed refrigerant R32 + R236fa', Applied Thermal Engineering, 84, pp. 268–275. doi: https://doi.org/10.1016/J.APPLTHERMALENG.2015.03.047.

Yan, G., He, C. and Yu, J. (2018) 'Theoretical investigation on the performance of a modified refrigeration cycle using binary zeotropic hydrocarbon mixture R170/R290', International Journal of Refrigeration, 94, pp. 111–117. doi: https://doi.org/10.1016/j.ijrefrig.2018.07.023.

Yilmaz, B., Mançuhan, E. and Yilmaz, D. (2020) 'Theoretical analysis of a cascade refrigeration system with natural and synthetic working fluid pairs for ultra low temperature applications', Journal of Thermal Science and Technology, 40(1), pp. 141–153.

Chapter 69
Hospital Energy Analysis in Turkey

Can Coskun, Zuhal Oktay, Hüseyin Özbek, and M. Ziya Sogut

Nomenclature

HECC	Hourly-electricity consumption coefficient
CDH	Cooling degree-hours
HDH	Heating degree-hours
COP	Coefficient of performance

69.1 Introduction

The high-energy intensity in the healthcare facilities, particularly in hospitals, along with energy costs and associated environmental concerns make energy analysis crucial for this type of facility. This is approximately 2.5 times higher than that of residential buildings. The proposed analysis shows that Turkish healthcare facilities have higher energy intensity than the ones in European facilities. This is which necessitates the adoption of more energy-efficient approaches to the infrastructure and the management of healthcare facilities in Turkey.

Lombard et al. (Lombard et al. 2008) have calculated the energy use intensity for buildings that serve different purposes in their study where they examine energy consumption in buildings. While energy intensity was 262 kWh/m^2 in schools, it

C. Coskun (✉) · Z. Oktay · H. Özbek
Engineering Faculty, Izmir Democracy University, Izmir, Turkey
e-mail: can.coskun@idu.edu.tr; zuhal.oktay@deu.edu.tr

M. Z. Sogut
Maritime Faculty, Piri Reis University, İstanbul, Türkiye
e-mail: mzsogut@pirireis.edu.tr

M. Z. Sogut et al. (eds.), *Proceedings of the 2022 International Symposium on Energy Management and Sustainability*, Springer Proceedings in Energy,
https://doi.org/10.1007/978-3-031-30171-1_69

was found to be 786 kWh/m^2 in hospitals. Chung and Park (Chung and Park 2015) surveyed energy demand characteristics of the three types of buildings in Korea: hotels, hospitals, and offices. They established load models for the three types of buildings. Electricity, heating, hot water, and cooling energy loads were analyzed by using collected electricity and fuel consumption. Annual, monthly, daily, and hourly energy consumption models are derived for the buildings examined. It is stated by the author that the determined load models can be used in simulations, design, optimization, and planning of building energy systems. Bawaneh et al. (Bawaneh et al. 2019) provided an analytical overview of end-use energy consumption data in healthcare systems for hospitals in the United States. The energy density of US hospitals varies between 640.7 kWh/m^2 (very hot region) and 781.1 kWh/m^2 (very cold region), reaching an average of 738.5 kWh/m^2. Energy consumed per square meter is approximately 2.6 times higher than other commercial buildings. In their analysis, they found that US healthcare facilities have higher energy density than healthcare facilities in many other countries. They suggested that more energy-efficient approaches should be adopted for the infrastructure and management of healthcare facilities in the United States. Teke and Timur (Teke and Timur 2014) researched energy-saving and efficiency strategies for hospitals in Turkey. They predicted that air-conditioning systems were responsible for approximately 70% of the total electricity consumption in hospitals. González et al. (González et al. 2018) studied and analyzed the energy consumption of 23 public hospitals in Germany between 2005 and 2015. Their analysis results show that the average annual energy consumption of a hospital is 0.27 MWh/m^2, 14.37 MWh/worker, and 23.41 MWh/bed. It has been shown that the indicators based on the number of beds are the most suitable as a reference for measuring the energy consumption of a hospital. Congradac et al. (Congradac et al. 2012) investigated the heating and cooling energy consumption in hospitals in order to increase energy efficiency in hospitals, using a variety of currently available technologies. They focused on the creation of a mathematical tool for the exact calculation of room/building energy demands. Bagnasco et al. (Bagnasco et al. 2015) presented a load estimation model and used the load estimation model in a real case study to estimate the electricity consumption of the Cellini medical clinic in Turin. Load estimation was carried out using the application of an artificial neural network (ANN). Their work focuses on providing a detailed analysis and an innovative formal procedure for the selection of all ANN parameters. Vaziri et al. (Vaziri et al. 2020) investigated the efficient use of renewable energy resources in hospitals by using demand dispatch. Since the main purpose of hospitals is to provide health services, and not to reduce the cost of energy, they offer recommendations on the use of renewable energy systems, taking into account the specific constraints and limitations of the hospital. The comfort of patients and doctors can be maintained while reducing energy costs in hospitals by using Vazari's proposed model. Reddy et al. (Reddy et al. 2019) investigated the energy-saving potential of Indian hospitals. They indicated that nearly 60% of healthcare services and hospitals do not meet the minimum of energy performance index (EPI) criteria. They have found 42% energy-saving potential by implementing energy-efficient measures.

The main aim of this study is to determine the energy consumption characterization and cost analyses of inpatient institution buildings in Turkey.

69.2 Methodology

In the analyses conducted for 2019, the hourly-electricity consumption amounts of the selected inpatient institution were measured. The main parameter taken to determine the hourly-distribution characteristic is annual-based hourly-average-electricity consumption amount ($E_{\text{ave, an.}}$). A general distribution was obtained and named the hourly-electricity consumption coefficient (HECC) (Coskun 2019). The HECC for each month is represented in this study as follows:

$$\text{HECC} = \frac{E_{\text{act.}}}{E_{\text{ave},an.}} 100 \quad (69.1)$$

Here, $E_{\text{act.}}$ and $E_{\text{ave, an.}}$ indicate the hourly-actual and annual-average electricity consumption. The average indoor temperatures during the heating and cooling period were analyzed for the hospital examined. The air velocity varies between 0.3 m/s (emergency unit) and 0.5 m/s (laboratories) and reaches a value of 0.388 m/s on average. The relative humidity ranges from 30% (laboratories) to 65% (emergency) and reaches an average of 52.4%. Different indoor temperatures occur in different rooms in the hospital. The indoor temperature in the hospital rooms for heating season varies between 20.7 °C (emergency) and 28.3 °C (general surgery), and on average is 24.5 °C. The indoor temperature in the hospital rooms for cooling season varies between 18.4 °C (operating room) and 25.7 °C (emergency), and on average is 22.2 °C.

The total heat transfer coefficient (L) for buildings is calculated by using Eq. 69.2 (Oktay et al. 2011):

$$L = \sum\nolimits_{j=1}^{M} U_j \cdot A_j + I \cdot (\rho . c_p) \frac{V}{3600} \quad (69.2)$$

Here, the term M indicates the areas where heat is lost to the outside. *U* indicates the heat transfer coefficient of each building element such as windows, exterior walls, ceiling, and roof. *A* is the area of each building element. Another term in the formula, I, refers to the hourly air change rate. *V* is the volume of the covered building studied. '$\rho .\ c_p$' is the volumetric thermal capacity. The volumetric thermal capacity of air could be taken as 1200 $\text{Jm}^{-3}\ \text{K}^{-1}$ (ASHRAE 1997). In general, the air exchange rate varies between 0.5 and 2 (Coskun et al. 2014). Since our calculations in this study are based on unit square meters, Eq. 69.3 has been modified as follows:

$$S = \frac{L}{A_{\text{total}}} \tag{69.3}$$

S is the heat transfer coefficient per unit area (W/m^2.°C). A_{total} is the total closed area in m^2. 43.2% of the surface area of the building where heat loss consists of windows and 56.8% consists of exterior walls and roofs. The heat transfer coefficient of windows and exterior walls is equal to 3.46 W/m^2.°C and 0.63 W/m^2.°C. While 67.4% of the total area is closed area, 32.6% is open area. For the hospital, the volume (V) in Eq. 69.2 was calculated as 3.6 m^3 per m^2 area. S was calculated as 6.86 W/m^2.°C.

The annual heating energy demand (Q_{heating}) can be calculated with the following equation:

$$Q_{\text{heating}} = N \cdot \text{HDH} - Q_{\text{solar}} - Q_{\text{employee}} - Q_{\text{patient}} - Q_{\text{equapment}} \tag{69.4}$$

Here, HDH is the heating degree-hours (°C–hours) and can be explained as in the following equation:

$$\text{HDH} = \sum\nolimits_{j=1}^{t} (T_{\text{indoor}} - T_{\text{amb}})_j \tag{69.5}$$

T_{indoor} and T_{amb} are indoor reference and ambient temperatures. The indoor reference temperature was measured as 24.5 °C for the investigated hospital. The annual cooling energy demand can be calculated with the following equation:

$$Q_{\text{cooling}} = N \cdot \text{CDH} + Q_{\text{solar}} + Q_{\text{employee}} + Q_{\text{patient}} + Q_{\text{equapment}} \tag{69.6}$$

Here, CDH is the cooling degree-hours (°C–hours) and can be explained as in the following equation:

$$\text{CDH} = \sum\nolimits_{j=1}^{t} (T_{\text{amb}} - T_{\text{indoor}})_j \tag{69.7}$$

The indoor reference temperature was measured as 22.2 °C for the investigated hospital. With reference to the outdoor temperature of the year under study, the heating degree-hour value was calculated as 48,546 °C-hour (24.5 °C indoor reference), and the cooling degree-hour value was 12,125 °C-hour (22.2 °C indoor reference). Heating and cooling degree-hour values are a very important parameter for determining heating and cooling loads. However, for buildings with very high heat gains, making a decision by looking at direct heating or cooling degree-hour values may present erroneous results. It has been observed that major mistakes will be made in hospitals if calculations are made only by looking at the heating or cooling degree-hour values. Although the hourly value of the heating is much higher than the cooling degree-hour value, as a result of the measurements, it is clear that the cooling load is higher than the heating load. The biggest reason for this is thermal gains. While thermal gains reduce the heating load in the winter, they increase the cooling load in the summer.

Thermal gains can be expressed in three main groups as gains consisting mainly of devices, gains from solar radiation, and gains from people. Generally, hospitals are places with a lot of human circulation. Thermal gains can be investigated in two parts as employees and patients. The average number of employees per unit closed area is predicted as 0.028742 employee/m^2. The average annual working time of employees for the hospital examined was determined as 2160 hours/year. During the study, it was determined that 240 W of thermal heat per employee was emitted. Thermal gains from people can be calculated from Eq. 69.8:

$$\begin{aligned} Q_{\text{employee}} &= \left[\frac{0.028742\ \text{employee}}{\text{m}^2}\right] \cdot \left[\frac{2160\ \text{hours}}{\text{year}}\right] \cdot \left[\frac{240\ \text{W}}{\text{employee}}\right] \\ &= 14.9 \text{kWh}/\text{m}^2.\text{year} \end{aligned} \tag{69.8}$$

The number of patients per unit area on an annual basis is calculated as 25.9 patient/ m^2.year. During the study, it was determined that 260 W of thermal heat per patient was emitted:

$$Q_{\text{patient}} = \left[\frac{25.9\ \text{patient}}{\text{m}^2}\right] \cdot \left[\frac{2.5\ \text{hours}}{\text{year}}\right] \cdot \left[\frac{260\ \text{W}}{\text{patient}}\right] = 16.835 \text{kWh}/\text{m}^2.\text{year} \tag{69.9}$$

The formula given below states the solar gain, Q_{solar}, through the glazing or the window to the building (Nielsen et al. 2001):

$$Q_{\text{solar}} = g \cdot \left[\sum\left(q_{\text{dir}} \cdot \left(1 - \tan^p\left(\frac{i}{2}\right)\right) \cdot \Delta t\right) + \sum((q_{\text{dif}} + q_{\text{ref}}) \cdot f \cdot \Delta t\right] \tag{69.10}$$

Here, g represents the total solar energy transmittance at an incidence angle of 0°; q_{dir} defines the average value of the direct solar radiation at the surface in the time step in W/m^2; q_{dif} is the mean value of the diffuse solar radiation on the surface in the time step in W/m^2; q_{ref} represents the mean value of the reflected solar radiation on the surface in the time step in W/m^2; i is the mean incidence angle of the direct solar radiation in the time step in degrees; f is a factor that adjusts for the total solar energy transmittance for diffuse solar radiation; Δt is the length of the time step in hours; and p represents the dependence of the incidence angle.

69.3 Results and Discussion

Using the 2019 data set, we determined hourly-electricity consumption profile. The hourly-electricity demand profile was calculated and given in Fig. 69.1. It was determined that the highest electricity consumption on an hourly basis occurred

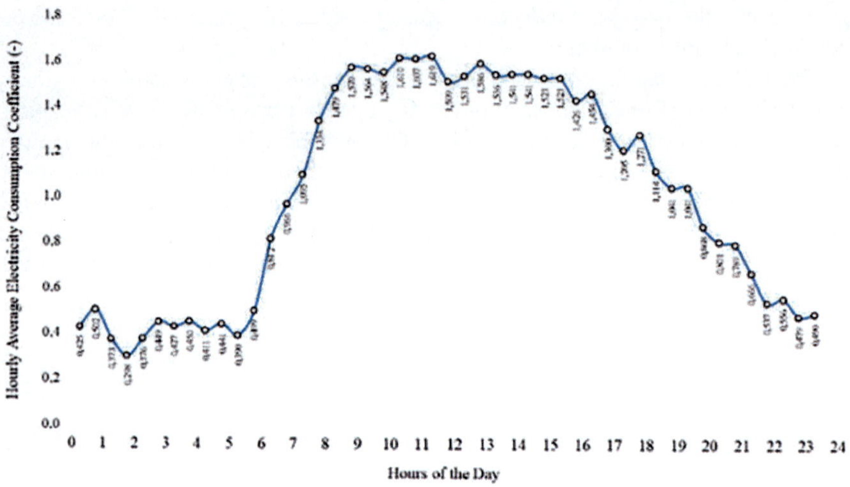

Fig. 69.1 Hourly-electricity consumption coefficient

Table 69.1 Installed power, average power, and capacity utilization rate of equipment

Equipment	Total installed power per unit area (W/m^2)	Average power equipment power per unit area (W/m^2)	Equipment capacity utilization rate (%)
Elevator	4.735	1.899	40.1
Lighting	3.940	1.773	45.0
Medical imaging	15.789	3.030	19.2
Others	0.684	0.450	65.8
Total	25.148	7.153	28.4

between 11:30 and 12:00. The lowest electricity consumption was between 01:30 and 02:00. As of 7 o'clock, the daily average electricity consumption value is reached. The electricity consumption value decreases gradually after 15:00. Equipment electricity consumption was collected in four main groups as elevators, lighting, medical imaging systems, and other equipment.

Air-condition electricity utilization is calculated in heating and cooling. Considering the hourly amount of electricity consumption, the capacity utilization rate of electrical equipment varies between 19.2% and 45% per hour, reaching an average of 28.4% (Table 69.1).

The total installed equipment power per unit closed area is predicted as 25.148 W. The highest installed power per unit closed area is due to the medical imaging system. Considering the average power demand of 7.153 W/m^2, the lighting is 24.79% (1.773 W/m^2), medical imaging 42.36% (3.03 W/m^2), elevator 26.56% (1.9 W/m^2), and other 6.29% (0.45 W/m^2). According to the data, hourly and daily equipment electricity load density for equipment utilization during the year is given in Fig. 69.2. Load density prediction is very suitable device for energy

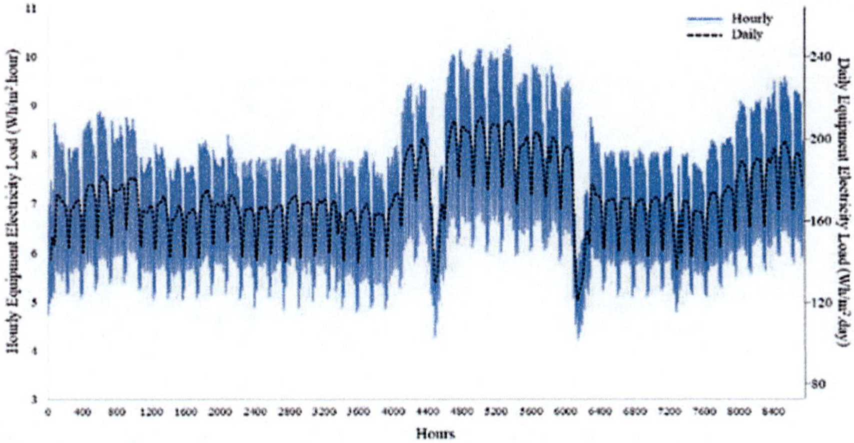

Fig. 69.2 Hourly and daily equipment electricity load density

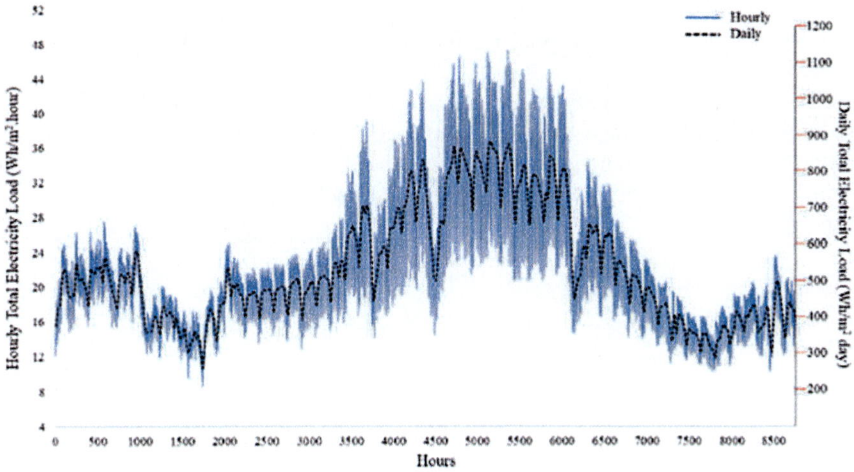

Fig. 69.3 Hourly and daily total electricity load density

analyses of different systems (Akyuz et al. 2013). As can be seen Fig. 69.2, hourly electricity load density changes between 4.22 and 10.23 W/m^2. The daily equipment electric load density changes between 121.4 Wh/m^2.day and 210.4 Wh/m^2.day, average 171.7 Wh/ m^2.day. The hourly and daily total electricity load density during the year is given in Fig. 69.3. The total electricity load includes equipment and heating and cooling electrical devices. As can be seen in Fig. 69.3, the hourly electricity load density changes between 8.65 and 47.34 Wh/m^2, average 22.14 Wh/m^2. The daily electric load density changes between 258.9 Wh/m^2.day

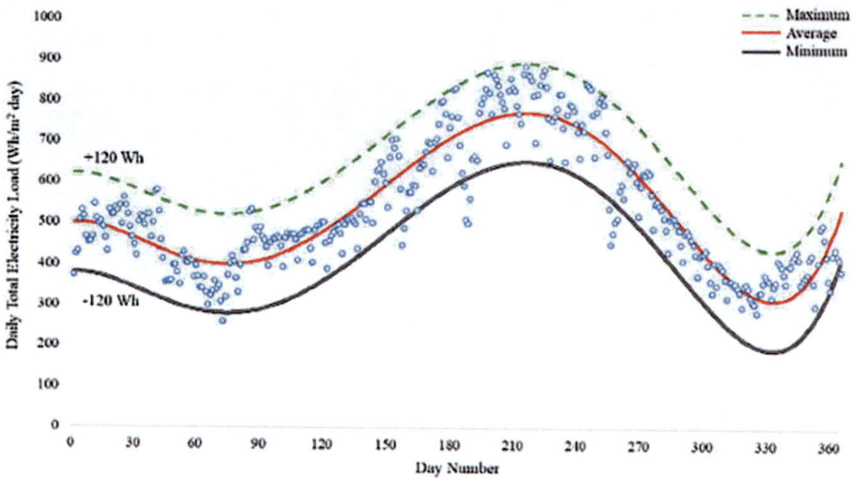

Fig. 69.4 Daily total electricity load distribution during a year

and 882.4 Wh/m^2.day, average 531.3 Wh/m^2.day. The hourly and daily load density increases in summer.

The daily total electricity load distribution during a year is determined and given in Fig. 69.4. The average daily trend in electricity consumption ($\mathrm{DTEC_{ave}}(dn)$) is determined, and the function is given in the equation below. The maximum and minimum daily electricity consumption function occurs above and below 120 W of the average trend:

$$\mathrm{DTEC_{ave}}(dn) = 500.6152 + 0.48167 \cdot dn - 8.h82345 \cdot 10^{-2} \cdot dn^2 \\ +1.182846 \cdot 10^{-3} \cdot dn^3 - 4.84666 \cdot 10^{-6} \cdot dn^4 + 6.19239 \cdot 10^{-9} \cdot dn^5 \tag{69.11}$$

$$\mathrm{DTEC_{max}}(dn) = \mathrm{DTEC_{ave}}(dn) + 120\ \mathrm{W} \tag{69.12}$$

$$\mathrm{DTEC_{min}}(dn) = \mathrm{DTEC_{ave}}(dn) - 120\ \mathrm{W} \tag{69.13}$$

The heat loss by conduction and infiltration was calculated as 333 kWh/m^2. The thermal gains reached 96.1 kWh/m^2. The average daily and seasonal equipment thermal load was assumed as 185 Wh/m^2 day and 33.3 kWh/m^2. The thermal gains were calculated as 33.3 kWh/m^2 (34.65%) from devices, 15.9 kWh/m^2 (16.54%) from people, and 46.9 kWh/m^2 (48.8%) from the sun. The energy requirement for the heating period was calculated as 236.9 kWh/m^2. The conduction and infiltration cooling requirement for the cooling season was calculated as 83.2 kWh/m^2. The thermal gains were calculated as 177.4 kWh. The thermal gain was 33.3 kWh/m^2 (18.77%) from devices, 15.9 kWh/m^2 (8.96%) from people, and 128.2 kWh/m^2 (72.26%) from the sun. The energy requirement for the cooling period was calculated as 260.6 kWh/m^2.

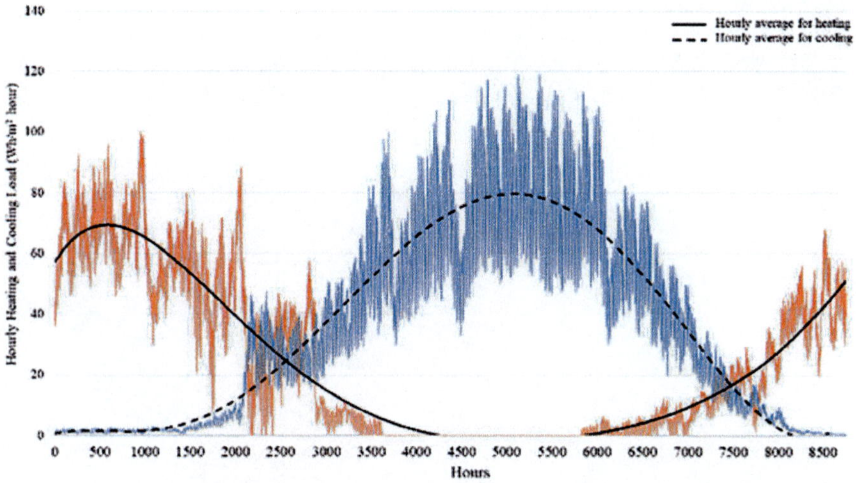

Fig. 69.5 Hourly heating and cooling load distribution

The hourly actual heating and cooling load distribution has been measured and is given in Fig. 69.5. There is a 3.4% difference between the predicted (236.9 kWh/m^2) and the measured (229.1 kWh/m^2) seasonal heating load. There is a 2.8% difference between the predicted (260.6 kWh/m^2) and the measured (268.1 kWh/m^2) seasonal cooling load. As expected, the heating load increases in the winter period, while the cooling load increases in the summer period. It has been observed that there is a need for both heating and cooling during the hours of the day, especially in the spring and autumn seasons. The measured maximum hourly heating and cooling load achieved is 100 and 119 Wh/m^2. Individual air-conditioning systems for the examined hospital have a higher installed capacity than the central air conditioning system. The central air-conditioning system has 38.72% of the installed capacity. When all air-conditioning systems are evaluated according to their capacities and COP values, the average COP value is calculated as 3.207.

Analyzing energy consumption data from the inpatient institution in Turkey resulted in an electricity intensity of 61.71 kWh/m^2 and a heat density (heating, cooling, and hot water) of 510.86 kWh/m^2 per year (Fig. 69.6). The cooling, heating, and hot water energy requirements constitute 46.83%, 40.01%, and 2.39% of the total energy requirement (572.6 kWh/m^2.year), respectively. The cooling energy requirement is the part with the highest energy demand.

Coşkun and Ertürk (Coşkun and Ertürk 2014) calculated the annual electricity consumption of residences as 1708 kWh in Turkey. It varies from 1344 to 1984 kWh/year in Turkey. The amount of electricity consumed in residences constitutes 23.715% of the total electricity consumption. Coskun et al. (Coskun et al. 2021), in another study, determined the average area of residences in Turkey as 119.7 m^2. In this context, the annual electricity consumption per m^2 in residences in Turkey can be calculated as 14.27 kWh/m^2.year. Coskun et al. (Coskun et al. 2021) calculated

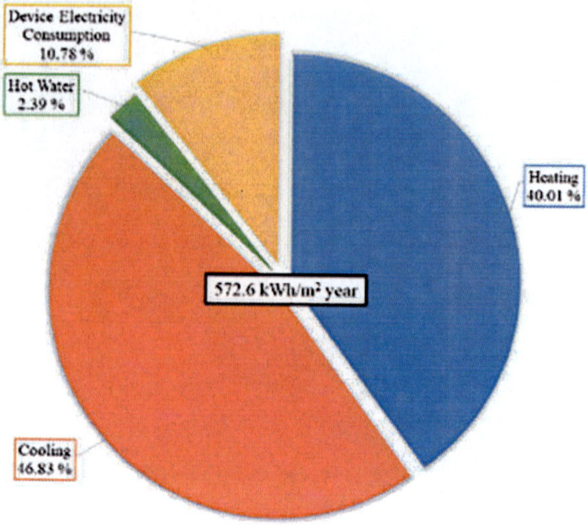

Fig. 69.6 Annual energy utilization distribution for the investigated heath facility

the annual heating and cooling energy amount as 93.5 kWh/m^2 and 1.876 kWh/m^2, respectively. The amount of energy consumed in residences is found as 109.65 kWh/m^2.year. The energy consumption in the hospital was calculated as 5.22 times higher than the average energy consumption in the residence (109.65 kWh/m^2.year).

69.4 Conclusion

In this study, energy analysis was performed for the selected inpatient health institutions in Turkey. The characteristics of the energy use in inpatient health institutions were determined. The results obtained in this study are presented below:

- The energy consumption in inpatient health institution was calculated as 5.22 times higher than the average energy consumption in the residential buildings in Turkey.
- The cooling, heating, and hot water energy requirements constitute 46.83%, 40.01%, and 2.39% of the total energy requirement (572.6 kWh/m^2.year), respectively. The cooling energy requirement is the part with the highest energy demand.
- The highest installed power per unit closed area is due to the medical imaging system. Medical imaging systems consume 42.36% of the average equipment power demand.
- Considering from the perspective of all hospitals, the device with the highest capacity utilization rate appears to be MRI with a rate of 53.41%. The average capacity utilization rate is estimated to be 15.18% (average of 3.6 h per day) for all medical imaging devices.

- Tomography and MRI consume 96.97% of the medical imaging system total energy consumption. The energy-efficient selection of tomography and MRI devices appears to be quite effective in reducing the electricity consumption at hospitals. In the context of tomography and MRI device, even at the level of 1%, more efficient device usage can provide an annual energy saving of 4.85 GWh for Turkey.
- The hourly electricity load density changes between 8.65 and 47.34 Wh/m^2, average 22.14 Wh/m^2. The daily electric load density changes between 258.9 Wh/m^2.day and 882.4 Wh/m^2.day, average 531.3 Wh/m^2.day. The hourly and daily load density increases in summer.
- Heating and cooling degree-hour values are a very important parameter for determining heating and cooling loads. However, for buildings with very high heat gains, making a decision by looking at direct heating or cooling degree-hour values may present erroneous results.

References

ASHRAE handbook fundamentals. (1997) Atlanta: American Society of Heating, Refrigerating and Air- Conditioning Engineers Inc.

Akyuz E, Demiral D, Coskun C, Oktay Z (2013) Estimation of the monthly based hourly wind speed characteristics and the generated power characteristics for developing bidding strategies in an actual wind farm: a case study. Arabian Journal for Science and Engineering 38(2) 263–275. https://doi.org/10.1007/s13369-012-0439-3

Bawaneh K, Ghazi NF, Rasheduzzaman M, Deken B (2019) Energy Consumption Analysis and Characterization of Healthcare Facilities in the United States. Energies 12(19):3775. https://doi.org/10.3390/en12193775

Bagnasco A, Fresi F, Saviozzi M, Silvestro F, Vinci A (2015) Electrical consumption forecasting in hospital facilities: An application case. Energy and Buildings. 103:261–270. https://doi.org/10.1016/j.enbuild.2015.05.056

Chung M, Park HC (2015) Comparison of building energy demand for hotels, hospitals, and offices in Korea. Energy 92:383–393 https://doi.org/10.1016/j.energy.2015.04.016

Coskun C (2019) A time-varying carbon intensity approach for demand-side management strategies with respect to CO2 emission reduction in the electricity grid. International Journal of Global Warming 18(1/2):1 https://doi.org/10.1504/IJGW.2019.101768

Coskun C, Ertürk M (2014) Electricity generation and geothermal energy based combined heat and power systems analysis. (in Turkish) Turkey Alim books publishing house ISBN-13:978-3-639-67018-9

Coskun C, Erturk M, Arcaklioğlu E, Balci K, Oktay Z (2021) The climate change impact projections on seasonal residential sector CO2 emissions and energy demand forecasting for Turkish provinces. Int. J. Global Warming 24(3/4):281–306 https://doi.org/10.1504/IJGW.2021.116710

Coskun C, Ertürk M, Oktay Z, Hepbasli A (2014) A new approach to determine the outdoor temperature distributions for building energy calculations. Energy conversion and management 78:165–172. https://doi.org/10.1016/j.enconman.2013.10.052

Čongradac V, Prebiračević B, Jorgovanović N, Stanišić D (2012) Assessing the energy consumption for heating and cooling in hospitals. Energy and Buildings 48: 146–154. https://doi.org/10.1016/j.enbuild.2012.01.022

González GA, García-Sanz-Calcedo J, Rodríguez SD (2018) Evaluation of Energy Consumption in German Hospitals: Benchmarking in the Public Sector. Energies 11(9):2279. https://doi.org/10.3390/en11092279

Lombard LP, Ortiz J, Pout C (2008) A review on buildings energy consumption information. Energy and Buildings 40:394–398 https://doi.org/10.1016/j.enbuild.2007.03.007

Nielsen TR, Duer K, Svendsen S (2001) Energy performance of glazings and windows. Solar Energy 69:137–143. https://doi.org/10.1016/S0038-092X(01)00062-7.

Reddy S, Sandbhor S, Dabir V (2019) Bringing Energy Efficiency for Hospital Building through the Conservative and Preventive Measures. International Journal of Innovative Technology and Exploring Engineering 8(12):3056–3060 https://doi.org/10.35940/ijitee.L2470.1081219

Teke A, Timur O (2014) Overview of Energy Savings and Efficiency Strategies at the Hospitals. International Journal of Economics and Management Engineering 8(1):242–248. https://doi.org/10.5281/zenodo.1090536

Oktay Z, Coskun C, Dincer I (2011) A new approach for predicting cooling degree-hours and energy requirements in buildings. Energy 36(8):4855–4863. https://doi.org/10.1016/j.energy.2011.05.022

Vaziri SM, Rezaee B, Monirian MA (2020) Utilizing renewable energy sources efficiently in hospitals using demand dispatch. Renewable Energy 151:551–562. https://doi.org/10.1016/j.renene.2019.11.053.

Chapter 70
Sustainability and Energy Efficiency of Passive Architecture for Modular Residences in Brazil

C. Gomide Sergio and A. R. Ismail Kamal

Nomenclature

A_{op}	Opaque surface area
A_{tr}	Transparent surface area
Re	Reynolds number
Pr	Prandtl number
Nu	Nusselt number
F_{st}	Solar factor
L	Characteristic dimension
Ig	Incident global radiation
U	Heat transfer coefficient
K	Conductivity
V	Hourly ventilation frequency
Q_{op}	Opaque wall relative load
Q_{tp}	Load relative to transparent wall
Q_t	Total heat load
Q	Total heat transfer
q_n	Heat flow
d_s	Vertical distance
h_{in}	Convective coefficient
R_{total}	Total thermal resistance

C. Gomide Sergio (✉)
Faculty of Mechanical Engineering, Energy Department, State University of Campinas, Campinas, São Paulo, Brazil

A. R. Ismail Kamal
UNICAMP, Campinas, São Paulo, Brazil
e-mail: kamal@fem.unicamp.br

M. Z. Sogut et al. (eds.), *Proceedings of the 2022 International Symposium on Energy Management and Sustainability*, Springer Proceedings in Energy,
https://doi.org/10.1007/978-3-031-30171-1_70

t_{emax} Maximum external temperature
T_{emin} Minimum external temperature
t_{imax} Maximum internal temperature

70.1 Introduction

Brazilian urbanization in the twentieth century, stimulated by the industrialization process, represented a decisive factor in the process of migration from rural to urban areas. It focused on meeting the demand for social housing. The architectural design patterns of single-family units absorbed outdated contours in terms of the biophysical profiles of their inhabitants, the constructive, environmental, and climatic requirements in a large part of the Brazilian housing stock built.

A determining factor for this work was the observation of the constructive characteristics from studies that pointed to both the bad use of electrical energy and the thermal performance, which invariably derive from similar problems. Related to the thermal performance of the houses appear the thermophysical characteristics of the fences with low level of use of ventilation and natural lighting, the use of materials' good conductors of heat, and bad orientation of the implantation. The thermal performance of these construction systems (opaque and transparent seals) focused on the demands of electricity consumption in the Brazilian residential sector, for air-conditioning and water heating.

From this analysis, a literature review points to the concept of building sustainability relating to the two categories. The first refers to the compatibility between construction and climate systems. The methods for determining the thermal performance establish recommendations for the project and relate the sensation of thermal comfort of its inhabitants, defined in the works of Olgyay (1963), Givoni (1963), Koenigsberger et al. (1971, 1974), and Evans and Shiller (1986). The second refers to the measurement and tools to support the design of sustainability of the construction system in the preliminary design and design phases of the buildings (LCA, Eco-Quantum, ENVEST, ATHENA). They support decision-making regarding the definition of the performance to be achieved for the final design solution. They are oriented toward life cycle analysis systems, products, performance, and environmental impacts (BREEAM, LEED, GBTool, and MARS-SC). The Brazilian experience resulted in legislation that linked the standardization of rules for labeling buildings, called the Brazilian Labeling Program (BLP), coordinated by the National Institute of Metrology, Standardization, and Industrial Quality (INMETRO 2010). It favors energy efficiency in homes through the Technical Quality Regulation (TQR), which has the methodology for obtaining the label.

The climatic complexity existing in Brazil imposed the challenge of proposing an architectural party combining elements of construction systems (opaque and transparent fences) in order to make them dynamic structures. These structures become solutions for the removal of radiation heat in hot climates and its retention in cold climates. Controlling the heat transfer resulting from fluctuations in solar radiation and temperature allows for air-conditioning of the internal environment by natural/passive means. In this work, the architectural design was applied to a standard single-family modular residence project aimed at the low-income population in three distinct climatic regions of Brazil. The compatibility of climatic data combined with the properties of the constituent elements of the construction systems allowed the methodology to assess the sustainability of the project. For the development of the architectural modeling of this type of project, a relation between area and volume of compact configuration was imposed, ensuring a more homogeneous exposure of the thermal areas of the opaque and transparent surfaces. The energy efficiency resulting from the thermal performance of this proposal is related to its most representative objective, reducing demand for electricity. The cities defined for the simulation are Belém (North Region), Recife (Northeast), and Curitiba (South) in climatic zones 2 (Curitiba) and 8 (Belém and Recife), according to Brazilian Standard (ABNT NBR 15220-3 2005).

70.2 Methodology

For opaque and transparent seals, the methodology called "Relative Assessment of the Sustainability of Building Solutions (MARS-SC), addresses three groups of environmental, functional, and economic indicators and their parameters. Once parameters to be considered in the evaluation are defined, quantification is carried out, allowing the comparison of solutions and the aggregation of these parameters for a precise evaluation of the final solution. Normalization is a mathematical procedure that avoids scaling effects in the aggregation of the parameters of each indicator. Pi represents the value of parameter i (sustainability) where P_i^* and P_{i*} are respectively the highest and lowest value of the sustainability parameter i. Values are dimensionless and scaled from 0 (worst value) to 1 (best value), according to Eq. 70.1:

$$P_i = \frac{P_i^* - P_i}{P_i - P_{i_*}} \tag{70.1}$$

By aggregation of indicators, according to Mateus and Bragança (2004), the assessment of the performance of each constructive solution involves parameters combined with each indicator depending on its importance (weight) to the requirements of the project. The value obtained in Eq. 70.2 is partial and represents the relative performance of the solution at the level of each indicator I_j:

$$Ij = \sum_{i=1}^{n} .Wi.P \tag{70.2}$$

The aggregation of the parameters, of the environmental, functional, and economic indicators to determine the sustainable solution, is obtained by Eqs. 70.3, 70.4, and 70.5.

Environmental performance:

$$I_E = \sum\nolimits_{i=1}^{n} .W_{Ei}.P_{Ei} \tag{70.3}$$

Functional performance:

$$I_F = \sum\nolimits_{i=1}^{n} .W_{Fi}.P_{Fi} \tag{70.4}$$

The quantified economic performance considers the value resulting from the sum of all costs involved in the construction life cycle for each proposed solution:

$$I_E = P_E \tag{70.5}$$

The global performance score of each constructive solution in terms of its environmental, functional, and economic indicators is obtained by Eq. 70.6:

$$NS = W_{G1}.I_E + W_{G2}.I_F + I_{G3}.I_{ec} \tag{70.6}$$

70.2.1 Analysis of the Behavior of Building Systems

The structural design of the opaque fence (walls and slab) of the proposed construction systems was characterized by a high mass per unit area. The classification of thermal inertia presented in this project is based on the concept of equivalent surface, according to Frota and Schiffer (2001). The reference wall received the highest sustainable score and wall 2 the second highest score. The configuration of this ventilated wall allows both the storage of heat in cold seasons and the flow of heat dissipating it, through the thermocirculation effect. Slab 4, with a double profile, contemplates the same parameters and weights of the walls with double profiles and will respond similarly to greater exposure to solar radiation. Figure 70.1 presents the results of performance evaluations and their parameters and sustainable note.

For the transparent seals (windows), according to Ismail (2010), they must behave as dynamic structures, in the control of heat transfer. The total heat transfer (Q) depends on the temperature fields of (∂T) at all points along the surface of the plate (ds), the heat flux (q_n”), according to Eq. (70.7). The heat flux is conditioned by the convective coefficient on the inner surface. The interaction between ambient temperature and the temperature of the inner surface of the glass plate, including the height of the window, conditions the convective coefficient, according to Eq. 70.8:

Constructive Solution	Sustainable profile	EPI KW.h/m²	GWP kgeqCO₂/m²	Performance Environmental Indicator I_E	D_{nw} dB	EP m	U W/m².°C	Performance Functional Indicator I_F	CC R$/m²	Performance Indicator Economic I_{EC}	Sustainable note SN
Wall ref		0.70	0.30	1.00	0.33	0.33	0.04	0.70	1.00	1.00	0.85
Wall 1		0.29	0.22	0.51	0.18	0.07	0.32	0.57	0.48	0.48	0.53
Wall 2		0.69	0.29	0.98	0.15	0.05	0.33	0.53	0.59	0.59	0.68
Wall 3		0.00	0.16	0.16	0.12	0.12	0.24	0.48	0.50	0.50	0.39
Wall 4		0.53	0.04	0.57	0.02	0.02	0.30	0.34	0.42	0.42	0.42
Wall 5		0.38	0.00	0.38	0.00	0.00	0.26	0.26	0.10	0.10	0.26
Wall 6		0.19	0.20	0.39	0.18	0.12	0.00	0.30	0.10	0.10	0.29
Wall 7		0.26	0.22	0.48	0.20	0.14	0.18	0.52	0.00	0.00	0.40
Wall 8		0.11	0.14	0.25	0.16	0.12	0.22	0.50	0.00	0.00	0.33

Constructive Solution	Sustainable profile	IPE KW.h/m²	GWP kgeqCO₂/m²	Performance Environmental Indicator I_E	D_{nw} dB	EP m	U W/m².°C	Performance Functional Indicator I_F	CC R$/m²	Performance Indicator Economic I_{EC}	Sustainable note SN
Slab ref		0.15	0.06	0.21	0.00	0.29	0.06	0.35	1.00	1.00	0.44
Slab 1		0.00	0.00	0.00	0.00	0.20	0.00	0.20	0.00	0.00	0.10
Slab 2		0.54	0.12	0.66	0.32	0.33	0.00	0.65	0.21	0.21	0.56
Slab 3		0.55	0.25	0.80	0.33	0.00	0.32	0.65	0.03	0.03	0.57
Slab 4		0.70	0.30	1.00	0.12	0.33	0.33	0.78	0.50	0.50	0.79

Fig. 70.1 Summary of results, sustainable note – wall and slab (by authors, 2022)

$$Q = \int_{O}^{S} = qn''.ds \tag{70.7}$$

$$q_{n''} = -K.\frac{\partial T}{\partial n} \tag{70.8}$$

The associated strategy considers exploring the characteristic dimension parameter (L), changing the thickness of the air blades as a way of comparing their effects in different domains. The comparison allows to verify that with an increase in the thickness of the air blade, the behavior of heat transfer (Q) of the regions of surface temperatures along its vertical extension changes. The boundary conditions related to domains (1) and (2) are shown in Fig. 70.2.

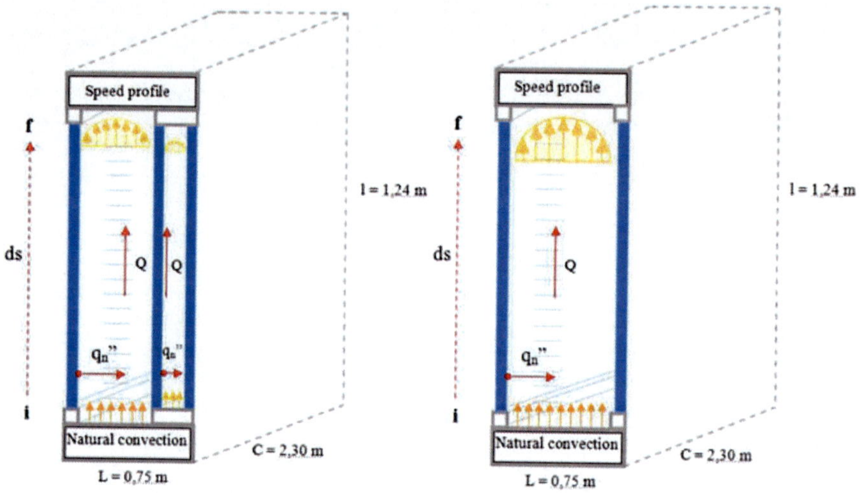

Fig. 70.2 Characterization of flows in domains (1) and (2) (by authors, 2022)

By reducing the effect of convective current and heat flux along the vertical axis, the dimension (L) is the original problem parameter for determining the Nusselt number (Nu), Eq. 70.9:

$$Nu = h.\frac{L}{K_f} \tag{70.9}$$

Considering the arbitrary part of Eq. (70.9), this strategy focuses on the efficiency of the window by increasing its thermal insulation. For climates characterized by reduced thermal gains, such as Curitiba in winter, the objective of retaining the heat absorbed from the inside to the outside of the environment is feasible. The behavior of fluid temperature and heat transfer (Q) throughout the day and in the four climatic contexts is shown in Fig. 70.3.

The structural design of the proposed windows (1) and (2) seeks to meet the strategy of controlling the convective effects inside the cavities of each proposed solution. The different air depths in the cavities made it possible to obtain values and compare them with each other and with a reference solution. The highest sustainable note was given to the window solution (1) and the reference window, the second largest note. The configuration of windows (1) and (2) allow heat storage in cold seasons, influencing the flow and temperature of the fluid as well as heat transfer. The performance and its parameters and sustainable note of each proposed solution for windows are shown in Fig. 70.4.

According to Frota and Schiffer (2001), the Centre Scientifique et Technique du Bâtiment (CSTB) (1958) method made it possible to simulate and evaluate the interaction of building systems under the direct influence of local climatic variables

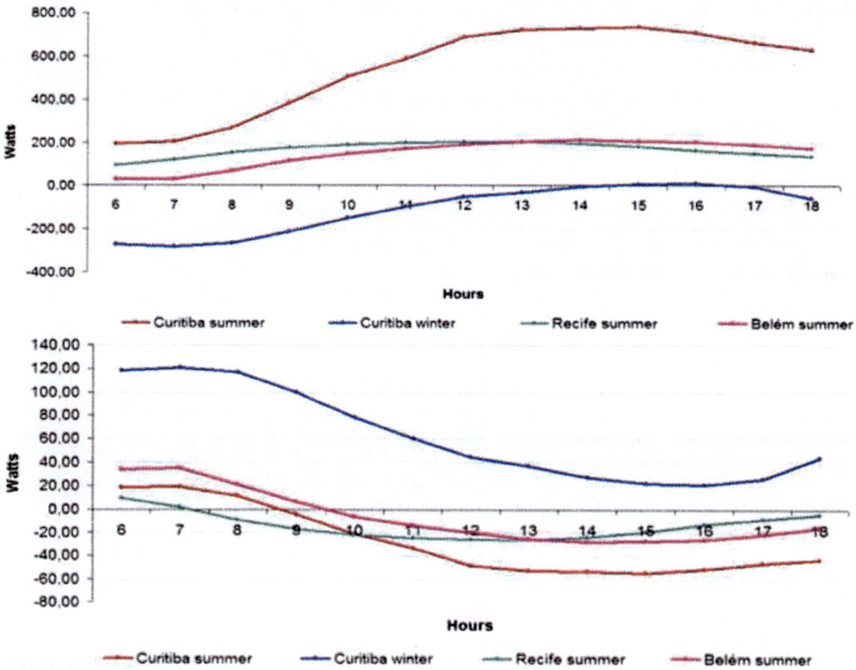

Fig. 70.3 Energy flow and heat transfer – domains (1) and (2) (by authors, 2022)

Constructive Solution	Sustainable profile	IPE KW.h/m²	GWP kgeqCO₂/m²	Performance Environmental Indicator I_E	$D_{n,w}$ dB	EP m	U W/m².°C	Performance Functional Indicator I_F	CC R$/m²	Performance Indicator Economic I_{EC}	Sustainable note SN
 Window ref		0.70	0.30	1.00	0.00	0.00	0.00	0.00	1.00	1.00	0.50
 Window 1	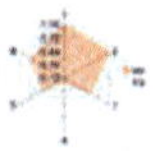	0.70	0.30	1.00	0.17	0.00	0.14	0.31	0.71	0.71	0.60
 Window 2		0.00	0.00	0.00	0.33	0.00	0.33	0.66	0.00	0.00	0.33

Fig. 70.4 Summary of results, sustainable note – window (by authors, 2022)

and the interaction of their residents. The thermal behavior of the construction system is established by NBR 15220-3 (2005). The values of these thermal gains as a function of solar radiation (I_g) were determined for a continuous period of 13 h a day. These values relate to each fence orientation. The adjustments for calculating these gains consider the thermal areas of opaque (Q_{op}) and transparent (Q_{tr}) surfaces, according to Eqs. 70.10 and 70.11:

$$Q_{op} = A_{op}. \propto .U.R_{se}.\Delta t \tag{70.10}$$

$$Q_{tr} = A_{tr}.F_{st.}\Delta t \tag{70.11}$$

The thermal gains generated as a function of temperature variation (Δt) consider the thermal areas of the opaque surfaces (Q_{op}) of the transparent surfaces (Q_{tr}), according to Eqs. 70.12 and 70.13:

$$Q_{op} = A_{op}. \propto .U.\Delta t \tag{70.12}$$

$$Q_{tr} = A_{tr}.\Delta t \tag{70.13}$$

The heat gains generated internally from the occupation and interaction of five people living in light/moderate activity and domestic equipment were considered. The relative wind load (Q_{win}) for heat removal is related to the thermal areas of opaque surfaces (10) and transparent surfaces (11):

$$Q_{wind} = 0.35.\Delta t \tag{70.14}$$

In situations where the removal of heat from the internal environment by ventilation is unfavorable, the relative wind load will be disregarded.

70.3 Result and Discussion

The combined action of the thermal properties of the sealing elements (opaque and transparent) made it possible to achieve 53.65% of thermal comfort, maintaining control of thermal fluctuations resulting from solar radiation and external temperatures, in the internal environment, as shown in Fig. 70.5. It is worth noting the limits as for natural air-conditioning, as demonstrated in Curitiba (winter) due to the relative severity of low temperatures and the reduced thermal gain by radiation.

Figure 70.6 presents a summary of the results obtained comparing the climatic characteristics of each region evaluated.

70.4 Conclusion

As for the objective proposed in this work, energy efficiency by natural/passive means was satisfactorily achieved. In hypothesis, these results focus on the reduction of consumption in 20% of the demand for electric energy destined to the

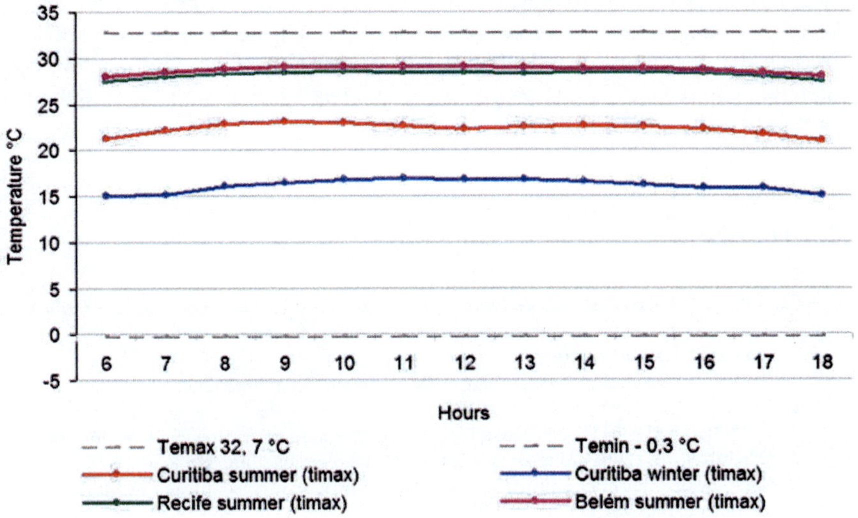

Fig. 70.5 Maximum internal temperature performance – t_{imax} (by authors, 2022)

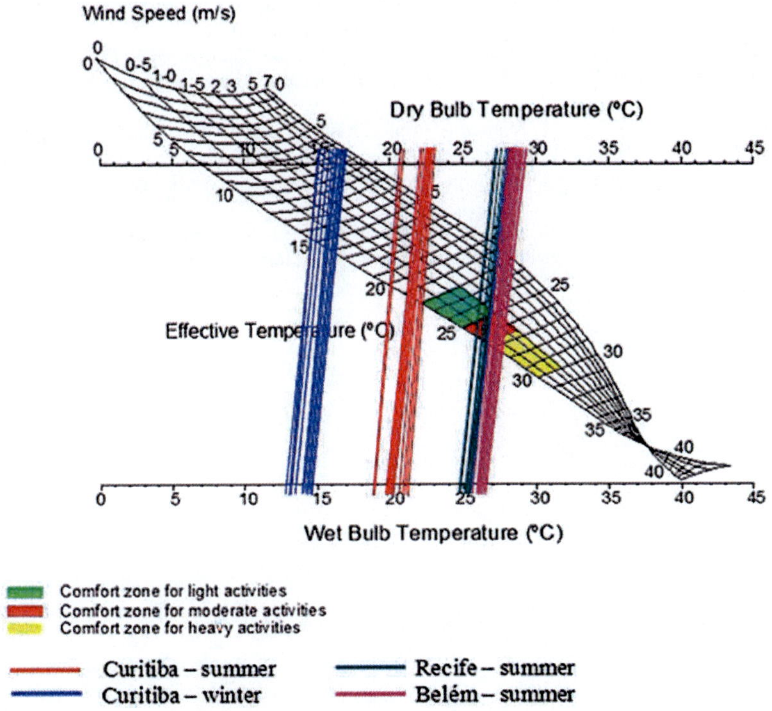

Fig. 70.6 Nomogram of effective temperature (by authors, 2022)

climatization and in approximately 50% of the 24% of the demand for energy destined to the heating of water, in Brazil. The contribution of this work is to envision the possibility of reducing the demand for electricity and meeting consumption, combining thermal properties of building systems with the resources of climate diversity.

References

ABNT (2005) Associação Nacional de Normas Técnicas. NBR 15.220-3 Desempenho térmico de edificações, Zoneamento bioclimático brasileiro e diretrizes construtivas para habitações de interesse social, Rio de Janeiro.

CSTB (1958) Centre Scientifique et Technique du Bâtiment. Hygrothermique et ventilation (D5). Paris

Evans J M, Shiller, S (1986) Diseño bioambiental y arquitectura solar, Buenos Aires, Eudeba

Frota A B, Schiffer S R (2001) Manual de conforto térmico: arquitetura, São Paulo, Studio Nobel. http://professor.pucgoias.edu.br/SiteDocente/admin/arquivosUpload/18350/material/ManualConfortoTERMICO.pdf

Givoni B (1963) Estimation of the effect of climate on man, development of a new thermal index. Haifa, Building Research Station

INMETRO (2010) Instituto Nacional de Metrologia Normalização e Qualidade Industrial. Programa Brasileiro de Etiquetagem (PBE) https://www.gov.br/inmetro/pt-br/assuntos/avaliacao-da-conformidade/programa-brasileiro-de-etiquetagem

Ismail K A R (2010) Janelas térmicas: modelagem e aplicação. Campinas, SP: Ed. autor.

Koenigsberger O H, Mahoney C, Evans J M (1971) Climate and house design. New York, United Nations

Koenigsberger O H et al (1974) Manual of tropical housing and building. London, Longmans

Mateus R, Bragança L (2004) Avaliação da sustentabilidade da construção: desenvolvimento de uma metodologia para a avaliação a sustentabilidade de soluções construtivashttps://hdl.handle.net/1822/7333

Olgyay V (1963) Design with climate. Bioclimate approach to architectural regionalism. New Jersey, Princeton University Press. https://doi.org/10.4236/ce.2018.912135

Chapter 71
Energy Retrofitting of a Restaurant Under Continental Climate Using TRNSYS Energy Simulation Tool

G. Uslu, H. U. Helvaci, and G. Gokcen Akkurt

Nomenclature

A	Area (m^2)
G	Total solar energy transmittance (–)
Q	Energy (kJ)
U	Overall heat transfer coefficient (W/m^2K)
WWR	Window-to-wall ratio (%)

71.1 Introduction

Buildings still have a significant share in total energy consumption worldwide. This causes a significant usage of fossil fuels that leads to increasing CO2 emissions into the atmosphere (Lebied et al. 2018). Therefore, it is important to note that buildings with low-energy consumptions are mandatory to reduce carbon dioxide (CO2) emissions (Basarir et al. 2012). Appropriate retrofitting strategies where the outdoor climate condition is taken into account can help reduce the energy consumption of buildings (Shandilya et al. 2020). For instance, thermal insulation is a commonly

G. Uslu (✉)
Izmir Institute of Technology, Izmir, Turkey
e-mail: gokceuslu@iyte.edu.tr

H. U. Helvaci
Department of Mechanical Engineering, Doğuş University, Istanbul, Turkey
e-mail: hhelvaci@dogus.edu.tr

G. G. Akkurt
Department of Energy Systems Engineering, Izmir Institute of Technology, Izmir, Turkey
e-mail: guldengokcen@iyte.edu.tr

M. Z. Sogut et al. (eds.), *Proceedings of the 2022 International Symposium on Energy Management and Sustainability*, Springer Proceedings in Energy,
https://doi.org/10.1007/978-3-031-30171-1_71

applied retrofitting measure for reducing the total energy consumption of buildings, and the thickness of the insulation layer is a significant parameter that affects the thermal condition inside a building (Kisilewicz 2019). With reference to the results of several investigations (Bojic et al. 2001; Liu et al. 2020; Zogou and Stamatelos 2011), increasing thermal insulation thickness could lead to a significant drop in the heating energy consumption of buildings.

However, it either has no effect or causes a slight increase in the cooling energy consumption in summer. Window-to-wall ratio (WWR) which is the ratio of the window area to the wall area determines the amount of solar gain of the building. Several studies have focused on the connection between window-to-wall ratio and building energy consumption (Su and Zhang 2010; Djamel and Noureddine 2017; Shaeri et al. 2019). These researchers conclude that an increase in the WWR could result in a decrease in the heating energy consumption and an increase in the cooling energy consumption. Another common and simple measure for decreasing the cooling energy consumption during summer is nighttime ventilation which is based on the circulation of outdoor air in the building (Zogou and Stamatelos 2011). The overall heat transfer coefficient (U-value) and the total solar energy transmittance (g-value) are the important parameters of window panes that affect the rate of heat transfer through a structure and portion of solar radiation entering the building, respectively (Bienvenido-Huertas et al. 2018; Moghaddam et al. 2021). In cold climates where the heating energy consumption is more pronounced, windows with a low U value and high g value can be utilized. On the other hand, using windows with low g value could help decrease the cooling energy use in hot climates (Sarihi et al. 2021).

The aim of this study is to apply and evaluate the effect of various retrofitting approaches mentioned above on total energy demand of a building in a cold climate. In this regard, a restaurant located in Ottawa, Canada, is modeled and simulated using the TRNSYS simulation tool. The impact of each retrofitting options on cooling and heating energy demand is investigated and the results are compared with the reference case.

71.2 Methodology

71.2.1 Thermal Energy Model

The thermal simulation model comprises building energy balances for each zone. In this balanced model, the same temperature, humidity, and other properties are considered for any air location in the zone. The energy balance is expressed as given in Eq. 71.1 (Klein 1988).

$$\frac{DQ_{air}}{dT} = \dot{Q}_{heat} - \dot{Q}_{cool} + \dot{Q}_{vent} + \dot{Q}_{inf} + \dot{Q}_{trans} + \dot{Q}_{gain} + \dot{Q}_{sol}\ [\mathrm{kJ/h}] \tag{71.1}$$

71.2.2 Building Description

The building investigated in this study is a restaurant consisting of three zones: a dining area, a storage, and a kitchen. The total exterior and interior wall areas of the restaurant are 205 m^2 and 67.5 m^2, respectively. While the external walls have an overall heat transfer coefficient as 0.501 W/m2K, consist of gypsum, insulation, stucco from inside to outside; internal walls, which have an overall heat transfer coefficient as 1.386 W/m^2K, consist of gypsum, wood, gypsum. The roof and floor areas are 225 m2 and include plastboard, airspace, insulation, concrete, roofing for roof; stone, insulating, concrete for floor, respectively. The overall heat transfer coefficients for roof and floor are 0.452 W/m^2K and 0.497 W/m^2K, respectively. The layout of the restaurant is given in Fig. 71.1.

There is a 10 m^2 double glazing window in the dining room. Heat gains are observed because of people, lighting, and stoves in the kitchen. The lighting schedule is 07:00–22:00. The frequency of customers in the restaurant varies depending on the weekend and weekdays. For the temperature for heating, the dining room and kitchen are set to 20 °C when the restaurant is occupied and 15 °C when it is unoccupied. In addition, cooling is applied in the kitchen when the temperature is above 25 °C. The infiltration rate of the restaurant is assumed as 0.5 ACH for occupied and unoccupied periods.

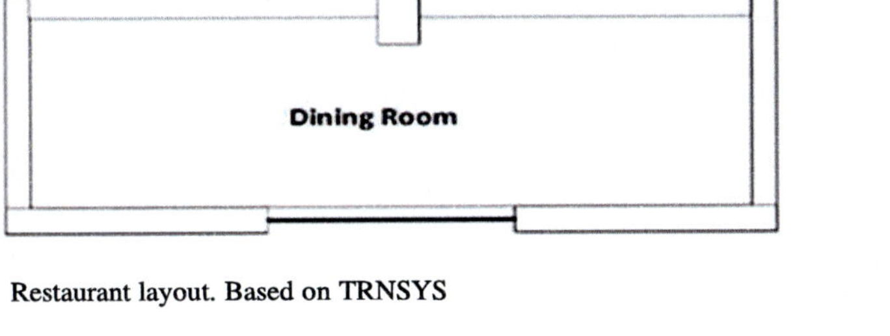

Fig. 71.1 Restaurant layout. Based on TRNSYS

71.2.3 Simulation Process

The simulations of the building model are carried out using TRNSYS 16 software to predict the thermal characteristics of the restaurant in terms of heating and cooling energy consumptions. The schematic diagram of the TRNSYS simulation model is presented in Fig. 71.2.

Type 56 is a building model component of TRNSYS where the geometry and the characteristics of the building envelope components such as walls, roof, and floor are inserted. Additionally, heat gains from people and lights and heating and cooling set point temperatures are defined in Type 56. For the dining room, people and lighting are determined to calculate the internal gains of the restaurant, whereas for the kitchen, people, lighting, and stoves are determined.

The details of the operating conditions of the restaurant is shown in Table 71.1.

Type 109 of TRNSYS is used to obtain weather data files in the standard typical meteorological year (TMY2) format. Hourly values of dry bulb temperature, relative humidity, wind direction and speed, and total and direct solar horizontal radiation for 1 year (8760 h) for the considered city are transferred to Type 56. In addition to this, Type 33e and Type 69b are used to calculate the psychrometric properties of moist air and effective sky temperature, respectively. Type 25c and Type 65d are used to print and plot the simulation results. The simulation period of this study is 1 year, from January 1 to December 31.

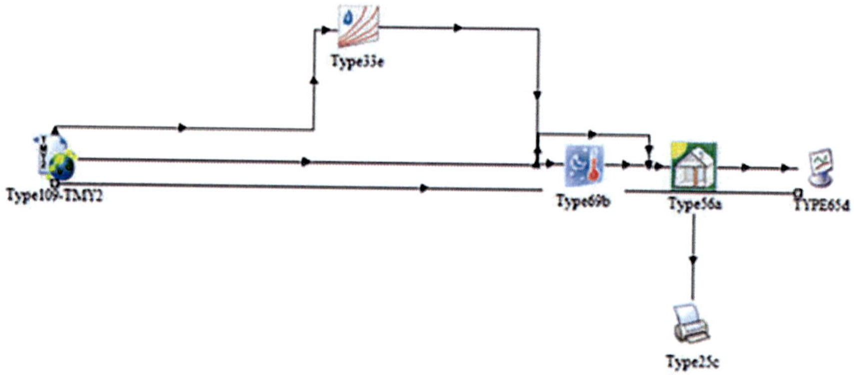

Fig. 71.2 Schematic diagram of the TRNSYS model

Table 71.1 Occupancy schedule and set temperatures

Time	Occupancy	Heating set temperature (°C)	Cooling set temperature (°C)
00:00–07:00	0	5*occupancy+15	26
07:00–22:00	1	5*occupancy+15	26
22:00–24:00	0	5*occupancy+15	26

Table 71.2 Applied retrofit scenarios

Retrofit scenarios	Retrofitting parameters	
Insulation	0.228 m, 0.38 m	
WWR	30%, 40%, 50%	
Window type	Solar control	U = 1.26 W/m2K g = 0.368
	Low-e	U = 1.24 W/m2K g = 0.642
Nighttime ventilation	4 ACH June 15–September 15 From 21:00 to 08:00	

71.2.4 *Retrofit Strategies*

In this study, various retrofitting scenarios such as insulation, WWR, window type, and nighttime ventilation are applied to decrease the cooling and heating energy requirements of the restaurant for the considered city. The data for retrofit actions are summarized in Table 71.2.

71.2.5 *Climatic Conditions*

In order to investigate the effect of each retrofitting on the continental climate, Ottawa, Canada, is considered as a case study. Ottawa has a continental climate that is warm in summer and very cold in winter; therefore, the total energy requirement is dominated by heating energy consumption. The need for heating and cooling was predicted as 12,960 kWh and 7320 kWh for the reference case. Reference case corresponds to the thermal simulation of the restaurant without any retrofitting applied.

71.3 Results and Discussion

Transient simulation results of each retrofit scenario along with the reference case are presented and discussed in this section. The heating and cooling energy requirements represent the summation of each zone, namely, the kitchen, the dining room, and the storage. Figure 71.3 illustrates the effect of increasing insulation thickness on energy requirement. It can be seen that the total heating requirement decreased from 12,960 kWh to 2388 kWh as the insulation thickness increased from 0.076 m to 0.228 m. However, the increase in insulation thickness led to an increase in the cooling energy requirement from 7320 kWh to 14,030 kWh.

This could be explained by the fact that an increased insulation layer reduces the cooling rate of the external walls, leading to a higher wall temperature (Kolaitis et al. 2013). Similar findings were reported by Aste et al. (2009) and Bojic et al. (2001).

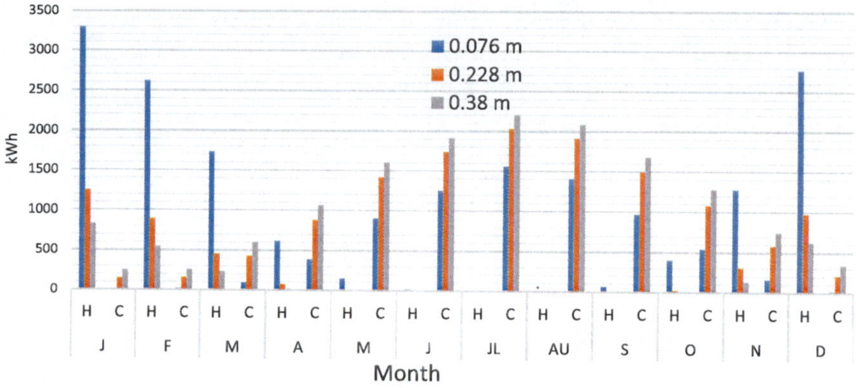

Fig. 71.3 Effect of insulation thickness on heating and cooling energy requirement

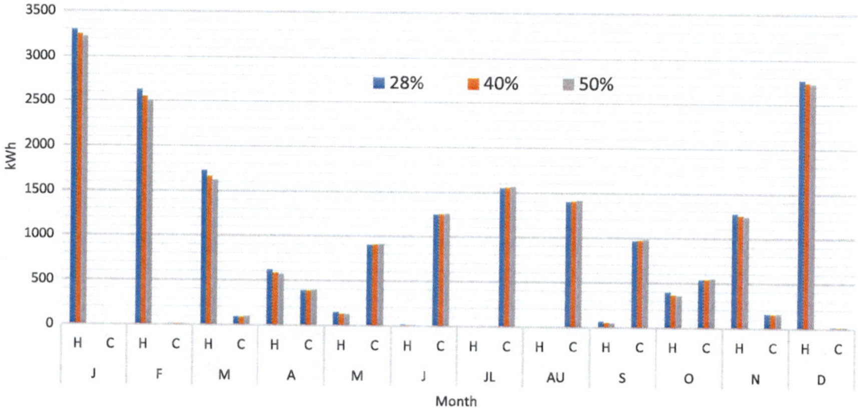

Fig. 71.4 Effect of WWR on heating and cooling energy requirement

The effects of WWR on the cooling and heating energy demand are illustrated in Fig. 71.4. It can be seen that increasing WWR with the reference window technology decreases the annual heating demand by 3.9%, yet it increases the cooling demand by 1.2%. The decrease in the heating energy demand is due to the increased amount of solar heat entering directly from the glass into the building. However, the greater window area also causes overheating of the building, especially in summer. This increase could be eliminated by introducing the overhangs or shading mechanisms (Zogou and Stamatelos 2011; Shaeri et al. 2019).

Figure 71.5 shows the impact of window type on the total heating and cooling energy demand. The low-emissivity (low-e) window panel decreased the heating requirement from 12,960 kWh to 12,640 kWh, yet the annual cooling demand rose from 7320 kWh to 7349 kWh. This is because the low-e panel has a lower U value that reduces the heat transmission between the restaurant and the environment.

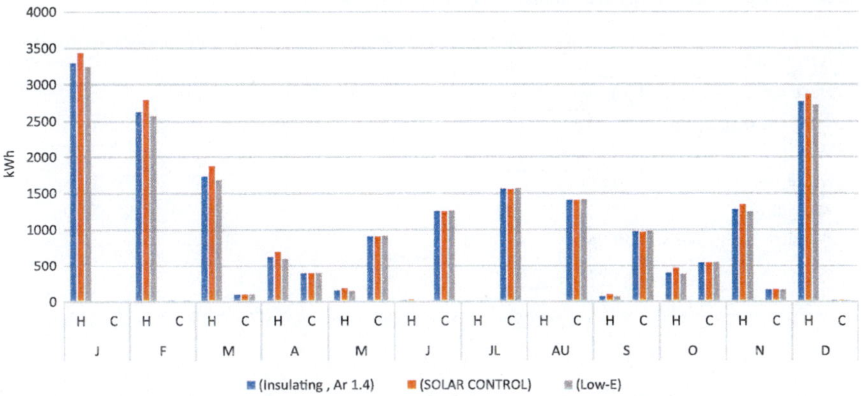

Fig. 71.5 Effect of window type on heating and cooling energy requirement

Fig. 71.6 Effect of nighttime ventilation on cooling energy requirement

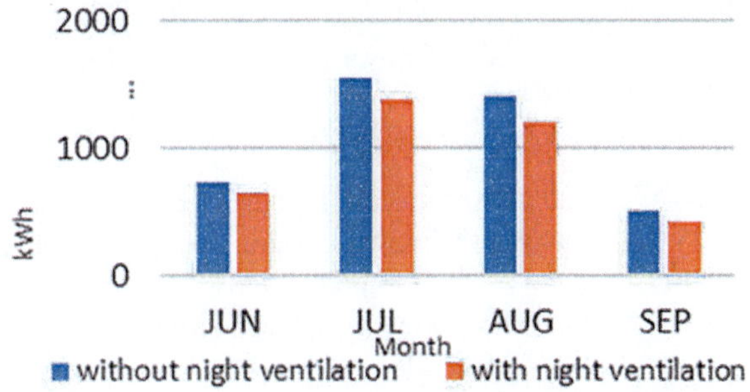

Therefore, this results in a smaller amount of heat loss. In addition, having a higher g value allows a higher portion of solar radiation to enter the building through the windows, reducing heating energy demand (Moghaddam et al. 2021). On the other hand, the solar control panel decreased the cooling requirement from 7320 kWh to 7246 kWh, while the heating requirement increased from 12,960 kWh to 13,720 kWh. Due to having a lower solar gain factor, solar control panels lead to a decrease in the cooling energy requirement (Zogou and Stamatelos 2011).

The effect of nighttime ventilation on the cooling energy demand is investigated and represented in Fig. 71.6. It can be seen that nighttime ventilation decreased the total cooling energy requirement for summer months by 17.9%. It is worth noting that nighttime ventilation requires a control strategy to prevent overcooling (Kolokotroni and Aronis 1999).

Finally, to determine the combined effect of thermal insulation, WWR, nighttime ventilation, and low-e window retrofit options, another simulation is performed, and the results are shown in Table 71.3 with single retrofit scenarios. The results show that the combined retrofitting scenario caused a 25.5% decrease in total energy requirement.

Table 71.3 Total energy requirements for all retrofit scenarios

Retrofitting	Total annual energy requirement (kWh)	Change in energy requirement (%)
Base case	20.280	–
Thermal insulation	16.418	−19.0
WWR	19.859	−2.1
Nighttime ventilation	19.809	−2.3
Window technology	20.966	3.4
	19.989	−1.4
Combination of all	15.110.9	−25.5

71.4 Conclusions

In this study, the effects of various retrofit scenarios, which are insulation thickness, WWR, window type, and night time ventilation, on the heating and cooling needs of a restaurant located in Ottawa were examined. Furthermore, the combined impact of all retrofit scenarios was also determined. As a result of the simulations, the following conclusions can be drawn:

- While insulation thickness is increased, the heating energy requirement significantly decreased, while the cooling energy need is increased. This increase can be offset by the additional precautions such as shading and nighttime ventilation. Nighttime ventilation provides an advantage in the summer months to reduce the cooling demand as a passive cooling technology.
- WWR retrofitting fairly decreases the heating energy requirement in the countries with high heating need in winter. On the other hand, cooling energy requirement slightly increases.
- Window type is the other option for improving the heating-cooling energy requirement. Heating energy need is decreased by low-e windows with low thermal transmissivity value.
- Finally, it can be suitable to consider all mentioned retrofits to balance increasing heating or cooling need obtained from single retrofit. For a feasible implementation of each retrofit options, economic analysis should be conducted.

References

Aste N, Angelotti A, Buzzetti M (2009) The influence of the external walls thermal inertia on the energy performance of well insulated buildings. Energy and buildings 41:1181–1187 https://doi.org/10.1016/j.enbuild.2009.06.005.

Basarir B, Diri BS, Diri C (2012) Energy efficient retrofit methods at the building envelopes of the school buildings. Retrieved 10:2016.

Bienvenido-Huertas D, Rodríguez-Álvaro R, Moyano JJ, et al. (2018) Determining the U-value of façades using the thermometric method: Potentials and limitations. Energies 11:360 https://doi.org/10.3390/en11020360.

Bojic M, Yik F, Sat P (2001) Influence of thermal insulation position in building envelope on the space cooling of high-rise residential buildings in Hong Kong. Energy and Buildings 33:569–581 DOI:https://doi.org/10.1016/S0378-7788(00)00125-0.

Djamel Z, Noureddine Z (2017) The impact of window configuration on the overall building energy consumption under specific climate conditions. Energy Procedia 115:162–172 https://doi.org/10.1016/j.egypro.2017.05.016.

Kisilewicz T (2019) On the role of external walls in the reduction of energy demand and the mitigation of human thermal discomfort. Sustainability 11:1061 https://doi.org/10.3390/su11041061.

Klein SA (1988) TRNSYS-A transient system simulation program. University of Wisconsin-Madison, Engineering Experiment Station Report 12–38.

Kolaitis DI, Malliotakis E, Kontogeorgos DA, et al. (2013) Comparative assessment of internal and external thermal insulation systems for energy efficient retrofitting of residential buildings. Energy and buildings 64:123–131 DOI: https://doi.org/10.1016/j.enbuild.2013.04.004.

Kolokotroni M, Aronis A (1999) Cooling-energy reduction in air-conditioned offices by using night ventilation. Applied Energy, Elsevier, vol. 63(4), pages 241–253, August. www.elsevier.com/locate/apenergy.

Lebied M, Sick F, Choulli Z, El Bouardi A (2018) Improving the passive building energy efficiency through numerical simulation–A case study for Tetouan climate in northern of Morocco. Case studies in thermal engineering 11:125–134 DOI:https://doi.org/10.1016/j.csite.2018.01.007.

Liu X, Chen X, Shahrestani M (2020) Optimization of insulation thickness of external walls of residential buildings in hot summer and cold winter zone of China. Sustainability 12:1574 DOI: https://doi.org/10.3390/su12041574.

Moghaddam SA, Mattsson M, Ameen A, et al. (2021) Low-Emissivity Window Films as an Energy Retrofit Option for a Historical Stone Building in Cold Climate. Energies 14:7584 https://doi.org/10.3390/en14227584.

Sarihi S, Saradj FM, Faizi M (2021) A critical review of façade retrofit measures for minimizing heating and cooling demand in existing buildings. Sustainable Cities and Society 64: 102525 DOI: https://doi.org/10.1016/j.scs.2020.102525.

Shaeri J, Habibi A, Yaghoubi M, Chokhachian A (2019) The optimum window-to-wall ratio in office buildings for hot–humid, hot–dry, and cold climates in Iran. Environments 6:45 https://doi.org/10.3390/environments6040045.

Shandilya A, Hauer M, Streicher W (2020) Optimization of thermal behavior and energy efficiency of a residential house using energy retrofitting in different climates. Civil Engineering and Architecture 8:335–349 DOI: https://doi.org/10.13189/cea.2020.080318.

Su X, Zhang X (2010) Environmental performance optimization of window–wall ratio for different window type in hot summer and cold winter zone in China based on life cycle assessment. Energy and buildings 42:198–202 DOI: https://doi.org/10.1016/j.enbuild.2009.08.015

Zogou O, Stamatelos A (2011) Application of building energy simulation in the sizing and design optimization of an office building and its HVAC equipment.

Chapter 72
An Energy Analysis of a New Biomass Gasification Integrated Geothermal System Design

Utku Seker, Muhammed H. Taheri, Gulden G. Akkurt, and Mousa Mohammadpourfard

Nomenclature

h	Enthalpy, *kJ/kg*
m	Mass flow rate, *kg*
Q	Heat, *kW*
W	Work, *kW*

72.1 Introduction

Growth in energy consumption is strongly linked to the increase in the world population, economic development, and human well-being. Because of the increasing energy demand and greenhouse gas emission, and global warming, renewable energy resources as well as novel energy conversion systems are getting more

U. Seker (✉)
Izmir Institute of Technology, Energy Engineering Programme, Izmir, Turkey
e-mail: utku.seker@tr.bosch.com

M. H. Taheri
Department of Mechanical Engineering, University of Tabriz, Tabriz, Iran
e-mail: mhaditaheri@tabrizu.ac.ir

G. G. Akkurt
Department of Energy Systems Engineering, Izmir Institute of Technology, Izmir, Turkey
e-mail: guldengokcen@iyte.edu.tr

M. Mohammadpourfard
Faculty of Chemical and Petroleum Engineering, University of Tabriz, Tabriz, Iran
e-mail: mohammadpour@tabrizu.ac.ir

M. Z. Sogut et al. (eds.), *Proceedings of the 2022 International Symposium on Energy Management and Sustainability*, Springer Proceedings in Energy,
https://doi.org/10.1007/978-3-031-30171-1_72

attention as global energy supplies. The novel processes for energy production must be more efficient, cost-beneficial, and environmentally benign. Biomass is recently receiving significant attention as a substitute to fossil fuels because of its lower cost in comparison with other energy sources, noticeable availability, and near-zero carbon or sulfur dioxide emissions. This energy source is currently recognized as the fourth source of energy in the world which can be gasified into different forms of syngas (or biogas). Biomass gasification is one of the thermochemical conversion process technology pathways that uses a controlled process involving heat, steam, and oxygen to convert biomass to hydrogen and other products, without combustion (Toklu 2017). Besides, geothermal resources have the highest availability since they are not dependent on weather conditions, and conversion technologies are available that allow electricity generation from geothermal fluids at low temperatures with a history of more than 100 years. It is obviously advantageous to implement these resources as a unit system to provide benefit from each resource along with the cogeneration benefits.

This study is aimed to show possible benefits of combining a biomass gasification with a geothermal power plant which operates on an organic Rankine cycle (ORC). Since both energy resources are renewable, this combined system offers low carbon solution compared to using oil or fossil fuels. Biomass gasification provides additional heat to the geothermal cycle and a better match of the heat range is possible. The present study highlights energy analysis of two different designs which are increasing geothermal fluid temperature and increasing working fluid temperature of the ORC by combustion gas leaving the biomass gasification-gas turbine cycle. Two different working fluids are used to compare the thermodynamic performance of both designs.

72.2 Material and Methods

72.2.1 System Description

The two different designs for a cogeneration system (Fig. 72.1) consist of a biomass gasification-gas turbine cycle and an ORC. Biomass and geothermal fluid from a well are the primary energy inputs. The main outputs of the proposed system are power and heat. Biomass enters into a gasifier with a flow rate of 5 kg/s. One of the streams of biomass gasification process is pressurized air which is compressed in an air compressor and directed to the combustion chamber. The final product of the process is synthesis gas (syngas) and needs to be cleaned at a cleaner. Synthesis gas and pressurized air are mixed at a combustion chamber. The combustion products pass through the gas turbine to generate power. In this study, two different designs are presented for the combustion gas that leaves the gas turbine.

In the first design, the combustion gas is used to increase geothermal fluid temperature through a heat exchanger (Fig. 72.1a), while working fluid temperature of the ORC is increased in the second design (Fig. 72.1b). In Design I, exhaust gases

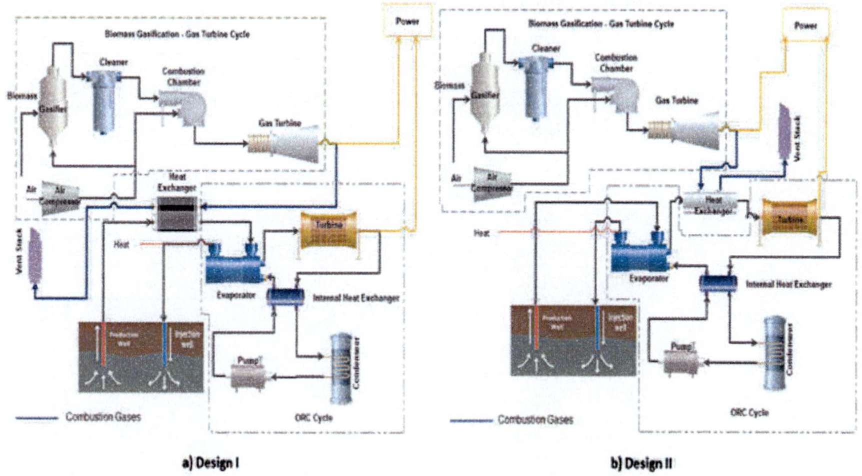

Fig. 72.1 System design

from the gas turbine enter the heat changer to increase geothermal fluid temperature. As a result, geothermal fluid transfers higher amount of heat to the working fluid of the ORC in the evaporator and more power is generated in the turbine of the ORC (Srinivas et al. 2014). This design is analyzed for two different working fluid: R123 and n-pentane (Fig. 72.1a).

In Design II, the combustion gas leaving the gas turbine is directed to the ORC instead of geothermal fluid. In this way, working fluid that leaves the evaporator is sent to a heat exchanger where the combustion gas transfers its heat to the working fluid. This also increases turbine output of the ORC. Again, R123 and n-pentane are used to evaluate the cycle performance and compare with the first design.

72.2.2 *Methods*

Energy analysis is conducted for both designs and two different working fluids by EES software (F-chart 2022). Then, the results are compared based on generated power, mass flow rate of the working fluid, and remaining available combustion gas temperature. The assumptions made for the calculations are the following:

1. All the processes are assumed to work under steady-state condition.
2. Combustion products and air are considered as ideal gases.
3. Pressure drops throughout the heat exchangers and pipelines are neglected.
4. The turbines and pumps have isentropic efficiencies.
5. The kinetic and potential energy changes are negligible.

The data used in the analyses are given in Table 72.1.

Table 72.1 Data for system analysis

Parameter	Value
Temperature (°C)	25
Pressure (kPa)	100
Geothermal fluid injection temperature (°C)	164.7
Biomass integrated gasification cycle	
Air mass flow rate (kg/s)	74.07
Pressure ratio (–)	12
Combustion gas flow rate (kg/s)	80.85
Combustion gas temperature (°C)	455.6
Binary cycle	
Pump inlet temperature (°C)	40
Pressure drop ratio (–)	7.7
Turbine efficiency (%)	85
Pump efficiency (%)	90
Compressor efficiency (%)	85

Table 72.2 Heat exchange data for Design I

	Heat exchanger inlet		Heat exchanger exit	
Streams	Temperature (°C)	Flow rate (kg/s)	Temperature (°C)	Flow rate (kg/s)
Geothermal fluid	168.2	450	188.7	450
Combustion gases	455.6	80.85	211.3	80.85

Mass and energy balance equations for calculations are given in Eqs. 72.1 and 72.2 (Yari 2010).

$$\Sigma m_{in} = \Sigma m_{out} \tag{72.1}$$

$$Q - W = \Sigma m_{out} h_{out} - \Sigma m_{in} h_{in} \tag{72.2}$$

72.3 Results and Discussion

An energy analysis is conducted for a new cogeneration system proposed in this study.

In Design I, geothermal fluid temperature increased from 168.2 °C to 188.7 °C, while combustion gas temperature decreased from 455.6 °C to 211.3 °C in the heat exchanger (Table 72.2). Injection temperature of the geothermal fluid is kept at 164.7 °C. Figure 72.2 exhibits the change in generated power of the ORC and working fluid flow rate with temperature difference between the evaporator and turbine for Design I. Although generated power is almost the same, the required mass flow rate of n-pentane is 50% lower than R123.

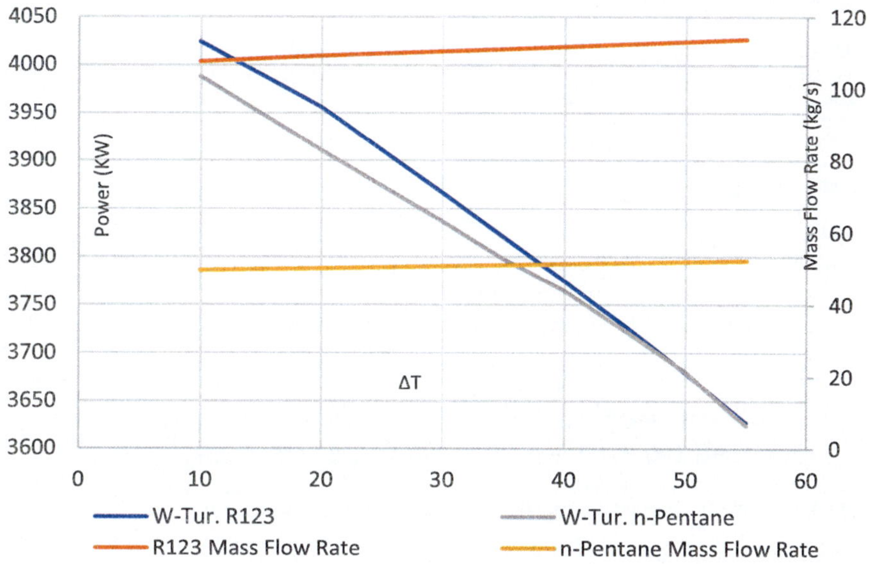

Fig. 72.2 Power output and working fluid mass flow rate change with ΔT (between evaporator and turbine) for Design I

Figure 72.3 exhibits the power output of ORC and working fluid flow rate change with temperature difference between the evaporator and heat exchanger for Design II. While R123 delivered 8.5% higher work output compared with that of n-pentane, mass flow rate requirement is 50% lower for n-pentane.

It should be noted that combustion gas temperature leaving the biomass gasification-gas turbine cycle is constant at 211.3 °C for Design I due to the stable input condition of geothermal fluid temperature. However, in Design II, the combustion gas temperature leaving the biomass gasification-gas turbine cycle changes depending on assumed temperature difference between the evaporator and heat exchanger. Combustion gas temperature leaving the heat exchanger for both working fluids as 173.1 °C at maximum generated power condition in Design II.

Table 72.3 compares both design options based on ORC work output, working fluid flow rate, and combustion gas temperature leaving the heat exchanger. The table indicates that increasing working fluid temperature of the ORC (Design II) instead of increasing geothermal fluid temperature (Design I) leads to an increase in the produced power which is encountered as 48% for R123 and 37% for n-pentane. Consider that the remaining available temperatures of combustion gases are 211.3 ° C and 173.1 °C for Design I and II, respectively. Design II absorbed 18% more heat from the combustion gases than Design I. Moreover, R123 shows 0.9% better performance than n-pentane based on power production. A 116% higher working fluid flow rate can be counted as a considerable drawback for R123. Higher flow rate leads to increase in possible maintenance cost of the system.

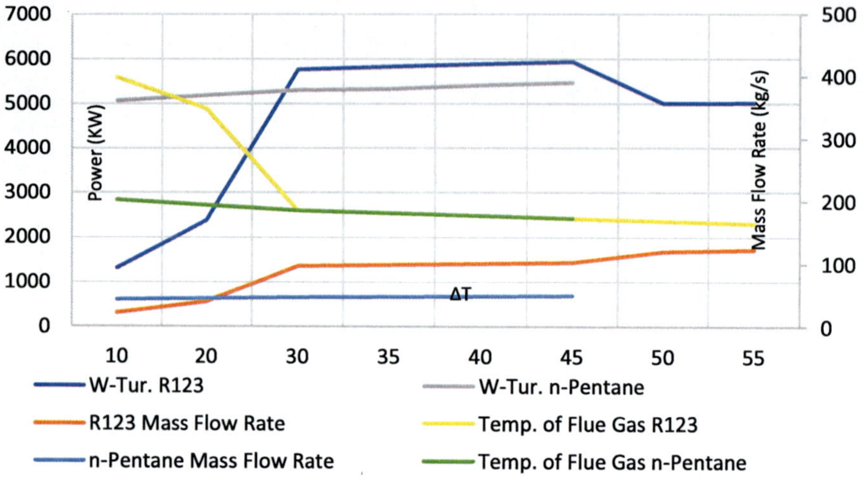

Fig. 72.3 Power output and working fluid mass flow rate change with ΔT (between evaporator and HEX) for Design II

Table 72.3 Comparison of Design I and II based on maximum produced power (kW) and working fluid flow rates

	Design I		Design II	
Parameters	R123	n-pentane	R123	n-pentane
Geothermal fluid temperature-HEX inlet (°C)	168.2	168.2	168.2	168.2
Geothermal fluid temperature-HEX exit (°C)	188.7	188.7	–	–
Geothermal fluid temperature-injection well (°C)	167.4	167.4	167.4	167.4
Combustion gases-HEX exit (°C)	211.3	211.3	173.1	173.1
Mass flow rate (kg/s)	107.5	49.6	103.4	49.3
Power output (kW)	4024	3988	5942	5474
Specific work output (kW/(kg/s))	37.4	80.4	57.5	111.0

72.4 Conclusions

A renewable energy-based biomass-geothermal cogeneration system is analyzed in this study. Supporting geothermal heat with available biomass resources is developed and demonstrated.

The overall energy analysis has been conducted using EES software (F-Chart 2022). Two different cogeneration designs and two different working fluids for the ORC based on generated power, working fluid mass flow rate, and combustion gas temperature leave the HEX.

Under given conditions, it can be concluded that Design II with n-pentane as the working fluid of ORC showed the highest performance based on produced power and flow rate. Specific work output data given in Table 72.3 indicates that n-pentane delivers 111.0 kW power for a unit flow rate, while R123 delivers only 57.5 kW. For a final decision, a feasibility study should be conducted.

References

F-Chart (2022). EES: Engineering Equation Solver I F-Chart Software: Engineering Software. Retrieved 23 February 2022, from https://fchartsoftware.com/ees/

Srinivas, S., Eisenberg, D., Seifkar, N., Leoni, P., Paci, M., & Field, R. (2014). Simulation-Based Study of a Novel Integration: Geothermal–Biomass Power Plant. Energy & Fuels, 28(12), 7632–7642. doi: https://doi.org/10.1021/ef501601b.

Toklu, E. (2017). Biomass energy potential and utilization in Turkey. Renewable Energy, 107, 235–244. doi: https://doi.org/10.1016/j.renene.2017.02.008.

Yari, M. (2010). Exergetic analysis of various types of geothermal power plants. Renewable Energy, 35(1), 112–121. doi: https://doi.org/10.1016/j.renene.2009.07.023.

Chapter 73
Drone Models in Urban Transport (New Concept Integration)

Dung D. Nguyen, Omar Alharasees, and Utku Kale

Nomenclature

UAV/UAS	Unmanned aerial vehicle/system
ATM	Air traffic management
GPS	Global Positioning System

73.1 Introduction

Currently, drones are the center of concern for development and production in air transport technology. The market for their civil use, driven by economic and social needs, is quickly expanding. On the other hand, a significant issue impedes the swift use of drones in urban uses and smart city mobility. The current air traffic management (ATM) system is unable to manage the expected and forecasted number of drones flying low in the metropolitan region between buildings in a difficult atmosphere (with, e.g., reflection), due to, e.g., (i) the restrictions in the system capability, (ii) the mandatory staff, (iii) the estimated expense, and (iv) the needed time of the system improvement.

Drones make processes more efficient and adaptable while boosting accuracy and cost-effectiveness (Kitonsa et al. 2018). Consequently, profitable-oriented drone employment is connected to a diverse range of economic potential. Drone use as a

D. D. Nguyen (✉) · O. Alharasees · U. Kale
Faculty of Aerospace Engineering, Le Quy Don Technical University, Hanoi, Vietnam

Department of Aeronautics and Naval Architecture, Budapest University of Technology and Economics, Budapest, Hungary
e-mail: dungnd@lqdtu.edu.vn; ddnguyen@vrht.bme.hu; oalharasees@edu.bme.hu

M. Z. Sogut et al. (eds.), *Proceedings of the 2022 International Symposium on Energy Management and Sustainability*, Springer Proceedings in Energy,
https://doi.org/10.1007/978-3-031-30171-1_73

mode of transportation is still in developing states. Nonetheless, delivery-based drones are now able to transport loads of up to 2–3 kg and perform aircraft operations within the urban area from a strictly technical standpoint (Joerss et al. 2016). Furthermore, passenger-based drones, sometimes known as "air taxis," have already shown their practical capabilities to transport passengers within and between cities (Kellermann et al. 2020).

Operational management concerns have begun to emerge with several studies looking into drone maintenance regimens (Martinetti et al. 2018), battery life management/charging, and cost-effective working characteristics (Goss et al. 2017; Pinto et al. 2019). Crucially, as the focus shifts to logistic operations, new concerns emerge, such as enhancing distribution techniques (El-Adle et al. 2019). According to a preliminary research (Ferrandez et al. 2016; Carlsson and Song 2018; Chung 2018; Liu et al. 2018), integrated automobile and air drone delivery procedures are more effective forms of logistics network distribution than present methodologies (Wang et al. 2019).

Given the exceedingly high number of drones and widely variable performance characteristics, traditional air traffic management (ATM) systems will be unable to provide drone services cost-effectively. Traditional ATM frameworks are designed for human-crewed aircraft. However, without a pilot on board, there will be a distinct set of administration concerns (Merkert and Bushell 2020), such as avoiding collisions, tracking trajectories, path planning, communication, and control, which are not observed in human-crewed aircraft operations.

As a result, incorporating drones into smart city transportation is a critical task requiring new, highly automated, self-driving solutions. The Department of Aeronautics and Naval Architecture at Budapest University of Technology and Economics conducted considerable research on developing new operational concepts (Rohacs and Rohacs 2016; Jankovics and Kale 2019) and the adaption of drones in smart-connected city transport systems (Nguyen et al. 2020; Nguyen and Nguyen 2021).

The potential of integrating drones into the smart city transportation system is investigated in this research. The proposed concept is built on a 3D airspace "road network." The drones' airway system is made up of typical components.

73.2 Methodologies and Materials

73.2.1 Airway Safety Rule Definitions

The authors examined and assessed numerous contemporary rules and related studies on drone safety and security to describe the airways. The statements below were written to explain the airway network features within the study's scope:

- Setting velocity limitations of 30 m/s in corridors due to the width restrictions, 20 m/s only for drones flying in preplanned designated trajectories at least 20 m from any facilities or any structure, and lastly 10 m/s specific to drones moving 20 m nearby any facility (but 5 m away).

- The suggested longitudinal spacing between drones in a preplanned designated trajectory is determined by the velocity, acceleration, deceleration, and strength of connectivity between the drones. For unconnected vehicles, the basic spacing "time" should be minimized to one second and another second for every 10 m/s of flight speed sec, which should be extended when the (i + 1) drone has a higher velocity than the (i-th) drone. The spacing time of connected drones could be reduced by 30–40% (depending on the actual air turbulence), and another 30% can reduce the spacing time of formation flight situations.
- Figures illustrating the proposed typical features of airways define the designed spacing (horizontally and vertically). As a general guideline, the horizontal and vertical spacing between the centers of gravity of a specific drone should be five to eight times their greatest measurements while flying in the same direction. If drones are flying in the other direction within the same corridor, a safe distance of the length of an empty lane might be used.
- The airways and the overall network should be formulated from the components listed above, and drones should only be able to change lanes horizontally or vertically.
- The designated channels and preplanned trajectories for the particular drone are defined and cannot intersect with any other channel.

73.2.2 Safe Airspace and Airway Network Design

Drone trajectories and, more broadly, airway networks are constructed and planned to utilize multidisciplinary and multidimensional objective optimization to reduce the total effect of drones while reducing the overall expense. Total refers to the total of all drones' impacts or expenses. Drone activities include a wide range of direct, short-term, or long-term consequences, reflections, and externalities, including the influence on nature, the built environment, health concerns caused by pollutants and mishaps, and economic repercussions. The total cost is calculated by taking into account all expenses, such as the operation, manufacture, improvement, and function of all essential assets, as well as any external expenditures incurred as a result of unwanted events.

The airway network is used in metropolitan zones, where precise location and traffic control necessitate unique regulations of accuracy and precision in a complicated environment. Partially implementing road traffic regulations and incorporating distinctive markers into urban infrastructure might lead to the development of rules. Passive, dynamic, and active approaches are used to run the airway network (like the sectorization). If safety and security issues are recognized, dedicated zones on the airway network are designated for an emergency landing.

This paper deals with the civil and commercial application of drones in urban areas – smart cities. Following trajectories/channels, the recommended urban drone traffic system and management might operate relatively autonomous vehicles in corridors and relatively small air vehicles (mostly with less than 60 kg take-off mass) following trajectories/channels. The corridors are far enough from the built environment to be able to react even when drones are unintentionally leaving the fixed corridors (e.g., due to malfunction). The smaller drones following the fixed trajectory may cause fewer damages and problems, limiting the possible unlawful actions.

73.2.3 Following Process

The rate of errors and accidents will grow and rise significantly with increasing the number of drones within the applied system. Drone-following models (one-by-one following mechanisms) in air traffic flows are necessary to analyze drone air traffic safety and the intelligent transport network. The drone-following models are built on the premise that each drone moves within the guidance of its leader. In other words, such an approach is described as a function of relative velocity or safety distance between two drones. In the case of three drones within the system in the same direction at the same time, two of them can follow the leader.

The drone's speed in the drone-following mechanisms is determined by the air traffic scenario, specifically the position of the drone prior to it and its speed, which revealed that a linear model assumes that the drone's operator stabilizes the drone's acceleration to maintain a relative sequence to the drone ahead.

The safe distance (SD) model is given as follows:

$$\ddot{X}_n(t+T) = \lambda \frac{\left[\dot{X}_n(t)\right]^p}{\left[X_{n-1}(t) - X_n(t)\right]^q} \left[\dot{X}_{n-1}(t) - \dot{X}_n(t)\right] \tag{73.1}$$

where $X_n(t + T)$ is the acceleration of n-th drone after a reaction.

$X_{n-1}(t) - X_n(t)$ is the relative distance between the (n-1)-th drone and the n-th drone, T is the delay time of a control device, λ is the weight coefficient related to the controlling device, and p and q are the parameters related to speed and position of the drone ahead.

It seems that this model is well applicable to the drones flying in the desired flight path. However, the air turbulences and wind flow separated from infrastructure cause rather stochastically disturbed motion of drones. With the types of advanced control appliances, the controller's relative position and reaction time also affect the control close-loop. This methodology indicates an improved model, called the Markov model.

The Markov model is constructed on the approximation of the stochastic estimation of decision-making speed, which shows data variation of the controller speed and in relative positions between the drones, which is explained below:

$$\ddot{X}_n[k+1] = c_v(\dot{X}_{n-1}[k] - \dot{X}_n[k]) + c_x[(X_{n-1}[k] - X_n[k]) - \Delta X_{\text{pdn}}] + \varepsilon[k] \quad (73.2)$$

where c_v and c_x are the coefficients depending on the time, given drone and controllers, $\Delta X_{\text{pdn}} = \dot{X}(t)$ is the predefined safety distance between the drones, k is the number of steps in a chain ($t = k.\ \Delta t$), and $\varepsilon[k]$ is the random value disturbing the process.

73.2.4 Obstacle Avoidance Method

Drone collisions with buildings, vehicles, and the surrounding environment are already a serious concern as drone usage grows more popular. To ensure airspace safety, drone operations require a collision avoidance system, particularly for autonomous drones' operations in crowded (urban) open airspace. For autonomous vehicles, collision avoidance and conflict detection are also useful techniques. Numerous simulation system algorithms are already used to investigate the important parameters in such an environment. One of the most important components of these systems is the obstacle model, which assumes that each obstacle is defined as a cylinder with a radius r_{Bl} and a center C_{Bl}, as illustrated in Fig. 73.1.

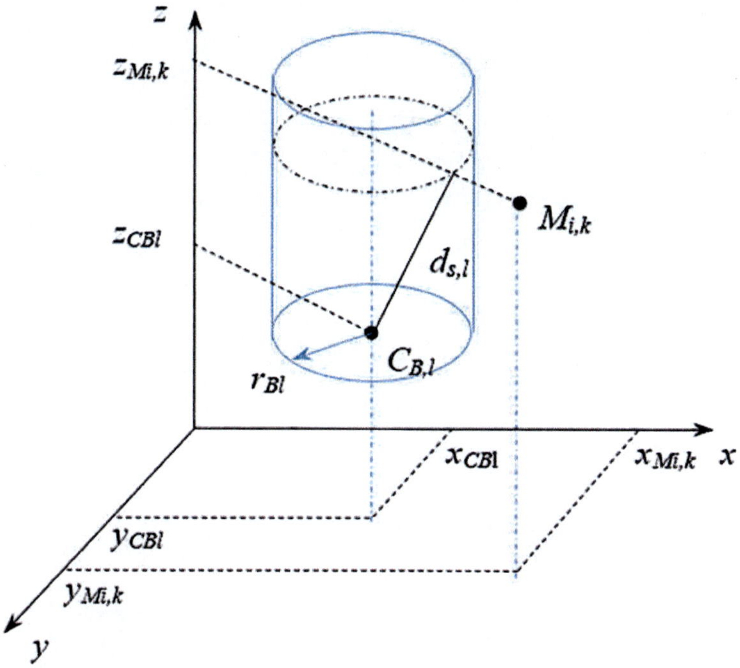

Fig. 73.1 Obstacle representation and safe distance estimation

73.2.5 Desired Landing Orbit for UAVs

The landing stage methodology is one of the most important aspects of any flying excursion in drone and UAV operations, as it allows the drones and UAVs to land at the desired location securely. The steps of general landing methods are as follows: (i) traveling against the wind, (ii) descending, and (iii) slowing down. Several elements, including wind disturbance, engine traction force, aerodynamic force, and propeller reaction moment, will impact this process.

By solving the aircraft's motion equations and using analytical approaches, methodologies for defining and analyzing landing regions are devised. The chosen landing route is predicted based on the landing regions, and the UAV subsequently lands precisely at the designated place. The UAV landing areas include the following three zones (Fig. 73.2).

Deceleration zone: this is the minimum circle on the horizontal plane covering the projection of the UAV's orbit, which flies straight with decreasing speed during the landing phase. Then, the deceleration zone's structure is a loop with center 0 and radius R_1.

Descending zone: this is the smallest circle on the horizontal plane covering the projection of the UAV's orbit, which flies in the process of elevation decline. This area is a circle with a center 0 and radius R_2.

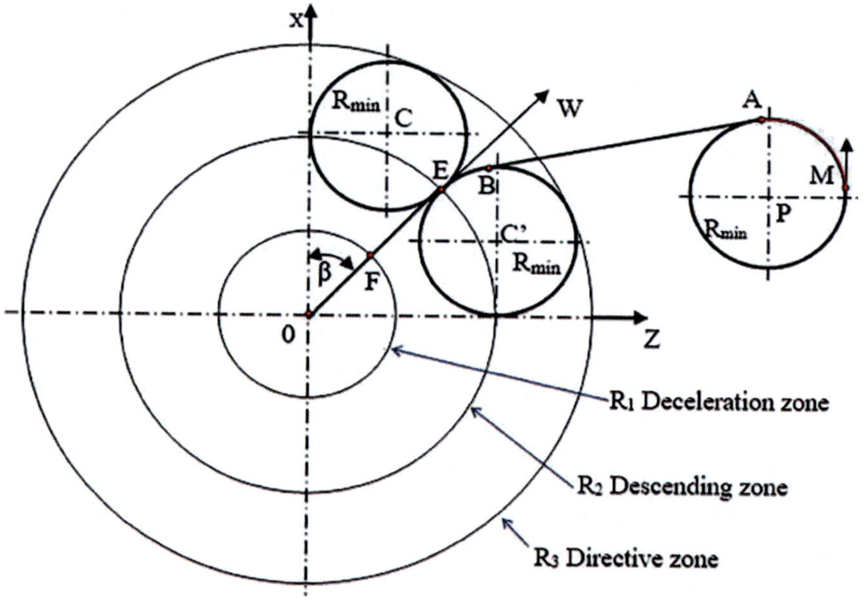

Fig. 73.2 The proposed UAV landing zones

Directive zone: this is the smallest circle in the horizontal plane, covering projections of two circles with radius R_{min}. Two circles are tangent to each other at the opposite of the wind direction.

R_{min} is the smallest rounding radius of the UAV. Thus, the directive zone is the circle with center 0 and radius R_3.

Approach, glide slope, and flare are the most common landing trajectories for UAVs. The landing trajectory chosen and implemented would determine the safety of the landing.

The directional phase, the descending phase, and the deceleration phase are the three phases of a UAV's landing operation. When the UAV enters each landing zone, the landing zones are calculated. The radius of each region will be used to establish landing zones. The approach entails solving the differential motion structure to explore the kinetic dynamics of UAVs. As a result, UAV dynamics will be utilized to estimate the deceleration zone, and then analytical methods will be employed to identify the remaining landing regions.

73.3 Results

73.3.1 Drone-Following Process in the Traffic Flow

Drone-following models for managing drones in smart cities' transportation management systems were introduced in previous studies (Dung 2020; Dung and Rohacs 2018). Such models were created on the primary concept that drones fly toward a leading drone. This approach has been a novel method for managing several drones in smart cities.

This subsection provides the main results achieved in the simulation research on the safe distance and Markov models (see Figs. 73.3 and 73.4).

It should be emphasized that there is neither an accident nor an implausible deceleration. The speed of a drone is adapted in accordance with the speed of the drone ahead of it. However, due to the variation in acceleration, the trailed drone can respond more swiftly than the leading drone.

Even though the safe distance and the Markov models are alike, the followed drone's response in the safe distance model is faster than in the Markov model. Moreover, the movement of the followed drone reveals a slower stabilized scenario

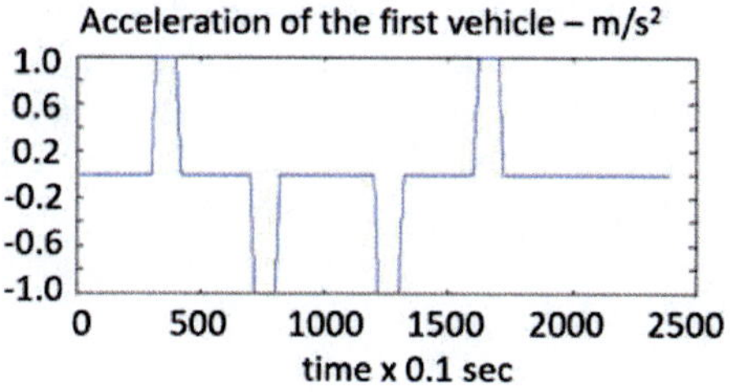

Fig. 73.3 First drone application verification tests (acceleration and deceleration)

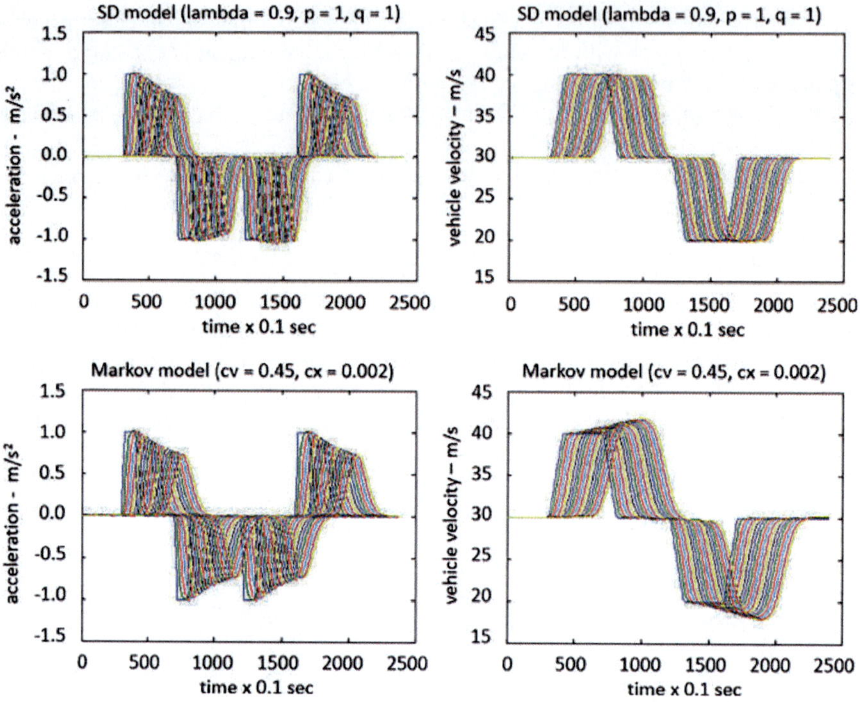

Fig. 73.4 Safe distance and Markov drone-following model results

in both models. In parallel with the safe distance model, the Markov model reflects the variations in space between consecutive drones. Additionally, the more significant the variation of the followed drone's speed in the Markov model, the lesser the spacing is between drone's sequence.

These results verified that the developed Markov model might perform the longitudinal safety separation of drones as the safe distance model. In general form, the results partly validated the proposed method that aims to adapt drones in the urban intelligent air transport system. The drone-following method needs to be verified by several immediate next steps, such as the safe distance being measured in direct sequence drones and multiple drones alongside designing and conducting an investigational experiment to accumulate quantitative data for drone operation in space.

As it seems, the developing Markov model might be more accurate in case of motion of drones in significant air turbulence and separated wind flow from the infrastructure, and it can be used in areas where problems with GPS positioning might have appeared, especially comparing and working together with the GPS techniques or acoustic sensors, etc.

73.3.2 Experiment Results of Drone Management System

In experimental studies, a cloud-based managing method had been applied to drones flying in a smart city environment (Nguyen 2021), including the physical, cloud, and control layers.

The result is illustrated in Fig. 73.5. Initially, the drone was at the start location. When a drone received the GCS order, it started flying, visited the established points, and completed the task. The results show that the chosen route and real route are correlated. The difference in the routes characterizes GPS location due to the fact that the drone receives the GPS location.

The proportional integral derivative (PID) control is completely satisfactory to follow set locations at small velocities (when the drone aerodynamics varies a little with no wind disturbance). For preliminary drones' operations, tracking was accomplished accurately, within a moderate error range.

Figure 73.6 shows that a linear controller realized the elevation flying control with the rise in the perpendicular velocity. The flight control unit (FCU) has operated exceptionally beneficial in altitude control, even with a minimal difference between the chosen and actual heights.

During conducting the experiment, the drone's video streamed downward, facing the camera, allowing the operator to supervise and manage the drone in a real-time environment.

The experimental results proved that the suggested and applied CbDMS (cloud-based drone managing system) is a cloud solution that allows drones' management and operation in a real-time environment. The observed effectiveness could be enhanced by increasing the regularity of refreshing GPS coordinates or adding filtering techniques (Kalman filters).

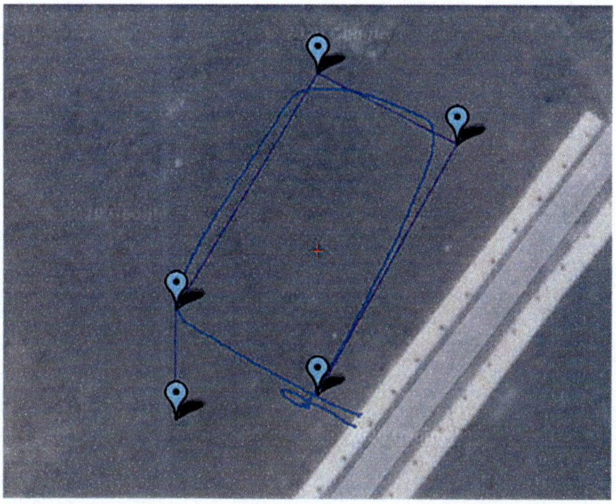

Fig. 73.5 The variation in routes – desired and real routes (pink line – desired route, blue line – real route)

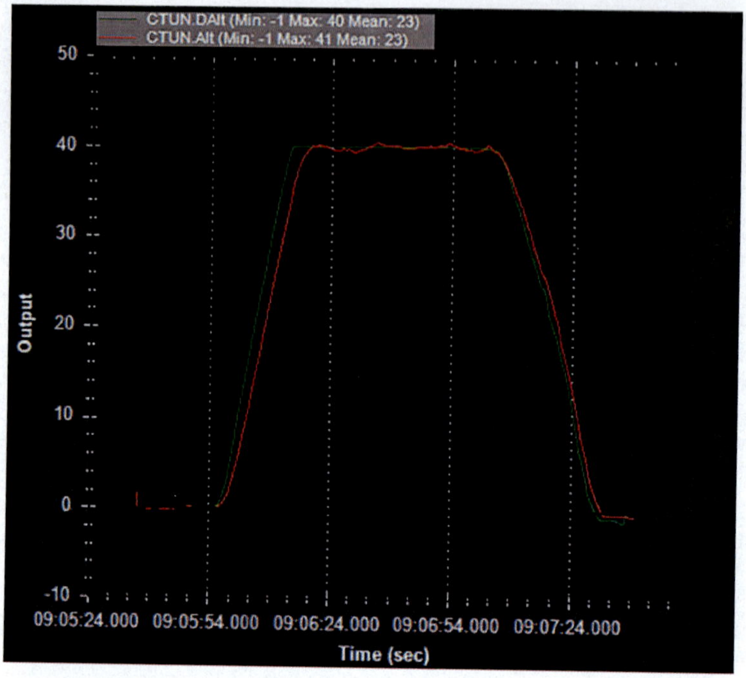

Fig. 73.6 The variation between desired and actual altitude of drone, green line – desired altitude, red line – actual altitude

73.3.3 Calculating the Desired Landing Orbits for UAVs

The previous research presented the methodologies used to determine and calculate the landing stages (Rohacs and Dung 2019). The intended landing orbit is predicted based on the landing regions, allowing the UAV to land precisely at the desired location. Figures 73.7 and 73.8 illustrate the simulation results for a UAV landing in the provided direction.

The planned landing orbit in this scenario consists of two shapes and two loops. The UAV completes two turns with the specified roll angle of $\gamma \leq 20^0$ at a height of H = 500 m. A straight flight with a speed of V = 40 m/s occurs between these two intervals. The drone then flies in the correct orbit in the provided direction, beginning the process of descending and finally directing the operation at a slower speed. The simulation result provided is sufficient and necessary for implementing control requests. The landing direction is defined as the direction from the UAV's present position to the target landing spot, and this landing distance is the shortest.

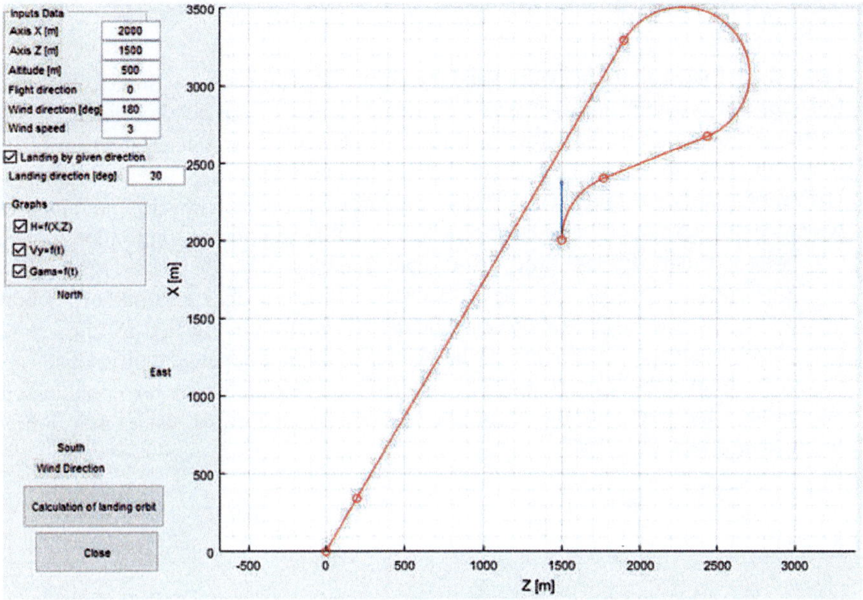

Fig. 73.7 The trajectory for UAV landing in the given direction

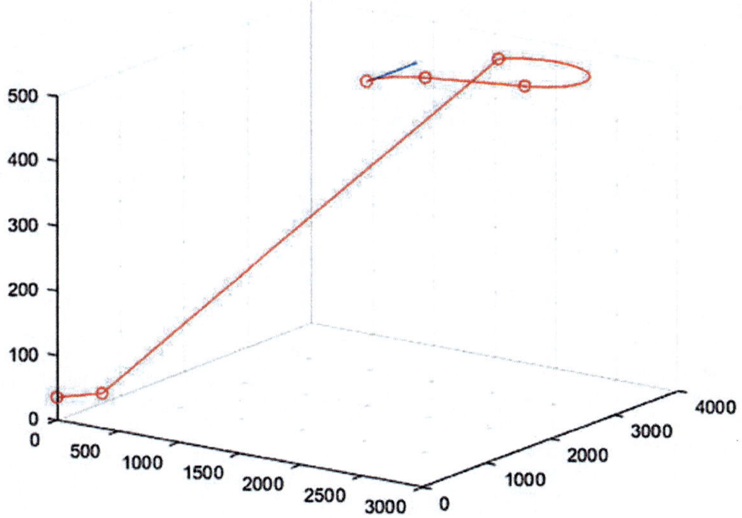

Fig. 73.8 The altitude of the UAV landing in the given direction

73.4 Conclusion

Several approaches for controlling and managing drone operations, such as intended trajectory following management, following process, obstacle avoidance, and desired land orbit, are discussed in this study. By presenting such a scenario, we intend to alleviate much of the existing work required to manage a large number of UAVs, particularly drones, with each new autonomous system development project. Researchers, developers, and hobbyists will be able to construct autonomous systems with substantially less time, effort, and money.

Professionals will also be able to exchange best practices and outcomes both inside and between organizations and fields. More study is needed to understand better how the suggested system may be planned, which is dependent on unanswered concerns about important qualities connected with the potential for considerable drone usage and how long they could be sustainable in terms of harsh environment and different aspects.

References

Carlsson, J. G. and Song, S. (2018). Coordinated logistics with a truck and a drone Management Science. INFORMS Inst.for Operations Res.and the Management Sciences, 64(9), pp. 4052–4069. doi: https://doi.org/10.1287/MNSC.2017.2824.

Chung, J. (2018). Heuristic method for collaborative parcel delivery with drone. Journal of Distribution Science. Korea Distribution Science Association (KODISA), 16(2), pp. 19–24. https://doi.org/10.15722/JDS.16.2.201802.19.

Dung, N. D. (2020). Developing Models for Managing Drones in the Transportation System in Smart Cities. Electrical, Control and Communication Engineering, 15(2), pp. 71–78. doi: https://doi.org/10.2478/ecce-2019-0010.

Dung, N. D. and Rohacs, J. (2018). The drone-following models in smart cities. In 2018 IEEE 59th Annual International Scientific Conference on Power and Electrical Engineering of Riga Technical University, RTUCON 2018 - Proceedings. Institute of Electrical and Electronics Engineers Inc. https://doi.org/10.1109/RTUCON.2018.8659813.

El-Adle, A. M., Ghoniem, A. and Haouari, M. (2019). Parcel delivery by vehicle and drone. Journal of the Operational Research Society. Taylor & Francis, 72(2), pp. 398–416. doi: https://doi.org/10.1080/01605682.2019.1671156.

Ferrandez, S. M. et al. (2016). Optimization of a truck-drone in tandem delivery network using k-means and genetic algorithm. Journal of Industrial Engineering and Management. Universitat Politecnica de Catalunya, 9(2), pp. 374–388. doi: https://doi.org/10.3926/JIEM.1929.

Goss, K., Musmeci, R. and Silvestri, S. (2017). Realistic models for characterizing the performance of unmanned aerial vehicles. The 26th International Conference on Computer Communications and Networks, ICCCN 2017. Institute of Electrical and Electronics Engineers Inc. https://doi.org/10.1109/ICCCN.2017.8038444.

Jankovics, I. and Kale, U. (2019). Developing the pilots' load measuring system. Aircraft Engineering and Aerospace Technology, 91(2). https://doi.org/10.1108/AEAT-01-2018-0080.

Joerss, M. et al. (2016). Parcel delivery: The future of last mile. McKinsey & Company, (September), p. 32. Available at: https://bdkep.de/files/bdkep-dateien/pdf/2016_the_future_of_last_mile.pdf.

Kellermann, R., Biehle, T. and Fischer, L. (2020). Drones for parcel and passenger transportation: A literature review. Transportation Research Interdisciplinary Perspectives. Elsevier, 4, p. 100088. https://doi.org/10.1016/J.TRIP.2019.100088.

Kitonsa, H. et al. (2018). Significance of drone technology for achievement of the United Nations sustainable development goals. Journals.urfu.ru, 4(3), pp. 115–120. https://doi.org/10.15826/recon.2018.4.3.016.

Liu, J., Guan, Z. and Xie, X. (2018). Truck and Drone in Tandem Route Scheduling under Sparse Demand Distribution. 8th International Conference on Logistics, Informatics and Service Sciences, LISS 2018 – Proceeding. Institute of Electrical and Electronics Engineers Inc. https://doi.org/10.1109/LISS.2018.8593233.

Martinetti, A., Schakel, E. J. and van Dongen, L. A. M. (2018). Flying asset: Framework for developing scalable maintenance program for unmanned aircraft systems (UAS). Journal of Quality in Maintenance Engineering, 24(2), pp. 152–169. https://doi.org/10.1108/JQME-12-2016-0073.

Merkert, R. and Bushell, J. (2020). Managing the drone revolution: A systematic literature review into the current use of airborne drones and future strategic directions for their effective control. Journal of Air Transport Management. Elsevier Ltd, 89(September), p. 101929. https://doi.org/10.1016/j.jairtraman.2020.101929.

Nguyen, D.-D. (2021). Cloud-Based Drone Management System in Smart Cities. In Krishnamurthi, R., Nayyar, A., and Hassanien, A. E. (eds) Development and Future of Internet of Drones (IoD): Insights, Trends and Road Ahead. Cham: Springer International Publishing, pp. 211–230. https://doi.org/10.1007/978-3-030-63339-4_8.

Nguyen, D. D. et al. (2020). Intelligent Total Transportation Management System for Future Smart Cities. Applied Sciences. https://doi.org/10.3390/app10248933.

Nguyen, H. P. D. and Nguyen, D. D. (2021). Drone Application in Smart Cities: The General Overview of Security Vulnerabilities and Countermeasures for Data Communication. In Krishnamurthi, R., Nayyar, A., and Hassanien, A. E. (eds) Development and Future of Internet of Drones (IoD): Insights, Trends and Road Ahead. Cham: Springer International Publishing, pp. 185–210. https://doi.org/10.1007/978-3-030-63339-4_7.

Pinto, R. et al. (2019). A network design model for a meal delivery service using drones. International Journal of Logistics Research and Applications. Taylor & Francis, 23(4), pp. 354–374. https://doi.org/10.1080/13675567.2019.1696290.

Rohacs, D. and Rohacs, J. (2016). Magnetic levitation assisted aircraft take-off and landing (feasibility study – GABRIEL concept). Progress in Aerospace Sciences, 85, pp. 33–50. https://doi.org/10.1016/j.paerosci.2016.06.001.

Rohacs, J. and Dung, N. D. (2019). Robust planning the landing process of unmanned aerial vehicles. International Journal of Sustainable Aviation. Inderscience Publishers, 5(1), p. 1. https://doi.org/10.1504/ijsa.2019.10021483.

Wang, K. et al. (2019). Cooperative route planning for the drone and truck in delivery services: A bi-objective optimisation approach. Journal of the Operational Research Society. Taylor & Francis, 71(10), pp. 1657–1674. https://doi.org/10.1080/01605682.2019.1621671.

Chapter 74
Impact Analysis for Improving Rational Entropy Management Regarding Container Ships

M. Koray

74.1 Introduction

Container ships are assumed to be the largest polluters among the world's merchant fleet, because of emitting 30% of total CO_2 emission. Despite constituting 13.2% of the merchant fleet, the levels of CO_2 emissions are also relatively higher due to their faster voyage speeds and circular sailing frequencies compared to other ship types.

In 2019, CO_2 emissions from container shipping amount to around 240 million tons. These CO_2 emissions represent approximately 0.6% of the total global CO_2 emissions (EU 2020). Although there have been significant industrial and technological improvements regarding energy efficiency of merchant ships in recent years, the operational option to reduce the average speed has come to the fore among the solutions found. Naturally, a few knots (nautical mile per hour) reductions in average speed will have a positive effect on energy efficiency and rational energy management, and their operational requirements will also be a limiting factor. The average speed of ships in the world merchant fleet is demonstrated in Fig. 74.1.

Considering that the average speed of container ships is 14.2 mph as shown in Fig. 74.1, it should also be taken into account that even one unit of speed reduction in order to increase energy efficiency will adversely affect the "just-in-time" cargo delivery process. In this case, container ships face two main problem areas regarding energy efficiency. The first is time pressure and the second is environmental protection factors. Container throughput at ports worldwide from 2012 to 2020 with a forecast for 2021 until 2024 is published by Martin Placek in 2021. Based on reference data, container throughput at ports worldwide from 2021 to 2030 have

M. Koray (✉)
Maritime Faculty, Piri Reis University, İstanbul, Türkiye
e-mail: nmkoray@pirireis.edu.tr

M. Z. Sogut et al. (eds.), *Proceedings of the 2022 International Symposium on Energy Management and Sustainability*, Springer Proceedings in Energy,
https://doi.org/10.1007/978-3-031-30171-1_74

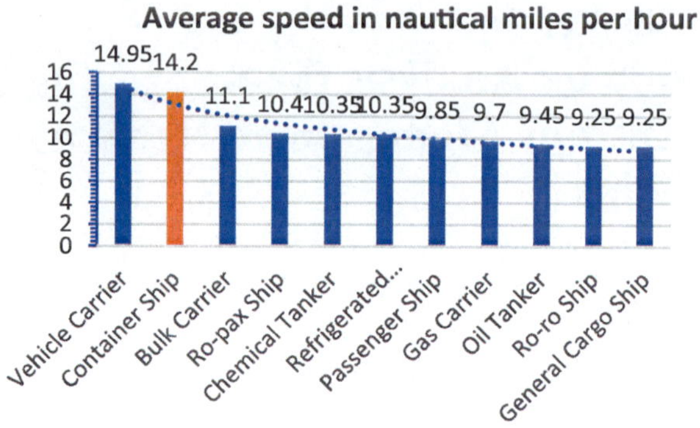

Fig. 74.1 Average speed of vessels in the world merchant fleet in 2018, by ship type (Placek 2021a, b)

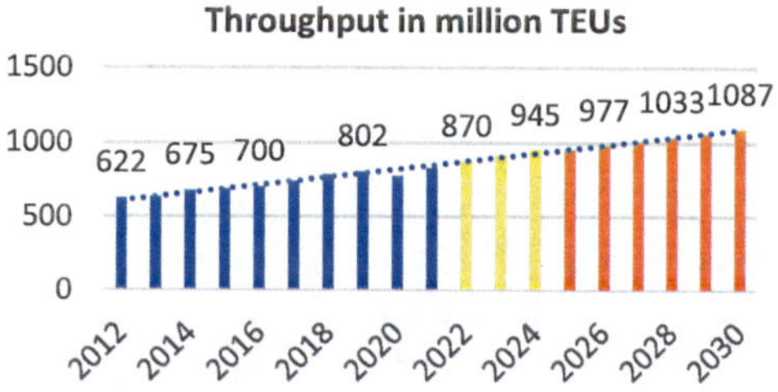

Fig. 74.2 Container throughput at ports worldwide from 2012 to 2020 with a forecast for 2021 until 2030 (Placek 2021a, b)

been predicted again considering pandemic and war risk. Predictions are demonstrated in Fig. 74.2.

Considering that 800 million TEU container volume was handled in 2019 and 240 million tons of CO_2 was emitted in this period, ceteris paribus, it is estimated that approximately 312 million tons of CO_2 for 1087 million TEU throughput will be emitted by container ships in 2030. In these circumstances, it is calculated that about 12 g of CO_2 is emitted for every 1 kg of cargo carried on container ships. The average carbon footprint of container ships in the United Kingdom in 2021 are calculated by Ian Tiseo as 16.14 g of CO_2 per metric ton of goods shipped per kilometer (Ian Tiseo 2021). Examining the reason for the 4.12 g difference between the world average and the emission value detected in the UK ports will also form the basis of rational energy management. It is considered that the average CO_2 emission volume of the global ocean fleet is moving away from the average CO_2 emission volume, especially due to the different age, tonnage, and fuel type of the ships calling at busy container ports. Container throughput at ports for the top ten countries worldwide from 2010 to 2020 in million TEUs is demonstrated in Fig. 74.3.

The relative superiority of the container trade of Chinese origin compared to other countries is clearly seen in Fig. 74.3. The ten-year average of the other nine countries is around 15.1–48.4 million tons of container throughput. The annual CO_2 emissions from the container trade of top ten countries have been calculated around 141,8 m tons. Countries with a container throughput in the range of 5–15 million tons in their ports from 2010 to 2020 in million TEUs are shown in Fig. 74.4. The annual CO_2

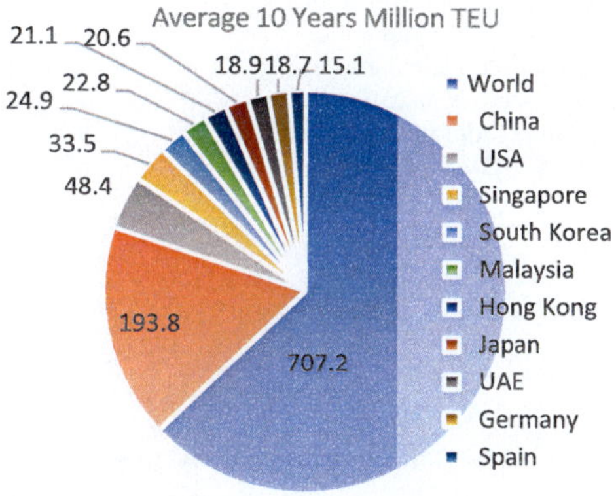

Fig. 74.3 Container throughput at ports for the top ten countries worldwide from 2010 to 2020 in million TEUs (UNCTADSTAT 2022)

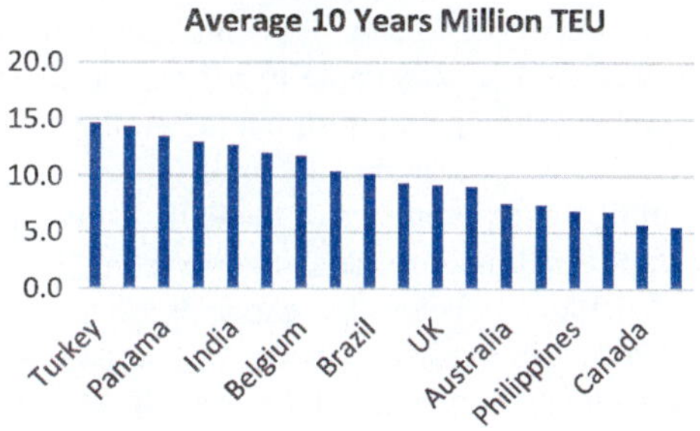

Fig. 74.4 Countries with a container throughput in the range of 5–15 m tons in their ports from 2010 to 2020 in million TEUs (UNCTADSTAT 2022)

emissions from the container trade of these countries have been calculated around 61 m tons.

Countries with a container throughput in the range of 5–5000 tons in their ports are shown in Fig. 74.4. The annual CO_2 emissions from the container trade of these countries have been calculated around 27,44 m tons.

As a result of the scrutinization of previous studies, it is seen that maritime transportation is the most efficient option among other freight transport modes (Bouman et al. 2017). However, the regulations to protect the marine environment are quite strict regarding the maritime transportation, and the industrial products developed for preventive measures are at exorbitant prices for modernization of secondhand ships (Fig. 74.5).

The Energy Efficiency Existing Ship Index (EEXI) and Carbon Intensity Indicator (CII), which contain the new carbon intensity standards developed by IMO, will enter into force in 2023. However, a common standard that can measure carbon emissions metrically has not yet been fully determined. Currently, the estimation of the carbon intensity of international shipping is based on the Energy Efficiency Operational Indicator (EEOI), Annual Efficiency Ratio (AER), Distance (DIST), and Time (TIME) (IMO 2021).

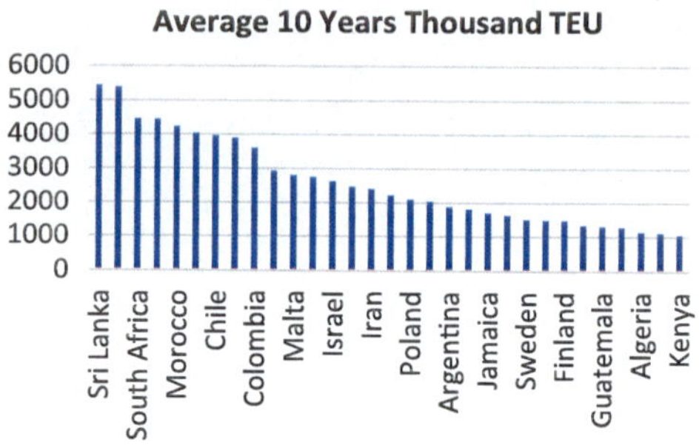

Fig. 74.5 Countries with a container throughput in the range of 5–15,000 tons in their ports from 2010 to 2020 in thousand TEUs (UNCTADSTAT 2022)

Faber et al. claimed that by reducing the amount of cargo carried per ship in relation to CO_2 reduction, one of the options of running the ships at idle or navigating at lower speeds would be possible. Although technically these suggestions seem reasonable, it is considered that other options will not be cost-effective, although it is acceptable to reduce the transfer speed to some extent, especially in terms of the operation of container ships (Jasper Faber et al. 2010). As a matter of fact, similar paradoxes are encountered in the decarbonization studies carried out to fulfill the requirements of the Paris Agreement. Since decarbonization processes and ship operating costs have inverse rating, it is not sufficient to reach the emission limit of 0.1% and below by choosing an alternative fuel, selecting a different power transmission system, or choosing operational options such as lower voyage speed, with today's technology by oneself. Paul Gilbert et al. have emphasized that with regard to decarbonization and carbon management, the complex nature of the industry and its interaction with other modes of transport make it difficult to develop policies that limit absolute emissions (Gilbert et al. 2015).

It is impossible to disagree with this view. However, there is an inverse relationship between the long-term extension of the process of adopting new technologies by experience at sea against the necessity of making money for the shipowners in the short term. Although air transport is not at a level to compete with the high cargo carrying potential of maritime shipping, air transport provides a competitive advantage compared to container ships in cases where fast service is required. Therefore, the option of average speed reduction needs to be rational to reduce CO_2 emissions from container ships (Haakon Lindstad et al. 2015). Considering the measures to reduce energy efficiency and CO_2 emissions and the potential impact, it is confirmed by previous studies that vessel size (Tillig et al. 2015; Wang et al. 2010; Miola et al. 2011; Stott and Wright 2011; Gucwa and Schäfer 2013; Halfdanarson and Snåre 2015; Pauli 2016; Zöllner 2009), biofuels (Eide et al. 2013; Wang et al. 2010; Bengtsson et al. 2013; Brynolf et al. 2014), speed optimization (Corbett et al. 2009; Miola et al. 2011; Wang and Lutsey 2013; Chatzinikolaou and Ventikos 2014; Norlund and Gribkovskaia 2013), and capacity utilization (Buhaug et al. 2009) are the prominent ones among the options of hull design power transmission system, alternative fuels, and alternative energy sources and operation.

74.2 Methodology

The methodology related to defining the research problem is constructed by collecting data, classifying data, multiregression analysis, defining weighted parameters, determining energy consumption range, calculating exergy for each operation hour, optimizing the handling process, and interpreting outcomes as shown in Fig. 74.6.

The objective of this methodology is to calculate operator errors more accurately than other methods. The problems related with "Rational Entropy Management Regarding Container Ships" can be listed as follows:

- Can this analysis determine the gap of the current container ship operation procedures?
- Can the existing procedures provide rational entropy management for container ship's operations?
- Can the determination of an exergy value for finding out the operation failures?

Firstly, fuel consumption data were obtained from daily fuel consumption log of a post-Panamax container vessel during cargo loading operation. However, the data consisted of discrete variables and they were not satisfied level. Therefore, it was necessary to classify the data and make it meaningful. In order to classify the data, harmonization, merging, clearing, and selection phases were applied to obtained fuel consumption data. After following these stages, multiple regression analysis was performed using appropriate data and then defined weighted parameters. Thus, the energy consumption range was determined. The value of energy consumption range provides calculation of the exergy value for each operation hour. The value shows operating errors considering the value as a reference. At the end of the study, the normal value, the tolerated value, and the extreme value of the energy consumption per hour during the loading operations on a post-Panamax container vessel were determined.

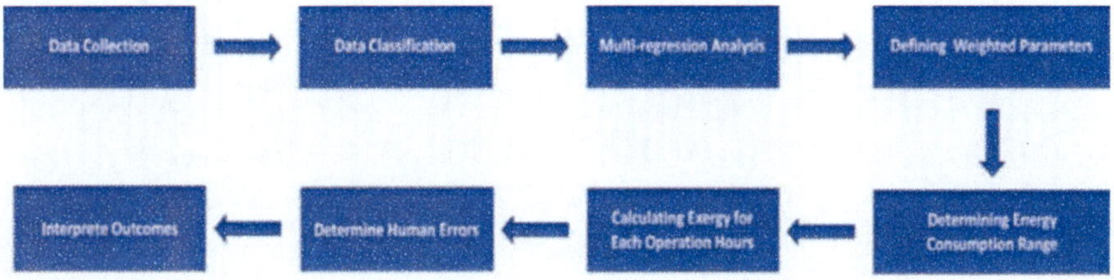

Fig. 74.6 The methodology related with defining the research problem

74.3 A Model for Operating Error Detection: A Case Study for Post-panamax Container Vessel Loading Operation

In the light of the methodology of the study, in 2021, loading operations are scrutinized for a post-Panamax container vessel within fuel consumption data. It is assumed that fuel consumption per hour or per day during loading operation is a fundamental factor to determine operating errors, and others are ceteris paribus unless otherwise stated as mutatis mutandis.

In this study, 145 loading operations for the post-Panamax container vessel are detected. Only 69 observations were found meaningful in 145 loading operations. Besides, operating errors were detected in 76 loading operations.

Sampled post-Panamax container vessel's actual fuel consumption is calculated as 0,17 tons per hour throughout 2021. According to oil record book's data and fixed or regular records, related with technical specifications, default fuel consumptions per hour are described as between 0,16 and 0,22 tons per hour.

In order to calculate the normal fuel consumption during loading operations, in 2021 initially actual fuel consumptions per 24 h are extracted from the oil record book, because other data were not continuous but intermittent. Somehow it was loaded for a short time, but it was interrupted due to malfunctions or various problems. These intermittent data were included in the calculation after linear regression analysis, as these data are important for comparing one hour of fuel consumption. Sixty-nine observations mentioned related with fuel consumptions are shown in Fig. 74.7.

As it is understood that 24-h consumption values are close to each other as seen in Fig. 74.3, initially linear regression analysis was performed with 24-h data. Linear regression analysis for calculating tolerated fuel consumptions are shown in Fig. 74.8.

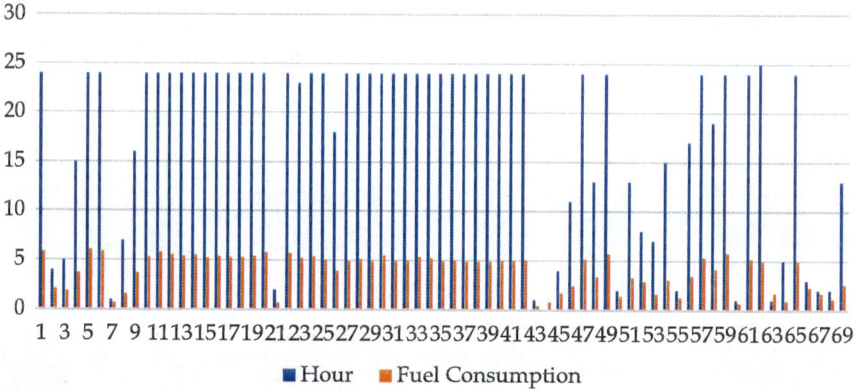

Fig. 74.7 Fuel consumptions during 69 loading operations of sampled post-Panamax container vessel

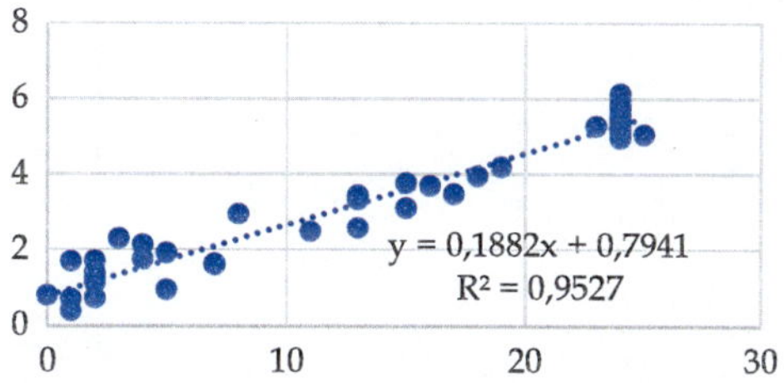

Fig. 74.8 Linear regression analysis for calculating tolerated fuel consumptions

Tolerated fuel consumptions were calculated by linear regression analysis for each of the 69 observations during loading operations considering variable.

$$y = -0,1882x + 0,7941.$$

Fuel consumptions range from −0,79 tons to 0,94 tons per hour. However, fuel consumption range is decreasing from −0,35 tons to 0,82 tons per 24 h. The average fuel consumption is calculated as 0,17 tons per hour. In addition to these results, linear regression analysis was performed to find out how much time was wasted as a result of operational errors. Linear regression analysis is shown in Table 74.1.

Wasted times were calculated by linear regression analysis for each of the 69 observations, and then readjusted consumption values are determined for the 145 observations including intermittent operations considering y variables as shown below.

$$y = -3,2025988 + (5,06140201 \times \text{Fuel Consumption})$$

$$ymin = -4,4135656 + (4,78638713 \times \text{Fuel Consumption})$$

$$ymax = -1,9916321 + (5,33641688 \times \text{Fuel Consumption})$$

When it is applied, the formulae of y variables for the 69 observations and fuel consumption ranges (differences between actual and calculated fuel consumption values) are calculated as from −5,44 tons to 7,28 tons. Anomalies are shown in Fig. 74.9.

When we considered the 145 observations, it can be seen in Fig. 74.10 that there will be differences between actual and calculated fuel consumption values. The differences (wasted time) are from −4,23 to 31.40 h.

If y variables' formulae are applied on the 145 observations, it can be seen in Fig. 74.11 that running hour ranges are elongated from −2,74 h to 171,82 h. If the reference is taken as 24 h, the reason of wasted time should be investigated. As can be seen, the extreme values obtained from the calculations made with the unsorted data are exaggerated.

When the remaining reasonable data is analyzed after extracting low reliability ones, the differences between current running hours and predicted hours identify operational errors.

Table 74.1 Multi-regression analysis for calculating tolerated running hours

Regression statistics						
Multiple R	0,97,606,268					
R square	0,95,269,836					
Adjusted R square	0,95,199,236					
Standard error	199,638,542					
Observations	69					
ANOVA	**Df**	**SS**	**MS**	**F**	**Significance F**	
Regression	1	5378,27,218	5378,27,218	1349,4413	4035E-46	
Residual	67	267,032168	398,555,474			
Total	68	5645,30,435				
	Coefficients	**Standard error**	**t stat**	**P-value**	**Lower 95%**	**Upper 95%**
Intercept	−32,025,988	0,60,669,423	−52,787,692	15124E-06	−44,135,656	−19,916,321
Fuel consumption	506,140,201	0,13,778,243	36,7,347,424	4035E-46	478,638,713	533,641,688

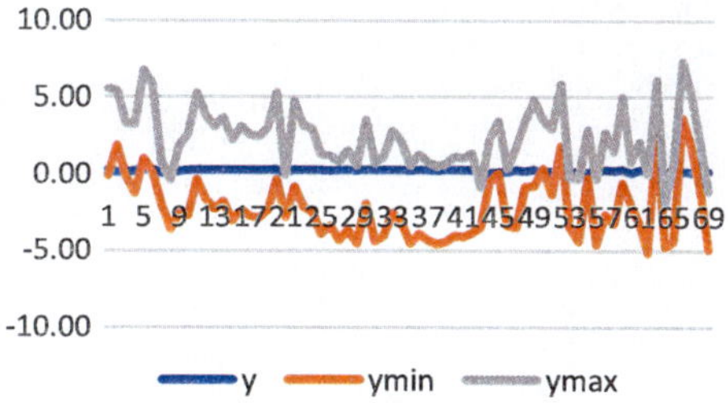

Fig. 74.9 Min, max, and average wasted time for the 69 observations

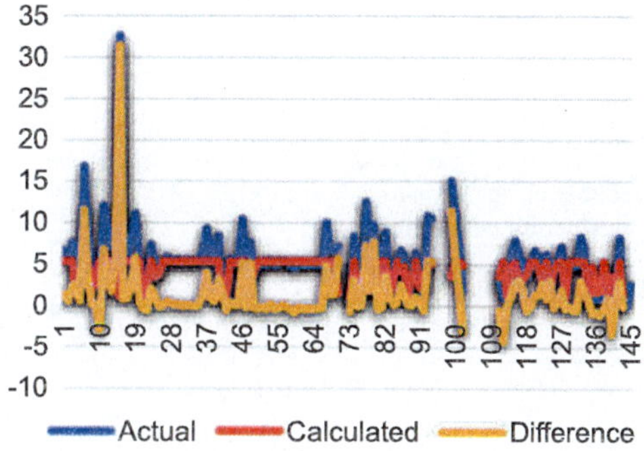

Fig. 74.10 Fuel consumptions for the 145 observations (actual, calculated, and difference)

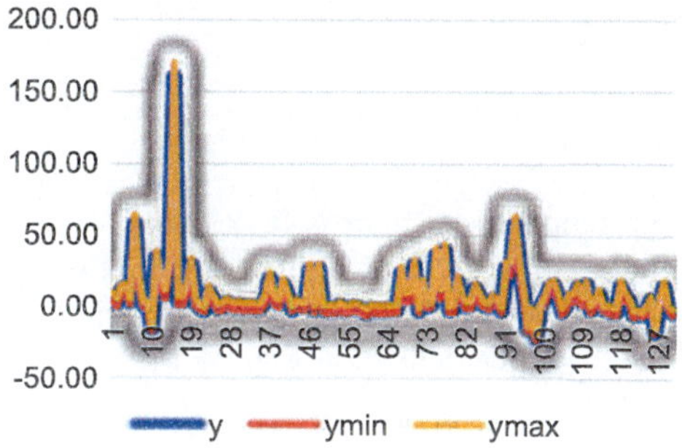

Fig. 74.11 Min, max, and average running hours for the 145 observations

The results are shown in figures below. In Fig. 74.12, actual running hours, average running hours, and minimum and maximum y variables are displayed together. When reliable observations are considered, it can be seen in Fig. 74.12 that there will be differences between actual and calculated fuel consumption values.

Calculated running hours are from 0,13 to 29.76 h.

The time differences between running hours are shown in Figs. 74.12 and 74.13.

Fuel consumptions ranges (differences between actual and calculated fuel consumption values) are calculated as from −5,24 tons to 5,80 tons. Anomalies are shown in Fig. 74.13.

Fuel consumption ranges per hour (differences between actual and calculated fuel consumption values) are calculated as from 0,01 tons to 0,70 tons. The average fuel

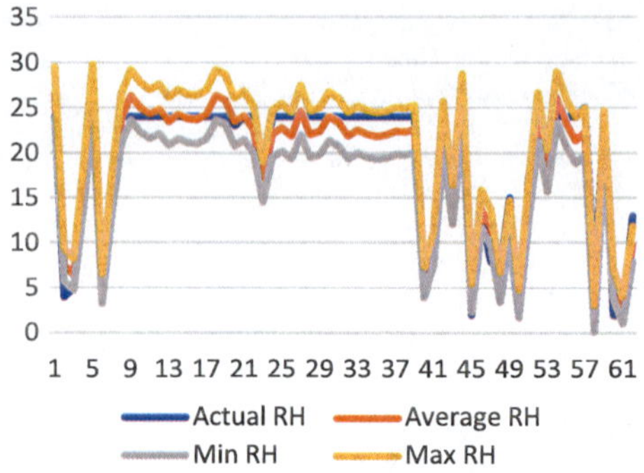

Fig. 74.12 Min, max, and average running hours of reliable observations

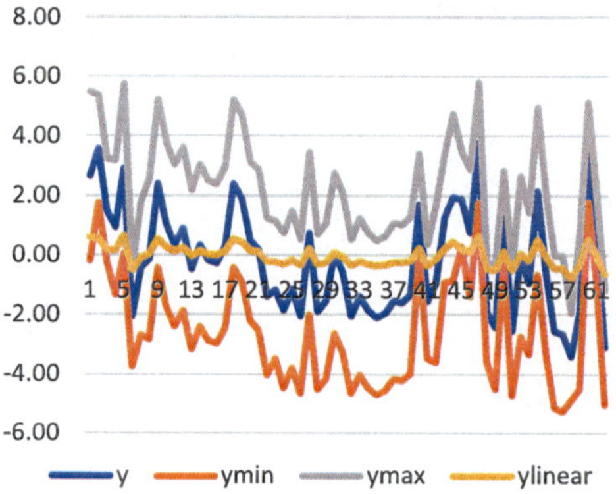

Fig. 74.13 Min, max, and average running hours of reliable observations considering actual time

consumption per hour is 0,18 tons. When calculated average fuel consumption is compared with SFC value as 0,15 tons per hour, wasted fuel consumption is calculated as 0,03 tons per hour. This value is determined as an operating error. Anomalies are shown in Fig. 74.14.

74.4 Evaluating Technical Efficiency Analysis of Container Ships

Within the scope of the study, the technical efficiency of 1441 container ships over 5000 GT was examined, and their fuel consumption and CO_2 emissions were analyzed on the basis of m tons. An evaluation was made by comparing the findings obtained as a result of the analysis of the data gathered by EMSA (2022). In this context, the Energy Efficiency Operational Indicator (EEOI), Annual Efficiency Ratio (AER), Distance (DIST), and Time (TIME) data were analyzed in light of the criteria determined by IMO. These criteria are defined in the following units:

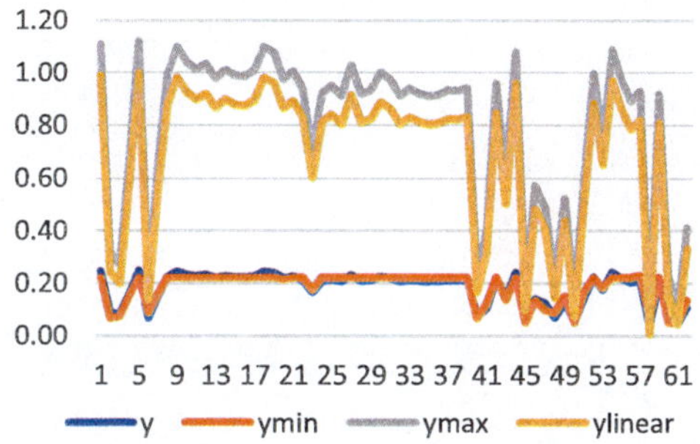

Fig. 74.14 Min, max, and average one-hour fuel consumption of reliable observations

$$\mathrm{EEOI} = \mathrm{gCO_2/ton/nautical\ mile} = \frac{\sum m_{\mathrm{CO_2}}}{\sum (m_{\mathrm{payload}} \times D_{\mathrm{payload}}} \tag{74.1}$$

$$\mathrm{AER} = \mathrm{gCO_2/dwt/nautical\ mile} = \frac{\sum m_{\mathrm{CO_2}}}{\mathrm{dwt} \times \sum D_{\mathrm{total}}} \tag{74.2}$$

$$\mathrm{DIST} = \mathrm{kgCO_2/nautical\ mile} = \frac{\sum m_{\mathrm{CO_2}}}{\sum D_{\mathrm{total}}} \tag{74.3}$$

$$\mathrm{TIME} = \mathrm{ton\ CO_2/h} = \frac{\sum m_{\mathrm{CO_2}}}{\sum T_{\mathrm{underway}}} \tag{74.4}$$

The CO_2 consumption of 715 container ships with a total fuel consumption of 6300 to 44,700 m tons ranges from 20,000 to 139,500 m tons in the calculation made, and the technical efficiency of these ships was found to be between 5.35 and 38.12. Technical efficiencies of selected 715 container ships are demonstrated in Fig. 74.15.

As ships age, the difference between fuel consumption and CO_2 emissions increases. It has been observed that the ships approaching the age of 40 and still used in maritime trade have released almost three times the total fuel consumption of CO_2, and it has been determined that their technical efficiency has increased to 38.12. It has been calculated that the technical efficiency of ships between 0 and 5 years of age decreased to 5.35.

The CO_2 consumption of 693 container ships with a total fuel consumption of 300–6400 m tons ranges from 1100 to 19,900 m tons in the calculation made, and the technical efficiency of these ships was found to be between 5.84 and 47.68. Technical efficiencies of selected 693 container ships are demonstrated in Fig. 74.16.

As can be seen in Fig. 74.16, it has been determined that technical efficiency increases negatively in short-range voyages where fuel consumption is relatively low, and total fuel consumption and CO_2 emissions rise up to three times in old ships over 25 years old.

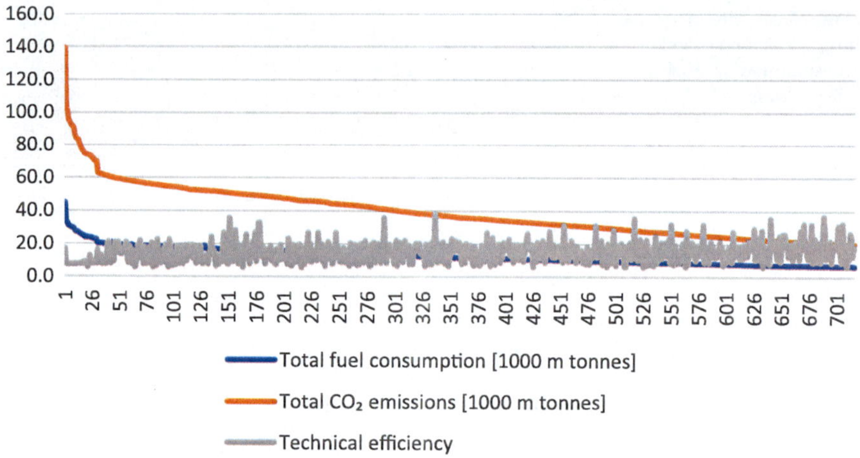

Fig. 74.15 Technical efficiencies of selected 715 container ships

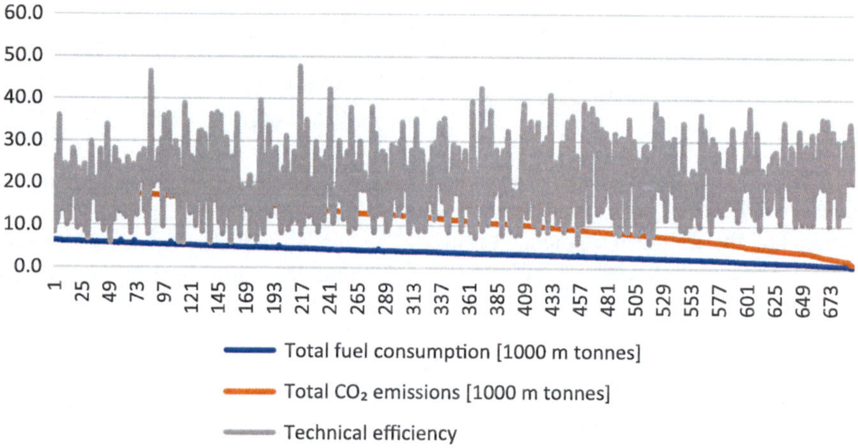

Fig. 74.16 Technical efficiencies of selected 693 container ships

74.5 Conclusions

In order to increase the efficiency of rational entropy management, the issues that should be done based on the findings obtained within the scope of the study are given in the following items:

- Considering the total fuel consumption consumed by the container ships during the navigational activities throughout the year, it has been determined that an average of 0.03 tons/h of fuel is wasted. In this case, it is foreseen that approximately 1341 m tons of wastage will occur on a ship with a fuel consumption of 44,700 m tons/year, and about 4023 m tons of additional CO_2 emission will be realized in ships over 20 years of age.

- For container ships, with a technical efficiency value of 10 or more, a 30–40 day's medium range voyages should be planned by reducing the voyage cycles. Besides, cruise speeds should be lower, i.e., 1–2 kts., than the average speed of 14.2 kts. Additionally, planning short shipping cycles to prevent the increase of CO_2 emissions should be avoided.
- Optimal ship sizes should be re-evaluated according to economies of scale. After the following of decommissioning small-size container ships, the focus should now be on the construction of 120,000 DWT (13,000 TEU)'s higher cargo carrying capacity vessels with a technical efficiency of 5 or less.

References

Bengtsson SK, Fridell E, Andersson KE (2013). Fuels for short sea shipping: A comparative assessment with focus on environmental impact, Proceedings of the Institution of Mechanical Engineers, Part M: Journal of Engineering for the Maritime Environment, 228(1), 44–54. https://journals.sagepub.com/doi/10.1177/1475090213480349

Bouman EA, Lindstad E, Riallandi AI, Strømman AH (2017) State-of-the-art technologies, measures, and potential for reducing GHG emissions from shipping – A review. Transportation Research An International Journal Part D; Transport and Environment Vol 52: 408–421. https://doi.org/10.1016/j.trd.2017.03.022

Brynolf S, Fridell E, Andersson K (2014) Environmental assessment of marine fuels: liquefied natural gas, liquefied biogas, methanol and bio-methanol, Journal of Cleaner Production, 74, 86–95. https://doi.org/10.1016/j.jclepro.2014.03.052

Buhaug Ø, Corbett JJ, Endresen Ø, Eyring V, Faber J, Hanayama S, Lee DS, Lee D, Lindstad H, Markowska AZ, Mjelde A, Nelissen D, Nilsen J, Pålsson C, Winebrake JJ, Wu W, Yoshida K (2009) Second IMO GHG Study 2009, International Maritime Organization London-UK. https://wwwcdn.imo.org/localresources/en/OurWork/Environment/Documents/SecondIMOGHGStudy2009.pdf

Chatzinikolaou S, Ventikos NP (2014) Assessment of ship emissions in a life cycle perspective, Proceedings of MARTECH 2014: 2nd International Conference on Maritime Technology and Engineering, Lisbon, Portugal. https://www.researchgate.net/publication/280313109_Assessing_Environmental_Impacts_of_Ships_from_a_Life_Cycle_Perspective

Corbett JJ, Wang H, Winebrake JJ (2009) The effectiveness and costs of speed reductions on emissions from international shipping, Transportation Research Part D: Transport and Environment, 14(8), 593–598. https://doi.org/10.1016/j.trd.2009.08.005

Eide, M.S., Chryssakis, C., Endresen, Ø., 2013. CO2 abatement potential towards 2050 for shipping, including alternative fuels. Carbon Manage. 4 (3), 275–289.

EMSA\THETIS-MRV (2022) CO_2 Emission Report, Reporting Period 2020, Version 127, https://mrv.emsa.europa.eu/#public/emission-report

EU (2020) 2019 Annual Report on CO2 Emissions from Maritime Transport, https://www.verifavia-shipping.com/uploads/files/eu-mrv-co2-emissions-report-2019.pdf

Faber J, Freund M, Köpke M, Nelissen D (2010) Going Slow to Reduce Emissions: Can the Current Surplus of Maritime Transport Capacity be Turned into an Opportunity to Reduce GHG Emissions? Seas at Risk, Delft, p. 26. https://seas-at-risk.org/wp-content/uploads/2021/03/Speed_Study_January_2010_final_-_SAR_version.pdf

Gucwa M, Schäfer A (2013) The impact of scale on energy intensity in freight transportation, Transportation Research Part D: Transport and Environment, Vol.23: 41–49, ISSN 1361-9209

Halfdanarson J, Snåre MW (2015) Implementation and application of an integrated framework for economic and environmental assessment of maritime transport vessels, Department of Energy and Process Engineering, MSc Dissertation, Norwegian University of Science and Technology, Trondheim, Norway https://ntnuopen.ntnu.no/ntnuxmlui/bitstream/handle/11250/2353104/13997_FULLTEXT.pdf?sequence=1&isAllowed=y

IMO (2021) Fourth IMO GHG Study 2020 Full Report https://wwwcdn.imo.org/localresources/en/OurWork/Environment/Documents/Fourth%20IMO%20GHG%20Study%202020%20-%20Full%20report%20and%20annexes.pdf

Gilbert P, Larkin AB, Mander S & Walsh C (2015) Technologies for the high seas: meeting the climate challenge, Carbon Management, https://www.tandfonline.com/doi/full/10.1080/17583004.2015.1013676

Norlund EK, Gribkovskaia I (2013) Reducing emissions through speed optimization in supply vessel operations, Transportation Research Part D: Transport and Environment, 23, 105–113, https://doi.org/10.1016/j.trd.2013.04.007

Lindstad H, Asbjørnslett BE & Strømman AH (2015) Opportunities for increased profit and reduced cost and emissions by service differentiation within container liner shipping. Maritime Policy & Management, 43(3), 280–294. https://doi.org/10.1080/03088839.2015.1038

Miola A, Marra M, Ciuffo B (2011) Designing a climate change policy for the international maritime transport sector: market-based measures and technological options for global and regional policy actions. Energy Policy 39 (9), 5490–5498. https://www.marleenmarra.nl/docs/publications/MiolaMarraCiuffo2011EnergyPolicy.pdf

Pauli G (2016), Emissions and Inland Navigation. In: Psaraftis, H. (eds) Green Transportation Logistics. International Series in Operations Research & Management Science, vol 226. Springer, Cham. https://doi.org/10.1007/978-3-319-17175-3_14

Placek M (2021a) Average Speed of Vessels in The World Merchant Fleet in 2018 by Ship Type, Statista, Transportation & Logistics, Water Transport, https://www.statista.com/statistics/1268217/average-speed-of-ships-by-ship-type/#statisticContainer

Placek M (2021b) Global container port throughput 2012–2024, Statista, Transportation & Logistics, Water Transport (in million TEUs), https://www.statista.com/statistics/913398/container-throughput-worldwide/#statisticContainer

Stott P, Wright P (2011) Opportunities for improved efficiency and reduced CO2 emissions in dry bulk shipping stemming from the relaxation of the the Panamax beam constraint. Int. J. Maritime Eng. 153 (A4), A215–A229.

Tillig F, Mao W, Ringsberg JW (2015) Systems modelling for energy-efficient shipping Chalmers University of Technology, Department of Shipping and Marine Technology, Division of Marine Technology, SE-412 96 Gothenburg, SWEDEN https://publications.lib.chalmers.se/records/fulltext/211480/211480.pdf

Tiseo I, (2021) Carbon footprint of cargo ship types in the UK 2021, Statista, Energy & Environment, Emissions (in grams of CO2e per metric ton of goods shipped per kilometer), https://www.statista.com/statistics/1233482/carbon-footprint-of-cargo-ships-by-type-uk/

Wang H, Lutsey N (2013) Long-term potential for increased shipping efficiency through the adoption of industry-leading practices, ICCT, September 30th. https://theicct.org/sites/default/files/publications/ICCT_ShipEfficiency_20130723.pdf

Wang, H, Faber, J, Nelissen, D, Russell, B, and St Amand, D. Reduction of GHG emissions from ships. Marginal Abatement Costs and Cost Effectiveness of Energy-Efficiency Measures. Netherlands: N. p., 2010. https://www.unccleam.org/wp-content/uploads/library/marginal_abatement_cost.pdf

UNCTADSTAT (2022) Maritime Transport, Data, Table, Container Port Throughput, Annual (in twenty-foot equivalent units – TEUs). https://unctadstat.unctad.org/wds/TableViewer/tableView.aspx?ReportId=13321

Zöllner, J., 2009. Strömungstechnische Möglichkeiten zur Reduzierung des Kraftstoffverbrauchs und der CO2-Emissionen von Binnenschiffen. Vortrag beim ZKR Kongress “Rheinschifffahrt und Klimawandel”. Bonn, June 24–25th.

Chapter 75
Reducing the Environmental Impact of Aviation by Minimizing Flight Delays

Ingrid Sekelová, Peter Korba, Simona Pjurová, Siva Marimuthu, and Utku Kale

75.1 Introduction

The need for regulations concerning the number of emissions that are created by aviation is evident. The aviation industry should try to minimize its effect on the environment as much as possible. The consequences of climate change are starting to influence air transportation. It is believed that these changes will become more and more common. Among various incentives to reduce aviation emissions, one of them is to create a Single European Sky, which would promote better air traffic management, more direct routes, and thus, fewer emissions. The longer the aircraft flies, the more emissions it will create, which is why reducing the number of flight delays and diversion is essential. There are many causes of flight delays and diversions, and some of these causes can be prevented from happening. This article aims at presenting data focusing on the number of delays in aviation.

I. Sekelová (✉) · P. Korba · S. Pjurová
Faculty of Aeronautics, Technical University of Košice, Košice, Slovakia
e-mail: ingrid.sekelova@tuke.sk; peter.korba@tuke.sk; simona.pjurova@tuke.sk

S. Marimuthu
School of Digital, Technologies and Arts, Staffordshire University, Staffordshire, UK
e-mail: siva.marimuthu@staffs.ac.uk

U. Kale
Faculty of Transportation Engineering and Vehicle Engineering, Budapest University of Technology and Economics, Budapest, Hungary
e-mail: kale.utku@kjk.bme.hu

M. Z. Sogut et al. (eds.), *Proceedings of the 2022 International Symposium on Energy Management and Sustainability*, Springer Proceedings in Energy,
https://doi.org/10.1007/978-3-031-30171-1_75

75.2 Aviation Transport, Energy, and Climate Change

According to the 2021 annual report published by the US Department of Transportation, transport, in general, account for a large share of energy and related emissions. For the last 70 years, transport has exceeded the annual uses of energy by other sectors (commercial and residential) (U.S. Energy Information Administration, 2022). Even though the COVID-19 pandemic temporarily decreased energy consumption and emissions, it has slowly started recovering and reaching its previous numbers (U.S. Department of Transportation 2021).

The major source of energy in the transport sector is petroleum, but in recent years, the use of renewable energy is increasing, which has been proven to be better for the environment, thanks to the reduced number of emissions that are released into the atmosphere. The imposed restrictions of many countries gravely influenced the aviation sector and, consequently, jet fuel consumption, which declined by 44% in 2020 when compared to the pre-pandemic year 2019 (U.S. Department of Transportation 2021). This decrease resulted in low emissions in the spring of 2020, resulting from the decline in jet fuel and motor gasoline use.

Air travel has experienced growth rates in passengers' demand for years, and even though the pandemic created a setback, this industry continues to grow rapidly. In 2021, the pandemic increased the number of passengers by 18% (Mazareanu 2021). Therefore, in recent years, there have been many incentives to decrease the impact of aviation on the environment. It is also in the interest of the aviation sector since climate change is continuously affecting air transport. Climate change is not only threatening the economic well-being of airlines and airports, but it can severely jeopardize the safety of flights.

International Civil Aviation Organization (ICAO) published a document that focused on this issue, and it exemplifies the possible effect of climate change on air transport (ICAO 2020). The increased intensity of storms can cause flight delays, cancelations, and closures of airports, or it can negatively affect jet engines' performance and maintenance requirements. The temperature changes can result in increased cooling or heating costs at terminals and reduce payload (passengers or cargo), which can have enormous economic consequences. The precipitation change can cause flood damage to runways or infrastructure. It can lead to flight delays or cancelations. The changes in the wind are a dangerous aspect since deviations from prevailing wind directions at airports can affect the use of the runway, reducing the airport's operability and safety. Changes in the wind can create extreme storms and winds that can damage the airport infrastructure, besides delays or cancelations. Changes to the jet stream can impact flight times or fuel costs. Changing icing conditions like freezing rain can also cause flight delays and cancelations; it may require de-icing methods with associated costs. Desertification is related to the increased occurrence of dust that affects engines and consequently increases maintenance requirements. Such sandstorms can damage airframes, fuselage, and engines and also disrupt the effective operation of airports (ICAO 2020).

From 2017 to 2020, US accident investigators recognize that 65% of severe injuries resulted from turbulence, which may worsen in the next few years due to climate change. Some of these effects are already observable. For example, wildfires have been recently happening in Greece, Turkey, Canada, and California. Concerning aviation, these wildfires result in canceled flights and closed airports. In the case of a wildfire in Canada, at least 40 flights were canceled over 24 h (Campbell 2021).

75.3 Initiatives for Reduction of Emissions

There are several initiatives to reduce the amount of aviation's contribution to emissions and greenhouse gases. One of these initiatives is the frequent flyer levy, which may be viewed as the only solution for meeting Europe's climate goals. If this policy were valid, it could prevent aviation from becoming the most significant source of emissions by 2050. This tax should discourage people from flying too much by leaving one return flight tax-free, and other flights would be taxed based on their frequency. This money would be used to implement mitigating measures to offset the impact of flight on the environment (Campbell 2021). This initiative is supported by studies that show that a minority of the country makes the majority of the flights. For example, according to the research done by the climate campaign group Possible, 22% of the population of Canada takes 73% of flights (Harrabin 2021).

Another possible approach is implementing more efficient aircraft, sustainable aviation fuels, or renewable energy carriers into aviation. For example, using sustainable aviation fuels should be possible to cut emissions up to 80%. There are ongoing projects that concentrate on electrification of aviation systems, research, and electric and hybrid aircraft investments. Implementing smaller electric aircraft is expected in the next decade by commercial aviation by 2050 (Schwab et al. 2021). However, this implementation can face numerous logistics, finances, or feasibility barriers.

The European Union introduced the initiative of making the airspace more efficient. It aims to improve air traffic management performance and subsequently reduce flight delays with their initiative called Single European Sky, which was launched back in 1999. Its expected completion is set around 2030–2050. By that time, it could triple the capacity of the European airspace, save costs for air traffic management, improve safety, and most importantly, reduce the environmental impact of aviation by 10%. This initiative should be reached by reduced fragmentation of the European airspace between, e.g., member states. The direct result would be shorter flight times, thanks to shorter paths and lower flight costs and emissions. The efficiency of air traffic management in Europe seems to be improving. In 2008, the average delay for each flight represented 1.43 min; this number became lower in 2016 as it was only 0.86 min of delay per flight, while the current goal is 0.5 min. Unfortunately, the full integration of a Single European Sky meets with issues because of its huge scope and will not be achieved earlier than in 2030 (Debyser 2021).

75.4 Results and Discussion

This paper addresses reducing emissions by focusing on the number of delays and diversions of flights that could have been prevented from happening. Delays are a complex issue that is influenced by many factors. Similarly, in aviation accidents, it is never only one factor that causes the accident; it is a series of subsequent contributing factors. One factor leads to the other. While each of these factors may seem insignificant, when they combine together, they may result in the worst-case scenario into an aircraft accident or with a certain amount of luck, for example, into a delay. If we try to eliminate these smaller contributing factors, it can be more feasible, and it can make a difference at the end of the day.

The current percentage of delayed flights is quite high, as is shown in Fig. 75.1. As published by the Office of Aviation Consumer Protection, in June 2021, delayed flights represented around 23% of all flights performed, accounting for 127,565 out of 546,124 total flights. It is clear from Fig. 75.1 that the least represented are diverted flights, with 0.33% coming up to 1806 flights. The third category is canceled flights that were not operated but were entered in the carrier's computer system within 7 days of scheduled departure. Canceled flights came up to 8850 flights. The biggest category represented is fortunately on-time flights, which occupy almost 75% of all flights; this percentage represents 407,903 flights (The Office of Aviation Consumer Protection 2021).

Delayed flights may have diverse causes. According to the Office of Aviation Consumer Protection report, there are five types of delays. Most delays (37%) were caused by the air carrier, meaning that the factors that contributed to the delay were in the airline's control, e.g., maintenance or crew problems. The second most common reason (35%) was caused by the late arrival of the previous flight, which created a chain reaction and impacted the flight in question to have a delay too. The national aviation system accounts for 20% of all delays. This system comprises many conditions, such as non-extreme weather conditions, airport operation, heavy traffic volume, etc. Extreme weather contributed only to 5% of all delays, but

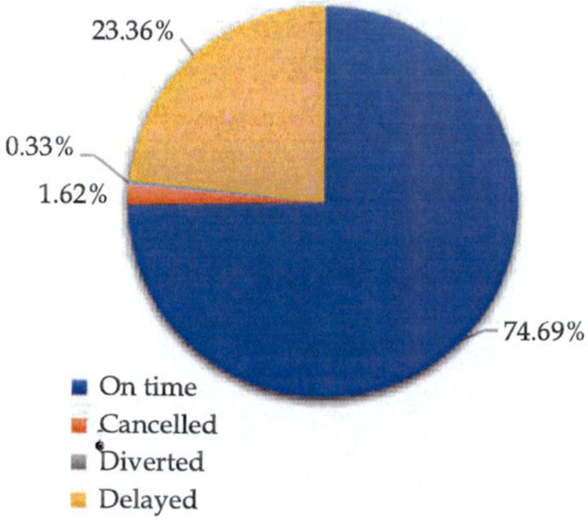

Fig. 75.1 Delayed, diverted, canceled, and on-time flights in June 2021 in the USA. (The Office of Aviation Consumer Protection 2021)

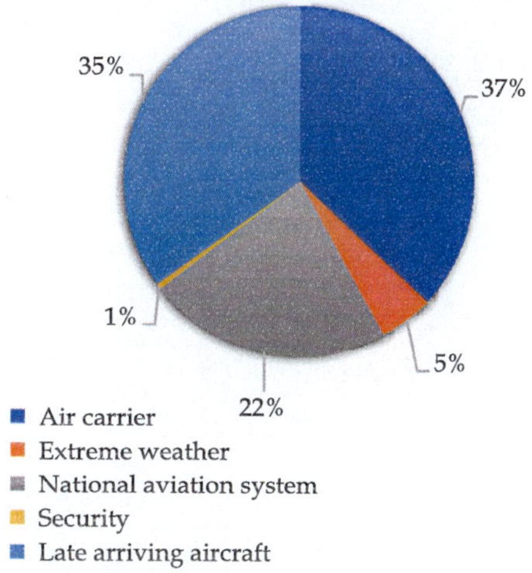

Fig. 75.2 Causes of delays in June 2021. (The Office of Aviation Consumer Protection 2021)

according to reports from ICAO mentioned above, this cause can occur more and more often in the future. The last reason that accounts only for 1% of delays is caused by security issues, such as the evacuation of the terminal, re-boarding of the aircraft because of a security breach, and long waiting lines for screening procedures (The Office of Aviation Consumer Protection 2021). We can see that the three most common causes can be influenced by human activity, which is not the case, for example, in the case of extreme weather conditions. It would be beneficial to have a closer and more detailed look at these causes that can be changed, such as new policies, company rules, and a better understanding of human factors.

The air carrier was denoted as the second most common reason for delays. According to the above-presented data, this category is quite broad, containing maintenance together with other factors that the air carrier can influence. In order to evaluate the factors which are represented in this category more closely, we focused our attention on the second study (Fig. 75.2).

A closer look into the causes of delays was provided in 2017 by a study performed in Europe, in which they studied the data about delays from 2008 to 2014 of a European airline. They focused on the causes of delays and came to the finding that sometimes during the year or even days, some reasons are more common than others. According to their study, it is hard to determine the cause of delays shorter than 15 min. It can be seen in Fig. 75.3 that delays caused by maintenance or aircraft defects, together with the delays caused by the delay of previous flights, are becoming more frequent as the length of the delay increases.

Air carriers try to eliminate long delays by effective management, which is then demonstrated in the lowering frequency of delays caused by air traffic control as the length of the delays increases. Figure 75.3 presents the data that was collected during this study. We can see that delays caused by another delayed flight are cited as the most numerous cause, while delays caused by maintenance are the second most common cause. In Fig. 75.3, it is clear that these two most common causes are more

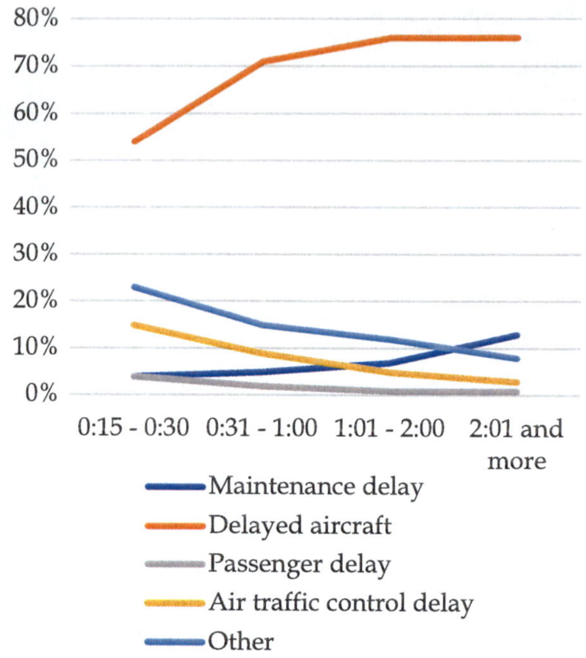

Fig. 75.3 Causes of delays of different times. (Zámková et al. 2017)

likely to cause long delays, which is the opposite tendency compared to passenger or air traffic control delays.

Delays caused by maintenance were more frequent during the summer season, which may be explained by the increased number of technical deficiencies that occur in summer as a consequence of more frequent flights. The peak season (July and August) is a widespread period of time for delays caused by air traffic control and delays caused by previous flights; during this peak time, the airspace is significantly more loaded than during the rest of the year, which allows for a chain reaction of delays. An interesting finding may be that delays caused by passengers' aircraft handling by suppliers (for handling, fuel, and catering) are more common at night. The reason behind this could be that it is harder to accumulate staff to solve unusual situations at night. Delays caused by maintenance are also happening more often at night because various repairs on aircraft must be done during the night before the majority of the aircraft take-off (from 5 AM to 7 AM). This study concluded that the airport staff and air traffic control together with supply and service companies are successful in reducing long delays. The most common cause of all delays is the previously delayed aircraft, which could be solved by having more aircraft available or the initiative Single European Sky. The recommendation of this study is to pay more attention to maintenance as it is the second most common cause of delays longer than 2 h. Better logistics could solve maintenance issues to ensure that necessary parts and continuous training are available as high-quality and timely maintenance is essential (Zámková et al. 2017).

75.5 Conclusion

Air transport contributes to the overall consumption of energy. Even though the COVID-19 pandemic decreased the number of emissions in 2020, it was only a temporary setback and in 2021 air transport continues to grow. In recent years, various incentives were created in order to prevent aviation from becoming the biggest polluter. Numerous delays and diversions can contribute among other causes to the number of emissions. Our proposition is to closely study the role of maintenance in the number of delays, as it can be seen that maintenance was determined as one of the most common causes of delays by the study performed in 2017. Maintenance is an integral part of air travel. The primary purpose is evident as it is impossible to maintain safe and efficient transportation without proper maintenance of its most essential elements, such as the aircraft. In our further study, we would like to focus on human factors present in aircraft maintenance, which contribute to ineffective communication that can result in diverse issues, one of them being delays or diversions.

References

Campbell, M (2021). Euronews. “Head slamming” turbulence and cancelled flights: turbulence climate crisis change travel? [Online] https://www.euronews.com/travel/2021/08/25/head-slamming-turbulence-cancelled-flights-how-will-climate-crisis-change-travel.

Debyser, A (2021). Air transport: Single European Sky. [Online] https://www.europarl.europa.eu/factsheets/en/sheet/133/air-transport-single-european-sky.

Harrabin, R (2021). A few frequent flyers ‘dominate air travel’. [Online] https://www.bbc.com/news/science-environment-56582094.

ICAO (2020). Effects of Climate Change on Aviation Business. [Online] https://www.icao.int/environmental-protection/Documents/Factsheet%20Business%20and%20Economics%20Final.pdf

Mazareanu, E (2021). Global air traffic – annual growth of passenger demand 2006-2022. [Online] https://www.statista.com/statistics/193533/growth-of-global-air-traffic-passenger-demand/.

Schwab et al. (2021). Electrification of Aircraft: Challenges, Barriers, and Potential Impacts. [Online] https://www.nrel.gov/docs/fy22osti/80220.pdf.

The Office of Aviation Consumer Protection (2021). Air Travel Consumer Report. [Online] https://www.transportation.gov/sites/dot.gov/files/2022-02/August%202021%20ATCR%20REV.pdf

U.S. Department of Transportation (2021). Transportation Statistics Annual Report 2021. [Online] https://www.bts.gov/sites/bts.dot.gov/files/2022-01/TSAR_FULL%20BOOK-12-31-2021.pdf.

U.S. Energy Information Administration (2022). February 2022 Monthly Energy Review. [Online] https://www.eia.gov/totalenergy/data/monthly/pdf/mer.pdf.

Zámková et al. (2017). Factors influencing flight delays of a European Airline. Acta Universitatis Agriculturae et Silviculturae Mendelianae Brunesis. 2017, 65 (5): 1800–1807. https://doi.org/10.11118/actaun201765051799

Chapter 76
CFD Investigation of Aircraft Preconditioned Air (PCA) Unit Flow Deflector Structures

Murat Ayar, Kerim Gumrukculer, Arif Hepbaşlı, and T. Hikmet Karakoc

76.1 Introduction

As a basic framework, UHŞÜ is a packaged air-conditioning unit. The fans that circulate the air inside the aircraft have a very high pressure of 10,000pa. In a normal packaged air conditioner, this pressure is in the range of 100–300pa. In the cooling section, by using the refrigerant through the compressor, the energy of the air passing through the evaporator is transferred to the refrigerant, and cooling is provided. The cooling unit consists of four main cooling cycle elements, namely the compressor, condenser, expander, and evaporator.

After the ambient air is conditioned, it is pressed into the duct system. The most basic feature of these devices is that they use 100% ambient air and are designed for rapid cooling. They are generally of two types. The most common are the types mounted under the bellows. The other type is the air-conditioning of aircraft that

M. Ayar (✉)
Eskisehir Technical University, Eskisehir, Turkey
e-mail: muratayar@eskisehir.edu.tr

K. Gumrukculer
IMBAT Air Conditioning and Refrigeraiton Systems, Eskisehir, Turkey
e-mail: kerim.gumrukculer@imbat.com

A. Hepbaşlı
Yasar University, Izmir, Turkey
e-mail: arif.hepbasli@yasar.edu.tr

T. H. Karakoc
Faculty of Aeronautics and Astronautics, Eskisehir Technical University, Eskisehir, Türkiye

Information Technology Research and Application Centre, Istanbul Ticaret University, İstanbul, Türkiye
e-mail: hkarakoc@eskisehir.edu.tr; thkarakoc@ticaret.edu.tr

M. Z. Sogut et al. (eds.), *Proceedings of the 2022 International Symposium on Energy Management and Sustainability*, Springer Proceedings in Energy,
https://doi.org/10.1007/978-3-031-30171-1_76

need cabin cooling with the device mounted on the ground. Users demand devices that consume less energy, and the limitations of the authorities force device manufacturers to work in this direction. As a result, the use of inverter compressors in all HVAC (heating, ventilating, and air-conditioning) systems has been increasing in recent years.

Studies on aircraft air-conditioning generally focus on the distribution of in-flight airflow. Besides experimental methods, computational fluid dynamics (CFD) methods are also used to examine the air distribution. There are studies on cabin air distribution and circulation related to aircraft air-conditioning. Aircraft cabin air distribution was investigated using the PIV flow imaging technique (Cao et al. 2014). In addition to the general air distribution, the effects of the structure of the holes providing the airflow on the air distribution performance were investigated (Fang et al. 2013). There are studies in which the vents are compared with experimental studies on a full-size cabinet and computational studies (Fang et al. 2015).

In addition to studies examining the air distribution provided by the ventilation system in aircraft, there are also studies on air quality. Fiser and Jícha evaluated three different air distribution systems on the model of an aircraft that can carry nine passengers in terms of air quality they offer (Fiser and Jícha 2013). As for the air quality in the aircraft cabin, a general evaluation study has been made that includes the new trends, effects, and costs related to the air quality in the aircraft passenger cabins (Hocking 2000). In addition to the studies in which air quality was investigated experimentally, different flights were simulated and a study was conducted by applying a questionnaire about the air quality feeling to the passengers (Park et al. 2011).

In this study, the effect of flow deflectors in the evaporator section of a PCA unit was investigated and the geometry with the most suitable form was determined with various geometric improvements.

76.2 Method

This study was carried out in order to examine the flow and thermal distribution in an air-conditioning unit with computational fluid dynamics (CFD) analysis and to make necessary improvements. The inlet of the unit, whose geometry is given in Fig. 76.1, has atmospheric pressure at a temperature of 350 °C, and evaporators in the unit and resistance are positioned behind the evaporators.

The hybrid mesh structure created by using the mesh process required for CFD analysis from tetrahedral and hexahedral elements is given in Fig. 76.2.

It has been diversified from 2 million to 20 million elements so that the number of elements can be determined.

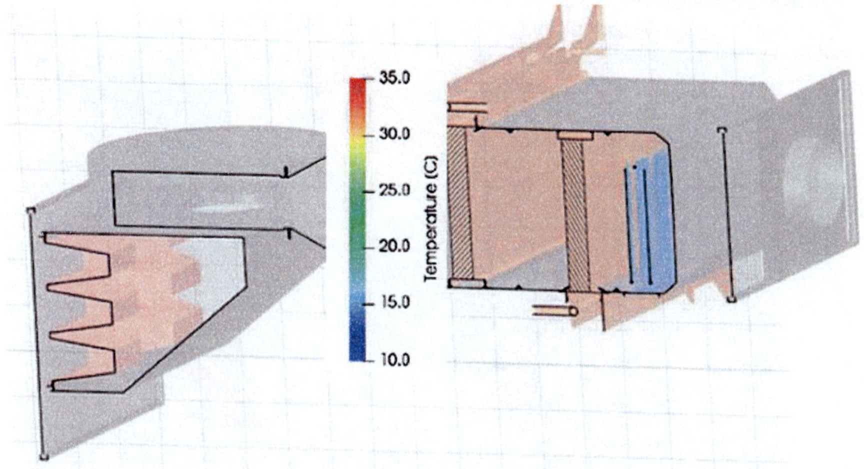

Fig. 76.1 CAD model of PCA

Fig. 76.2 Mesh structure

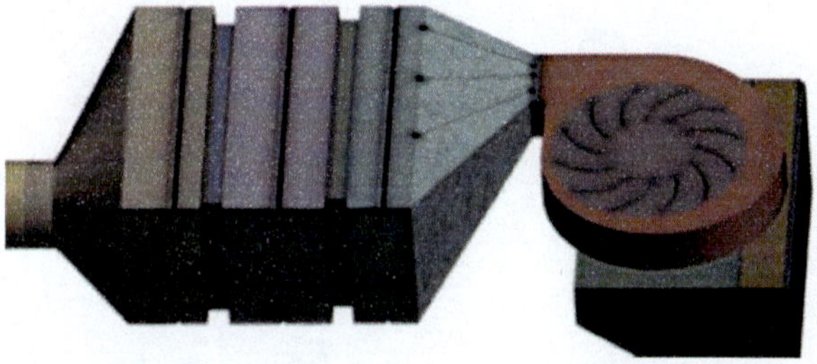

76.3 Results and Discussion

The effect of the routers at the fan outlet has priority in operation. The flow leaving these diverters enters the evaporator at equal rates, providing a more effective heat transfer in the evaporator, thus increasing the accuracy of the evaporator model in operation. The current situation and ten improvement attempts for the 2D model are created in this direction (Figs. 76.3 and 76.4).

When the CFD results are examined, it is seen that the revisions made the flow in the evaporator laminar, allowing for more suitable velocity and thermal distributions.

Another benefit of the revised model was seen in the pressure drop in Fig. 76.5. It is seen that the pressure loss is reduced by approximately 14% compared to the current model.

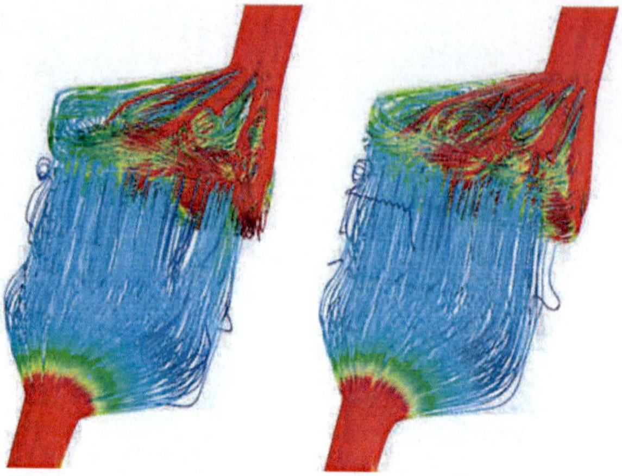

Fig. 76.3 Flow velocity distribution for initial design and final revision

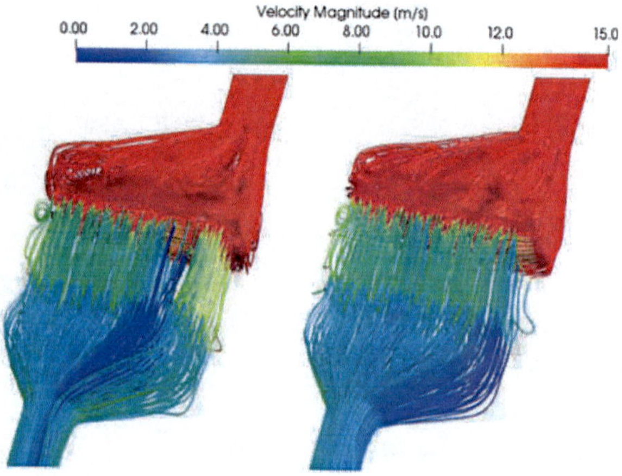

Fig. 76.4 Thermal distribution for initial design and final revision

76.4 Conclusion

The task of air-conditioning units is to provide filtered clean air at the desired temperature and humidity values to a closed space. However, during conditioning, the air encounters differently shaped components of the unit, so every obstacle it faces creates a pressure drop. High-pressure drops will lead to selecting a motor that requires greater power. Therefore, when internal losses are reduced, capacities will be reduced, and air-handling units with lower energy consumption can be designed. For this reason, it is essential to calculate these pressure losses accurately and in detail since the PCA units, which are the subject of the study, work especially at high pressure.

Fig. 76.5 Pressure distribution for initial design and final revision

By modeling the deflectors in the evaporator section of the PCA unit with the CFD method, it was possible to determine the structures and parts of the deflectors that resist the flow. By making use of these results, it was possible to reduce the pressure loss by making changes in the structure and placement of the deflectors.

CFD studies and revisions were iterated in parallel, resulting in a pressure gain of 14% compared to the initial design. With pressure gain, energy savings can be achieved by using a lower-power motor during the design phase.

Acknowledgments This study was funded by the Scientific and Technological Research Council of Turkey (TUBITAK) ARDEB 1501 (Grant No 3191527).

References

Cao, X., Liu, J., Pei, J., Zhang, Y., Li, J., & Zhu, X. (2014). 2D-PIV measurement of aircraft cabin air distribution with a high spatial resolution. Building and Environment, 82, 9–19.

Fang, Z., Liu, H., Li, B., Baldwin, A., Wang, J., & Xia, K. (2015). Experimental investigation of personal air supply nozzle use in aircraft cabins. Applied ergonomics, 47, 193–202.

Fang, Z., Li, N., Li, B., Liu, H., Dong, Y., Liu, F., ... & Kong, F. (2013). Experimental research on the attenuation rules of personalized air-conditioning nozzle jet flow in aircraft cabins. International Journal of Ventilation, 12(3), 285–296.

Fišer, J., & Jícha, M. (2013). Impact of air distribution system on quality of ventilation in small aircraft cabin. Building and Environment, 69, 171–182.

Hocking, M. B. (2000). Passenger aircraft cabin air quality: trends, effects, societal costs, proposals. Chemosphere, 41(4), 603–615.

Park, S., Hellwig, R. T., Grün, G., & Holm, A. (2011). Local and overall thermal comfort in an aircraft cabin and their interrelations. Building and Environment, 46(5), 1056–1064.

Chapter 77
Effects of the Covid-19 Pandemic in the Natural Gas Sector: A Situation Evaluation on Supply and Demand

Yonca Özğan and Selçuk Sarıkoç

Nomenclature

bcm	Billion Cubic Metres
CO_2	Carbon Dioxide Gas
OPEC	Organization of Petroleum Exporting Countries
EMRA	Energy Market Regulatory Authority

77.1 Introduction

The concept of energy, which has direct or indirect connections with transportation, trade, tourism, automotive, and many other sectors, has increased its importance even more with the increasing population and technological developments that we cannot keep up with. It is thought that the consumption of resources such as coal and oil will continue for decades due to the fact that nonrenewable energy resources have become traditional and have been used for years, the amount of world reserves are at a satisfactory level, and the tools and equipment used in industry and transportation and the machines are integrated according to nonrenewable energy resources. Natural gas, which is the most used energy source after oil and coal among

Y. Özğan (✉)
Technology and Innovation Management/Graduate School Of Natural And Applıed Scıences, Amasya Unıversıty, Amasya, Turkey

S. Sarıkoç (✉)
Department of Motor Vehicles and Transportation Technologies, Amasya University, Amasya, Turkey

Tasova Yuksel Akin Vocational School, Amasya, Turkey

M. Z. Sogut et al. (eds.), *Proceedings of the 2022 International Symposium on Energy Management and Sustainability*, Springer Proceedings in Energy,
https://doi.org/10.1007/978-3-031-30171-1_77

nonrenewable energy sources, is a bridge between nonrenewable energy sources and renewable energy sources due to its high combustion efficiency, easy and economical use, and because it emits less emissions and burns cleaner than coal and oil in the state.

77.2 Natural Gas in the World

Despite its greenhouse gas emissions, natural gas is the cleanest burning, fastest growing, and the least CO_2 amount compared to other fossil fuels. Natural gas currently accounts for about a quarter of global production (International Energy Agency (IEA) 2022a, b, c, d, e, f).

77.3 Natural Gas Reserved in the World

According to OPEC data, there are proven 206.2 trillion cubic meters of natural gas reserves in the world as of 2019. Nearly half of the natural gas reserves (40.9%) are in the Middle East countries with 79.1 trillion cubic meters (Botas 2022). This order is followed by European and Asian countries with 62.2 trillion cubic meters and African/Asia Pacific countries with 33.1 trillion cubic meters. On a country basis, Russia ranks first with 50,279 bcm. Russia, which has a share of approximately 25%, is followed by Iran, Qatar, and the USA, respectively (British Petrol 2022a, b).

77.4 Natural Gas Production in the World

Although Russia is the country with the richest natural gas reserves in the world, the USA ranks first in natural gas production with 33,899 billion cubic feet. This ranking is followed by Russia, Iran, and China. Turkey ranks 71st in the world natural gas production with 17 billion cubic feet (International Energy Agency (IEA) 2022a, b, c, d, e, f). World dry natural gas production ranking in 2019 is given in Table 77.1.

With the pandemic, natural gas production in the world decreased by 123 billion cubic meters in 2020, while the largest decreases were experienced in Russia with 41 billion cubic meters and in the USA with 15 billion cubic meters (British Petrol 2022a, b). Natural gas production by region is shown in Fig. 77.1.

Table 77.1 World dry natural gas production ranking in 2019. (International Energy Agency (IEA) 2022a, b, c, d, e, f)

Ranking	Country	Billion cubic meters
1	USA	960
2	Russia	678
3	Iran	238
4	China	179
5	Canada	179
6	Qatar	167
7	Australia	143
8	Norway	115
9	Saudi Arabia	113
10	Algeria	88

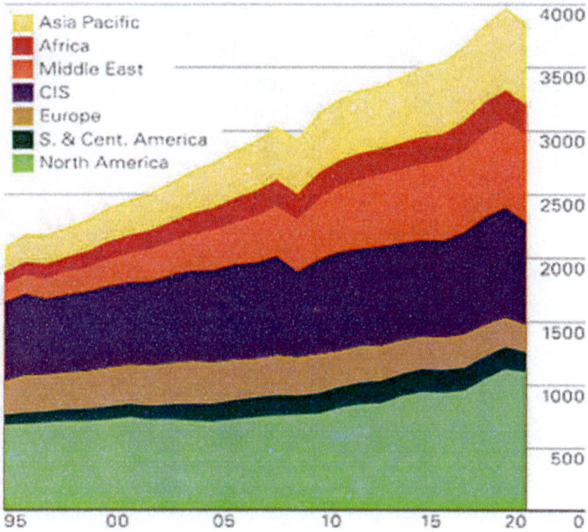

Fig. 77.1 Natural gas production by regions. (British Petrol 2022a, b)

77.5 Natural Gas Consumption in the World

According to OPEC data, there are proven 206.2 trillion cubic meters of natural gas reserves in the world as of 2019. Nearly half of the natural gas reserves (40.9%) are in the Middle East countries with 79.1 trillion cubic meters. This order is followed by European and Asian countries with 62.2 trillion cubic meters and African/Asia Pacific countries with 33.1 trillion cubic (Botas 2022). World natural gas consumption ranking in 2019 is presented in Table 77.2.

With the Covid-19 pandemic, natural gas consumption decreased by 81 billion cubic meters. While gas consumption fell in most countries, demand increased by an exceptional 6.9% in China and Iran. There was a decrease in gas demand, but the share of gas in primary energy continued to increase. This is due to the general decline in primary energy demand (British Petrol 2022a, b) (Fig. 77.2).

Table 77.2 World natural gas consumption ranking in 2019. (International Energy Agency (IEA) 2022a, b, c, d, e, f)

Ranking	Country	Billion cubic meters
1	USA	882
2	Russia	479
3	China	307
4	Iran	221
5	Canada	125
6	Saudi Arabia	113
7	Japan	102
8	Germany	96
9	Mexico	86
10	UK	77

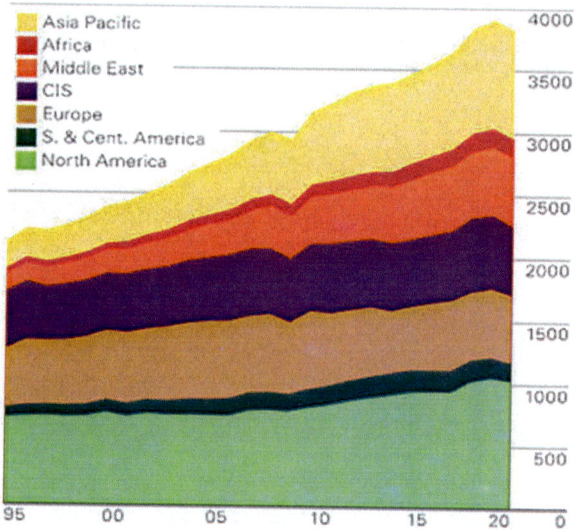

Fig. 77.2 Natural gas consumption by regions. (British Petrol 2022a, b)

77.6 Natural Gas Export in the World

Russia meets about 40% of the natural gas needs of European countries. According to the EMRA natural gas sector report, Turkey met 33.6% of its natural gas needs from Russia in 2020. Russia, the natural gas export giant, was followed by the USA, Qatar, and Norway, respectively (International Energy Agency (IEA) 2022a, b, c, d, e, f) (Table 77.3).

Table 77.3 World natural gas export ranking in 2019. (International Energy Agency (IEA) 2022a, b, c, d, e, f)

Ranking	Country	Billion cubic meters
1	Russia	257
2	USA	132
3	Qatar	127
4	Norway	110
5	Australia	100
6	Canada	76
7	Netherlands	43
8	Algeria	43
9	Turkmenistan	38
10	Malaysia	36

Table 77.4 World natural gas import ranking in 2019. (International Energy Agency (IEA) 2022a, b, c, d, e, f)

Ranking	Country	Billion cubic meters
1	China	132
2	Japan	105
3	Germany	95
4	USA	78
5	Italy	71
6	Mexico	59
7	Netherlands	56
8	South Korea	55
9	France	55
10	UK	47

77.7 Natural Gas Import in the World

Due to its increasing energy need, China ranks first in oil and natural gas imports. While Japan, German, and USA countries follow this ranking, and Turkey, which imports approximately 98% of the natural gas it uses, ranks 11th in the world natural gas imports ranking (International Energy Agency (IEA) 2022a, b, c, d, e, f) (Table 77.4).

77.8 Summary and Conclusions

With the Covid-19 pandemic, as of 02.03.2022, 5,960,972 people around the world lost their lives, and nearly 438 million people went through the quarantine process (World Health Organization (WHO) 2022). Despite the availability of a vaccine, the increase in the number of deaths and cases still remains uncertain as to when the Covid-19 pandemic will end. The pandemic, which we feel strongly in every aspect of our lives, has also seriously affected the energy sector. With the onset of the pandemic, there was a decrease in all energy sources, while an increase in renewable energy sources was observed as an exception.

The use of natural gas caused a global excess supply in 2019 due to the decrease in annual growth, a mild winter, the decrease in production due to the pandemic, and the recession of the world economy due to this, and natural gas prices began to decline.

Although the beginning of 2020 was negative for natural gas, it experienced great recovery in 2021. With the removal of the restrictions and the spread of the vaccine, workplaces were opened, production continued, community areas such as cafes, restaurants, and shopping malls were opened to full-time use by people, and the demand for heating and natural gas increased.

In 2021–2022, there has been an increase of more than 400% in natural gas prices, especially in Europe. The most important reason for this is the rapidly increasing energy demand after the pandemic. With the economic recovery, the demand for natural gas increased, but the supply remained stable. Due to the inability of the supply to meet the demand and the inability to obtain sufficient energy from renewable energy sources such as water and wind due to drought due to global climate change, natural gas instead of coal is preferred because it is cleaner in Europe. Due to reasons such as rapid consumption, natural gas prices have increased significantly (Euronews 2022).

Although the demand for natural gas, which has been increasing continuously for years, has stagnated with the pandemic, this situation has been temporary and the supply has come to a level that cannot keep up with the demand in a short time. The war between Russia and Ukraine, which started on Tuesday, February 24, 2022, has negatively affected and continues to affect the warring countries and the countries that have commercial relations with the warring countries. Russia, which is a pioneer in natural gas exports, continues to export natural gas to Europe without slowing down, even though there has been an increase in this export. However, this situation could not prevent the natural gas price increase, and it is predicted that the price increase will continue to increase in the future. Although it is stated that natural gas will continue to increase in the next 15 years, its consumption will be minimized until 2050 in accordance with the Paris Agreement, and what will happen in the long term remains uncertain (International Energy Agency (IEA) 2022a, b, c, d, e, f).

Acknowledgments The authors would like to thank Amasya University for its support. This paper was produced from the Master Thesis of Yonca Özğan.

References

Botas. (2022). "Home, Contact, Frequently Asked Questions Where and how much are oil and natural gas reserves in the world?" Retrieved 23 March 2022, 2022, from https://www.botas.gov.tr/Sayfa/dunyadaki-petrol-ve-dogal-gaz-rezervleri-nerelerdedir-ve-ne-kadardir/251.

British Petrol. (2022a). "Home, Energy economics, Statistical Review of World Energy, Natural gas, Natural Gas Production. British Petrol." Retrieved 20 March 2022, 2022, from https://

www.bp.com/en/global/corporate/energy-economics/statistical-review-of-world-energy/natural-gas.html.

British Petrol. (2022b). "Home, Energy economics, Statistical Review of World Energy, Natural gas, Natural Gas Reserves. British Petrol." Retrieved 22 March 2022, from https://www.bp.com/en/global/corporate/energy-economics/statistical-review-of-world-energy/natural-gas.html.html#natural-gas-reserves.

Euronews. (2022). "Home, My Europe, Europe News, Why Europe's energy prices are soaring and could get much worse." Retrieved 20 March 2022, 2022, from https://www.euronews.com/my-europe/2021/10/28/why-europe-s-energy-prices-are-soaring-and-could-get-much-worse.

International Energy Agency (IEA). (2022a). "Energy efficiency, The first fuel of a sustainable global energy system." Retrieved 24 March 2022 2022, from https://www.iea.org/topics/energy-efficiency.

International Energy Agency (IEA). (2022b). "Gas." Retrieved 24 March 2022 2022, from https://www.iea.org/fuels-and-technologies/gas.

International Energy Agency (IEA). (2022c). "International, Rankings, Dry Natural Gas Consumption 2019." Retrieved 26 March 2022 2022, from https://www.eia.gov/international/rankings/world?pa=35&u=0&f=A&v=none&y=01%2F01%2F2019&ev=false.

International Energy Agency (IEA). (2022d). "International, Rankings, Dry Natural Gas Exports 2019." Retrieved 21 March 2022 2022, from https://www.eia.gov/international/rankings/world?%20pa=64&u=0&f=A&v=none&y=01%2F01%2F2019&ev=false&pa=89.

International Energy Agency (IEA). (2022e). "International, Rankings, Dry Natural Gas Imports 2019." Retrieved 20 March 2022 2022, from https://www.eia.gov/international/rankings/world?%20pa=64&u=3&f=A&v=none&y=01%2F01%2F2019&ev=false&pa=64

International Energy Agency (IEA). (2022f). "International, Rankings, Dry Natural Gas Production 2019." Retrieved 24 March 2022 2022, from https://www.eia.gov/international/rankings/world?pa=10&u=0&f=A&v=none&y=01%2F01%2F2019&ev=false.

World Health Organization (WHO). (2022). "Coronavirus (COVID-19) Dashboard, Overview." Retrieved 20 March 2022, 2022, from https://covid19.who.int/

Chapter 78
"Drones GIS System" in Urban Transport

Dung D. Nguyen, Omar Alharasees, Utku Kale, Munevver Ugur, and Tahir Hikmet Karakoc

Nomenclature

UAV/UAS	Unmanned Aerial Vehicle/System
ATM	Air Traffic Management
GPS	Global Positioning System

78.1 Introduction

Since the eighteenth century industrial revolution, urban planning and urbanization problems have occurred due to the rapid increase in the global population and the concentration of the population in certain centers (Akintunde et al. 2016). This unnatural pace of urbanization has created significant social and environmental challenges for decision-makers (Muslim et al. 2014). Modeling and simulation techniques are effective tools to explore urban development mechanisms and enable

D. D. Nguyen (✉) · O. Alharasees · U. Kale · M. Ugur · T. H. Karakoc
Faculty of Aerospace Engineering, Le Quy Don Technical University, Hanoi, Vietnam

Department of Aeronautics and Naval Architecture, Budapest University of Technology and Economics, Budapest, Hungary

Department of Airframe and Powerplant Maintenance, Faculty of Aeronautics and Astronautics, Eskişehir Technical University—ESTU, Eskişehir, Turkey

Information Technology Research and Application Center, Istanbul Ticaret University, Istanbul, Turkey
e-mail: dungnd@lqdtu.edu.vn; oalharasees@edu.bme.hu; kale.utku@kjk.bme.hu; munevver929@eskisehir.edu.tr; hkarakoc@eskisehir.edu.tr

M. Z. Sogut et al. (eds.), *Proceedings of the 2022 International Symposium on Energy Management and Sustainability*, Springer Proceedings in Energy,
https://doi.org/10.1007/978-3-031-30171-1_78

planning in growth management. Therefore, monitoring and modeling cities' urban sprawl is a key parameter to avoid potential problems (Dereli et al. 2017; Kıvılcım et al. 2013).

Every city is a complex system. Geographical information systems combine all spatial and nonspatial information in a single system. While providing consistency in the analysis of geographical data, it establishes various connections and relationships between events depending on geographical proximity and ensures that all stakeholders access this data and analysis under a single platform. GIS can integrate spatial data with other data sources. Web scenes provide a realistic perception by showing geographic data and events in three dimensions. All geographic layers can be mapped with three-dimensional symbology, performing the generation and presentation of GIS maps on the computer.

Urban life requires a high level of understanding and solutions to the challenges faced by the society in that city. It is one of the high-tech application areas where large amounts of data are needed. Big data means that complex urban systems face challenges. They are voluminous and complex data with different qualities that have the potential to generate new hypotheses and new methods for understanding interactions between social, biophysical, and infrastructure domains (Herold et al. 2002; McPhearson et al. 2016).

In GIS, area data are usually represented by mosaicking and object data by topological vector data. Mosaic, formed by surrounding cells' coming together, is of three types: square cells, hexagonal cells, and triangular cells. However, all cells have the same shape and size within themselves. Each cell is assigned a value, and these values are related to all other cell values. The tessellation process occurs in many GIS software with different names, such as raster or raster map. The size of a single raster cell is called raster resolution. The structure formed by these cells is called a grid.

Drones can follow fixed trajectories or predefined corridors. For this, real-time GIS support is one of the requirements. Thus, active conflict/obstacle detection and problem resolution can be achieved. GIS support can be provided by 3D modeling.

The digital terrain model represents the geographical structure or the real terrain in the area. In other words, it enables the structure of the land to be expressed digitally in all aspects, such as elevation, slope, aspect direction, drainage, etc. In this way, sustainable GIS mapping support can be obtained to integrate transportation systems in smart cities.

High-resolution maps are the cornerstone of evaluating urban footprints by integrating them into global settlement models. In the last decade, rapid progress has been made in preparing such maps. The following is a list of some other datasets that provide GIS-related information in the field of urbanization:

- Global Map: It is a series of digital maps covering the whole world to accurately express the state of the environment on a global scale. It was developed in collaboration with the National Geospatial Information Authorities (NGIAs) around the world. The World Map provides eight major map themes with a nominal ground resolution of 1 km for raster data and a scale of 1:1,000,000 for vector data.

These themes are (i) transportation, (ii) boundary, (iii) drainage, (iv) population centers, (v) height, (vi) flora, (vii) land cover, and (viii) land use.

- World's Grid Population (GPW): A dataset from NASA's socioeconomic data and application center includes population density of past, present, and future raw population estimates. The goal of GPW is to provide a spatially disaggregated population layer compatible with datasets from the social, economic, and Earth science disciplines and satellites. These data are globally consistent and open for research, spatial decision-making, and communication.
- World Bank Geodata: In this data, a wide variety of World Bank datasets have been converted to KML format, including education and financial data.
- Global Management Fields: These are the database sections of low-level information belonging to administrative fields like countries and provinces. Version 3.6 was released in 2018. Version 4 will be released in December 2021. It covers 386,735 administrative areas and scientists can download spatial data by country.
- Armed Conflict Location and Incident Dataset: This data includes all reported conflict incidents in 50 developing world countries from 1997 to the present.
- Global Rural–Urban Mapping Project (GRUMP): It is a data set containing information about rural and urban population balances, taken from NASA's socioeconomic data and application center.
- Open Street Map (OSM): It contains information about worldwide landmarks, buildings, roads and road names, ferry routes, etc. It is crowdsourced data that contains a lot of important information.
- Geohive8: The initiative is made available by Ordnance Survey Ireland for easy access to "public spatial data" and includes population and county statistics. It is not in GIS data format but can be easily converted from CSV format (Tohidi and Rustamov 2020).

Following the Singapore guideline, the most promising options for urban air transport management must use specifically designed airspaces with predetermined fixed routes or fixed corridors (Fig. 78.1) (Pathiyil et al. 2016). Corridors refer to multilane "highway" channel routes in this context.

78.2 Operational Concept

The fundamental flight data recommended flight tunnel, and side wing profiles are depicted in Fig. 78.2 of the created cockpit equipment to support the EU-supported so-called Gabriel idea (magnetic levitation assisted take-off and landing concept of an undercarriage-less aircraft).

In theory, a variety of critical services and control mechanisms are required to achieve the goal of safely managed UAV traffic. As a result, this study aims to aid in the development and administration of UAV operations for civil use, particularly for bigger drones operating at low altitudes.

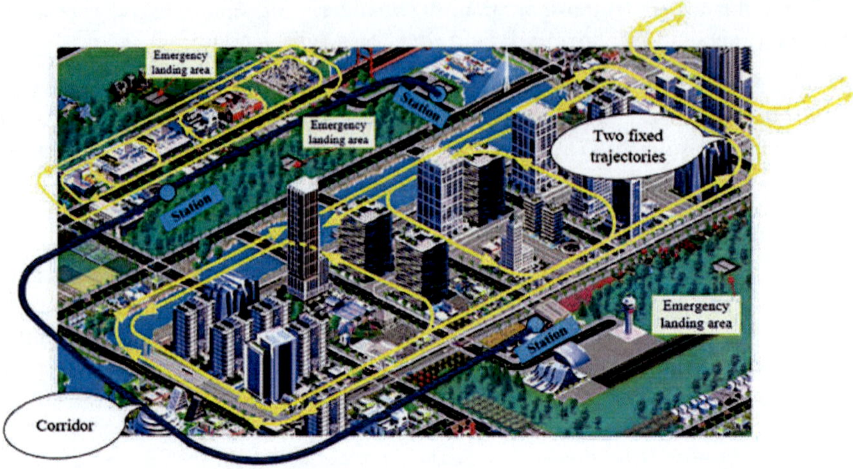

Fig. 78.1 Recommended concepts

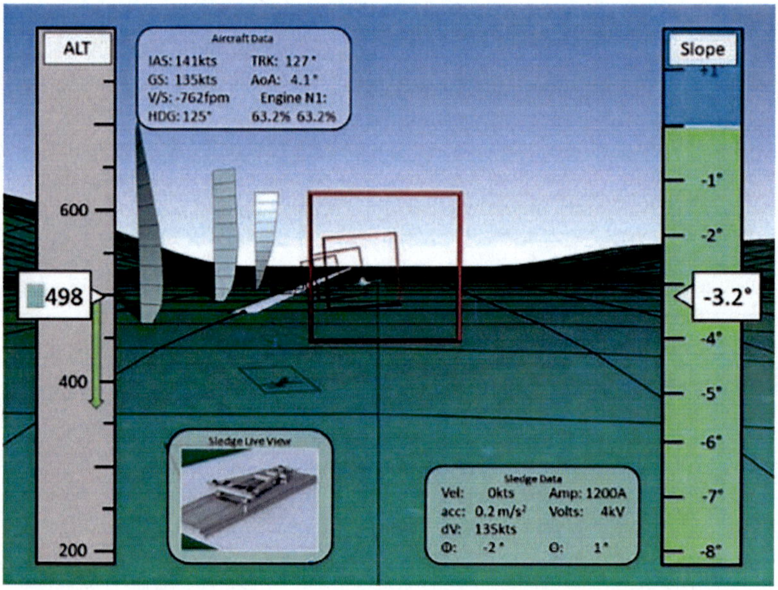

Fig. 78.2 Advanced precision landing assist cockpit instruments

The role of the drone operator is in transition from active controlling to passive monitoring. This means that the drone operators will have a more passive role, such as observation or inactive supervision, particularly in abnormal situations and emergencies. Such a control environment should be supported with adequate data to permit the UTM to support the smart city vision.

The operational concepts describe the usage of drones by the given assigned users (IEEE 1998; Cloutier et al. 2009). The operators use drone applications. In this paper, the common purpose of drones is to diversify air freight transportation, which involves collecting and ultimately distributing tiny packages directly from senders to receivers. Hence, this paper covers flights in city areas as well.

Drones operate via the following prearranged trajectory (Fig. 78.1). Each drone usually follows its own prearranged corridor such as speed, altitude, heading, or altering "lanes." Thanks to this feature, drones cannot bump into other drones.

Drones fly by following a predefined trajectory or corridor (Fig. 78.1). Each drone flies on its trajectory that might be part or follow the generally predefined trajectories with, for example, changing "lanes," heading, altitude, or speed. Drones can never meet other drones and drones moving in opposite directions on their trajectory.

According to the similarities of road networks, the settled trajectories and corridors have been arranged. The corridors involving a few lanes in a two-way direction are called "highways." There should be minimum 30 m between the corridors and any surrounding structures/obstacles. Main airways have settled trajectories.

High-speed delivery drones will operate in a corridor attached by nodes in the harbor area, one node in the factory area, and another in the cargo air terminal (Fig. 78.1). Thanks to the prearranged trajectory, drones can avoid obstacles and each other. Moreover, a safety puffer is established, which assures that it is unlikely that any drones can come across each other.

Operators must inform the air traffic management center about the planned flight and the expected target points before the flight. Utilizing other users' trajectories and surrounding data like minimum safe altitude and static obstacles, the automated center locates the trajectory for the flight in a 3D virtual channel being optimized by using the GIS map and opening a slot. When the drone misses the open slot, this process should be restarted.

The flight is entirely autonomous; nevertheless, the drone regularly predicts its position and any potential collision and accommodates its motion to the real flight. The position of drones depends on GPS and GIS mapping (Gazis et al. 1961; Pang et al. 2020) using fixed-point markers in the infrastructure and active, intelligent surveillance (Fig. 78.3). GIS enables the definition of optimal corridors depending on diminishing the habitants to be affected by any kind of environmental threat or potential collision situations.

Both the general concept and system framework are shown in Fig. 78.3.

The prearranged trajectory may have a return, round flights, and stops. When the drone arrives at its destination, it could follow the next part of a prearranged trajectory depending on the next target destination. The operational center naturally defines and initiates the slots for the following parts of the flight.

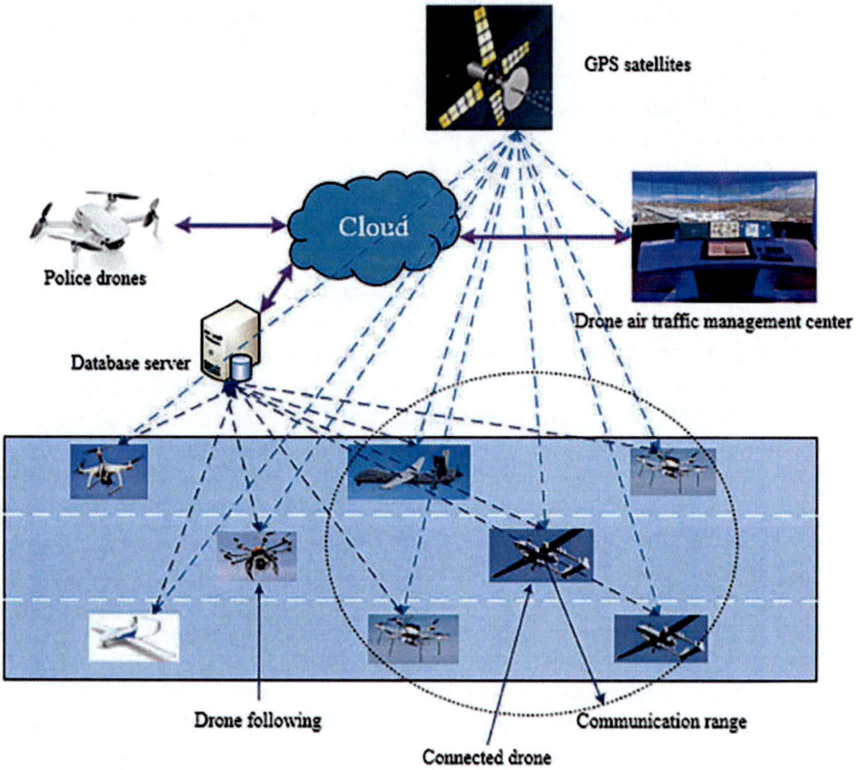

Fig. 78.3 General system layout of the suggested autonomous/Connected UAV management system

78.3 Airway Network

The airway network is a complex system that subdivides traffic flow and might reduce consistency. Aircraft have significant timeline schedules and routes. Therefore, it is very critical to manage the airway network. It is recommended to use four different sectors: geographical sector, sectors for restricted areas, sectors in vertical separation (between the large buildings), and sectors for vertical motion (climb/descent) between structures.

Trajectories and lanes are simple elements of the motions network. Lanes are significant elements for the aircraft that allow motions and flight modes such as descent, climb, coordinated turn, or just a straight flight (see Table 78.1).

Table 78.1 The typical elements of the airway network

Element type	Description
One direction: a) Vertical direction b) Three-dimensional direction	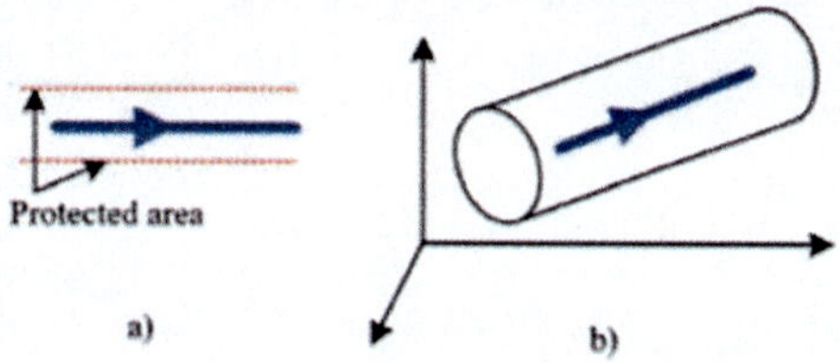
Two directions: a) Vertical direction b) Three-dimensional direction	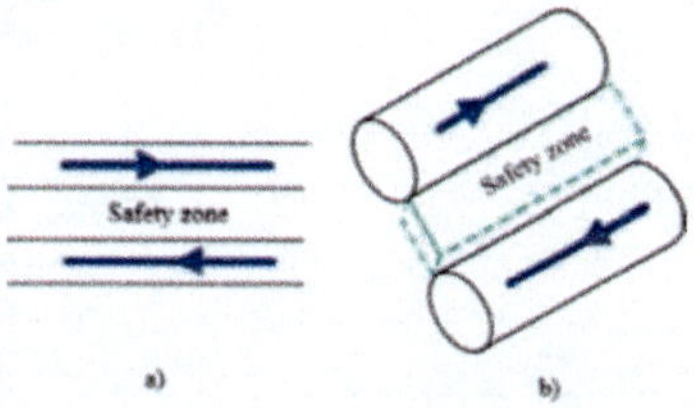
One direction, multiple lanes: a) Vertical direction b) Three-dimensional direction: v- vertical safe distance h- horizontal safe distance	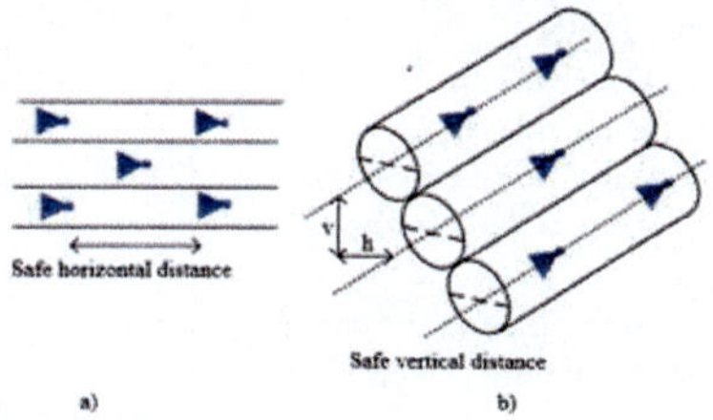
Two directions, multiple lanes: a) Vertical direction b) Three-dimensional direction	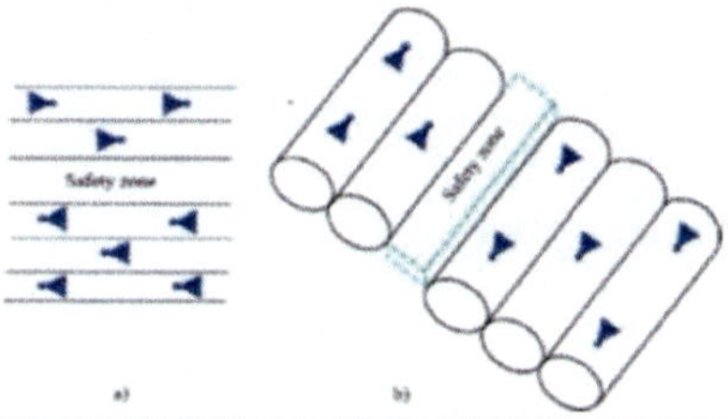
Turning: a) One direction at the same elevation b) Two directions at the same elevation	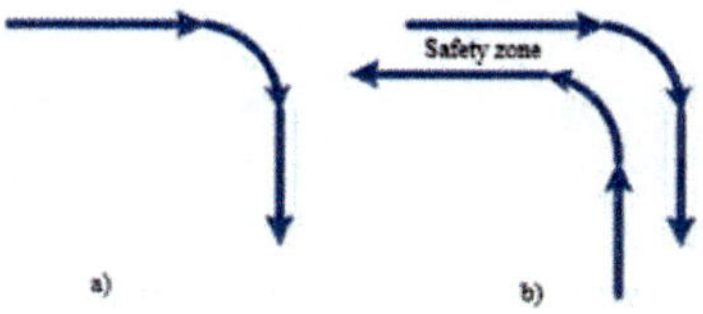

(continued)

Table 78.1 (continued)

Element type	Description
Changing altitude in the same direction	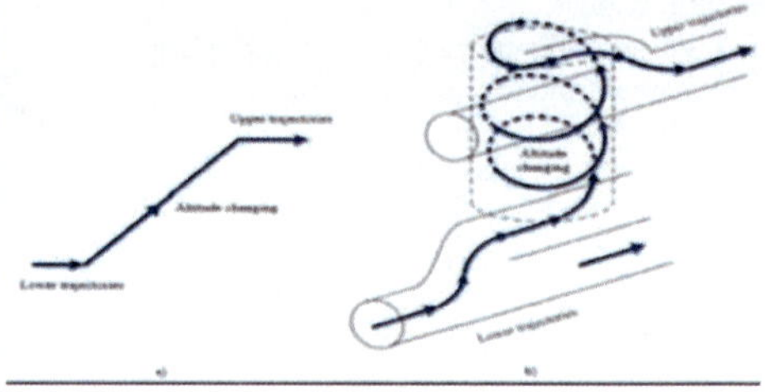
Crossing: a) Changing lane b) Changing direction (in top view)	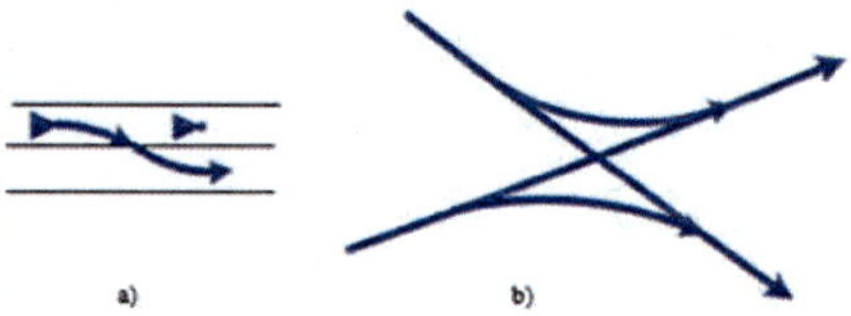
Changing direction (different elevation): 1- new lane transfer 2- flying in the new lane 3- change the elevation 4- turning on the same elevation 5- flying at the new lane in the chosen direction 6- merging in the lane at the same elevation and in the new direction selected	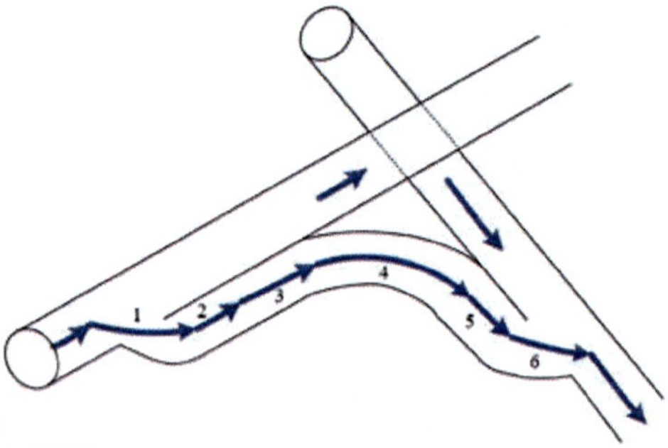

78.4 Conclusion

Integrating drone flights into air traffic management is complex. This study presents operational concepts for drone operations in urban areas, including airspace design, recommended construction of the airways, and essential safety requirements.

The rules given in this study were recommended to terminate or lessen causal factors and any unsafe situations. Analysis of the control actions, their possible hazardous situations, casual factors, and scenarios should be examined and improved to develop the effectiveness of these requirements. Obviously, the implementation of this concept needs theoretical and practical investigation, applying a wide range of methods, techniques, and solutions that were studied, developed, and improved by the current authors.

References

Akintunde, J. A., E. A. Adzandeh, and O. O. Fabiyi. 2016. "Spatio-Temporal Pattern of Urban Growth in Jos Metropolis, Nigeria." Remote Sensing Applications: Society and Environment 4 (October): 44–54. https://doi.org/10.1016/J.RSASE.2016.04.003.

Cloutier, R, A Mostashari, S McComb, A Deshmukh, J Wade, D Kennedy, and Peter Korfiatis. 2009. "Investigation of a Graphical CONOPS Development Environment for Agile Systems Engineering." In .

Dereli, MA, MA Uğur, N Polat - SGEM2017; Albena, undefined Bulgaria, and undefined 2017. 2017. "Spatio-Temporal Analysis of Urban Expansion Using Remote Sensing Data." Researchgate.Net. https://doi.org/10.5593/sgem2017/23/S10.029.

Gazis, Denos C, Robert Herman, and Richard W Rothery. 1961. "Nonlinear Follow-the-Leader Models of Traffic Flow." Operations Research 9 (4): 545–67. https://doi.org/10.1287/opre.9.4.545.

Herold, Martin, Joseph Scepan, and Keith C. Clarke. 2002. "The Use of Remote Sensing and Landscape Metrics to Describe Structures and Changes in Urban Land Uses." Environment and Planning A: Economy and Space 34 (8): 1443–58. https://doi.org/10.1068/A3496.

IEEE. 1998. "IEEE Guide for Information Technology - System Definition - Concept of Operations (ConOps) Document." IEEE Std 1362–1998. https://doi.org/10.1109/IEEESTD.1998.89424.

Kıvılcım, Cemal Özgür, Z Duran, Cemal Ö Kivilcim, and Zaide Duran. 2013. "İmpact of Rapid Urbanization on the Morphology of Historical İstanbul: Üsküdar District Case Study." In International Symposium on Environmental Pollution and Its Impact on Life in the Mediterranean Region (MESAEP), 14–18. Istanbul, Turkey.

McPhearson, Timon, Steward T.A. Pickett, Nancy B. Grimm, Jari Niemelä, Marina Alberti, Thomas Elmqvist, Christiane Weber, Dagmar Haase, Jürgen Breuste, and Salman Qureshi. 2016. "Advancing Urban Ecology toward a Science of Cities." BioScience 66 (3): 198–212. https://doi.org/10.1093/BIOSCI/BIW002.

Muslim, Mohammad, Suraj Kumar Singh, Mariella Aquilino, Eufemia Tarantino, and Majid Farooq. 2014. "Dynamics and Forecasting of Population Growth and Urban Expansion in Srinagar City-A Geospatial Approach." Academia.Edu. https://doi.org/10.5194/isprsarchives-XL-8-709-2014.

Pang, B, W Dai, T Ra, and K H Low. 2020. "A Concept of Airspace Configuration and Operational Rules for UAS in Current Airspace." In AIAA/IEEE 39th Digital Avionics Systems Conference (DASC), 1–9. https://doi.org/10.1109/DASC50938.2020.9256627.

Pathiyil, L., Low, K. H., Soon, B. H., & Mao, S. (2016). Enabling safe operations of unmanned aircraft systems in an urban environment: a preliminary study. In The International Symposium on Enhanced Solutions for Aircraft and Vehicle Surveillance Applications (ESAVS 2016) (p. 10). Berlin, Germany: German Institute of Navigation and the German Aerospace Center (DLR).

Tohidi, Nasim, and Rustam B. Rustamov. 2020. "A Review of the Machine Learning in GIS for Megacities Application." Geographic Information Systems in Geospatial Intelligence, November. https://doi.org/10.5772/INTECHOPEN.94033.

Chapter 79
Enhancing the Performance of an Active Greenhouse Dryer by Using Copper Oxide and Zinc Oxide Nano-enhanced Absorber Coating

Ceylin Şirin, Fatih Selimefendigil, and Hakan F. Öztop

Nomenclature

E	Consumed electrical energy (kWh)
M_d	Final dry mass (g)
M_i	Initial wet mass (g)
m_w	Extracted water mass (kg)
MC_{db}	Moisture content on dry basis (gwater/gdry matter)
MPGD	Modified parabolic greenhouse dryer
PGD	Parabolic greenhouse dryer
SEM	Scanning electron microscope
SMER	Specific moisture extraction rate (kg/kWh)
UGD	Uneven-span greenhouse dryer

79.1 Introduction

Solar energy plays an important role in reducing harmful environmental emissions. Solar energy can be used in different types of applications such as electrical energy generation, hot water production, space heating, and drying.

C. Şirin (✉) · F. Selimefendigil
Department of Mechanical Engineering, Manisa Celal Bayar University, Manisa, Turkey
e-mail: fatih.selimefendigil@cbu.edu.tr

H. F. Öztop
Department of Mechanical Engineering, Fırat University, Elazığ, Turkey
e-mail: hakanfoztop@firat.edu.tr

M. Z. Sogut et al. (eds.), *Proceedings of the 2022 International Symposium on Energy Management and Sustainability*, Springer Proceedings in Energy,
https://doi.org/10.1007/978-3-031-30171-1_79

Solar dryers are divided into two groups including direct and indirect dryers. Greenhouse drying systems are one of the direct solar drying systems, and they are widely utilized worldwide. Greenhouse drying systems are mostly employed in agricultural product drying applications. Therefore, the shelf lives of fresh products can be extended. There are some efficiency enhancement methods in greenhouse drying systems. These methods mostly include utilizing auxiliary heating systems (Tuncer et al. 2020), applying thermal energy storage systems (Selimefendigil and Şirin 2022), north-wall modifications (Chauhan and Kumar 2016), and flooring modifications (Ahmad and Prakash 2019).

In recent years, utilizing nanoparticles in solar-thermal systems has become a popular performance enhancement technique. Nanoparticles with high thermal conductivity values can be applied in absorber surface materials to improve the performance of solar-thermal systems (Abdelkader et al. 2020; Sivakumar et al. 2020). In a study, Kumar et al. (2020) utilized nano-embedded absorber coating in a solar heater. Graphene nanomaterials were used in the study and thermal performance was improved extensively. In another work, Selimefendigil et al. (2022) used nanoplatelets in black paint to improve the performance of a greenhouse dehumidifier. The exergy efficiency of the system was improved by 4.87% by employing nano-embedded absorber coating. There are also some studies that analyzed nano-enhanced absorber coating applied to solar desalination systems (Thakur et al. 2018; Kabeel et al. 2019).

In this study, the impact of applying nano-embedded black paint on the absorber surface of a greenhouse dryer has been analyzed experimentally. The main aim of this work is to enhance the thermal performance of a greenhouse dryer without utilizing complex additional components. In the first part of the work, different geometrical modifications have been tested. The most successful greenhouse geometry (parabolic) was selected to be tested in the second part of the experimental process. The parabolic greenhouse dryers with the same dimensions have been manufactured. One of the greenhouse dryers was designed as a conventional type. Zinc oxide (ZnO) and copper oxide (CuO) nano-integrated black paints have been applied to the second and third greenhouse dryers, respectively. Designed three parabolic greenhouse dryers have been tested in the same environmental conditions, and the results have been discussed.

79.2 Materials and Methods

79.2.1 Preparation of Nano-enhanced Absorber Coating Materials

In this study, two of the parabolic greenhouse dryers have been modified with nano-enhanced matt black paint. In this regard, two different nano-embedded absorber coating materials were prepared. Copper oxide (CuO) and zinc oxide (ZnO)

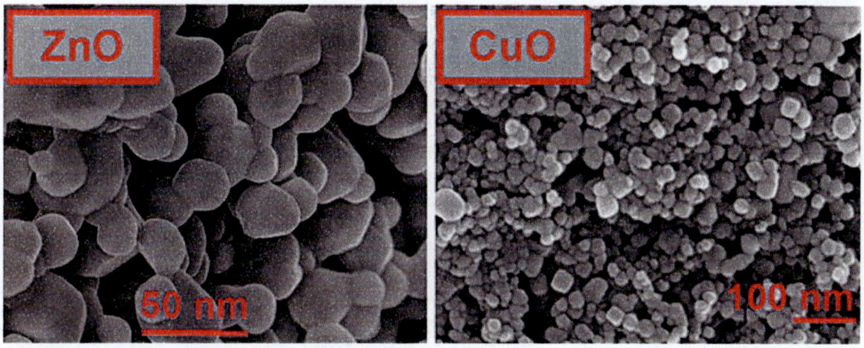

Fig. 79.1 SEM view of the used nanoparticles

nanoparticles (Nanografi Co, Turkey) were selected to be utilized in matt black paint in the current study. Nanoparticles have been integrated into industrial matt black paint in the initial stage of the preparation process. Then, obtained mixtures were stirred for 2 h with a mechanical mixer and applied to 1-mm thick aluminum sheet bottom plates of two greenhouse dryers. Both concentration ratios are 2% (wt/wt). Specific surface area and mean particle size of the CuO nanoparticles are >20 m^2/g and <77 nm, respectively. The same parameters for ZnO nanoparticles are 20.0–65.0 m^2/g and 18 nm, respectively. Scanning electron microscope (SEM) images of employed nanomaterials are shown in Fig. 79.1. Similar surveys were considered in the preparation process of the nano-enhanced black paint (Sivakumar et al. 2020; Selimefendigil et al. 2022; Abdelkader et al. 2020).

79.2.2 *Experimental Setup*

Three different greenhouse dryer geometries including and uneven-span (UGD), parabolic (PGD-I), and modified parabolic (MPGD) types have been analyzed for selecting the system with the highest performance. According to the result of the first test, the parabolic greenhouse gave the best performance result. Therefore, three same-sized parabolic greenhouse dryers have been manufactured to observe the effect of adding nano-enhanced absorber coating on the drying performance. One of the parabolic greenhouse dryers was designed as a conventional type (PGD-I). Zinc oxide (ZnO) and copper oxide (CuO) nano-integrated black paints have been applied to the second (PGD-II) and third (PGD-III) parabolic greenhouse dryers, respectively. Frames and transparent cover of the greenhouse dryers were made from wood and polyethylene greenhouse cover with 91% transmissivity, respectively. Also, 50-W powered AC fans have been placed in each greenhouse to provide airflow. Dimensions of the developed greenhouse dryers are shown in Fig. 79.2. Also, photographs of the experimental setups are given in Fig. 79.3.

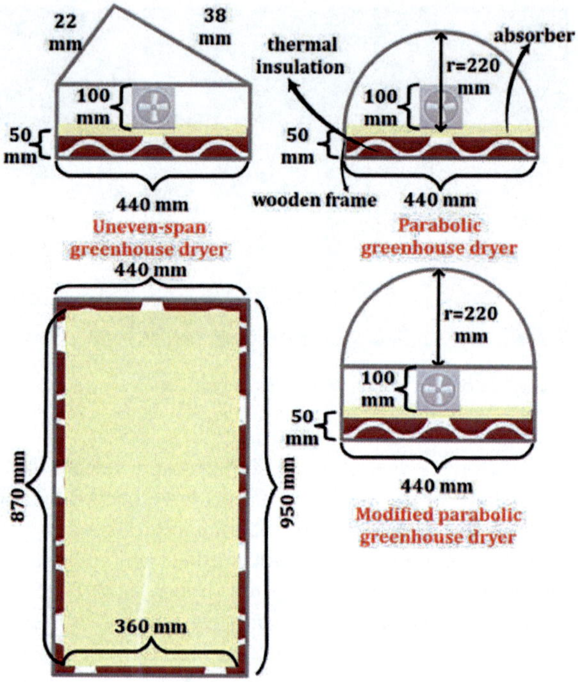

Fig. 79.2 Dimensions of the developed systems

79.2.3 Experimental Procedure

Two experiments have been performed within the scope of this work. Experiments were done in Manisa, Turkey, in July 2021. In the first experiment (Exp. 1), uneven-span, parabolic, and modified parabolic types of greenhouse dryers have been tested to determine the system geometry with the best performance. In the second experiment (Exp. 2), the effect of integrating nano-enhanced absorber coating on the drying performance of a parabolic greenhouse dryer has been analyzed. Both experiments have been performed at a constant 0.014 kg/s flow rate. Moreover, 100 g of fresh red pepper samples (sliced in 5-mm pieces) have been utilized for each system. Initial moisture content of the samples is 5.21 ± 0.22 gwater/gdry matter. Tests were started at 10:00 and finished when the difference among two mass measurements was lower than 1%. Temperature values were measured and recorded 10 every 10 s while other parameters were observed 20 min intervals. Accuracy values of utilized thermocouples, dataloggers, solarmeter, anemometer, and digital balance are ±0.5 °C, ±0.3 °C, ±10 W/m^2, ±2%, and ±0.02 g, respectively.

Fig. 79.3 A photograph of the experimental setup

79.3 Theoretical Calculations

Moisture content (dry basis) parameter is employed to model the loss of the drying product mass in the drying process, and it could be obtained by using the equation below (Aghbashlo et al. 2012):

$$MC_{db} = \left(\frac{M_i - M_d}{M_d} \right) 100 \tag{79.1}$$

Here, M_i and M_d depict the initial wet mass (g) and final dry mass (g), respectively.

Specific moisture extraction rate is a crucial performance indicator in evaluating drying systems, and it can be expressed as (Ismaeel and Yumrutaş 2020)

$$\mathrm{SMER} = \frac{m_w}{E} \tag{79.2}$$

In Eq. (79.2), m_w and E show extracted water mass (kg) and consumed electrical energy (kWh), respectively.

79.4 Results and Discussion

In this section, experimentally obtained results have been presented and discussed. Figure 79.4 presents the change in environmental conditions via drying time. Mean solar radiation values in Exp. 1 and Exp. 2 were obtained as 862 and 885 W/m^2, respectively. Also, obtained mean ambient temperatures in Exp. 1 and Exp. 2 are 31.5 °C and 31.8 °C, respectively. It should be indicated that in each experiment, all greenhouse dryers were operated till the drying process of the last greenhouse dryer terminated to observe the thermal behavior of the greenhouse dryers. In Exp. 1, average outlet air temperature values for UGD, MPGD, and PGD-I were attained as 42.87 °C, 41.96 °C, and 43.81 °C, respectively. In Exp. 2, the obtained mean outlet temperatures for PGD-I, PGD-II, and PGD-III are 44.01 °C, 45.11 °C, and 45.95 °C, respectively. As can be seen, applying ZnO and CuO nanoparticles in absorber coating material enhances the outlet temperature of the greenhouse dryer as 2.49% and 4.40%, respectively.

Figure 79.5 presents a change in moisture content values via drying time. In the first test, the drying time for UGD, MPGD, and PGD-I were obtained as 320, 360, and 300 min, respectively. In Exp. 2, the obtained drying times for PGD-I, PGD-II, and PGD-III are 280, 260, and 220 min, respectively. Modifying greenhouse dryer geometry and applying CuO nano-enhanced black paint decreased drying time significantly.

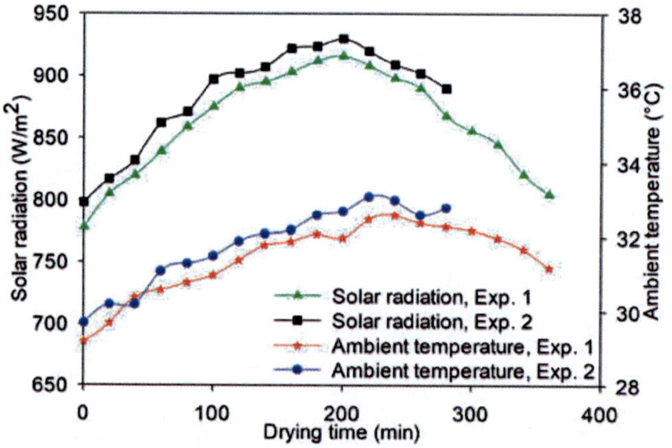

Fig. 79.4 Change in environmental conditions via time

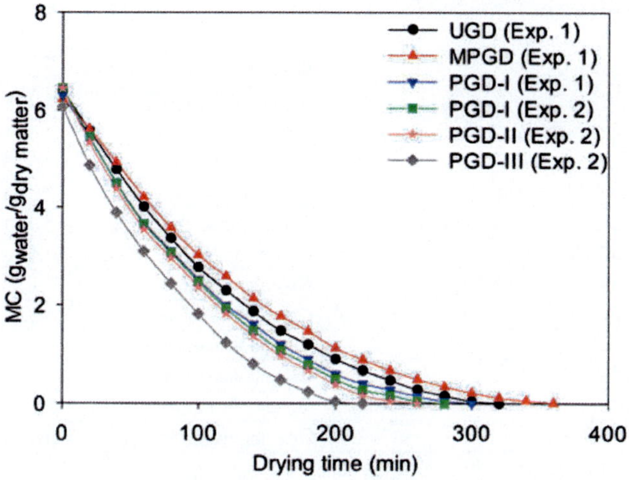

Fig. 79.5 Change in moisture content values via time

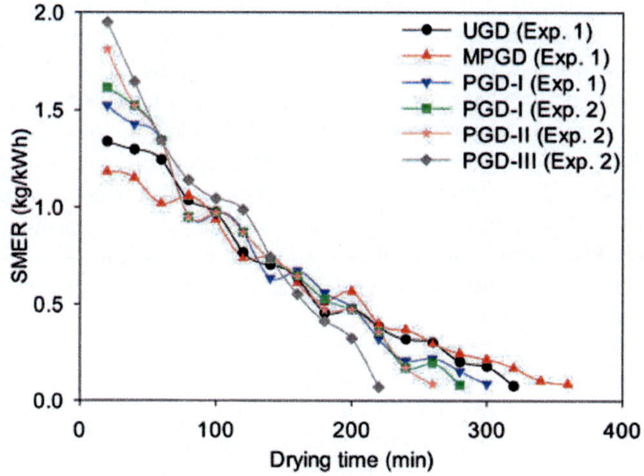

Fig. 79.6 Change in SMER values via time

Figure 79.6 presents a change in SMER values via drying time. In the first experiment (Exp. 1), SMER values for UGD, MPGD, and PGD-I were obtained as 0.65, 0.57, and 0.69 kg/kWh, respectively. In Exp. 2, attained SMER metrics for PGD-I, PGD-II, and PGD-III are 0.74, 0.80, and 0.94, respectively. As can be seen, drying time and SMER results of PGD-I in Exp.1 and Exp.2 varied in small amounts. As is known, many parameters impact drying performance such as environmental conditions and moisture content of the specific drying product. SMER results of this

Table 79.1 Comparison of obtained SMER values with similar works available in the scientific literature

Reference	Dryer type	SMER (kg/kWh)
Fudholi et al. (2015)	Indirect solar dryer	0.29
Dorouzi et al. (2018)	Indirect solar dryer	0.274
Sözen et al. (2020)	Indirect solar dryer	0.73
Chapchaimoh et al. (2016)	Heat pump dryer	0.47
Mohanraj and Chandrasekar (2009)	Indirect solar dryer	0.84
Tuncer et al. (2020)	Greenhouse dryer	0.36
Selimefendigil and Şirin (2022)	Greenhouse dryer	1.29
This study (UGD)	Greenhouse dryer	0.65
This study (MPGD)	Greenhouse dryer	0.57
This study (PGD-I)	Greenhouse dryer	0.69–0.74
This study (PGD-II)	Greenhouse dryer	0.80
This study (PGD-III)	Greenhouse dryer	0.94

study showed the successful utilization of nanoparticles in absorber coating to enhance the performance of an active greenhouse dryer.

Table 79.1 presents a comparison of obtained SMER values in the current research with similar studies that investigated various types of drying applications available in the scientific literature. As can be seen, the parabolic-formed greenhouse dryer had better performance than the uneven-span and modified parabolic greenhouse drying systems. Also, integrating ZnO and CuO nanoparticles in the absorber coating of the parabolic greenhouse dryer improved SMER values by 8.10% and 27.02%, respectively, in comparison to the conventional forced convection parabolic greenhouse dehumidification system. In addition, the designed, manufactured, and tested greenhouse drying systems in this study are in good line with some studies about drying systems, including indirect solar drying applications, heat pump dryers, and greenhouse dryers.

It should be stated that the developed greenhouse dryers in this work are pilot-scale systems. In further studies, large-scale applications can be tested. Also, in this study, only the drying performance of the systems was analyzed. The main goal of this work is to show the applicability of these nanoparticles for improving drying performance. It should be said that sample quality investigation and analysis of the toxic impact on the sample have not been surveyed within the scope of this work. To overcome this bottleneck, various nano-additives can be chosen and additional coating applications can be considered. Moreover, tested systems can be utilized to dry products other than food materials, including timber, manure, and municipal sewage sludge.

79.5 Conclusion

In this study, the effect of integrating ZnO and CuO nano-additives in black paint on the drying performance of parabolic greenhouse dryers has been investigated. Main outcomes of this work can be given as follows:

- In the initial stage of the experimental process, parabolic greenhouse dryers gave better performance in comparison to uneven-span and modified parabolic greenhouse dryers.
- Geometrical modifications and applying nanoparticles to the system significantly decreased drying time.
- SMER value of the parabolic greenhouse dryer was improved by 8.10% and 27.02%, respectively, by adding ZnO and CuO nanoparticles to the absorber coating.

Acknowledgments This work is supported by the Scientific Research Project Coordination Unit (BAP) of Manisa Celal Bayar University under Grant no 2020-088, whose support is gratefully acknowledged.

References

Abdelkader TK, Zhang Y, Gaballah ES, Wang S, Wan Q, Fan Q (2020) Energy and exergy analysis of a flat-plate solar air heater coated with carbon nanotubes and cupric oxide nanoparticles embedded in black paint. Journal of Cleaner Production 250:119501. https://doi.org/10.1016/j.jclepro.2019.119501

Aghbashlo M, Mobli H, Rafiee S, Madadlou A (2012) Energy and exergy analyses of the spray drying process of fish oil microencapsulation. Biosyst. Eng. 111: 229–241. https://doi.org/10.1016/j.biosystemseng.2011.12.001

Ahmad A, Prakash O (2019) Thermal analysis of north wall insulated greenhouse dryer at different bed conditions operating under natural convection mode. Environmental Progress & Sustainable Energy 38: e13257. https://doi.org/10.1002/ep.132572

Chapchaimoh K, Poomsa-ad N, Wiset L, Morris J (2016) Thermal characteristics of heat pump dryer for ginger drying. Applied Thermal Engineering 95:491–498. https://doi.org/10.1016/j.applthermaleng.2015.09.025

Chauhan P S, Kumar A (2016) Performance analysis of greenhouse dryer by using insulated north-wall under natural convection mode. Energy Reports 2:107–116. https://doi.org/10.1016/j.egyr.2016.05.004

Dorouzi M, Mortezapour H, Akhavan HR, Moghaddam AG (2018) Tomato slices drying in a liquid desiccant-assisted solar dryer coupled with a photovoltaic-thermal regeneration system. Solar Energy 162:364–371. https://doi.org/10.1016/j.solener.2018.01.025

Fudholi A, Sopian K, Alghoul MA, Ruslan MH, Othman MY (2015) Performances and improvement potential of solar drying system for palm oil fronds. Renewable Energy 78:561–565. https://doi.org/10.1016/j.renene.2015.01.050

Ismaeel HH, Yumrutaş R (2020) Thermal performance of a solar-assisted heat pump drying system with thermal energy storage tank and heat recovery unit. International Journal of Energy Research 44:3426–3445. https://doi.org/10.1002/er.4966

Kabeel AE, Sathyamurthy R, Sharshir SW, Muthumanokar A, Panchal H, Prakash N, Prasad C, Nandakumar S, El Kady MS (2019) Effect of water depth on a novel absorber plate of pyramid solar still coated with TiO2 nano black paint. Journal of Cleaner Production 213:185–191. https://doi.org/10.1016/j.jclepro.2018.12.185

Kumar R, Kumar Verma S, Kumar Sharma V (2020) Performance enhancement analysis of triangular solar air heater coated with nanomaterial embedded in black paint. Materials Today: Proceedings 26:2528–2532. https://doi.org/10.1016/j.matpr.2020.02.538

Mohanraj M, Chandrasekar P (2009) Performance of a forced convection solar drier integrated with gravel as heat storage material for chili drying. J Eng Sci Technol. 4:305–314.

Selimefendigil F, Şirin C (2022) Experimental investigation of a parabolic greenhouse dryer improved with CuO nano-enhanced latent heat thermal energy storage unit. International Journal of Energy Research 46:3647–3662. https://doi.org/10.1002/er.7412

Selimefendigil F, Şirin C, Öztop HF (2022) Improving the performance of an active greenhouse dryer by integrating a solar absorber north wall coated with graphene nanoplatelet-embedded black paint. Solar Energy 231:140–148. https://doi.org/10.1016/j.solener.2021.10.082

Sivakumar S, Velmurugan C, Jacob Dhas DSE, Solomon AB, Dev Wins KL (2020) Effect of nano cupric oxide coating on the forced convection performance of a mixed-mode flat plate solar dryer. Renewable Energy 155:1165–1172. https://doi.org/10.1016/j.renene.2020.04.027

Sözen A, Şirin C, Khanlari A, Tuncer AD, Gürbüz EY (2020) Thermal performance enhancement of tube-type alternative indirect solar dryer with iron mesh modification. Solar Energy 207: 1269–1281. https://doi.org/10.1016/j.solener.2020.07.072

Thakur AK, Agarwal D, Khandelwal P, Dev S (2018) Comparative Study and Yield Productivity of Nano-paint and Nano-fluid Used in a Passive-Type Single Basin Solar Still. Advances in Smart Grid and Renewable Energy. Lecture Notes in Electrical Engineering. 435. Springer, Singapore. https://doi.org/10.1007/978-981-10-4286-7_70

Tuncer AD, Sözen A, Khanlari A, Amini A, Şirin C (2020) Thermal performance analysis of a quadruple-pass solar air collector assisted pilot-scale greenhouse dryer. Solar Energy 203:304–316. https://doi.org/10.1016/j.solener.2020.04.030

Chapter 80
Investigation of Light Transmittance of Coatings Containing SiO_2 and TiO_2 Nanoparticles Made by Electrospinning Technique

Ali Samet Sarkın, Şafak Sağlam, and Nazmi Ekren

Nomenclature

PLA Polylactic acid
PMMA Polymethyl methacrylate

80.1 Introduction

The electrospinning technique is to obtain coating by transferring material between positive and negative poles by using high DC voltage and magnetic fields (Zheng 2019). DC source is the transmitting (positive) side and the surface to be coated is the receiving (negative) side. In this technique, a solution is prepared and the substances to be transferred is added. Generally, nanosized materials are used. It is used to obtain polymer, metal, ceramic, and composite materials (Varghese et al. 2019). 3D coatings can be obtained. In this way, it is used in the biomedical field for self-healing and drug release, such as skin dressing and wound dressing (Gendelberg et al. 2019). Its usage area has increased with the studies carried out in the last 20 years.

In an electrospinning application, a high voltage DC power supply, needle, spinneret, and a grounded conductive collector are used (Chronakis 2005). A typical

A. S. Sarkın (✉) · Ş. Sağlam · N. Ekren
Kadirli Vocational High School of Higher Education, Osmaniye Korkut Ata University, Osmaniye, Türkiye

Faculty of Technology, Marmara University, Istanbul, Türkiye
e-mail: sametsarkin@osmaniye.edu.tr; ssaglam@marmara.edu.tr; nazmiekren@marmara.edu.tr

M. Z. Sogut et al. (eds.), *Proceedings of the 2022 International Symposium on Energy Management and Sustainability*, Springer Proceedings in Energy,
https://doi.org/10.1007/978-3-031-30171-1_80

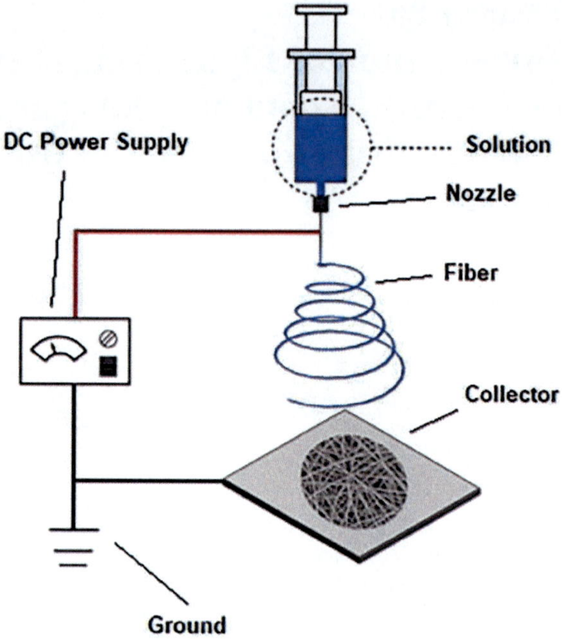

Fig. 80.1 Basic electrospinning setup. (Long et al. 2019)

electrospinning application can be seen in Fig. 80.1 (Long et al. 2019). It is known that the collector is applied with a rotating drum in electrospinning applications, but there are also collector types as static collectors and precision-deposited collectors (Matthews et al. 2002; Li et al. 2003, 2004; Deitzel et al. 2001; Buttafoco et al. 2006). The nozzle used as the solution output source determines the physical properties of the fiber, like the thickness of the fiber and the cone output (Kızıldağ et al. 2012; Thompson et al. 2007). Nozzle types can be coaxial, gas-assistant, a two-component nozzle, pointed-tip, and multi-needle (Sun et al. 2003; Larsen et al. 2004; Lin et al. 2005; Sun et al. 2006).

While nozzle type and width are effective parameters in the physical properties of the fibers seen in Fig. 80.1, the distance between the nozzle and collector, voltage level, solution content, flow rate, and collector type are the basic parameters of the electrospinning technique. The applied voltage varies with solution content, distance, and type of collector. It is generally applied in the range of 10–24 kV. The higher voltage increases the rate of formation and output of the fiber, but the flow rate and the voltage must be balanced. If the flow rate cannot meet the high flowing due to voltage, the balance of the fiber is disturbed, the conical output is not obtained, and thinner fibers can be observed and even interruptions in the fiber flow occur. Multi-fiber outputs are observed under high flow rate and high voltage (Zhong et al. 2002; Lee et al. 2004). The distance between the nozzle and the collector is usually set between 10 and 20 cm in applications. The distance determines the spread of the fiber over the collector. It depends on the flow rate and voltage. Increasing the distance reduces the thickness of the fiber, and this effect can also cause the fiber to break. If the distance is close, it may disrupt the homogeneous distribution and lead to thicker fiber formation. The last of the basic parameters is the flow rate. The flow rate (ml/h) should be adjusted in accordance with the solution content. The flow rate can be applied in the range of 0.1–1 ml/h. The flow rate should be adjusted considering the voltage level and the distance. In cases of high flow rate, if the voltage level is insufficient, dripping from the nozzle and clogging over time can be observed. Moreover, the droplets carried on the fibers may form (wet fibers) and disrupt the surface distribution (Chowdhury and Stylios 2010). At an insufficient flow rate, thinning and breaking of the fibers may occur. In case of an insufficient flow rate, the voltage can be reduced. In addition to the basic parameters, conductive particles in the solution content affect the main parameters. The high or low humidity of the room increases the fiber thickness, while high room temperature leads to thin fiber (Yuya et al. 2010; Içoğlu and Ogulata 2013).

One of the uses of the electrospinning technique is to obtain a coating. Our study aims to obtain coating on glass by electrospinning technique. It is aimed to increase the light transmittance of the glass, reduce the reflection from the surface of the glass, and provide a self-cleaning via a hydrophobic effect. Problems of coatings are adhesion, durability on the surface, and homogeneous distribution on the surface. When the literature was investigated, it was seen that TiO_2/SnO_2 and photocatalyst nanofiber were obtained as coating materials in the electrospinning method. Titania nanofiber with a porous structure was obtained by using a TiO_2/PVP mixture solution with a 1-kV/cm electric field at a flow rate of 0.1 ml/h and thinner fibers with a thickness of 52–374 nm (Li and Xia 2003). A coating was prepared 50–50% SiO_2/TiO_2 and PVAc (polyvinyl/acetate) polymer, and fiber thicknesses of 50–400 nm were obtained by calcining at 600–1400 °C for 2 h (Ding et al. 2003). The application was made using CdS and ZnS doped TiO_2 and PVP (polyvinyl/pyrrolidone) polymer with 15 cm and a voltage of 10 kV (Li et al. 2016). The nano-coating obtained with $BaTiO_3$/PVP composite material was applied at a voltage level of 12–15–18 kV to obtain an average fiber thickness of 170–379 nm (Ghalambor and Noroozpoor 2014).

80.2 Methodology

The study aims to make a coating with nanostructures that increase light transmission and reduce reflection on glass surfaces with the electrospinning technique. Coating materials are composite solutions containing nano-sized (10–30 nm particle size, Rutile, 99.5% pure) SiO_2/TiO_2 and containing PLA (polylactic acid) or PMMA (polymethyl methacrylate) as polymer. The polymers and nanoparticles were weighed on precision balances and mixed in a magnetic stirrer. Chloroform ($CHCl_3$) was used as the main solvent at the rate of 20 ml. Polymers and nanoparticles were prepared in different ratios and contents, as shown in Table 80.1.

Table 80.1 Solution contents

Sol. No.	PLA (g)	PMMA (g)	$CHCl_3$ (ml)	SiO_2 (g)	TiO_2 (g)
1	1.6	–	20	0.2	0.2
2	2	–	20	0.1	–
3	2	–	20+ eth + Ast.A.	–	0.1
4	2	–	20	–	0.05
5	1.6	–	20	0.05	0.05
6	2	–	20	0.05	–
7	1.6	–	20	0.025	0.025
8	–	1.6	20	0.025	–
9	–	1.6	20	0.025	0.025
10	2	–	20	0.05	0.05

The solutions were made at ambient temperature and humidity by turning off the temperature of the magnetic stirrer and covering them with parafilm at a speed of 1000 rpm. In solutions, number 1 was obtained by mixing them all for 4 h. In other solutions, the polymer was first dissolved in 20 ml of chloroform ($CHCl_3$) and then nanoparticles were added. Solution-2 was dissolved in polymer for 2 h and mixed with nanoparticles for another 2 h. In solution-3, the polymer was mixed for another 2.5 h with the added nanoparticle after 4 h of dissolving time. In solution-4, after the polymer was dissolved in 2 h, nanoparticles were added and mixed for 18 h. In solution-6, the polymer was mixed for 2 h and for another 3 h after the nanoparticle was added. In solution-7, the polymer was dissolved for 2.5 h and homogenized, and then nanoparticles were added and mixed for 4 h. After solution-8 was mixed for 2 h to dissolve the polymer, SiO_2 was added and mixed for another 18 h. Solution-9 has 1 h of polymer dissolution and 3 h of mixing time until it becomes homogeneous after adding nanoparticles. Solution-10 has a polymer dissolution time of 2 h and a nanoparticle mixing time of 2.5 h.

After the prepared solutions were put into the injector, they were placed in the electrospinning (Inovenso-Basic electrospinner) seen in Fig. 80.2 with an automatic injector pump (Inovenso-IPS12). The automatic injector pump controls the flow rate, and the value is entered in ml/h. Voltage (kV) and distance (cm) adjustments are made in the electrospinning device. The application time is determined as second. Four electrospinning parameters were used in different values in applications. All parameters can be seen in Table 80.2.

Applications were made on soda-lime laboratory glass slides (25 × 75 × 1 mm, pre-cleaned) using different parameters.

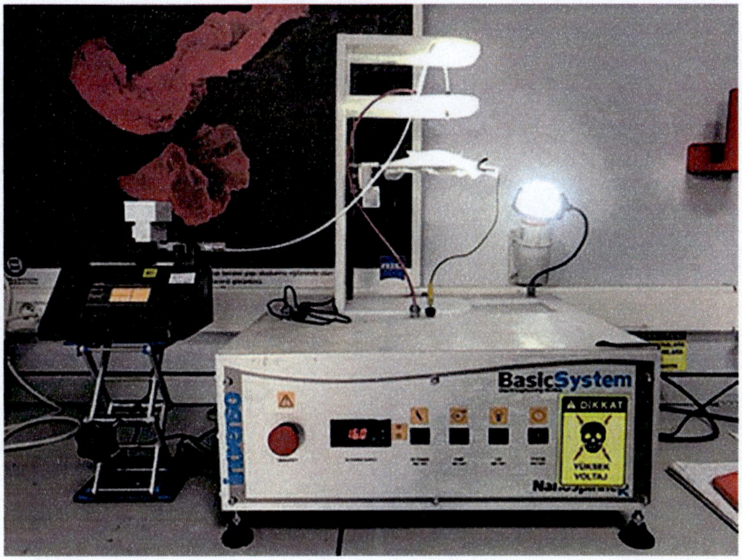

Fig. 80.2 Electrospinning application setup

Table 80.2 Parameters

Flow rate (ml/h)	Voltage (kV)	Duration (sec)	Distance (cm)
0.1	16	60	10
0.2	20	120	15
0.3	22	150	
0.5	24	180	
		210	
		240	
		300	

80.3 Results and Discussion

The fiber structures of the obtained coatings were imaged under an optical microscope (Olympus – Colorview) by zooming in 5× –10× –20×. The physical structure of the fibers, surface adhesion, homogeneous spread on the surface, and the distribution of nanoparticles on the fibers were investigated by optical microscope imaging. The prepared coatings, which can be seen in Fig. 80.3, were obtained with different solution contents and parameters.

Figure 80.3a coating was prepared with solution-5. The parameters are as follows: flow rate, 0.2 ml/h; voltage, 20 kV; distance, 10 cm; and duration, 120 s. Figure 80.3b was also prepared with solution-5. Distance and voltage were not changed. The flow rate was set as 0.3 ml/h and the duration was set as 180 s. Although the surface distribution is good in Fig. 80.3a, it is seen that the surface

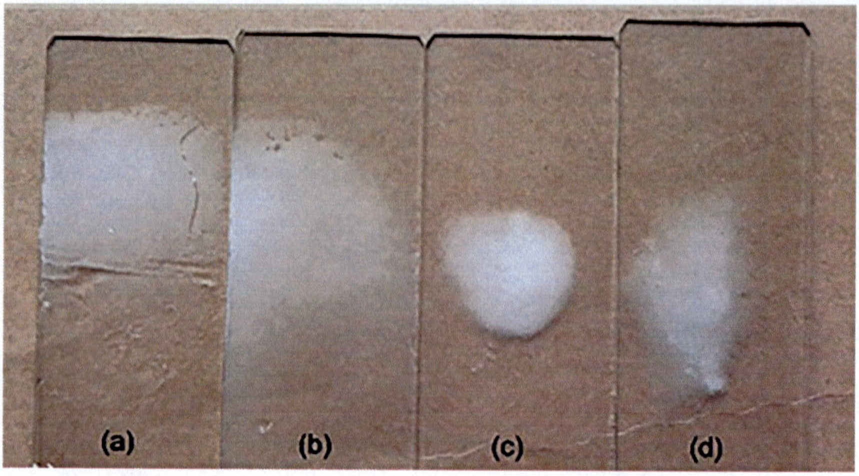

Fig. 80.3 Samples

adhesion is insufficient. Figure 80.3b was found to be more successful in terms of surface distribution and adhesion. Figure 80.3c coating was prepared with solution-7. Electrospinning parameters are as follows: flow rate, 0.2 ml/h; voltage, 20 kV; distance, 15 cm; and duration, 120 s. In Fig. 80.3c, it was observed that the surface distribution was poor, but the adhesion to the surface was good. The coating seen in Fig. 80.3d was prepared with solution-4. In this coating, the parameters were set as follows: flow rate, 0.2 ml/h; voltage, 24 kV; distance, 15 cm; and duration, 210 s. With these parameters, there was no homogeneous distribution on the surface, and deposition was observed in some areas.

When the electrospinning samples prepared with different solutions and parameters were investigated, deposition was observed on the surfaces since the voltage increased the flow rate and there was not enough flowing. It has been observed that the high-set voltage negatively affects the surface distribution. It has been seen that the flow rate of 0.2 or 0.3 for all solutions is enough. Distance, which is the distance between the receiver and the source, was applied as 10–15 cm. It was observed that the dispersion on the surface was more homogeneous in coatings with 10 cm. The duration was applied in the range of 60–300 s. In this parameter, sufficient coating was achieved in the range of 120–180 s. It was analyzed that coating thickness increased and light transmission decreased in coatings with more duration time.

After the coatings were physically investigated, fiber structures were analyzed with an optical microscope at 5× −10× −20× zoom. The image of the fiber formations and the distribution of nanomaterials of the samples given in Fig. 80.3 can be seen in Fig. 80.4.

Homogeneous distribution and nanomaterials on the fiber can be seen in Fig. 80.4a, b in optical microscope imaging, which is similar to the physical appearance of the coatings. In Fig. 80.4c, which was prepared with solution-7, the

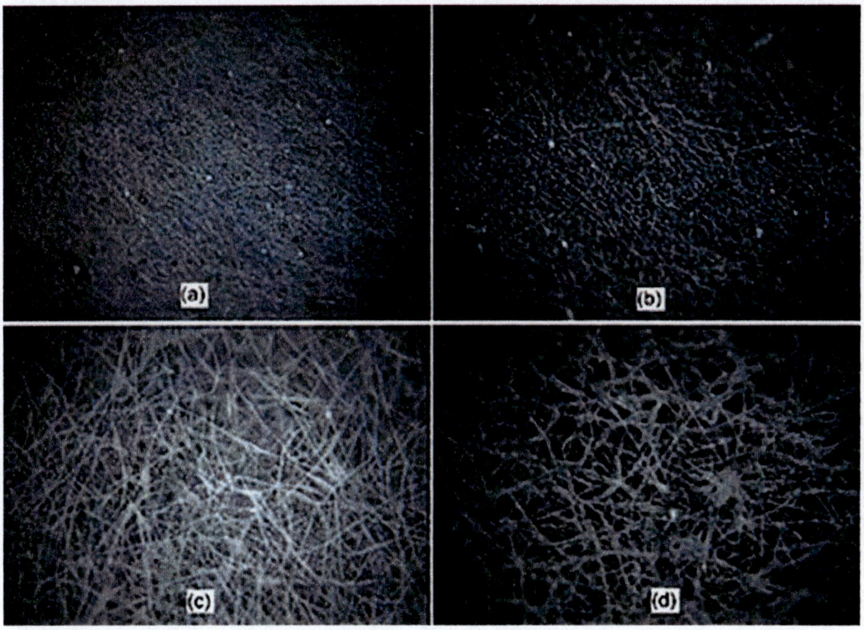

Fig. 80.4 Samples under the optical microscope

application parameters and the solution preparation time optimized the viscosity. In this way, it was observed that the polymer density and nanoparticle distribution were compatible. In Fig. 80.4d, similar to the image in Fig. 80.3d, low nanoparticle distribution was observed among the dense polymer due to the high duration parameter.

After the coatings were obtained, the light transmittance was measured, and the result of the study was achieved. Light transmittance was measured by scanning 300–800 nm wavelength with a PG Instrument-T60 UV-VIS Spectrophotometer. The light transmittance of the blank glass slide was 91%. All light transmittances can be seen in Fig. 80.5.

The light transmittance at 300–800 nm wavelength of the coatings of all solutions is given in Fig. 80.5. Peak values of light transmittance at 550 nm are given in Table 80.3.

80.4 Conclusion

Anti-reflection and self-cleaning coatings were made on the laboratory slide glass with the electrospinning technique. PLA and PMMA polymers and SiO_2 and TiO_2 nanoparticles were used in a literature survey. Solutions prepared at different ratios

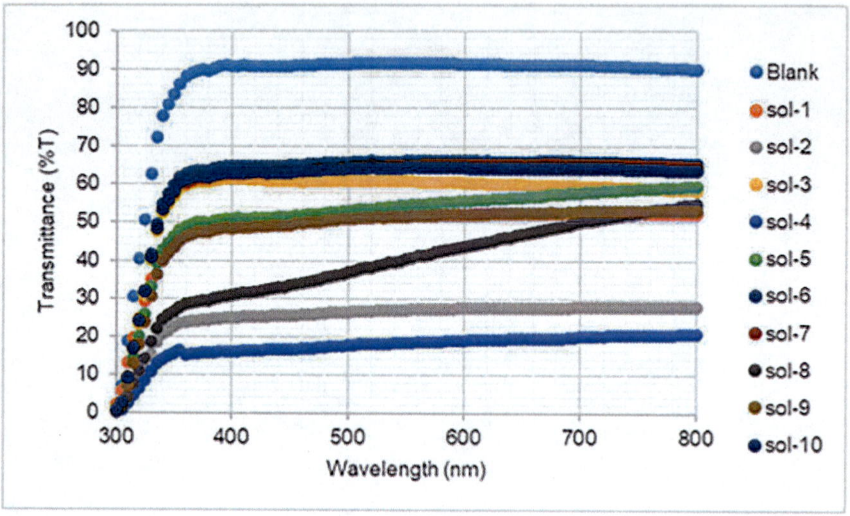

Fig. 80.5 The transmittance of coatings (%T)

Table 80.3 The peak of % Transmittance

Solution number	Peak %T
1	52
2	27.1
3	60.9
4	18.5
5	54.5
6	66.2
7	64.8
8	40.4
9	51.7
10	64.4

were applied with different parameters in electrospinning. It was measured that solution-7 containing 1.6 g of PLA, 0.025 g of SiO_2, and 0.025 g of TiO_2 in coatings achieved the highest light transmittance. Although 64.4% light transmission was achieved in the coatings made with solution-10 with lower viscosity, it failed in terms of surface adhesion and coating thickness. In coatings, solutions containing 0.025 g of SiO_2 and 0.025 g of TiO_2 were successfully evaluated in terms of surface adhesion, formation of fibers, and nanoparticles on the fiber. All coatings were considered unsuccessful, with light transmittance of less than 91% of blank glass. Solution contents and electrospinning parameters affect the failure of coatings. Reducing the polymer ratio of the solutions will increase the viscosity and become more fluid, making coating easier. In electrospinning parameters, it is thought that by decreasing the voltage, slower flowing will be achieved, and controlled flow will be provided.

Acknowledgments This work is derived from the doctoral Ph.D. thesis titled "Fotovoltaik Panellerde Verim Artırıcı Nano Yapılı Kaplamalar" (in English "Nanostructures Efficiency-Enhanced Coatings on Photovoltaic Panels") by student Ali Samet Sarkın and thesis advisors Nazmi Ekren and Şafak Sağlam.

References

Buttafoco L, Kolkman NG, Engbers-Buijtenhuijs P, Poot AA et al. (2006) Electrospinning of collagen and elastin for tissue engineering applications. Biomaterials 27(5):724–734. https://doi.org/10.1016/j.biomaterials.2005.06.024

Chowdhury M, Stylios G (2010) Effect of experimental parameters on the morphology of electrospun nylon 6 fibres. International Journal of Basic & Applied Sciences 10(6):70–79. http://www.ijens.org/100906-2323%20IJBAS-IJENS.pdf

Chronakis IS (2005) Novel nanocomposites and nanoceramics based on polymer nanofibers using electrospinning process—A review. Journal of Materials Processing Technology 167:283–293. https://doi.org/10.1016/j.jmatprotec.2005.06.053

Deitzel JM, Kleinmeyer J, Harris D, Beck Tan NC (2001) The effect of processing variables on the morphology of electrospun nanofibres and textiles. Polymer 42(1):261–272. https://doi.org/10.1016/S0032-3861(00)00250-0

Ding B, Kim H, Kim C, Khil M et al. (2003) Morphology and crystalline phase study of electrospun TiO_2–SiO_2 nanofibres. Nanotechnology 14:532–537. https://doi.org/10.1088/0957-4484/14/5/309

Long YZ, Yan X, Wang XX, Zhang J, Yu M (authors). Ding B, Wang X, Yu J (eds) (2019). Electrospinning: Nanofabrication and Applications, A volume in Micro and Nano Technologies. Elsevier – William Andrew.

Ghalambor Dezfuli AM, Noroozpoor A (2014) Synthesis and characterization of electrospun BaTiO3/PVP composite nanofibers, in Proceedings of the 14th IEEE International Conference on Nanotechnology. https://doi.org/10.1109/NANO.2014.6968017

Gendelberg N, Bird D, Ravindra NM (authors). Guo JZ, Song K, and Liu C (eds) (2019). Polymer-Based Multifunctional Nanocomposites and Their Applications. Elsevier

Icoglu HI, Ogulata RT (2013) Effect of ambient parameters on morphology of electrospun polyetherimide (PEI) fibers, Tekstil ve Konfeksiyon 23:313. https://dergipark.org.tr/tr/download/article-file/345725

Kızıldağ N, Beceren Y, Kazanci M, Cukul D (2012) Effect of needle diameter on diameter of electrospun silk fibroin nanofibers, in RMUTP International Conference: Textiles and Fashion, p. 17. http://textileconference.rmutp.ac.th/wp-content/uploads/2011/09/Proceeding-book_The-4-RMUTP-International-Conference.pdf

Larsen G, Spretz R, Velarde-Ortiz R (2004) Use of coaxial gas jackets to stabilize taylor cones of volatile solutions and to induce particle-to-fiber transitions. Advanced Materials 16(2):166-169. https://doi.org/10.1002/adma.200306021

Lee JS, Choi KH, Ghim HD, Kim SS et al. (2004) Role of molecular weight of atactic poly(vinyl alcohol) (PVA) in the structure and properties of PVA nanofabric prepared by electrospinning. Journal of Applied Polymer Science 93(4):1638–1646. https://doi.org/10.1002/app.20602

Li D, Wang Y, Xia Y (2003) Electrospinning of polymeric and ceramic nanofibers as uniaxially aligned arrays. Nano Letters 3(8):1167–1171. https://doi.org/10.1021/nl0344256

Li D, Xia Y (2003) Fabrication of Titania Nanofibers by Electrospinning. Nano Letters 3(4): 555–560. https://doi.org/10.1021/nl034039o

Li D, Wang Y, Xia Y (2004) Electrospinning nanofibers as uniaxially aligned arrays and layer-by-layer stacked films. Advanced Materials 16(4):361–366. https://doi.org/10.1002/adma.200306226

Li Y, Gao F, Zhao L, Ye Y et al. (2016) Reversing CdS and ZnS preparation order on electrospun TiO2 and its effects on photoelectrochemical property. Micro & Nano Letters 11(11):731–733. https://doi.org/10.1049/mnl.2016.0134

Lin T, Wang H, Wang X (2005) Self-crimping bicomponent nanofibers electrospun from polyacrylonitrile and elastomeric polyurethane. Advanced Materials 17(22):2699–2703. https://doi.org/10.1002/adma.200500901

Matthews JA, Wnek GE, Simpson DG, Bowlin GL (2002) Electrospinning of collagen nanofibers. Biomacromolecules 3(2):232–238. https://doi.org/10.1021/bm015533u

Sun D, Chang C, Li S, Lin L (2006) Near-field electrospinning. Nano Letters 6(4):839–842. https://doi.org/10.1021/nl0602701

Sun Z, Zussman E, Yarin AL, Wendorff JH, Greiner A (2003) Compound core-shell polymer nanofibers by co-electrospinning. Advanced Materials 15(22):1929–1932. https://doi.org/10.1002/adma.200305136

Varghese RJ, Sakho EHM, Parani S, Thomas S, Oluwafemi SO, Wu J (authors). Thomas S, Sakho EHM, Kalarikkal N, Oluwafemi SO et al. (eds) (2019). Nanomaterials for Solar Cell Applications. Elsevier

Thompson CJ, Chase GG, Yarin AL, Reneker DH (2007) Effects of parameters on nanofiber diameter determined from electrospinning model. Polymer 48(23): 6913–6922. https://doi.org/10.1016/j.polymer.2007.09.017

Yuya N, Kai W, Kim BS, Kim IS (2010) Morphology controlled electrospun poly (vinyl pyrrolidone) fibers: effects of organic solvent and relative humidity. Journal of Materials Science and Engineering with Advanced Technology, 2:97.

Zheng Y (2019). Bioinspired Design of Materials Surfaces - Materials Today. Elsevier

Zhong XH, Kim KS, Fang DF, Ran SF et al. (2002) Structure and process relationship of electrospun bioabsorbable nanofiber membranes. Polymer 43(16):4403–4412. https://doi.org/10.1016/S0032-3861(02)00275-6

Chapter 81
An Assessment of Sustainable Waste Management Strategies in Airports

Orhan Yucel, Alper Dalkiran, Seval Kardes Selimoglu, and T. Hikmet Karakoc

Nomenclature

ATAG	Air Transport Action Group
LCC	Lockheed California Company
IATA	International Air Transport Association
ACI	Airport Council International

81.1 Introduction

Air transportation is a fast-growing industry that serves 4.5 billion passengers and handles 62 million tonnes of cargo annually (ATAG 2021). Environmental effects create undesirable problems for airports and surrounding areas, and 6.1 million tonnes of waste are generated annually from air transport operations (IATA 2021).

O. Yucel (✉)
Cappadocia University, Vocational Collage, Nevşehir, Turkey
e-mail: orhan.yucel@kapadokya.edu.tr

A. Dalkiran
School of Aviation, Süleyman Demirel University, Keciborlu, Isparta, Türkiye

S. K. Selimoglu
Anadolu University, Eskisehir, Turkey

T. H. Karakoc
Faculty of Aeronautics and Astronautics, Eskisehir Technical University, Eskisehir, Türkiye

Information Technology Research and Application Centre, Istanbul Ticaret University, İstanbul, Türkiye

M. Z. Sogut et al. (eds.), *Proceedings of the 2022 International Symposium on Energy Management and Sustainability*, Springer Proceedings in Energy,
https://doi.org/10.1007/978-3-031-30171-1_81

Despite the common need, there exist few studies in the literature on finding solutions to the problems. This study, by bridging the gap in the field, investigates the most commonly used waste management strategies and alternatives to create and improve a sustainable green economy in airports.

The study assesses the sustainable solutions used by the world's leading airports to the various problems caused by airport wastes. For this purpose, an in-depth literature review on the subject was conducted. Additionally, Sustainability Reports, Annual Reports, and Company Web Pages of airports are studied using the Content Analysis Method.

Fast growth in aviation has raised concerns about its environmental footprint on the issues like energy consumption, materials use, and waste generation.

The airport waste management system usually encompasses the identification of the waste sources, equipment, facility, and infrastructure requirements to process various types and amount of waste and the effectiveness of waste collection, storage, recycling, and waste disposal (Janić 2011). It requires segregation of waste in waste collection areas, reduction in waste generation, increased reuse and recycling of wastes, and permanent improvement in airport waste management applications (Eriksson 2017; Arushanyan et al. 2017; Baxter et al. 2018).

Waste Management Strategy in the airports aims to eliminate or minimize the negative effects of the wastes and creates additional economic value through effective approaches like source segregation, reuse, recycle, etc.

The type and properties of the waste define both the proper waste management program and treatment and disposal technics (Shi et al. 2016).

Airports can use incentives and advertising programs as part of their waste management strategy. Participation and support from the businesses and airport stakeholders are crucial to achieving sustainable waste management goals for the airports.

81.2 Airport Waste Types and Handling

Airport Waste means unwanted or unused products/materials/substances that are generated or brought at the airport site and needs handling with special methods (Baxter et al. 2018).

Airports, especially large ones with heavy traffic, produce a large amount of waste like small cities. If the type and characteristics of the waste are known, proper waste management programs and treatment and disposal technics can be established (Shi et al. 2016).

Various types of waste are generated from airport operations, including Municipal Solid Waste (MSW); Construction and Demolition Debris (CDD); waste from flights (deplaned waste); compostable waste; hazardous and industrial waste; and lavatory waste. The management of waste in a proper and useful way leads to improved airport operations and decreases environmental impacts (ICAO 2022).

MSWs are wastes from daily operations in an airport terminal. They are produced by tenants, airlines, and cargo operators in the form of product packages, furniture, clothing, bottles, leftover foods, and newspapers (Mehta 2015; Parameshwar 2011).

Deplaned waste is a specific type of MSW that is unloaded from aircraft after flights. It constitutes almost 20% of an airport's total MSW and includes "galley waste"—materials collected as part of the de-catering process, including compactor boxes, waste bags, food carts, and bonded carts (ICAO 2022). The length of the flight affects the amount and type of waste generated on an aircraft (Chandrappa and Das 2012).

Cleaning wastes also arises from the left-over wastes of passengers, lavatory waste bins, and medical items. Compostable wastes are generated from unused or trash food in restaurants and food stores in an airport.

Research shows that around 25% of cabin waste worth US $ 3.9 billion comprised leftover food and beverages in 2017. Since these wastes are landfilled or incinerated, minimization would lead to significant cost savings for airlines. This practice is included in the Sustainable Development Goals (SDGs), which require cutting global food waste per capita in half by 2030 (IATA 2019).

Green wastes are generated from landscaping activities like tree, bush, and grass cutting, leaves, weeds, tree branches, etc. They can be composted in airport sites or regional facilities (Sebastian and Louis 2021).

Activities like land clearing, maintenance of runways and taxiways, and new terminal construction generate C & D wastes. These generate wastes like pieces of concrete, asphalt, wood, metal, soil, bricks, rock, stone, gravel, sand, roofing materials, drywall, carpet, plastic, pipe, etc. (ICAO 2022). Airports need to handle C & D wastes with special waste streams and facilities.

Wastes from airplane lavatories contain cleaning agents, bacteria, virus, or other harmful microorganisms, which can be harmful to health. These wastes in the airplane are kept in special tanks and carried away to the cleaning sites with special trucks, undergo a disinfection treatment, and after that they can be released for drainage (Sebastian and Louis 2021).

Spill clean-up and remediation waste arises during the cleaning of spills and the remediation of contamination. They can be generated as a result of spills from storage tanks, vehicles, and maintenance activities and during oil and gas production (Mehta 2015; FAA 2013; Baxter et al. 2018).

Hazardous wastes are generated from aircraft refueling and maintenance, vehicle maintenance, emergency power generation, and other operations. Oil leakages, chemicals used for cleaning, and substances having combustible, toxic, corrosive, or reactive effects are the primary type of hazardous waste and should be handled in accordance with special regulations (ICAO 2022).

Hazardous waste disposal should be controlled and supported by the necessary funding at local, regional, and national levels. In the USA, it is regulated by the Environmental Protection Agency (EPA), the Organization for Health and Safety (OSHA), and the Department of Transportation (DOT) (Wieher 2002).

81.3 Airport Waste Management

Waste management is the process of handling the waste from aircraft, concessioners, maintenance operations, ground vehicles, offices, construction, and other activities at the airport as well as compliance with the different requirements of the various types of waste (ICAO 2022).

The Waste Management Strategy in the airports aims to minimize the negative effects of the waste streams and generate economic advantages by focusing on the alternative usages of the wastes.

The airport waste management system includes the identification of the waste sources, facility, equipment, and infrastructure requirements for different types and volumes of waste streams and the efficiency and effectiveness of the waste collection, storage, recycling, and disposal practices (Janić 2011).

Waste management system incorporates source separation, recycling, and minimization of waste streams, a charging regime for airport waste generators, reuse of material and equipment as much as possible, and the promotion and usage of renewable products that cause minimum harm to the environment (Dimitriou and Voskaki 2010).

Three main waste treatment options at airports are as follows (Baxter et al. 2018):

- Recycling: the waste is broken down to get ingredients for the manufacturing of new items or waste-to-energy conversion (Spilsbury 2010).
- Incineration: "an established combustion process to reduce the wastes into CO_2, water vapor, gases and a relatively small, non-combustible residue" (Guyer 2011).
- Landfill: depositing the waste that cannot be recycled or incinerated in a special area. Airports eliminate the cost of transportation if their landfilling sites are close to the airport, but these sites can attract pests and birds (ICAO 2021).

81.4 Results and Discussion

Waste Management Strategy and Practices of World leading international airports with high volumes of passengers are studied from the Company Reports. The result of the study is presented in Table 81.1. Table 81.1 also includes the amount of waste and recycling rates of the airports, which can provide benchmarking information among the airports.

The practice of the world's leading airports indicates several waste disposal streams. Wastes from the terminals and airport services are processed at the central container station, where they are separated for incineration, disposal, or landfilling (Baxter et al. 2018).

Source segregation significantly contributes to the success of waste management programs. Multiple-bin collection system is used at airports to collect paper, mixed recyclables, organics, etc. (ADR 2019).

Table 81.1 Waste data and management practices in busiest airports

Airport/reference	Pax/year (millions)	Waste/year (tons)	Recycle rate	Waste management practices
Istanbul Airport (IGA 2020)	52.9	26.7	31%	All stakeholders signed the Zero Waste Declaration, which is a prerequisite for the purchasing processes. Zero Waste Program trainings are performed. The waste-related KPIs of all the firms are monitored regularly. Records of all sorts of waste, from its source to disposal, are kept by a database and software running 24/7. Wastes are mainly sorted at the source; mixed wastes are handled at the Solid Waste Collection and Sorting Center. A Waste to Art workshop was held to re-evaluate the construction wastes. The workshop aimed to create artwork and design products from waste, reduce waste, and raise waste awareness. A photo art competition in collaboration with a social organization was organized. The award-winning photos and videos were published in airport advertising areas and social media channels. Management of 240 tons of waste in a day is planned.
Mumbai Airport (Mumbai Airport 2020)	45	5.4	43%	Source separation and recycling of nonhazardous waste streams and processing of organic waste at OWC (Organic Waste Compactor) facility. E-Waste bins are set up at terminals. SRA (Strategic Research Agenda) for disposal mechanism. Universal storage facility for hazardous waste. Single-use plastic-free airport status. Single-use plastic items are banned. All waste is handled by a third-party contractor at the airport using national standards.
Heathrow Airport (Heathrow 2020)	80.9	23.9	48.9%	New passenger bins were installed to better segregate waste at the source. Waste contracts are awarded based on circular economy principles. Trialed new technology to help turn unrecyclable plastic waste into useful products, including low-carbon fuels.

(continued)

Table 81.1 (continued)

Airport/reference	Pax/year (millions)	Waste/year (tons)	Recycle rate	Waste management practices
Hongkong Airport (Hong Kong Airport 2020)	60.9	33.9	12.4%	Built-in hand dryers to reduce the use of paper towels in washrooms were installed in the passenger terminals. 108 washrooms will be renovated by mid-2021. Trial of an Artificial Intelligence (AI) robotic sorter system to recover recyclables from different waste streams. Food waste was converted into fishmeal and biogas at a facility.
Incheon Airport (Incheon Airport 2019)	78.39	53.8	71%	Use of the resources recovery facility, and resources classified treatment facility based on the integrated waste operation system. The ice removal solution wastes are treated through an outside treatment agency after piled up in storage tanks. Process improvement to reduce the amount of vinyl packaging and use of plastic boxes and rolltainers instead of pouches for duty-free item transportation.
Kuala Lumpur Airport (Malaysia Airports 2019)	62.3	14.5	14%	A Recycling Awareness Program is conducted at landside and airside areas, and offices. Collected recyclables are weighed, and monetary incentives are given based on the weight. The DHES Material Recovery Facility is used to collect, sort, and recycle the wastes.
Frankfurt Airport (Fraport 2019)	70.6	28.21	68.8%	An advanced sorting systems is used for separating nonhazardous waste either for recycling or sending into regional waste-to-energy plants. Hazardous waste is disposed of in incinerators or in physical/chemical treatment plants where recycling facilities are not available. The hazardous waste is collected separately and channeled away for recycling, where recycling facilities are not available. Management of wastes is performed in the Integrated Facility Management service unit. Working together with stakeholders like disposal companies, contractors, universities, and authorities. Up-to-date knowledge acquired from benchmark activities is added into established airport practices.

JFK Int Airport (JFKIAT 2019)	62.5	6757	57%	Implement K-Cup recycling program for airport terminal offices. A K-Cup recycling program in the JFKIAT office has been implemented. K-Cups are collected and converted into waste-to-energy. The electricity is used to power homes locally. Collaborate with food and cleaning partners to implement waste and recycling strategies. All staff are trained on source segregation, composting, waste, and recycling practices. Work with food and beverage partners for composting all food. Processed food waste is taken to local farms.
Leonardo da Vinci Airport (ADR 2019)	73	12,775	82.7%	Separate waste collection in the terminals and car parking areas. An organic waste composting plant is built, enabling 150 tons of reduction. Planned to build a composting plant capable of eliminating 1000 tons of organic waste per year. Special equipment is installed near the security checks for compacting PET bottles to optimize the collection of plastic and reduced waste by around 20 tons. All temporary waste storage areas are fenced to prevent unauthorized third parties from abandoning waste without control, enabling 200 tons of waste reduction. Regular meetings with the F & B operators served by door-to-door collection to raise their awareness. Increase separate waste collection process, elimination of nonseparated waste containers from the Food & beverage areas and kitchens, and launching a special promotional campaign for passengers. Prohibition of nonbiodegradable and/or noncompostable items in food and beverage items through the contracts; these biodegradable materials and food waste will be delivered to the airport's self-composting plant and the compost produced will be reused in the airport's green areas.

Some airports have set up counters for the disposal of liquids before the security checks as a way of reducing energy usage in waste management processes. With the "Refill" program, around 140 million tons of beverage bottles were collected at security checkpoints of Manchester Airport Group airports, which serve 42 million passengers annually (Winter 2019).

Food wastes can be donated, sent out for bioenergy production, and often composted. Processed compost can be used for soil enrichment in the airport. Wastes from landscaping, coffee grounds, paper towels, etc., are recycled at the composting units (ADR 2019; USEPA 2009).

Waste compaction reduces the volume of wastes and is helpful during the transportation of the wastes to treatment or disposal sites. Furthermore, 80% of the waste from cabins can be recycled, which is a common practice in large airports (ICAO 2021).

Since wastes from international flights can contain contaminants, diseases, or pest infestations, some countries have special regulations and handle them separately from the domestic ones. These wastes are either burned in the site or landfilled. Biohazardous cabin wastes and hospital or healthcare wastes are usually incinerated (Sebastian and Louis 2021).

Some airports use incinerators for diverting food wastes, paper cups, meal plates, glasses, etc., and produce energy to heat terminal buildings (Fraport 2019).

Experiences showed that regional policies contribute to the success of airport waste management programs. In this context, EU's banning of single-use plastic improved Europe airports' waste reduction stream significantly (Chua 2019).

According to Sebastian and Louis (2021), airport's approach to waste management varies in a wide range; while some airports with larger passenger numbers reach zero waste targets, others continue to landfill 75% of their wastes.

Airport tenants and businesses can be incentivized to use biodegradable packages and sustainable alternatives. Using recycled food packages and serve ware, choosing sustainable materials instead of single-use plastics, eliminating the Styrofoam usage, and using biodegradable trash bags are among the sustainable measures of waste management programs (ADR 2019).

Airports also use financial incentives, advertising programs, and nonfinancial incentives like recognition awards for employees and businesses to reduce waste and landfilling as part of their sustainable waste management. Banning single-use plastics by authorities can create further supporting effects on the issue (ADR 2019; Mumbai International Airport 2020).

Stakeholder engagement, training programs, and campaigns constitute an important part of a sustainable waste management program in achieving collaboration and active involvement of the businesses, and other stakeholders in an airport (FAA 2013).

Continuous interest of research for the greening of airports is growing in aviation circles. There is a gap between studies and practices carried out by airports (Greer et al. 2020). Research on water management, climate change, and waste management specific to airport operations is limited, as evidenced that environmental accounting of airports needs further analysis.

81.5 Conclusion

The study assessed the sustainable solutions used by the world's leading airports to the actual problems caused by airport wastes.

Airport Waste Management Strategy aims to eliminate the negative effects of the wastes and creates additional economic values with the implementation of source segregation, reuse, and recycle practices.

The airport waste management system is based on the identification of the waste sources, equipment, facility and infrastructure requirements and effectiveness of the waste collection, storage, recycling, and waste disposal (Janić 2011).

The system requires segregation of waste in waste collection areas, reduction in waste generation, increased reuse and recycling of wastes, and permanent improvement in airport waste management applications (Eriksson 2017; Arushanyan et al. 2017; Baxter et al. 2018).

The type and properties of the waste define both the proper waste management program and the treatment and disposal technics as well (Shi et al. 2016).

Airports can use incentives and advertising programs as part of their waste management strategy. Participation and support from the businesses and airport stakeholders are crucial to achieve sustainable waste management goals for the airports.

Airports should benchmark the well-performing airports and have their own waste management strategies uniquely customized for their conditions.

The amount of resources used, waste generated, sources of waste generation, and recycle rates should be tracked continuously, and process and technology improvements should be made to improve waste management performance.

Route maps for waste generation, including the recycling processes, should be drawn starting from the material and equipment acquisition to the end of the life cycle.

Landfilling of wastes should be avoided as much as possible. If not, it should be performed in accordance with environmental standards.

Depending on the airport size and capabilities, setting up of waste processing facilities for recycling and energy generation can be made a priority; if not feasible, facilities outside of the airport can be used.

Airports can use incentives and advertising programs as part of their waste management strategy.

Participation and support from the businesses and airport stakeholders are crucial to achieve sustainable waste management goals for the airports.

References

ADR (2019) Sustainability Report 2019. Aeroporti di Roma. https://www.adr.it/documents/17615/18251773/BILANCIO+SOSTENIBILITA+2019+EN_DEF_2020.09.01_pubblicato.pdf/ba98b5e1-3f52-4023-8318-d103df30ff58.

Airport Authority Hong Kong (2020). Sustainability Report 2019/2020. https://www.hongkongairport.com/iwov-resources/file/sustainability/ sustainability-report/AA_Sustainability_Report_201920_ENG.pdf.

Arushanyan Y, Björklund A, Eriksson O, Finnveden G et al (2017) Environmental assessment of possible future waste management scenarios. Energies, 10, 247. https://doi.org/10.3390/en10020247

ATAG (2021) Fact & Figures. Air Transport Action Group: https://www.atag.org/component/factfigures/?Itemid=

Baxter G, Srisaeng P, Wild G (2018) Sustainable airport waste management: the case of Kansai international airport. Recycling; 3:6. https://doi.org/10.3390/recycling3010006.

Chandrappa R, Das DB (2012) Solid Waste Management: Principles and Practice. Springer Science & Business Media.

Changi Airport (2020) Resilience & Sustainability. Changi Airport Group. https://www.changiairport.com/content/dam/cacorp/sustainability/sustainable-changi/sustainability-report/CAG-Sustainability%20Report%20FY1920.pdf.

Dimitriou DJ, Voskaki AJ (2010) Regional airports' environmental management: Key messages from the evaluation of ten European airports. International Journal of Sustainable Development and Planning, 5(2), 150-162. https://doi.org/10.2495/SDP-V5-N2-150-162

Eriksson O (2017) Energy and waste management. Energies, 10, 1072. https://doi.org/10.3390/en10071072

FAA (2013) Recycling, reuse, and waste reduction at airports: a synthesis document. https://www.faa.gov/airports/resources/publications/reports/environmental/media/recyclingsynthesis2013.pdf 2013.

Fraport (2019) GRI Report 2019. Frankfurt Airport Services Worldwide. https://www.fraport.com/en/newsroom/media-center/publications.html.

Greer F, Rakas J, Horvath A (2020) Airports and environmental sustainability: a comprehensive review. Environmental Research Letters, 15(10), 103007. https://doi.org/10.1088/1748-9326/abb42a

Guyer, JP (2011) Engineering Soundbite: Solid Waste Incineration; Amazon Digital Services, Inc.: Seattle, WA, USA.

Heathrow (2020) Heathrow 2.0 2020 Sustainability Progress. Heathrow Airport Limited. https://www.heathrow.com/content/dam/heathrow/web/common/documents/company/heathrow-2-0-sustainability/futher-reading/ Heathrow-Sustainability-Report-2020.pdf.

ICAO (2022) Waste Management at Airports, Eco Airport Toolkit. ICAO. https://www.icao.int/environmental-protection/documents/waste_management_at_airports_booklet.pdf

IATA (2021) IATA Workshop on Environment and Sustainability. https://www.iata.org/contentassets/15d7b7ae73cc497da8c65d209a9e4f89/iata-workshop-env%2D%2Daspac-21-april-final.pdf

IATA (2019) IATA Cabin Waste Handbook. https://www.iata.org/contentassets/821b593dd8cd4f4aa33b63ab9e35368b/iata-cabin-waste-handbook---final-resized.pdf [Accessed 15 March 2022].

ICAO (2021) Environment. Waste management at airports. Eco Airport Toolkit. https://www.icao.int/environmental-protection/documents/waste_management_at_airports_booklet.pdf.

IGA (2020) A New Airport Experience: Sustainability Report 2020. https://www.igairport.aero/sites/sustainability/en/Documents/%c4%b0GA-SRD%20Raporu_2020_ENG.pdf.

Incheon Airport (2019) Incheon International Airport Corporation Green Report 2019. https://www.airport.kr/co_file/ko/file01/2019_Green_report_ENG.pdf.

Chua JM (2019) Basically everything you're given on an airplane is wrapped in plastic. That's a huge environmental problem. Goods by Vox. https://www.vox.com/the-goods/2019/7/9/20680969/airplanes-plastic-zero-waste-flights#:~:text=Basically%20everything%20you're%20given,are%20trying%20 to%20change%20that.

Janić M (2011) Greening airports: Advanced technology and operations. Springer Science & Business Media.

JFKIAT (2019) JFK International Air Terminal 2019 Sustainability Report. https://s3.amazonaws.com/takeshape-api.prod.assets/068c892d-2bca-468f-8ba0-7ca9d5f19862/dev/33b400fb-cb83-47af-8977-de4aa661d100/JFKIAT_2019_sustainability_report_FINAL.pdf.

Malaysia Airports (2019) Gaining Momentum: Sustainability Report 2019. Malaysia Airports Holdings. https://www.malaysiaairports.com.my/sites/corporate/files/2020-06/MAHB-Sustainability-Report-2019_0.pdf.

Mumbai Airport (2020) Sustainability Report 2020. Mumbai International Airport Ltd. https://csmia.adaniairports.com/sustainabilityreport.pdf.

Mehta P (2015) Aviation waste management: An insight. International Journal of Environmental Sciences, 6(1), 179.

Parameshwar HK (2011) Solid waste management in airports: A case study of Bangalore International Airport. In Proceedings of the International Conference on Green Technology and Environmental Conservation (GTEC-2011), Chennai, India, 15–17 December 2011; pp. 152–158. https://doi.org/10.1109/GTEC.2011.6167661

Sebastian RM, Louis J (2021) Understanding waste management at airports: A study on current practices and challenges based on literature review. Renewable and Sustainable Energy Reviews, 147, 111229. https://doi.org/10.1016/j.rser.2021.111229

Shi H, Mahinpey N, Aqsha A, Silbermann R (2016) Characterization, thermochemical conversion studies, and heating value modeling of municipal solid waste. Waste Manag.; 48:34–47. https://doi.org/10.1016/j.wasman.2015.09.036.

Spilsbury L (2010) Waste and Recycling Challenges; The Rosen Publishing Group: New York, NY, USA.

USEPA (2009) Developing and implementing an airport recycling program. National Service Center for Environmental Publications. https://nepis.epa.gov/Exe/ZyPDF.cgi/P1005N5Q.PDF?Dockey=P1005N5Q.PDF

Wieher CR (2002) Hazardous Waste Curriculum for Aviation Maintenance. Florida Department of Environmental Protection Contract No. SW130. https://www.broward.org/EnvironmentAndGrowth/EnvironmentalProgramsResources/Publications/Documents/CurriculumAviationMaint.pdf

Winter M (2019) Sustainability Series: How can airports reduce their plastic waste? International Airport Review. https://www.internationalairportreview.com/article/91476/sustainability-series-plastic-waste/

Chapter 82
Assessment of Engine Characteristics of Diesel Engine Fuelled with Graphene Nano Additive Doped *Syzygium cumini* Biodiesel Blends

Murugu Nachippan N., Parthasarathy M., Selçuk Sarıkoç, Senthilkumar P. B., P. V. Elumalai, Backiyaraj A., and O. D. Samuel

Nomenclature

B100	*Syzygium cumini* biodiesel
D100	Diesel
BSFC	Brake-specific fuel consumption
CO	Carbon monoxide
HC	Hydrocarbon
NO_x	Nitrogen oxide
PM	Particulate matters

Murugu Nachippan N. (✉) · Senthilkumar P. B.
School of Mechanical and Construction, Vel Tech Rangarajan Dr. Sagunthala R & D Institute of Science and Technology, Chennai, Tamilnadu, India

Parthasarathy M.
SankarAdvanced Mobility Solutions, Chennai, India

S. Sarıkoç
Department of Motor Vehicles and Transportation Technologies, Tasova Yuksel Akin Vocational School, Amasya University, Amasya, Turkey

P. V. Elumalai
Department of Mechanical Engineering, Aditya Engineering College, Surampalem, Andhra Pradesh, India

Backiyaraj A.
Departmentof Mechatronics Engineering, Agni College of Technology, Chennai, India

O. D. Samuel
Department of Mechanical Engineering, Federal University of Petroleum, Resources, Effurun, Nigeria

M. Z. Sogut et al. (eds.), *Proceedings of the 2022 International Symposium on Energy Management and Sustainability*, Springer Proceedings in Energy,
https://doi.org/10.1007/978-3-031-30171-1_82

AB	Argemone biodiesel
NaOH	Sodium hydroxide

82.1 Introduction

In the last decade, the energy consumptions by industrial growth have posed massive issues for energy production and the environment. Urbanization and the increase in population have made the scarce of conventional fossil fuels. Apart from that, greenhouse gas emission is another main reason for limiting the use of fossil fuels to avoid global warming. Greenhouse gases have therefore contributed to a 0.8 °C rise in global mean temperature during the last century.

Biodiesel share of 10% and 20% with diesel shows better engine running characteristics (Sanjid et al. 2015). Biodiesel combustion produces fewer HC, CO, and PM in exhausts than fossil diesel combustion in diesel engines (Sharma et al. 2020). It's attributed to biodiesel's intrinsic lubricity and lower soot production, consistent with the fuel's lower PM emissions (Dwivedi et al. 2013). Slow oil pyrolysis from highly yielding seeds with anhydrous sodium hydroxide catalyst produced castor biodiesel. Performance metrics and biodiesel emissions at a 10% concentration were comparable to those of gas oil. It is suggested as the suitable percentage (Attai et al. 2020). Overall performance showed that, without engine modifications, 5% of biodiesel blends outperformed others. The maximum torque has been noticed being at this blend, while the BSFC was minimal due to the low fuel consumption. It is recognized that HC emissions have slightly increased (Eiman Ali Eh Sheet et al. 2020). Result proved that waste cooking oil biodiesel blends would be an excellent replacement for diesel (Chiatti et al. 2018). It was discovered that blends of B10 and B20 for apricot and several biodiesels performed better than diesel at 80% load.

An exergy study is also conducted to study the effectiveness and efficiency of the biodiesel blend (Singh et al. 2017). Biodiesel used in the diesel engine has an excellent reduction in emission except for nitrogen oxide. But in the case of fuel consumption, it was poor when compared to diesel because of its high viscous characteristics (Narendranathan et al. 2014) Furthermore, Argemone biodiesel/diesel blends reduced exhaust pollutants. Up to 30% blending of Argemone biodiesel/diesel improved engine performance and reduced exhaust pollutants such as HC, CO, and smoke. Compared to AB30, a greater blend (AB40) demonstrated a reduction in performance attributes (Singh et al. 2018). Canaki et al. (2003) proposed that thermal efficiency is increased for biodiesel owing to earlier injection and earlier start of combustion. Nantha Gopal et al. (2015) studied the engine characteristics of pongamia methyl ester (PME 100) in several blends, including 20%, 40%, 60%, and 80% of biodiesel share with diesel. Dwivedi et al. (2011) observed experimentally that engine power is decreased when diesel content is increased. Engine power is lost due to biodiesel's low heating value when compared to diesel, but consumption of fuel is increased as a result of the low heating value, as stated by Xue et al. (2011).

A bioethanol emulsion with respect to diesel was prepared with a ratio of 15% and 84% of bioethanol and diesel along with 1% of surfactant. Combustion and emission parameters were calculated; the NOx emissions were found to be very low compared to pure biodiesel (Murugan Sivalingam et al. 2015). CNT-emulsified fuel has shown the advantage of improved engine performance, combustion characteristics, and reduced emissions (Sadhik Basha et al. 2014; Narinder Singh et al. 2015). For the aluminum oxide nanoparticles blended Mahua biodiesel, the data from experiments exposed a significant improvement in thermal efficiency and a reduction in exhaust emissions (Syed Aalam et al. 2015). According to Hoque et al. (2015), Karanja outperformed the other three biodiesels in terms of performance. Compared to neat diesel, Mahua emits the least amount of CO and NOx, 44.44% and 38.3%, respectively. The motive of the study is to determine the diesel engine working characteristics of graphene-blended *Syzygium cumini* biodiesel.

82.2 Materials and Methods

82.2.1 Syzygium Cumini Seeds

Syzygium cumini, also known as jambul, black plum, Indian blackberry, or black cumin, is a plant in the Myrtaceae family. Biodiesel from *Syzygium cumini* seeds was extracted, and oil conversion into biodiesel was done through transesterification. Figure 82.1a, b shows the *Syzygium cumini* fruit and *Syzygium cumini* seeds.

Fig. 82.1 (**a**) *Syzygium cumini* fruit; (**b**) *Syzygium cumini* seeds

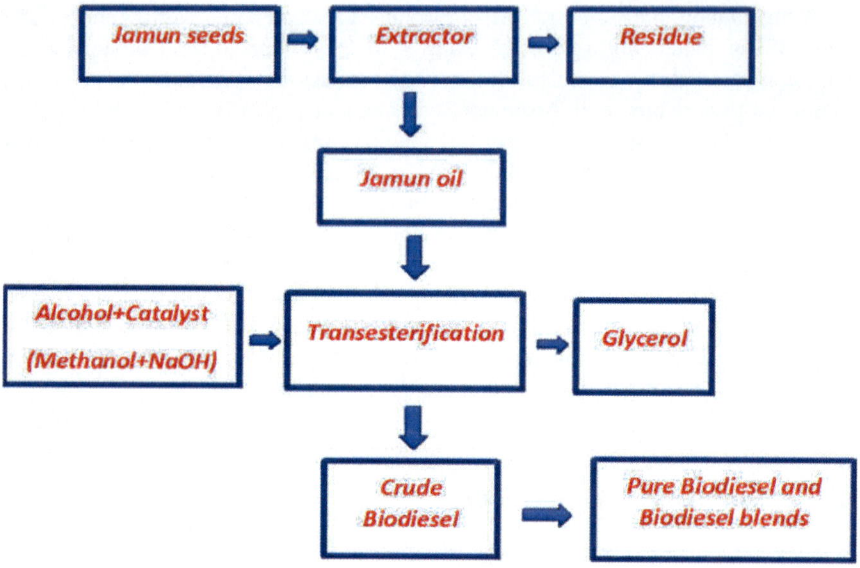

Fig. 82.2 Steps involved in the transesterification process

82.2.2 Biodiesel Preparation Process

Mutual exchange of an organic group of ester and an organic group of alcohol in the existence of an acid or base catalyst is called the transesterification process. The suitable biodiesel production process was an alkaline homogeneous catalysis process. Strong catalyst NaOH is used as a base catalyst for the production process. Continuously stirred methanol and NaOH (sodium hydroxide) are heated at around 65 °C with 400 rpm. After 2 h of heating, the glycerol settled over the flask was removed by density separation. Produced crude biodiesel should be purified, and the water purification method should be employed. After water washing, the crude biodiesel is heated up to 120 °C to remove water and excess glycerol. Figure 82.2 shows the steps involved in the biodiesel production process.

82.2.3 Graphene

A single-layer (monolayer) sheet of carbon atoms linked together in a hexagonal arrangement is known as graphene. It is purchased from Sigma-Aldrich. Fig. 82.3a, b shows the microscopic graphene image at 50 μm and 2 μm. Figure 82.4 shows the EDAX spectra of graphene. Table 82.1 displays the elemental composition of graphene.

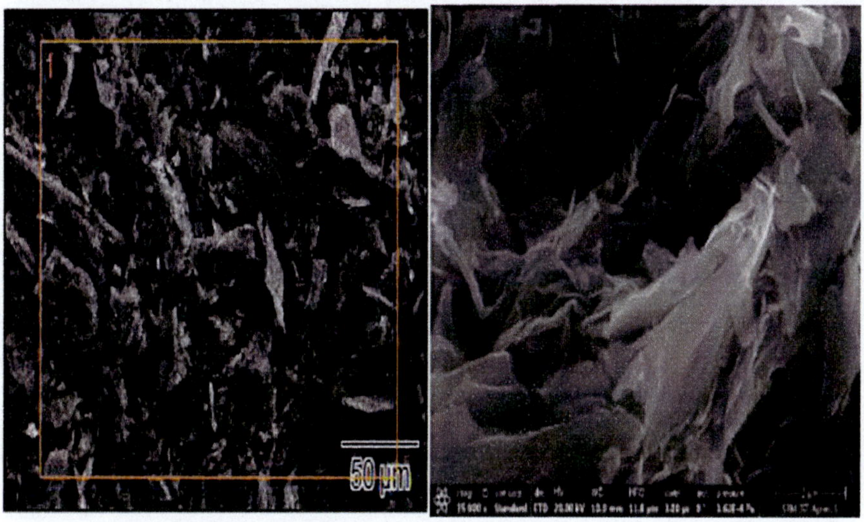

Fig. 82.3 (a) and (b) SEM image of graphene.

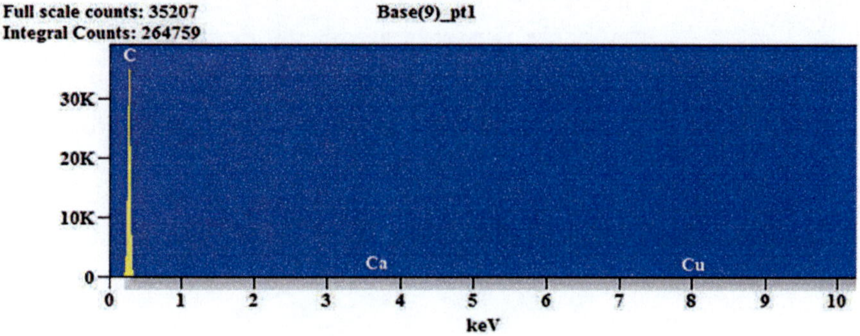

Fig. 82.4 EDAX spectra of graphene

Table 82.1 Elemental composition of graphene

Element	Elemental composition (%)
C	99.01
Ca	0.30
Cu	0.69

82.3 Experimental Setup

At a constant speed of 1500 rpm, tests were conducted in a 3-cylinder, 4-stroke, inline, water-cooled, and direct injection using diesel (D100) and several biodiesel blends (B10, B20, B30, B40, and B100) at varied load conditions. Figure 82.5

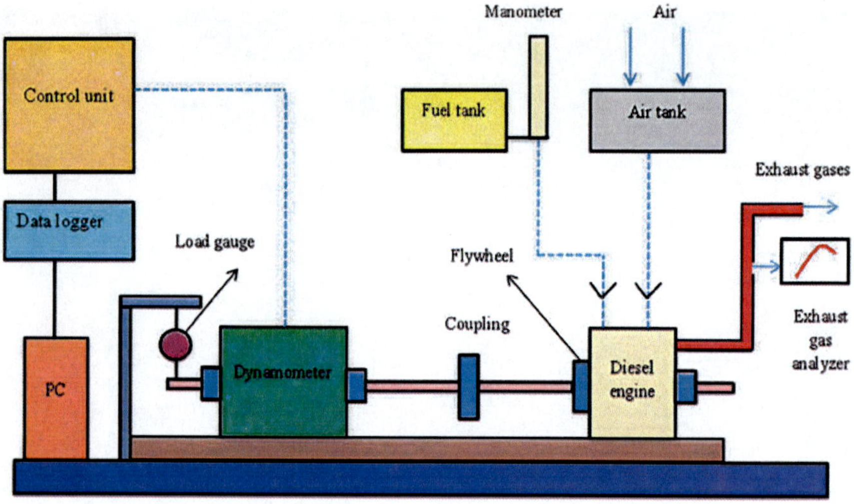

Fig. 82.5 Schematic representation of the experimental setup

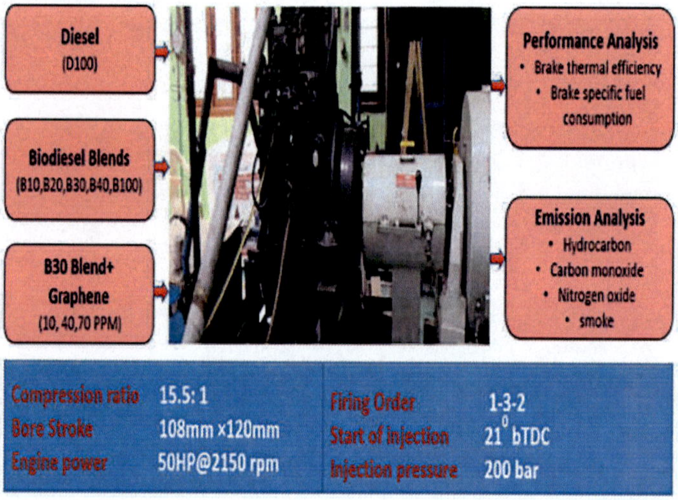

Fig. 82.6 Experimental methodology

displays the schematic representation of experimental setup and Fig. 82.6 displays the experimental methodology. The smoke from exhaust gases is measured using an AVL 437C smoke meter. The exhaust gas composition is determined using an AVL DI GAS 444 gas analyzer. The diesel (D100) and *Syzygium cumini* biodiesel (B100)

Table 82.2 Properties of biodiesel blend

	Diesel	*Syzygium cumini* biodiesel blends				
Properties	D100	B10	B20	B30	B40	B100
Calorific value (MJ/kg)	43.4	42.5	40.3	39.7	38.9	38.4
Kinematic viscosity 40 °C (mm^2/sec)	2.5	2.6	2.8	3.2	3.5	4.1
Flash point (°C)	50	62	75	88	112	120
Density 15 °C (kg/m^3)	837	842	858	673	794	882

are used to evaluate the diesel engine's working characteristics. In addition, four biodiesel blend at blending ratios of 10, 20, 30, and 40 (on volume basis) were prepared and tested. Table 82.2 shows the properties of the biodiesel blend.

82.4 Results and Discussion

82.4.1 Brake Thermal Efficiency

For all tested fuels, brake thermal efficiency increases from no load to full load. This is due to more fuel being fed into the combustion chamber, which increases combustion efficiency from no load to full load. At B30, the inherent availability of oxygen and optimum energy values makes the combustion efficiency high resulting in brake thermal efficiency slightly equivalent to diesel compared to other biodiesel blends. The addition of graphene in three different concentrations of 10 ppm, 40 ppm, and 70 ppm to the B30 blend shows better thermal efficiency due to the reactivity of nanoparticles in the blend (Fig. 82.7).

The brake thermal efficiency found for graphene nano additive doped in the B30 blend of 70 ppm is 28.41, compared to 27.05 for the B30 blend. Compared to B30, a 5% increase in brake thermal efficiency is observed for graphene nano additive of 70 ppm doped B30 blend.

82.4.2 Brake-Specific Fuel Consumption

Brake-specific fuel consumption decreases from no load to full load for all tested fuels due to the high amount of fuel supplied for engine combustion as load increases. Biodiesel blends have higher fuel consumption than diesel. Because of the low calorific value of the *Syzygium cumini* biodiesel blend in comparison to diesel, more fuel is required to maintain the same power output. However, B30 blend shows better brake thermal efficiency due to the domination of oxygen concentration. Better combustion makes the brake-specific fuel consumption value closer to diesel than other biodiesel blends (Fig. 82.8).

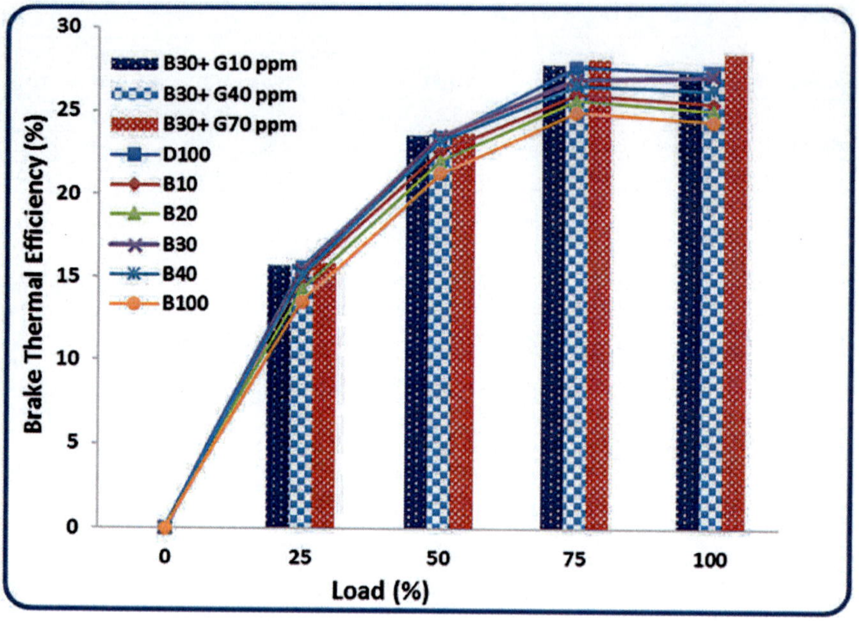

Fig. 82.7 Influence of graphene on brake thermal efficiency in diesel engine

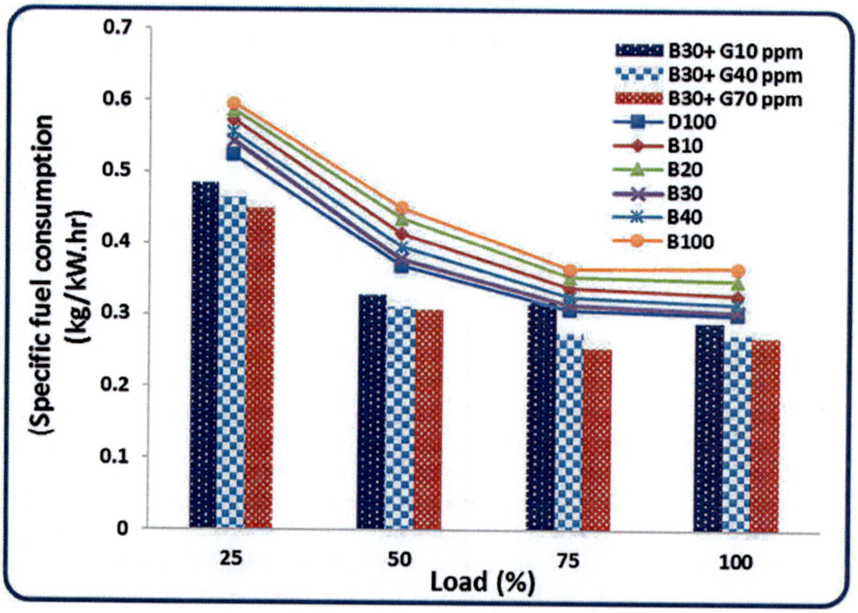

Fig. 82.8 Influence of graphene on brake-specific fuel consumption in diesel engine

In addition to graphene at three different concentrations of 10 ppm, 40 ppm, and 70 ppm in the B30 blend, nanoparticles present in the blend reduce brake-specific fuel consumption. The addition of nanoparticles improves the surface-to-volume ratio. This results in a more homogeneous mixing of fuel and air, as well as improved combustion efficiency. The brake-specific fuel consumption obtained for graphene nano additive of 70 ppm doped in the B30 blend is 0.269 kg/kW.hr. compared to 0.305 kg/kW.hr. for the B30 blend. Compared to B30, a 12% decrease in brake-specific fuel consumption is observed for the graphene nano additive of 70 ppm doped B30 blend.

82.4.3 Hydrocarbon Emissions

Hydrocarbon emissions increase from no load to full load for all tested fuels. When the engine is under heavy load, a rich mixture is supplied to power it. When the biodiesel blend share is increased, hydrocarbon emissions in biodiesel are reduced compared to diesel. This is primarily due to the higher oxygen content of biodiesel, which accelerates the combustion reaction. Addition of graphene of three different concentrations of 10 ppm, 40 ppm, and 70 ppm in the B30 blend reduced hydrocarbon emissions due to the catalytic activity of nanoparticles present in the biodiesel. Graphene nanoparticles strengthen the interaction of fuel and air and improve combustion efficiency. The hydrocarbon emissions obtained for graphene nano additive of 70 ppm doped in the B30 blend is 25 ppm compared to 42 ppm for the B30 blend. Compared to B30, a 22% decrease in hydrocarbon emission is observed for the graphene nano additive of 70 ppm doped B30 blend (Fig. 82.9).

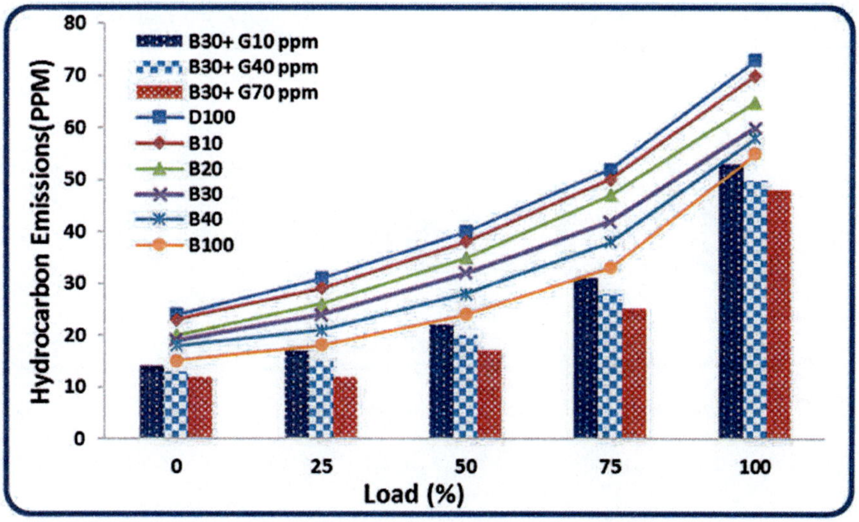

Fig. 82.9 Influence of graphene on hydrocarbon emissions in diesel engine

82.4.4 Carbon Monoxide Emissions

Carbon monoxide emissions increase from no load to full load for all tested fuels. At full load, a high air–fuel ratio results in higher carbon monoxide emissions. When the proportion of biodiesel in a blend increases, carbon monoxide emissions in biodiesel are reduced compared to diesel. In biodiesel blends, oxygen availability is important, and it is responsible for complete fuel combustion when compared to diesel (Fig. 82.10).

Addition of graphene of three different concentrations of 10 ppm, 40 ppm, and 70 ppm in the B30 blend reduced carbon monoxide emissions due to the catalytic activity of nanoparticles present in the biodiesel. Graphene nanoparticles improve the surface–volume ratio of fuel that improves atomization, oxidization, and combustion efficiency. The carbon monoxide emission obtained for graphene nano additive of 70 ppm doped in the B30 blend is 0.121% vol compared to 0.2% vol for the B30 blend. Compared to B30, a 16% decrease in carbon monoxide emission is observed for the graphene nano additive of 70 ppm doped B30 blend.

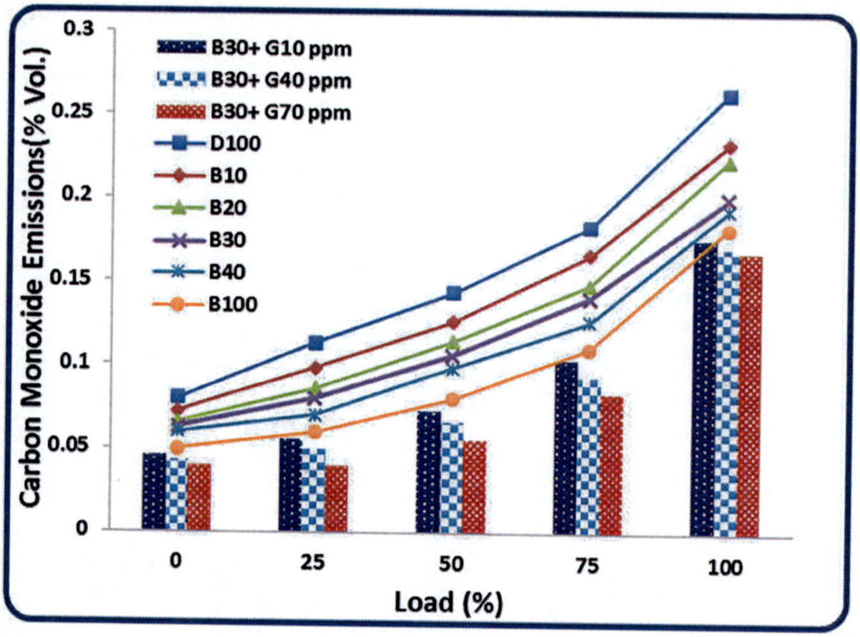

Fig. 82.10 Influence of graphene on carbon monoxide emissions in diesel engine

82.4.5 Nitrogen Oxide Emissions

Nitrogen oxide emissions rise with increasing load for all tested fuels. The primary cause of NOx emissions at high loads is increased in-cylinder temperature caused by combustion. When compared to diesel, NOx emissions from all biodiesel blends are high. NOx emission increases when the biodiesel blend share increases. This is because biodiesel has high oxygen availability, particularly at higher loads, the combustion temperature rises steeply, and NOx generation will be more. The addition of graphene of three concentrations of 10 ppm, 40 ppm, and 70 ppm in B30 blend nitrogen oxide emissions was reduced due to better heat transfer characteristics in the cylinder. Graphene nanoparticle absorbs oxygen and acts as a reducing agent during combustion; it also improves the heat transfer rate and reduces the peak temperature inside the cylinder. The nitrogen oxide emissions obtained for graphene nano additive of 70 ppm doped in the B30 blend is 550 ppm compared to 792 ppm for the B30 blend. Compared to B30, a 31% decrease in nitrogen oxide emissions is observed for the graphene nano additive of 70 ppm doped B30 blend (Fig. 82.11).

82.4.6 Smoke Emissions

Smoke emissions increase when the load increases for all tested fuels. The foremost reasons for smoke emission at high loads are rich fuel mixture zone and lack of oxygen availability in fuel. Compared to diesel, smoke emissions observed for all biodiesel blends are low (Fig. 82.12).

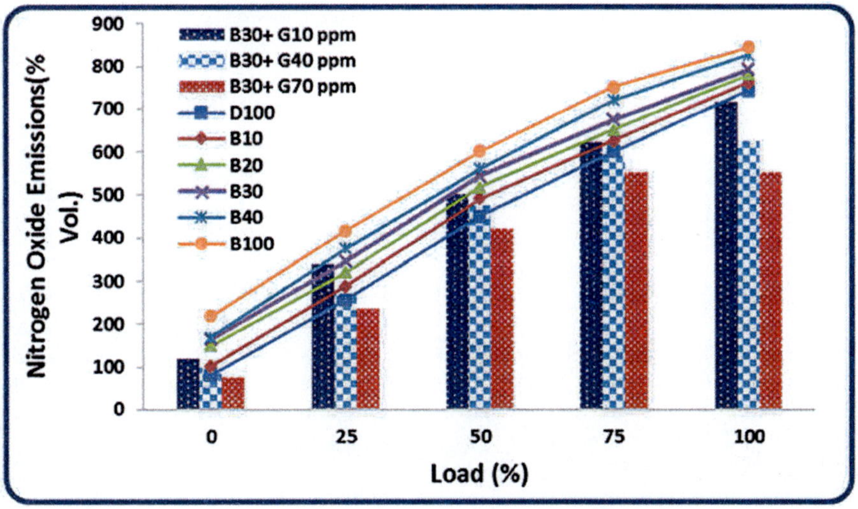

Fig. 82.11 Influence of graphene on nitrogen oxide emissions in diesel engine

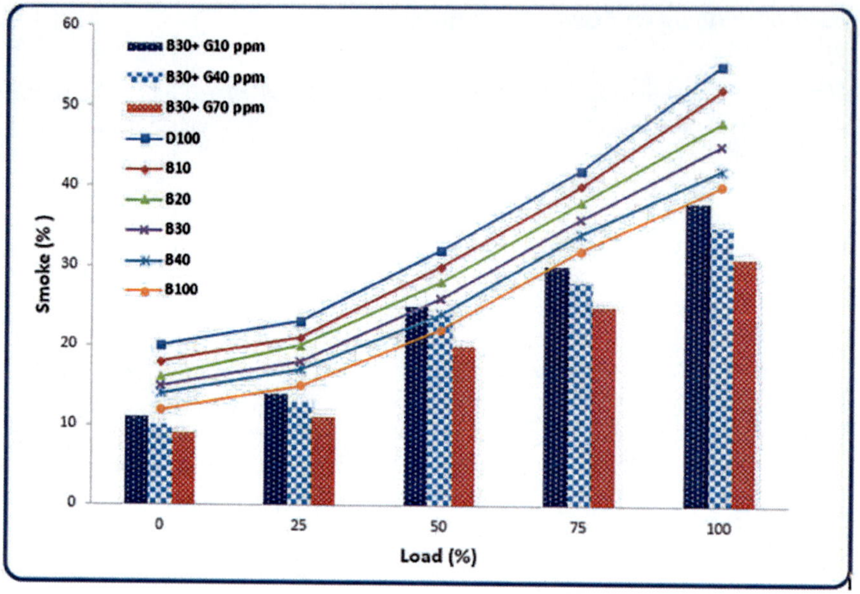

Fig. 82.12 Influence of graphene on smoke emissions in diesel engine

The smoke emission decreases when the biodiesel blend share increases. This is because biodiesel has high oxygen availability and particularly at lower loads; smoke emission is not significant due to the lower fuel-rich zone and better combustion. However, at higher loads, enhanced fuel-rich zone makes smoke emission high. Addition of graphene of three different concentrations of 10 ppm, 40 ppm, and 70 ppm in the B30 blend reduced smoke emissions due to improved air-fuel mixing, better evaporation, and fuel ignition characteristics. The smoke emissions obtained for graphene nano additive of 70 ppm doped in the B30 blend is 31% compared to 42% for the B30 blend. Compared to B30, a 32% decrease in smoke emissions is observed for the graphene nano additive of 70 ppm doped B30 blend.

82.5 Conclusion

In this work, *Syzygium cumini* biodiesel is produced using single-step transesterification. Biodiesel blends are prepared at blending ratios of 10, 20, 30, 40 (in volume basis) along with diesel, and these blends are evaluated for performance and emission characteristics. Based on the evaluation, B30 blends show the optimal result, and its performance is equivalent to diesel with lesser hydrocarbon, carbon monoxide, and smoke emissions. The influence of graphene nanoparticles on three concentrations of 10 ppm, 40 ppm, and 70 ppm is dispersed in B30 blend, and its performance and emission characteristics are evaluated. It is observed that a 5% increase in brake thermal efficiency and 12% reduction in brake-specific fuel consumptions are observed for graphene nano additive of 70 ppm doped B30 blend. Hydrocarbon, carbon monoxide, nitrogen oxide, and smoke emissions are reduced due to the presence of graphene nano additive of 70 ppm doped B30 blend. Compared to B30, a 22% decrease in hydrocarbon emission, a 16% decrease in carbon monoxide emission, a 31% decrease in nitrogen oxide emissions, and a 32% decrease in smoke emissions are observed.

References

Attai, Y. A., Abu-Elyazeed, O. S., ElBeshbeshy, M. R., Ramadan, M. A., Gad, M. S. (2020) Diesel Engine Performance, Emissions and Combustion Characteristics of Castor Oil Blends Using Pyrolysis. Advances in Mechanical Engineering 12 (12), https://doi.org/10.1177/1687814020975527.

Canaki M., Van Garpen J. H., (2003) Comparison of engine performance and emissions for petroleum diesel fuel, yellow-grease biodiesel and soybean-biodiesel biodiesel. Transactions of the ASAE, 46(4):937–944.

Chiatti, G., Chiavola, O., Recco, E. (2018) Effect of Waste Cooking Oil Biodiesel Blends on Performance and Emissions from a CRDI Diesel Engine. In Improvement Trends for Internal Combustion Engines, Ceper, B. A., M., Eds., InTech, https://doi.org/10.5772/intechopen.69740.

C. Syed Aalam, C.G. Saravanan (2015). Effects of nano metal oxide blended Mahua biodiesel on CRDI diesel engine. Ain Shams Engineering Journal

Dwivedi, G., Jain, S., Sharma, M. P. (2013) Diesel Engine Performance and Emission Analysis Using Biodiesel from Various Oil Sources. Review. 14.

Dwivedi G., Jain S., Sharma M. P., 2011, Impact of Biodiesel and its Blends with Diesel and Methanol on Engine Performance, International Journal of Energy Science, 1(2): 105–109.

Eiman Ali Eh Sheet, J. A. A. Y., 2020, Effect of Biodiesel Blending with Diesel Fuel on CI Engine Performance and Energy Availability. 32 (2), 14.

J. Sadhik Basha, R.B. Anand b, (2014) Performance, emission and combustion characteristics of a diesel engine using Carbon Nanotubes blended Jatropha Methyl Ester Emulsions. Alexandria Engineering Journal 53: 259–273.

K. Nantha Gopal, R. Thundil Karupparaj (2015) Effect of pongamia biodiesel on emission and combustion characteristics of DI compression ignition engine. Ain Shams Engineering Journal, 6: 297–305.

Murugan Sivalingam, Subranshu Sekhar Mahapatra, Dulari Hansdah, Bohumil Horak (2015) Validation of some engine combustion and emission parameters of a bioethanol fuelled DI diesel engine using theoretical modelling. Alexandria Engineering Journal 54: 993–1002.

Narendranathan, S. K., Sudhagar, K. (2014) Study on Performance & Emission Characteristic of CI Engine Using Biodiesel. AMR: 984–985. https://doi.org/10.4028/www.scientific.net/AMR.984-985.885.

Narinder Singh and R S Bharj (2015) Effect of CNT-Emulsified Fuel on Performance Emission and Combustion Characteristics of Four Stroke Diesel Engine. International Journal of Current Engineering and Technology, Vol. 5, No. 1.

N. Hoque, M. Mourshed and B.K. Das (2015). Performance and emission comparison of Karanja (pongamia pinnata), Pithraj (aphanamixis polystachya), Neem (azadira chtaindica) and Mahua (madhuca longofolia) seed biodiesel as a potential feedstock for biodiesel production in Bangladesh. International Journal of Automotive and Mechanical Engineering, Volume 12, p. 2967–2982, July–December.

Sanjid, A., Masjuki, H. H., Kalam, M. A., Rahman, S. M. A., Abedin, M. J., Fattah, I. M. R. (2015) Characterization and Prediction of Blend Properties and Evaluation of Engine Performance and Emission Parameters of a CI Engine Operated with Various Biodiesel Blends. RSC Adv. 5 (17): 13246–13255. https://doi.org/10.1039/C4RA14748K.

Sharma, A. K., Sharma, P. K., Chintala, V., Khatri, N., Patel, A. (2020) Environment-Friendly Biodiesel/Diesel Blends for Improving the Exhaust Emission and Engine Performance to Reduce the Pollutants Emitted from Transportation Fleets. IJERPH 17 (11):3896. https://doi.org/10.3390/ijerph17113896.

Singh, N., Kumar, D., Sarma, A. K., Jha, M. K. (2017) Exergy Conceptual-Based Study for Comparative Thermodynamic Performance of CI Engine Fueled with Petroleum Diesel and Biodiesel Blends. J. Energy Eng. 143 (1), 04016030. https://doi.org/10.1061/(ASCE)EY.1943-7897.0000380.

Singh, M., Gandhi, S. K., Mahla, S. K., Sandhu, (2018) S. S. Experimental Investigations on Performance and Emission Characteristics of Variable Speed Multi-Cylinder Compression Ignition Engine Using Diesel/Argemone Biodiesel Blends. Energy Exploration & Exploitation 2018, 36 (3): 535–555. https://doi.org/10.1177/0144598717738573.

Xue J., Grift T. E., Hansena A. C. (2011) Effect of biodiesel on engine performances and emissions. Renewable and Sustainable Energy Reviews, Vol. 15,: 1098–1116.

Chapter 83
Experimental Studies on Compression Ignition Engine Operated with Blends of Tamanu Biodiesel and Diesel

P. B. Senthilkumar, M. Parthasarathy, Selçuk Sarıkoç, N. M. Nachhippan, A. Backiyaraj, P. V. Elumalai, and O. D. Samuel

Nomenclature

CI	Compression ignition
HC	Hydrocarbon
CO	Carbon monoxide
NO_x	Oxides of nitrogen
ASTM	American Society for Testing and Materials
BTE	Brake thermal efficiency

P. B. Senthilkumar (✉) · N. M. Nachhippan
Department of Automobile Engineering, Vel Tech Rangarajan Dr. Sagunthala R & D Institute of Science and Technology, Chennai, India

M. Parthasarathy
Sankar Advanced Mobility Solutions, Chennai, India

S. Sarıkoç (✉)
Department of Motor Vehicles and Transportation Technologies, Amasya University, Tasova Yuksel Akin Vocational School, Amasya, Turkey

A. Backiyaraj
Department of Mechatronics Engineering, Agni College of Technology, Chennai, India

P. V. Elumalai
Department of Mechanical Engineering, Aditya Engineering College, Surampalem, Andhra Pradesh, India

O. D. Samuel
Department of Mechanical Engineering, Federal University of Petroleum, Resources, Effurun, Delta State, Nigeria

M. Z. Sogut et al. (eds.), *Proceedings of the 2022 International Symposium on Energy Management and Sustainability*, Springer Proceedings in Energy,
https://doi.org/10.1007/978-3-031-30171-1_83

BP	Brake power
SEC	Specific energy consumption
FTIR	Fourier transform infrared spectroscopy
GC-MS	Gas chromatography mass spectrometry

83.1 Introduction

Nonedible oils are becoming more and more popular due to their obtainability in many areas of the world. When compared to edible oils, non-edible oils could be more efficient and economical. Its biodiesel can be used as a promising substitute for diesel because other plant/seed raw materials have been used or are being considered (Niju et al. 2019). Tamanu oil seems to be a best alternative biodiesel source, with characteristics like diesel fuel. The characteristics of tamanu biodiesel comply with ASTM standards. Catalytic esterification of tamanu gives high yields, usually over 95%, making it an economically attractive feedstock. Investigated the possibility of using Donax deltoides shells (DDS) as a heterogeneous alkaline catalyst suitable to produce tamanu methyl esters (biodiesel) from Calophyllum inophyllum oil (CIO)-waste cooking oil (WCO) mixture. Through the calcination process, DDS transformed into the active form of the CaO catalyst. Dewang et al. (2017) analyzed that the transesterification process does not have to take an hour to reach the standard. The efficiency of the transesterification process was also measured. During the 2-h transesterification reaction, the highest yield was 98.2%. The output in 1 h and in 3 h was 94.8% and 97.7%, respectively. Fauzan et al. (2019) reported that degumming, esterification, and transesterification are used in a three-stage process to convert tamanu oil, which contains a large amount of free FAME. Naveenkumar and Baskar (2019) reported that most researchers focus on the biodiesel production from tamanu oil. The maximum biodiesel yield, 89.0% (v/v), was obtained at 55 °C. Krishnaji Bawane et al. (2014) reported that compression ratio zone, the brake thermal efficiency of biodiesel and its mixtures is slightly lower than diesel fuel. This is because biodiesel made from Calophyllum inophyllum is less volatile and reduces atomization characteristics, air-fuel mixture uniformity, and het releases rat and brake thermal efficiency. Gopinath et al. (2015) analyzed that the engine characteristics results are compared with the parameters of mineral diesel fuel combustion. In that, the results of combustion characteristics show that polanga biodiesel is nearer to diesel fuel. The injection and combustion of polanga biodiesel starts much faster than diesel fuel (Hwai Chyuan Ong et al. 2019). Calophyllum inophyllum biodiesel blends are tested at full load from 1000 to 2400 rpm, and the results are compared with diesel. The ranges of biodiesel blends are PB10, PB20, PB30, CIB10, CIB20, and CIB30. The four-ball testing machine is used for friction and wear tests at various temperatures under 30 and 70 kg load at rated speed. The average SFC increased from 6.15% to 9.2% (Bridjesh et al. 2016).

Optimization strategies such as artificial neural network and ant colony optimization were utilized to optimize process variable used for production of CI40CP60 oil combination (40% by weight Calophyllum inophyllum oil blended in with 60% Ceiba) Catalytic transesterification measure variable pentandra oil to amplify the production of biodiesel (Kurniati et al. 2019). The composition of tamanu methyl ester crude oil is present in palmitic acid, oleic acid, stearic acid, and linoleic acid.

The biodiesel processing technology utilized three methods, which are conventional, microwave, and electromagnetic induction. Electromagnetic induction methods were fast and easy to produce biodiesel, and energy consumption was low. Using gas chromatography and mass spectrometry, the study evaluated the characteristics of biodiesel. Viscosity of biodiesel (Raj and Kandasamy 2012). The characteristics of a variable compression ignition engine were evaluated using a blend of diesel and tamanu methyl esters. The engine is tested at 1500 rpm. The compression ratios vary from 14 to 18. It shows the results of low HC and CO emissions at higher loads. As compression ratios increased, HC, CO, and NOx emissions decreased while NOx emissions increased. Rahul Krishnaji Bawane et al. (2020) analyzed that four-stroke compression ignition engine is used to test the performance and emissions of diesel and tamanu-based biodiesel blends of 25%, 50%, and 75%. Blends of 25% give better performance and emission characteristics compared with other blend ratios. Dinesh et al. (2019) reported that a small amount of ethanol (10–20%) can improve the fuel's characteristics, performance, and emissions. However, when ethanol is added to a mixture of diesel and biodiesel, it is discovered that the 20% ethanol mixture reduces smoke density better than the 40% Tamanu biodiesel mixture in the current investigation. Fuel and exhaust gas drop under certain load circumstances. Gautam (2019) revealed that the use of 1D/3D modelling and simulation methodologies adds to the benefits and information available regarding the flow through the intake system, as well as the impact of geometric characteristics on its efficiency and engine performance. Furthermore, the structural integrity of the system is crucial due to the enormous suction pressure imposed on the intake manifold, as an explosion in the booster chamber is a distinct possibility. Parthasarathy et al. (2020) compared the performance analysis of conventional diesel fuel engine and homogeneous charge compression ignition (HCCI) engine with tamanu methyl ester.

HCCI engine operates on various intake air temperatures like 800 °C, 900 °C, and 1000 °C. Mineral diesel fuel engine has high brake thermal efficiency (6.5%), low emission of carbon monoxide (1.6%), and hydrocarbon (31.2%) than HCCI engine. The purpose of this work is to investigate the fuel characterization of test fuel using GC-MS and FT-IR. The theoretical and experimental evaluation of the engine's characteristics was conducted using a tamanu biodiesel and diesel blend.

Fig. 83.1 A pictorial view of Tamanu tree, fruit, and seed

83.2 Materials and Methods

Tamanu oil is derived from dried and peeled tamanu seed. Tamanu seeds are removed the shell and dried at 36 °C. The particle size is 1–4 mm. Tamanu oil is extracted by mechanical presses and further processed by filtering the neat tamanu oil. Figure 83.1 shows a pictorial view of tamanu tree, fruit, and seed.

In Asian countries, tamanu is a very popular tree. An important use of tamanu oil is for many medicinal purposes and cosmetics. Long chain alkyl esters like methyl, propyl, or ethyl are present in the vegetable oil. Tamanu oil is extracted by two processes: acid esterification and alkali esterification process. The initial test was conducted to find out the maximum yield of tamanu biodiesel. The molar ratio of methyl alcohol to neat tamanu oil ratio was 4:1, 6:1, 8:1, and 10:1, and the yield was 81.9%, 86%, 81.2%, and 76%, respectively. The neat tamanu oil has high free fatty acid and contains triglyceride mixture. This mixture converts into diacylglycerol using acid esterification.

Sulphuric acid (H_2SO_4) catalyst was added to preheated oil at 56–600 °C in a flask for 30 min. The mixture content was heated up to 2 h and cooled at room temperature. After the transesterification process, the solution would separate into two layers, with the glycerol at the bottom being removed and the biodiesel at the top accumulating. Potassium hydroxide (KOH) catalyst added to preheat oil at 600 °C. Then it was stirred for 90 min. The mixture content was transferred to a separation flask. The glycerol was settled in the separation flask for 10 h. After this process, the soap and excess methanol were removed using hot water. For removing moisture content, the tamanu biodiesel was heated at desired temperature. Finally, tamanu biodiesel was obtained. Figure 83.2 shows the transesterification process. Based on ASTM standards, the fuel properties were tested in fuel testing laboratory. Table 83.1 shows the properties of test sample fuels.

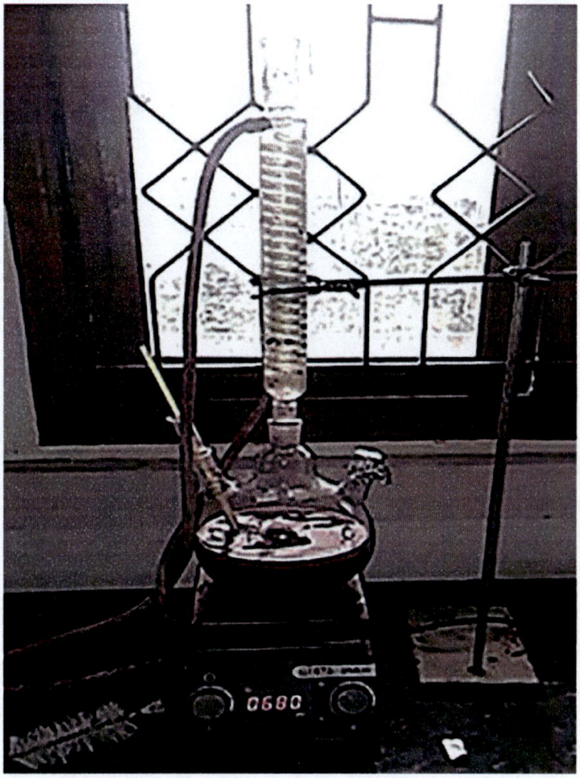

Fig. 83.2 Transesterification process

83.3 Experimental Setup

The actual experiment was conducted on a mono cylinder, four strokes, liquid-cooled, and naturally aspirated diesel engine. The blended fuels, neat biodiesel, and diesel fuel were tested in single cylinder Kirloskar engine at constant speed of 1500 rpm. The actual test engine was loaded with eddy current dynamometer. The experimental test engine was connected to different instruments, like burette, pressure transducer, and exhaust gas analyzer. The data acquisition system received cylinder pressure and stored the data in the computer. Experimental test engine fuel consumption was measured by the burette. The exhaust gas of HC, CO and NO_x was measured by the AVL Di gas analyzer. Table 83.2 shows the specification of actual test engine. Figure 83.3 shows a pictorial view of the experimental test engine setup.

Table 83.1 Summary study for the different biodiesel blend and diesel

Biofuel	Performance characteristics		Emission characteristics				Combustion characteristics		References
	BSFC ▲▼ (%)	BTE ▲▼ (%)	HC ▲▼ (%)	CO ▲▼ (%)	NOX ▲▼ (%)	Smoke ▲▼ (%)	P ▲▼ (%)	HRR ▲▼ (%)	
Garcinia gummi-gutta methyl ester (B10, B20, B30, B40, B100)	25,8▲	5,5▼	14▼	15,1▼	22,7▲	–	4,1▼	9,7▼	Lingesan Subramani et al. (2018)
Sapindus seed biodiesel (B10, B20, B30)	19,5▲	16,6▲	29,4▲	8,7▼	11,6▲	–	–	–	Padmanabhan et al. (2020)
Linseed biodiesel (B20)	4,2▼	1,5▲	9,3▼	20,8▼	1,3▼	4,4▼	–	–	Elumali et al. (2020)
Ethanol (B10, B20, B30)	3,1▲	5,7▼	5,8▲	3,7▲	17,2▼	9,8▲	4,8▼	3,8▼	Shanmugam et al. (2021)
Cymbopog on flexuosus biodiesel (B20, B40, B60, B80, B100)	5▲	12,5▼	14,7▼	15,7▼	4,1▲	15▼	8,8▼	13,2▼	Dhinesh et al. (2017)
Spirulina biodiesel (B20, B40, B60, B80)	5,2▲	4,2▲	34,4▼	14,6▼	6,8▼	12,3▼	1,9▼	5,9▼	Upendra Rajak et al. (2018)
Mahua biodiesel (B5, B10, B15, B20)	4,9▲	3,5▲	16,3▼	42,1▼	21,2▲	37,5▼	6,3▼	3▼	Raman and Kumar (2020)
Tamrind seed methyl ester (B20)	7,4▲	8▼	15,6▼	10,2▼	9,7▲	2,51▼	2,1▼	13,2▼	Dhana Raju et al. (2018)
Calophyllum inophyllum biodiesel (B10, B20, B30, B100)	6,5▲	3,5▼	8,6▼	19,6▼	19,1▼	7,3▼	3,1▼	8,1▼	Tamilvanan et al. (2019)
Calophyllum inophyllum methyl ester (B100)	12,5▲	3,1▼	25▼	30▼	13,3▲	20▲	6,2▼	16,9▼	Ashok et al. (2017)

Table 83.2 Engine specification

Description	Details
Make	Kirloskar
Max.BP@ 1500 rpm	5.2 KW
Bore × stroke	87.5 mm × 110 mm
Compression ratio of engine	17.5:1
Injection pressure	220 bar
Timing of injection	21ob TDC (during end of the compression)
Type of cooling	Liquid
Governor type	Mechanical
Injection nozzle type	3-hole nozzle

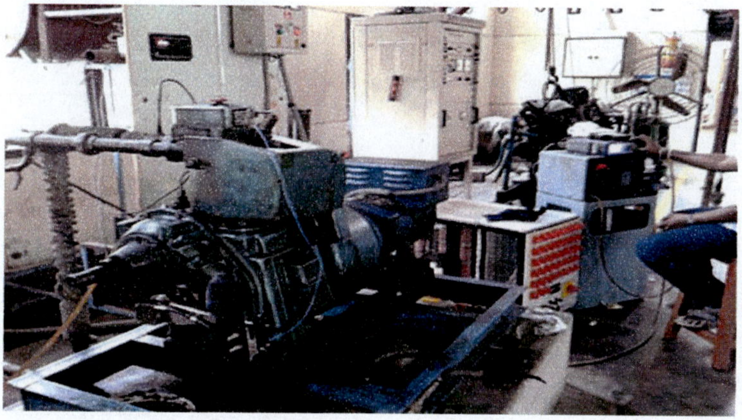

Fig. 83.3 Pictorial view of experimental test engine setup

Table 83.3 FT-IR band value of tamanu biodiesel

Sl. no.	Peak value	Band value
1	419.52	431.10–406.99
2	721.39	797.58–655.81
3	1163.10	1220–1127.41
4	1742.71	1784.19–1706.07
5	2923.17	2991.64–2881.70

83.4 Results and Discussion

83.4.1 FTIR for Tamanu Biodiesel

In this test, the model shimadzu IR tracer 100 was employed to fourier transform infrared spectroscopy analysis. The test sample of tamanu biodiesel reached the high peaks found in the wave number range from 500 to 4000 cm^{-1}. Based on the results, the liquid sample presents the acids, alcohols, and alkynes. Table 83.3 shows the FT-IR band value of tamanu biodiesel. Figure 83.4 shows the FT-IR analysis of tamanu biodiesel.

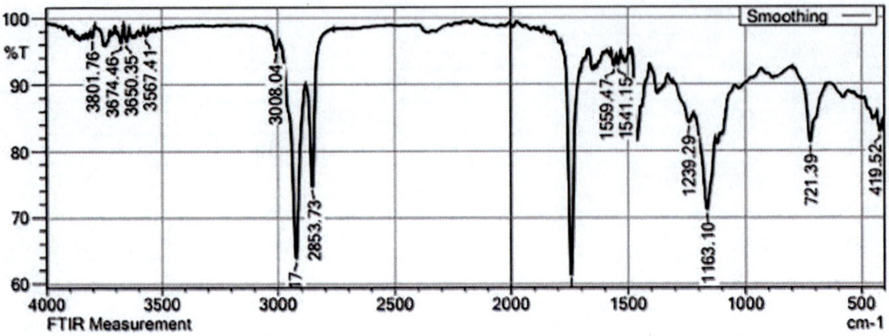

Fig. 83.4 FTIR analysis of biodiesel

Table 83.4 GC-MS peak value of tamanu biodiesel

Sl. no.	Peak no.	Run time	Chemical name
1	9	6.909	Hexanoic acid, methyl ester
2	19	8.520	Decane
3	37	12.463	N-octanal dimethylacetal
4	45	14.272	Octadec-9-en-1-al dimethyl acetal
5	61	16.821	2-propenoic acid, 3-phenyl-, methyl
6	69	18.608	Tetradecane
7	72	19.532	Nonanedioic acid, dimethyl ester
8	81	21.197	Decanedioic acid, dimethyl ester
9	89	23.423	Heptadecane
10	95	25.578	Hexadecanoic acid, methyl ester
11	100	26.676	Octadecanoic acid, methyl ester
12	103	28.644	9,12-octadecadienoic acid (z,z)
13	106	29.077	9-octadecenoic acid, methyl ester
14	112	30.232	Methyl 9-octadecenoate
15	113	30.518	Triacontanoic acid, methyl ester
16	125	32.743	Heptacosanoic acid, methyl ester
17	136	34.964	Octadecanoic acid, 2,3-dihydroxypropyl ester
18	139	35.876	Geranyl linalool isomer B

83.4.2 GCMS for Tamanu Biodiesel

In present study, fuel characteristics were tested by using shimadzu gas chromatography. The GC-MS data showed that the fuel was composed primarily of a compound with a high peak between 2.2 and 36.0 min. Table 83.4 exposes the GC-MS peak value of tamanu biodiesel.

The test sample presents chemical contents like olefins and paraffin. Figure 83.5 shows the GC-MS analysis of tamanu biodiesel.

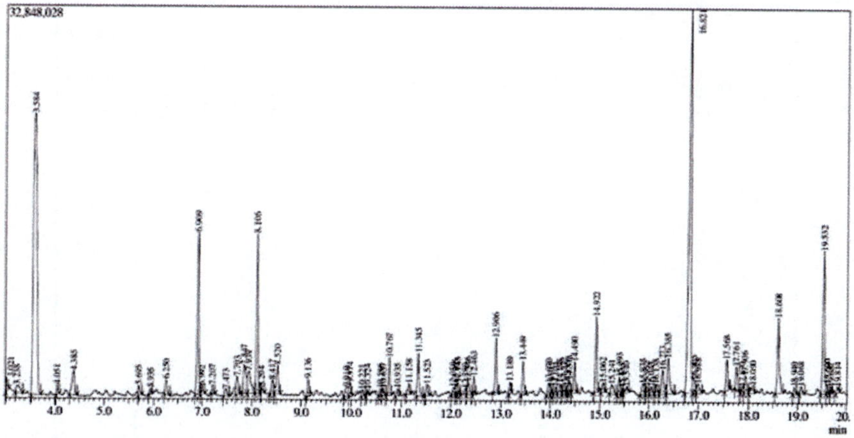

Fig. 83.5 GC-MS analysis

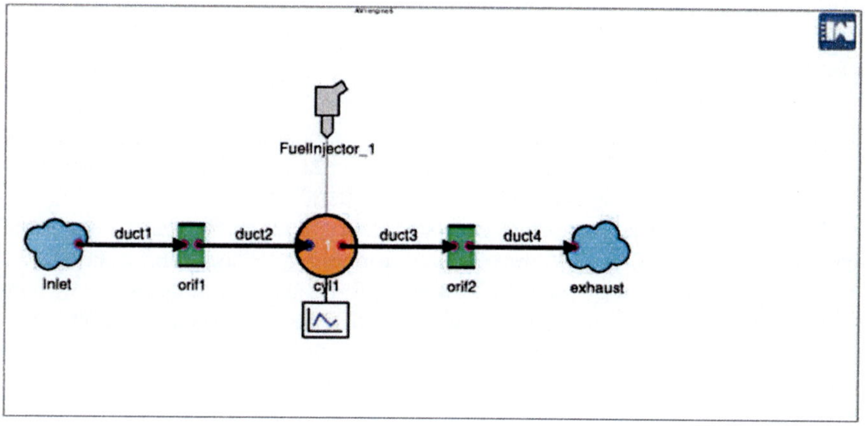

Fig. 83.6 One-dimensional engine model setup

83.4.3 One-Dimensional Ricardo Wave Engine Simulation

The 1D diesel engine simulation was performed on Ricardo-wave. Figure 83.6 shows the one-dimensional simulation Kirloskar engine setup. The first step was to measure the physical parameters, like bore, stroke, bowl design, the temperature on piston, connecting rod length, valve lift, valve diameter, cam profile, mass flow rate, injection timing, injection pressure, injection hole, and combustion duration. From the data, a comprehensive 1D modelling of the engine was built. The blend was created by a combination of bio-oil, ethanol, and diesel fuel. Ricardo would automatically blend the fuel according to the input by the user. At first, the

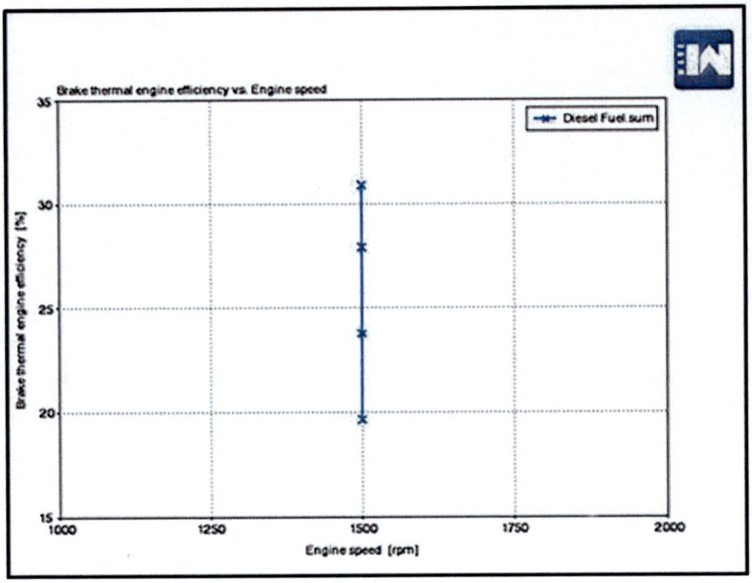

Fig. 83.7 BTE of diesel fuel from WAVE software

simulation was conducted for typical diesel fuel to acquire the same engine performance denoted in the specification. After obtaining the same result in the specification, the same process was repeated for B20 and B100.

B20, B100, and neat diesel fuel were simulating in mono cylinder CI engine at constant speed of 1500 rpm. It was observed from Fig. 83.7, the BTE for diesel was 31.6%, whereas BTE of B20 and B100 were 29.7% and 27.8% from Figs. 83.8 and 83.9, respectively. The one-dimensional simulation result showed that B20 blend portrayed better BTE than B100 at rated engine power. Based on this motivation, experimental work was done with mentioned test fuels.

83.4.4 Specific Energy Consumption

Figure 83.10 represents the variation in SEC of B20 blend, B100, and diesel concerning brake power. As brake power is increased, the results showed that the specific energy consumption (SEC) of all fuels tested decreased. Among all the test fuel, diesel fuel has least specific energy consumption at peak load. The results of SEC for Tamanu biodiesel, B20, and diesel were 12.03, 12.4, and 13.3 MJ/kW.hr., respectively. The energy value and viscosity of fuel determined the fuel consumption rate during the combustion. The results revealed that B100 found higher SEC than other test fuels diesel and B20. Neat B100 operated CI engine at full load was 6.8% higher SEC than B20 blend owing to low calorific value and poor vaporization

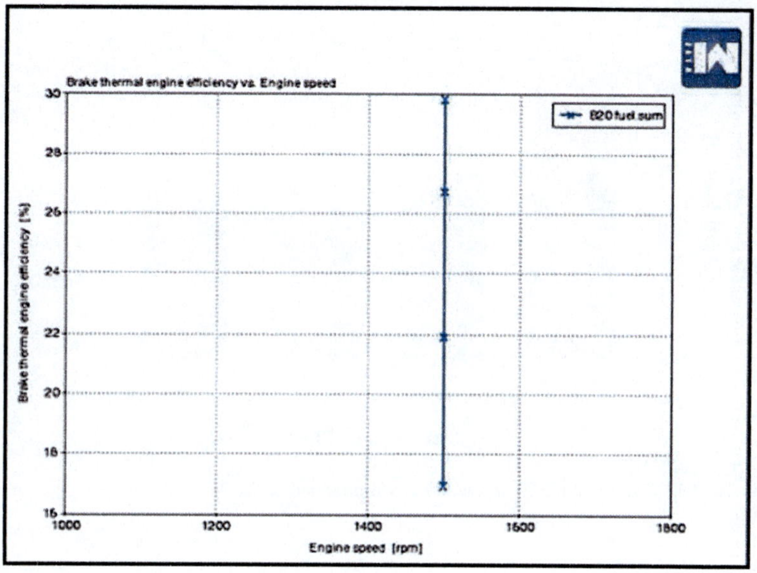

Fig. 83.8 BTE of B20 fuel from WAVE software

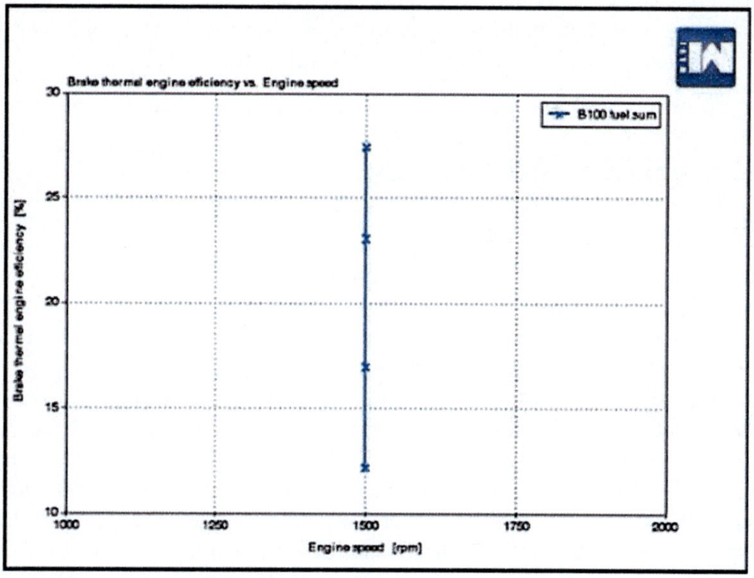

Fig. 83.9 BTE of B100 fuel from WAVE software

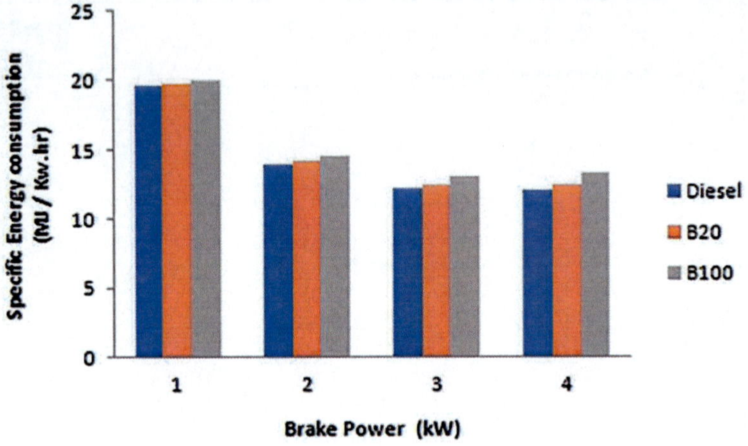

Fig. 83.10 BSEC of diesel, B20 and B100 at various loads

characteristics of raw biodiesel, which results in inferior combustion. In addition, B20 blend also showed similar trend with diesel fuel at all BP condition. This result might be due to reasonable viscosity and density of B20 blend, which enhance the oxidation rate at the end of compression stroke. The cylinder temperature of biodiesel was more than conventional fuel; therefore, it causes more heat loss from chamber, which results in decline in the combustion efficiency.

83.4.5 Brake Thermal Efficiency

From Fig. 83.11, it was detected that trend of BTE with respect to BP for different test fuel. It was noted that the BTE was linearly increased with all BP conditions for all test fuels. The thermal efficiency of neat diesel, B20, and B100 were 30.49%, 29.92, % and 26.82%, respectively. The result exhibited that the standard CI engine with neat diesel fuel was higher BTE as related to other test fuels at the same operating conditions. Due to its higher heat content and greater air utilisation in the combustion area, B20 was shown to have a higher BTE than B100. The brake thermal efficiency of neat diesel was 1.77% and 9.98% greater than B20 and B100 owing to better combustion and oxidation of diesel. The B100 fuel was showed 8.3% lower BTE than B20 blend due to combined benefits of volatility and vaporization characteristics of diesel domination blend shows better combustion process. However, the brake thermal efficiency of B20 blend was slightly dropped than diesel fuel at rated BP condition.

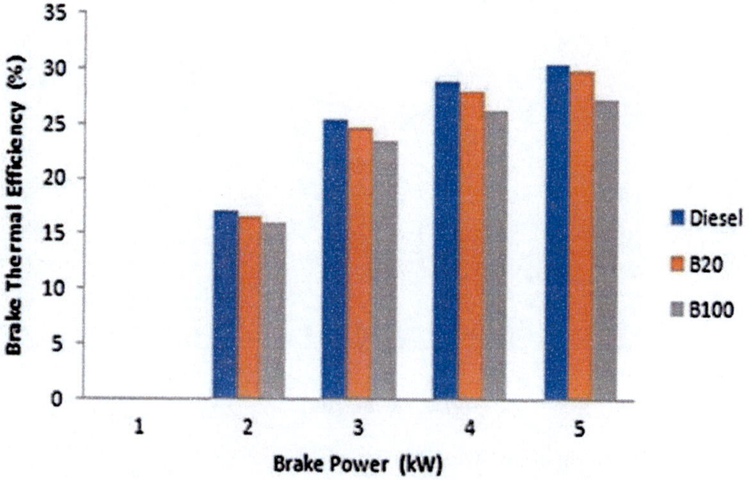

Fig. 83.11 BTE of diesel, B20 and B100 at various loads

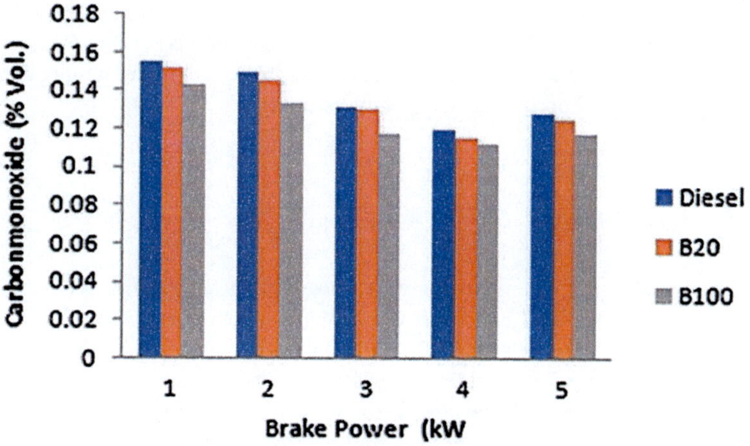

Fig. 83.12 CO emission of diesel, B20 and B100 at various loads

83.4.6 Carbon Monoxide

Figure 83.12 illustrates the effect of tamanu biodiesel blend with diesel on carbon monoxide (CO) formation. The figure found that B100 noticed lower CO than normal diesel due to existence of inherent O_2 content in the tamanu fuel. The CO emission of neat diesel, B20, and B100 at standard CI engine was 0.129% vol., 0.124% vol., and 0.119% vol., respectively. It was showed that CO emission was lower for biodiesel

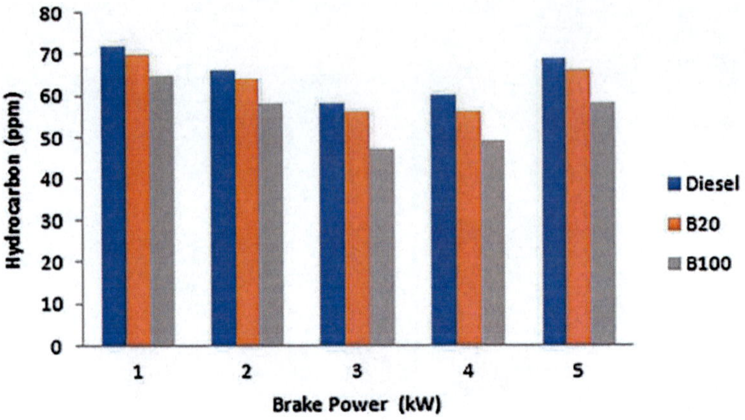

Fig. 83.13 HC emission of diesel, B20 and B100 at various loads

blend because of low C/H ratio of biodiesel. The CO emission of neat diesel was 3.87% and 7.75% higher than B20 and B100 fuel at rated BP condition owing to low carbon weightage in the tamanu biodiesel. B20 blend was found to have higher CO emission than B100 due to lack of O_2 availability in the chamber. Biodiesel fuel has a lower CO generation rate than diesel fuel because its high cylinder temperature and high oxygen content improve the CO_2 conversion rate.

83.4.7 Hydrocarbon

Figure 83.13 exhibits the comparison of hydrocarbon (HC) formation for diesel, B20 and B100 for various load condition. From the result, it was noted that HC level gradually decreased with rise in BP condition up to mid-load condition, beyond which it reaches the peak condition at peak load. It was noticed that B100 emits lower HC emission than diesel and B20. It might be ascribed to more oxygen presence and high cetane rating of biodiesel partially burnt to form a CO emission. B20 emits higher HC emission than B100 at all load range. It might be due to reduction of O_2 content and poor air utilization, which leads to unburnt particle presence in the exhaust. The result of HC emission for diesel was 68 ppm, which was 2.94% and 14.7% higher than B20 and B100, respectively. It was mainly due to low carbon and hydrogen ratio in biodiesel and excess quantity of O_2 in the chamber, which exhibits partial combustion of hydrocarbon, therefore lower HC formation.

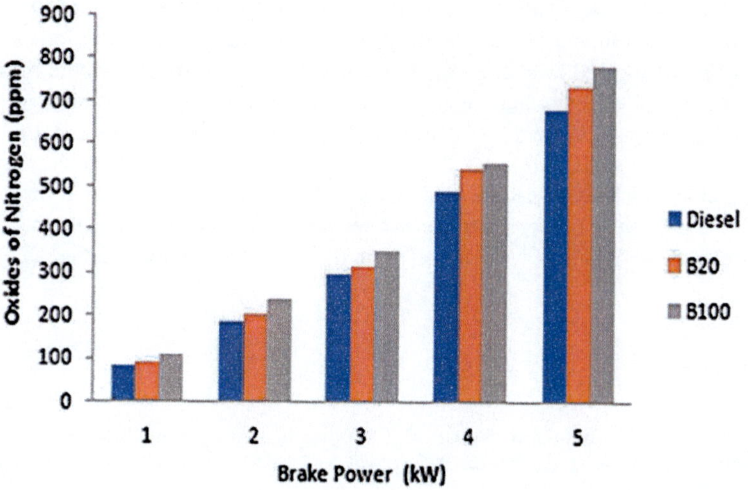

Fig. 83.14 NOx emission of diesel, B20 and B100 at various loads

83.4.8 Oxides of Nitrogen

Figure 83.14 displays the impact of tamanu biodiesel blend with diesel on oxides of nitrogen (NO_x) emission formation. The figure shows that the NO_x emission of diesel fuel emits lower emission than B20 and B100 owing to low temperature in the combustion when increase the load. The results is found that increases the load and percentage of tamanu blends with diesel fuel to drastically increase the NO_x emission. The result of NO_x emission of diesel was 678 ppm, for B20 it was 734 ppm and for B100 it was 782 ppm. It was found that NO_x emission of biodiesel was more than diesel. It might be due to less aromatic compounds and sufficient oxygen presence in tamanu biodiesel blends, which leads to increased combustion temperature and NO_x emission.

83.5 Conclusion

In the present investigation, the experimental results was found to be the qualities of biodiesel produced from non-edible oil. Test fuel properties were tested in fuel laboratory. The viscosity and density of tamanu biodiesel were more than that of conventional fuel. In contrast, the heating value of tamanu biodiesel is slightly less than the diesel fuel. One-dimensional theoretical simulations result was compared to real-time experimental result. The same operating conditions featured in theoretical simulation and experimental analysis. Simulation results were verified with actual experimental results for fuels B20, B100, and neat diesel fuel. The Ricardo wave simulation results were very close to experimental results. The SEC of diesel was less than tamanu biodiesel because of high calorific value of tamanu biodiesel. The BTE of B100 showed least at all operating condition. This result was the combined effect of higher viscosity and density of tamanu biodiesel, which is poor in atomization characteristics. Better volatility and enhanced calorific value of the blend result in higher BTE than neat B100, which leads to greater thermal efficiency. Biodiesel blend (B20) was showed lower HC and CO emission than diesel but penalty that slightly higher NOx emissions. Based on engine characteristics, it was concluded that B20 blend was elected as an optimum fuel for unmodified CI engine.

References

A. Gopinath, K. Sairam, and R. Velraj, 2015, Combustion Analysis of Polanga (Calophyllum inophyllum) Biodiesel, Appl. Mech. Mater., vol. 812, pp. 51–59, https://doi.org/10.4028/www.scientific.net/amm.812.51.

B. Ashok, K. Nanthagopal, R. Thundil Karuppa Raj, J. Pradeep Bhasker, D. Sakthi Vignesh, 2017, Influence of injection timing and exhaust gas recirculation of a Calophyllum inophyllum methyl ester fuelled CI engine, Fuel Processing Technology, Volume 167, Pages 18–30, ISSN 0378-3820, https://doi.org/10.1016/j.fuproc.2017.06.024.

B. Dhinesh, R. Niruban Bharathi, J. Isaac Joshua Ramesh Lalvani, M. Parthasarathy, K. Annamalai, 2017, An experimental analysis on the influence of fuel borne additives on the single cylinder diesel engine powered by Cymbopogon flexuosus biofuel, Journal of the Energy Institute, Volume 90, Issue 4, Pages 634–645, ISSN-1743-9671, https://doi.org/10.1016/j.joei.2016.04.010.

Elumalai PV, Dhinesh B, Parthasarathy M, Pradeepkumar AR, Mohamed Iqbal S, Jayakar J, Nambiraj M, 2020, An experimental study on harmful pollution reduction technique in low heat rejection engine fuelled with blends of pre-heated linseed oil and nano additive, Journal of Cleaner Production, https://doi.org/10.1016/j.jclepro.2020.124617.

Hwai Chyuan Ong, Jassinnee Milano, Arridina Susan Silitonga, Masjuki Haji Hassan, Abd Halim Shamsuddin, Chin-Tsan Wang, Teuku Meurah Indra Mahlia, Joko Siswantoro, Fitranto Kusumo, Joko Sutrisno, 2019, Biodiesel production from Calophyllum inophyllum-Ceiba pentandra oil mixture: Optimization and characterization, J. Clean. Prod., vol. 219, pp. 183–198, https://doi.org/10.1016/j.jclepro.2019.02.048.

K. Dinesh, A. Tamilvanan, S. Vaishnavi, M. Gopinath, and K. S. R. Mohan, 2019, Biodiesel production using Calophyllum inophyllum (Tamanu) seed oil and its compatibility test in a CI engine, Biofuels, vol. 10, no. 3, pp. 347–353, https://doi.org/10.1080/17597269.2016.1187543.

Lingesan Subramani, M. Parthasarathy, Dhinesh Balasubramanian, KrishnaMoorthy Ramalingam, 2018, Novel Garcinia gummi-gutta Methyl Ester (GGME) As a Potential Alternative Feedstock for Existing unmodified DI Diesel Engine, Renewable Energy. https://doi.org/10.1016/j.renene.2018.02.134

M. Parthasarathy, S. Ramkumar, J. Isaac Joshua Ramesh Lalvanib, P.V. Elumalai, B. Dhinesh, R. Krishnamoorthy, S. Thiyagarajan, 2020, Performance analysis of HCCI engine powered by tamanu methyl ester with various inlet air temperature and exhaust gas recirculation ratios, Fuel Volume 282, 118833. https://doi.org/10.1016/j.fuel.2020.118833

M. T. Raj and M. K. K. Kandasamy, 2012, Tamanu oil – an alternative fuel for variable compression ratio engine, Int. J. Energy Environ. Eng., vol. 3, no. 1, pp. 1–8, https://doi.org/10.1186/2251-6832-3-18.

N. A. Fauzan, E. S. Tan, F. L. Pua, and G. Muthaiyah, 2019, Physiochemical properties evaluation of Calophyllum inophyllum biodiesel for gas turbine application, South African J. Chem. Eng., vol. 32, pp. 56–61, 2020, https://doi.org/10.1016/j.sajce.2020.02.001.

P. Bridjesh, N. Prabhu Kishore, M. V. Mallikarjuna, and N. Alekhya, 2016, Performance analysis of variable compression ratio diesel engine using Calophyllum inophyllum biodiesel, Indian J. Sci. Technol., vol. 9, no. 35, pp. 6–10, https://doi.org/10.17485/ijst/2016/v9i35/95577.

P. S. Gautam, 2019, Experimental Analysis of CI Engine Fuelled with Calophyllum Inophyllum Biodiesel and Diesel with Ethanol as an Additive, Vol. 14, no. 9, pp. 76–81.

Rahul Krishnaji Bawane, A. Muthu raja, Gajanan N. Shelke & Anshul Gangele, 2020, Impact Analysis of Calophyllum Inophyllum Oil Biodiesel on Performance and Emission Characteristic of Diesel Engine under Variation in Compression Ratio, Engine Load, and Blend Proportion, International Journal of Ambient Energy, https://doi.org/10.1080/01430750.2020.1730955.

Roshan Raman, Naveen Kumar, 2020, Performance and emission characteristics of twin cylinder diesel engine fueled with mahua biodiesel and DEE, Transportation Engineering, Vol.2, 100024, ISSN2666-691X, https://doi.org/10.1016/j.treng.2020.100024.

R. Krishnaji Bawane, S. V Channapattana, N. B. Gadge, and S. M. Ingole, 2014, Experimental Investigation of Performance Characteristics of Calophyllum Inophyllum Biodiesel in CI Engine by Varying Compression Ratio, Int. J. Eng. Adv. Technol., no. 3, pp. 2249–8958.

R. Naveenkumar and G. Baskar, 2019, Biodiesel production from Calophyllum inophyllum oil using zinc doped calcium oxide (Plaster of Paris) nanocatalyst, Bioresour. Technol., https://doi.org/10.1016/j.biortech.2019.02.078.

Shanmugam, R., Murugesan, P., Guye, G.G. et al., 2021, Effect of additives on the stability of ethanol-diesel blends for IC engine application. Environ Sci Pollut Res 28, 12153–12167 (2021). https://doi.org/10.1007/s11356-020-10934-6

S. Dewang, Suriani, S. Hadriani, Bannu, and B. Abdullah, 2017, Physical Properties of Nyamplung Oil (Calophyllum inophyllum L) for Biodiesel Production, IOP Conf. Series: Journal of Physics: Conf. Series 846 (2017) 012009. https://doi.org/10.1088/1742-6596/846/1/012009.

S. Kurniati, S. Soeparman, S. S. Yuwono, L. Hakim, and S. Syam, 2019, A novel process for production of Calophyllum inophyllum biodiesel with electromagnetic induction, Energies, vol. 12, no. 3, https://doi.org/10.3390/en12030383.

S. Niju, G. Vishnupriya, and M. Balajii, 2019, Process optimization of Calophyllum inophyllum-waste cooking oil mixture for biodiesel production using Donax deltoides shells as heterogeneous catalyst, Sustain. Environ. Res., vol. 1, no. 1, pp. 1–12. https://doi.org/10.1186/s42834-019-0015-6.

S. Padmanabhan, T. Vinod Kumar, M. Chandrasekaran & S. Ganesan, 2020, Investigation of Sapindus seed biodiesel with nano additive on single cylinder diesel engine, International Journal of Ambient Energy, 41:10, 1106–1109, https://doi.org/10.1080/01430750.2018.1501755.

Tamilvanan, A., Balamurugan, K. & Vijayakumar, M., 2019, Effects of nano-copper additive on performance, combustion and emission characteristics of Calophyllum inophyllum biodiesel in CI engine. Journal of Thermal Analysis and Calorimetry, 136(1), 317–330 – September 2018. https://doi.org/10.1007/s10973-018-7743-4

Upendra Rajak, Prerana Nashine, Tikendra Nath Verma, 2018, Assessment of diesel engine performance using spirulina microalgae biodiesel, Energy, https://doi.org/10.1016/j.energy.2018.10.098

V. Dhana Raju, P.S. Kishore, K. Nanthagopal, B. Ashok, 2018, An experimental study on the effect of nanoparticles with novel tamarind seed methyl ester for diesel engine applications, Energy Conversion and Management, Volume 164, Pages 655–666, ISSN 0196-8904, https://doi.org/10.1016/j.enconman.2018.03.032

Chapter 84
Optimization of the High-Speed De-Laval Nozzle to Reduce the Acoustic Energy by Using the Truncated Nozzle

T. Kumaran, Selçuk Sarıkoç, C. Sarathkumar, A. Backiyaraj, and M. Parthasarathy

Nomenclature

MLN	Minimum length Nozzle
CFD	Computational fluid dynamics
C-D	Convergent divergent
FW-H	Ffowcs Williams and Hawkings
3D	Three dimensional
dB	Decibel

T. Kumaran (✉) · C. Sarathkumar
Department of Aeronautical Engineering, Vel Tech Rangarajan Dr Sagunthala R & D Institute of Science and Technology, Chennai, India
e-mail: tkumaran@veltech.edu.in

S. Sarıkoç (✉)
Department of Motor Vehicles and Transportation Technologies, Tasova Yuksel Akin Vocational School, Amasya University, Amasya, Turkey

A. Backiyaraj
Departmentof Mechatronics Engineering, Agni College of Technology, Chennai, India

M. Parthasarathy
SankarAdvanced Mobility Solutions, Chennai, India

M. Z. Sogut et al. (eds.), *Proceedings of the 2022 International Symposium on Energy Management and Sustainability*, Springer Proceedings in Energy,
https://doi.org/10.1007/978-3-031-30171-1_84

84.1 Introduction

In recent decades, the high speed aircraft engine performance was improved in much research on various flight operating conditions. Engines like Scramjet and Ramjet are operating at maximum velocities and also used in multiple aircraft high speed operations. The inlet isolator, combustor, and nozzle, as well as other fuel injections, flame holders, and cooling mechanisms, are major parts of a scramjet engine. The nozzle is one of essential components of this type of high speed air breathing engines which is used to expand the high temperature gas from the combustion chamber. The combustion chamber of the Supersonic vehicle is in Supersonic speed at the Mach number of above 1.2, so the exhaust gas from the combustion chamber of the scramjet having high thermal energy and high enthalpy to initiate a proper nozzle efficiency (Kumar et al. 2021; Jose et al. 2021). For the nozzle, the divergent length is slightly difficult to satisfy the ideal requirement of the high speed air breathing engine nozzle efficiency. To improve the requirement, a truncated nozzle is designed to raise the aerodynamic characteristics and performance, then velocity streamline method with an offset circular geometric entrance was considered to implement 3D asymmetric truncated nozzle (Zhu et al. 2018a). The optimistic performance of the convergent divergent nozzle is improvised by use of supercritical jets having high pressure energy; especially in the ground effect, the supercritical jet provides diverted shock and expansion waves in high speeds (Margason et al. 1996). The analysis model's nozzle expansion pattern is found to be highly accurate with the FLUENT simulation process. All of the relevant parameters throughout the scramjet were evaluated when the evaluation programme was completed. Each component of nozzle inlet and output was measured for static and total pressures, temperature, enthalpy, and velocity. In addition, mixed gas contents such as oxygen, hydrogen, and nitrogen were obtained. In the exergy analysis, these physical parameters were used (Zhu et al. 2018a, b, c). The asymmetric truncated is modelled to increase the efficiency and reduce the acoustic effects in the de-laval nozzle. Based on the nozzle exit area, the acoustic and pressure energy is highly influenced (Quadros et al. 2021). In the present research, we have to investigate the pressure energy and acoustic effects expanding from the de-laval nozzle by using ANSYS FLUENT by simulating the nozzle in various velocities and various offset truncated shape. The results of various cases were compared and the performance characteristics of this modified nozzle discussed.

84.2 Methodology

The method of optimizing the nozzle in the divergent portion is offset with variable angle variations based on the energy transmission of hot gas from the throat. The regions which we consider for the asymmetric area by three regions of the de-laval nozzle follow the transition region, kernel region, and the uniform region, as shown in Fig. 84.1.

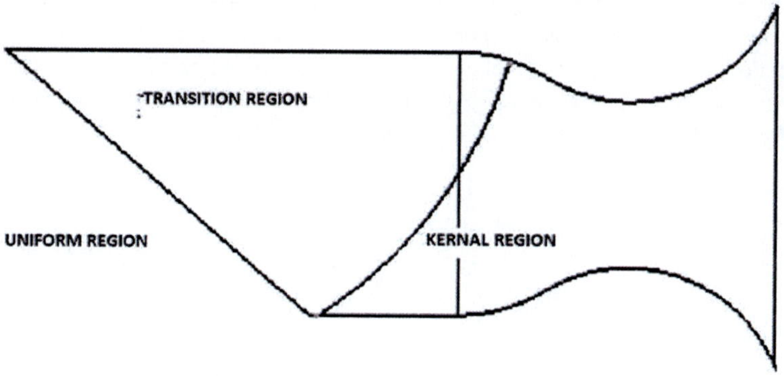

Fig. 84.1 Schematic representation of the nozzle based on various regions

The transition region is a section of simple wave region which is influenced by the upper wall of the divergent duct, so the geometry of the transition region is essentially considered in the performance of adiabatic expansion of gas in deliver nozzle in high speed. The kernel region is a downward region in which the expansion wave is influenced by the nozzle based on the lower wall, and, finally, the uniform region is where the exit of the gas is exhaust and the Mach number of the nozzle is influenced by this region (Fu et al. 2016). In this research, we will design the modified kernel region and transition region so that we can estimate the influence of expansion waves and simple waves in the performance of the nozzle and also evaluate the sound waves based on the upper and lower wall of the nozzle. 3D modelling is done for various offset angles, like 40, 50, and 60 degrees, and simulated using computational fluid dynamics (CFD) to calculate the performance characteristics and acoustics energy.

84.2.1 Governing Equations

The equation to solve numerical solution for a nozzle considered is hyperbolic derived by the using the characteristics inviscid, irrotational, and isentropic the following equations obtained from (Clarke 1978) for the obtaining the asymmetric flow and plane flow for the nozzle (Turkyilmazoglu 2015) in Eq. 84.2:

$$\{\{(u^2 - a^2)u_x + (v^2\_a^2)v_x + 2uvu_y - \delta a^2 v/y = 0, u_y - v_x = 0 \tag{84.1}$$

$$\left\{\left(\frac{dy}{dx}\right) = \Lambda\pm = \tan(\theta \pm \alpha),\ (u^2 - a^2)du_\pm + (2uv - (u^2 - a^2)\Lambda\pm)dv_\pm - \left(\frac{\delta a^2 v}{y}\right)dx_\pm = 0\right. \tag{84.2}$$

where $\delta = \text{n}$ and $\delta = 0$ indicate the flow which we consider for the evaluation; and θ and α are the flow angle and Mach angle, respectively; then u and v are the x and y direction velocities, respectively. Other parameters were calculated by the following equations in finite difference method based on characteristic lines which are used to calculate the pressure, specific heat, velocity and temperature.

$$Ma_4 = \frac{V_4}{\sqrt{\gamma R T_{\text{tin}}}} - \frac{\gamma - 1}{2} V_4^2, \quad (84.3)$$

$$P_{\text{tin}} = \left(1 + \frac{\gamma - 1}{2} Ma_4^2\right) 1^{\frac{\gamma}{\gamma - 1}} P_4, \quad (84.4)$$

$$T_{\text{tin}} = \left(1 + \frac{\gamma - 1}{2} Ma_4^2\right) T_4. \quad (84.5)$$

After that, while solving the computation fluid dynamic problem in these types of nozzles, we have to use (Zhu et al. 2018c):

$$C_p = a_1 + a_2 T + a_3 T^2 + a_4 T^3 + a_5 T^4 \quad (84.6)$$

A polynomial characterizing the thermos physical parameters of the gas is included with varying specific heat in the above equation.

84.3 Results and Discussion

The minimum length nozzle or MLN is modelled based on various offset angle; the performance characteristics, such as pressure (P), velocity at outlet (V), turbulence kinetic energy (k), and acoustic energy based on decibel (dB), were evaluated in computational fluid dynamics by use of ANSYS and CATIA V5. The modelling of the convergent divergent nozzle is shown in Fig. 84.2.

The model was meshed and converted into a finite element model having the element size of 0.02 mm to increase the accuracy of the CFD. The meshed model is shown in Fig. 84.3 and the boundary conditions, such as inlet, outlet, and wall, were set in the meshing for solution initialization.

The solution for the computational model is set as turbulent flow because of high Mach number so that it considers K-omega model, initializes the energy equation for the evaluation, and activates the acoustic model for the analysis, and three different cases were considered for the validation of result comparison.

Based on the requirement for solving the numerical model the elements and nodes were created on this configuration of 40, 50, and 60-degree offset configurations were solve the turbulence omega model and the pressure distribution over the nozzle at subsonic inlet velocity of 0.8 Mach no is shown in Figs. 84.4, 84.5, and 84.6. From this analysis, the pressure is varying with respect to the different configurations and the nozzle exit has uniform flow for all three cases.

Fig. 84.2 Modelling of minimum length nozzle based on various regions

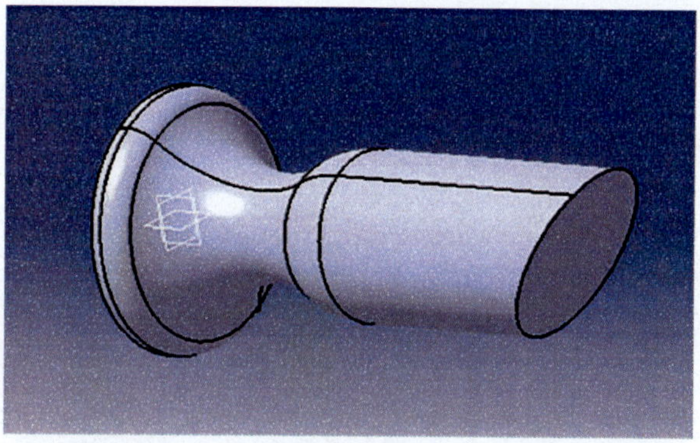

Fig. 84.3 Meshing of minimum length nozzle based on various regions

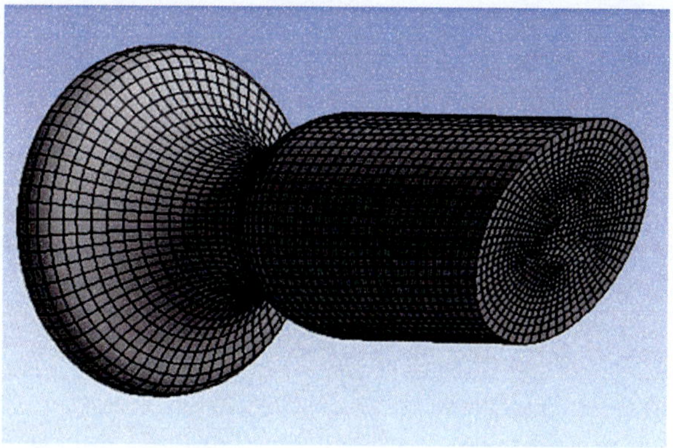

Fig. 84.4 Pressure contour at 40 degrees

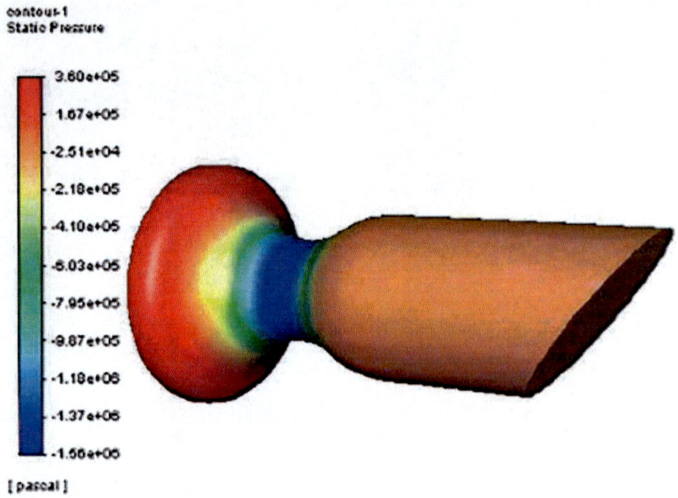

Fig. 84.5 Pressure contour at 50 degrees

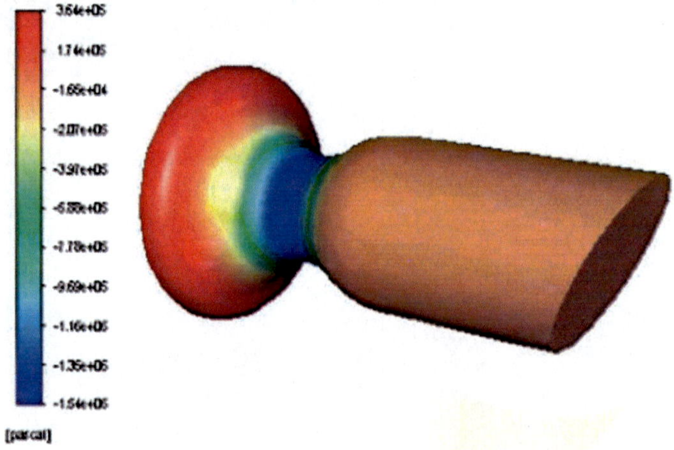

Fig. 84.6 Pressure contour at 60 degrees

Similarly, the turbulence created on the throat and the exit of the nozzle is evaluated based on the turbulence model solution, and it is shown in Figs. 84.7, 84.8, and 84.9; it explains that in all three cases, the turbulent kinetic energy with respect to the geometry of this nozzle is initially very low, but when it flows in the convergent section, it is very high, and it finally ends with the turbulent kinetic energy level related to 2250 m^2/s^2.

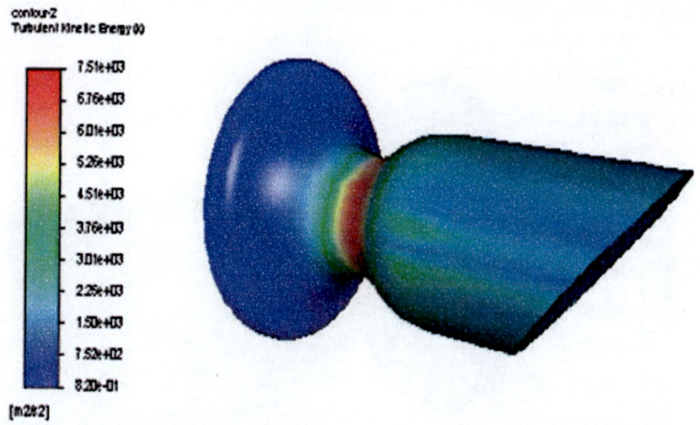

Fig. 84.7 Turbulence K contour at 40 degrees

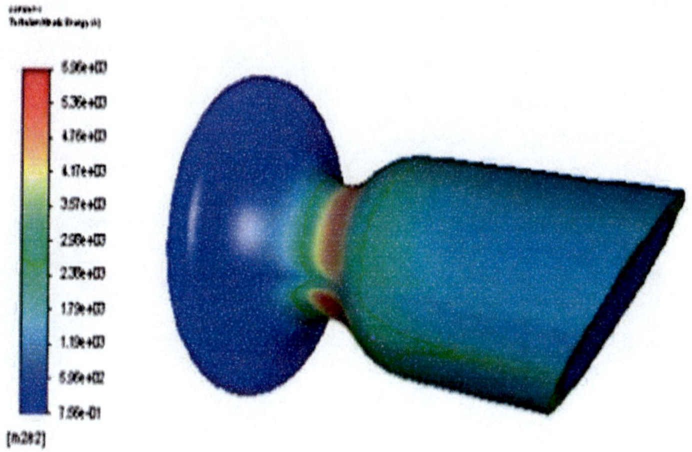

Fig. 84.8 Turbulence K contour at 50 degrees

After the turbulent analysis the solution is considered one acoustic model activating the FW-H acoustics model to solve with acoustic broad band noise considerations to evolve the acoustic power which is generated in the output boundaries and wall boundaries of the C-D nozzle shown in Fig. 84.10 for the initial case for 40-degree offset design; the dB of this model starts with 137.

Compared to the results from the figures for turbulent kinetic energy, the configuration with a 60-degree offset has a low level of turbulence at the exit, so it shows that we can suggest this case for good engine performance to increase the efficiency while also having low turbulent effects.

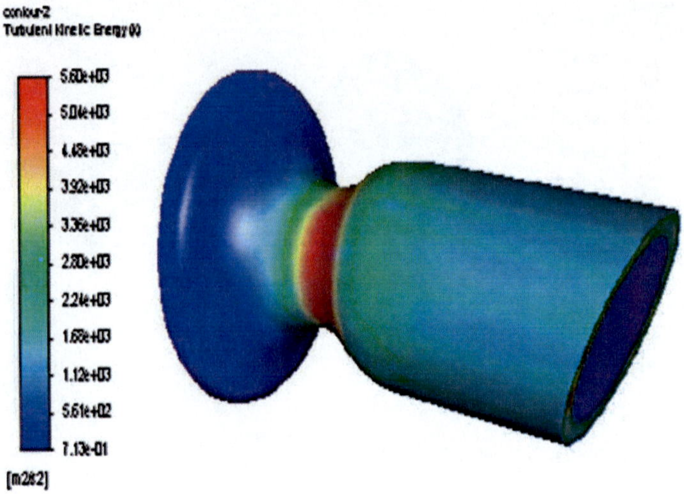

Fig. 84.9 Turbulence K contour at 60 degrees

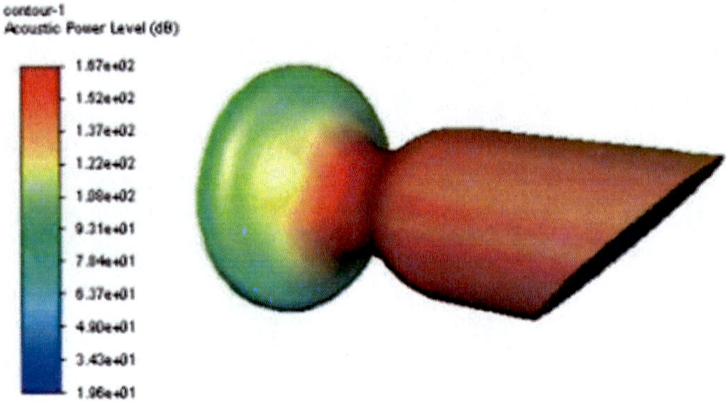

Fig. 84.10 Acoustic power at 40 degrees

The remaining case which consider for the evaluation shown in Figs. 84.11 and 84.12 explains the variation of the acoustic power based on the design as follows 149 and 133 dB. From this acoustic analysis, the sound power level based on these various offset designs is varying with respect to the area and the geometry and the 60-degree offset nozzle configuration is producing low acoustic power level compared to the other configurations.

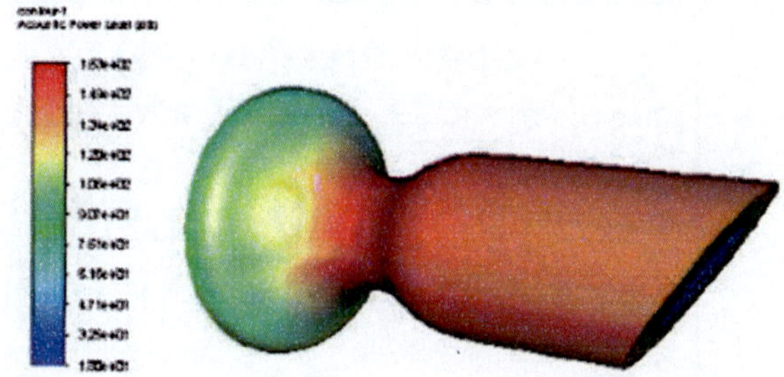

Fig. 84.11 Acoustic power at 50 degrees

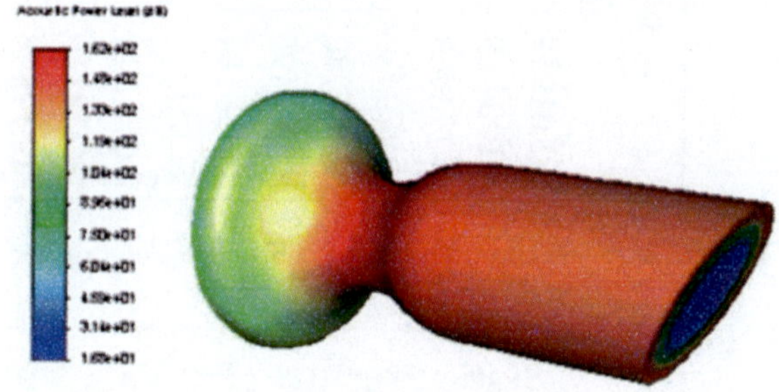

Fig. 84.12 Acoustic power in MLN at 60 degrees

The results from the contour of pressure generated in wall conditions in turbulent flow at subsonic speed the energy which is exerted from the wall respect to the Reynolds no the graph between the static pressure level on the 40 degree offset configuration is plot related to the acoustic energy is shown in Fig. 84.13. It shows that the variation of acoustic energy fully depends on the pressure energy of the exhaust gas. The variation of dB in each node is demonstrated in the plot, and the decibel value is increasing at the throat, but after the expansion in the gas of this deliver nozzle slightly reduced to 137.

Similarly, the expansion of the pressure energy of the hot gas exhaust from the nozzle for the 50-degree offset truncated nozzle is shown in Fig. 84.14, which explains the variation of expansion and the efficiency with respect to pressure on the exit wall boundary, having the decibel level reduced gradually from the throat. The throat area depends on the operating Mach number of the nozzle so that the modified offset geometry is creating the acoustic energy level of 149 dB for this case.

Finally, the geometry of this 60-degree offset minimum length nozzle also has linear variation with respect to the pressure energy, but also has adequate performance in nozzle expansion energy and the minimum acoustic power level at the exit wall conditions with respect to the Supersonic speed, as shown in Fig. 84.15.

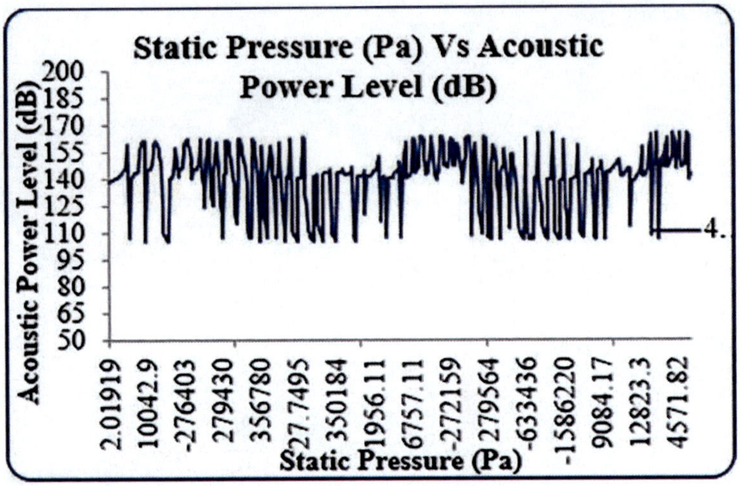

Fig. 84.13 Acoustic power vs. static pressure on a 40-degree offset configuration

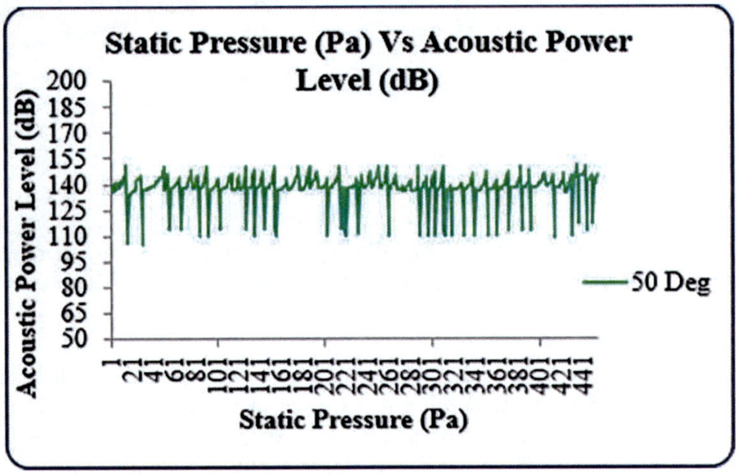

Fig. 84.14 Acoustic power vs. static pressure on a 50-degree offset configuration

84.4 Conclusion

The development of research in energy conservation on a system is utilised in this high-speed convergent divergent nozzle to propose a design of high performance, and noise control of the C-D nozzle is analysed based on the variable minimum length nozzle. Based on the current research on this type of nozzle, the computational fluid dynamic analysis is done using two models. Using K-Omega model the variable geometry cases evaluating the pressure energy and initiate the FW-H

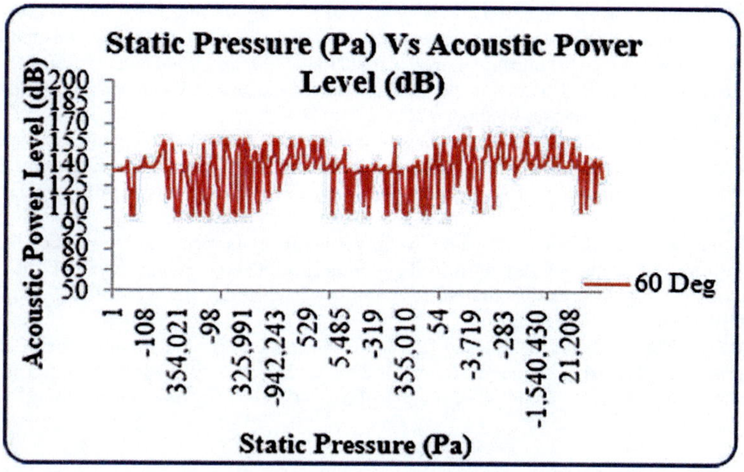

Fig. 84.15 Acoustic power vs. static pressure on a 60-degree offset configuration

equations for the acoustic analysis to predict the dB created on the exit wall boundary conditions. The results of the three cases solved on this method the 60-degree offset configuration shows the improved validation compared to the other cases and the normal convergent divergent nozzle which used in the aircraft. The gas pressure energy is one of the major factors influencing the acoustic energy and also the noise with respect to the frequency on normal mode of the nozzle.

References

Clarke, J. (1978). Gas Dynamics, Volumes 1 and 2. By M. J. ZUCROW and J. D. HOFFMAN. Wiley, 1976, 772 pp. 219.00 or $35.00 and 1977, 480 pp. S18.95 or $37. One-Dimensional Compressible Flow. By H. DANESHYAR. Pergamon Press, 1976. 179 pp. E7.50 or $15.00 (hardback), 23.75 or $7.50 (flexicover). Journal of Fluid Mechanics, 87(4), 789–792. https://doi.org/10.1017/S0022112078211901

Fu, Lei; Zhang, Shuai; Zheng, Yao (2016). Design and Verification of Minimum Length Nozzles with Specific/Variable Heat Ratio Based on Method of Characteristics. International Journal of Computational Methods, (), 1650034–. https://doi.org/10.1142/S0219876216500341

Jose, Riyan & Raj, Rhitik & Dewang, Yogesh & Sharma, Vipin. (2021). A Review on Scramjet Engine. https://doi.org/10.1007/978-981-16-0159-0_48.

Kumar, S., Murari Pandey, K., & Kumar Sharma, K. (2021). Recent developments in technological innovations in scramjet engines: A review. Materials Today: Proceedings, 45, 6874–6881. https://doi.org/10.1016/j.matpr.2020.12.1086

Margason, Richard J.; Arledge, Thomas K.; Wardwell, Douglas A.; Hange, Craig; Naumowicz, Tim (1996). SAE Technical Paper Series [SAE International SAE International Powered Lift Conference and Exposition – (NOV. 18, 1996)] SAE Technical Paper Series – Jet Efflux Characteristics and their Influence on STOVL Aircraft Propulsion-Induced Effects. , 1(), –. https://doi.org/10.4271/962250

Quadros, Jaimon & Khan, Sher & T., Prashanth. (2021). Study of buzz phenomenon in suddenly expanded flow structure using experiments and visualization. Numerical Heat Transfer, Part B: Fundamentals. 80. 1–17. https://doi.org/10.1080/10407790.2021.1944784.

Turkyilmazoglu, M. [2015] "Parabolic partial differential equations with nonlocal initialand boundary values," Int. J. Comput. Methods 12(5), 1550024.

Zhu M, Fu L, Zhang S, Zheng Y (2018a). Design and optimization of three-dimensional supersonic asymmetric truncated nozzle. Proceedings of the Institution of Mechanical Engineers, Part G: Journal of Aerospace Engineering.;232(15):2923–2935. https://doi.org/10.1177/0954410017718567

Zhu, Meijun & Fu, Lei & Zhang, Shuai & Zheng, Yao. (2018b). Design and Optimization of Three-Dimensional Supersonic Asymmetric Truncated Nozzle. Proceedings of the Institution of Mechanical Engineers Part G Journal of Aerospace Engineering. 232. 2923–2935. https://doi.org/10.1177/0954410017718567

Zhu, Meijun; Zhang, Shuai; Zheng, Yao (2018c). Conceptual design and optimization of scramjet engines using the exergy method. Journal of the Brazilian Society of Mechanical Sciences and Engineering, 40(12), 553–. https://doi.org/10.1007/s40430-018-1468-y

Chapter 85
Energy Analysis of a Solar Air Heater with a Fin-Added Absorber

Abid Ustaoglu, Mustafa Sabri Gok, Alaattin Metin Kaya, Zeyad Amjed, Tayfun Altiok, and Fatih Kocyigit

Nomenclature

SAH Solar Air Heater
FASAH Fin Absorber Solar Air Heater

85.1 Introduction

Today, most of the energy consumption is supplied by coal, natural gas, and oil. These generate several difficulties, such as a rise in greenhouse gas levels, global warming, and environmental damage.

For more nations to satisfy their obligations to cut carbon dioxide emissions, the percentage of renewable energy has to expand to account for the majority of the world's energy supply. Solar energy offers considerable promise as an alternative to

A. Ustaoglu (✉) · M. S. Gok
Department of Mechanical Engineering, Bartin University, Bartin, Turkey
e-mail: msabrigok@bartin.edu.tr

A. M. Kaya
Department of Mechanical Engineering, Bursa Uludag University, Bursa, Turkey
e-mail: alaattinkaya@uludag.edu.tr

Z. Amjed · T. Altiok
Graduate School, Bartin University, Bartin, Turkey

F. Kocyigit
Department of Aviation Management, Dicle University, Diyarbakir, Turkey
e-mail: fatih.kocyigit@dicle.edu.tr

M. Z. Sogut et al. (eds.), *Proceedings of the 2022 International Symposium on Energy Management and Sustainability*, Springer Proceedings in Energy,
https://doi.org/10.1007/978-3-031-30171-1_85

these fuels. An hour of solar radiation arriving on the globe produces sufficient energy to cover the world's energy consumption for a complete year.

Turkey heavily employs solar collectors that create hot water from solar energy. However, its usage in space heating is quite restricted. Air solar collectors are a form of heat exchanger that warms the air flowing through it with the rays originating from the sun. Since the heat transfer rate to the air is lower than the ratio to the liquid in these systems, it is required to exceed the heat transfer surface territory to alleviate this shortfall.

For their simplicity and cheap cost, solar air heaters are significant methods for using solar radiation to heat space, dry agricultural items such as vegetables, fruits, and seeds, and processors in industries in buildings. The air solar collector design will be created in the study, and the design will be altered to provide airflow. A radiator will be used to increase the surface area. The radiator will be laser or water jet cut to size in a way that does not obstruct airflow and is compatible with the solar collector design. The obtained system's design will be tailored to the previously estimated air solar collector geometry.

There were many studies on the solar air heater system. This research seeks to conduct an empirical investigation of the impact of adding geometry of radiator consisting of big surface area fins on the performance of solar collector into improving the amount of heat transfer in the solar air collector. By reviewing the literature, it turns out that this is the first time that work has been done on an extended surface solar collector with radiator geometry. Priyam and Chand (2016) examined the system comprised of two undulating blades on an absorber plate and a thermally insulated bottom side. It has been shown that the usage of wavy fins enhances the heat transfer surface area and also the heat transfer rate. Increasing the airflow rate from the solar air heater has been shown to result in not only improved efficiency but also a rise in pressure loss. Bhushan and Singh (2010) undertook experiments to boost the rate of heat transmission to the air moving in the duct of a solar-powered air heater. In the findings, the artificially roughened absorbent plate surface was deemed to be an effective strategy. Various roughness geometries were tested to analyze the heat transmission and friction qualities of an intentionally roughened duct in solar air heaters. It has been noted that artificial roughness is a suitable way to boost the thermal efficiency of solar air heaters. Abo-Elfadl et al. (2020) prepared the tubular type of a solar air heater in an experimental setting (SAH). A flatness solar air heater with the same size and materials was compared against a tubular solar air heater. The tubular one is more efficient, has more thermal force, and has less heat loss. Heydari and Mesgarpour (2018) studied a solar air heater with a spiral influx track and a triangular crossing channel numerically. With two distinct flow rates, the heat transfer coefficient and thermal efficiency were determined. When compared to simple safety and roughly 8.6% of dual double painting, the average thermal efficiency of dual passage with a greater spiral design was 14.7%. Hassan and Abo-Elfadl (2018) analyzed the recirculation of the air beneath the plate of the absorber to pull part of the air over the absorbent surface and then remove additional heat from the absorbent plate. The findings revealed that the temperature of the absorbent plate lowers with increasing the rate of top airflow. M. Sabzpooshani et al.

(Sabzpooshani et al. 2014) showed the influence of varying fin coefficients, glass cover thickness, segregation bottom thickness, & intake air temperature at different mass influx rates on energy efficiency. It was discovered in the research that placing blades and baffles at low mass influx had a favorable outcome in energy efficiency but demonstrated an opposing consequence at larger mass flow rates.

The geometry of the produced radiator will be situated in particular portions of the collector along with the airflow. A CFD evaluation of the solar air collector will be undertaken with the enhanced surface area. In these investigations, the effect of numerous parameters such as the angle of placement of the coolant and the number of radiators on system performance will be investigated. The intended solar air collector will be produced. Basic materials will be obtained, and the system will be created. The air will be flowing with the aid of a fan. The system will be evaluated in varied weather situations at varying radiator placements, angles, and numbers. Finally, the tilt angle of the system will be adjusted to 30 degrees. Each system was aimed at the south. The inlet, exit, and surface of the temperature of the solar air heater and ambient temperatures were monitored to assess system performance. The system will be updated, and the solar air collector with enlarged surface area and coolant form will be economically evaluated. Thus, it will be ideal for single usage. Solar energy may be utilized for house or space heating. When solar energy is adequate, it eliminates the expense of fuel, and when it is insufficient, it will help to decrease the cost of fuel.

85.2 Energy Analysis

Irradiation is evaluated utilizing a pyranometer. The horizontal surface comprises direct, Hb, and diffuse solar radiation Hd. It is mentioned below Dummy paragraph.

$$H = H_b + H_d = I_b \sin \alpha_s + H_d \tag{85.1}$$

where I_b and α_s indicate direct sun radiation and solar altitude, respectively. The sun height is computed as

$$\sin \alpha_s = \sin \delta \sin \phi + \cos \delta \cos \phi \cos \omega \tag{85.2}$$

The general energy balance equation is stated as

$$\sum \dot{E}_{in} - \sum \dot{E}_{out} = \Delta \dot{E}_{system} \tag{85.3}$$

The exergy balance equation for a steady-state condition is shown as

$$\sum \dot{E}x_{in} - \sum \dot{E}x_{out} = \Delta \dot{E}x_{dest} \tag{85.4}$$

The exergy balance equation could be defined as

$$\sum\left(1-\frac{T_{amb}}{T_s}\right)\eta_{opt}\dot{Q}_S - W + \sum \dot{m}_{in}\psi_{in} - \sum \dot{m}_{in}\psi_{in} = \dot{E}x_{dest} \tag{85.5}$$

The exergy input and output are specified as

$$\psi_{in} = (h_{in} - h_{amb}) - T_{amb}(s_{in} - s_{amb}) \tag{85.6}$$

$$\psi_{out} = (h_{out} - h_{amb}) - T_{amb}(s_{out} - s_{amb}) \tag{85.7}$$

85.2.1 Experimental Setup

Figure 85.1 shows the solar air heater together with a schematic of its layout and its dimension, and Fig. 85.2. shows the experimental setup. Solar radiation measurements are made with the help of a thermometer. An anemometer was used to measure incoming and outgoing air velocities. The air mass flow rate was determined by measuring air velocity, cross-sectional area, temperature, and corresponding density. Use thermocouples to determine the air temperature, inlet and outlet temperatures, and surface temperatures. The temperature and solar radiation measurements will be transmitted to the computer with the help of the datalogger. An experimental analysis was performed on a clear sky day. The tilt angle of the system is set to 30 degrees. Each system was directed to the south. The inlet, outlet, and surface temperature of the solar air heater and ambient temperatures were measured to report system performance.

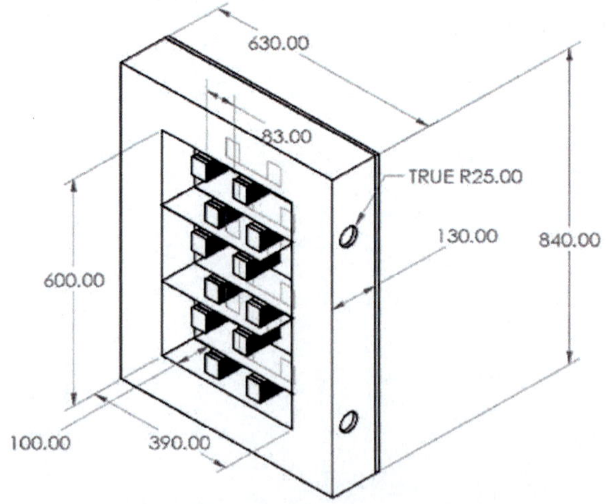

Fig. 85.1 Solar air heater from the side and its dimensions

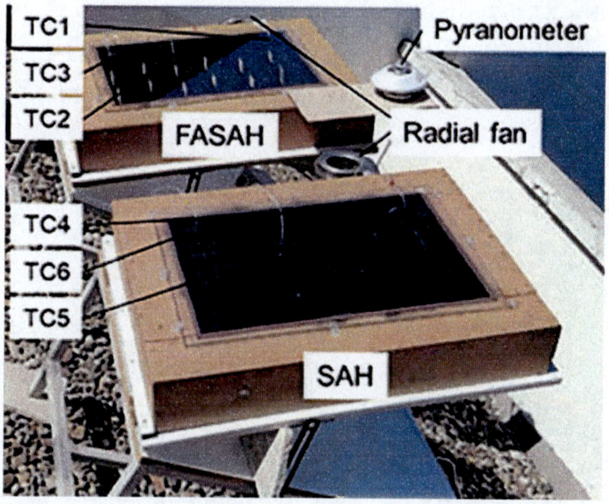

Fig. 85.2 Experimental setup of SAH systems

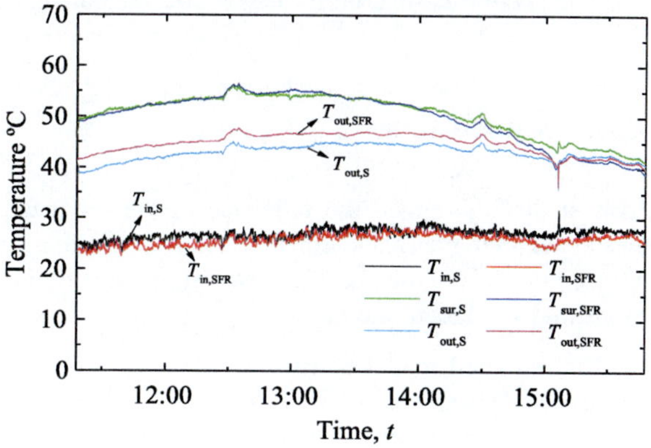

Fig. 85.3 Time and temperature-transformation diagram for a solar air heater

85.3 Results and Discussion

Figure 85.3 shows the solar air heater's time and temperature conversion diagram. Temperature measurement for the entry and exit areas for fins and without fins is shown. The table also shows six different temperatures at the top, indicating the surface temperature and the others like entry (Wijeysundera et al. 1982) and exit temperatures. It is also noted that the device that contains fins have higher temperatures than those without fins by 5–8 °C.

Figure 85.4 shows the thermal efficiencies of conventional and fin-added solar air heaters. The advantage of the FASAH system appears after 14:00 hours. The average thermal efficiency of the FASAH system was around 71.3%, while that of the SAH

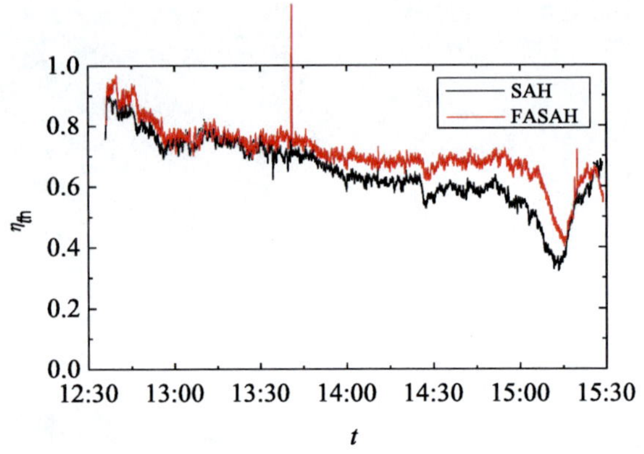

Fig. 85.4 Thermal efficiency of conventional and fin-added solar air heaters

system was 65.2%. The maximum difference in the thermal efficiencies was observed between 14:30 and 15:10. The results show that the proposed system provided better thermal performance than the conventional one.

85.4 Conclusion

This study highlighted the influence of the thermal performance of the solar air heater without fins and with fins. Active efficiency and thermal efficiency were utilized to analyze the heat transfer performance of the solar air heater. The conclusions may be drawn like this:

- The designed model delivers satisfactory performance results for plate and fin solar air heaters.
- Various performance aspects of the solar air heater were tested.
- Boosting the airflow rate via the solar air heater leads to increased collector effectiveness.
- It is desirable to employ fins to improve the surface zone of heat transmission and the heat transfer coefficient to raise the thermal performance of solar air heaters.
- Under identical circumstances of work, it was observed that a solar air heater with fins produces greater values of higher temperature and thermal effectiveness than the flat plate collector.
- It can be concluded that the air entry temperature is 24 °C and the exit temperature is 38 °C, which is an increase at a rate of 36.84% for a solar heater without fins, and for a solar heater with fins, the air entry is at a temperature of 24 °C and the exit temperature is 42 °C, which is an increase of 42.85%.

References

Abo-Elfadl, S., Hassan, H., & El-Dosoky, M. F. (2020). Study of the performance of double pass solar air heater of a new designed absorber: An experimental work. Solar Energy, 198(October 2019), 479–489. https://doi.org/10.1016/j.solener.2020.01.091

Bhushan, B., & Singh, R. (2010). A review on methodology of artificial roughness used in duct of solar air heaters. Energy, 35(1), 202–212. https://doi.org/10.1016/j.energy.2009.09.010

Hassan, H., & Abo-Elfadl, S. (2018). Experimental study on the performance of double pass and two inlet ports solar air heater (SAH) at different configurations of the absorber plate. Renewable Energy, 116, 728–740. https://doi.org/10.1016/j.renene.2017.09.047

Heydari, A., & Mesgarpour, M. (2018). Experimental analysis and numerical modeling of solar air heater with helical flow path. Solar Energy, 162(January), 278–288. https://doi.org/10.1016/j.solener.2018.01.030

Priyam, A., & Chand, P. (2016). Thermal and thermohydraulic performance of wavy finned absorber solar air heater. Solar Energy, 130, 250–259. https://doi.org/10.1016/j.solener.2016.02.030

Sabzpooshani, M., Mohammadi, K., & Khorasanizadeh, H. (2014). Exergetic performance evaluation of a single pass baffled solar air heater. Energy, 64, 697–706. https://doi.org/10.1016/j.energy.2013.11.046

Wijeysundera, N. E., Ah, L. L., & Tjioe, L. E. (1982). Thermal performance study of two-pass solar air heaters. Solar Energy, 28(5), 363–370. https://doi.org/10.1016/0038-092X(82)90253-5.

Chapter 86
Energy Performance Evaluation of PCM-Impregnated Vermiculite in Brick

Ertugrul Erdogmus, Osman Gencel, Abid Ustaoglu, Hande Torlaklı, Ahmet Sarı, Gokhan Hekimoglu, Mucahit Sutcu, and Fatih Ergur

Nomenclature

EPS	Expanded PolyStyrene
HVAC	Heating, Ventilation and Air Conditioning
LHS	Latent Heat Storage
PCM	Phase Change Material

E. Erdogmus
Department of Environmental Engineering, Bartin University, Bartin, Turkey
e-mail: eerdogmus@bartin.edu.tr

O. Gencel
Civil Engineering Department, Bartin University, Bartin, Turkey
e-mail: ogencel@bartin.edu.tr

A. Ustaoglu (✉)
Department of Mechanical Engineering, Bartin University, Bartin, Turkey
e-mail: austaoglu@bartin.edu.tr

H. Torlaklı
Graduate School, Bartin University, Bartin, Turkey

A. Sarı · G. Hekimoglu
Department of Metallurgical and Material Engineering, Karadeniz Technical University, Trabzon, Turkey
e-mail: ahmet.sari@ktu.edu.tr; ghekimoglu@ktu.edu.tr

M. Sutcu
Department of Metallurgical and Material Engineering, Manisa Celal Bayar University, Manisa, Turkey
e-mail: mucahit.sutcu@cbu.edu.tr

F. Ergur
Iki Nokta Architecture Const. Cont. Ind. Trade Co. Ltd., 67100, Zonguldak, Turkey
e-mail: fatih@ikinoktamimarlik.com

M. Z. Sogut et al. (eds.), *Proceedings of the 2022 International Symposium on Energy Management and Sustainability*, Springer Proceedings in Energy,
https://doi.org/10.1007/978-3-031-30171-1_86

SHS	Sensible Heat Storage
TES	Thermal Energy Storage

86.1 Introduction

Today, depending on the developments in living standards, the share of the building sector in total energy consumption is increasing day by day. Inconsistency and intermittency in environmentally-friendly renewable energy sources, which are of great importance in reducing the consumption and effects of fossil fuels, can create some problems in terms of availability and reliability in applications (Aqib et al. 2020). Thermal energy storage (TES) systems are an alternative solution for these problems. Energy storage, which has a wide application area today, is defined as the storage of energy for future use. TES systems are classified into three groups: chemical energy storage, sensible energy storage (SHS), and latent energy storage (LHS). These have various advantages and limitations in relation to the purpose of use (Yang et al. 2020). Compared to others, LHS is seen as an important method in increasing energy efficiency, as it provides various advantages such as high heat storage density, constant temperature during phase changes, and small volume change (Huang et al. 2020).

Approximately 35% of global energy consumption is due to thermal requirements in buildings (Moreno et al. 2020). Heating, ventilation, and air conditioning (HVAC) systems, which are used to meet the changing thermal comfort demands in the building sector, constitute half of the total consumption in buildings and have an important share of one-fifth in total energy use (Khan et al. 2020).

TES with PCM has become one of the topics that are emphasized today in terms of the various benefits it provides for energy conservation (Khan et al. 2020). PCMs are substances that absorb heat at a certain phase change temperature storing in the form of latent heat or releasing energy (Gencel et al. 2021). This heat is consumed or stored; it is used for solid, liquid, and gas phase changes. LHS applications with PCM can provide 14 times more heat storage per volume unit than SHS (Javadi et al. 2020). At the same time, the integrated use of PCMs with LHS systems provides various advantages over other TES applications due to their chemical stability, economic, energy storage density, and noncorrosiveness effects (Alehosseini and Jafari 2020).

PCM with TES applications for use in buildings can be classified into three groups: active, free cooling, and passive systems used in heating and cooling (Ručevskis et al. 2020). Within the scope of passive systems, solar and heat protection, heat modulation, and heat dissipation technologies are evaluated (Moreno et al. 2020). For passive applications, PCMs are integrated into building components (Ručevskis et al. 2020). Integrating PCMs into building components such as windows, floors, and walls helps reduce heat transfer rates during peak hours, while also enabling temperature control and increased durability (Moreno et al. 2020). In this way, advantages can be provided both in terms of reduction in energy consumption and thermal comfort (Ručevskis et al. 2020).

In the literature, studies on the use of PCM for building applications are generally related to the effect of integrating PCM into building elements, as well as the type of PCM used, layer thickness, and its location in the building elements on thermal performance. Jamil et al. (2016) evaluated the applicability of PCM in terms of both air temperature reduction and thermal comfort for a house in Melbourne conditions with an experimental and numerical approach. PCM has been found to provide a temperature reduction of up to 1.1 °C in daytime indoor air temperature and up to 34% reduction in thermal discomfort hours. Alawadhi (2008) investigated the thermal performance of PCM-integrated brick with cylindrical holes by numerical simulation. In the results, it has been seen that the increase in the amount of PCM has a positive effect on energy saving by affecting heat gains. It has been found that the best performance is obtained when the PCM is in the centerline location of the brick. Castell et al. (2010) investigated the effect of using PCM in conventional and alveolar brick with experimental analyzes in real conditions. The addition of PCM was found to provide up to 1 °C reduction for the peak temperatures and was effective in reducing daily temperature fluctuations.

Tunçbilek et al. (2020) carried out a comparative analysis based on a numerical model established and verified under the conditions of the Marmara Region, Turkey, on the effects of the integration of PCM into conventional brick on seasonal and annual thermal performance. In the results obtained considering the fusion temperature, quantity, and location parameters, it was seen that the addition of PCM provided a 17.6% reduction in annual thermal load and higher energy savings could be achieved with PCM when near the indoor ambient. Pirasaci (2020) evaluated the thermal performance and savings potential of a residential apartment with a PCM layer integrated on its exterior walls for the heating season by creating a numerical model. It has been observed that with the integration of the RT21HC layer, a reduction of approximately 6% in the annual heating energy requirement is achieved.

86.2 Material and Methods

86.2.1 System Description

Two building models using 24 hollow cavity bricks are shown in Fig. 86.1a to examine the effect of PCM in improving the thermal performance of hollow brick. Models were insulated with 5-cm expanded polystyrene (EPS). The exterior of the models formed in the form of tunnels is coated with MDF. For comparative analyses, the hollow brick was filled one-third using vermiculite impregnated with 45% PCM. The phase change temperature of the PCM is 27 °C. Tunnel models in the experimental setup were created with dimensions of 100 cm (length) × 30 cm (width) × 30 cm (height). The bricks used as wall parts in both models are 30 cm

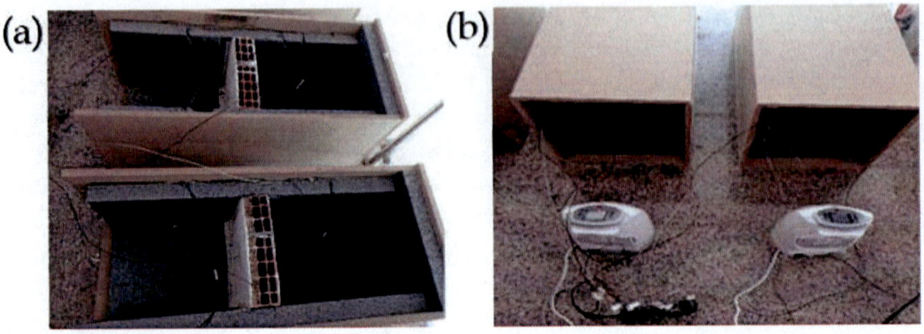

Fig. 86.1 Models of the building wall: (**a**) top view of building wall fragment and (**b**) experimental setup

(length) × 8.6 cm (width) × 30 cm (height). Also, 2-kW fans were used as a heat source. Temperature monitoring at different points of reference and PCM building models was carried out with K- and T-type thermocouples.

86.2.2 Experimental Setup

The effect of PCM on heat transfer rate and thermal performance was evaluated for the building wall part. PCM-impregnated vermiculite with a phase change temperature of 27 °C was used in the study. In building models, inlet air temperature, exterior and interior wall temperatures, room temperature, PCM, and cavity temperatures are the measured parameters. Temperatures were recorded throughout the day at regular 2-second intervals (Fig. 86.1b). The operating temperature of the fans was determined as 37 °C. When the interior wall temperatures reached the phase change temperature, the temperature readings were continued by turning off the fans.

86.3 Results and Discussion

In this study, PCM is integrated into a standard hollow brick and analyzed to improve the heat transfer rate and thermal performance of the exterior walls of a building. Two simplified building models were developed to be used in experimental analysis, and their energy performance was evaluated. The temperature measurements are recorded, and the T type thermocouple is used; the error in the measurements is $\pm$ 0.1%, which corresponds to $\pm$ 0.5 in the reading for $-$ 100°C < TS $\leq$ 400 °C.

In Fig. 86.2, the changes in room center temperatures of the reference and PCM building models during the day are given. While peak temperatures are observed at the moment of turning off the fans for both rooms, a rapid decrease in temperatures is observed afterward. The PCM stores heat energy up to the phase change temperature

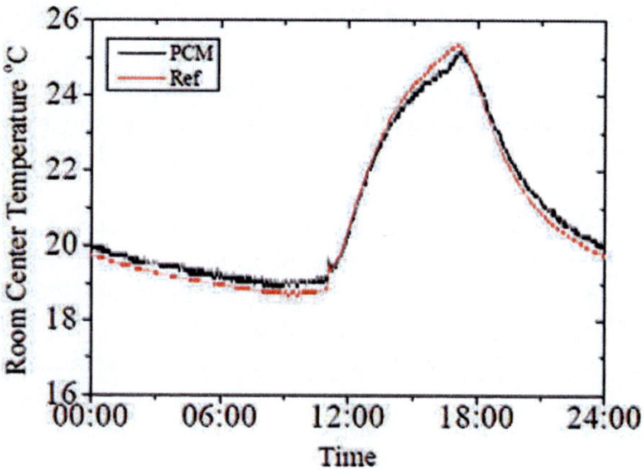

Fig. 86.2 Temperature difference of room center in the reference and PCM

during the day. At night when temperatures drop, PCM releases this stored heat, returns it to the room, and solidifies. Accordingly, when compared to the reference room, the PCM's ability to store heat during the day, prevent overheating, and reach higher temperatures with the heat it releases during the night provides a significant advantage in terms of thermal comfort and energy savings.

From 00:00 to approximately 11:05, higher temperatures are observed in the test room with PCM compared to the reference. From approximately 11:05, the temperature difference between the PCM and the reference rooms, as well as the effect of the PCM, decreases. Sudden increases in the temperatures of the test chambers were observed from 11:00 to 12:00. From 12:00 to 16:44 when the fans were turned off, the temperature difference between the PCM and the reference rooms increases due to the cooling of the PCM-impregnated vermiculite integrated room temperature. The test rooms reach maximum temperature after about 20 minutes. The maximum temperature for the reference and PCM rooms was observed at 17:04 as 25.4 °C and 25.2 °C, respectively. From these values, there is a decrease in room temperatures and the difference between these temperatures. At 17:25, the room temperatures take the same value. For 18:00–24:00, the PCM releases the absorbed heat and reaches higher temperatures compared to the reference room.

86.4 Conclusion

An experimental study was carried out on standard brick filled with PCM to evaluate the effect of PCM on improving the energy performance of a building wall. The results can be summarized as follows:

- It showed that the room with PCM could provide up to 0.5 °C reduction in daily indoor air temperature compared to the reference room.
- For maximum temperature during the day, the PCM provides an advantage of about 0.2 °C over the reference room.

- The test room with PCM is cooler during the day with high temperatures and warmer at night compared to the reference room.
- It has been found that integrating PCM into the exterior wall is beneficial in terms of reducing energy consumption during the day by preventing heat transfer from the outdoor to the indoor environment and is effective in reducing heat losses at night.
- An improvement in the thermal performance of the hollow standard brick was observed with the integration of PCM.
- PCM was found to be effective in reducing daily peak thermal loads and smoothing temperature fluctuations.
- It has been observed that the use of PCM in building applications can contribute to energy efficiency due to its positive effects on the improvement of heat transfer rate and thermal comfort.

Acknowledgments The authors gratefully acknowledge the financial support from the Bartın University Scientific Research Projects Coordinator (BAP) (2021-FEN-B-005).

References

Alawadhi, EM (2008) Thermal Analysis Of A Building Brick Containing Phase ChangeMaterial. Energy and Buildings 40(3):351–357. https://doi.org/10.1016/j.enbuild.2007.03.001

Alehosseini E, Jafari SM (2020) Nanoencapsulation of phase change materials (PCMs) and their applications in various fields for energy storage and management. Advances in Colloid and Interface Science 283:102226. https://doi.org/10.1016/j.cis.2020.102226

Aqib M, Hussain A, Ali HM, Naseer A, Jamil F (2020) Experimental Case Studies of The Effect of Al2O3 and MWCNTs Nanoparticles on Heating and Cooling of PCM, Case Studies in Thermal Engineering 22:100753. https://doi.org/10.1016/j.csite.2020.100753

Castell A, Martorell I, Medrano M, Pérez G, Cabeza LF (2010) Experimental Study of Using PCM in Brick Constructive Solutions For Passive Cooling, Energy and Buildings 42(4):534–540. https://doi.org/10.1016/j.enbuild.2009.10.022

Gencel O, Sarı A, Ustaoglu A, Hekimoglu G, Erdogmus E, Yaras A, Sutcu M, Cay V.V (2021) Eco-friendly Building Materials Containing Micronized Expanded Vermiculite and Phase Change Material For Solar Based Thermo-regulation Applications, Construction and Building Materials 308:125062. https://doi.org/10.1016/j.conbuildmat.2021.125062

Huang S, Lu J, Li Y, Xie L, Yang L, Cheng Y, Chen L, Zheng L, Li W, Zhang Y, Wang L (2020) Experimental Study on The Influence of PCM Container Height on Heat Transfer Characteristics Under Constant Heat Flux Condition, Applied Thermal Engineering, 172:115159. https://doi.org/10.1016/j.applthermaleng.2020.115159

Jamil H, Alam M, Sanjayan J, Wilson J (2016) Investigation of PCM as retrofitting option to enhance occupant thermal comfort in a modern residential building. Energy and Buildings. 133: 217–229. https://doi.org/10.1016/j.enbuild.2016.09.064

Javadi FS, Metselaar, HSC, Ganesan P (2020) Performance improvement of solar thermal systems integrated with phase change materials (PCM), a review. Solar Energy. 206:330–352. https://doi.org/10.1016/j.solener.2020.05.106

Khan RJ, Bhuiyan MZH, Ahmed DH (2020) Investigation of heat transfer of a building wall in the presence of phase change material (PCM). Energy and Built Environment. 1:199–206. https://doi.org/10.1016/j.enbenv.2020.01.002

Moreno S, Hinojosa JF, Hernández-López I, Xaman J (2020) Numerical and Experimental Study of Heat Transfer in A Cubic Cavity With A PCM in A Vertical Heated Wall, Applied Thermal Engineering 178:115647. https://doi.org/10.1016/j.applthermaleng.2020.115647

Pirasaci T (2020) Investigation Of Phase State And Heat Storage Form Of The Phase Change Material (PCM) Layer Integrated Into The Exterior Walls of the residential-apartment during heating season. Energy 207:118176. https://doi.org/10.1016/j.energy.2020.118176

Ručevskis S, Akishin P, Korjakins A (2020) Parametric analysis and design optimisation of PCM thermal energy storage system for space cooling of buildings. Energy and Buildings. 224: 110288. https://doi.org/10.1016/j.enbuild.2020.110288

Tunçbilek E, Arıcı M, Bouadila S, Wonorahardjo S (2020) Seasonal and Annual Performance Analysis of PCM-Integrated Building Brick Under The Climatic Conditions of Marmara Region, Journal of Thermal Analysis and Calorimetry 141(1): 613–624. https://doi.org/10.1007/s10973-020-09320-8

Yang L, Huang JN, Zhou F (2020) Thermophysical properties and applications of nano-enhanced-PCMs: An update review. Energy Conversion and Management. 214:112876. https://doi.org/10.1016/j.enconman.2020.112876

Chapter 87
Exergetic Performance Assessment of a Two–Stage Compression Transcritical CO_2 Refrigeration Cycle

Ibrahim Karacayli and Ozay Akdemir

Nomenclature

$\dot{E}$	Rate of energy, W or kW
c_p	Specific heat, J/kg°C
h	Specific enthalpy, kJ/kgK
$\dot{m}$	Mass flow rate, kg/s
T	Temperature, °C or K
$\dot{W}$	Power, W or kW
$\dot{X}$	Rate of energy, W or kW
RI	Relative irreversibility
σ	Pressure ratio
η	Efficiency, %
φ	Rational exergy efficiency, %
ε	Specific flow exergy, kJ/kg
γ	Improvement in total exergy destruction

87.1 Introduction

In recent years, increasing energy needs and the importance of global warming have led scientists to develop alternative energy sources. A considerable amount of research has been carried out to solve environmental problems such as developing and using environmentally friendly refrigerants like CO_2 (Wang et al. 2022). CO_2 as

I. Karacayli · O. Akdemir (✉)
Ege University, Faculty of Engineering, Department of Mechanical Engineering, Izmir, Turkey
e-mail: ibrahim.karacayli@ege.edu.tr; ozay.akdemir@ege.edu.tr

M. Z. Sogut et al. (eds.), *Proceedings of the 2022 International Symposium on Energy Management and Sustainability*, Springer Proceedings in Energy,
https://doi.org/10.1007/978-3-031-30171-1_87

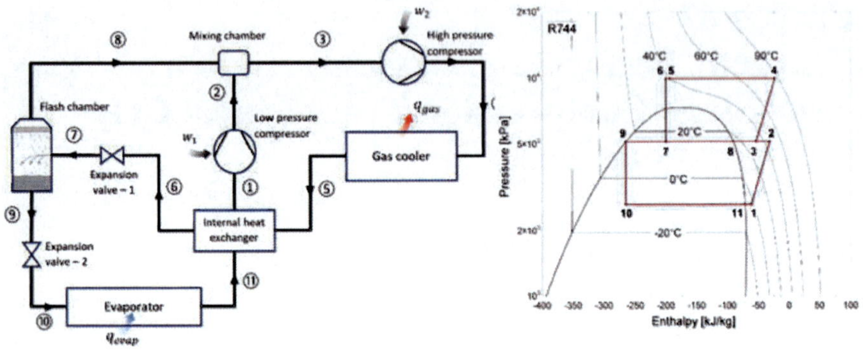

Fig. 87.1 Schematic of the two–stage compression transcritical CO_2 refrigeration cycle with internal heat exchanger and logP–h diagram of that cycle

a refrigerant has several advantages such as high specific heat, high thermal conductivity, low viscosity, zero ozone-depletion potential (ODP), and negligible global warming potential (GWP). Since the critical temperature and pressure of CO_2 are quite low as 31.1 °C and 73.8 kPa, respectively, CO_2 refrigeration cycle takes place above the critical point and it has a lower COP and high irreversibility (Ozcan et al. 2013; Elbarghthi et al. 2021).

Figure 87.1 presents a schematic and the corresponding logP–h diagram of the TCRC with an internal heat exchanger. The cycle consists of two compressors, a mixing chamber, a gas cooler, an internal heat exchanger, two expansion valves, a flash chamber (separator), and an evaporator. The saturated vapor CO_2 enters the low-pressure compressor at state (1) and is compressed to the intermediate pressure at state (2) with an isentropic efficiency, $\eta_{is,\ lpc}$. The mixed fluid (3) is compressed with an isentropic efficiency, $\eta_{is,\ hpc}$ to high-side pressure at state (4) by the high-pressure compressor. Next, the supercritical CO_2 cools to state (5) at constant pressure in the gas cooler and then passes through the internal heat exchanger. The high-pressure refrigerant expands to intermediate pressure at state (7) in the throttle valve and then enters to flash chamber to be split into two streams such as saturated vapor (8) and saturated liquid (9). Subsequently, the saturated liquid part of the CO_2 expands again to low pressure at state (10) and enters the evaporator to provide a cooling effect and saturated vapor part (8) passes through the mixing chamber.

In the present study, the exergetic performance based on energy and exergy analyses of the TCRC with an internal heat exchanger and flash chamber using R744 was theoretically investigated at different design conditions such as gas cooler pressure and gas cooler exit temperature. Furthermore, performance comparisons in terms of total exergy destruction, relative irreversibility, and second-law efficiency with single-stage TRC cycle were presented.

87.2 Thermodynamics Analysis

In order to simplify the evaluation of the coefficient of performance of that refrigeration system, exergy destructions, and exergy efficiencies of each component and whole refrigeration cycle, the following assumptions were made:

- All processes are steady state.
- There are not any chemical, magnetic, or nuclear reactions, and the change of kinetic and potential energies are negligible.
- Effectiveness of all heat exchangers is equal to 1, and compression and expansion processes are adiabatic.
- In the expansion valves, in the flash chamber and in the mixing chamber, there is no work transfer.
- The dead state temperature and pressure are taken to be 27 °C and 101.325 kPa.

Isentropic efficiency for the high- and low-pressure compressors is obtained from (Robinson and Groll 1998)

$$\eta_{is} = 0.815 + 0.022\sigma - 0.041\sigma^2 + 0.0001\sigma^3 \tag{87.1}$$

The intermediate pressure for two-stage compression refrigeration cycles is estimated as the geometric mean of the gas cooler and evaporator pressure:

$$P_{\text{int}} = \sqrt{P_e \cdot P_{\text{gas}}} \tag{87.2}$$

The general energy equation can be defined as

$$E_{\text{heat,in}} + E_{\text{mass,in}} - E_{\text{work,out}} = \Delta E_{\text{system}} \tag{87.3}$$

The general exergy balance can be written as

$$X_{\text{heat,in}} + X_{\text{mass,in}} - X_{\text{work,out}} - X_{\text{dest}} = \Delta X_{\text{system}} \tag{87.4}$$

$$\left(1 - \frac{T_0}{T_k}\right) q_{k,\text{in}} + \varepsilon_{\text{in}} - w_{\text{out}} - X_{\text{dest}} = \Delta X_{\text{system}} \tag{87.5}$$

where flow exergy is $\varepsilon = (h - h_0) - T_0(s - s_0)$. For cycling processes that operate at a steady state, the terms of the change in energy and exergy of the system in Eqs. (87.3)–(87.5) are zero.

The rational exergy for all components is expressed as (Hepbasli and Akdemir 2004)

$$\varphi = \frac{\dot{X}_{\text{dest,out}}}{\dot{X}_{\text{cons,in}}} \tag{87.6}$$

The relative irreversibility is the ratio of exergy destruction of a component to total irreversibility:

$$RI = \frac{X_{\text{dest,k}}}{X_{\text{dest,total}}} \tag{87.7}$$

The improvement in total exergy destruction is

$$\gamma = \frac{X_{\text{dest,TCRC}} - X_{\text{dest,TRC}}}{X_{\text{dest,TRC}}} \tag{87.8}$$

87.3 Results and Discussion

Pressure, temperature, vapor quality, specific enthalpy, specific entropy, and flow exergy data for $T_{evap} =- 10$ °C, $P_{cool} = 10\ MPa$, and $T_5 = 40$ °C are listed in Table 87.1.

In the energy and exergy analyses, the energy transfer rate, the exergy destruction rate in the EERHX, the relative irreversibility of all components, and the functional exergy efficiency of the TCRC were determined. The results of the energy and exergy analyses are illustrated in Table 87.2.

The effects of the gas cooler temperature on the relative irreversibility of all components are indicated in Fig. 87.2. While the highest relative irreversibility was recorded in the gas cooler, the relative irreversibility decreased abruptly after 43 °C.

Table 87.1 Property data for the TCRC

Point	P (kPa)	T (°C)	h (kJ/kg)	s (kJ/kgK)	ε (kJ/kg)
0	101,3	35.0	7.618	0.02603	
1	2649	−3.0	−61.68	−0.8031	186.2
2	5146	54.4	−21.42	−0.7662	215.1
3	5146	36.39	−47.31	−0.8476	214.3
4	10,000	99.02	−4.141	−0.8128	246.7
5	10,000	40.0	−194	−1.384	232.7
6	10,000	38.84	−200.3	−1.404	232.7
7	5146	15.48	−200.3	−1.373	223.2
8	5146	15.48	−90.86	−0.9938	215.8
9	5146	15.48	−265.3	−1.598	227.6
10	2649	−10	−265.3	−1.577	220.9
11	2649	−10	−71.64	−0.8405	187.8

Table 87.2 Energy and exergy balance results of the TCRC

Component	$\dot{E}$ (*kW*)	$\dot{X}_{dest}(W)$	*RI*	ψ
LP compressor	25.25	7.135	13.76	71.74
Mixing chamber	0.0	1.080	2.083	99.5
HP compressor	43.17	10.71	20.66	75.19
Gas cooler	−189.9	13.98	26.97	94.33
Internal HX	0.0	1.062	2.049	99.7
Expansion valve 1	0.0	9.470	18.27	97.05
Expansion valve 2	0.0	4.205	8.111	79.79
Evaporator	121.5	4.197	8.097	24.23

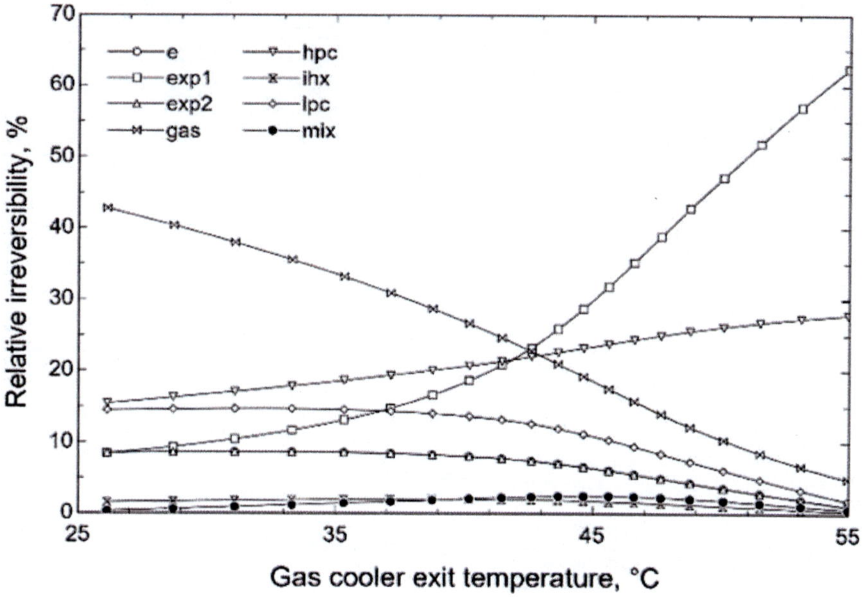

Fig. 87.2 Variation of relative irreversibility of components with gas cooler exit temperature

After this temperature, the relative irreversibility of the expansion valve-1 and high-pressure compressor increases significantly. However, there was no significant change in relative irreversibility with the increase of gas cooler exit temperature in other components.

The gas cooler pressure was gradually increased from 7 MPa to 12 MPa while the gas cooler exit temperature increased from 10 °C to 70 °C. The effects of the gas cooler pressure and the gas cooler exit temperature on the exergy destruction of expansion valve-1, high-pressure compressor, and gas cooler are indicated in Fig. 87.3. Exergy destruction of the gas cooler was increased with gas cooler pressure, but in the expansion valve and high-pressure compressor, exergy destruction almost didn't change until a certain gas cooler exit temperature, but then it was raised. Exergy destruction of the gas cooler is minimum (1.749 kW) for a gas cooler

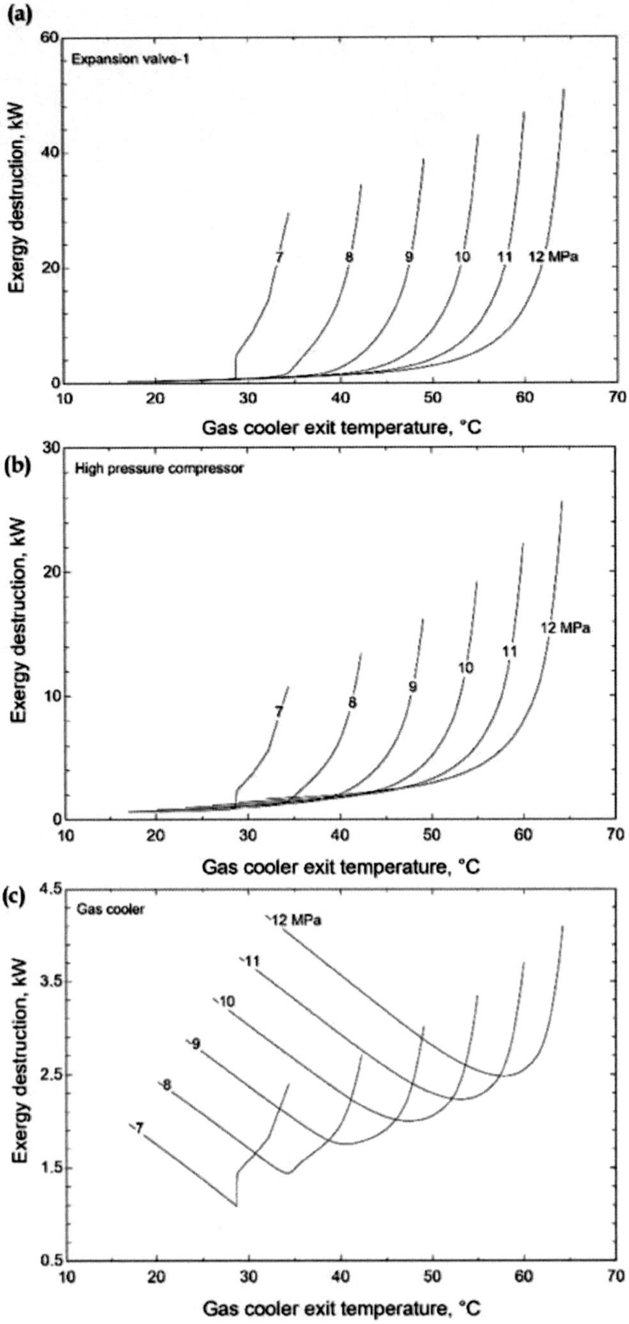

Fig. 87.3 Variation of exergy destruction of (**a**) expansion valve-1, (**b**) high-pressure compressor, and (**c**) gas cooler variations with gas cooler exit temperature for different gas cooler pressure

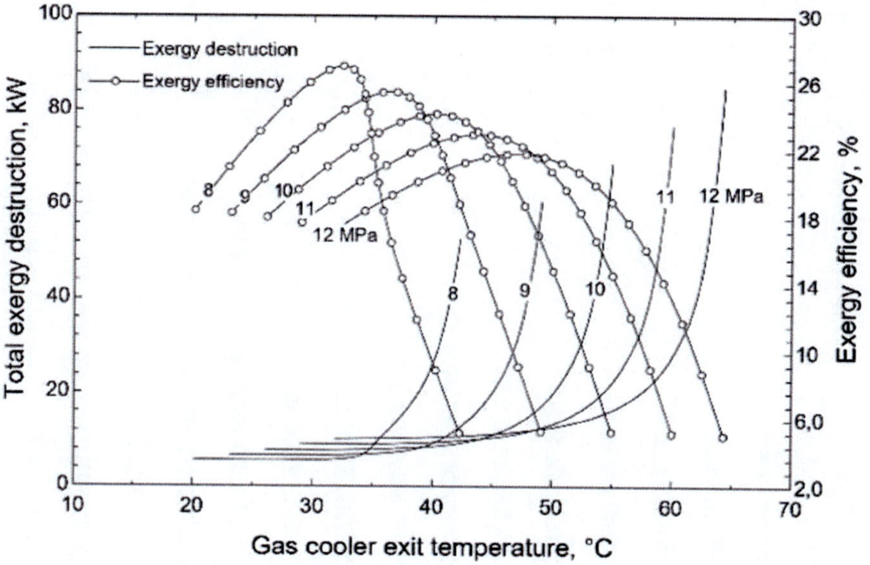

Fig. 87.4 Variation of total exergy destruction and exergy efficiency of the TCRC with gas cooler exit temperature for different gas cooler pressure

exit temperature of 39.9 °C at a gas cooler pressure of 9 MPa. At this point, the exergy destruction of the expansion valve and high-pressure compressor were obtained as 2.288 and 1.885 kW, respectively, as the lowest values.

As can be seen in Fig. 87.4, the exergy efficiency of the TCRC was negatively affected by gas cooler pressure. Total exergy destruction increased with gas cooler pressure, but the exergy efficiency was increased until a certain gas cooler exit temperature, then it was decreased. The maximum exergy efficiency was obtained at 37.9 °C at a gas cooler pressure of 9 MPa, and the total exergy destruction was 7.634 kPa.

Figure 87.5 (**a**) and (**b**) shows the variation of total exergy destruction and exergy efficiency of conventional (single-stage compression) TRC and TCRC with gas cooler pressure, respectively.

As can be seen in Fig. 87.6, improvements in both exergy destruction and exergy efficiency became more significant when the gas cooler pressure was increased.

87.4 Conclusion

The effect of gas cooler pressure and gas cooler exit temperature on the exergetic performance of the TCRC using R744 is investigated. The gas cooler pressure was gradually increased from 7 MPa to 12 MPa while the gas cooler exit temperature

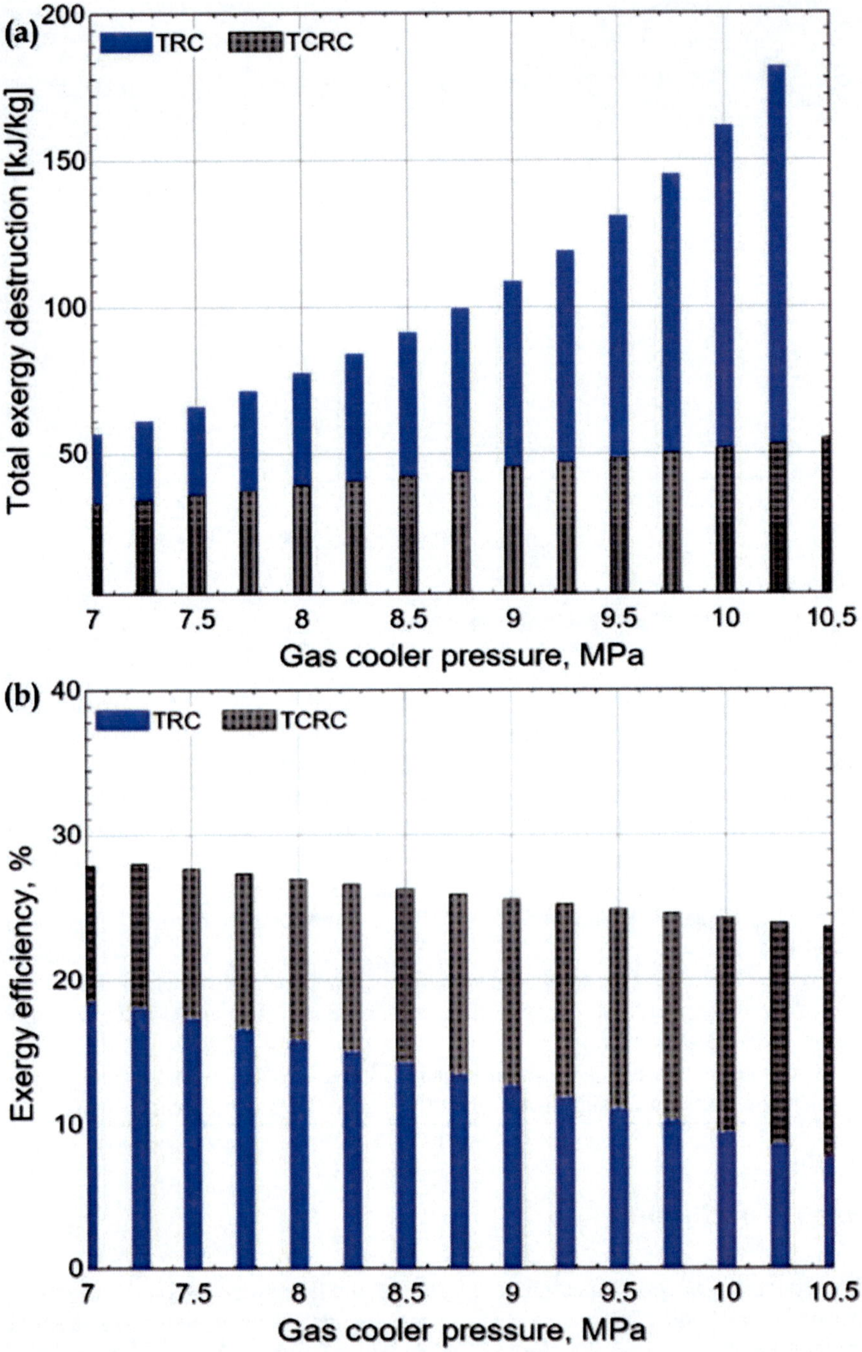

Fig. 87.5 (**a**) COP and (**b**) exergy efficiency of TCRC variations with gas cooler exit temperature for different gas cooler pressure

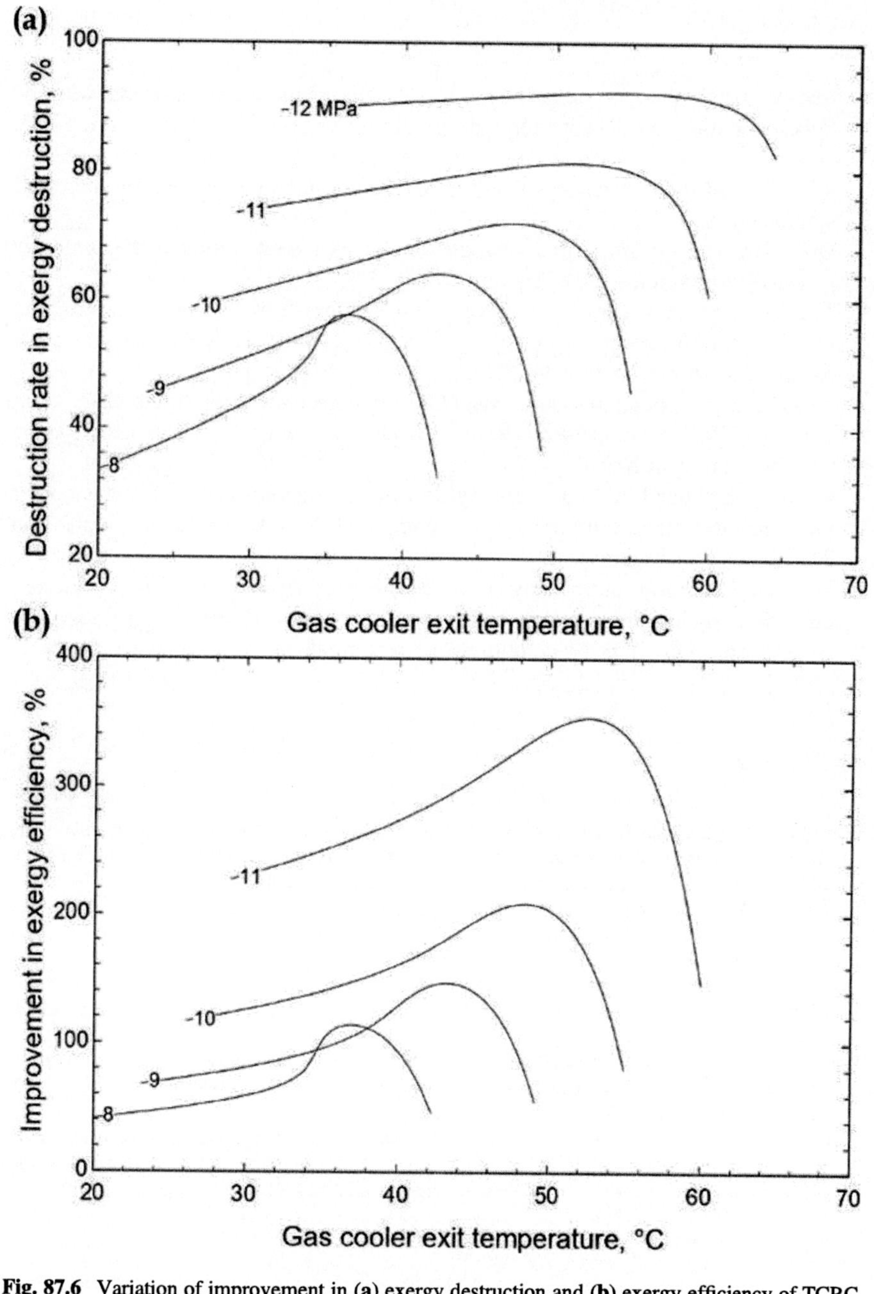

Fig. 87.6 Variation of improvement in (**a**) exergy destruction and (**b**) exergy efficiency of TCRC with gas cooler exit temperature for different gas cooler pressure

increased from 10 °C to 70 °C. The concluding remarks achieved from this study are listed below:

- In cases where the gas cooler outlet temperature is low, the components with the lowest and highest ratios of exergy destruction to total exergy destruction are IHX and the gas cooler, respectively. With the rise in temperature, the relative irreversibility of the expansion valve-1 and the high-pressure compressor increases significantly.
- Operating the system with lower gas cooler exit temperature and gas cooler pressure will increase the exergy efficiency.
- The optimum gas cooler exit temperature to obtain the lowest exergy destruction and the highest exergy efficiency for 8, 9, 10, 11, and 12 MPa pressures were recorded as approximately 34, 39, 46, 54, and 59 °C, respectively.
- At 9 MPa gas cooler pressure, the TCRC system instead of single-stage compression TRC system caused a 59.6% reduction in total exergy destruction and a 110.2% increment in exergy efficiency.
- On the other hand, although exergy destruction increases at higher gas cooler pressures, the amount of improvement compared to the single-stage compression TRC system also increases.
- For a further study, performing (advanced) exergy analyses for different design conditions such as intermediate pressure, subcooling, and superheating degree to achieve optimum working conditions are recommended.

References

Elbarghthi AFA, Dvorak V, Hafner A, Banasiak K (2021) The potential impact of the small-scale ejector on the R744 transcritical refrigeration system. Energy Conversion and Management. 249(114860):1–13. https://doi.org/10.1016/j.enconman.2021.114860.

Hepbasli A, Akdemir O (2004) Energy and exergy analysis of a ground source (geothermal) heat pump system. Energy Conversion and Management. 45(5):737–753. https://doi.org/10.1016/S0196-8904(03)00185-7

Ozcan HG, Gunerhan H, Yaldirak H (2013) Alternatif Soğutkan Karbondioksit ve Otobüs Klimalarında Uygulaması. 11th National Plumbing Engineering Congress, April 17-20, 2013. Izmir, Turkey.

Robinson DM, Groll EA (1998) Efficiencies of transcritical CO_2 cycles with and without an expansion turbine, International Journal of Refrigeration, 21(7): 577–589. https://doi.org/10.1016/S0140-7007(98)00024-3

Wang M, Zhang J, Liu H (2022) Thermodynamic analysis and optimization of two low-grade energy driven transcritical CO2 combined cooling, heating and power systems, Energy. 249(123765):1–13. https://doi.org/10.1016/j.energy.2022.123765

Chapter 88
Research Status and Trends of Life Cycle Analysis of Ethanol- and Biodiesel-Fueled Engines: A Bibliometric Analysis in 2000–2021

Selçuk Sarıkoç and Jeffrey Dankwa Ampah

Nomenclature

CO_2	Carbon dioxide
GHG	Greenhouse gas
ICE	Internal combustion engine
LCA	Life cycle analysis
NP	Number of publications
PY	Publication year
TC	Total citations
WoS	Web of Science

88.1 Introduction

Using alternative fuel is an effective means to alleviate the petroleum shortage and environmental problems from the overreliance and combustion of fossil fuels in engines (Jin et al. 2020, 2021; Liu et al. 2019). Due to their biological background and renewability, biodiesels and ethanol have been acknowledged to be very promising substitutes for conventional fuels such as diesel and gasoline in the internal combustion engine (ICE). The focus of most existing studies on these alternative

S. Sarıkoç (✉)
Department of Motor Vehicles and Transportation Technologies, Amasya University, Tasova Yuksel Akin Vocational School, Amasya, Turkey

J. D. Ampah (✉)
School of Environmental Science and Engineering, Tianjin University, Tianjin, China

M. Z. Sogut et al. (eds.), *Proceedings of the 2022 International Symposium on Energy Management and Sustainability*, Springer Proceedings in Energy,
https://doi.org/10.1007/978-3-031-30171-1_88

fuels has for a long time relatively been limited only to the ICE's operational phase, i.e., tank-to-wheel tailpipe emissions. For biodiesel and ethanol to be highly feasible in the context of zero carbon future (carbon neutrality), research into the other phases (well-to-wheel) of their use in ICE needs to be investigated extensively. Against this background, life cycle analysis (LCA) becomes a relevant tool to ascertain the energy consumption and environmental burdens of ethanol- and biodiesel-fueled engines. Fortunately, emerging studies in the twenty-first century have made conscious effort in this direction. Ou et al. (2009) investigated the energy consumption and greenhouse gas (GHG) emissions of six biofuel pathways by LCA in China. A limited LCA comparing large- and small-scale production of ethanol for heavy engines under Swedish conditions was investigated by Bernesson et al. (2006). The LCA of mid-range passenger cars powered by liquid and gaseous biofuels was conducted by Ternel et al. (2021), and the results were compared with greenhouse gas emissions of electric vehicles and forecast to 2030.

The above-mentioned studies and many others have provided significant contribution to literature, but there is a general limitation with these groups of studies. It is difficult to deduce from these studies the key actors of development, research hotspots, evolutionary nuances, and future direction of the field of ethanol and biodiesels in ICE from the perspective of their lifetime energy consumptions and associated environmental impacts.

In order to address this research gap, a useful tool currently on the rise is bibliometric analysis. It is a type of research study that provides the basis for what has been done and what needs to be investigated (Schneider and Borlund 2004). The methodology consists of descriptive and exploratory techniques; these are deemed worthy tools to analyze the relevant literature's recent trends (Donthu et al. 2021). In recent years, some studies in the field of environmental science and energy have successfully used the tool to map the research output of LCA of buildings (Geng et al. 2017), LCA of ecosystem services (VanderWilde and Newell 2021), and LCA of electric vehicles (Ayodele and Mustapa 2020).

Considering all over, this study will enhance our understanding of life cycle analyses. In this respect, the current study thus seeks to contribute to the existing studies by revealing the publication output per year, key actors (authors and countries/ regions), research trends, and development (keywords) in the field of LCA on biodiesel- and ethanol-fueled ICE.

88.2 Materials and Methods

On January 7, 2022, the bibliometric analysis was conducted to reveal the extant literature on LCA of ethanol- and biodiesel-fueled ICE published from 2000 to date. Two most powerful and popular databases for bibliometric analysis in the likes of Scopus and Web of Science Core Collection (WoS) were used to increase the reliability of the obtained results. The complete search strategy adopted from the identification of relevant papers to their extraction has been outlined in Table 88.1.

Table 88.1 Search strategy and dataset used in the current study

	Scopus	Web of science
Search strategy	("life cycle analysis" OR "life cycle assessment") AND (Biodiesel OR ethanol) AND ("internal combustion engine" OR "compression ignition engine" OR "spark ignition engine" OR "diesel engine" OR "gasoline engine")	
Timespan	2000 to present	
Initial documents (English)	124	212
Retrieved documents (removing irrelevant documents)	83	118
Removed duplicates	53	
Final documents	148	

NB: During search, double-inverted commas (") were used in Scopus, while single-inverted commas (') were used in WoS. Assessment in keywords was written as 'assess*' in WoS

Table 88.2 Main information on the dataset

Information	Record
Timespan	2000:2021
Sources (journals, books, etc.)	70
Documents	148
References	9089
Authors	544
Institutions	260
Countries/regions	43

The respective documents from Scopus (83) and WoS (118) were merged with programming codes in Rstudio, where 53 duplicated materials were deleted, bringing the final dataset used in the current analysis to 148.

The R-bibliometrix tool called biblioshiny (Aria and Cuccurullo, 2017) was used in obtaining the bibliometric results, including publication output per year, productivity of key actors, and word cloud of keywords. Table 88.2 describes the main characteristics of the final dataset.

88.3 Results and Discussion

The results in Fig. 88.1. show that the research field is growing at an annual growth rate of 12.25%, with the highest contribution recorded in 2020, whereas no publication was recorded in 2001, 2002, and 2007. From the start of 2015, a total of 82 papers have been published, which is 16 papers greater than what was published in the preceding 15 years. The results signify a growing trend, and an exponential increase in publication output could be expected in the near distant future.

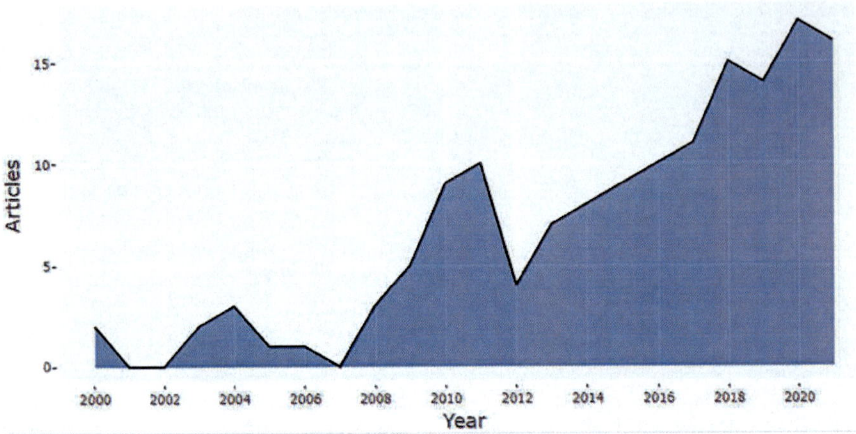

Fig. 88.1 Annual publication output

Table 88.3 Productivity of top authors in the field

Author	NP	H_Index	TC	PY_Start
Tabatabaei M	12	8	444	2016
Aghbashlo M	9	5	338	2017
Bandbafha H	8	5	237	2018
Maclean H	5	5	522	2000
Mohammadi P	5	3	60	2019
Rajaeifar M	5	5	217	2016
Lave L	4	4	502	2000
Boretti A	3	1	3	2011
Khalife E	3	2	54	2019
Khanali M	3	2	177	2018

The top 10 authors driving the research output in the field are shown in Table 88.3. Tabatabaei M from Agricultural Biotechnology Research Institute in Iran has been the most productive author with 12 publications. These publications have amassed 444 citations, and hence the h-index of this author in this research area is 8. It is worth noting that the first publication of this author on the subject was recorded only five years ago. This shows the level of commitment of this author in developing the research field. However, based on citations, the author with the most cited articles in the field is Maclean H from the University of Toronto in Canada. His five articles have accumulated 522 citations, at least 20 times higher than that of any other author. This implies that the works of this author are the most influential in the field.

The country's scientific production of the research field is displayed in Fig. 88.2a. USA, Iran, and China occupy the helm of affairs with 52, 44, and 27 research publications, respectively. Relatively, research output from Africa, South America, and Oceania are heavily underrepresented. In Fig. 88.2b, the research collaborations

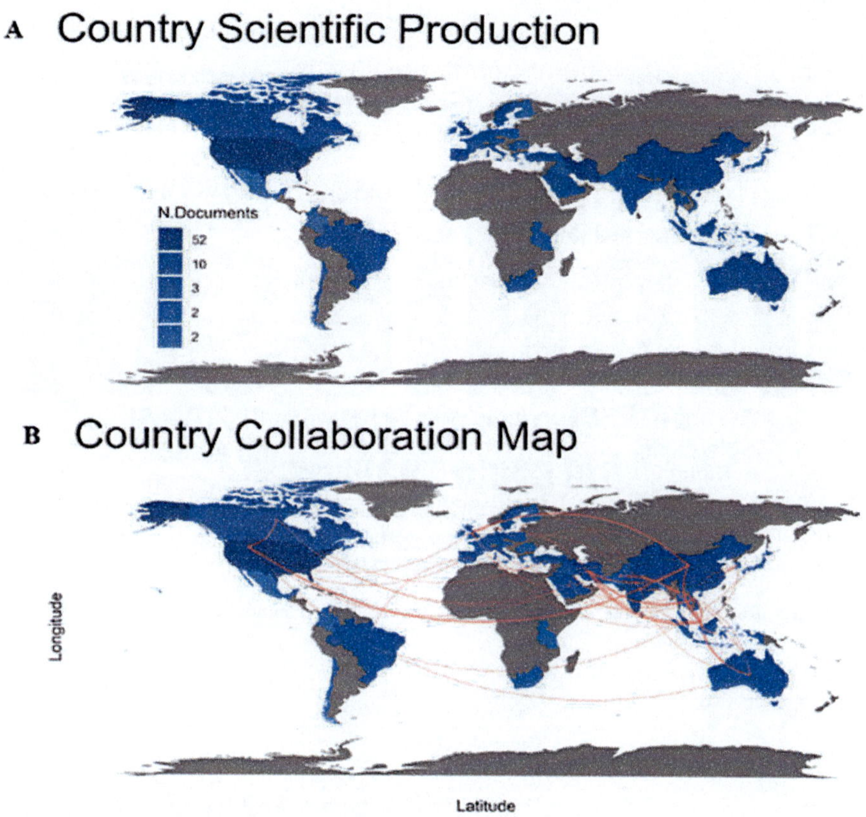

Fig. 88.2 Country productivity and collaboration map

between the various countries/regions are shown. It can be seen that the development of the field is mostly from internal ties with few external collaborations. The most collaboration between any two countries exists between Iran and Malaysia (4 documents). By strengthening the culture of internal and external collaborations, the research field can be developed significantly through knowledge and resource sharing.

Through keyword analysis, the research hotspots of the field have been identified in Fig. 88.3. It can be seen that with climate change being a keen topic of interest in the research field, the research community has for the last two decades focused on the environmental impact of these alternative fuels from its production to the end use in ICE. Relatively, most of the studies were dedicated to ascertaining the impacts of biodiesel in contrast to ethanol. Of the various biodiesel feedstocks available as fuel for ICE, Jatropha and Rapeseed methyl esters have been the most investigated. The greenhouse gas emissions from the life cycle of these low-carbon fuels are of paramount interest to the researchers. Besides LCA, techno-economic analysis has also been a minor methodology of assessment by various investigators.

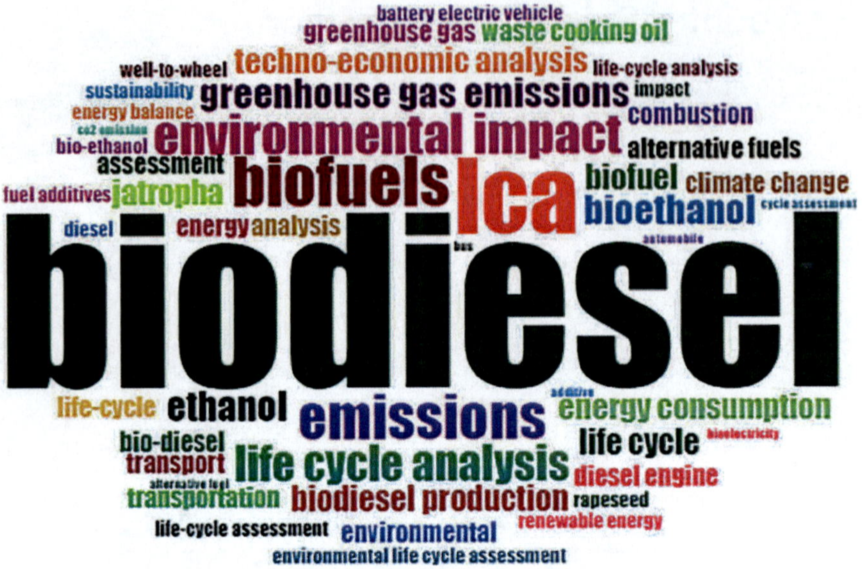

Fig. 88.3 Research hotspots according to most frequently used author keywords

Table 88.4 Most relevant papers in the field

Paper	Total citations
1. Carraretto et al. (2004)	459
2. Ou et al. (2009)	271
3. MacLean and Lave (2003)	264
4. Hosseinzadeh-Bandbafha et al. (2018)	157
5. Hossain and Davies (2010)	156

Finally, the five most cited documents in the research field have been listed in Table 88.4. The most relevant paper has so far received 459 citations and was a work done by Carraretto et al. (2004). As part of an extensive work, the authors assess some environmental aspects of bio-diesel, in particular the benefit of biodiesel to the total net emission of CO_2 during the whole life cycle and the evaluation of its net energy requirement. It has been found that the CO_2 emissions from biodiesel are around 0.21 $kgCO_2/kg$. Furthermore, if the full amount of CO_2 produced by the processes is totally assigned to biodiesel, the CO_2 output is about 0.72 $kgCO_2/kg$ biodiesel. The results also show that the total energy required to produce 1 MJ of biofuel is 1.398 MJ of primary energy, with fossil fuels accounting for 35% of the total (Carraretto et al. 2004).

88.4 Conclusion

The research on LCA of ethanol- and biodiesel-fueled internal combustion engines is increasing. Currently, there is no study that reveals the key actors and evolutionary nuances or hotspots in this growing research area. Against this backdrop, the current study uses bibliometric analysis to map the development of the research field in the last 21 years. Some of the key findings from the current work have been summarized below:

- The research field is growing at an annual growth rate of 12.25%, with the peak of contribution recorded in 2020.
- The most productive author in the field is Tabatabaei M, and Maclean H has the most influential papers. USA, Iran, and China are the main contributors in the field. Studies from Africa, South America, and Oceania are heavily underrepresented, while external collaborations in the field among countries are very low.
- The main goal of most studies is to ascertain the environmental impacts of these low-carbon fuels in the context of climate change.
- Relatively, there are more studies on biodiesel-fueled ICE than its ethanol counterpart.
- Only a few studies have made efforts to assess the techno-economic impacts of the fuels and their engines from an LCA approach.
- Per the number of citations, the most relevant work in the field is titled "Biodiesel as alternative fuel: Experimental analysis and energetic evaluations" published in 2004.

In conclusion, the research field is still immature, and more work is needed to further drive its development. Researchers must make a conscious effort to establish collaborations with researchers outside their countries. Environmental impacts can go beyond greenhouse gases, while the technical and economic feasibility studies of the biodiesel- and ethanol-fueled ICE could be further explored. Future works can make an attempt to map the research field of the LCA on other clean alternative fuels or powertrains like methanol, hydrogen, ammonia, and electric vehicles.

Acknowledgments The authors are grateful to Amasya University and Tianjin University for their support.

References

Aria M, Cuccurullo C (2017) bibliometrix: An R-tool for comprehensive science mapping analysis. Journal of informetrics 11 (4):959-975

Ayodele BV, Mustapa SI (2020) Life Cycle Cost Assessment of Electric Vehicles: A Review and Bibliometric Analysis. Sustainability 12 (6):2387

Bernesson S, Nilsson D, Hansson P-A (2006) A limited LCA comparing large- and small-scale production of ethanol for heavy engines under Swedish conditions. Biomass and Bioenergy 30 (1):46-57. https://doi.org/10.1016/j.biombioe.2005.10.002

Carraretto C, Macor A, Mirandola A, Stoppato A, Tonon S (2004) Biodiesel as alternative fuel: Experimental analysis and energetic evaluations. Energy 29 (12):2195-2211. https://doi.org/10.1016/j.energy.2004.03.042

Donthu N, Kumar S, Mukherjee D, Pandey N, Lim WM (2021) How to conduct a bibliometric analysis: An overview and guidelines. Journal of Business Research 133:285-296. https://doi.org/10.1016/j.jbusres.2021.04.070

Geng S, Wang Y, Zuo J, Zhou Z, Du H, Mao G (2017) Building life cycle assessment research: A review by bibliometric analysis. Renewable and Sustainable Energy Reviews 76:176-184. https://doi.org/10.1016/j.rser.2017.03.068

Hossain AK, Davies PA (2010) Plant oils as fuels for compression ignition engines: A technical review and life-cycle analysis. Renewable Energy 35 (1):1-13. https://doi.org/10.1016/j.renene.2009.05.009

Hosseinzadeh-Bandbafha H, Tabatabaei M, Aghbashlo M, Khanali M, Demirbas A (2018) A comprehensive review on the environmental impacts of diesel/biodiesel additives. Energy Conversion and Management 174:579-614. https://doi.org/10.1016/j.enconman.2018.08.050

Jin C, Ampah JD, Afrane S, Yin Z, Liu X, Sun T, Geng Z, Ikram M, Liu H (2021) Low-carbon alcohol fuels for decarbonizing the road transportation industry: A bibliometric analysis 2000-2021. Environ Sci Pollut Res. doi:https://doi.org/10.1007/s11356-021-15539-1

Jin C, Zhang X, Han W, Geng Z, Tessa Margaret Thomas M, Dankwa Jeffrey A, Wang G, Ji J, Liu H (2020) Macro and micro solubility between low-carbon alcohols and rapeseed oil using different co-solvents. Fuel 270. doi:https://doi.org/10.1016/j.fuel.2020.117511

Liu H, Wang X, Wu Y, Zhang X, Jin C, Zheng Z (2019) Effect of diesel/PODE/ethanol blends on combustion and emissions of a heavy duty diesel engine. Fuel 257. doi:https://doi.org/10.1016/j.fuel.2019.116064

MacLean HL, Lave LB (2003) Evaluating automobile fuel/propulsion system technologies. Progress in Energy and Combustion Science 29 (1):1-69. https://doi.org/10.1016/S0360-1285(02)00032-1

Ou X, Zhang X, Chang S, Guo Q (2009) Energy consumption and GHG emissions of six biofuel pathways by LCA in (the) People's Republic of China. Applied Energy 86:S197-S208. https://doi.org/10.1016/j.apenergy.2009.04.045

Schneider JW, Borlund P (2004) Introduction to bibliometrics for construction and maintenance of thesauri. Journal of Documentation

Ternel C, Bouter A, Melgar J (2021) Life cycle assessment of mid-range passenger cars powered by liquid and gaseous biofuels: Comparison with greenhouse gas emissions of electric vehicles and forecast to 2030. Transportation Research Part D: Transport and Environment 97:102897. https://doi.org/10.1016/j.trd.2021.102897

VanderWilde CP, Newell JP (2021) Ecosystem services and life cycle assessment: A bibliometric review. Resources, Conservation and Recycling 169:105461. https://doi.org/10.1016/j.resconrec.2021.105461

Chapter 89
Investigation of the Relationship Between the Body Mass Index (BMI) of the Human Body and the Thermal Sensation Vote (TSV): A Case Study

Aydın Ege Çeter, Mehmet Furkan Özbey, Şevval Örfioğlu, and Cihan Turhan

Nomenclature

ASHRAE	American Society of Heating, Refrigerating, and Air-Conditioning Engineers
BMI	Body Mass Index, kg/m^2
I	Insulation, $K.m^2/W$
ISO	International Organization for Standardization
M	Metabolic rate, met, W/m^2
PMV	Predicted Mean Vote
RH	Relative Humidity, %
T	Temperature, °C
TC	Thermal Comfort
TSV	Thermal Sensation Vote

A. E. Çeter
School of Natural and Applied Science, Mechanical Engineering, Atılım University, Ankara, Turkey

M. F. Özbey
Faculty of Engineering, Mechanical Engineering, Atılım University, Ankara, Turkey

Faculty of Engineering, Energy Systems Engineering, Atılım University, Ankara, Turkey
e-mail: furkan.ozbey@atilim.edu.tr

Ş. Örfioğlu · C. Turhan (✉)
Faculty of Engineering, Energy Systems Engineering, Atılım University, Ankara, Turkey
e-mail: cihan.turhan@atilim.edu.tr

M. Z. Sogut et al. (eds.), *Proceedings of the 2022 International Symposium on Energy Management and Sustainability*, Springer Proceedings in Energy,
https://doi.org/10.1007/978-3-031-30171-1_89

v	Velocity
ΔTV	Thermal Vote Difference

Subscripts

a	air
cl	clothing
g	globe

89.1 Introduction

Providing thermal comfort (TC) of the occupants has great importance in indoor environments since most of the people spend approximately 85% of their time in interior zones (Salata et al. 2018). For the calculation of the TC, Predicted Mean Vote (PMV), one of the most common TC indices, was developed by Fanger (1970), and the PMV method uses four environmental parameters (indoor air temperature (T_a), relative humidity (RH%), air velocity (v_a), and mean radiant temperature (T_r)) and two personal parameters (clothing insulation (I_{cl}) and metabolic rate (M)); furthermore, the method is accepted by international standards such as ASHRAE 55 and ISO 7730 (ASHRAE 2017; ISO 2005). On the other hand, while it is known that the anthropometric parameters (gender, age, body mass index) of the human body have an influence on thermal sensation, traditional TC indices remain insufficient for accurate predictions of thermal sensation (Kruger and Drach 2017). Therefore, the importance of examination of the gap between the PMV and thermal sensation becomes a vital issue. To this aim, anthropometric parameters should be investigated to estimate Thermal Sensation Vote (TSV) in detail (Kruger and Drach 2017).

Body Mass Index (BMI) is one of the anthropometric parameters that could affect the gap between PMV and TSV (Majewski et al. 2020). According to the World Health Organization (WHO), BMI is a measure for indicating nutritional status and defined as the ratio of a person's weight in kilograms to the square of person's height in meters (World Health Organization 2022). In the literature, there are a few studies that investigate the relationship between BMI and TSV. For instance, Majewski et al. (2020) investigated the difference between PMV and TSV by applying questionnaires to 1369 participants. The outcomes of the study showed that the BMI might affect the difference, and participants that have large BMI felt warmer than small BMI ones. In another study conducted by Forcada et al. (2021), experiments were performed in nursing homes with 894 participants in a temperate climate. According to the results of the study, researchers revealed that the BMI had no correlation with the TC of the occupants. Indraganti et al. (2015) conducted of series of experiments with 2787 participants in office buildings in India. The results of the study showed that participants who had low BMI values required higher comfort temperature.

However, the gap between the PMV and TSV in relation to the BMI should be investigated in detail, especially for different cultures and occupants in different climate zones. For this reason, this study aims to investigate the BMI of young male and female students in a university study hall which is located in a Csb type-climate zone.

89.2 Methodology

Experiments were conducted in a part of the study hall of Atılım University, which is located in the Csb climate zone (Kottek et al. 2006) and has a total capacity of approximately 900 occupants. The study hall is a mixed-mode building and has only a heating system via radiators. On the other hand, ventilation is provided by windows and doors.

Before the experiments, permissions were taken from both the students and the Ethical committee of the university.

Measurement campaigns were conducted between August 16 to December 21, 2021. Environmental data (indoor air temperature, relative humidity, air velocity, and mean radiant temperature) and the PMV values were collected with Delta Ohm HD32.3TCA (DeltaOhm 2021) every 1 min from 1.1 m height (ASHRAE 2017) during the experiments. The measurement probes of the device have an accuracy for indoor air temperature (T_a), relative humidity (RH%), air velocity (v_a), and globe temperature (T_g) as $<\pm$ 0.5 °C, ±3% (Max ±5%), and ±0.1 °C, respectively. Figure 89.1 shows the Delta Ohm HD32.3TCA microclimate device and its location in the study hall.

A total number of 549 male and female students without any health problems participated in the experiments with an age interval of 20 and 28 years old. A 30-min resting time was conducted in order to avoid the effect of high metabolic rate by the activities to arrive at the study hall. Hereafter, students were requested to fill out an online questionnaire (Fig. 89.2). To this aim, a mobile application was developed for collecting the TSV, personal information (gender, age, height, and weight), and clothing insulation data. Detailed information on the mobile application can be found in Turhan and Özbey (2021).

Data analyses were performed according to the BMI definitions of the World Health Organization (2022). In the following equation (Eq. 89.1), the calculation method of the BMI can be found.

$$\mathrm{BMI}\left(\mathrm{kg/m^2}\right) = \mathrm{Weight\ (kg)/Height^2}\left(\mathrm{m^2}\right) \tag{89.1}$$

Table 89.1 depicts the classification of the BMI values. In this study, obesity classes are gathered in one group as an obesity group.

On the other hand, the difference between thermal votes (ΔTV) was calculated with respect to Eq. 89.2.

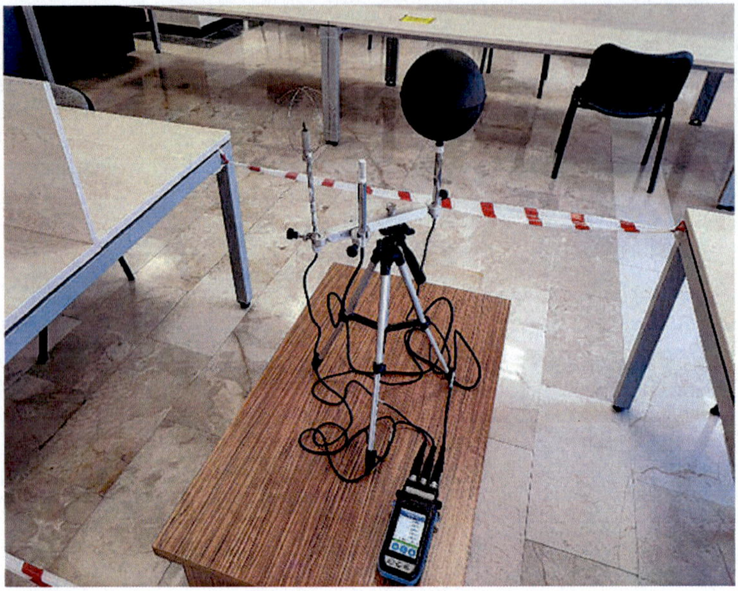

Fig. 89.1 Delta Ohm HD32.3TCA microclimate device

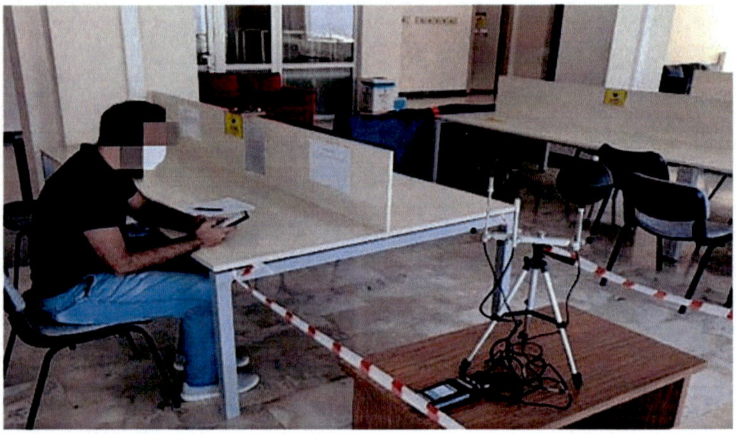

Fig. 89.2 Snapshot from the experiments

$$\Delta TV = PMV - TSV \tag{89.2}$$

For analyzing data and investigating the relationships, statistical analysis was applied to the BMI classes. However, it is worth reminding that obesity classes I, II, and III were assumed in the same class and not analyzed individually.

Table 89.1 Classification of the body mass index

BMI	Class
<18.5	Underweight
18.5–24.9	Normal weight
25.0–29.9	Pre-obesity
30.0–34.9	Obesity class I
35.0–39.9	Obesity class II
>40	Obesity class III

WHO (2022)

Table 89.2 The mean values and standard deviation of data

	N	Mean	Standard deviation
BMI	549	24.21	4.48
ΔTV	549	1.15	0.81

Table 89.3 Correlation between body mass index and difference of thermal vote

	T-value	*p*-value
BMI – ΔTV	2.9	0.001*

Note: *significance at 1%

Application of the statistical analyses was based on the null hypothesis (Ho: τi = 0) as follows:

> There is not a significant and positive relationship between the BMI of the human body and the ΔTV.

The null hypothesis was evaluated for the BMI classes of Underweight, Normal weight, Pre-obesity, and Obesity (Classes I, II, and III) and the difference between thermal votes (ΔTV) at a significance level of 5% ($\alpha = 0.05$). One-tailed t-test was performed to determine the relationship between BMI and ΔTV. If the *p*-value was lower than 0.05, the null hypothesis was rejected (Kim 2015).

89.3 Results and Discussion

The one-tailed t-test was performed on the MINITAB (2022) data analysis tool in order to determine the relationship between the BMI and ΔTV and, thus, evaluate the null hypothesis.

Table 89.2 shows the number of samples, mean of the data, and standard deviation of the BMI and the ΔTV.

The relevant data was evaluated with a one-tailed t-test with a significance level of 0.05 ($\alpha = 0.05$), as discussed in the methodology part. The outcomes of the evaluation can be found in Table 89.3.

Therefore, the null hypothesis was rejected as a result of the t-test at a significance level of 5%.

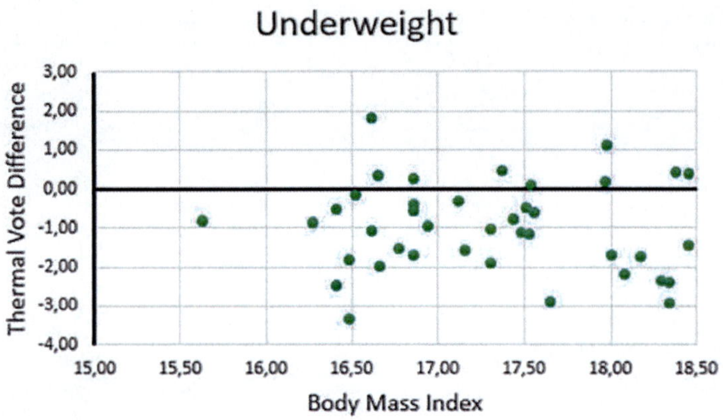

Fig. 89.3 Thermal vote difference of underweight class

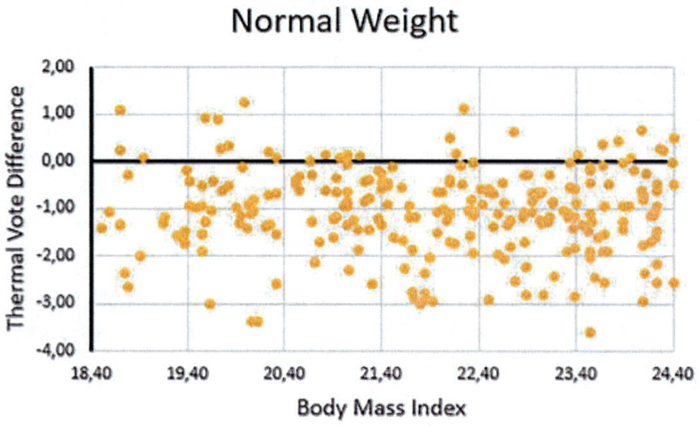

Fig. 89.4 Thermal vote difference of normal weight class

On the other hand, a further study was conducted in order to examine each Body Mass Index separately. However, due to the low sample size of Obesity Classes I–II and III, both classes were considered and regrouped in one class and named as Obesity class (World Health Organization 2022).

Figures 89.3, 89.4, 89.5, and 89.6 depict the difference between the thermal vote and the body mass index. In the figures, the positive ΔTV values represent the situation where the occupant feels colder than the PMV value. In contrast, negative values represent the situation where the occupant feels warmer than the PMV.

Figure 89.3 shows the thermal vote difference distribution of occupants included in the Underweight BMI class. The thermal sensation characteristic of occupants was concentrated more intensely in the warm region. In other words, occupants feel slightly warmer than the measured PMV value. However, there were occupants that felt cooler than the measured value. This is the result of the climate adaptation for the

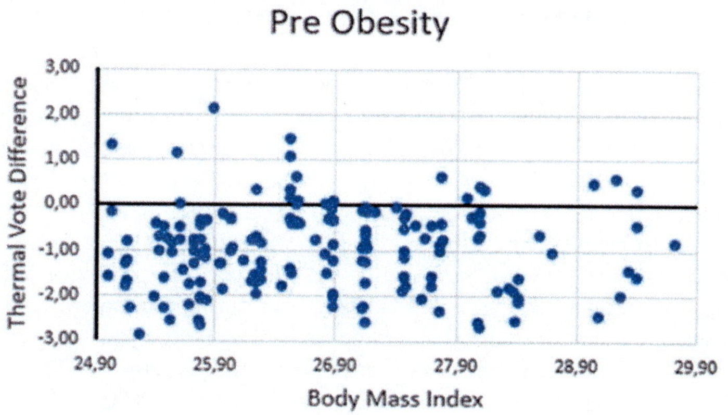

Fig. 89.5 Thermal vote difference of pre-obesity class

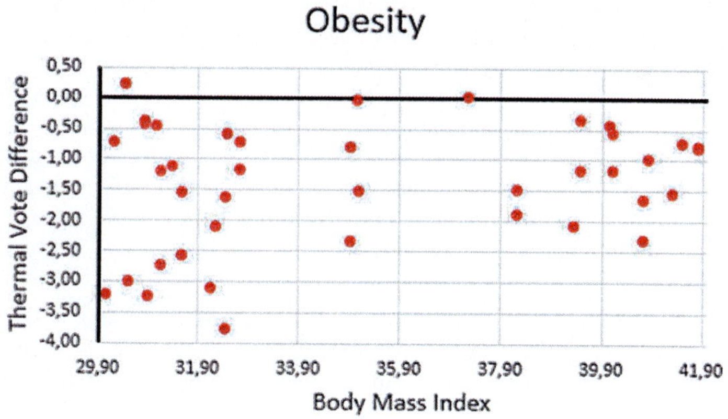

Fig. 89.6 Thermal vote difference of obesity class

occupants in the Csb-type climate zone. Occupants in warmer regions generally feel warmer than the measured PMV value. However, the study found that the difference was under −3. Only 18% of the data exceeded the −2 value of ΔTV.

A similar situation is valid to the Normal weight and Pre-obesity BMI classes shown in Figs. 89.4 and 89.5. Occupants in Normal weight and Pre-obesity BMI classes are predominantly concentrated in the warmer region and feel warmer than the measured PMV value. However, there were occupants that felt colder than the measured PMV in these BMI classes as well. Only 17% of the data exceeded the −2 value of ΔTV.

On the other hand, the effect of BMI on the thermal vote difference could be observed more sharply in the Obesity class shown in Fig. 89.6. According to the figure, the occupants were clustered in the warmer zone, and all of the occupants felt warmer than the measured PMV value or felt equal to the PMV. The difference exceeded even −4 for the Obesity class. Furthermore, 31% of the data exceeded the −2 value of ΔTV.

The fact that occupants feel warmer than the measured PMV value is considered a normal situation for a temperate climate zone. In addition, various personal factors can affect the thermal sensation vote, and it is usual for occupants to be clustered in the warmer and colder zones for different climate zones (Turhan and Akkurt 2019; Turhan and Özbey 2021). However, in the Obesity class, the effect of high BMI values on the thermal vote difference was observed significantly.

In addition, there were some limitations of the study. For instance, due to the COVID-19 pandemic, occupants used face masks during the experiments. Moreover, the stress factor was also ignored in the study. The experiments were conducted in a study hall, which could increase the stress level of the occupants. Therefore, the occupants could feel warmer due to the high-stress level and face masks. Further studies will be able to examine these limitations in detail and eliminate their impact on the current results.

89.4 Conclusions

This study was conducted with 549 young adult female and male occupants in a study hall of the university. Occupants were requested to fill out online surveys to collect their personal information and Thermal Sensation Vote. Additionally, environmental data and PMV values were collected while the occupants filled out the surveys.

For investigating the relationship between the Body Mass Index and the thermal vote difference, a one-tailed t-test was applied to the obtained data at a significance level of 0.05 ($\alpha = 0.05$). The results showed that there was a positive relationship between BMI and ΔTV at a significance level of 0.01.

Additionally, a further study was conducted with the aim of evaluating each BMI class individually. The outcomes of the examination showed that the characteristic of the Underweight, Normal weight, and Pre-obesity BMI classes were fitted to the usual characteristics of occupants who live in the temperate climate zone. However, the characteristics of the obesity BMI class revealed the effect of the Body Mass Index on the thermal vote difference. This study showed that all occupants in the Obesity class feel warmer than the measured PMV value.

Acknowledgments The Scientific and Technological Research Council of Turkey (TÜBİTAK) funded this research, and their contribution is gratefully acknowledged (Project Number: **120M890**).

References

American Society of Heating, Refrigerating and Air-Conditioning Engineers. (2017) Thermal Environment Conditions for Human Occupancy (Standard 55). Atlanta, GA: ASHRAE.

DELTA OHM HD32.3TCA, Microclimate Measurement Device, Delta OHM, Italy.

Fanger PO (1970) Thermal Comfort: Analysis and Applications in Environmental Engineering. Danish Technical Press, Copenhagen.

Forcada N, Gangolells M, Casals M, Tejedor B, Macarulla M, Gaspar K (2021) Field study on thermal comfort in nursing homes in heated environments. Energy and Buildings 244: 111032. https://doi.org/10.1016/j.enbuild.2021.111032

Indraganti M, Ooka R, Rijal H. B (2015) Thermal comfort in offices in India: Behavioral adaptation and the effect of age and gender. Energy and Buildings 103: 284-295. https://doi.org/10.1016/j.enbuild.2015.05.042

ISO 7730 (2005) Ergonomics of the thermal environment – Analytical determination and interpretation of thermal comfort using calculation of the PMV and PPD indices and local thermal comfort criteria.

Kim TK (2015) T-test as a parametric statistic. Korean journal of anaesthesiology 68(6), 540. https://doi.org/10.4097/kjae.2015.68.6.540

Kottek M, Grieser J, Beck C, Rudolf B, Rubel F (2006) World map of the Köppen-Geiger climate classification updated. Meteorologische Zeitschrift. 15(3): 259 – 263. https://doi.org/10.1127/0941-2948/2006/0130

Kruger EL, Drach P (2017) Identifying potential effects from anthropometric variables on outdoor thermal comfort. Building and Environment 117: 230-237. https://doi.org/10.1016/j.buildenv.2017.03.020

Majewski G, Orman ŁJ, Telejko M, Radek N, Pietraszek J, Dudek A (2020) Assessment of thermal comfort in the intelligent buildings in view of providing high quality indoor environment. Energies 13(8): 19. https://doi.org/10.3390/en13081973

Minitab, Minitab LLC., 2022

Salata, F, Golasi I, Verrusio W, de Lieto Vollaro E, Cacciafesta M, de Lieto Vollaro A (2018) On the necessities to analyse the thermohygrometric perception in aged people. A review about indoor thermal comfort, health and energetic aspects and a perspective for future studies. Sustainable Cities and Society 41: 469-480. https://doi.org/10.1016/j.scs.2018.06.003

Turhan C, Akkurt GG (2019) The relation between thermal comfort and human-body exergy consumption in a temperate climate zone. Energy and Buildings 205: 109548. https://doi.org/10.1016/j.enbuild.2019.109548

Turhan C, Özbey MF (2021) Effect of pre-and post-exam stress levels on thermal sensation of students. Energy and Buildings 231: 110595. https://doi.org/10.1016/j.enbuild.2020.110595

World Health Organization, Body Mass Index, WHO, https://www.euro.who.int/en/health-topics/disease-prevention/nutrition/a-healthy-lifestyle/body-mass-index-bmi, accessed on January 30, 2022.

Chapter 90
Numerical Investigation of Reactant Channel Design on PEM Fuel Cell Performance

Hüseyin Kahraman and Muzaffer Furkan Sün

90.1 Introduction

Today, various types of petroleum-based fuels are used. The use of fossil fuels causes irreversible damage to the environment. Therefore, the search for alternative energy sources has been accelerated. In recent years, hydrogen has stood out as an alternative energy carrier to fossil fuels (Abdul Rasheed et al. 2017). PEM fuel cells generate electricity using hydrogen and only release water and heat as the reaction product. Due to the fast startup, flexibility in input fuel, compact design, lightweight, low cost, and solidity of electrolyte, PEMFCs are the most suitable candidate for mobile and transportation applications and is regarded as a promising power source for electric vehicles (Tan et al. 2022). One of the most important components in PEMFC is the bipolar plate (BP). The main functions of BPs can be listed as follows:

- Connect cells electrically in-series.
- Transfer heat from the active cells to the cooling cells.
- Separate gasses in adjacent cells.
- Provide structural support for the stack.
- House the flow field channels (Kahraman and Orhan 2017).

In PEM fuel cells, homogeneity of the reactant distribution on the active surface is an important phenomenon to obtain optimum performance from the fuel cell stack. The study of channel design in PEMFC has been studied for many years by many researchers. In the previous studies, it was tried to obtain a homogeneous flow density distribution by changing the width and depth of the flow channels and the

H. Kahraman (✉) · M. F. Sün
Department of Mechanical Engineering, Faculty of Technology, Sakarya University of Applied Sciences, Sakarya, Turkey
e-mail: huseyink@subu.edu.tr; y205009029@subu.edu.tr

M. Z. Sogut et al. (eds.), *Proceedings of the 2022 International Symposium on Energy Management and Sustainability*, Springer Proceedings in Energy,
https://doi.org/10.1007/978-3-031-30171-1_90

channel profiles (Houreh et al. 2019). Designs such as parallel, serpentine, Pin-type, interdigitated, and modified versions of these designs have also been frequently investigated. Some researchers have adapted designs inspired by nature to flow channels. Leaf venation patterns, plant veins, and lung-inspired approaches can be seen in the literature (Kjelstrup et al. 2010; Ghadhban et al. 2021; Kang et al. 2019; Saripella et al. 2015; Liu et al. 2019).

In this study, the snail spiral and phi vortex flow channel design inspired by nature are modeled in the ANSYS program. In addition, it has been observed that the spiral structure is shaped according to the Fibonacci series in the snail and phi vortex. While designing the flow channels, the Fibonacci series expansion was also taken into consideration. The results were compared with the serpentine channel design, and the graphs showed that the Fibonacci spiral model performed better.

90.2 Numerical Simulation

The flow field numerical simulations of the 3D, laminar, unsteady, and two-phase flow are modeled by using ANSYS FLUENT 15.0. Assumptions used in fuel cell modeling are:

- Ideal gas properties.
- Incompressible flow.
- Laminar flow.
- Isotropic and homogeneous electrolyte, electrode, and bipolar material structures.
- A negligible ohmic potential drop in components.
- Mass and energy transport are modeled from a macroperspective using volume-averaged conservation equations (Kahraman and Coban 2020).

The conservation of mass and momentum equations has been applied to the model. The continuity equation for the flow of a fluid with density ρ (kg/m^3), velocity v (ms^{-1}), the operator ∇, $\frac{d}{dx}+\frac{d}{dy}+\frac{d}{dz}$, and S_m representing the additional mass sources and no source or sink terms may be written as

$$\frac{\partial \rho}{\partial t}+\nabla\cdot(\rho v)=\mathrm{S}_m \tag{90.1}$$

The conservation of momentum equation is given as (Eq. 90.2)

$$\frac{d(\rho v)}{dt}+\nabla\cdot(\rho v)=-\nabla \mathrm{p}+\nabla\cdot(\mu_{\mathrm{mix}}\nabla \mathrm{v})+\mathrm{S}_m \tag{90.2}$$

Here, p is the fluid pressure (Pa), μ is the viscosity (kgm^{-1} s^{-1}), and S_M is the momentum source. This momentum source term varies in different parts of the fuel cell. The reaction rate is an important function of temperature. Thus, it is important to account for temperature variations within the cell by solving the energy conservation equation. Energy conversation for any domain in an FC is described by Eq. (90.3):

$$(\rho c_p)\frac{dT}{dt} + (\rho c_p)(v \cdot \nabla T) = \nabla \cdot (k_{\text{eff}} \nabla T) + S_e \tag{90.3}$$

where c_p is the mixture-averaged specific heat capacity, J kg^{-1} K^{-1}, T is the temperature, K, k is the thermal conductivity, W m^{-1} K^{-1}, and S_e is the energy source term. S_e includes the heat from reactions, ohmic heating, and heat associated with a phase change. The species balance equation represents mass conservation for each reactant. Species conservation for the gas phase is

$$\frac{\partial(\varepsilon \rho x_i)}{\partial t} + \nabla \cdot (\varepsilon \rho x_i) = \left(\nabla \cdot \rho D_i^{\text{eff}} \nabla x_i\right) + S_{s,i} \tag{90.4}$$

where x_i is the mass fraction of gas species, D_{eff} is a function of porosity, and $S_{s,i}$ represents additional species sources. In the Nafion membrane, two types of water flux are present: back diffusion and electroosmotic drag. Both fluxes can be accounted for by the following equation:

$$J_{H_2O}^{M} = 2n_{\text{drag}}\frac{i}{2F}\frac{\lambda}{22} - \frac{\rho_{\text{dry}}}{M_m} D_\lambda \frac{d\lambda}{dz} \tag{90.5}$$

The water content is not constant in Eq. (90.5). By obtaining the water content, the resistance of the electrolyte can be estimated.

90.3 Geometric Model Design and Boundary Conditions

The designs of the flow channel located on the double-sided bipolar plates in the fuel cell were first drawn by the SOLIDWORKS CAD 17 program. These emerging geometric flow designs consist of the serpentine design, which is frequently encountered in the literature and reveals various results by making comparisons, and the spiral design, which is created with the help of geometric objects using daily life and revealed by taking the snail modeling as an example. The geometry of the flow channel of the spiral design is shown in Fig. 90.1 due to the snail-shaped folds.

In order to understand the logic of this spiral layer, the model of a simple (single) channel fuel cell is shown schematically, avoiding complexity and the components on it are indicated. In addition to the technical information of the geometry that emerged while creating the model, its values in mm are given in Tables 90.1 and 90.2 for both serpentine and spiral models, respectively.

Figure 90.1 shows the detailed settings about the materials, operating conditions, and boundary conditions of the PEMFC model. The preliminary values of the average current density of 1 Acm^{-2} and the O_2 flow rate of 10 sccm cm^{-2} are

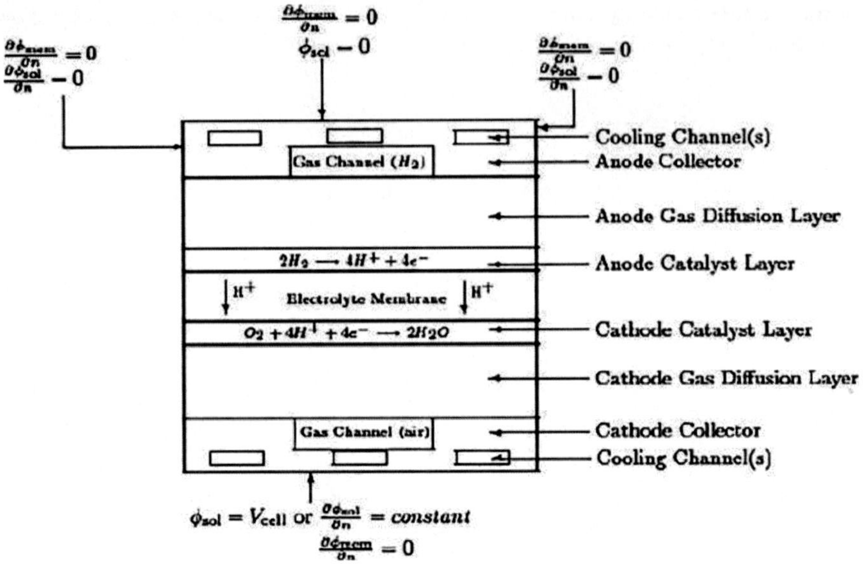

Fig. 90.1 Boundary conditions for the electric potential (solid and membrane)

Table 90.1 Fuel cell dimensions for single channel geometry

Component	Value	Unit
Current collector plate thickness	2	Mm
Gas diffusion layer thickness	0.3	Mm
Catalyst layer thickness	0.01	Mm
Membrane thickness	0.049	Mm
Flow channel width	1	Mm
Flow channel depth	1	Mm
Fuel cell total thickness	4.26	Mm
Active area (serpentine design)	4	cm^2
Active area (spiral design)	3.887	cm^2

Table 90.2 Dimensioning for spiral flow design

Component	Value	Unit
Current collector plate thickness	2	Mm
Gas diffusion layer thickness	0.3	Mm
Catalyst layer thickness	0.01	Mm
Membrane thickness	0.05	Mm
Spiral duct width	1	Mm
Spiral channel depth	1	Mm
Fuel cell total thickness	4.67	Mm
Active area (spiral design)	4	cm^2
Spiral wall thickness	1	Mm

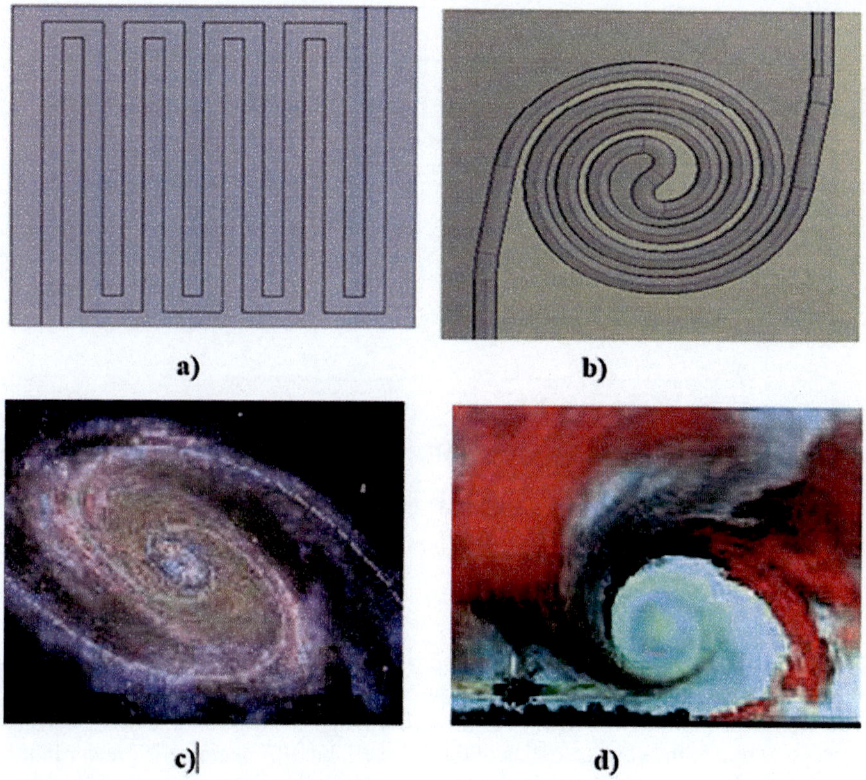

Fig. 90.2 Flow channel geometries performed by solid works program: (**a**) serpentine design, (**b**) spiral design, (**c**) cosmetic universe representation, and (**d**) phi vortex

used. The stoichiometric value was 2.5. The water transport mechanism inside the flow field is studied using the simulation of a three-dimensional, transient, two-phase flow. Air and water are treated as a gaseous phase and a liquid phase, respectively, and no mixture phase. Because of the small sizes of channels and low flow rate, the fluid flow is considered laminar. For simplifying the model calculation, heat transfer, electrochemistry reactions, and phase change are not considered. The simulation is carried out using the commercial software package ANSYS Fluent. Figure 90.2 shows the standard serpentine and spiral designs inspired from natural shapes (Fig. 90.2c, d).

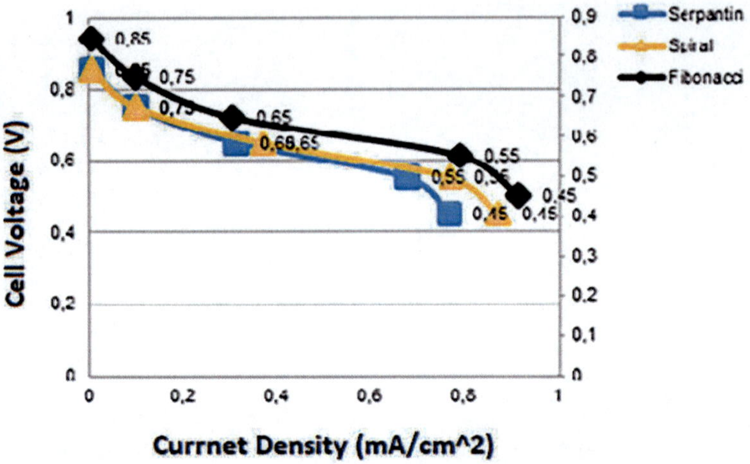

Fig. 90.3 Polarization data of designs created according to serpentine, spiral, and Fibonacci law

90.4 Results and Discussion

Figure 90.3 demonstrates the polarization curves of the PEMFC with two different flow fields. As could be observed, the spiral model shows better performance than that of the serpentine flow field. As the cell voltage increases, the current density decreases due to the characteristics of the fuel cell, but this decrease is greater in the serpentine design. In addition, in the spiral model designed based on the Fibonacci series expansion, the current density gives better results in response to the falling cell voltage. To explain in numbers, when the cell current decreased from 1 V to 0.6 V, the current densities reached 0.6, 0.63, and 0.8 mA/cm^2, for serpentine, spiral, and Fibonacci spiral models, respectively.

The current density distribution behaviors of the flow channels can be observed in Fig. 90.4. In a serpentine flow field, fluids flow through a continuous path from start to end. Serpentine design has good water removal due to high flow velocity, coverage of the entire active area, and most common for small active areas. However, reactants in the serpentine flow field depletion along the length of the channels lead to uneven reactant distribution. In addition, because of the high-pressure drop, flooding danger can be seen at high current densities and water can build up in bends, causing local current density reductions. These disadvantages are also seen in the results of the simulation study. While the current density in the entrance region of the serpentine design is high, it decreases as it approaches the exit region. This is due to the high current produced by the high reactant concentration in the inlet region. In the exit region, since the reactants are depleted and decreased, the reaction rate is low in these regions, and low current density locations are formed. In the spiral model designed to eliminate this inhomogeneous current density distribution, the current

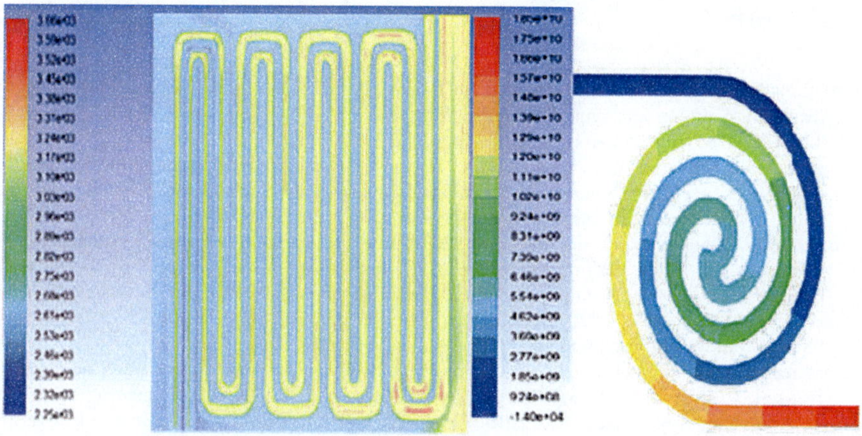

a) b)

Fig. 90.4 The current density distributions of the surface between gdl and cathode catalyst as a result of the current given at 0.75 V: (**a**) serpentine design and (**b**) spiral design

density is relatively more homogeneous. However, it was observed that the reaction rates decreased in the exit regions.

90.5 Conclusions

It can be inferred that the flow field geometry and structure of PEMFC are extremely important to internal and external heat and mass transfer processes, and the optimization of flow field configuration is a certain research direction of PEMFC for further large-scale commercial development. Under the conditions of high current density and high humidification, the amount of reactant transportation and water production has a significant impact on the flow field design methodology. In this study, different designs are used to optimize the inhomogeneous current density caused by the weaknesses of the serpentine design. The current density distribution should also be simulated for computers with more powerful processors and lower cell voltages. Also, as stream channel designs become more complex, more powerful computers are needed. As a result, in this study, a design that includes the mathematical balance in nature was used to be an alternative to the serpentine design, and it was observed that the cell performance increased.

References

Abdul Rasheed, R.K., Liao, Q., Caizhi, Z., Chan, S.H., 2017. A review on modelling of high temperature proton exchange membrane fuel cells (HT-PEMFCs). Int. J. Hydrog. Energy 42, 3142–3165. https://doi.org/10.1016/j.ijhydene.2016.10.078

Ghadhban, S.A., Alawee, W.H., Dhahad, H.A., 2021. Study effects of bio-inspired flow filed design on Polymer Electrolyte Membrane fuel cell performance. Case Stud. Therm. Eng. 24, 100841. https://doi.org/10.1016/j.csite.2021.100841

Houreh, N.B., Shokouhmand, H., Afshari, E., 2019. Effect of inserting obstacles in flow field on a membrane humidifier performance for PEMFC application: A CFD model. Int. J. Hydrog. Energy 44, 30420–30439. https://doi.org/10.1016/j.ijhydene.2019.09.189

Kahraman, H., Coban, A., 2020. Performance improvement of a single pem fuel cell using an innovative flow field design methodology. Arab. J. Sci. Eng. 45, 5143–5152.

Kahraman, H., Orhan, M.F., 2017. Flow field bipolar plates in a proton exchange membrane fuel cell: Analysis & modeling. Energy Convers. Manag. 133, 363–384.

Kang, H.C., Jum, K.M., Sohn, Y.J., 2019. Performance of unit PEM fuel cells with a leaf-vein-simulating flow field-patterned bipolar plate. Int. J. Hydrog. Energy 44, 24036–24042. https://doi.org/10.1016/j.ijhydene.2019.07.120

Kjelstrup, S., Coppens, M.-O., Pharoah, J.G., Pfeifer, P., 2010. Nature-Inspired Energy- and Material-Efficient Design of a Polymer Electrolyte Membrane Fuel Cell. Energy Fuels 24, 5097–5108. https://doi.org/10.1021/ef100610w

Liu, S., Chen, T., Xie, Y., Zhang, J., Wu, C., 2019. Numerical simulation and experimental study on the effect of symmetric and asymmetric bionic flow channels on PEMFC performance under gravity. Int. J. Hydrog. Energy, The 2nd International Symposium on Hydrogen Energy and Energy Technologies (HEET-2018) 44, 29618–29630. https://doi.org/10.1016/j.ijhydene.2019.06.046

Saripella, B.P., Koylu, U.O., Leu, M.C., 2015. Experimental and Computational Evaluation of Performance and Water Management Characteristics of a Bio-Inspired Proton Exchange Membrane Fuel Cell. J. Fuel Cell Sci. Technol. 12. https://doi.org/10.1115/1.4032041

Tan, Q., Lei, H., Liu, Z., 2022. Numerical simulation analysis of the performance on the PEMFC with a new flow field designed based on constructal-theory. Int. J. Hydrog. Energy 47, 11975–11990. https://doi.org/10.1016/j.ijhydene.2022.01.243

Chapter 91
Modeling Thermal Performance in the Uruguayan Residential Sector

Sofía Gervaz, Federico Favre, and Pedro Curto-Risso

91.1 Introduction

Residential buildings contribute to the world's total final energy demand in a relevant proportion. As a consequence, in recent years different building energy models have been developed with the purpose of effectively understanding, assessing, and managing energy consumption in the residential building sector (Aydinalp et al. 2002; Cerezo-Davila et al. 2016 and Wilson et al. 2017).

When aiming at characterizing building energy requirements for whole cities or countries, two main modeling approaches are distinguished. The top-down approach utilizes the estimate of total building sector energy consumption and relates it with major drivers such as gross domestic product (GDP), energy price, population, weather conditions, etc. (Li et al. 2017). On the contrary, the bottom-up approach is based on determining the energy consumption of individual buildings and then extrapolating these results to represent a region or nation (Swan and Ugursal 2009). Each technique thus requires different levels of detail for the input data and produces results with a different applicability.

Unlike in top-down models, the bottom-up technique is used to assess the contribution of each end-use toward the aggregate demand of the building sector. Most of these models rely on the definition of archetypes, which consist of building definitions that are used to represent a group of buildings with similar properties. The strengths of the bottom-up approach are that the resulting models allow identifying areas of improvement by analyzing the demand of each end-use and also to evaluate

S. Gervaz (✉) · F. Favre · P. Curto-Risso
Universidad de la República, Facultad de Ingeniería, Montevideo, Uruguay
e-mail: sofiag@fing.edu.uy; ffavre@fing.edu.uy; pcurto@fing.edu.uy

M. Z. Sogut et al. (eds.), *Proceedings of the 2022 International Symposium on Energy Management and Sustainability*, Springer Proceedings in Energy,
https://doi.org/10.1007/978-3-031-30171-1_91

the impact of technology or design changes. However, models of this type require a large set of input information and great effort is required to develop and maintain them.

In Uruguay, though an important consumer, there are yet no developments that model the energy requirements of the residential sector in detail, nor for thermal requirements for space conditioning in particular. In this regard, the objective of this work is to develop a bottom-up model to analyze thermal performance of the Uruguayan residential sector disaggregated according to relevant characteristics of the buildings and their local climate. The model is focused on modeling thermal energy requirements as they are the most complicated to characterize due to the high level of interdependence they have on many variables (local weather, construction materials, building design, occupants' behavior, etc.). The analysis includes the study of the residential sector as it is and also the evaluation of the impact of different energy efficiency strategies.

91.2 Methodology

A physics-based bottom-up approach was followed based on buildings identified as representative of the residential sector. Being a physics-based model, the thermal performances of the representative buildings (or archetypes) were obtained by simulations performed focusing on their physical characteristics and the thermodynamic principles that govern the interaction between a building and its surroundings. Then, the results obtained for these archetypes were extrapolated for the whole residential sector based on their prevalence among the housing stock.

The development of this bottom-up model is based on the Uruguayan housing stock characterization performed in Curto-Risso and Picción (2017), where the archetypes were defined as well as their prevalence in the residential sector. Then, important parameters for the simulations were defined such as the weather information to use, the occupation schedules in the archetypes, the behavior of the occupants regarding ventilation and use of solar protections, the thermal comfort criteria, etc. Finally, there was the process automation stage, in which a tool was developed to automatically generate and simulate each archetype model and also process its results.

During the housing stock characterization, the attributes of the residential buildings that are closely related to their thermal performances were identified and divided into categories. Those are as follows:

- Geographic location: the country was divided into four regions according to their climatic characteristics (MVD, N-NW, SW-Ctr, NE, and S-SE; see Fig. 91.1). The city of Montevideo defined a region on its own (MVD) as it holds 41% of the total residential buildings.
- Household type: house or apartment.
- Size: less than 40 m^2, between 40 and 70 m^2, or more than 70 m^2.

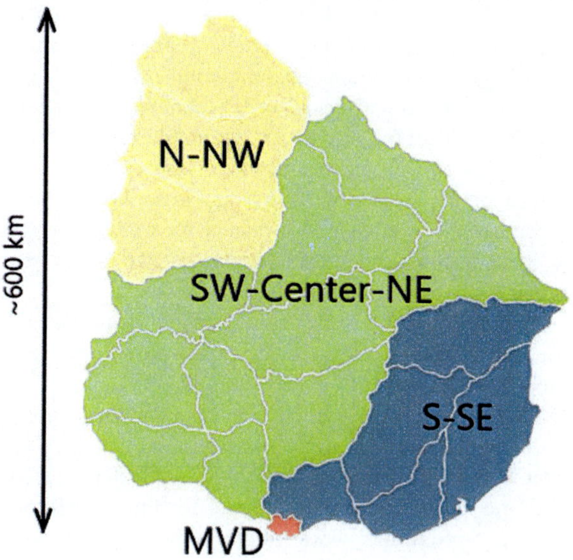

Fig. 91.1 Regions considered

- Socioeconomic status: income deciles 1–4, 5–7, or 8–10.
- Vintage: less than 10 years, between 10 and 30 years, and more than 30 years.

Each possible combination of type, size, socioeconomic status, and vintage defined an archetype to which a typical geometry, construction materials, and a number of occupants were assigned. Then, the weight of each archetype along the four geographic regions was determined for the whole housing stock.

During the next stage, weather files based on the Typical Meteorological Year (Alonso-Suárez et al. 2016) were selected for the simulations as they proved to be the most representative or Uruguayan weather. The occupation schedules as well as internal gains due to occupancy, lighting, and equipment were defined as established in NBR15575 technical rule (Associação Brasileira de Normas Técnicas 2013). The thermal comfort model considered was the ASHRAE55 adaptive model (ANSI/ASHRAE Standard 2017), which relates indoor acceptable temperature ranges to outside meteorological parameters based on the idea that humans can adapt to different conditions during different times of the year. The thermal performances of the archetypes were measured as the quantity of thermal energy required to maintain comfortable conditions during the occupied hours. Occupants were defined to use natural ventilation and operate solar protections in order to reduce thermal requirements when they are at home, and windows remained closed and protections inactive (open) during unoccupied hours.

Besides, each archetype represents a large quantity of different households with similar characteristics in terms of typology and construction materials. However, these households may be located in different geographic regions and have different orientations and surroundings, all of which will affect the thermal performance of the buildings. Therefore, for the results to be representative, every archetype in every geographic region should be simulated as many times as possible considering these variations, so as to obtain averaged results for each. This led to automation as the final stage of the model development.

A tool developed based on Python scripts (and in particular its Eppy library) enables the user to perform a large number of simulations as it automates the whole process by generating several models for each archetype varying its orientation and surroundings. It then performs the EnergyPlus simulations for every model and processes the results. Then, results for each archetype are extrapolated according to the housing stock characterization in order to represent the whole residential sector.

EnergyPlus v.8.7 is the simulation engine in which the energy calculation method for each archetype relies on. Regarding the surface heat balances, EnergyPlus default models were used for the convection coefficients calculations: TARP for inside and DOE-2 for outside coefficients. For the case of walls and roofs, the ConductionTransferFunction was used as the algorithm for solving the conduction heat fluxes. For the ground, the model selected was GroundDomain:Slab combined with the KusudaAchenbach model for the undisturbed ground temperatures. Heat transfer through the windows was solved by the layer-by-layer approach, whereas infiltration and ventilation loads were calculated by means of the AirflowNetwork model. The IdealLoads HVAC system was used to calculate thermal energy requirements for achieving thermal comfort. Details regarding all these models can be found in U.S. Department of Energy (2021).

91.3 Results and Discussion

Results for a simulation of 500 models show that, as expected, when distinguishing between the type of household, apartments have better performances than houses on average (see Table 91.1).

Apartments performing better than houses is mainly due to the difference in the percentage of the exposed area. Whereas in houses it is 87% on average, in apartments it is 32%. Besides, and based on the impact socioeconomic status have on the construction qualities, the income decile prevalence in each type of household also contributes to apartments requiring less thermal energy than houses (45% of houses correspond to income deciles 1–4 while 66% of apartments correspond to income deciles 8–10).

Energy requirements obtained when disaggregating the results by the geographic regions are shown in Fig. 91.2. MVD and S-SE total demand is less than 60 kWh/m^2,

Table 91.1 Annual thermal requirements

Type	Cooling	Heating
	kWh/m^2	kWh/m^2
House	32.6	33.4
Apartment	20.5	14.2

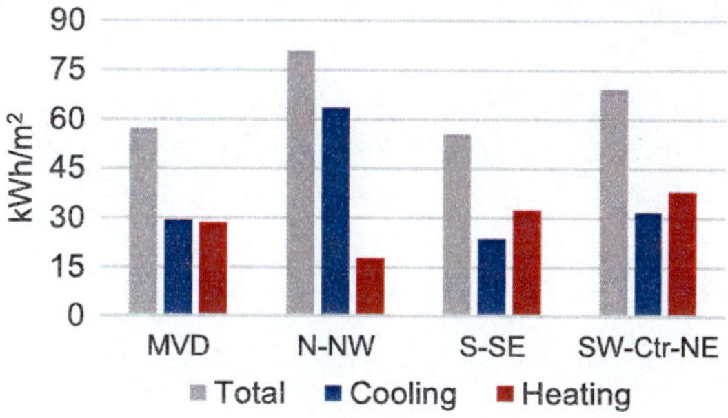

Fig. 91.2 Thermal energy requirements in regions

Table 91.2 Regions' mean temperatures

	MVD (°C)	N-NW (°C)	S-SE (°C)	SW-Ctr-NE (°C)
Cooling period	21.7	25.4	21.5	23.1
Heating period	10.8	13.7	11.0	10.7

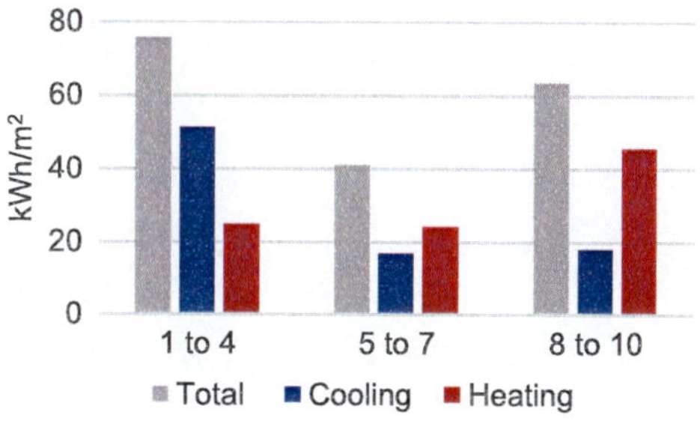

Fig. 91.3 Thermal energy requirements in income deciles

whereas N-NW is around 80 kWh/m^2 and SW-Center-NE is nearly 70 kWh/m^2. For the case of N-NW, the difference is due to the cooling requirements, which are double of those in the other regions. SW-Center-NE, on the other hand, has both higher cooling and heating requirements than MVD and S-SE; yet, the differences are less evident than those for N-NW.

Heating and cooling loads of Fig. 91.2 are in accordance with the differences in mean temperatures for each region shown in Table 91.2 and also with Uruguayan solar and wind maps (Alonso-Suárez et al. 2016; MIEM et al. 2009). The N-NW's higher irradiance and temperatures and lower wind speeds result in higher heat gains and lower losses than in the rest of the regions.

When distinguishing between income decile ranks, the differences between energy requirements are related to different construction materials, number of occupants, and household area. While deciles 1–4 have the highest demands, deciles 5–7 have the lowest. Moreover, when distinguishing between cooling and heating, it remains clear that the reason for deciles 1–4's high requirements are the cooling loads, whereas the opposite is true for deciles 8–10 (see Fig. 91.3).

Fig. 91.4 Thermal energy requirements in income deciles per occupant

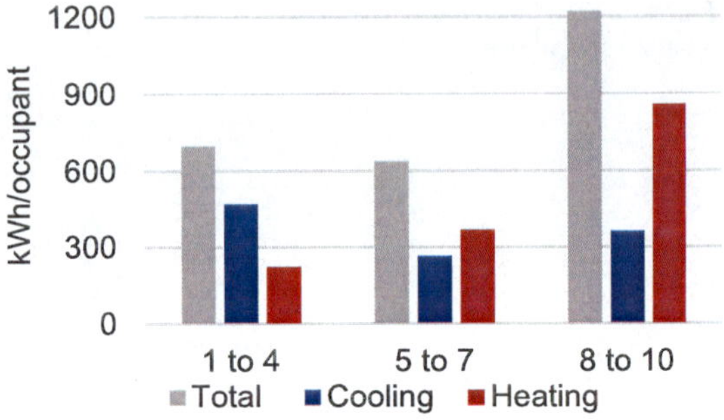

Fig. 91.5 Energy savings per efficiency strategy

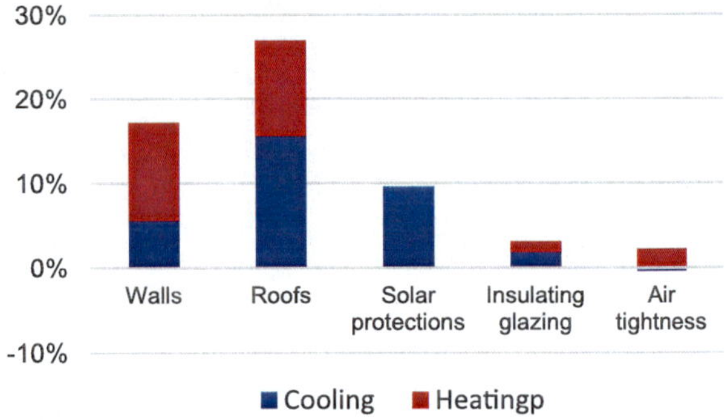

However, if energy requirements for the different socioeconomic categories are analyzed relative to the number of occupants in the households rather than relative to their areas, the results are different (see Fig. 91.4). A person in the highest income category thus requires, on average, 1200 kWh each year so as to maintain thermal comfort whereas a person in deciles 1–4 requires 700 kWh, and this is despite the higher deciles living in houses with better quality constructions. These results are very different than those from Fig.91.3, where deciles 1–4 have the highest requirements. Hence, the basis on which the results are expressed should be conscientiously selected, as the best alternative would depend on the aim of the analysis.

Five energy efficiency measures were proposed and modeled with the aim of analyzing their impact on the residential sector's thermal performance. Those measures are the improvement of construction materials in the exterior walls and in the roof, the incorporation of solar protections in every window, changing the windows constructions to insulating glazing, and improving the buildings' air tightness. Five new simulations (of 500 models each) were performed, incorporating one efficiency measure in each.

Results show that the strategy with the highest impact on thermal performance when applying it to the whole residential sector is improving the construction materials of the roofs (see Fig. 91.5). The annual benefits of this strategy are 27% of thermal energy requirements and are both for cooling and heating loads in similar proportions.

Table 91.3 Strategy cost relative to energy savings

Walls	Roofs	Solar protections	Insulating glazing
$/kWh	$/kWh	$/kWh	$/kWh
962.4	205.2	97.0	764.9

Apart from the energy saved, the efficiency measures were also analyzed in terms of their cost-effectiveness. In order to do so, the extra cost associated with each measure is estimated based on INCA (2019) and is expressed relative to the energy savings. This was not done for the improvement of the air tightness given that it would be very difficult to estimate its cost and it has a very low impact on energy requirements.

Results show that although improving the roofs' construction is the strategy with the highest annual savings, the most cost-effective is the incorporation of solar protections in every window of the households (see Table 91.3). However, the energy savings obtained when incorporating solar protections—and consequently its cost-effectiveness—will be highly dependent on how they are operated by the occupants.

91.4 Conclusions

A physics-based bottom-up model was developed to characterize the thermal performance of the Uruguayan residential sector. Efforts in making the model as accurate as possible led to automation as the strategy to generate and simulate a large number of buildings models. A platform was thus developed based on Python scripts and Eppy library, as well as on the outcomes of a housing stock characterization process performed in Curto-Risso and Picción (2017). This platform relies on EnergyPlus as the simulation engine, and it is capable of carrying out the whole simulation process, including the characterization, generation, and simulation of the models as well as the results processing.

As it is, the tool developed is capable of quantifying annual or seasonal energy requirements for thermal comfort in the Uruguayan housing stock as it was characterized in Curto-Risso and Picción (2017). Also, evaluating the impact of energy efficiency retrofits in the enclosures of the buildings or the usage patterns. Besides, the effect climate change could produce in the energy required for heating and cooling could also be determined. All of these studies could be performed either for the whole residential sector or disaggregated according to certain relevant characteristics of the buildings modeled. Analyzing the results distinguishing among different building categories allows identifying major areas of improvement, which might in turn lead to the design of targeted energy policies and of cost-effective retrofit measures.

Nevertheless, the tool developed has its limitations, and further work should be devoted to improving the model. Given the high impact they have on results, housing stock characterization and the hypotheses considered (such as the definition of the usage patterns) should be improved. Also, considering the EnergyPlus models used, latent heat ought to be accounted for in the energy balance equations and pressure coefficients might be more accurately determined when solving infiltration and ventilation loads.

Future work will incorporate the remaining end-uses apart from space conditioning and the transformation of energy requirements into final energy demand. By doing so, it would be possible to develop a forecasting model, which can be used to obtain projections of energy demand in the residential sector. Moreover, further development could produce results on an hourly scale, hence providing power demand curves.

Acknowledgments We thank the Comisión Académica de Posgrado of Universidad de la República for the economic support throughout this work.

References

Alonso-Suárez R., Bidegain M., Abal G. and Modernell P., 2016. Año Meteorológico Típico para Aplicaciones de Energía Solar (AMTUes): series horarias típicas para 5 sitios del Uruguay (versión 2.4). Publicación de la Comisión Sectorial de Investigación Científica de la Udelar. ISBN: 978-9974-0-1647-7.

ANSI/ASHRAE Standard 55-2017, Thermal Environmental Conditions for Human Occupancy.

Associação Brasileira de Normas Técnicas, Comitê Brasileiro da Construção Civil, 2013. Edificações habitacionais – Desempenho.

Aydinalp M., Ismet Ugursal V., and Fung A. S., 2002. Modeling of the appliance, lighting, and space-cooling energy consumptions in the residential sector using neural networks. Applied Energy 71.2, pages 87-110.

Cerezo-Davila C., Reinhart C. F. and Bemis J. L., 2016. Modeling Boston: A workflow for the efficient generation and maintenance of urban building energy models from existing geospatial datasets. Energy, Elsevier, vol. 117(P1), pages 237-250.

Curto-Risso P. and Picción A., 2017. FSE_1_2017_1_14479 – Eficiencia energética en el sector residencial. Situación actual y evaluación de estrategias de mejoramiento para distintas condiciones climáticas en el Uruguay. Montevideo-Salto, Uruguay. https://doi.org/10.4000/books.cemca.5418.

INCA, 2019. Costo de Componentes de Obra. URL: https://costos.todouy.com/

Li W., Zhou Y., Cetin K., Eom J., Wang Y., Chen G and Zhang X., 2017. Modeling urban building energy use: A review of modeling approaches and procedures. Energy, Elsevier, vol. 141, pages 2445-2457.

MIEM, IIE and IMFIA, 2009. Proyecto de Energía Eólica – Mapa eólico de Uruguay. URL: http://www.energiaeolica.gub.uy

Swan L. and Ugursal V., 2009. Modeling of end-use energy consumption in the residential sector: A review of modeling techniques. Renewable and Sustainable Energy Reviews, Elsevier, vol. 13, pages 1819-1835.

U.S. Department of Energy, 2021. Engineering Reference. EnergyPlus Documentation.

Wilson E., Christensen C., Horowitz S., Robertson J. and Maguire J., 2017. Energy Efficiency Potential in the U.S. Single-Family Housing Stock. Technical Report, NREL, TP-5500-68670.

Chapter 92
Thermal Analysis of a Social Interest Household in Uruguay

Federico Favre, Gabriel Pena, Sofía Gervaz, Juan Romero, María López, and Lucía Pereira

Nomenclature

HVAC	Heat Ventilation Air Conditioning
MEVIR	Movement for the Eradication of Unhealthy Rural Housing
COP	Coefficient of Performance
NPV	Net Present Value

92.1 Introduction

In recent years, social inclusion policies have been paying more attention to the energetic demand for low-income socio-economic sectors in Uruguay. The energy required for thermal comfort usually cannot be affordable for families in this context. In the present work, the thermal behavior of a household of a social-interest typology is studied in order to analyze possible improvements for its design. The house is the archetype designed by the "Movement for the Eradication of Unhealthy Rural Housing" (MEVIR).

F. Favre (✉) · G. Pena · S. Gervaz · J. Romero
Faculty of Engineering, UdelaR, Montevideo, Uruguay
e-mail: ffavre@fing.edu.uy; gabpena@fing.edu.uy

M. López · L. Pereira
Faculty of Architecture, UdelaR, Montevideo, Uruguay

M. Z. Sogut et al. (eds.), *Proceedings of the 2022 International Symposium on Energy Management and Sustainability*, Springer Proceedings in Energy,
https://doi.org/10.1007/978-3-031-30171-1_92

92.2 Methodology

The study was carried out by means of numerical simulations using EnergyPlus software developed by Crawley et al. (2000); a sketch of the geometry is presented in Fig. 92.1.

EnergyPlus default models were used for the surface heat balances and convection coefficients calculations. For solving the conduction heat fluxes at walls and roofs, the *ConductionTransferFunction* was used. To determine conduction through the ground, the model selected was *GroundDomain:Slab*. For windows heat transfer, the layer-by-layer approach was followed, whereas infiltration and ventilation loads were solved by means of the *AirflowNetwork* model. Finally, energy requirements were calculated using an ideal HVAC system.

The internal and external temperature of one particular household was monitored during a period of 2 weeks. The obtained measures were compared with a simulation for that particular period in order to verify the accuracy of the model, obtaining a very good agreement between numerical and experimental data, according to the ASHRAE Guideline 14.

Using the verified model, a base case is simulated for an entire year to analyze the thermal behavior of the house. To do that, the Typical Meteorological Year of Montevideo from Alonso-Suarez et al. (2016) was used. The Uruguayan climate is mild and wet according to the Köppen classification. It must be noted that Uruguay is located in the south hemisphere, and because of that, winter is from June to August and the sun path is through the north. The energy required to achieve the adaptive thermal comfort ASHRAE-55 was determined by varying the orientation with a step of 30°. Cooling and heating requirements were differentiated. Then, variations of the base case were simulated, such as (i) reduction of the roof thermal transmittance, (ii) inclusion of solar protection in every window, and (iii) a more efficient user regarding the solar protections and natural ventilation operation. These variations

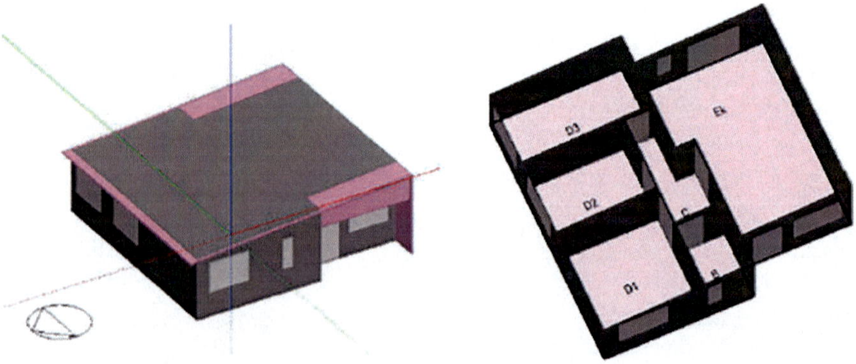

Fig. 92.1 Model of the household

represent 48 simulations. Therefore, in order to automate the simulation process, a python script based on Eppy libraries was implemented.

For each energy efficiency measure, an economic analysis was performed in order to determine which are the more attractive measures. The cost for implementing each measure is calculated considering additional materials and labor costs. The economic benefits due to energy saving were estimated considering that the thermal requirements are covered with electric systems with a COP of 2. Finally, the repayment time (as the time to achieve positive NPV) is calculated according to these estimations for each measure.

92.3 Results and Discussion

92.3.1 *Base Case*

First, a standard case, named Base Case, was simulated in order to characterize the household thermal behavior for the original design. The geometry is presented in Fig. 92.1. The envelope materials are summarized in Table 92.1 and correspond to the one reported by MEVIR.

To determine the air tightness, a blower door test was carried out in a particular MEVIR household. A value of $n_{50} = 7.77ACH$ was measured. With this value, local tightness was estimated according to the windows' and doors' locations in order to define the AirflowNetwork parameters.

The occupants' schedules and behaviors (regarding ventilation and solar protections) were defined, trying to represent as accurately as possible the real occupants. To do so, several interviews were carried on with families living in MEVIR houses. A family with two adults and two minors was defined as standard. The schedule and user behavior according to solar protection and ventilation are summarized in Table 92.2.

Impact of Orientation

The base case was simulated for different orientations. The 0° orientation corresponds to the front facade oriented to the north, as shown in Fig. 92.2, which in the south hemisphere is the one facing the sun. The cooling and heating requirements obtained are presented in Fig. 92.3. As can be observed, this house performs better in

Table 92.1 Typology specifications

Item	Composition (cm)	Thickness	U (W/m2K)
External wall	Brick (12) + Basecoat (1) + Air cam (2) + EPS (3) + Brick (12)	30	0.67
Roof	Sheet steel (0.05) + glass wool (5 cm) + OSB wood (1.5)	6.55	0.7
Floor	Ceramic (1.5) + Mortero (2) + concrete (12)	15.5	–
Windows	Simple glass and aluminum frame	0.4	5.8

Table 92.2 Users' schedule main point

Item	From 16/03 up to 15/11	From 16/11 up to 15/03
Bedroom occupation	22–07 (all); 17–20 (D2 & D3)	
Living room and kitchen occupation	07–09 (3p); 09–18 (1p); 18–20 (2p); 20–22 (4p). Weekends: 09–18 (3p).	
Ventilation (open windows)	12–13	08–13 and 18–24
Solar protection (open)	08–18	08–14 and 18–22

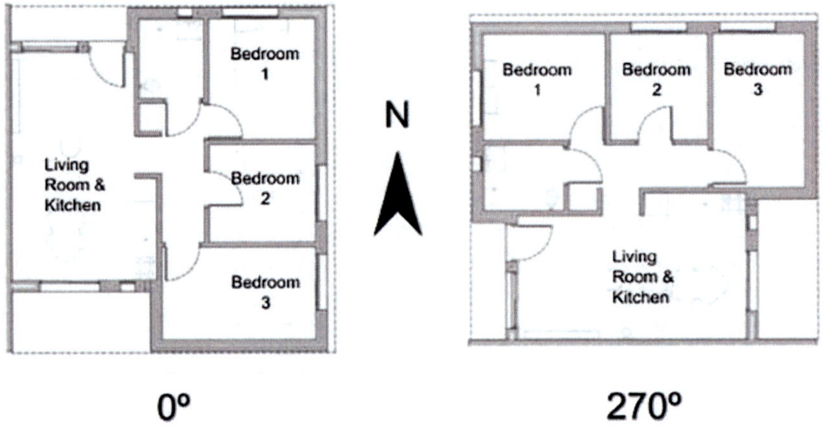

Fig. 92.2 Model of the household

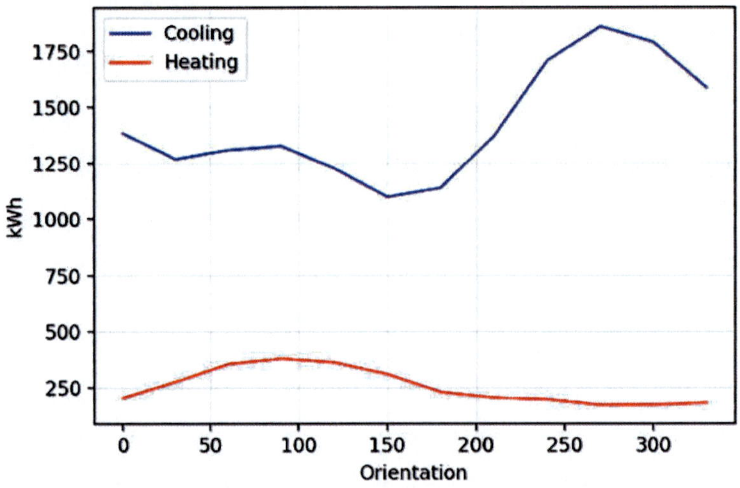

Fig. 92.3 Base case HVAC requirements

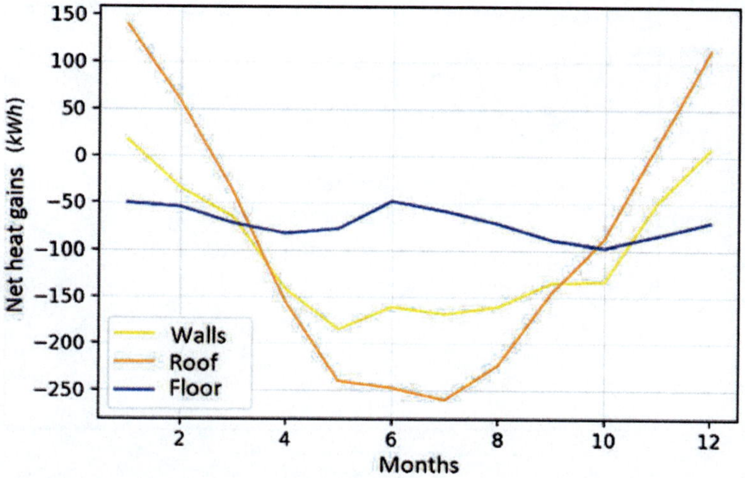

Fig. 92.4 Envelope net heat gains; house orientation 0°

winter than in summer for all orientations, being cooling demand much higher than heating demand. This is due to the low transmittance of the envelope. Because of that, in winter with the internal gains (people and equipment), the solar gains, and a small energy demand, the comfort temperature can be achieved. However, in summer, this internal and solar gains, and also the gains through the envelope and air infiltrations, must be compensated by de HVAC systems, generating an important cooling demand.

As was expected, an important dependency between the energy requirement and the house orientation was observed. Cooling requirements are maximized for the 270° orientation, for which the bedroom windows (which include sun protection) face north, while the kitchen and living room windows (both without sun protection) face east and west, respectively. On the other hand, the most critical orientation for heating is 90°, for which no opening faces north.

Envelope Net Heat Gains

The difference between gains and losses for the roof, exterior walls, and floor are presented in Fig. 92.4. First, the floor heat flow is outgoing during all-year round. This is because the underground temperature is always lower than the inside temperature. These heat losses range from 50 up to 100 kWh per month, being significant (and negative) during the cold period. Second, the net balance for the exterior walls only generates heat gains in the summer months, although they are very low. Heat losses are maximized between April and August, overtaking 150 kWh per month. Lastly, according to the results of the simulations, roof is the opaque envelope surface that causes greater heat gains in summer and greater heat losses in winter for this typology. This is the usual behavior of houses exposed to Uruguayan climate, with similar walls and roof thermal transmittances. A reduction of the roof transmittance could reduce energy demand.

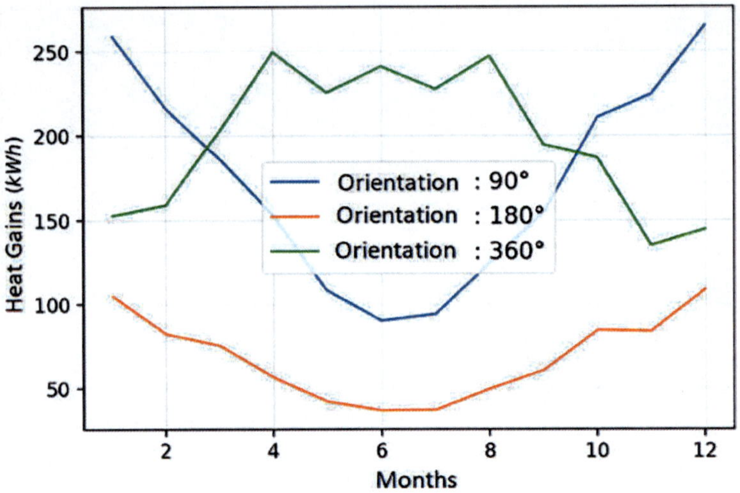

Fig. 92.5 Windows' solar gains; house orientation 0°

Windows' Solar Gains

Glazed surfaces are the ones that generate the greatest net heat gains, so their orientation is central to the house thermal behavior. The net heat gains for each facade for the house at orientation 0° are presented in Fig. 92.5. As was expected, north facade windows (green curve) present a positive oscillation according to energy requirements. This means higher gains in winter (around 240 kWh/month) than in summer (150 kWh/month).

On the other hand, windows oriented to the east (blue curve in Fig. 92.5), as well as west oriented, present high net gains in summer, with values over 250 kWh/month in December and January. This is due to the high direct solar radiation during the morning. Therefore, including solar protections with an adequate operation could reduce cooling energy demand.

Air Infiltration and Natural Ventilation

The annual evolution of infiltrations and ventilation heat losses are presented together in Fig. 92.6 since the EnergyPlus version used in this study does not allow discriminating it. Important heat losses are observed in summer, over 400 kWh/month, mainly due to ventilation. These heat losses are good since they reduce cooling demand. Then, an adequate operation of the windows can be very beneficial in the final energy requirements.

In winter months, the occupants maintain the windows closed for more time. Because of that, low heat losses are obtained in that period (less than 100 kWh/month).

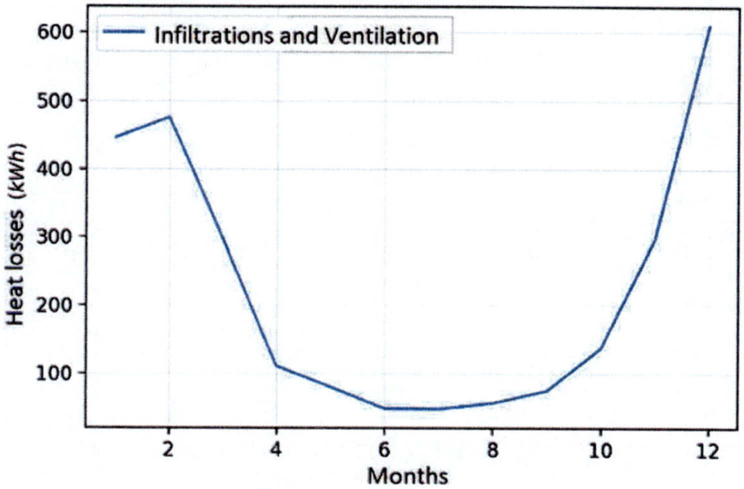

Fig. 92.6 Infiltration and natural ventilation losses. Base case, orientation 0°

92.3.2 Energy Efficiency Measures

Following, three energy efficiency measures are studied in order to improve even more the thermal behavior of the house. As was noted before, solar protection is expected to reduce cooling demand. Also, a more efficient operation of solar protections and openings can save energy demand. Moreover, a reduction of the roof transmittance can reduce both cooling and heating requirements.

Impact of the User

In order to estimate the occupants' behavior influence on the energy demand, an "efficient user" was defined. It consists of the same occupancy schedule but varies the way it operates the solar protections and the openings.

The annual evolution of cooling and heating demand for the efficient user are presented in Fig. 92.7. An important cooling demand reduction is obtained for all orientations. For orientations between 270° and 30°, the cooling demand is less than half for the efficient user; in particular, at 330° the cooling demand is less than a third of the original user one. This important reduction is due to efficient use of solar protections, which highly reduces the net heat gain through glazed areas. Moreover, the efficient user takes more advantage of natural ventilation, which also contributes to the cooling demand reduction.

The heating demand is less affected since for winter the original user also takes advantage of windows gains and does not ventilate much.

According to local energy costs, the savings for the efficient user are equivalent to 94.3 US$ per year. Considering a training program for the occupants of the houses with a cost of 240 US$, the repayment period is less than 3 years.

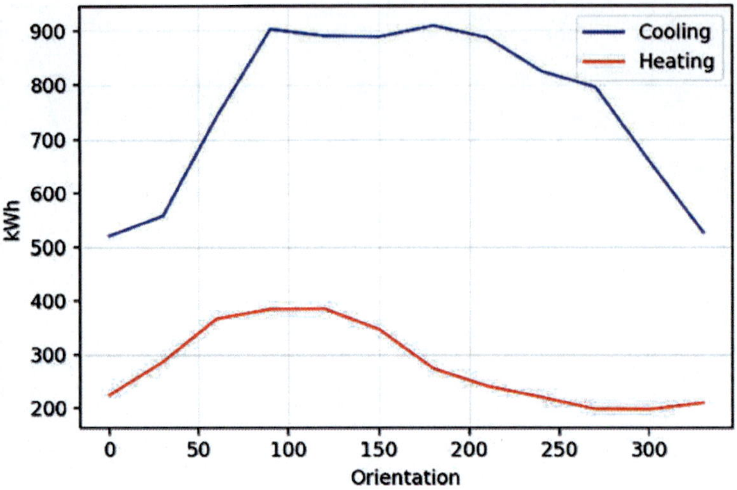

Fig. 92.7 Efficient user HVAC requirements

Solar Protections

When using blinds instead of curtains as solar protections for the kitchen and the living room, an important reduction of energy demand was estimated. For cooling demand, the reduction is thanks to the lower heat gains through the windows. Then, the most notorious reduction is for the orientation 90°, for which the front facade is oriented to the east. For the heating demand, the reduction is due to the lower transmittance of the windows when the blinds are down.

With this reduction in the energy demand, again considering the local costs of electricity, the inclusion of blinds can produce savings of 18.5 US$ per year. The cost of including this in the house design is estimated at 300 US$, meaning a repayment period of more than 20 years.

Roof with Less Transmittance

Considering a roof with less transmittance than the original (0.33 W/m^2K against 0.70 W/m^2K), a saving of around 60% is obtained for heating demand for an entire year for all orientations and savings of 30% for cooling demand. That means a saving of 25.8 US$ per year.

The proposed roof has a lower cost than the original. Therefore, the inversion for this measure is considered negative, estimated at −90US$.

92.4 Conclusion

The thermal behavior of a social interest household in Uruguay (MEVIR) was studied by means of numerical simulations using EnergyPlus. To define and calibrate the model, experimental measures were used. Moreover, interviews with families living in this type of house were performed to determine the schedules and behavior of the occupants in order to define this in the model. The results of the simulations for the base case show that the MEVIR archetype has a good thermal design because of the low thermal transmittance of its opaque envelope. Principally for winter, a low heating demand is estimated.

Thereafter, three energy efficiency measures were analyzed. All of them show a clear reduction in energy demand. The reduction of the roof transmittance is the most attractive because it has no cost and reduces energy demand. Actually, the proposed roof is cheaper than the original. Moreover, a training program for occupants to take advantage of solar protections and natural ventilation could have an important effect on reducing energy demand with a moderate cost. Finally, including blinds in the living room and kitchen windows also reduces energy requirements but with a high repayment period.

Acknowledgments This research was supported by "Administración Nacional de Usinas y Transmisiones (UTE)—Uruguay." The authors also thank the "Dirección Nacional de Energía (DNE)" for assistance with the Blower door measurements and MEVIR for facilitating information regarding the house archetype.

References

Alonso-Suarez R, Bidegain M, Abal G, and Modernell P. 2016. Año Meteorológico Típico para Aplicaciones de Energía Solar (AMTUes): series horarias típicas para 5 sitios del Uruguay.

Crawley D., Lawrie L., Pedersen C. and Winkelmann F., 2000. EnergyPlus: Energy Simulation Program. ASHRAE Journal 42, pages 49–56.

Chapter 93
Analysis of Isentropic Efficiency in the Utilization of Alternative Low GWP Refrigerants in a Hermetic Reciprocating Compressor

Ümit İşkan, Mahmut Cüneyt Kahraman, and Mehmet Direk

Nomenclature

HFC	Hydrofluorocarbon
GWP	General Warming Potential
Te	Evaporator Temperature (°C)
Tc	Condensing Temperature (°C)
W	Work of per unit refrigerant (kJ/kg)
Comp	Compressor

93.1 Introduction

HFC-R134a is one of the most preferred refrigerants due to its physical and chemical properties. However, the GWP value of R134a is 1300 (Lemmon et al. 2013); therefore, restrictions and gradual bans on its use are imposed. In this context, the research for alternative refrigerants to R134a continues. Several refrigerants such as R1234yf, R1234ze(E), R516a, ND, and R456a (Heredia-Aricapa et al. 2020; Direk and Soylu 2018; Işkan et al. 2021a, b) come to the table as an alternative to R134a.

Conventional mechanical vapor compression refrigeration cycles consist of a condenser, evaporator, expansion valve, and compressor. Energy consumption

Ü. İşkan · M. C. Kahraman (✉) · M. Direk
Faculty of Engineering, Department of Energy Systems Engineering, Yalova University, Yalova, Türkiye
e-mail: umit.iskan@yalova.edu.tr; mcuneyt.kahraman@yalova.edu.tr; mehmet.direk@yalova.edu.tr

M. Z. Sogut et al. (eds.), *Proceedings of the 2022 International Symposium on Energy Management and Sustainability*, Springer Proceedings in Energy,
https://doi.org/10.1007/978-3-031-30171-1_93

depending on the operating conditions of the compressor significantly affects system efficiency values. The efficiency values of the compressor are mostly considered constant in previous studies. In the literature, theoretical and experimental studies have been carried out to determine the performance of the compressor. In one of these studies, Tian et al. (2004) have developed a correlation that determines the volumetric efficiency depending on the superheat temperatures and compression ratios of the compressor. It was determined that the mathematical models, which they developed, had high similarity to the results obtained from the experimental work. İşkan et al. (2021a), in their study, have determined that the volumetric efficiency values obtained by the correlation, which has been developed by Tian et al. (2004), has showed 10% similarity with the yield values obtained from the experiments. When the literature was carefully analyzed, it was seen that the effect of the compressor on the performance of the vapor compression refrigeration system was examined with several theoretical and experimental studies (Alkan and Hoşöz 2010a, b).

İşkan et al. (2021b) analyzed the volumetric efficiency of the compressor in the utilization of low GWP refrigerants for air conditioners of automobiles. Efficiency analysis was based on compression ratios and speed change. As a result, they have determined that the volumetric efficiency values of R134a, R1234yf, and R1234ze (E) varied between 0.5 and 0.75. İşkan et al. (2021a) in another paper compared the performances of R134a, R445a, R515a, R456a, ND, and R516a refrigerants with each other when the compressor speed and volumetric efficiency values were constant. As a result, when the volumetric efficiency value was constant 0.7, the highest COP value was obtained from the system, which was operated with R445a.

The isentropic efficiency value of the compressor is another parameter affecting the performance of the compressor in mechanical vapor compression refrigeration cycles. In this study, the isentropic efficiency of the compressor was calculated theoretically. Isentropic efficiency is analyzed comparatively for R134a and R1234yf and R1234ze(E) considered as alternatives. It is aimed to find the effect of isentropic efficiency and deviations from network while calculating the compressor network at constant evaporation temperatures and different condensing temperatures. Compressor isentropic efficiency was calculated theoretically using the technical parameters obtained from the manufacturer according to the compressor model. The isentropic efficiency of the hermetic compressor was determined when alternative refrigerants were utilized without making any changes in the cooling system and compared with the value found with R134a (Table 93.1).

93.2 Material and Methods

The evaporation temperature in the evaporator for the applied refrigerants is introduced as 5 °C, and the superheat value is introduced as 10 °C while calculating the isentropic efficiency. Compressor work and flow rate are calculated by applying the evaporation temperature (T_e) and condensing temperature (T_c) in the polynomial

Table 93.1 Properties of refrigerants. (Lemmon et al. 2013)

Refrigerant	R134a	R1234ze(E)	R1234yf
Boiling point (K)	247.1	254.18	243.7
Critical pressure (kPa)	4059.3	3634.9	3382.2
Critical temperature (K)	374.2	382.5	367.85
Liquid density (kg.m^{-3})	1206.7	1163.1	1091.9
Steam density (kg.m^{-3})	32.35	26.32	37.925
Latent heat value (kJ.Kg^{-1})	216.9	166.92	180.25
GWP	1300	4	6
ASHRAE security class	A1	A2L	A2L

Table 93.2 Compressor power and flow rates

C	P (kW)	$\dot{m}$ (kg/s)
0	263.673791384226	0.0249577536540033
1	−23.462482917541	0.00102319771980924
2	26.4312560156009	−0.000113176114630227
3	−0.735462122173667	0.0000145397723806345
4	1.07779116746245	−0.000001740744771439
5	−0.299204398954573	−0.0000010971554743365
6	−0.045827013942256	7.2271289544845E-08
7	0.0092312066324335	−1.89962944650491E-08
8	−0.0051991491642368	−3.51577530392532E-08
9	0.00166030078844383	−3.51577530392532E-08

function given in Eq. 93.1. Finally, the isentropic efficiency for R134a was calculated by applying the steps of the basic flow diagram. Since the compressor is designed for R134a, first, the thermos-physical properties of the refrigerants are emulated to R134a, and then the alternative flow diagram is applied for the utilization of other refrigerants (Table 93.2).

$$Y = C_0 + C_1T_e + C_2T_c + C_3T_e^2 + C_4T_eT_c + C_5T_c^2 + C_6T_e^3 + C_7T_cT_e^2 + C_8T_eT_c^2 + C_9T_c^3 \quad (93.1)$$

Steps of isentropic efficiency calculation (R134a):

- Select the evaporation temperature (Te) for R134a. Evaporation pressure is calculated.
- Compressor input values are found by adding superheating value Te at equal evaporation pressure.
- Enthalpy and entropy values at compressor input are calculated.
- Enthalpy value of the isentropic compressor output is found at the desired compressor output pressure.
- Tc at the compressor outlet pressure is calculated.
- Te and Tc are applied in the equation.

- Wcomp and flow rate are calculated.
- The energy consumed by the compressor per unit refrigerant mass is found by using the Wcomp and flow rate value.
- An initial isentropic efficiency value of 0.5 is estimated.
- At the estimated isentropic efficiency value, the work consumed on the compressor per unit mass of refrigerant is found.
- The values found in the eighth and tenth steps are proportional to each other and the isentropic efficiency value is found.
- Find the difference between the isentropic efficiency value found in Step 11 and the estimated isentropic efficiency value.
- If the difference is greater than 0.01, the estimation is increased by 0.005 and the steps are repeated from Step 9.
- If the difference calculated in Step 12 is less than 0.01, the isentropic efficiency value found in Step 11 is accepted as the isentropic efficiency value of the compressor.

Calculation steps for alternative refrigerants:

- Compressor input values of the refrigerant to be desired to use in the compressor are calculated.
- Density and enthalpy values of the refrigerant at the compressor inlet are calculated.
- The same density and enthalpy values are calculated for R134a, and other physical properties are found.
- The temperature value of R134a corresponding to the compressor inlet pressure is calculated and Te is found.
- Compressor output values of the alternative refrigerant at the compressor input value are calculated isentropically.
- Density and enthalpy for R134a at compressor output values are applied and other physical properties are found.
- The temperature value of R134a corresponding to the compressor outlet pressure is calculated and Tc is found.
- The basic flow diagram steps are applied from Step 6 to calculate the isentropic efficiency of alternative refrigerants.

93.3 Results and Discussion

The condensing temperatures were applied between 30 °C and 70 °C by increasing 4 °C during the experiments. The alteration of the isentropic efficiency of the hermetic reciprocating compressor depending on the compression ratio and condensing temperature is given in Fig. 93.1. When Fig. 93.1 is analyzed, it is seen that the isentropic efficiency increases up to the 3.5 value of the compression ratio, but after this value, it starts to decrease.

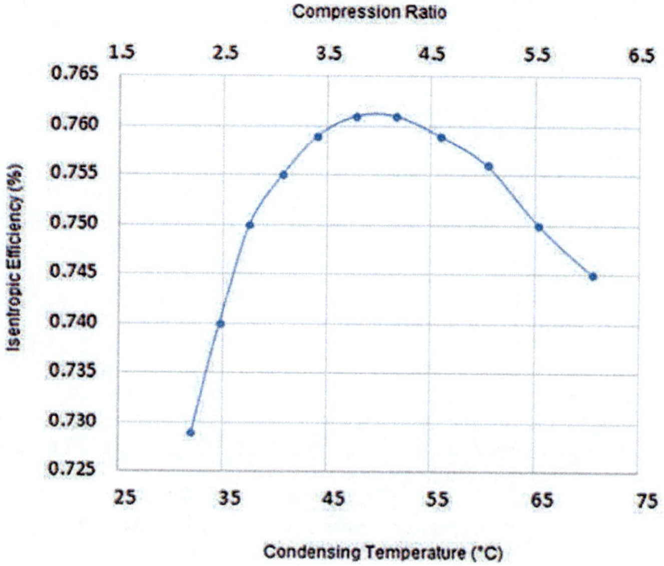

Fig. 93.1 Alteration of isentropic efficiency with compression ratio condensing temperature

Congruently, it is seen that the isentropic efficiency depending on the condensing temperature decreases after the condensing temperature reaches 50 °C.

In Fig. 93.2, the alteration of the power supplied to per unit refrigerant in the compressor depending on the condensing temperature for the isentropic and real case is given. It is seen that the difference between the isentropic power and the actual power value given to the refrigerant in the compressor per unit refrigerant rises with the increase in the condensing temperature.

In Fig. 93.3 when R1234yf and R1234ze(E) as low GWP alternative refrigerants are used instead of R134a, the values obtained from the mathematical model to find the isentropic efficiency of the selected semi-hermetic compressor are given comparatively. Isentropic efficiency values were compared depending on both the condensing temperature and compression ratios. In the case of utilization of R1234yf and R1234ze(E) refrigerants, it is seen that the isentropic efficiency value alters similarly but is approximately 7% lower than R134a.

93.4 Conclusion

In this study, the isentropic efficiency of the compressor was calculated theoretically and it was evaluated comparatively with utilization to the use cases of different refrigerants. In the case of utilization of R1234yf and R1234ze(E) refrigerants as an alternative to R134a in the refrigeration cycle, performance evaluation has been made depending on compressor compression ratios and condensing temperatures.

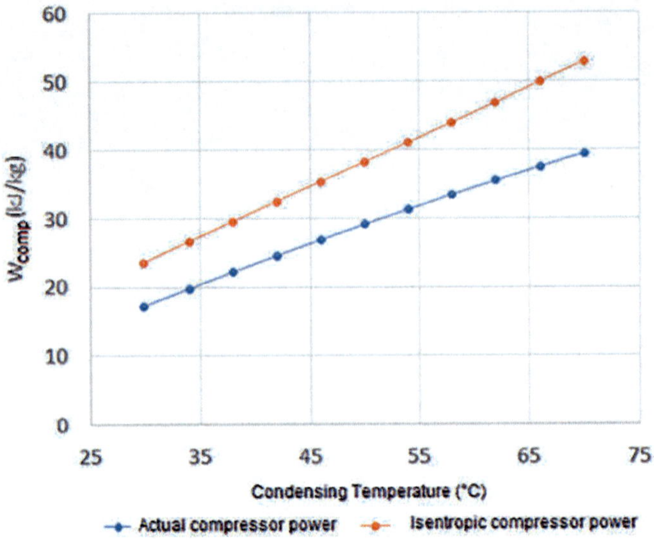

Fig. 93.2 Alteration of the power (Wcomp) given to the refrigerant in the compressor per unit refrigerant with condensing temperature

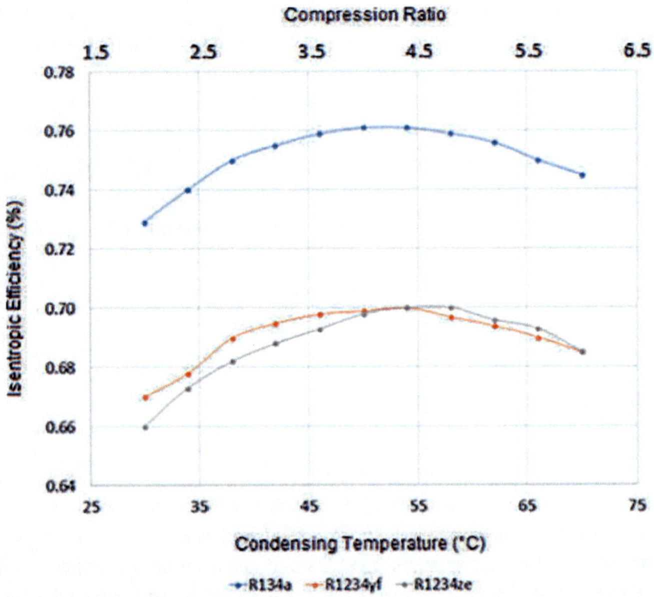

Fig. 93.3 Alteration of isentropic efficiency value with compression ratio and condensing temperature for R134a, R1234yf, and R1234ze(E) refrigerants

It is seen that the isentropic efficiency increased up to 3.5 compression ratio and started to decrease after this value from the analysis of the study. It was concluded that the isentropic efficiency dropped at condensing temperatures above 50 °C. When R1234yf and R1234ze(E) refrigerants are used instead of R134a in the system, it has been observed that the isentropic efficiency curves depending on the condensing temperature and compressor compression ratios are similar. It has been determined that the isentropic efficiency value of alternative refrigerants is approximately 7% lower than R134a. It was concluded that the selected refrigerants as an alternative to R134a could be used in terms of isentropic efficiency evaluation for the selected compressor.

References

Alkan, A., Hosoz, M., (2010a). Experimental performance of an automobile air conditioning system using a variable capacity compressor for two different types of expansion device. International Journal of Vehicle Design. 52: 160–176. https://doi.org/10.1504/IJVD.2010.029642

Alkan, A., Hosoz, M., (2010b). Comparative performance of an automotive air conditioning system using fixed and variable capacity compressors. International Journal of Refrigeration, 33: 487–495. https://doi.org/10.1016/j.ijrefrig.2009.12.018

Direk, M., Soylu, E. (2018). "The Effect of internal heat exchanger using R1234ze(E) as an alternative refrigerant in a mobile air-conditioning system", Strojniški vestnik – Journal of Mechanical Engineering, vol. 64, 2, pp. 114–120. https://10.5545/sv-jme.2017.5093

Heredia-Aricapa, Y., Belman-Flores, J. M., Mota-Babiloni, A., Serrano-Arellano, J., & García-Pabón, J. J. (2020). Overview of low GWP mixtures for the replacement of HFC refrigerants: R134a, R404A and R410A. International Journal of Refrigeration, 111, 113–123. https://doi.org/10.1016/j.ijrefrig.2019.11.012

Işkan, Ü., Direk, M., Koyunoğlu, C., Yüksel, F., (2021a). Volumetric efficiency evaluation of thu use of refrigerants R445a, R456a, ND and R516a with low GWP instead of R134a. Iğdır Üniversitesi Fen Bilimleri Enstitüsü Dergisi Vol 11 (1): 500–511. https://doi.org/10.21597/jist.734720

Işkan, Ü., Direk, M., Yüksel, F. Soylu E., (2021b) Otomobil iklimlendirme sistemlerinde Farkli Soğutucu Akişkan Kullanilmasinda kompresörün volumetrik verim analizi," Int. J. Pure Appl. Sci. 7(1), 41–50. https:// doi.org/10.29132/ijpas.881952

Lemmon, E.W., Huber, M.L., McLinden, M.O. (2013). NIST Standard Reference Database 23, NIST Reference Fluid Thermodynamic and Transport Properties. REFPROP version 10.0.

Tian, C., Dou, C., Yang, X., Li, X., (2004). A mathematical model of variable displacement wobbleb plate compressor for automotive air conditioning system. Applied Thermal Engineering, 24: 2467–2486. https://doi.org/10.1016/j.applthermaleng.2004.04.011

Chapter 94
Assessment and Prediction of Thermodynamic Metrics of Small Turbojet Engine at Different Working Points

Hakan Aygun and Mohammad Rauf Sheikhi

Nomenclature

AC	Air compressor
EcoEF	Ecological effect factor
ExD	Exergy destruction (MW)
EDF	Exergy destruction factor
ExEFF	Exergy efficiency
Ex	Exergy
FAR	Fuel-to air ratio
GT	Gas turbine
LHV	Lower heat value (kJ/kg)
$\dot{m}$	Mass flow rate (kg/s)
P	Pressure (kPa)
R	Specific gas constant (kJ/kg.K)
RPM	Revolutions per minute
SFC	Specific fuel consumption (kg/kNs)
SIP	Specific irreversibility production
T	Temperature (K)
TE	Turbojet engine
τ	Thrust (kN)

H. Aygun (✉)
Department of Aircraft Air-Frame and Power-Plant, Firat University, Elazığ, Turkey
e-mail: haygun@firat.edu.tr

M. R. Sheikhi
Faculty of Aeronautics and Astronautics, Eskisehir Technical University, Eskisehir, Turkey
e-mail: mohammadraufsheikhi@ogr.eskisehir.edu.tr

M. Z. Sogut et al. (eds.), *Proceedings of the 2022 International Symposium on Energy Management and Sustainability*, Springer Proceedings in Energy,
https://doi.org/10.1007/978-3-031-30171-1_94

94.1 Introduction

The aviation society has put environmental concerns related to aircraft engines on its agenda since the usage of gas turbine engines in the aviation sector has expanded day by day due to rising urgent demands such as military and civilian needs (Turgut et al. 2015). Meeting this energy demand could require progress in energy systems, and this case indirectly abates dependence on hydrocarbon fuels (Musharavati et al. 2021). Also, the aims of carbon dioxide emission reduction were supported by the Kyoto protocol in 1997. Since environmental issues could be related to system performance, both topics are of high significance for sustainable aviation (Balli and Caliskan 2021). In this framework, the investigations on the thermodynamic performance of gas turbine engines have drawn the attention of the related people.

This study differs from antecedent studies in that thermodynamics analysis is the first time implemented to this engine at 14 different points. Moreover, to estimate exergetic indicators regarding the engine, modeling is established using least square methods. The major objective of this study is to assess RPM effects using several well-known thermodynamics metrics.

94.2 System Description

The kind of gas turbine engine used in this study is a turbojet engine producing a thrust of 5.27 kN at maximum take-off power. Its specifications are very similar to those of TRI-60 manufactured by Microturbo Inc. Moreover, the specific fuel consumption of this engine is measured as 24.77 g/kNs whereas air and fuel mass flow passed in it are found at 8.10 kg/s and 0.13 kg/s, respectively (Fig. 94.1).

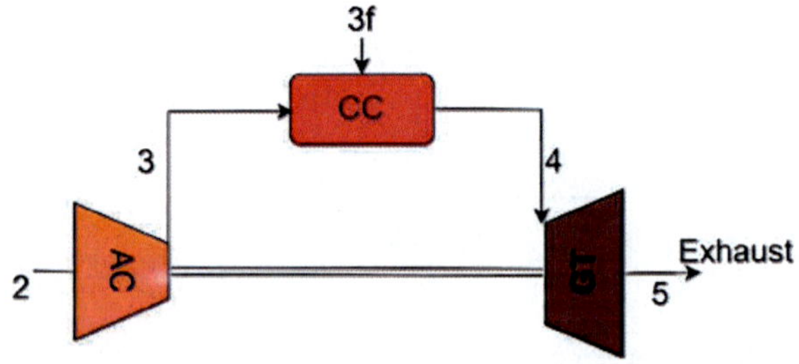

Fig. 94.1 Cross section of a conceptual turbojet engine

94.3 Methodology and Background

Having two well-known laws, thermodynamics analysis has been implemented in many thermal systems so far. The first law means variation of energy form, as well as its total amount, is constant, whereas the second law provides insight into irreversibility that occurred in the process (Şöhret et al. 2016). Moreover, the thermal process performs a definite direction where the quality of energy diminishes. The term exergy means the maximum useful work obtained from a system when it comes into equilibrium with its environment. Several exergetic formulas pertinent to the engine are presented in Table 94.1 (Aydin et al. 2015; Aygun 2022) Further information about these indicators can be found elsewhere.

94.4 Results and Discussion

This study incorporates the findings related to exergy analyses performed at 14 different revolutions per minute (RPM). These outcomes also address how the RPM affects exergetic indicators for the turbojet engine and its components. As can be seen, generally, exergy destruction tends to increase with engine RPM. One of the main factors leading to relatively high irreversibility might be attributed to being higher mass flow passed through components. At first glance, it could be mistaken to conclude that the engine becomes more inefficient at higher powers due to increased exergy destruction. Namely, an important reality is how much exergy is destructed per unit input exergy. In this context, ExD of the air compressor varies from 0.065 MW to 0.188 MW whilst that of the combustor changes from 0.74 MW to 1.76 MW. Moreover, it is observed that ExD increases from 0.029 MW to 0.213 MW since RPM is close to the maximum value (Fig. 94.2).

As can be observed, the changing RPM leads to having a different pattern of ExEFF for the components. There are directly proportional between ExEFF and RPM for AC and CC units, whereas that of GT could be negatively correlated with RPM. In this sense, the exergy efficiency of air compressors ranges from 83.14% to 90.22%, whereas that of gas turbines diminishes from 93.95% to 90.94%. Furthermore, ExEFF of the combustor increases from 60.4% to 77.1% because the combustion process improves at higher RPMs (Fig. 94.3).

Any attempt to abate environmental pollution also helps to attain zero-emission goals. Managing to measure exergetic metrics at operation points more than one point facilitated the investigation of the relationship between these metrics and power settings. These figures could help in deciding which points the engine should operate for the highest sustainability. In this framework, ExDF of the whole engine decreases from 53.92% to 36.31%. It means that at the lowest point, more than half of fuel exergy is subjected to destruction whereas this ratio experiences decrement to an important degree as the RPM increases. Similarly, the SIP of the engine diminishes from 0.83 MW/kN to 0.41 MW/kN. It means that the engine produces lower

Table 94.1 Several exergetic formulations

Exergy rate	$\dot{E}x = \dot{m}\left[c_{p(T)}\left[T - T_0 - T_0 \ln\left(\frac{T}{T_0}\right)\right] + RT_0 \ln\left(\frac{P}{P_0}\right)\right]$	(1)
Fuel exergy	$\dot{E}x_F = \dot{m}_F \dot{Q}_{\text{LHV}} \xi_F$	(2)
ExD	$\sum \dot{E}x_{\text{in}} - \sum \dot{E}x_{\text{out}} = \sum \dot{E}x_D$	(3)
ExDR	$X_k = \frac{\dot{E}x_{\text{dest},k}}{\dot{E}x_{\text{dest, tot}}}$	(4)
Exergy efficiency	$\eta_{ex} = \frac{\dot{E}x_{\text{out}}}{\dot{E}x_{\text{in}}}$	(5)
ExDF	$X_k = \frac{\dot{E}x_{\text{dest, tot}}}{\dot{E}x_F}$	(6)
SIP	$SIP_{\text{engine}} = \frac{\sum \dot{E}x_D}{\tau}$	(7)

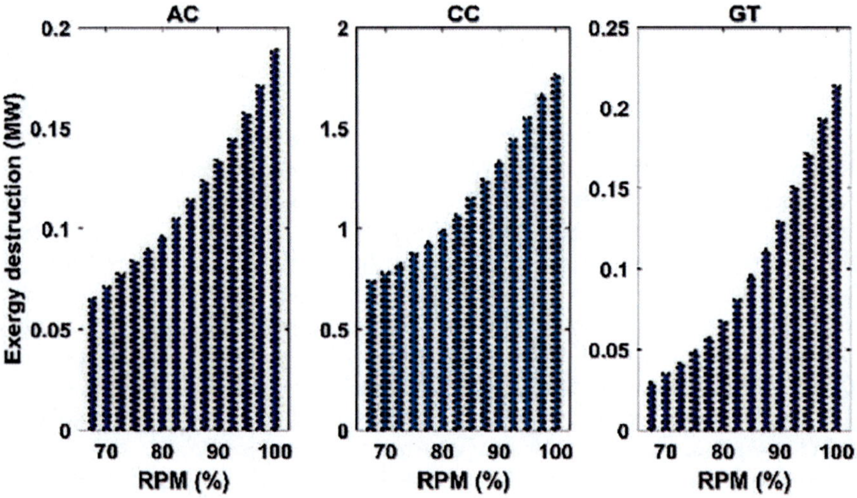

Fig. 94.2 Exergy destruction of three components versus 14 RPMs

irreversibility per unit thrust due to elevated RPM. Moreover, ExEFF of the turbojet increases from 1.97% to 14.13% as the power settings rise (Fig. 94.4).

Regarding exergetic indicators, it has been detected that RPM negatively correlates with exergy destruction factor and specific irreversibility production, whereas it positively correlates with exergy efficiency. An interesting phenomenon is observed, in which the model accuracy of ExDF and SIP are obtained higher with quadratic functions whereas that of ExEFF is achieved to high correctness with first-order function (Fig. 94.5 and Table 94.2).

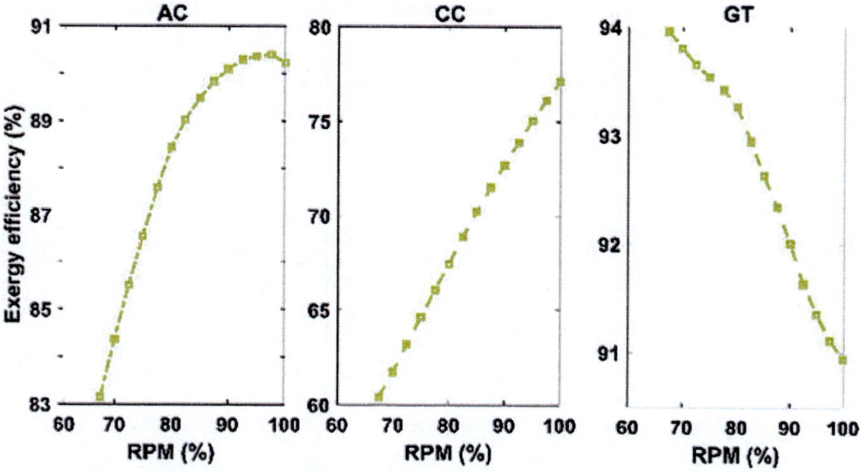

Fig. 94.3 Exergy efficiency of three components versus 14 RPMs

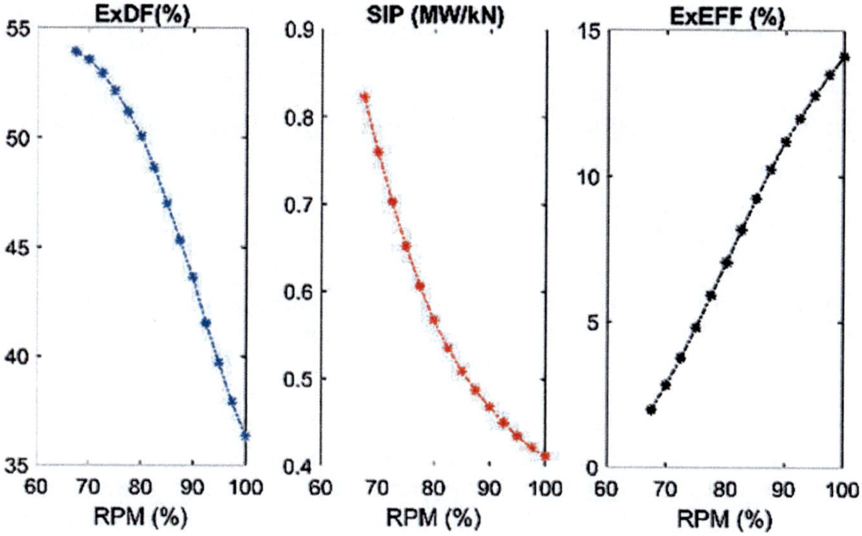

Fig. 94.4 ExDF, SIP, and ExEFF of the whole turbojet engine at different RPMs

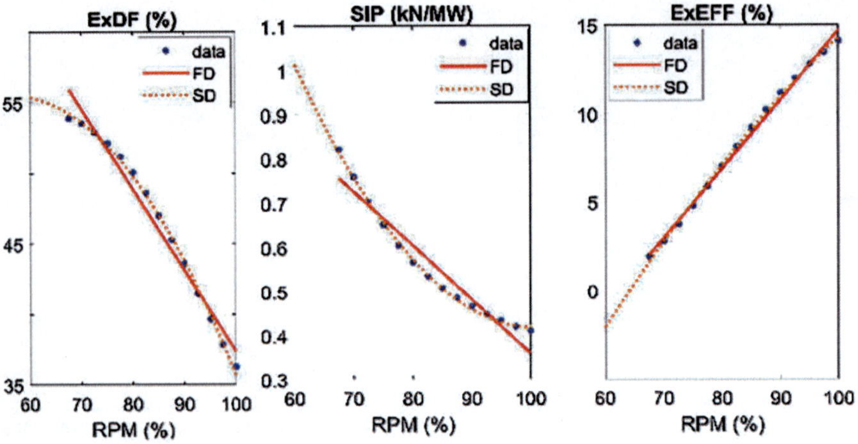

Fig. 94.5 Models of exergetic indicators at first and second degrees

Table 94.2 Coefficient of determination (R2) and the errors regarding several exergetic functions

	First-degree			Second-degree		
	ExDF	SIP	ExEFF	ExDF	SIP	ExEFF
SSE	0.014	0.0152	1.1 × 10–4	1.11 × 10–4	3.6 × 10–4	3.89 × 10–5
RMSE	0.0107	0.0356	0.0031	0.0032	0.0057	0.0019
R2	0.9711	0.933	0.9944	0.9972	0.9984	0.9982

94.5 Conclusion

Turbojet engines have been more popular as time progressed. In this framework, assessments of these systems in terms of thermodynamic and environmental performance are of high importance to improve system operation. This study emphasizes RPM effects highlighting important relationships with exergetic metrics. Although there have been many papers devoted to comprehending and quantifying exergetic behaviors of the engine, modeling of these parameters is the main gap in the aero-derivative field. It is thought that this study can shine a light on establishing exergetic functions that help in detecting the engine performance to an important extent. The variation of exergy efficiency of the engine with RPM appears to be almost linear, yet the analyses indicate that quadratic functions result in better conclusions for specific irreversibility production and exergy destruction factor models.

References

Aydin, H., Turan, O., Karakoc, T.H., Midilli, A., (2015). Exergetic sustainability indicators as a tool in commercial aircraft: a case study for a turbofan engine. International journal of green energy 12, 28–40.

Aygun, H., (2022). Thermodynamic, environmental and sustainability calculations of a conceptual turboshaft engine under several power settings. Energy 245, 123251.

Balli, O., Caliskan, H., (2021). On-design and off-design operation performance assessmentsof an aero turboprop engine used on unmanned aerial vehicles (UAVs) in terms of aviation, thermodynamic, environmental and sustainability perspectives. Energy Conversion and Management 243, 114403.

Musharavati, F., Khanmohammadi, S., Pakseresht, A., Khanmohammadi, S., (2021). Waste heat recovery in an intercooled gas turbine system: Exergo-economic analysis, triple objective optimization, and optimum state selection. Journal of Cleaner Production 279, 123428.

Şöhret, Y., Ekici, S., Altuntaş, Ö., Hepbasli, A., Karakoç, T.H., (2016). Exergy as a useful tool for the performance assessment of aircraft gas turbine engines: a key review. Progress in Aerospace Sciences 83, 57–69.

Turgut, E.T., Cavcar, M., Yay, O.D., Ucarsu, M., Yilmaz, E., Usanmaz, O., Armutlu, K., Dogeroglu, T., Miake-Lye, R.C., (2015). Analysis of test-cell emission measurements of newly overhauled turbofan engines. Journal of Propulsion and Power 31, 559–572.

Chapter 95
Techno-Economic Comparison of Wet Cooling Towers and Air Cooled Condensers for Combined Cycle Power Plant in Ankara Region

Görkem Zengin and Şaban Pusat

Nomenclature

M	Mass flow rate, kg/s
x	Humidity ratio, kg/kg
Q	Heat transfer capacity, kW
U	Overall heat transfer coefficient, W/m^{2}K
A	Heat transfer surface, m^2
LMTD	Log mean temperature diff., K
K	Mass transfer coefficient, kg/m^{2}s
a	Surface area density, m^2/m^3
V	Volume, m^3
L	Mass flow rate of water, kg/s
T	Temperature, °C
cp	Specific heat capacity, kJ/kgK
i	Enthalpy, kJ/kg
W	Electricity production output, kWe

Greek Letters

η	Efficiency

G. Zengin (✉) · Ş. Pusat
Yildiz Technical University, Istanbul, Turkey
e-mail: spusat@yildiz.edu.tr

M. Z. Sogut et al. (eds.), *Proceedings of the 2022 International Symposium on Energy Management and Sustainability*, Springer Proceedings in Energy,
https://doi.org/10.1007/978-3-031-30171-1_95

Subscripts

evap	Evaporation
blo	Blowdown
a	Air
o	Outlet
i	Inlet
tot	Total
wi	Water inlet
wo	Water outlet
w	Water
iso	Isentropic
tra	Transmission
HP	High pressure
LP	Low pressure
cond	Condensation

95.1 Introduction

Energy need is increasing day by day at the present time. Almost each new device needs energy to operate, so energy consumption is constantly increasing. On the other hand, new energy production methods are being invented and former methods' efficiencies are improving.

Combined cycle is the one of these energy production methods. It is based on Brayton and Rankine Cycles to produce energy. Brayton cycle side is named by Simple Cycle in this system. Electricity is produced by a gas turbine and generator couple in simple cycle power plants. Rankine Cycle is added to standard Brayton Cycle to improve efficiency of the overall plant (Tiwari et al. 2010). Simple Cycle and Combined Cycle are depicted in Figs. 95.1 and 95.2, respectively.

In this study, combined cycle power plants among energy production methods were examined and two major cooling methods' advantages and disadvantages were compared in terms of water consumption and total electricity production. There are many different cooling equipment and methods used in power plants; wet cooling towers and air cooled condensers are the most common ones.

95.1.1 Wet Cooling Towers

Wet cooling towers are one of the most economic heat rejection units which operate according to direct evaporative cooling principle. Hot circulation water sprays above fill section, and cold and dry air flows from air inlet louvers to fan stack. There are

Fig. 95.1 Simple cycle (Brayton). (Kachwaha 2014)

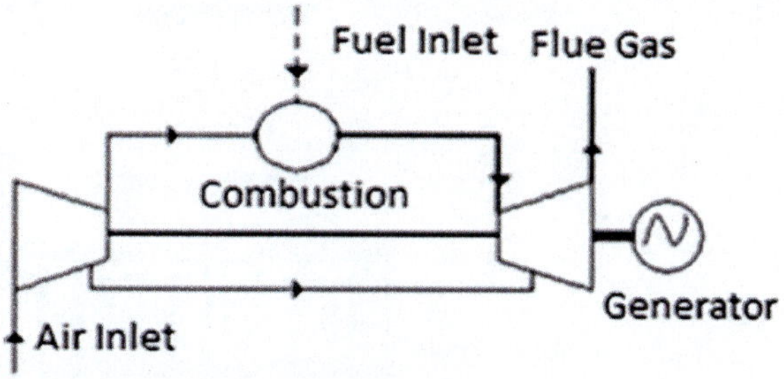

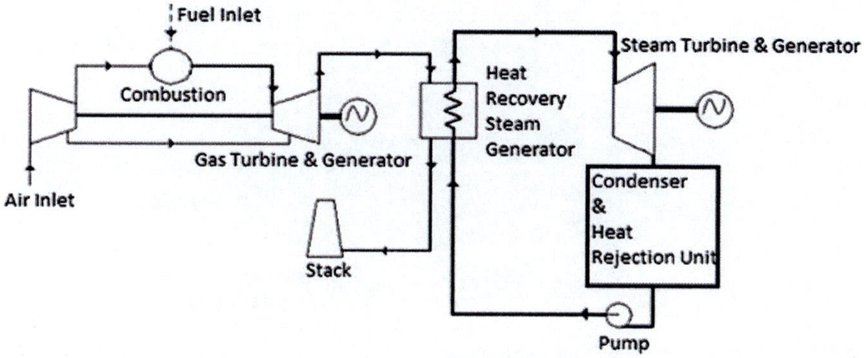

Fig. 95.2 Combined cycle (Brayton + Rankine). (Kachwaha 2014)

many different cooling tower types in the market, such as natural draft, mechanical draft, cross flow, counter flow, etc. (Kröger 2004). Mechanical draft counter flow cooling tower type is examined in this study, and its working principle is given in Fig. 95.3.

Hot circulation water and cold atmosphere air are in direct contact in the wet cooling tower. Heat and mass transfer occurs between water and air. Enthalpy of air increases, and enthalpy and temperature of circulation water decrease at the end of the operation. During this heat and mass transfer operation, dry and cold air absorb a little portion of the water from circulation water. This is the main evaporation loss which results in water consumption in the cooling tower.

There are three main reasons for water consumptions in wet cooling towers, and these are evaporation loss, blowdown loss, and drift loss.

In the wet cooling towers, generally, the amount of evaporation loss is about 1% of the total circulation water flow (Kröger 2004). This loss mainly depends on air's psychometric and water cycle's thermophysical parameters. Its equation is given in Eq. 95.1:

$$m_{\text{evap}} = m_a (x_{\text{outlet}} - x_{\text{inlet}}) \quad (95.1)$$

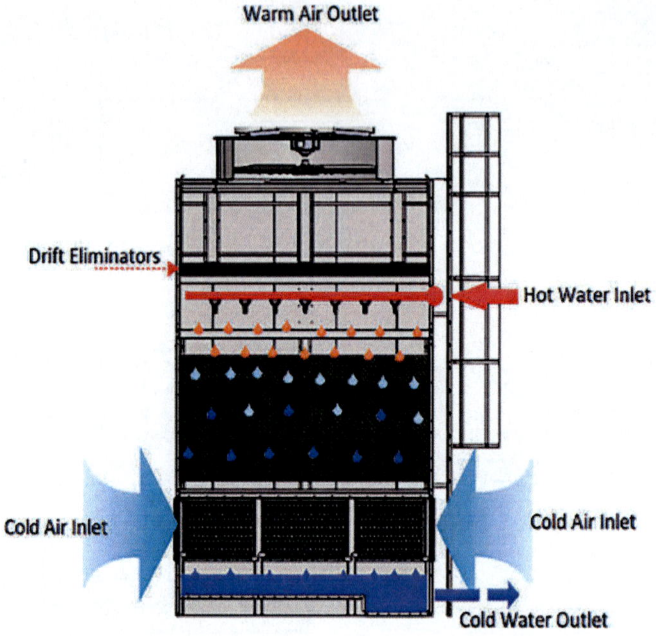

Fig. 95.3 Wet cooling tower working principle

Blow down water consumption is related to circulation water chemistry. Water is continuously evaporating in the cooling tower, but other elements or minerals are not going out (Savari et al. 2015). Circulation water's electrical conductivity is increasing constantly, and it's one of the major corrosion determinants, whereas water electrical conductivity should be lower than 2000 μs/cm (U.S. Environmental Protection Agency (EPA) 2017). Equation for blowdown water consumption is given in Eq. 95.2:

$$m_{\text{blo}} = \frac{m_{\text{evap}}}{\text{Concentration Cycles} - 1} \tag{95.2}$$

Blowdown water consumption levels can be reduced by increasing the cycle of concentration rate, but it results extra chemicals cost to the facility. In this study, cycle of concentration (CoC) is assumed to be 2.5.

Drift losses are related to air velocity, water loading, and the drift eliminator type used in the cooling tower. Water loading is the ratio of the circulation water flow to the cross sectional area of the cooling tower. In this study, drift loss ratio is assumed as %0.001.

95.1.2 Air Cooled Condensers

Air cooled condensers are another type of heat rejection units where high pressure and high temperature steam enter into the steam turbine in the Rankine Cycle, and the turbine outlet steam contains the lower energy that produces the electricity. And that outlet steam condenses in the air cooled condensers (Bansal et al. 2017).

Exhaust steam condenses inside the tubes and never directly contacts with the atmosphere air, and this is the second big advantage compared to the wet cooling towers. Water consumption is almost zero in air cooled condenser systems (Nazar 2020).

Where the air cooled condensers are used, surface condensers and secondary circulation pumps are not required as they are used in cooling tower systems. Secondary water circuits are not required, and this is the one of the major advantages of air cooled condenser systems.

An air cooled steam condenser's working principle is shown in Fig. 95.4.

Because of the dry cooling principles, required heat transfer surface and airflow are higher than cooling towers. Consequently, electricity consumption is higher than the cooling towers. These consequences are all depending on ambient air temperatures and exhaust steam's condensation temperature while they affect logarithmic mean temperature difference (LMTD). The main heat transfer equation of the air cooled condenser is given in Eq. 95.3:

$$Q_{\text{total}} = U * A * \text{LMTD} \tag{95.3}$$

When the temperature difference of the condensation point of the exhaust steam and the ambient air increases, the required airflow reduces (Lau et al. 1987). Effects of these parameters on net electricity production and water consumption are both examined in this study in detail.

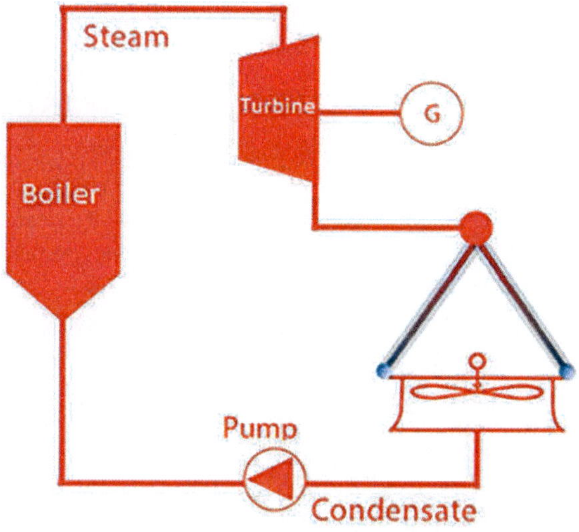

Fig. 95.4 Air cooled condenser working principle (SPG dry cooling)

95.2 Calculation Algorithms

95.2.1 Wet Cooling Towers

The Merkel Method is used in wet cooling tower performance calculations. It is the most common calculation method in the literature. Its main formulation is given in Eq. 95.4:

$$\frac{KaV}{L} \approx \frac{(T_{\mathrm{wi}} - T_{\mathrm{wo}})}{4}\left[\frac{c_{pw1}}{\Delta i_1} + \frac{c_{pw2}}{\Delta i_2} + \frac{c_{pw3}}{\Delta i_3} + \frac{c_{pw4}}{\Delta i_4}\right] \approx \frac{c_{pwm}(T_{\mathrm{wi}} - T_{\mathrm{wo}})}{4} \times \left[\frac{1}{\Delta i_1} + \frac{1}{\Delta i_2} + \frac{1}{\Delta i_3} + \frac{1}{\Delta i_4}\right] \quad (95.4)$$

This method was improved by four-point Chebyshev theorem. It is based on differential analysis of air and water enthalpy values (British Standard 4485 1988).

95.2.2 Air Cooled Condensers

Fundamentally, an air cooled condenser is a finned tube heat exchanger, and it can be evaluated by the ε-NTU method. Inside and outside, heat transfer coefficients are calculated according to the heat transfer equations (Owen 2013).

95.2.3 Steam Cycle

Steam cycle's energy balance calculations are made based on thermodynamic equations. The Power output of the steam turbine is given in Eq. 95.5 (Njoku and Diemuodeke 2020):

$$Q_{\mathrm{total}} = U * A * \mathrm{LMTD}$$

$$W_{st} = \eta_{\mathrm{iso}} * \eta_{\mathrm{tra}} * 8[mHP(iHP - iLP) + mLP(iLP - i_{\mathrm{cond}})] \quad (95.5)$$

Inlet enthalpy values are taken from water/steam thermophysical tables. The exhaust steam dryness factors and its enthalpy values are calculated by isentropic and thermodynamic equations.

95.3 Comparison Definition and Parameters

This comparison is prepared between air cooled condenser and wet cooling tower used combined cycle power plant in Ankara Region, Turkey. Only the steam cycle's performance and heat rejection unit's electricity consumption are taken into consideration, while other electricity production or consumption items are accepted as constant. Other operation and design parameters are given in Table 95.1.

Commercial parameters were obtained from local public institutions.

Ankara region's climatic conditions are used in thermo-economic feasibility analysis. Wet bulb temperature table is used for wet cooling towers, whereas dry bulb temperature table is used for air cooled condenser calculations. Related data are given in Tables 95.2 and 95.3.

The main meteorological data are graphed in hourly intervals. The data are divided into 5K differential temperature groups for simplifying the thermo-economic feasibility analysis.

95.4 Results

In this section, annual electricity production, water consumption and the net profit are examined according to the values in Tables 95.1, 95.2, and 95.3.

Water/steam cycle's energy electricity production capability and water consumption when wet cooling tower is used are given in Tables 95.4 and 95.5, respectively.

Water/steam cycle's energy electricity production capability when air cooled condenser is used is given in Table 95.6. Water consumption table is not given for air cooled condenser case study as it was given above for the cooling towers (Table 95.7).

Annual net electricity production values are calculated by the values taken from Tables 95.2 and 95.3. First, output values are calculated according to equations, and then they are multiplied by duration values.

Commercial comparison of the two systems is given in Table 95.8.

Table 95.1 Operation and design parameters

Item	Value	Unit
Power plant location	Ankara	–
Operation load	Full load	–
HP turbine inlet steam mass flow	100	kg/s
HP turbine inlet steam pressure	100	Bar(a)
HP turbine inlet steam temperature	540	°C
LP turbine inlet steam mass flow	125	kg/s
LP turbine inlet steam pressure	7,5	Bar(a)
LP turbine inlet steam temperature	250	°C

Table 95.2 Operation and design parameters (cont.)

	Value	Unit
Item		
Steam turbine isentropic efficiency (LP & HP)	90	%
Pump efficiency	90	%
Fan efficiency	65	%
Motor efficiency	90	%
Design range	10	K
Design air velocity	2.8	m/s
Surface condenser pressure drop	300	kPa
Air cooled condenser parameters		
Design dry bulb temperature	29.5	°C
Design approach temperature	20	K
Design air velocity	2.7	m/s
Commercial parameters		
Water price	1.20	Euro/m^3
Electricity price	0.15	Euro/kWh

Table 95.3 Wet bulb temperatures for Ankara Region

No.	Temperature range (°C)	Duration (h)
1	20–25	34
2	15–20	887
3	10–15	1992
4	5–10	2022
5	0–5	1717
6	−5 to 0	1594
7	−10 to −5	490
8	−15 to −10	24

Table 95.4 Dry bulb temperatures

No	Temperature range (°C)	Duration (h)
1	35–40	18
2	30–35	214
3	25–30	603
4	20–25	1076
5	15–20	1300
6	10–15	1401
7	5–10	1183
8	0–5	1613
9	−5 to 0	1035
10	−10 to −5	299
11	−15 to −10	18

Table 95.5 Wet cooling tower used in power plant's electricity production

Design wet bulb temperature (°C)	Total electricity consumption (fan + pump) (MW_e)	Steam turbine electricity production (MW_e)	Annual net electricity production (GW_h)
22.5	3.06	109.9	3.63
17.5	3.08	110.5	95.25
12.5	3.08	111.0	214.90
7.5	3.09	111.4	219.02
2.5	3.09	111.8	186.65
−2.5	3.10	112.2	173.86
−7.5	3.11	112.6	53.63
−12.5	3.12	112.9	2.63
Total			949.6

Table 95.6 Wet cooling tower used in power plant's water consumption

Design wet bulb temperature (°C)	Total water consumption (evaporation + blowdown + drift) (kg/s)	Annual water consumption (m^3)
22.5	156.27	19,126.9
17.5	145.06	463,214.9
12.5	136.94	982,047.7
7.5	127.85	930,656.2
2.5	122.75	758,729.8
−2.5	115.71	663,994.2
−7.5	100.03	176,446.1
−12.5	90.95	7857.7
Total		4,002,073.5

Table 95.7 Air cooled condenser used in power plant's electricity production

Design dry bulb temperature (°C)	Total electricity consumption (fan) (MW_e)	Steam turbine electricity production (MW_e)	Annual net electricity production (GW_h)
37.5	2.96	107.1	1.9
32.5	2.98	108.2	22.6
27.5	2.99	109.2	64.2
22.5	3.00	110.3	115.8
17.5	3.01	111.4	141.3
12.5	3.01	112.4	153.7
7.5	3.01	113.5	131.1
2.5	3.00	115.0	181.1
−2.5	3.00	115.8	117.0
−7.5	1.44	115.8	34.2
−12.5	0.76	115.8	2.1
Total			964.95

Table 95.8 Commercial comparison

Equipment type	Wet cooling tower	Air cooled condenser
Annual water price	–4,8 Million €	N/A
Annual electricity price	1424 Million €	1447 Million €
Total	1376 Million €	1447 Million €

95.5 Conclusions

According to the result of this study, an air cooled condenser has major advantage against a wet cooling tower in electricity production and water consumption. Temperature decrease is increasing the electricity production output but, the increase rate when the wet cooling tower is used is less than the other. Dry cooling principle is more stable than wet cooling. Wet cooling principle is based on psychometric values of the air, and the air's water containment capability is reducing by the wet bulb temperature decrease.

An air cooled condenser's main disadvantages are required installation area and its capital cost. The capital cost of the air cooled condenser is almost seven times higher than a cooling tower for the same case (California Energy Commission 2002).

References

A.K.Tiwari, Islam Mohd and M.N.Khan Thermodynamic Analysis of Combined Cycle Power Plant [Journal]. – [s.l.]: International Journal of Engineering Science and Technology, 2010. – 4: Vol. 2.

Bansal Mehul, Halamchi Behnam and Sopanen Jussi Theoretical Investigation into Air Cooled Condenser Performance Optimization Through Parameterization for a 10 MW Power Plant [Journal] // ASME 2017 International Mechanical Engineering Congress and Exposition (IMECE2017). – 2017. – pp. 1–10.

British Standard 4485 Water Cooling Towers, Part2: Methods for Performance Testing [Journal]. – 1988.

California Energy Commission Comparison of Alternate Cooling Technologies for California Power Plants Economic, Environmental and Other Tradeoffs [Report]. – California: PIER/EPRI, 2002.

Kachwaha Surendra Singh GTA modeling of combined cycle power plant efficiency [Journal]. – [s.l.]: Ain Shams Engineering Journal, 2014.

Kröger Detlev G. Air Cooled Heat Exchangers and Cooling Towers [Book]. – Oklohama: PennWell Corporation, 2004.

Lau S. C., Annamalai K. and Shelton S. V. Optimization of Air-Cooled Condensers [Journal]. – [s.l.]: Journal of Energy Resources Technology, 1987. – 91: Vol. 109.

Nazar Roshna Water-reduction potential of air-cooled condensers in coal power plants in India and anticipated trade-offs [Journal]. – [s.l.]: Applied Water Science, 2020. – 162: Vol. 10.

Njoku Ifeanyi Henry and Diemuodeke Ogheneruona E. Techno-economic comparison of wet and dry cooling systems for combined cycle power plants in different climatic zones [Journal]. – [s.l.] : Elsevier, 2020. – 227: Vol. 113610.

Owen Michael Air-cooled condenser steam flow distribution and related dephlegmator design considerations. – Stellenbosch: Faculty of Engineering at Stellenbosch University, 2013.
Savari Farzaneh [et al.] Prevent Corrosion in Cooling Towers: Finding the Optimum Amount of Makeup Water and the Outlet Water Stream, Experimentally [Journal] // European Journal of Technology and Design. – 2015. – pp. 4–11.
SPG Dry Cooling [https://spgdrycooling.com/products/air-cooled-condensers/]. – [s.l.]: SPG Dry Cooling.
U.S. Environmental Protection Agency (EPA) Water Efficiency Management Guide – Mechanical Systems [Report]. – [s.l.]: EPA-WaterSense, 2017.

Chapter 96
Numerical Analysis of the Ejector Performance with a Hydrofluoroolefin Group Refrigerant

Okan Gök and Aytunç Erek

Nomenclature

D	Diameter, m
L	Length, m
Ma	Mach number [−]
$\dot{m}$	Mass flow rate, kg/s
P	Pressure, Pa
T	Temperature, C
w	Entrainment ratio [−]
x	Location on the x-coordinate, m
x'	The ratio of x-coordinate to D_{const} [−]
y	Location on the y-coordinate, m
y'	The ratio of y-coordinate to D_{const} [−]
NXP	Nozzle exit position, m
NXP'	The ratio of NXP to D_{const} [−]
θ	Angle, °
conv	Converging duct of the ejector
const	Constant area mixing section of ejector
throat	Throat of the primary nozzle
diff	Diffuser
in	Inlet
out	Outlet
p	Primary (motive) fluid/nozzle
s	Secondary fluid

O. Gök (✉) · A. Erek
Faculty of Engineering, Mechanical Engineering Department, Dokuz Eylül University, İzmir, Turkey
e-mail: gok.okan@ogr.deu.edu.tr; aytunc.erek@deu.edu.tr

M. Z. Sogut et al. (eds.), *Proceedings of the 2022 International Symposium on Energy Management and Sustainability*, Springer Proceedings in Energy,
https://doi.org/10.1007/978-3-031-30171-1_96

96.1 Introduction

Considering the current environmental and technological perspective, it is crucial to develop and use components that enable the efficiency and performance of any system to be sustained or improved by reducing the amount of energy consumed for useful output. In this context, minimizing the energy consumed, especially for the cooling effect, is a priority for the spaces, components and various devices where cooling systems are used. For example, the expansion valve used in conventional refrigeration cycles causes throttling losses throughout the system. Therefore, the addition of an ejector to a conventional refrigeration cycle aims to improve the performance of the refrigeration cycle.

Ejectors are classified according to the phase of the fluid mixture exiting the ejector and the primary and secondary fluids entering the ejector (Elbel and Hrnjak 2008). In this study, the performance of the ejector in Heat-Driven Ejector Refrigeration Cycle (HDERC) was analyzed by Computational Fluid Dynamics (CFD) method. The type of ejector in this cycle, when the high-energy primary fluid is in the vapor phase and the secondary fluid is in the vapor phase, is called a vapor jet ejector. The outlet phase of this ejector type is a vapor. In the vapor jet type ejector, the two-phase flow may occur, with shock waves possible in its flow area. It is advantageous to use ejectors in cooling applications in terms of being resistant to multi-phase flow conditions, low production and maintenance costs, and no moving parts (Scott et al. 2008). For this reason, many researchers have focused on the physical mechanism of supersonic ejectors in recent years (Little and Garimella 2016; Besagni 2019).

The ejector is added to a conventional cooling cycle and takes place in a cooling cycle originating from renewable energy or waste heat etc. In general, the name given to this cycle is the Heat-Driven Ejector Refrigeration Cycle (HDERC). Although low Coefficient of Performance (COP) values are obtained compared to other ejector cooling cycles, studies are being carried out as they can work with renewable or waste heat sources. The Heat-Driven Ejector Refrigeration Cycle (HDERC) is shown schematically in Fig. 96.1.

The full of the refrigeration cycle is a combination of two loops. The lower cycle (1–2–6–7–8) is similar to a conventional refrigeration cycle, and the compressor stage is replaced by the upper cycle (1–2–3–4–5). This configuration is particularly attractive for air-conditioning applications because the entire cycle is mostly driven by low-grade energy (heat input into the generator). The overheater is added to the evaporator and condenser outlet, and it is ensured that the fluid is in the gas phase at the inlet of the ejector. This is a well-performing supersonic ejector, the critical parameter for an efficient refrigerant cycle.

The working principle of the ejector is based on converting the internal and flow energy of a high-pressure primary fluid into kinetic energy and increasing the pressure of a low-pressure secondary fluid by momentum transfer. The ejector consists of basically four sections as follows: (i) primary convergent–divergent

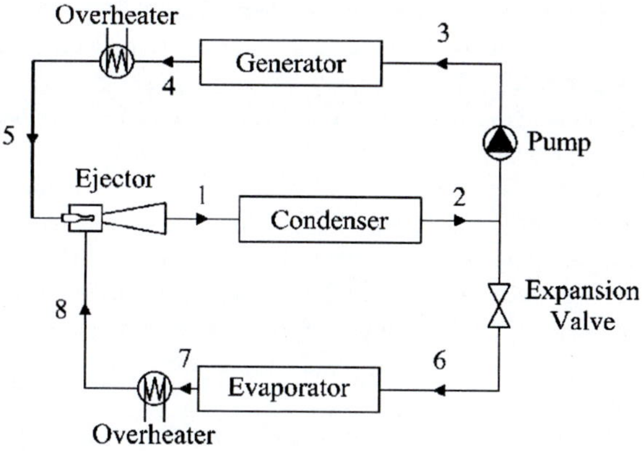

Fig. 96.1 Schematic representation of the heat-driven ejector refrigeration cycle (HDERC)

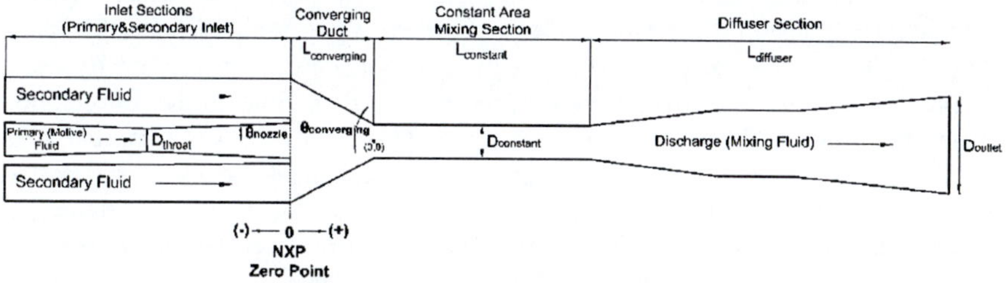

Fig. 96.2 Schematic representation of the computational domain for a typical ejector. (Metin et al. 2019)

nozzle, (ii) suction section where primary and secondary fluid flows enter, respectively, (iii) mixing section, and (iv) diffuser. The schematic view of the ejector was displayed, as shown in Fig. 96.2.

In many studies in the literature, CFC, HCFC, and HFC refrigerants have been used as the working fluid in refrigeration and air-conditioning applications, which include an ejector. These refrigerants are currently regulated due to environmental concerns about global warming (E. U. Regulation – No 517/2014). In line with the Montreal Protocol in 1987 and the Kigali Amendment in 2019, restrictions were applied on the using of CFCs, HCFCs, and HFCs. The types of these refrigerants have a direct effect in the case of leakage and release to the environment due to their high Ozone Depletion Potential (ODP) and Global Warming Potential (GWP) values, although they are stable, nontoxic, and flammable. For this reason, as an alternative to the commonly used R134a refrigerant, environmentally friendly fluids such as R1234yf with low GWP value, zero ODP, and excellent life cycle are preferred. Furthermore, R1234yf refrigerant, proposed as an alternative to R134a, has similar thermophysical properties. Therefore, when necessary, arrangements are made in the system, and it can easily be used as an alternative.

Many cooling applications, including the ejector refrigeration cycle, have been carried out using new-generation refrigerants as an alternative to HFC refrigerants. Özgür et al. (2014) carried out an exergy analysis of vapor-compressed cooling systems using R1234yf as an alternative to R1334a. As a result of this change, it is stated that while R1234yf provides a lower exergy removal rate, there is no significant difference in terms of energy efficiency. Zilio et al. (2011) obtained experimental results by using R1234yf instead of R134a for the European automotive air conditioning system. Using R1234yf, experimental and numerical studies were carried out to obtain similar cooling capacities obtained using R134a. Moreover, in general cooling application studies, new-generation environmentally friendly refrigerants instead of HFC fluids are used, even in cooling applications with ejectors.

For refrigeration systems including supersonic ejectors, the use of HFO refrigerants' drop-in replacement of R134a, etc., from HFC refrigerants has become a widespread phenomenon. Lawrence and Elbel (2014) carried out experimentally for Ejector Expansion Refrigeration Cycle using the refrigerants such as R134a and R1234yf. They obtained the Coefficient of Performance (COP) increase by using R134a and R1234yf for two different evaporator combinations in the refrigeration cycle. According to their results, the evaporator design provides COP improvement for EERC. Navarro-Esbrí et al. (2013) present an experimental analysis of R1234yf as a drop-in replacement for R134a. From their experimental results, cooling capacity and COP reductions of 6–13% were observed as a result of refrigerant replacement. However, the presence of an Internal Heat Exchanger (IHX) has been to reduce these reductions by 2–6%.

In addition to experimental studies, there are numerical studies in the literature. Fang et al. (2017) obtained numerical results by using R1234yf and R1234ze (E) from HFO refrigerants as an alternative to r134a for the supersonic ejector geometry. According to the numerical results, it was concluded that when using R1234yf, there was 9.6% and 19.8% reduction in COP and cooling power, and 4.2% and 26.6% reduction for R1234ze(E), respectively. Li et al. (2014) studied thermodynamic analysis in EERC using R1234yf as the refrigerant. According to the results, the cycle performances of the R1234yf and R134a were compared. Although it is insufficient compared to R134a in terms of performance, it is stated that R1234yf will be a good alternative.

In this study, the effects of using the new-generation environmentally friendly refrigerant R1234yf instead of R134a and the selected basic geometric parameter changes of the supersonic ejector were investigated numerically. The results obtained for the present study were based on the evaluation of the flow structures in the ejector and the global parameters for the studied ejector geometry and refrigerants. The entrainment ratio, one of the most important global performance parameters for ejector performance, is compared in terms of geometric changes depending on the primary nozzle exit position (*NXP*). The entrainment ratio is explained as the ratio of the secondary mass flow rate to the primary mass flow rate for the supersonic ejector. Also, Mach number distributions are compared on the symmetry axis of the ejector for the same cases and the secondary fluid. The strength of the motive jet is analyzed according to the selected cases, and the relationship between the secondary flow entrainment is discussed.

96.2 Numerical Method and Computational Model

The numerical solution steps described in this study were created in line with the validated numerical study presented in Metin et al. (2019). In this numerical study, as the next step study of Metin et al. (2019), the ejector used in the refrigeration systems was modeled with the use of Computational Fluid Dynamics (CFD) under single-phase flow assumption for R134a and R1234yf as the working fluid. The computational model was performed with a commercial finite volume solver, ANSYS Fluent v19.2. The computational domain was established for two-dimensional axisymmetric flow assumption neglecting three-dimensional effects of the flow features (Rusly et al. 2005; Zhu et al. 2009; Lamberts et al. 2017). The Root-Mean-Square (RMS) normalized residual value is set below 10^{-6} as the numerical convergence criteria of the governing equations. Besides, it was ensured that convergence was achieved by controlling the mass and energy conservation values at the inlet and outlet of the ejector geometry.

The schematic representation of the ejector is given in Fig. 96.2, and it was selected from the study of Del Valle et al. (2014) as the geometric dimensions. The main dimensions of the ejector geometry are presented in Table 96.1 following the reference study. It also represents the main dimensions of Case 1 in the results presented in this paper. The density, specific heat, thermal conductivity, viscosity, and molecular weight of the refrigerants used in the numerical study were determined using the REFPROP v9.1 database under the real gas approach (Lemmon et al. 2013).

In this paper, geometric parameters are presented using dimensionless parameters. The name given to the section between the primary nozzle outlet and the constant area mixing section is the primary nozzle exit position (NXP). The primary nozzle exit position (NXP) is divided by the nozzle throat diameter (D_{throat}) to obtain the NXP' value as a dimensionless parameter, as shown in Eq. (96.1). In addition to the geometric parameters, in the results of the ejector geometry positioned on the

Table 96.1 The main dimensions of the ejector. (Del Valle et al. 2014)

Geometric dimension	Value
Nozzle diverging angle (θ_p)	1.40°
Converging duct angle (θ_{conv})	29.95
Primary nozzle throat diameter (D_{throat})	2.00 mm
Converging duct length (L_{conv})	5.38 mm
Constant area mixing section diameter (D_{const})	4.80 mm
Constant area mixing section length (L_{const})	41.39 mm
Diffuser outlet diameter (D_{out})	10.00 mm
Diffuser length (L_{diff})	120.15 mm

Table 96.2 Boundary conditions of the ejector numerical model

Refrigerant Name	P_{gen} [kPa]	T_{gen} [°C]	T_{gen}[a] [°C]	P_{evap} [kPa]	T_{evap} [°C]	T_{evap}[a] [°C]	P_{ratio}[b] [–]	P_{cond} [kPa]	T_{cond} [°C]
R134a	2888.8	84.39	94.39	414.61	10	20	6.97	826.6	32.48
R1234yf	2753.9	84.39	94.39	395.25	6.8	16.8	6.97	787.9	30.21

[a]Represents +10 °C overheated cases
[b]The value given as P_{ratio} represents the ratio of the pressure at the primary inlet to the pressure at the secondary inlet of the ejector ($P_{primary}/P_{secondary}$)

x-axis and y-axis, the expressions x' and y' refer to the dimensionless parameters using D_{const}, as shown in Eq. (96.2) and Eq. (96.3), respectively.

$$NXP' = \frac{NXP}{D_{throat}} \tag{96.1}$$

$$x' = \frac{x}{D_{const}} \tag{96.2}$$

$$y' = \frac{y}{D_{const}} \tag{96.3}$$

The nozzle exit positions are designated NXP' as dimensionless and results are obtained for 0, −0.208, −0.417, −0.625, −0.833, −1.042, and −1.25, respectively.

The entrainment ratio was obtained depending on the seven different primary nozzle exit positions (NXP). The entrainment ratio (w), the name given to the ratio of the secondary mass flow to the primary mass flow, is expressed by Eq. (96.4).

$$w = \frac{\dot{m}_s}{\dot{m}_p} \tag{96.4}$$

The inlet boundary conditions for the primary and secondary inlets of the ejector have specified the pressure-inlet type. In addition, the boundary condition of the ejector outlet is defined as a pressure-outlet type. The pressure boundary conditions for R1234yf were determined to provide the primary and secondary fluid pressure ratios selected for R134a. As the temperature value, it was determined by finding the evaporation pressure of the R1234yf fluid at the determined pressure values. Also, it was examined for the R1234yf fluid in two different cases, with and without overheating. The boundary conditions specified for both fluids are presented in Table 96.2.

The governing equations are solved using the COUPLED algorithm for the steady-state compressible flow. The advection and diffusion terms of the governing equations were discretized with the second-ordered upwind and second-ordered central difference scheme, respectively. For the study of the turbulence model used in this study, it was determined as a result of a sensitive study by Metin et al. (2019). The turbulence model used for this numerical model is the SST k-ω turbulence model.

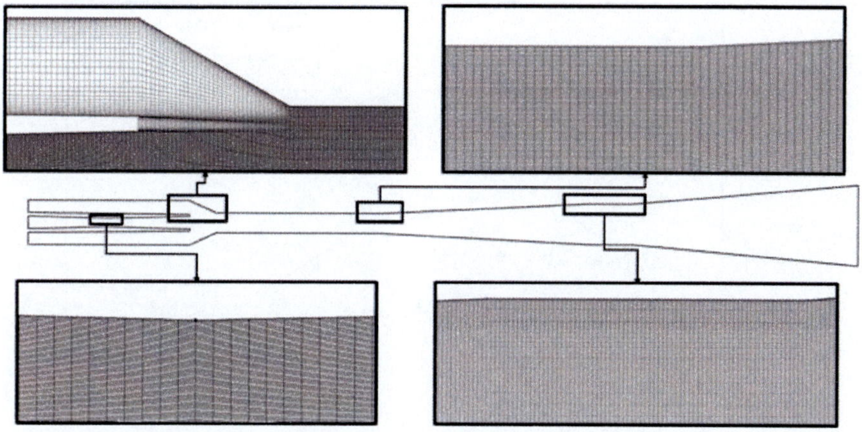

Fig. 96.3 Detailed view of the mesh structure of the ejector. (Metin et al. 2019)

ANSYS ICEM CFD v19.2 was used to create the structured mesh grid of the ejector used in the numerical study. In order to demonstrate the accuracy and precision of the numerical model, the element type and the number of elements are significant in the meshing step. It also plays a major role in determining complex structures such as shock series occurring in the flow area. In line with the mentioned work, the number of elements of the structured mesh grid, whose independence from the number of elements and the sensitivity of the mesh grid has been determined, is 200,000, approximately (Metin et al. 2019). The quadrilateral structure was chosen as the element type. A detailed view of the network structure is given in Fig. 96.3.

96.3 Results and Discussion

The effects of the distance between the nozzle exit and the inlet of the converging duct (*NXP*) is investigated using R1234yf instead of R134a through numerical simulations for supersonic ejectors. The effects of geometric parameters related to the ejector mixing section, entrainment ratio variations, Mach number distributions along the symmetry axis, and Mach number contours are presented in detail for both refrigerants.

Depending on the primary nozzle exit position (NXP), the entrainment ratio (w) change for different refrigerants can be seen in Fig. 96.4. When the primary nozzle exit position (NXP) is selected in accordance with the selected refrigerant, improvement to the entrainment ratio value is observed. The results obtained using the basic ejector geometry, in which Del Valle et al. (2014) carried out experimental studies directly from the literature, were named Case 1. It is named by using suffixes "a, b,

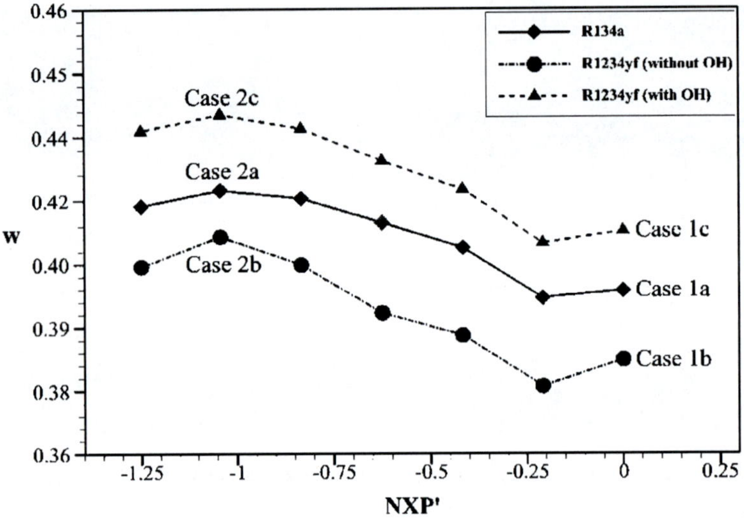

Fig. 96.4 Effect of the primary nozzle exit position of the ejector on the entrainment ratio (*w*)

and c" according to different refrigerants and overheating conditions: the suffix "a" for results using R134a, suffix "b" for results using R1234yf and without overheating, and suffix "c" for results using R1234yf and with overheating.

In the entrainment ratio results obtained depending on the different primary nozzle exit position (NXP), the improvement for R134a was 5.95% compared to base geometry. For R1234yf, the improvement compared to the base geometry was calculated as 7.65% and 6.65% without and with overheating, respectively. When the results in Fig. 96.5 are evaluated in general, when R1234yf is used without overheating instead of R134, a decrease in performance is observed in terms of entrainment ratio (*w*). This reveals the necessity to achieve similar performance by making improvements in terms of geometric and operating conditions when R1234yf fluid is used instead of R134a. When the R1234yf fluid is used with overheating, it is seen in the case indicated by Case 2c, where the best performance in terms of entrainment ratio (*w*) is achieved.

As shown in Fig. 96.5, the Mach number distribution across the x-axis line of the ejector is described and the three important cases for NXP changes, Case 2a, Case 2b, and Case 2c, are compared with each other. For the three different cases, the cases that give the best performance in terms of entrainment ratio are examined. NXP', which gives the best performance for all three different cases, corresponds to approximately -1.042. Considering the distribution of Mach Number on the x-axis line, it is seen that there are shock series in similar positions. Also, it is seen that the mixed fluid decreases to the subsonic level when x' is approximately 9.5.

In addition to the x-axis Mach number distribution, contour graphics are also presented for different cases. Contour graphs are given in Fig. 96.6 for all cases.

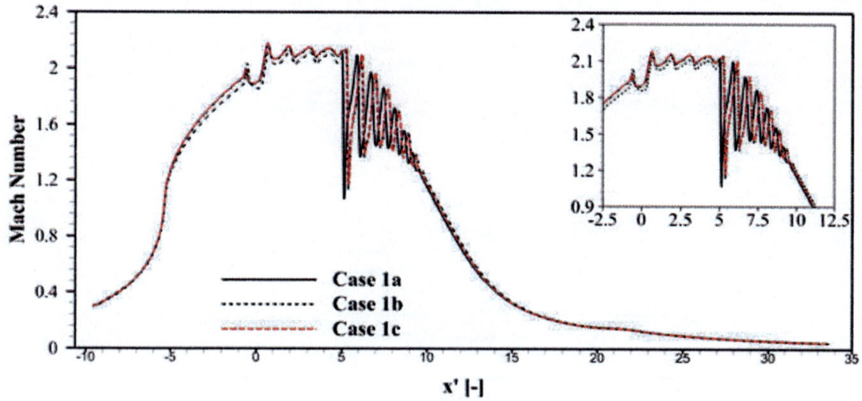

a)

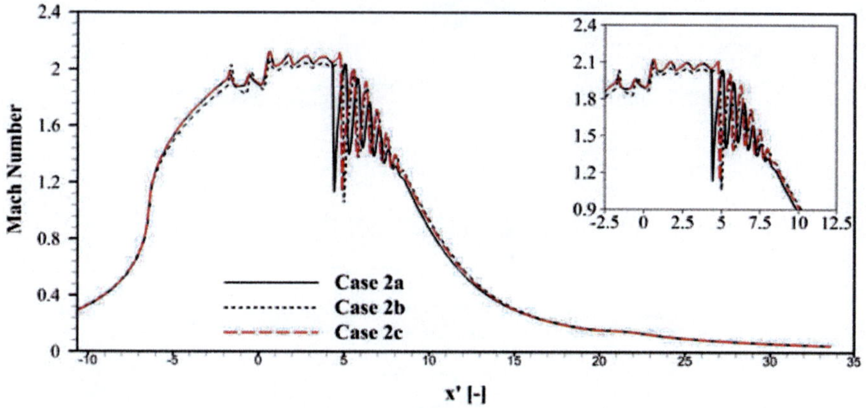

b)

Fig. 96.5 Mach number distribution throughout the ejector at symmetry axis for a) base geometry cases (Case 1a, Case 1b, and Case 1c) and b) best performance cases relative to base geometry cases (Case 2a, Case 2b, and Case 2c)

Figure 96.6a, c, and e represent the base geometry cases, and Fig. 96.6b, d, and e represent the best performance results (relative to base geometry cases). By means of the contour graphics of the Mach number, it is possible to have information about the structure of the flow that takes place in the ejector. Considering all working conditions, it is seen that the primary fluid reaches supersonic condition after the primary nozzle exit in all cases. Although the shock series seen from Mach contours are similar (there are differences in shock position), different values were obtained as the entrainment ratio. The reason for this situation is that the area called the effective area, which the secondary fluid can pass in the ejector, is changing depending on different NXP values.

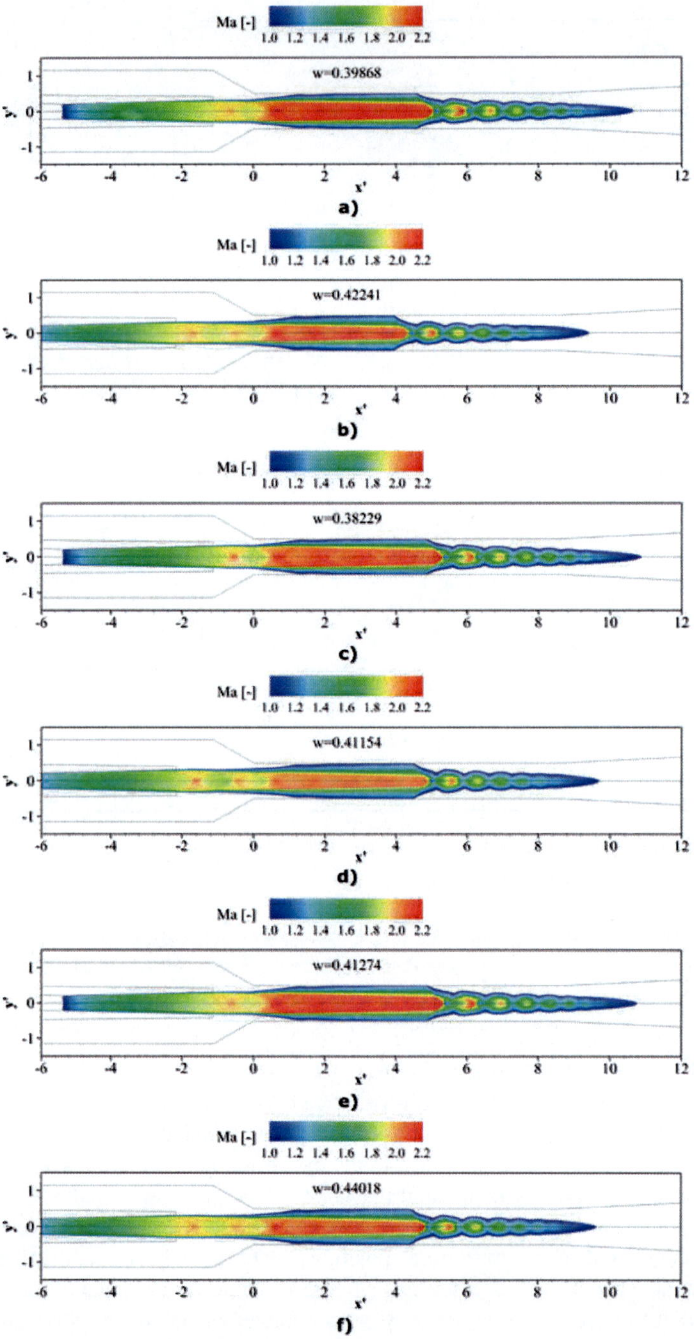

Fig. 96.6 Interaction of primary and secondary fluid flows for the cases studied: Mach number contour of Case 1a (**a**), Case 1b (**b**), Case 1c (**c**), Case 2a (**d**), Case 2b (**e**), and Case 2c (**f**)

96.4 Conclusion

In this study, a numerical model was constructed to examine the performance of an ejector operating with R134a fluid when using R1234yf under the geometric effects of the nozzle exit position (NXP). The results are shown by the ejector, the Mach number distributions along the axis of symmetry, and the Mach number contours. The main results can be summarized as follows:

- Using R1234yf instead of R134a fluid: When using R1234yf, which is one of the new-generation environmentally friendly HFO refrigerants instead of the R134a, which is in the process of being out of use, a decrease in ejector performance has been observed in terms of the entrainment ratio for similar operating conditions. Therefore, when it comes to using R1234yf fluid instead of R134a, it is very important to work toward an increased performance in terms of geometric and operating conditions.
- Primary nozzle position (NXP): When the performance of the ejector is evaluated in terms of the entrainment ratio, it is concluded that it can be increased by changing the primary nozzle exit position for fluid R134a and R1234yf. The entrainment ratio increase obtained for the R134a fluid was 5.95%, and the performance increase obtained as a result of the NXP change was 7.65% in the analyses carried out without overheating using R1234yf instead of the R134a fluid and was obtained as 6.65% with overheating.

In light of all this information, it is concluded that when the refrigerant is changed, the performance can be increased depending on the geometrical parameters and operating conditions. In the investigation of geometric and operating conditions using different refrigerants, detailed information can be obtained about the flow structures inside the ejector by using numerical models.

Acknowledgments The authors would like to acknowledge the support of the Scientific and Technological Research Council of Turkey (TUBITAK) under Grant No: 116 M367.

References

Besagni, G., 2019, Ejectors on the cutting edge: The past, the present and the perspective. Energy, 170, 998–1003. https://doi.org/10.1016/j.energy.2018.12.214

Del Valle, J. G., Jabardo, J. S., Ruiz, F. C., and Alonso, J. S. J., 2014, An experimental investigation of a R-134a ejector refrigeration system. International Journal of Refrigeration, 46, 105–113. https://doi.org/10.1016/j.ijrefrig.2014.05.028

Elbel, S., and Hrnjak, P., 2008, Ejector refrigeration: an overview of historical and present developments with an emphasis on air-conditioning applications. International Refrigeration and Air Conditioning Conference, July 14-17, 2008. Purdue University, West Lafayette, Indiana. https://docs.lib.purdue.edu/cgi/viewcontent.cgi?article=1883 & context=iracc

Fang, Y., Croquer, S., Poncet, S., Aidoun, Z., and Bartosiewicz, Y., 2017, Drop-in replacement in a R134 ejector refrigeration cycle by HFO refrigerants. International Journal of Refrigeration, 77, 87–98. https://doi.org/10.1016/j.ijrefrig.2017.02.028

Metin, C., Gök, O., Atmaca, A. U., and Erek, A., 2019, Numerical investigation of the flow structures inside mixing section of the ejector. Energy, 166, 1216–1228. https://doi.org/10.1016/j.energy.2018.10.095

Lamberts, O., Chatelain, P., and Bartosiewicz, Y., 2017, New methods for analyzing transport phenomena in supersonic ejectors. International Journal of Heat and Fluid Flow, 64, 23–40. https://doi.org/10.1016/j.ijheatfluidflow.2017.01.009

Lawrence, N., and Elbel, S., 2014, Experimental investigation of a two-phase ejector cycle suitable for use with low-pressure refrigerants R134a and R1234yf. International Journal of Refrigeration, 38, 310–322. https://doi.org/10.1016/j.ijrefrig.2013.08.009

Lemmon, E. W., Huber, M. L., and McLinden, M. O., 2013, NIST Standard Reference Database 23: Reference Fluid Thermodynamic and Transport Properties-REFPROP, Version 9.1, Standard Reference Data Program. National Institute of Standards and Technology, Gaithersburg, MD. https://doi.org/10.18434/T4/1502528

Li, H., Cao, F., Bu, X., Wang, L., and Wang, X., 2014, Performance characteristics of R1234yf ejector-expansion refrigeration cycle. Applied energy, 121, 96–103. https://doi.org/10.1016/j.apenergy.2014.01.079

Little, A. B., and Garimella, S., 2016, A critical review linking ejector flow phenomena with component-and system-level performance. International Journal of Refrigeration, 70, 243–268. https://doi.org/10.1016/j.ijrefrig.2016.05.015

Navarro-Esbrí, J., Moles, F., and Barragán-Cervera, Á., Experimental Analysis of the Internal Heat Exchanger Influence on A Vapour Compression System Performance Working with R1234yf as a Drop-in Replacement for R134a. Applied Thermal Engineering, 59(1–2), 153–161, 2013. https://doi.org/10.1016/j.applthermaleng.2013.05.028

Özgür, A. E., Kabul, A., and Kizilkan, Ö., 2014, Exergy analysis of refrigeration systems using an alternative refrigerant (hfo-1234yf) to R-134a. International Journal of Low-Carbon Technologies, 9(1), 56–62. https://doi.org/10.1093/ijlct/cts054

Regulation, E. U., 2014, No 517/2014 of the European parliament and the council of 16 april 2014 on fluorinated greenhouse gases and repealing regulation (EC) No 842/2006. J. Eur. Union. https://eur-lex.europa.eu/legal-content/EN/TXT/?uri=uriserv:OJ.L_.2014.150.01.0195.01.ENG

Rusly, E., Aye, L., Charters, W. W. S., and Ooi, A., 2005, CFD analysis of ejector in a combined ejector cooling system. International Journal of Refrigeration, 28(7), 1092–1101. https://doi.org/10.1016/j.ijrefrig.2005.02.005

Scott, D., Aidoun, Z., Bellache, O., and Ouzzane, M., 2008, CFD simulations of a supersonic ejector for use in refrigeration applications, International Refrigeration and Air Conditioning Conference, July 14–17, 2008. Purdue University, West Lafayette, Indiana. https://docs.lib.purdue.edu/cgi/viewcontent.cgi?article=1926&context=iracc

Zhu, Y., Cai, W., Wen, C., and Li, Y., 2009, Numerical investigation of geometry parameters for design of high performance ejectors. Applied Thermal Engineering, 29(5–6), 898–905. https://doi.org/10.1016/j.applthermaleng.2008.04.025

Zilio, C., Brown, J. S., Schiochet, G., and Cavallini, A., 2011, The refrigerant R1234yf in air conditioning systems. Energy, 36(10), 6110–6120. https://doi.org/10.1016/j.energy.2011.08.002

Chapter 97
Piezoelectric Energy Harvester in IoT: Recent Highlights of Bibliometric Analysis

N. R. Dhineshbabu, P. V. Elumalai, and Selçuk Sarıkoç

Nomenclature

IoT Internet of Things
EH Energy Harvesting

97.1 Introduction

Because of the development of smart cities and the continual augmentation of big data, the Internet of Things (IoT) has experienced tremendous expansion in recent years. Agriculture monitoring, environmental protection, transportation operations, logistics tracking and security, and other infrastructure industries are just a few of the areas where IoT is now being deployed. More importantly, IoT has tremendous potential and use in a variety of rapidly expanding services across all industries, including financial, home, education, healthcare, and so on (Garg and Garg 2017; He et al. 2017; Zhang et al. 2020). The powerful emergence of IoT effectively stimulates

N. R. Dhineshbabu (✉)
Department of Electronics and Communication Engineering, Aditya Engineering College, Surampalem, Andhra Pradesh, India

P. V. Elumalai
Department of Mechanical Engineering, Aditya Engineering College, Surampalem, Andhra Pradesh, India

S. Sarıkoç
Department of Motor Vehicles and Transportation Technologies, Amasya University, Amasya, Turkey

Tasova Yuksel Akin Vocational School, Amasya, Turkey

M. Z. Sogut et al. (eds.), *Proceedings of the 2022 International Symposium on Energy Management and Sustainability*, Springer Proceedings in Energy,
https://doi.org/10.1007/978-3-031-30171-1_97

the intelligent growth of these elements and makes the use and distribution of limited resources more equitable, increasing the efficiency and benefit of the industry. Perception, network, and application layers make up the basic core of IoT. The various layers are mostly self-contained, yet they all work together to bring IoT to life. As hardware and software implementation technology develops in wisdom medical, intelligent agricultural, autonomous drive, and smart houses, IoT will become more widely employed in more application scenarios (Rana et al. 2020).

Sensor miniaturization is the cornerstone and assumption of IoT implementation since sensors are a critical component of the perception layer. The basic role of sensors is to collect data about the environment or objects and send it to the background for processing across the network. Essentially, the major job of the perception layer is to provide power and technology that allows human personal wearable gadgets to run autonomously and continuously (He et al. 2017). In the electrical industry these days, innovative technologies, notably wearable electronic gadgets and IoTs, are gaining traction (Elvira-Hernández et al. 2020; Jornet, and Akyildiz 2012). The basic purpose of electrical energy boosting technology is to eliminate the need for batteries. It has operated autonomously, employing cutting-edge technology to transform mechanical, vibrational, and hydraulic energy into electrical energy, which may be utilized to power nanodevices and nanosystems for energy harvesting (Elvira-Hernández et al. 2020; Lee et al. 2020; Zhang et al. 2020).

The piezoelectric nanogenerator (PENG) was first presented by Wang's group in 2006, and it employs a zinc oxide nanowire array to convert nanoscale mechanical energy such as bending or compressing into electrical energy via the piezo potential (Cao et al. 2021; Whiter et al. 2014; Wu 2016). Nanogenerators can absorb energy from the environment and generate electrical signals directly under external stimuli, making them suitable for self-powered sensors, unlike traditional sensors that require an external power supply. From 2012 to 2021, a bibliometric survey of piezoelectric nanogenerators through IoT is presented.

When mechanical stress is applied to a piezoelectric material, an electric charge is generated, and the material's dimensions change when an electric field is applied across it. Direct and inverse piezoelectric effects are the terms used to describe the direct and inverse piezoelectric effects, respectively (Xue et al. 2014; Nyamayoka et al. 2018; Zhang et al. 2020). Elevating the applied stress or strain on the piezoelectric material, which releases more mechanical energy, is one of the key ways for improving power production. Piezoelectric materials include single crystalline minerals (quartz), piezoceramics (such as lead zirconate titanate (PZT), barium titanate, lead titanate, potassium niobate, lithium niobate, lithium titanate, sodium tungstate, and zinc oxide), piezoelectric semiconductors, polymers (such as polyvinylidene fluoride), piezoelectric composites, glass ceramics, and piezoelectric semiconductors (such as $Li_2Si_2O_5$ and Ba_2TiSiO_6) (Luo et al. 2017; Rovisco et al. 2020; Vivekananthan et al. 2019; Rana et al. 2020; Li et al. 2015 Lee et al. 2020; Chandran et al. 2021; Chung et al. 2019; Nyamayoka et al. 2018). This paper is a comprehensive look of the past decade's worth of research on piezoelectric nanogenerators from the scopus database. This article assists in assessing the effects of several articles in the subject of piezoelectric nanogenerators.

97.2 Materials Used for Analysis

97.2.1 Search Method

To combine and obtain as many relevant records as possible, Scopus' core collection field (which covers over 107 peer-reviewed journals published internationally) was chosen, as well as the Scopus core collection database's "Advanced Search" function. From 2012 until the present, we have been able to find 107 records.

97.2.2 Advanced Searching

The search terms applied to identify the closest matching publication included "Internet of things" AND "Piezoelectric Nanogenerator," which was used as the keyword in the title.

97.2.3 Data Analysis

Finally, using bibliometric tools like MS Excel and VOS viewer, 107 records on Piezoelectric Nanogenerator with the Internet of Things were chosen for data analysis and visualization. The link between the author and the keywords was investigated using VOSviewer (Yu et al. 2020; Van Eck and Waltman 2010), a software program for creating and visualizing bibliographic networks. The capabilities of VOSviewer are quite useful for studying large directory maps in an easy-to-understand format.

97.3 Results and Discussion

97.3.1 Statistical Analysis

On the topic of "Piezoelectric nanogenerator with IoT based," we found 107 papers. Scopus is a database that collects information on publications. On the same database, the following statistical analysis is performed.

- Documents by type.
- Documents by country.
- Documents by subject area.

Documents are organized by country/territory in Fig. 97.1. China has published 15 papers, with Singapore, India, the United States, and South Korea publishing 6, 4,

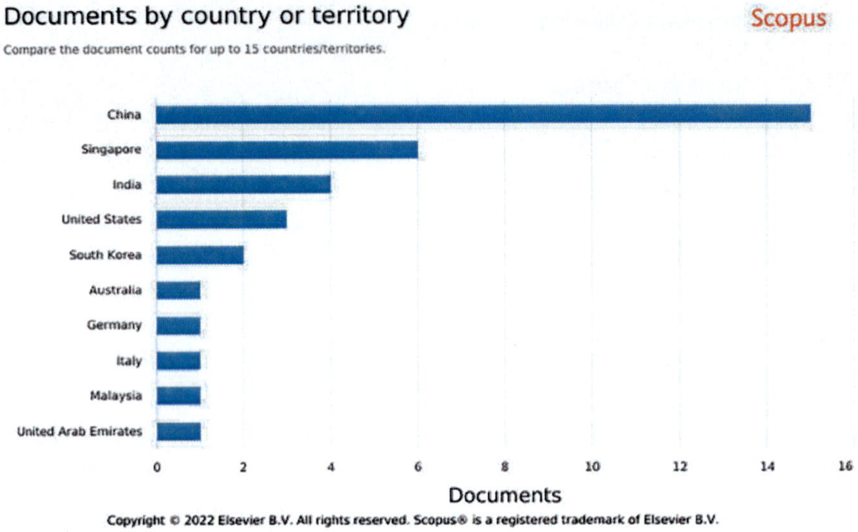

Fig. 97.1 Documents by country

3, and 2 documents, respectively which is clearly noticed from the Fig. 97.1. Other nations, such as Australia, Germany, Italy, Malaysia, the United Arab Emirates, and the United Kingdom, have issued each one between 2012 and 2022.

Review, Article, Conference Paper, and Book Chapters are the types of publications that are distributed depending on the year 2012–2022 (see Fig. 97.2). Review Papers account for 34.6% of all published documents, whereas Articles contribute for 57.9%.

The documents are organized by subject area in Fig. 97.3. The Materials science field has the highest percentage of documents published (28.6%), followed by Engineering (23.7%) and Energy (12.6%). Materials Engineering is the rationale for having the most articles on the subject of Materials Science, which includes blends of polymer, ceramics, or composites.

97.3.2 Network Analysis of Bibliographic Coupling

97.3.2.1 Bibliometric Analysis of the Keywords

The final analysis included keywords submitted by the authors of the publication that there were occurrences of a keyword more than twice in the Scopus core database. Only 244 keywords out of 1158 met the threshold. "Nanogenerator" (total link

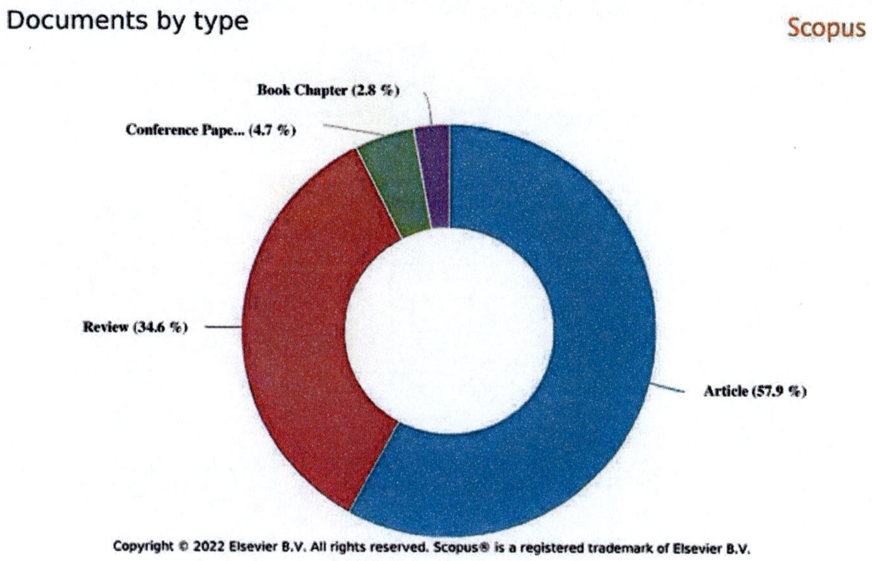

Fig. 97.2 Documents by type

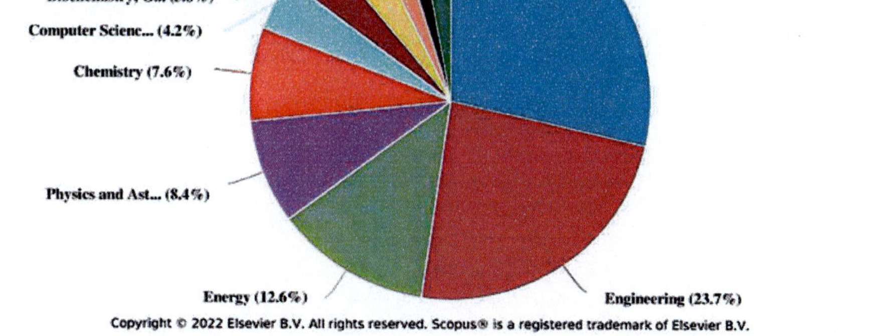

Fig. 97.3 Documents by subject area

strength 845) and "Piezoelectricity" (total link strength 709) were the most common terms, both of which had a strong relationship to "energy harvesting" and "self-powered." Another term was Piezoelectricity comparisons, with total link strength of more than 100 (see Fig. 97.4).

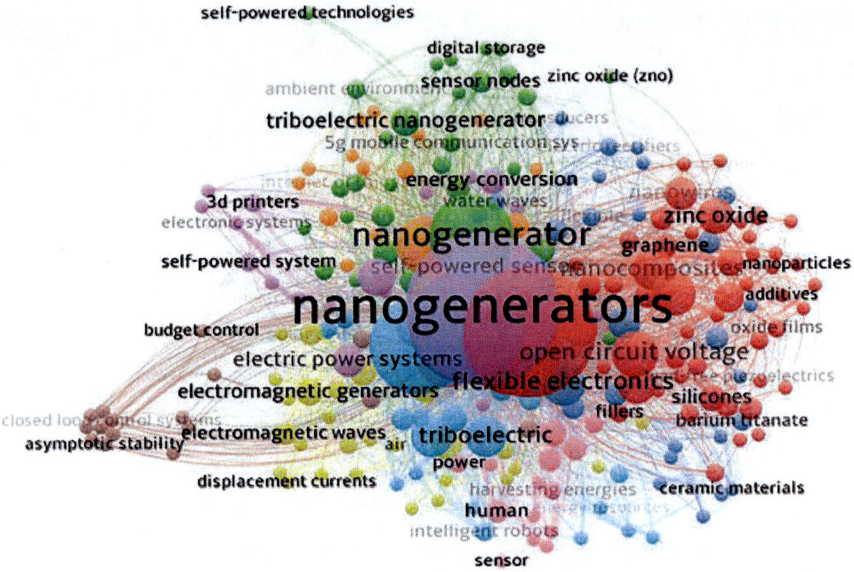

Fig. 97.4 Bibliometric analysis of the co-occurrence of keywords

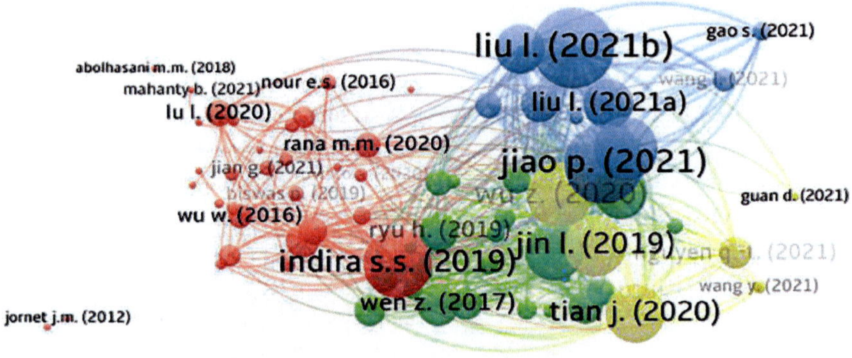

Fig. 97.5 Bibliographic coupling of documents

97.3.2.2 Bibliographic Coupling of Documents

The overall strength of bibliographic coupling relationships with other texts was estimated, and 72 documents out of a total of 107 satisfied the requirements. Documents containing at least three citations are taken into account. With 20 citations, Liu et al. (2021) has the strongest bibliographic coupling (Fig. 97.5) with total link strength of 709.

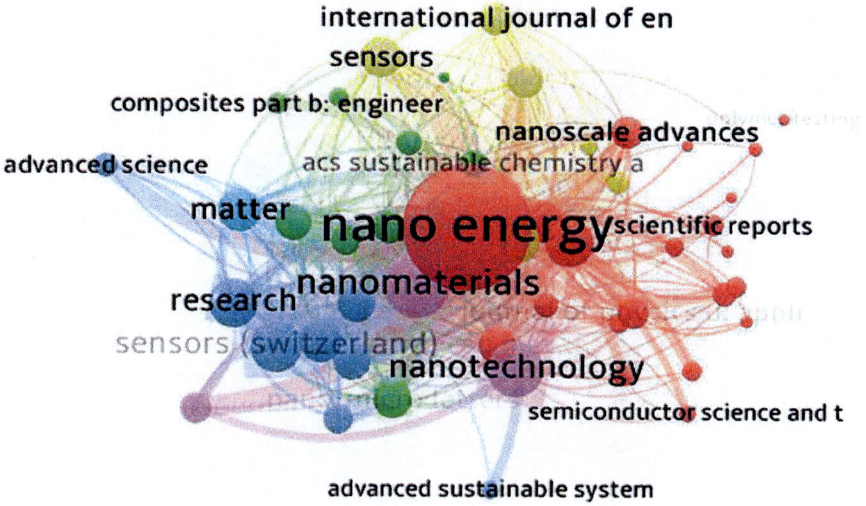

Fig. 97.6 Bibliographic coupling of sources

97.3.2.3 Bibliographic Coupling of Sources

In this analysis, 52 documents out of 64 sources matched the criteria, having at least one document and one source citation. With other sources and documents, Nano Energy has the greatest bibliographic coupling strength of 4297 with 850 citations (see Fig. 97.6).

97.3.2.4 Bibliographic Coupling of Authors

Considering a minimum threshold value of one document and three citations for an author, 45 authors out of a total of 74 satisfied the threshold criterion. With 4 documents and 1091 citations, He et al. (2017) has a maximum bibliographic coupling strength of 2050. The bibliographic coupling by authors is depicted in Fig. 97.7.

97.3.2.5 Bibliographic Analysis of the Citations

In this analysis, ignore the documents with a large number of authors per document of 5, and 121 documents out of a total of 159 authors met the requirements. Documents containing at least one citation were taken into account. In Fig. 97.8, with 1129 citations, Cao et al. (2021) has the strongest Bibliographic coupling with a total link strength of 54. Eight hundred and fifty-six papers from China have been cited 2038 times, and the total link strength is 116 (see Fig. 97.9).

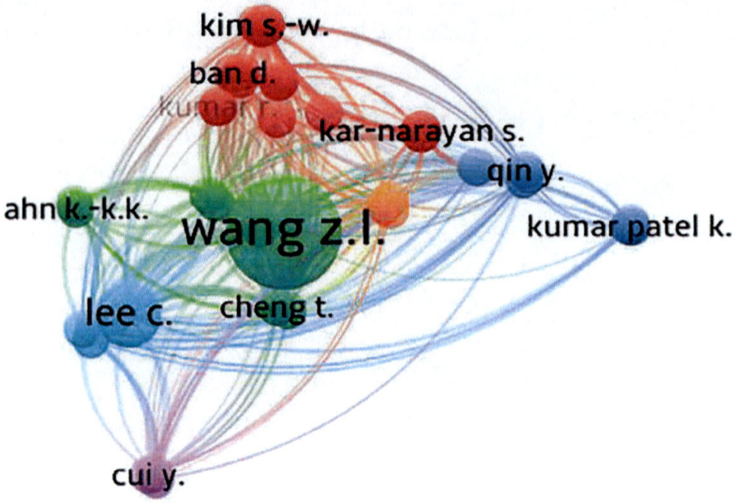

Fig. 97.7 Bibliographic coupling of authors

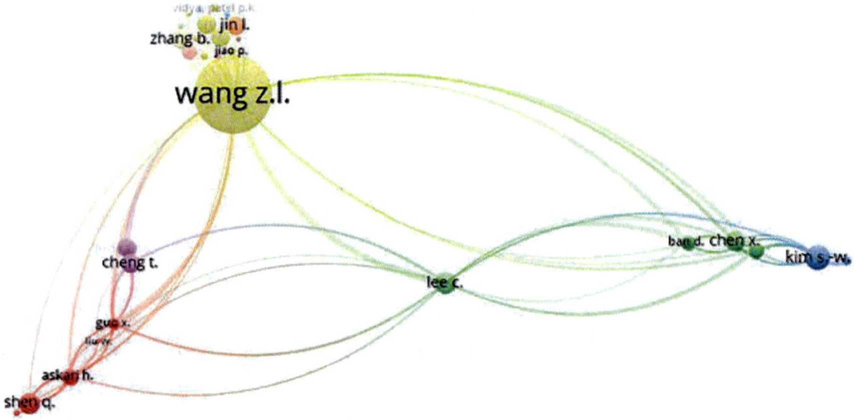

Fig. 97.8 Bibliometric analysis of the citations of authors

97.3.3 Discussion

In this study, 107 papers listed in the Scopus core database regarding piezoelectric nanogenerators with the Internet of Things were objectively and fully examined. Aspects of the Internet of Things are covered in published literature. "Nanogenerator" is the most often used term, and it is closely associated with "energy harvesting," "wearable technology," and "nanotechnology." China has made the most significant contribution to this subject. In the future, energy harvesting via piezoelectric nanogenerators might be a hotspot.

Fig. 97.9 Bibliometric analysis of the citations of countries

Despite this, some limits are unavoidable. To begin with, even though a high number of new research articles are added to Scopus every day, only a small portion of them can be indexed in the core database. As a result, the majority of non-English language articles were overlooked or ignored.

97.4 Conclusions

Relevant articles on nanogenerators have expanded dramatically in Scopus during the last 5 years, notably in the last 3 years; nevertheless, relevant publications on nanotechnology have only exhibited modest yearly growth. A unique technique, phased biographic coupling, is used to study the evolutionary trend of nanogenerator research in light of the varying timeframes of development of typical nanogenerators (piezo, tribo, thermal, and pyroelectric).

Since 2010, more and more publications introducing the potential use of nanogenerators have been published in broadly diverse Scopus categories in all subject areas, indicating that relevant research on nanogenerators has entered a phase of rapid expansion.

Emerging applications of nanogenerators might play a promising role in the future, especially with the growth of pervasive computing technologies, the Internet of Things, wearable electronic devices, renewable energy, and so on. The research process of implementing nanogenerators in numerous sectors has progressed from research to commercialization in recent years. Without a doubt, its expansion will continue in all emerging fields. Relevant publications are still few in possible interdisciplinary domains such as communication and renewable energy; on the other hand, these interdisciplinary topics might give scope for future improvements, leading to more viable applications.

References

Cao X, Xiong Y, Sun J, Zhu X, Sun Q, Wang ZL (2021) Piezoelectric Nanogenerators Derived Self-Powered Sensors for Multifunctional Applications and Artificial Intelligence.Adv. Funct. Mater. 31(33):2102983. https://doi.org/10.1002/adfm.202102983

Chandran AM, Varun S, Karumuthil SC, Varghese S, Mural PKS (2021)Zinc oxide nanoparticles coated with (3-aminopropyl)triethoxysilane as additives for boosting the dielectric, ferroelectric, and piezoelectric properties of poly(vinylidene fluoride) films for energy harvesting. ACS Applied Nano Materials. 4(2): 1798–1809. https://doi.org/10.1021/acsanm.0c03214

Chung MH, Yoo S, Kim H, Yoo J, Han S, Yoo K, Jeong H (2019) Enhanced output performance on LbL multilayer PVDF-TrFE piezoelectric films for charging supercapacitor. Scientific Reports, 9(1): 6581. https://doi.org/s41598-019-43098-6

Elvira-Hernández EA, Anaya-Zavaleta J C, Martínez-Cisneros E, López-Huerta F, Aguilera-Cortés LA, Herrera-May AL (2020) Electromechanical modeling of vibration-based piezoelectric nanogenerator with multilayered cross-section for low-power consumption devices. Micromachines, 11(9): 860. https://doi.org/10.3390/mi11090860

Garg N, Garg R (2017) Energy harvesting in IoT devices: A survey, 2017 International Conference on Intelligent Sustainable Systems (ICISS), Palladam, 7–8 Dec. 2017: 127–131. https://doi.org/10.1109/ISS1.2017.8389371

He X, Zi Y, Yu H, Zhang SL, Wang J, Ding W, Wang ZL (2017) An ultrathin paper-based self-powered system for portable electronics and wireless human-machine interaction. Nano Energy, 39: 328–336. https://doi.org/10.1016/j.nanoen.2017.06.046

Jornet JM, Akyildiz IF (2012) Joint energy harvesting and communication analysis for perpetual wireless nanosensor networks in the terahertz band. IEEE Transactions on Nanotechnology, 11(3): 570–580. https://doi.org/10.1109/TNANO.2012.2186313

Lee C, Park H, Lee J (2020) Recent structure development of poly(vinylidene fluoride)-based piezoelectric nanogenerator for self-powered sensor. Actuators, 9(3): 57. https://doi.org/10.3390/act9030057

Li X, Chen Y, Kumar A, Mahmoud A, Nychka JA, Chung H(2015) Sponge-templated macroporous graphene network for piezoelectric ZnO nanogenerator. ACS Applied Materials and Interfaces, 7(37): 20753–20760. https://doi.org/10.1021/acsami.5b05702

Liu L, Guo X, Liu W, Lee C(2021) Recent progress in the energy harvesting technology from self-powered sensors to self-sustained IoT, and new applications. Nanomaterials 11(11): 2975. https://doi.org/10.3390/nano11112975

Luo C, Hu S, Xia M, Li P, Hu J, Li G, Jiang H, Zhang W (2017)A Flexible Lead-Free BaTiO3/PDMS/C Composite Nanogenerators as Piezoelectric Energy Harvester. Energy Technology, 6(5):922–927. https://doi.org/10.1002/ente.201700756

Nyamayoka LTE, Zhang L, Xia X (2018)Feasibility study of embedded piezoelectric generator system on a highway for street lights electrification, Energy Procedia, 152:1015–1020. https://doi.org/10.1016/j.egypro.2018.09.110

Rana MM, Khan AA, Huang G, Mei N, Saritas R, Wen B, Ban D (2020)Porosity modulated high-performance piezoelectric nanogenerator based on Organic/Inorganic nanomaterials for self-powered structural health monitoring. ACS Applied Materials and Interfaces, 12(42): 47503–47512. https://doi.org/10.1021/acsami.0c12874

Rovisco A, Dos Santos A, Cramer T, Martins J, Branquinho R, Águas H,Barquinha P (2020) Piezoelectricity enhancement of nanogenerators based on PDMS and ZnSnO3 Nanowires through microstructuration. ACS Applied Materials and Interfaces, 12(16): 18421–18430. https://doi.org/10.1021/acsami.9b21636

Van Eck NJ, Waltman L (2010) Software survey: VOSviewer, a computer program for bibliometric mapping. Scientometrics 84: 523–538. https://doi.org/10.1007/s11192-009-0146-3

Vivekananthan V, Alluri NR, Chandrasekhar A, Purusothaman Y, Gupta A, Kim S (2019) Zero-power consuming intruder identification system by enhanced piezoelectricity of K0.5Na0.5NbO3 using substitutional doping of BTO NPs. Journal of Materials Chemistry C, 7(25): 7563–7571. https://doi.org/10.1039/C8TC06626D.

Whiter RA, Narayan V, Kar-Narayan S(2014) A Scalable Nanogenerator Based on Self-Poled Piezoelectric Polymer Nanowires with High Energy Conversion Efficiency. Advanced Energy Materials 4 (18): 1400519. https://doi.org/10.1002/aenm.201400519.

Wu W. (2016) High-performance piezoelectric nanogenerators for self-powered nanosystems: quantitative standards and figures of merit. Nanotechnology, 27, 112503. https://doi.org/10.1088/0957-4484/27/11/112503

Xue CJ, Li Q, Zhang Z, Zhang Z, Hai L, Gao R, Feng (2014) A Novel Arch-Shape Nanogenerator Based on Piezoelectric and Triboelectric Mechanism for Mechanical Energy Harvesting. Nanomaterials 5 (1): 36–46. https://doi.org//10.3390/nano5010036

Yu Y, Li Y, Zhang Z, Gu Z, Zhong H, Zha Q, Yang L, Zhu C, Chen E (2020) A bibliometric analysis using VOSviewer of publications on COVID-19. Ann Transl Med 8(13):816. https://doi.org/10.21037/atm-20-4235

Zhang T, Wen Z, Liu Y, Zhang Z, Xie Y, Sun X (2020) Hybridized nanogenerators for multifunctional self-powered sensing: Principles, prototypes, and perspectives. IScience, 23(12), 101813. https://doi.org/10.1016/j.isci.2020.101813

Chapter 98
The Effect of Energy Consumed at Airports on Sustainability and Examination of Society Education

Betul Kacar and Alper Dalkiran

98.1 Introduction

Sustainability is a concept that can be defined in many different ways from various sources. In order to fully understand the concept of sustainability, it is necessary to know the background that paved the way for the emergence of this concept.

Undoubtedly, the Industrial Revolution was the most momentous event in the emergence of this concept. The Industrial Revolution, which represents the transition from production based on human and animal power to production dominated by machine power, emerged in England in the eighteenth century. With this revolution, radical changes have occurred that will significantly affect human life, such as raw materials, mechanical power, and transportation activities. People who use animals to go from one place to another have started collectively using transportation vehicles such as trains. As a result of the increasing demand, the establishment of factories and the spread of this service worldwide very quickly became inevitable. These events, which accelerate the formation of civilization, also have a background that has gained importance in recent years and has not been given much attention. Globalization and industrialization, which have an effect that makes human life more difficult in the long run while making it more accessible, should be considered in their entirety in terms of their effects. The mentioned adverse effects can be evaluated within the concept of sustainability.

As more pressing global and ecological problems have come to the fore, the concept of sustainability has also come to the fore. Sustainability can be defined as

B. Kacar
Turk Hava Yollari, Istanbul, Turkiye

A. Dalkiran (✉)
School of Aviation, Süleyman Demirel University, Keciborlu, Isparta, Türkiye
e-mail: alperdalkiran@sdu.edu.tr

M. Z. Sogut et al. (eds.), *Proceedings of the 2022 International Symposium on Energy Management and Sustainability*, Springer Proceedings in Energy,
https://doi.org/10.1007/978-3-031-30171-1_98

"Development that meets the needs of the present without compromising the ability of future generations to meet their needs" (Brundtland Commission 1987). Although it has been learned through bad experiences, it is hopeful that people are starting to gain awareness of this issue.

This study aims to investigate how educated the society is on this subject and contribute to the examination of the level of consciousness. In order to add specificity to the study, articles in which international airports measure the level of awareness and education of staff and passengers on sustainability were examined and evaluated.

98.2 The Concept of Sustainability

Sustainability as a policy concept originates in the Brundtland Report of 1987 (Kuhlman and Farrington 2010). Although there are a few ideas about where the word "sustainability" comes from etymologically, a generally accepted idea is expected in the sources. According to Wiersum, the concept of sustainability first emerged in forestry as a "sustainable yield principle" (Wiersum 1995). It has been understood that the concept of sustainability, which has expanded its scope in the field of forestry, has been included in more different fields over time. The reason for this can be shown as the dependence of most industries on forest products.

When we examine the root of the word, it can be said that it emerged from the combination of the verb "sustain" and -able. The verb "sustain," which has been in use since the fourteenth century, means to continue; −able means doable ("Sustainable" 2022). In this case, the word sustainability means using natural resources without ending the use of natural resources, both when considering the area where it is first used and when it is analyzed in terms of origin.

The following parts of the study examine the factors that effectively expand the concept of sustainability and society's awareness and education level about the subject.

98.3 The Importance of the Concept of Sustainability in Airports

The current state of the planet's natural resources makes it clear that businesses are not doing. There is growing support for sustainability. However, there is a considerable divergence between rhetoric and reality (Barr 2012).

It is also possible to say that they are not the only ones responsible for the low number of businesses engaged in sustainability-related activities. As a result, in a working environment with deadly competition conditions for businesses, every expenditure made will be against the company. Therefore, it is essential to determine what large or small-scale enterprises can do about sustainability, to make them compulsory by law, to be supported by the state, and to accelerate these studies.

Air transportation, one of the businesses where sustainability is vital, is a sector that has negative contributions to environmental sustainability with its maintenance activities, terminal building activities, and hangars, especially emissions from aircraft. Due to its reputation as the most challenging industry to adopt sustainability (Nijkamp 1999; Gossling and Upham 2012; Daley 2016; Fagan-Watson et al. 2014), the aviation industry was chosen as a case study to investigate the procedure of creating sustainable policy.

Airports on a global scale must comply with government-enforced laws at various levels. Apart from these, some airports take the initiative and contribute to the environment. However, it should not be forgotten that it can be complicated for airline companies to achieve this goal. Although it is perceived to be the opposite by customers due to the high-ticket prices, it can be said that the profit margin is relatively low in airline transportation due to the high fuel and labor costs. Therefore, keeping the balance without forgetting the competitive conditions between the enterprises is necessary. Some of the leading environmental and sustainable development concerns raised concerning the operation of airports are emissions, noise issues, land use by airports, and energy consumption (Greening Airports 2003).

The answer to the question of what exactly includes the concepts of greenhouse gas and carbon emission, which we frequently encounter in the literature and whose names we frequently hear in environmental actions, is an essential subject for understanding the study.

This subject, which has been extensively explained in academic studies, is the subject of Bill Gates, who works on environmental sustainability, in the recently published "How to Avoid a Climate Disaster?" The book is explained in a simplicity that readers of all education levels can understand. Gates, who started his work by drawing attention to the importance of two numbers regarding sustainability and the environment, says that the first of these numbers is 51 billion, and the other is 0. While 51 billion expresses how many tons of greenhouse gases are released into the atmosphere each year, the number 0 indicates the base that this number should be reduced.

It is remarkable that the measures taken are insufficient for the environment and that it is impossible to completely correct the destruction even if much larger measures are taken from this stage. Therefore, after this point, it will be vital for individuals to take every precaution they can, especially businesses that are very harmful to the environment, such as airports. These precautions will only be possible by educating the communities, instilling environmental awareness in people at a very early age, and providing more training to the employees and decision-makers of the enterprises that have a more significant impact on increasing environmental sustainability. On a topic basis, some environmental activities in various airlines are reviewed in the next section.

98.4 Example Airports with "Sustainability" Activities

When we look at the airports that work for a sustainable environment, it is seen that Hong Kong Airport carries out significant activities. According to the information conveyed in the sustainability reports, the percentage allocated to the "Environmental and Staff Wellbeing" category in the distance personnel training was 8.7. In addition, there are exemplary practices such as activities that will minimize operational emissions and energy consumption, assessment of risks arising from climate change, increasing the airport's environmental performance, and dissemination of the use of wastes for different purposes (Sustainability Report 2021).

When the sustainability report of Gatwick Airport is examined, it is seen that issues such as noise, personnel training, and waste management can be evaluated under the sustainability title. However, it has been noticed that this airport places less emphasis on staff training than Hong Kong Airport. It is stated in the report that the person received training on sustainability only in the field of biodiversity (Gatwick Airport Performance Summary 2020).

98.5 Results and Discussion

Although there have been countless studies on the harm of carbon emissions to the environment, there is no invention that can improve the world in the current situation, and there are only studies that will prevent the situation from worsening.

Klaus, S. Lackner, Director of the Lenfest Center for Sustainable Energy, is trying to develop a system that it describes as "Goal is to take a process that takes 100,000 years and compress it into 30 minutes" in order to reduce the carbon emissions that have increased since the industrial revolution. In the system they developed with their team starting from a straightforward idea, they are developing an instrument they call an air extractor, which is modeled on the leaves of a tree. This instrument, which is too large to be installed in an aircraft, contains white leaves from a tree that take on the task of catching air from carbon dioxide. This tool has a strong affinity for carbon dioxide when dry and returns carbon dioxide to us when wet. Therefore, this air is carried out from the air where it can dry and thus take the carbon dioxide. Although this cycle changes depending on humidity, temperature, and wind, when it is finished in about an hour, the box is wetted, and after it gets wet, that air is removed from the box (Klaus S. Lackner 2022).

In this respect, it is up to decision-makers to get the word out about the need for training and integration of this gadget, which has data that may be entered into the engineering department's workspace.

98.6 Conclusion

We can say some things with certainty based on the study, other academic studies, and bad experiences. First, we must be aware of the environmental disasters that may occur in the present and the future as soon as possible. We can achieve this awareness by trying to instill environmental awareness in individuals from a young age. Small changes in size can cause considerable distances to be covered. Second, studies should be supported to obtain the energy required to operate technological devices sustainably used in the modern world. Third, setting targets such as minimizing carbon emissions instead of unrealistic targets such as zeroing will lead to faster progress. Finally, all aviation personnel, decision-makers, and authorities must take bold and firm steps for air transport, which contributes greatly to human life, to be the most reliable and environmentally friendly transportation method.

References

Airport Authority Hong Kong, 2021, "Sustainability Report 2020/21"
Barr, S., 2012. "Environment and Society: Sustainability policy and the citizen."
Brundtland Commission, 1987, Our Common Future. Oxford University Press.
Daley, B., 2016. Air transport and the environment. Routledge.
"Energy Consumption, Wastage and Water Contamination are Among Other Problems." 2003. Janić, Greening Airports, page at 20–21, 35.
Fagan-Watson B, Watson T, Elliott B, Vanner R, Shaw B. and Morris S 2014 What range of methods is available to evaluate the implementation and impact of complex environmental policies on complex systems?
Gatwick Airport, 2020, "10 Decade of Change"
https://www.earth.columbia.edu/articles/view/2523, "Klaus S. Lackner", accessed on 21.02.2022
Gossling S and Upham P. eds., 2012 Climate change and aviation: Issues, challenges and solutions.
Kuhlman, T., Farrington, J., 2010, "What is Sustainability?", sustainability, 3436–3448
Nijkamp P 1999 Sustainable transport: new research and policy challenge for the next millennium. European Review, 7(4), pp.551–563.
"Sustainable", https://www.etymonline.com/word/sustainable, Access Time: 12.02.2022
Wiersum, K. F., 1995, "200 Years of Sustainability in Forestry: Lessons from History", Environmental Management, 321–329

Index

M. Z. Sogut et al. (eds.), *Proceedings of the 2022 International Symposium on Energy Management and Sustainability*, Springer Proceedings in Energy,
https://doi.org/10.1007/978-3-031-30171-1

F